MOUNT ROYAL COLLEGE
3 2047 00623 3723

D0742502

Internet Technology Handbook

Internet Technology Handbook

Optimizing the IP Network

Mark A. Miller

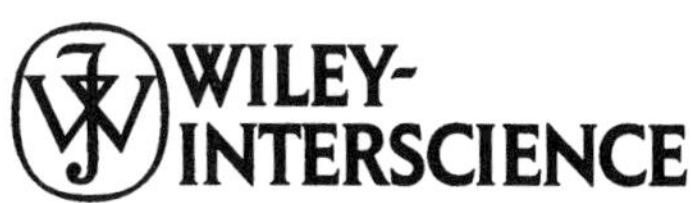

A JOHN WILEY & SONS, INC., PUBLICATION

Published by John Wiley & Sons, Inc., Hoboken, New Jersey.
Published simultaneously in Canada.

For general information on our other products and services please contact our Customer Care Department within the U.S. at 877-762-2974, outside the U.S. at 317-572-3993 or fax 317-572-4002.

Wiley also publishes its books in a variety of electronic formats. Some content that appears in print, however, may not be available in electronic format.

Library of Congress Cataloging-in-Publication Data is available.

ISBN 0-471-48050-9

Printed in the United States of America.

10 9 8 7 6 5 4 3 2 1

To Boomer, our faithful sentry

Contents at a Glance

Contents

Table of Illustrations

Preface

The Internet is an amazing communication medium. Its predecessor network, the ARPANET, was developed over three decades ago, yet many of the fundamental algorithms and protocols are still in use today. Granted, we don't have to run these protocols on a refrigerator-sized computer; our laptop or palmtop PC will do quite well. But think of it — how many other worldwide systems do we have today that have been serving us for several decades, and in much the same form? This is a true credit to the original architects of the Internet, as they clearly took the time to do their homework and designed a very solid system that benefits millions of us everyday.

But all systems require a tuneup or remodeling job over time, and the Internet is no exception. In the last few years we have seen the explosive growth of the World Wide Web, and the transport of multimedia signals, such as voice and video, over the Internet. As a result, new protocols to support real-time traffic have been developed. We have also seen the growth of electronic commerce, which has brought issues of network security to the forefront. And perhaps the biggest issue of all — the overall growth of the Internet — has inspired the development of a new protocol called Internet Protocol version 6, or IPv6, that addresses all three of the above challenges: multimedia support, enhanced security, and addressing shortages.

These and many other issues are addressed in this book. My objective has been to provide a comprehensive foundation of information that explains the underlying protocols of the Internet, such as the Transmission Control Protocol (TCP) and the Internet Protocol (IP), while still looking forward to newer technologies such as Voice over Internet Protocol (VoIP).

In working towards that objective, I had the support of a great team. My technical editor, Dr. John Thompson, made many constructive comments on the manuscript, and my copy editor, Annette Devlin, made sure that no grammatical rules were violated in the process. My assistant, Donna Mullen, produced all of the figures in the book, researched the appendices, and handled the final layout. My colleagues at John Wiley & Sons, George Telecki and Brendan Codey, provided great logistical support. And I would be remiss not to mention my support from the home front: Holly, Nathan, and Nicholas, with assistance from Boomer and Baron.

I trust that you will find the information in this text to be helpful. Drop me a note if you have any comments or suggestions for the next edition.

mark@diginet.com
March 2004

Credits

TECHNICAL EDITOR
Dr. John S. Thompson

COPY EDITOR AND PROOFREADER
Annette S. Devlin

ILLUSTRATOR AND BOOK DESIGNER
Donna M. Mullen

About the Author

Mark A. Miller, P.E., is the President of DigiNet Corporation, a Denver-based data communications engineering firm specializing in the design of local area and wide area networks. He is the author of numerous texts, including *LAN Troubleshooting Handbook,* 2nd Edition; *LAN Protocol Handbook; Internetworking,* 2nd Edition; *Troubleshooting Internetworks; Troubleshooting TCP/IP,* 3rd Edition; *Managing Internetworks with SNMP,* 3rd Edition; *Implementing IPv6,* 2nd Edition and *Voice of IP Technologies.* Mr. Miller is a frequent speaker at industry events, and has taught many tutorials on internetwork design and analysis at ComNet, Comdex, Network+Interop, Next Generation Networks, and other conferences. He is a member of the IEEE and a registered professional engineer in four states. For information on his many tutorials, including one that is based upon this text, contact him via `mark@diginet.com`, or visit `www.diginet.com`.

1

The Challenge of the Internet

To say that the Internet, with its millions of end users operating 24 hours per day, is a complex system would be one of the greatest understatements of the 21st century. However, if we take a step back and look at the Internet from a systems engineering perspective, we can observe a number of subsystems that are much easier to understand. Let's begin our journey into this fascinating architecture by exploring its origins.

1.1 A Brief History of the Internet

The Internet with a capital I is one of the world's most interesting achievements in computer science and networking technology. It provides a worldwide mechanism for user-to-user, computer-to-computer communication that spans corporate and national boundaries. This achievement is even more amazing because the Internet is self-governing, run by committees comprised largely of volunteers. In the past, many government organizations, such as the U.S. Department of Defense and individual states, have subsidized the basic expenses. Much of the research into the Internet protocols is conducted at major U.S. research universities, such as the University of California, the University of Colorado, the University of Illinois, and the University of Texas. As the Internet has evolved, however, the expenses for Internet connectivity have been passed down to the individual users — those who benefit from access to this public resource. Let's see how this unique system evolved.

Today's Internet was born in 1969 as the Advanced Research Projects Agency Network (ARPANET) and was sponsored by the U.S. Defense Advanced Research Projects Agency (DARPA), now known as ARPA. The purpose of the ARPANET was to test and determine the viability of a communication technology known as packet switching. The contract to build the original ARPANET was awarded to a firm known as Bolt, Baranek, and Newman (now BBN Technologies, Cambridge, Massachusetts). ARPANET went online in September 1969 at four locations: Stanford Research Institute (SRI), the

University of California at Santa Barbara (UCSB), the University of California at Los Angeles (UCLA), and the University of Utah. The original hosts were Honeywell minicomputers, known as Interface Message Processors (IMPs).

The initial test was successful, and the ARPANET grew quickly. At the same time, it became apparent that nonmilitary researchers could also benefit from access to a network of this type, so leaders in the university and industrial research communities made proposals to the National Science Foundation (NSF) for a cooperative network of computer science researchers. The NSF approved funding for the Computer Science Network (CSNET) in 1981.

In 1984, the ARPANET was split into two different networks: MILNET (for unclassified military traffic) and ARPANET (for nonmilitary traffic and research). In 1984, the NSF established the Office of Advanced Scientific Computing (OASC) to further the development of supercomputers and to make access to them more widely available. The OASC developed the NSFNET to connect six supercomputing centers across the United States, using T-1 lines operating at 1.544 Mbps in 1987, and subsequently upgrading to the T-3 rate (44.736 Mbps) in 1990. NSFNET, with its higher transmission rates, was a resounding success; as a result, the U.S. Department of Defense declared the ARPANET obsolete and dismantled it in June 1990.

In the meantime, NSFNET connections encompassed a system of regional and state networks. The New England Academic and Research Network (NEARNET), the Southeastern Universities Research Association Network (SURAnet), and the California Education and Research Federation Network (CERFnet) were among the family of NSFNET-connected networks. Since these networks were designed, built, and operated, in part, with government funds, regulations called Acceptable Use Policies, or AUPs, governed the types of traffic that could traverse these networks. In general, traffic that was for "research or educational purposes" was deemed acceptable; other traffic was either discouraged or prohibited. Few, if any, accounts of the "Internet Police" apprehending an AUP violator were recorded, however, again testifying to the self-governing nature of the Internet.

The business community, seeing the new opportunity for electronic commerce, began looking for ways to support general business traffic on the Internet without violating these regulations. As a result of this opportunity, the Commercial Internet Exchange Association (CIX) was formed in 1991. CIX was a nonprofit trade organization of Public Data Network service providers that promoted and encouraged the development of the public data communications internetworking services. Membership in CIX was open to organizations that offered TCP/IP or OSI public data internetworking services to the general public. CIX gave these service providers a neutral forum for the discussion and development of legislative, policy, and technology issues. Member networks agreed to interconnect with all other CIX members, and to exchange traffic. There were no restrictions placed on the traffic routed between member networks. Nor were there "settlements," or traffic-based charges, as a result of these interconnections. CIX has since disbanded, and has been replaced by the United States Internet Service Provider Association (US ISPA).

Another outgrowth of the Internet expansion was the founding of the nonprofit Internet Society (ISOC) in 1992. The Internet Society, headquartered in Reston, Virginia, is an international organization that strives for global cooperation and coordination for the Internet. Members of the ISOC include government agencies, nonprofit research and educational organizations, and for-profit corporations. The charter of the ISOC emphasizes support for the technical evolution of the Internet, educating the user community in the use and application of the Internet, and promoting the benefits of Internet technology for education at all grade levels.

In 1993, the NSF announced that it would no longer provide the traditional backbone architecture, but instead would specify a number of locations, called *Network Access Points* (NAPs), where various ISPs could interconnect and exchange traffic. This concept was based on the CIX concept and specified four NAPs, located at San Francisco, Chicago, New York, and Washington, DC. In addition to these four NAPs, two Federal Internet Exchange (FIX) points, one on each coast, were developed: FIX-East and FIX-West. Other NAPs, operated as commercial entities, also exist today.

In April 1995, the existing NSFNET backbone was retired and replaced by a new architecture that provides for very high-speed connectivity. This new network is called the very high-speed Backbone Network Service (vBNS). The vBNS is based on both Asynchronous Transfer Mode (ATM) and Synchronous Optical Network (SONET) technologies, and is designed as a noncommercial research network to develop and test high-speed applications, routing, and switching technologies.

Figure 1-1 illustrates a timeline of significant events in the history of the Internet. Shown at the top of the figure are the various government organizations, starting with DARPA, that have participated in the development of the Internet. On the second line are the various organizations, starting with the ARPANET Working Group (WG) and then proceeding to the Internet

Society (ISOC) and the World Wide Web Consortium (W3C), that have provided technical oversight into the development. At the lower portion of the figure are some of the significant milestones and their associated dates [1-1].

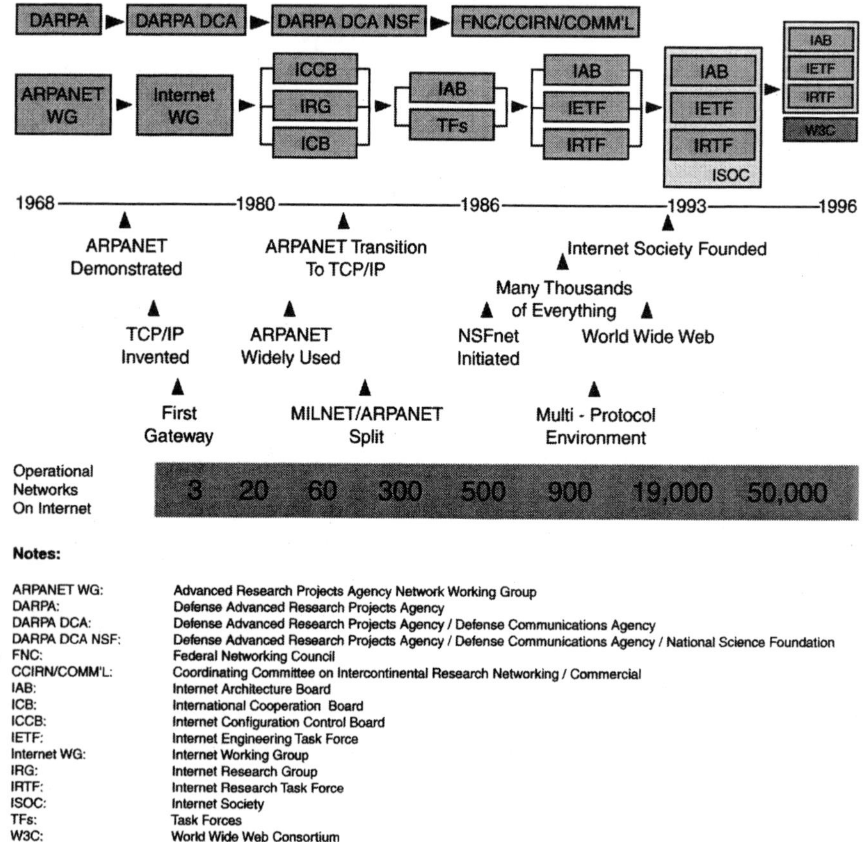

Figure 1-1. Internet Development Timeline.
(Source: www.isoc.org/internet/history/brief.shtml.)

Perhaps the most important aspect of the Internet history is the Internet's astounding growth in the last few years. Figures 1-2 and 1-3, taken from Reference [1-2], illustrate the growth in Internet Hosts and World Wide Web Servers, respectively. Note that in Figure 1-2 a new host counting mechanism, believed to be more accurate than the previous method, is shown for host statistics beginning in 1995.

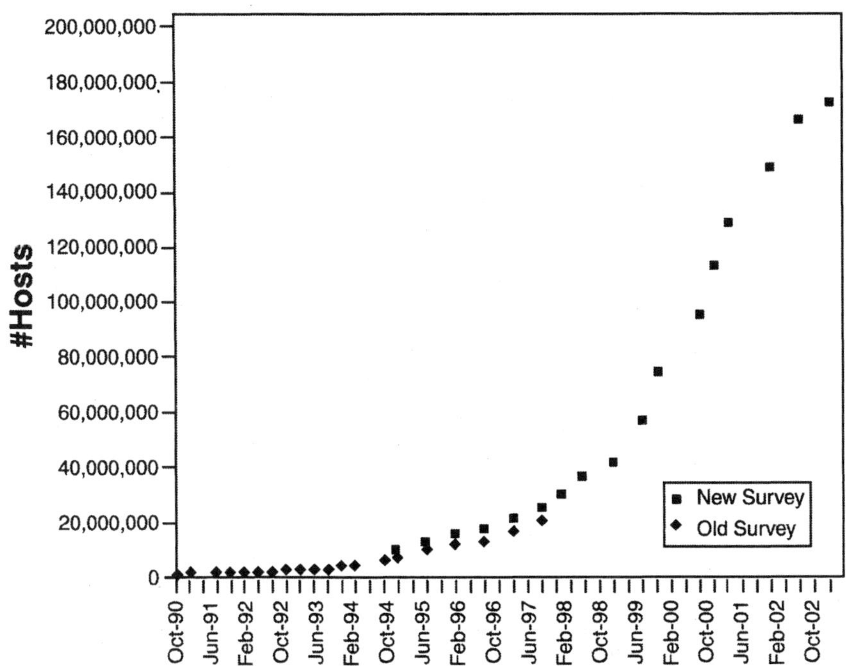

Figure 1-2. Internet Hosts.
(Source: Hobbes' Internet Timeline, www.zakon.org/robert/internet timeline/
Copyright 2003, Robert H. Zakon. Reprinted with permission.)

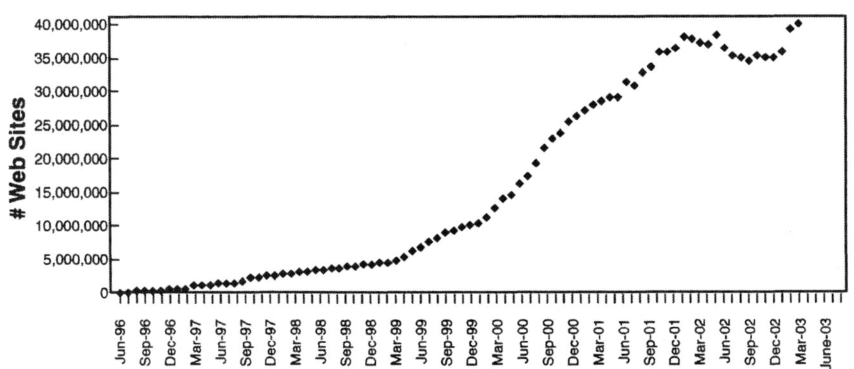

Figure 1-3. WWW Networks Growth.
(Source: Hobbes' Internet Timeline, www.zakon.org/robert/internet/timeline/
Copyright 2003, Robert H. Zakon. Reprinted with permission.)

Lessons learned from the ARPANET have had a significant effect on a number of data communication technologies, such as LANs and packet switching.

1.2 Governing and Documenting the Internet

An amazing characteristic of the largely volunteer Internet community is how smoothly this rather avant-garde organization operates. To quote from RFC 1726 [1-3]: "A major contributor to the Internet's success is the fact that there is no single, centralized point of control or promulgator of policy for the entire network. This allows individual constituents of the network to tailor their own networks, environments and policies to suit their own needs. The individual constituents must cooperate only to the degree necessary to ensure that they interoperate."

But no organization, avant-garde or not, can operate without some degree of structure. The Internet Society provides some of that structure, and one of the Internet Society's components is the Internet Architecture Board (IAB), chartered in 1992. The IAB consists of 13 members: 12 full members plus the chair of the Internet Engineering Task Force (IETF). The IAB's responsibilities include [1-4]:

- Appointing a chair of the IETF and its subsidiary Internet Engineering Steering Group (IESG)
- Oversight of the architecture for the protocols and procedures used by the Internet
- Oversight of the process used to create Internet standards
- Editorial management and publication of the Request for Comment (RFC) document series and administration of the various Internet assigned numbers
- Representing the interests of the Internet Society to other organizations
- Providing guidance to the Internet Society regarding Internet technologies

Two task forces report to the IAB. The IETF coordinates the technical aspects of the Internet and its protocols and ensures that it functions effectively. The Internet Research Task Force (IRTF) researches new technologies.

The IAB produces numerous protocol standards and operational procedures that require dissemination and archiving. Most of these documents are known as Requests for Comments documents (RFCs) and are reviewed by the appropriate IETF or IRTF members. Many of the Internet protocols have become U.S. military standards and are assigned a MIL-STD number. For example, the Internet Protocol is described in both RFC 791 and MIL-STD-1777. Internet documents are published according to two tracks. Off-track specifications are labeled with one of three categories: experimental, informational, or historic. These documents contain information that is useful but not appropriate for an Internet standard. Specifications that are destined to become Internet standards evolve through several levels of testing and revision, known as the standards track, described in RFC 2026 [1-5]. Three "maturity levels" are defined. The first level of maturity is called a Proposed Standard. This standard is stable, well understood, has been reviewed by the community, and has sufficient interest to be considered valuable. A Draft Standard is one from which at least two independent and interoperable implementations have been developed and for which sufficient operational experience has been obtained. After significant implementation and operational experience is obtained, a Draft Standard may be elevated to an Internet Standard. The various organizations that are involved in this process are described in RFC 2028 [1-6]. An interesting history of the RFC process is found in RFC 2555 [1-7].

Two sources of information on Internet standards and parameters are updated on a periodic basis. The Internet Assigned Numbers Authority (IANA) documents protocol parameters, assigned addresses such as port numbers, and many others [1-8]. The Internet Official Protocol Standards document (currently RFC 3300 [1-9]) describes the standards track process and lists recently published RFCs and the current standardization status of the various protocols. Information on how to obtain Internet documentation is given in Appendix B.

With the explosive growth of the Internet within the last few years, a wealth of information is available. Throughout this text, we will make extensive use of the RFCs and other documents in our study. Readers needing details on the Internet itself could benefit from the following: RFC 1207, "FYI on Questions and Answers — Answers to Commonly Asked 'Experienced Internet User' Questions" [1-10]; RFC 1402, "There's Gold in Them Thar Networks! Or Searching for Treasure in All the Wrong Places" [1-11]; RFC 1580, "Guide to Network Resource Tools" [1-12]; RFC 2664, "FYI on Questions and Answers — Answers to Commonly Asked 'New Internet User' Questions" [1-13]; RFC 1738, "Uniform Resource Locators" [1-14]; RFC

1935, "FYI on 'What is the Internet'" [1-15]; RFC 1983, "Internet Users' Glossary" [1-16]; and RFC 2151, "A Primer on Internet and TCP/IP Tools" [1-17].

1.3 The Protocols of the Internet

The Internet is based on an architecture that was developed by the Advanced Research Projects Agency (ARPA), which is funded by the United States government. There are literally hundreds of protocols that fit into the ARPA architecture and provide various functions within the Internet. In this text, we will concentrate our efforts on three broad categories of protocols. The first category could be called the *ARPA Core Protocols*, because they provide the underlying infrastructure that facilitates packet transport and end-user communications. These core protocols are shown in Figure 1-4, which will be explored in depth in Chapter 2. In general, these protocols would be visible (and known) to the end users of an Internet-connected network. Note that the associated RFC or MIL-STD reference is included with each protocol listed.

ARPA Layer	Protocol Implementation								OSI Layer
Process / Application	Hypertext Transfer	File Transfer	Electronic Mail	Terminal Emulation	Domain Names	File Transfer	Client / Server	Network Management	Application
	Hypertext Transfer Protocol (HTTP) RFC 2616	File Transfer Protocol (FTP) MIL-STD-1780 RFC 959	Simple Mail Transfer Protocol (SMTP) MIL-STD-1781 RFC 2821	TELNET Protocol MIL-STD-1782 RFC 854	Domain Name System (DNS) RFC 1034, 1035	Trivial File Transfer Protocol (TFTP) RFC 1350	Sun Microsystems Network File System Protocols (NFS) RFC 3530	Simple Network Management Protocol (SNMP) v1: RFC 1157 v2: RFC 1901-10 v3: RFC 3411-18	Presentation / Session
Host-to-Host	Transmission Control Protocol (TCP) MIL-STD-1778 RFC 793				User Datagram Protocol (UDP) RFC 768				Transport
Internet	Address Resolution ARP RFC 826 RARP RFC 903		Internet Protocol (IP) MIL-STD-1777 RFC 791			Internet Control Message Protocol (ICMP) RFC 792			Network
Network Interface	Network Interface Cards: Ethernet, Token Ring, MAN and WAN RFC 894, RFC 1042 and others								Data Link
	Transmission Media: Twisted Pair, Coax, Fiber Optics, Wireless Media, etc.								Physical

Figure 1-4. ARPA Core Protocols.

The second broad category of protocols could be called the *ARPA Control, Routing, and Address Resolution Protocols.* These protocols work behind the scenes to make sure that the packet transport functions operate properly and are kept up to date. The control, routing, and address resolution protocols are shown in Figure 1-5, and will be studied in greater detail in Chapter 3. In general, these protocols would not be visible to the end users of an Internet-connected network.

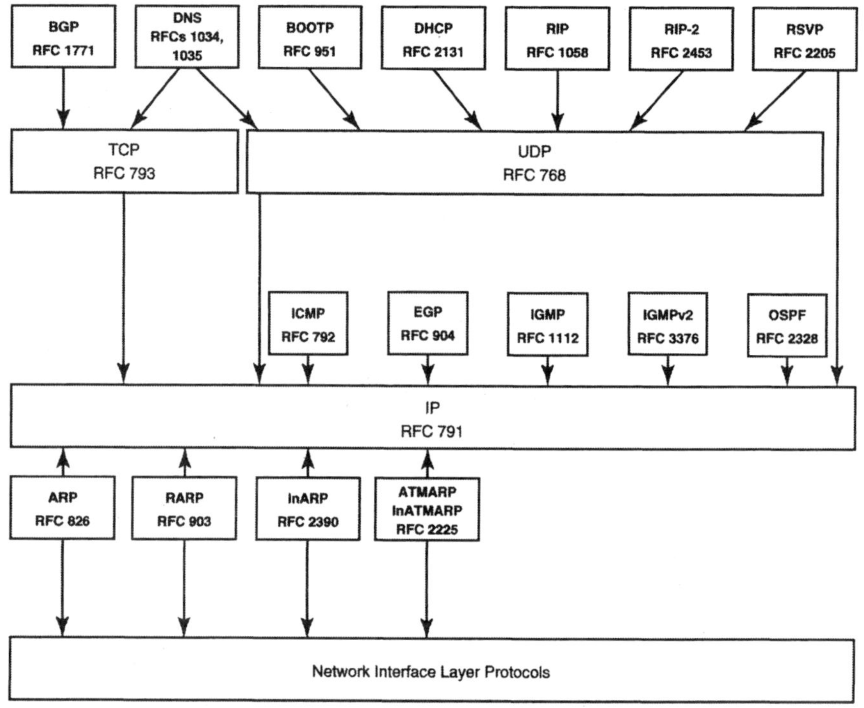

Figure 1-5. ARPA Control, Routing, and Address Resolution Protocols.

The third broad category of protocols could be called the *ARPA Multimedia Protocols.* These protocols facilitate the transmission of voice and video information over the Internet. The multimedia protocols are shown in Figure 1-6, and will be studied in greater detail in Chapter 9. In general, these protocols would also not be visible to the end users of an Internet-connected network.

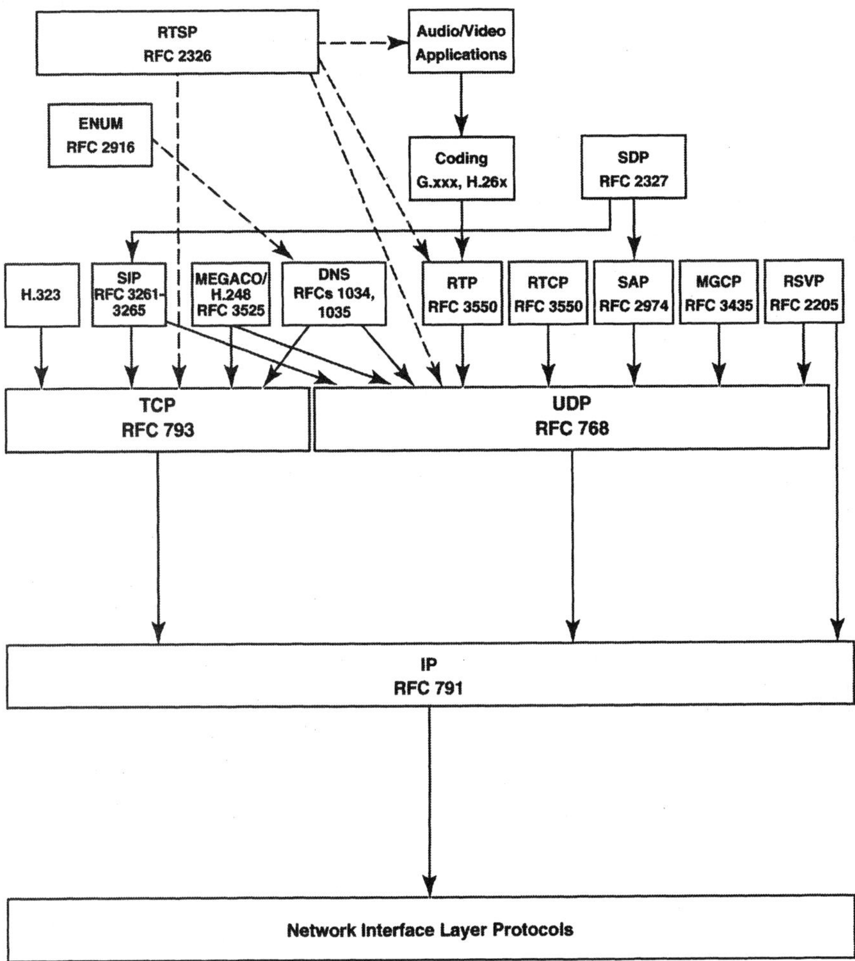

Figure 1-6. ARPA Multimedia Protocols.

1.4 Outline of This Book

With a brief glance at Figures 1-4, 1-5, and 1-6, you can see a large number of protocols that are required to support various Internet-based user applications. One of the challenges, therefore, is to sort out the various protocol functions and understand how they interrelate and work together. We will address this challenge by dividing this text into five major parts:

Part I — Introduction

Chapter 1: The Challenge of the Internet

Part II — Packet Transport

Chapter 2: Analyzing the IP Network
Chapter 3: Datagram Addressing and Delivery
Chapter 4: Routing and Intranetwork Communication
Chapter 5: End-to-End Reliability
Chapter 6: The Next Generation: IPv6
Chapter 7: Case Studies in Packet Transport

Part III — Application Support

Chapter 8: Data Transport
Chapter 9: Converged Networks and Multimedia Transport
Chapter 10: Case Studies in Application Support

Part IV — Network and Performance Management

Chapter 11: Network Management Architectures
Chapter 12: Network Management System Components
Chapter 13: Case Studies in Network and Performance Management

Part V — Reference Appendices

Appendix A: Acronyms and Abbreviations
Appendix B: Sources of Internet Information
Appendix C: Addresses of Standards Organizations
Appendix D: Trademarks

Of special note are Chapters 7, 10, and 13, which provide case studies, captured from live internetworks, that illustrate problems that can arise when using the Internet protocols and solutions that can address these problems. All of these case studies are illustrated with trace listings from the *Sniffer*® network analyzer from Network Associates, Inc.

1.5 Looking Ahead

In this chapter, we have considered a brief history of the Internet and examined some of the challenges that we will encounter from the study of the Internet protocols. In the next chapter, we will begin our work in earnest by considering the architecture of an Internet-connected network.

1.6 References

[1-1] Leiner, Barry M., et al. A Brief History of the Internet. Available from: http://www.isoc.org/internet/history/brief.shtml.

[1-2] Zakon, Robert Hobbes. Hobbes' Internet Timeline. RFC 2235. Current information is available from: http://www.zakon.org/robert/internet/timeline/.

[1-3] Kastenholz, F., and C. Partridge. "Technical Criteria for Choosing IP: The Next Generation." RFC 1726, December 1994.

[1-4] Carpenter, B., Ed. "Charter of the Internet Architecture Board." RFC 2850, May 2000. The IAB can be reached at www.iab.org.

[1-5] Bradner, S. "The Internet Standards Process — Revision 3." RFC 2026, October 1996.

[1-6] Hovey, R., and S. Bradner. "The Organizations Involved in the IETF Standards Process." RFC 2028, October 1996.

[1-7] RFC Editor, et al. "30 Years of RFCs." RFC 2555, April 1999.

[1-8] The Internet Assigned Numbers Authority (IANA) maintains an online database of protocol numbers and parameters at www.iana.org/numbers.html. This replaces the former "Assigned Numbers" document, which was last published as RFC 1700 in October 1994.

[1-9] Reynolds, J., et al. "Internet Official Protocol Standards." RFC 3300, November 2002.

[1-10] Malkin, G., et al. "FYI on Questions and Answers — Answers to Commonly Asked 'Experienced Internet User' Questions." RFC 1207, February 1991.

[1-11] Martin, J. "There's Gold in Them Thar Networks! Or Searching for Treasure in All the Wrong Places." RFC 1402, January 1993.

[1-12] EARN Staff. "Guide to Network Resource Tools." RFC 1580, March 1994.

[1-13] Plzak, A., et al. "FYI on Questions and Answers — Answers to Commonly Asked 'New Internet User' Questions." RFC 2664, August 1999.

[1-14] Berners-Lee, T., et al. "Uniform Resource Locators (URL)." RFC 1738, December 1994.

[1-15] Krol, E., and E. Hoffman. "FYI on 'What is the Internet.'" RFC 1935, April 1996.

[1-16] Malkin, G., Editor. "Internet Users' Glossary." RFC 1983, August 1996.

[1-17] Kessler, G., and S. Shepard. "A Primer on Internet and TCP/IP Tools." RFC 2151, June 1997.

2

Analyzing the IP Network

The 1960s could be described as the era of the vendor-proprietary, host-centric network. Telecommunications facilities were predominantly analog leased lines. Internetwork maintenance was dominated by the major players, the vendors that supplied the host system and the communication lines. Third-party maintenance of computer systems was not readily available because of the proprietary nature of those systems (and the vendors' desires to keep those systems proprietary). For most installations, maintenance of communication lines was the responsibility of the then dominant Bell System: AT&T and its various telephone company subsidiaries. AT&T published many standards for their analog lines, modems, etc., that could have enabled third parties to enter the communications business. Not until the divestiture of AT&T in 1984 did that industry experience a significant growth, however.

The 1970s brought about growth in distributed processing that came from the newfound popularity of minicomputers. For multi-location organizations, it made sense to move the computing power close to the end-user process. Thus, if the engineering center, manufacturing facility, and distribution warehouse were in separate locations, three computer centers would be established. Data communication between those distinct locations would invariably be required. This created the melding of the computer and communications technologies. As the network became distributed, managers had to extend their span of control. Modem and multiplexer-based networks incorporated in-hand diagnostics to thus extend the reach of the network control center.

The 1980s could be identified by a major revolution in both computer and data communication technologies. Price/performance improvements in computers, plus VLSI (very large-scale integration) components, helped fuel the LAN explosion. The advent of digital transmission lines such as T1/T3 and ISDN created radical changes in testing methods and practices for WAN facilities.

The 1990s have been described as the decade of internetworking. Three significant issues impacted the networks of this era. First, distributed computing replaced the host-centric architectures in most industries. Processing power moved closer to the end user, and that end user demanded stronger communication capabilities with his or her peers, irrespective of geography. Connectivity devices, such as bridges and routers, became more prevalent as a result. Second, the bandwidth required to connect these devices increased as well. Third, the move toward distributed computing brought an associated move away from the single vendor network. Open Systems, as we have come to know them, demanded a number of multivendor protocol solutions, instead of the proprietary solution that had been the norm until that time.

In the new millennium of 2000 and beyond, networks based on the Internet Protocol (IP) became the norm. Support for both voice and data services over IP gave rise to the term *converged networks*, which can support both real-time (voice and video) and non-real-time (data) applications. IP-based applications such as email, the World Wide Web, and electronic commerce are commonplace, to the extent that these technologies are impacting other business operations as well. For example, is your network secure enough to support new electronic commerce applications? Will electronic mail render post offices and courier services obsolete? Will conferences and trade shows be replaced by webcasts and product demonstrations that you can view from your workstation? Will business travel be displaced by low-cost, high-quality video conferencing? The effect of these technologies on our overall economic system has yet to be determined.

In this chapter, we will consider the architectures and infrastructures of IP-based internetworks, and will also introduce the concepts of protocol analysis that can assist us in managing these complex systems.

2.1 Standardizing Internetwork Architectures

There are two old axioms that have floated around the data communications industry for years. The first says something like "computer users are voracious consumers of memory, and once introduced to more, will never be satisfied with a lesser amount." The second axiom addresses data transmission speeds: "users are forever complaining of poor response times, and will revolt if the transmission speed is ever reduced."

The last decade has produced complex internetworks that consist of multiple LAN and WAN connections that may span the globe across multiple time zones. As a result, those of us who must analyze those internetworks could coin a third axiom:

"Internetworks, by nature, are exceedingly complex. When they work, they are marvels of engineering. When they fail, you are in big trouble."

Keeping out of "big trouble" (we all have our own definition of that term) is certainly a laudable goal. Being able to recover from those inevitable periods of network downtime is an underlying theme of this book. No matter how complex the internetwork problem, a thorough analysis should restore it to health. But analysis is getting ahead of the story. Let's see why IP-based internetworks have become so complicated to begin with.

2.1.1 A Historical Perspective on Connectivity

The more things change, the more they remain the same. At least it seems that way when the connectivity of computing resources is considered. The traditional centralized architecture (circa 1970) consisted of a large, centralized host processor with dumb terminals connected via terminal controllers. Typical I/O devices were card readers and magnetic tape drives. The early releases of IBM's System Network Architecture (SNA) are good examples of centralized host processing.

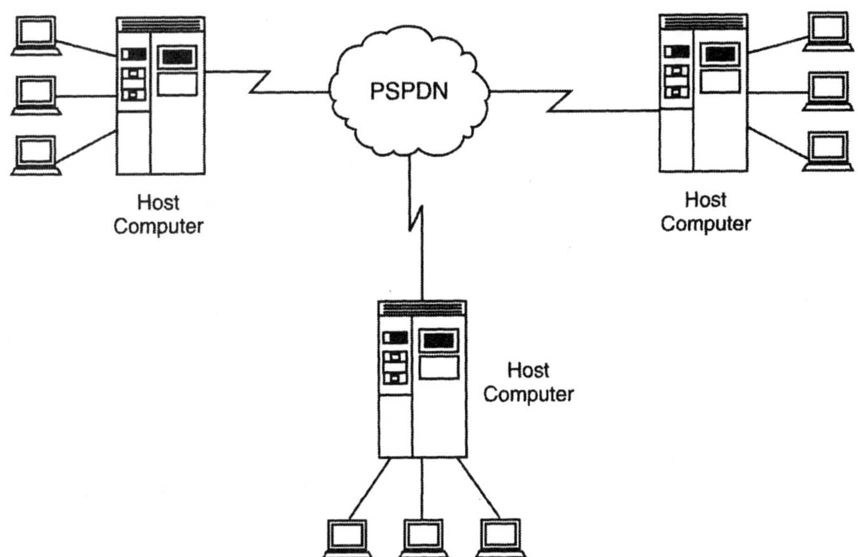

Figure 2-1. Traditional Distributed Processing.

In the late 1970s, distributed host processing, typical of Digital Equipment Corporation's (DEC) DECnet, provided connectivity between geographically dispersed hosts (Figure 2-1). The Wide Area Network (WAN) element was thus introduced into the configuration using either Point-to-Point (PTP) or Packet Switched Public Data Network (PSPDN) connections. These transmission links between hosts were slow by today's standards, and typically consisted of leased lines transmitting at 56 Kbps or less.

The entry of the PC in 1981 provided an opportunity for actual processing to be done on the desktop, instead of miles away at the host. With host-based systems, the speed of user applications was limited by the host's ability to handle input/output (I/O) and information processing requests. If a WAN link was between the user and host, another limiting factor (the speed of that link) was added.Thus, the previous user complaints of slow host response times and not enough I/O ports were replaced by new challenges — inadequate memory and slow processing at the PC.

The user's requirements did not change, however. If the power of a mainframe database was the user's expectation, he or she was probably disappointed with the early database programs available for PCs. As a partial compromise to the requirement for mainframe power with PC simplicity, PCs became emulators of host terminals (such as the IBM 3278 or DEC VT-100), melding local processing with remote host access. Two options for the PC to host were developed: a local connection via coax or EIA-232 interfaces (Figure 2-2a), or a remote connection via modems and the Public Switched Telephone Network, or PSTN (Figure 2-2b). The local connection facilitated access to a host within the same building or campus environment. Remote operation, such as Remote Job Entry (RJE), required a connection via modems. Both configurations brought increased productivity to the desktop. The PC provided local processing and data storage that was not available with proprietary terminals. As long as the user was content to maintain the physical and logical connection to the host information, terminal emulation provided a very cost-effective solution.

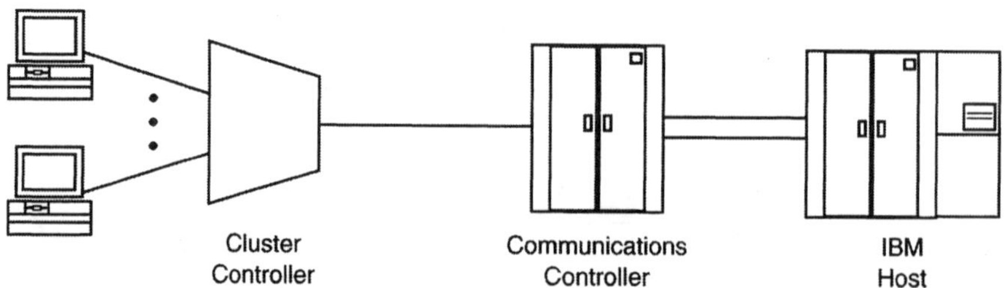

Figure 2-2a. Local PC to Host Connection.

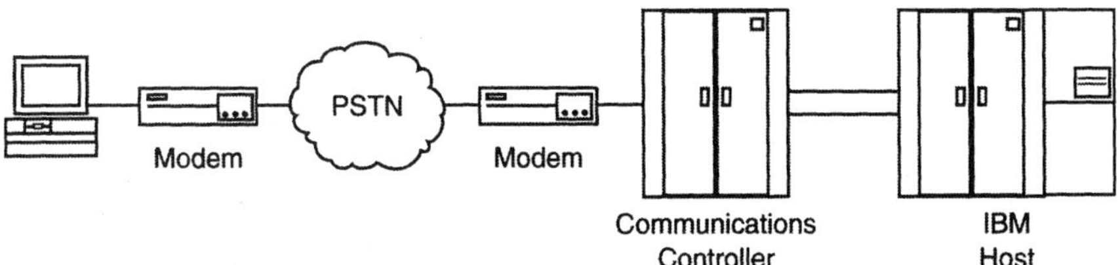

Figure 2-2b. Remote PC to Host Connection.

Each year of the late 1980s was routinely proclaimed as the "Year of the LAN," with printer and file sharing being the initial technical objectives. Access to host applications, such as financial databases, remained a user requirement, however. This requirement produced a merger of technologies, with the LAN as the "glue" connecting the host, peripherals, and workstations together. Two major brands of "glue" emerged. The first, Ethernet, was developed by DEC, Intel, and Xerox beginning in the early 1970s. It originally consisted of a coaxial cable bus topology network (Figure 2-3), but evolved into a number of options for both transmission medium (twisted pairs, coax, or fiber) and architecture (bus or star). In 1980, the Ethernet standard was turned over to the Institute of Electrical and Electronics Engineers (IEEE), which enhanced the standard and produced a successor technology, designated IEEE 802.3. The IEEE 802.3 standard is similar, but not identical, to the original Ethernet, as Section 2.4.1 will discuss. One of the advantages of IEEE 802.3 is that it has evolved from the original 10 Mbps transmission rate into the 1–10 Gbps rates as application requirements have dictated. In addition, this technology has migrated from the venue of the local area networks (LANs) into metropolitan area networks (MANs), thus solving some critical interoperability challenges over broader geographic areas.

The second type, token ring/IEEE 802.5 (Figure 2-4), was heavily influenced by IBM. This topology supported a variety of transmission media (shielded and unshielded twisted pairs, and fiber optic cables) in a distributed star architecture. Token ring networks were standardized at two transmission rates, 4 and 16 Mbps, with some experimental development at higher rates. For a variety of technical and economic reasons, however, Ethernet/IEEE 802.3 networks have achieved a considerably higher market share than token ring/IEEE 802.5 networks, relegating the latter to the "legacy network" designation for most applications.

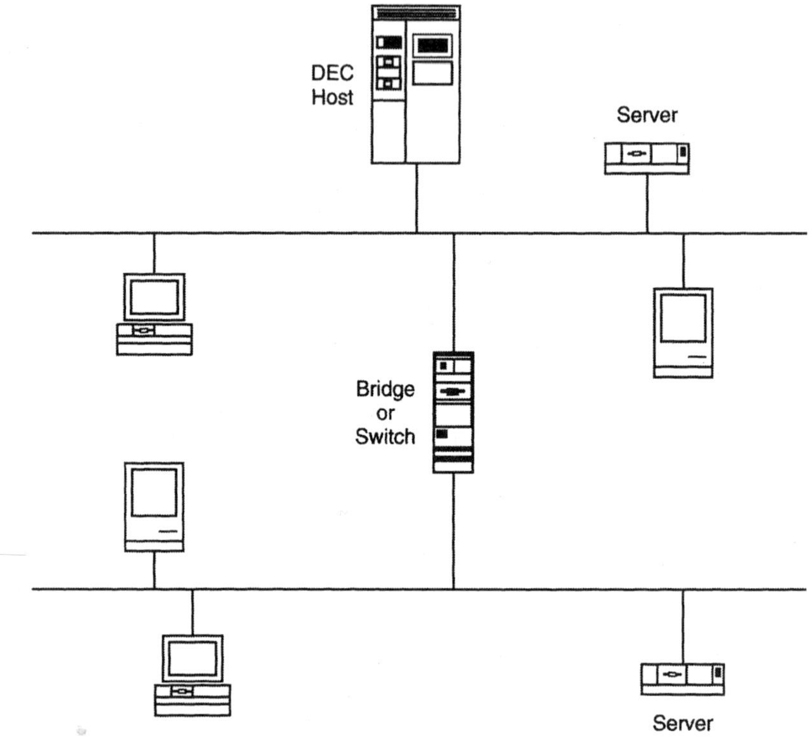

Figure 2-3. Connectivity with Ethernet / IEEE 802.3

The type of data processing architecture, centralized or distributed, has an impact on the design and support of the internetwork in a dramatic way. Centralized systems become host-centric, with access required to a small number of data sources. This architecture requires the connectivity design to be driven by the gateway required to access that host. Thus, the protocols, transmission speed, physical cabling, etc., of the gateway constrain the host-centric internetwork.

2.1.2 Open Systems Interconnection Principles

It is one issue for two workstations to exchange a sequence of bits, and an entirely different issue for those two workstations to exchange meaningful data. The difference is in the level of understanding as defined by the protocols (or rules of communication) employed.

In the 1970s, most network architectures were de facto standards from a specific vendor, instead of open standards designed by committees of international experts. Thus, IBM's software would not interoperate with DEC's (which in turn could not communicate directly with either Hewlett-Packard or Wang) unless a gateway having significant complexity was employed. The gateway provided protocol translations at all layers of the proprietary architecture. With a gateway in between, an SNA host connected to a token ring network could exchange an electronic mail message with a DECnet host connected to an Ethernet.

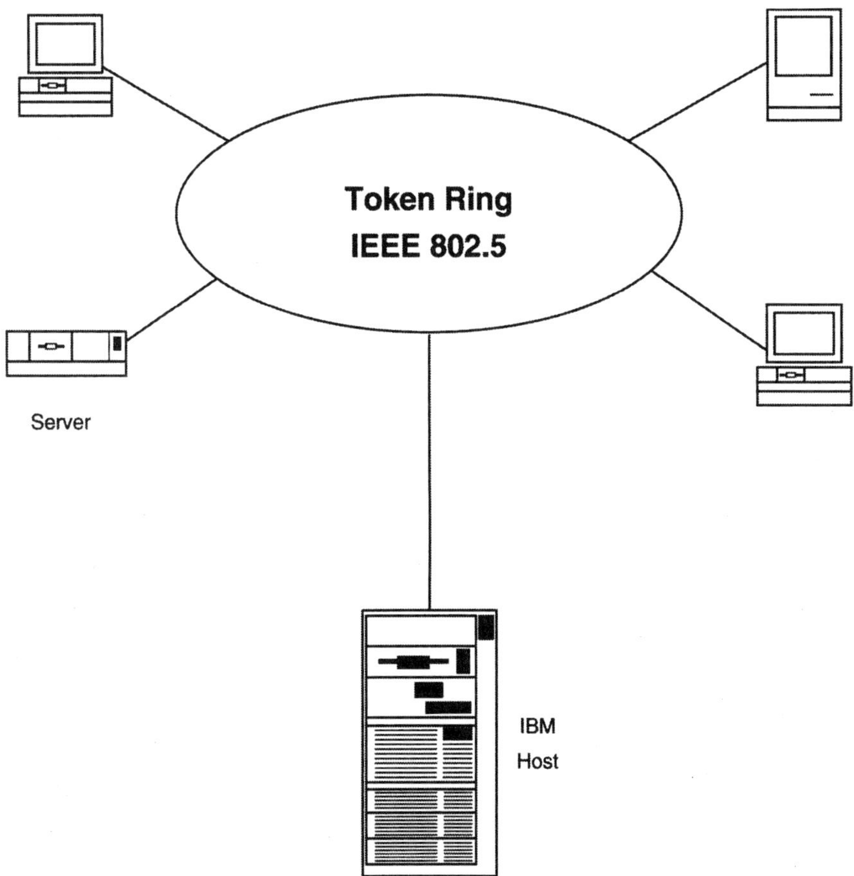

Figure 2-4. Connectivity with Token Ring / IEEE 802.5.

One objective driving the open systems concept was protocol interoperability between dissimilar vendors' systems. In other words, if all vendors adhered to a common architecture, communication problems between those systems would (in theory) go away. The basis for that theory was the concept of Open Systems Interconnection (OSI) and the seven-layer

OSI Reference Model that was developed by the International Organization for Standardization (ISO) [2-1].

The OSI Reference Model was designed as a standard to allow various "open" systems to communicate. A system that complies with standards (specifically OSI standards) — also referred to as protocols — for communication with other systems is defined as being open. An open system is standards-based instead of proprietary-based; my system can thus communicate and cooperate with your system using interfaces and protocols that both systems understand. In creating the OSI model, the ISO divided the network communication functions into seven layers. In general terms, the lower five layers provide connectivity functions, while the upper two layers provide interoperability functions (Figure 2-5).

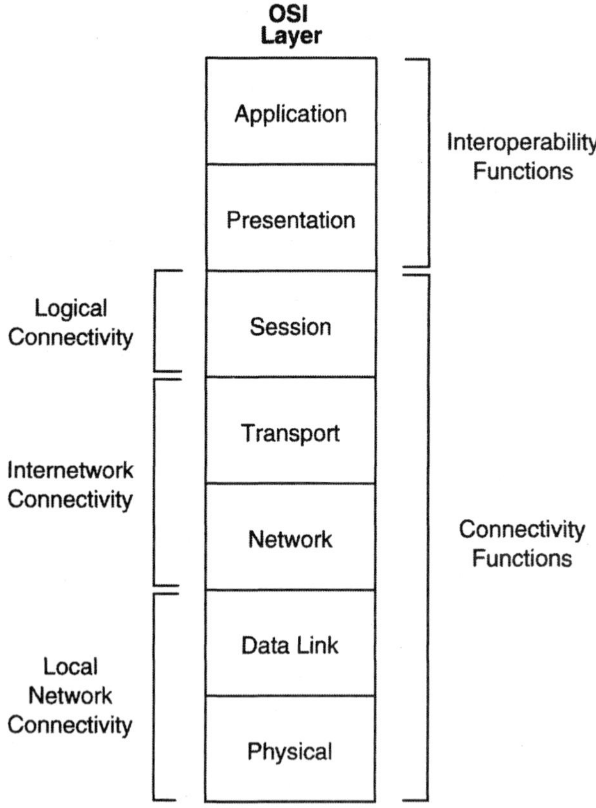

Figure 2-5. Internetworking and Interoperability within the OSI Framework.

The *Physical Layer* handles bit-transmission between one node (e.g., host, workstation) and the next. The Physical Layer functions include interfacing with the transmission media; encoding the data signal; defining the range of the voltage or current magnitudes; defining the connector sizes, shapes, and pinouts; and anything generally associated with the physical transmission of the bit stream.

The *Data Link Layer* maintains a reliable communication link between adjacent nodes. As such, it assumes that the Physical Layer is noisy or prone to errors. Data Link provides a reliable delivery mechanism to transmit a frame (or package) of data bits to the next node. The Data Link Layer inserts addresses in the data frame (including source and destination) and provides error control for the data — usually implemented with a Cyclic Redundancy Check (CRC).

The *Network Layer* establishes a path for the traveling data packet along the communication subnet from the source node to the destination node. The Network Layer switches, routes, and controls the congestion of these information packets within the subnet.

The *Transport Layer* provides reliable delivery of host messages originating at Layer 7, the Application Layer, in the same way that the Data Link Layer assures reliable delivery of frames between adjacent nodes. The major difference between the Data Link and Transport Layers is that the Data Link domain lies between adjacent nodes, whereas the Transport Layer's domain extends from the source to the destination (or end-to-end) within the communication subnet. Issues concerning source-to-destination messages are important in the Transport Layer. For example, the Transport Layer segments a long message into smaller units (packets) prior to transmission, and assures the reassembly of those packets into the original message at the receiver's end.

The *Session Layer* establishes and terminates process-to-process communication sessions between hosts. Translation between name and address databases, as well as synchronization between the two hosts, may be required to manage the sessions.

The *Presentation Layer* establishes the syntax (or form) in which data is exchanged between the two hosts. As such, the Presentation Layer provides a data manipulation function, not a communication function. Data compression and data encryption are two examples of Presentation Layer services.

The *Application Layer* provides end-user services, such as Application Layer file transfers, electronic messages, virtual terminal emulation, remote database access, and network management. The end user interacts with the Application Layer.

The seven layers divide into two important subsets (Figure 2-6). The first is comprised of the lower three layers (the Physical, Data Link, and Network Layers) and is termed the *communications subnetwork*, subnet, or the carrier portion of the system. The upper three layers (the Session, Presentation, and Application Layers) are collectively known as the *host process*, sometimes called the customer portion of the system. The middle layer (Transport) is the first end-to-end layer, and acts as a buffer between the two subsets. As such, the Transport Layer is often grouped with the upper layers as part of the host process.

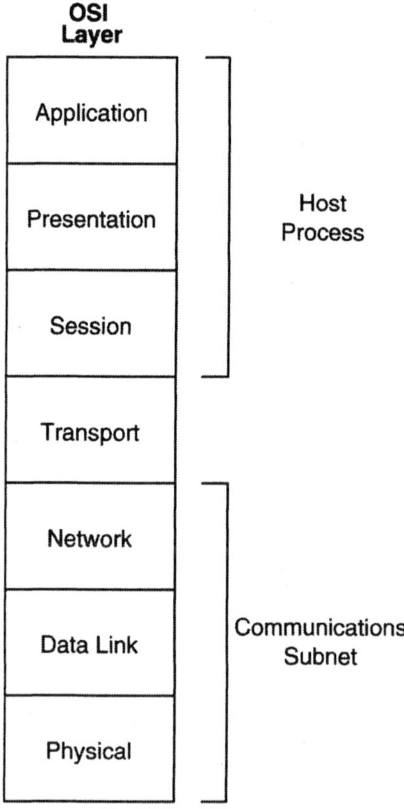

Figure 2-6. Communications and Host Functions within the OSI Framework.

A specific internetwork architecture results when two open systems are linked directly with a bi-directional communication channel such as a cable. The interface between the layers within the same system is a vertical relationship, whereas the protocol is a horizontal relationship between peer layers of the adjacent systems. The actual communication path originates in Open System A as an input to its Application Layer. The message then proceeds down through the seven layers (7 through 1) of System A, across the physical media (cable), and up the seven layers (1 through 7) of System B. Details of this process are shown in Figure 2-7, taken from Reference [2-2]. Data from Application process X is passed to the Application Layer protocol, which adds its Application Header (AH). The header contains protocol control information (PCI) necessary for the peer process (Y) to interpret the data. The AH plus Application Data (AD) is then passed down to the Presentation Layer. The Presentation Layer treats the AH and AD as its own data, appends the Presentation Header (PH), and passes the data unit down to the Session, Transport, and Network Layers in turn. Note that the accumulated headers and data from one layer are treated as data when it gets to the next lowest layer.

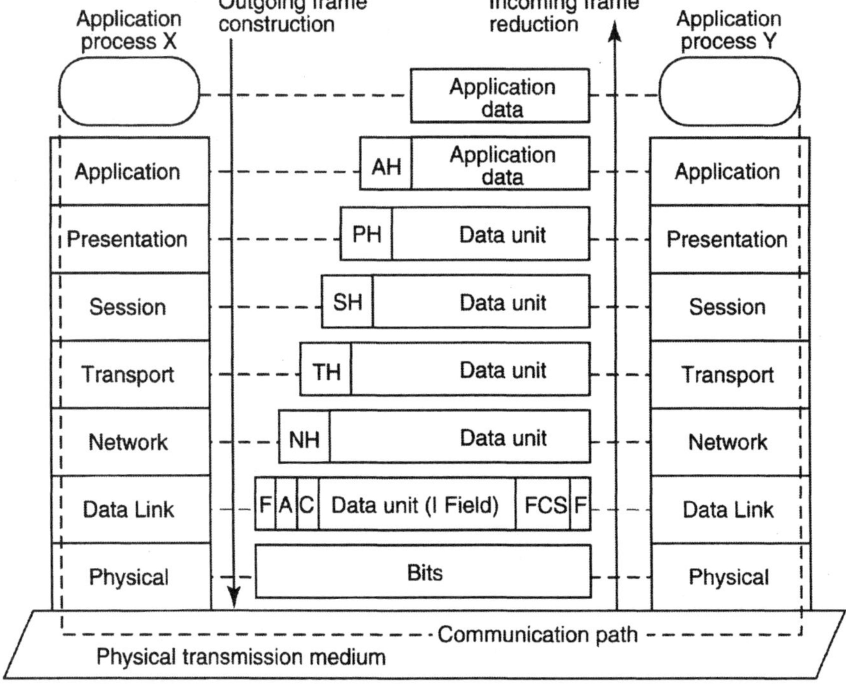

Figure 2-7. Building a Frame for Transmission (© 1986 IEEE).

When the encapsulated message reaches the Data Link Layer, Framing (F), Address (A), and Control (C) information is added as the Data Link Layer header. The Frame Check Sequence (FCS) and possibly additional Framing (F) characters are appended as the Data Link Layer trailer. The assembled frame is then passed to the Physical Layer. The Physical Layer encodes the data for transmission, accesses the transmission medium, and monitors the serial bit transmission. At the destination node, the reverse of this process occurs. The Physical Layer hands its bits to the Data Link Layer, which decodes and then strips off the Data Link Layer header and trailer. The Data Link Layer data unit is then passed to the Network, Transport, Session, and higher layers in turn. The process is completed when the electronic message (shown again as Application data) is delivered to Application Y.

Unfortunately, the economics of market share and stock price enter into the technical arena. These perturb the theoretical OSI model with proprietary hooks designed to gain, regain, or maintain market presence. The existence of these multiple architectures results in internetworks that require additional devices (such as gateways) to assure the internetworking and interoperability that users require for reliable communication.

2.1.3 Applying Connectivity Devices to the OSI Model

Let's explore how connectivity devices work with regard to OSI protocols. We'll assume that two open systems are connected with a physical transmission medium such as a twisted pair or fiber optic cable. What happens if the cable length is so long that the signal loses power? We can solve this by adding a repeater (Figure 2-8) that will amplify (or regenerate) the physical signal. Repeaters function at the Physical Layer and operate between like networks, such as token ring to token ring, or Ethernet to Ethernet. A repeater can be added to the internetwork to extend the range of the network; connected segments behave physically (and logically) as a single network. As technology has evolved, however, more intelligent connectivity devices such as switches and routers have been deployed, rendering the simple repeater to the legacy device category.

As more users are added to a LAN, the traffic on that LAN will correspondingly increase. When this occurs, it is advantageous to separate the one large network into smaller segments for traffic management purposes. The device for that function is called a bridge, or Layer 2 switch, which adds the functionality of the Data Link Layer (Figure 2-9). The bridge or switch logically separates two network segments by operating on the ad-

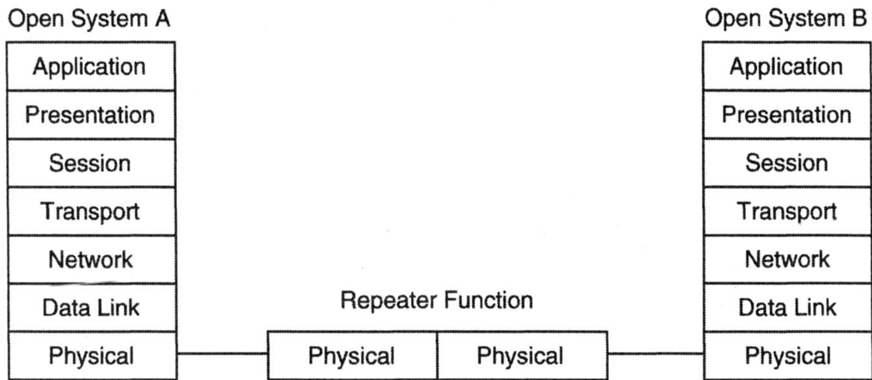

Figure 2-8. Comparing a Repeater to OSI.

dress within the Data Link Layer (or IEEE Medium Access Control [MAC] Layer) frame. Information that is either stored at the bridge or provided within the transmitted frame assists the bridge or switch in making a rather simple decision: pass the frame to the next segment (known as forwarding) or do not pass the frame to the next segment (known as filtering). Bridges or Layer 2 switches operate on networks having compatible Data Link Layer addressing schemes (such as IEEE 802.3 to 802.3, or 802.3 to 802.5), but are transparent to the protocols of the Network and higher layers. Layer 2 switches have very attractive price points, and have therefore displaced bridges for almost all applications that require the logical segmentation of a high traffic network.

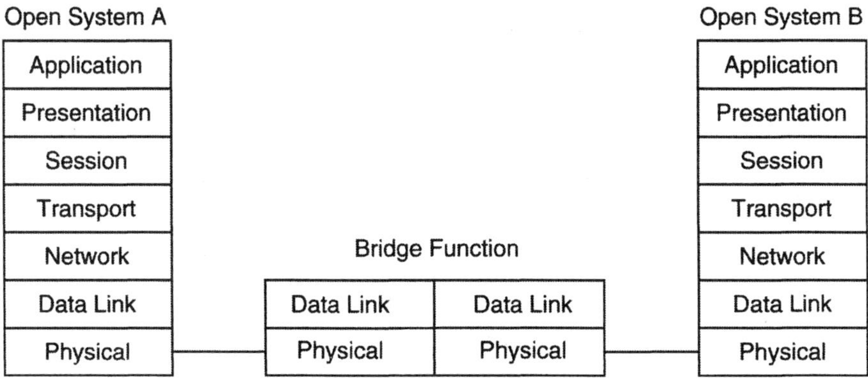

Figure 2-9. Comparing a Bridge to OSI.

Routers operate at the Network Layer, and may interpret either one or more protocols at that layer (Figure 2-10). Recall that the Network Layer makes a choice between available paths within the communication subnet, eventually connecting the source and destination hosts. A router performs similarly, reading information about the destination network address and forwarding that packet to the appropriate destination network. (Bridges, as discussed above, make a simple binary decision to forward or not to forward a frame after examining the Data Link Layer address.) The router thus serves a network-wide connectivity function. Routers may operate on one Network Layer protocol, such as the Internet Protocol (IP), or multiple protocols such as IP, DECnet, and Novell's IPX (Internetwork Packet Exchange).

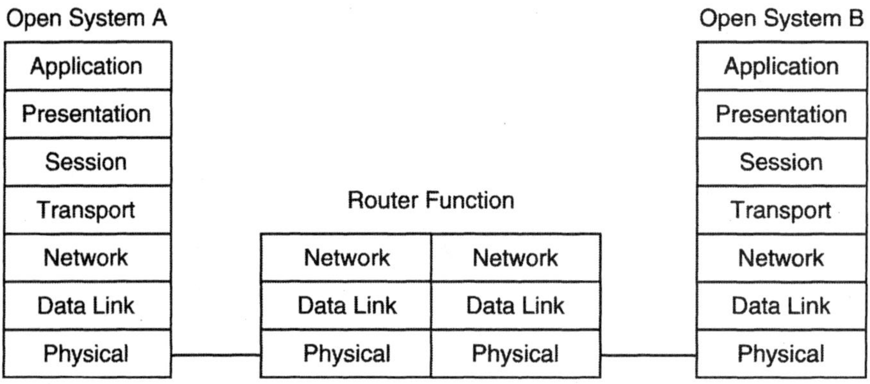

Figure 2-10. Comparing a Router to OSI.

Finally, gateways may operate at all seven OSI layers (Figure 2-11). Gateways are application-oriented, and may be responsible for connecting incompatible electronic mail systems, converting and transferring files from one system to another, or enabling interoperability between dissimilar operating systems.

2.1.4 The Advanced Research Projects Agency Network Architecture

Today's Internet was born in 1969 as the Advanced Research Projects Agency Network (ARPANET) and was sponsored by the U.S. Defense Advanced Research Projects Agency (DARPA), now known as ARPA. The purpose of the ARPANET was to test and determine the viability of a communication technology known as *packet switching*. The contract to build the original

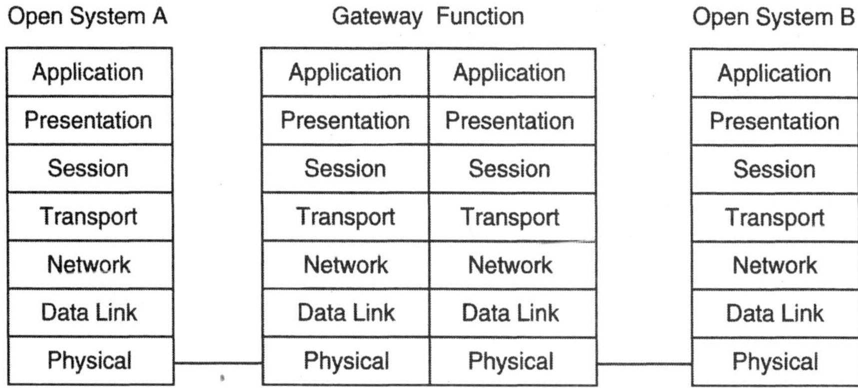

Figure 2-11. Comparing a Gateway to OSI.

ARPANET was awarded to a firm known as Bolt, Baranek, and Newman (now known as BBN Technologies, Cambridge, Massachusetts). ARPANET went online in September 1969 at four locations: Stanford Research Institute (SRI), the University of California at Santa Barbara (UCSB), the University of California at Los Angeles (UCLA), and the University of Utah. The original hosts were Honeywell minicomputers, known as Interface Message Processors (IMPs).

The initial test was successful, and the ARPANET grew quickly. At the same time, it became apparent that nonmilitary researchers could also benefit from access to a network of this type, so leaders in the university and industrial research communities made proposals to the National Science Foundation (NSF) for a cooperative network of computer science researchers. The NSF approved funding for the Computer Science Network (CSNET) in 1981.

In 1984, the ARPANET was split into two different networks: MILNET (for unclassified military traffic) and ARPANET (for nonmilitary traffic and research). In 1984, the NSF established the Office of Advanced Scientific Computing (OASC) to further the development of supercomputers and to make access to them more widely available. The OASC developed the NSFNET to connect six supercomputing centers across the United States, using T-1 lines operating at 1.544 Mbps in 1987, and subsequently upgrading to the T-3 rate (44.736 Mbps) in 1990. NSFNET, with its higher transmission rates, was a resounding success; as a result, the U.S. Department of Defense declared the ARPANET obsolete and dismantled it in June 1990.

For its first decade, the ARPANET grew quickly, adding an average of one new host computer every 20 days. The original protocol for internal network communications was known as the Network Control Program (NCP). When compared with the seven-layer OSI Model, the NCP provided the functions of the third and fourth layers (Network and Transport), managing the flow of messages between host computers and intermediate packet switches. NCP was designed with the assumption that the underlying communication subnetwork (i.e., OSI Physical, Data Link, and Network Layers) provided a perfect communication channel. Given ARPANET's mission to support government and military networks, which could include radio links under battlefield conditions, the assumption of a reliable communication channel needed reconsideration. In January 1973, ARPA made the Transmission Control Protocol (TCP) a standard for the Internet because of its proven performance. The ARPA internetwork architecture (Figure 2-12) consisted of networks connected by gateways [2-3]. (Note that in the OSI sense of the word, these devices were actually routers, operating at the OSI Network Layer.) The ARPA model assumed that each network used packet switching technology and could connect to a variety of transmission media (LAN, WAN, radio, and so on).

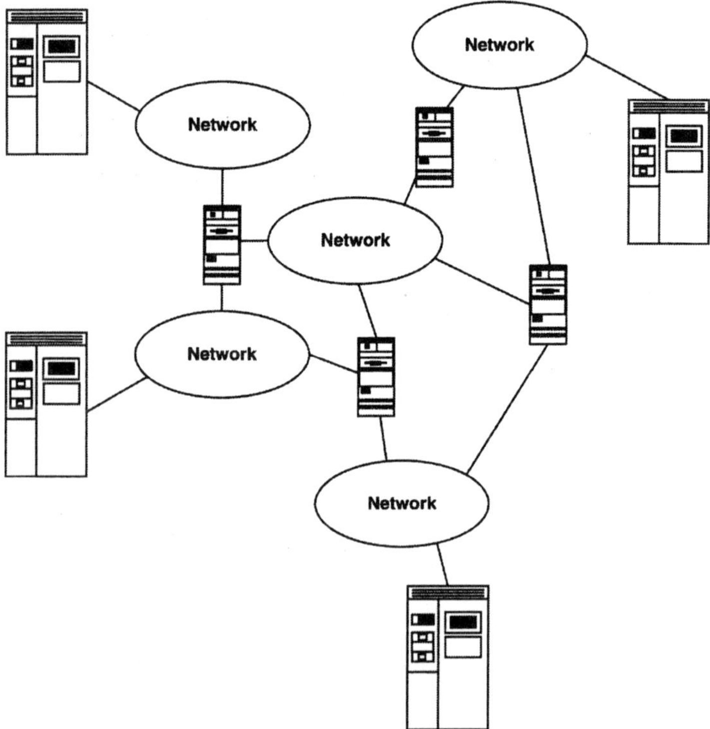

Figure 2-12. Networks Connected with Gateways to Form an Internetwork.

The ARPA Internet architecture consisted of four layers (Figure 2-13). The lowest layer was called the *Network Interface Layer* (it was also referred to as the Local Network or Network Access Layer) and comprised the physical link (e.g., LAN) between devices. The Network Interface Layer existed in all devices, including hosts and gateways.

OSI Layer	ARPA Architecture
Application	Process / Application Layer
Presentation	
Session	
Transport	Host-to-Host Layer
Network	Internet Layer
Data Link	Network Interface or Local Network Layer
Physical	

Figure 2-13. Comparing OSI and ARPA Models.

The *Internet Layer* insulated the hosts from network-specific details, such as addressing. The Internet Protocol (IP) was developed to provide end-to-end datagram service for this layer. (Datagram service is analogous to a telegram in which the information is sent as a package.) The Internet Layer (and, therefore, IP) existed only in hosts and gateways.

While the Internet Layer provided end-to-end delivery of datagrams, it did not guarantee their delivery. Therefore, a third layer, known as the *Host-to-Host Layer* (formerly called the Service Layer), was provided within the hosts. As its name implies, the Host-to-Host Layer defined the level of service and support that the host applications required. Two protocols were created for the Host-to-Host Layer: the Transmission Control Protocol (TCP) for applications needing reliable end-to-end service, and the User Datagram Protocol (UDP) for applications with less stringent reliability requirements. A third protocol, the Internet Control Message Protocol (ICMP), allowed hosts and gateways to exchange monitoring and control information.

The highest ARPA layer, the *Process/Application Layer*, resided only in hosts and supported user-to-host and host-to-host processing or applications. A variety of standard applications were developed. These included the Telecommunications Network (TELNET) for remote terminal access, the File Transfer Protocol (FTP) for file transfer, the Simple Mail Transfer Protocol (SMTP) for electronic mail, and many others.

Figure 2-13 also compares the OSI and ARPA architectures. Note that the OSI Physical and Data Link Layers represent the ARPA Network Interface (or Local Network) Layer; the OSI Network Layer corresponds to the Internet Layer; the OSI Transport Layer is functionally equivalent to the Host-to-Host (Service) Layer; and the OSI Session, Presentation, and Application Layers comprise the ARPA Process/Application Layer. References [2-4] and [2-5] describe the development of the ARPANET Reference Model and protocols.

2.2 Connectionless vs. Connection-Oriented Network Architectures

In general, communications networks can provide one of two different types of network services: connectionless (CON) network service or connection-oriented (CO) network service. More specifically, an individual layer within that network architecture can also be defined by the service that it provides to the adjacent layer above: either CO or CON service. Taken as a whole, the communication architecture may incorporate some combination of these services in support of a particular application.

The typical model for a connection-oriented network service is the Public Switched Telephone Network (PSTN). When the end user takes the telephone off-hook, they notify the network that service is requested. The network then returns dial tone, and the end user dials the destination number. When the destination party answers, the end-to-end connection is confirmed through the various switching offices along that path. A conversation (or more typically a voice mail message) can then ensue, and when the conversation is completed, the parties say goodbye and hang up. The network then disconnects the call, terminates the billing process, and makes the network resources available for another conversation.

For the duration of the connection (or conversation), the communication path may be modeled like a very long pipeline: information is inserted into one end of the pipeline and taken out at the other end of the pipeline (Figure 2-14a). During the time that the connection is active, certain statements can be made about that pipeline. For example, the data will arrive in the order in which it was sent (sequentiality). The data will follow the same path (either a physical or logical path, depending on the network architecture in use) from the source to the destination. Finally, the relative delay of the data will be constant along that path. That is, if there is a 20 millisecond delay on the first word that is transmitted through the telephone network, there should be the same relative delay for the second word, the third word, and so on. Since these characteristics, such as sequentiality, delay, and so on, positively impact the quality and reliability of the transmission, a connection-oriented network is often referred to as a reliable network. The Transmission Control Protocol (TCP) is an example of a connection-oriented protocol.

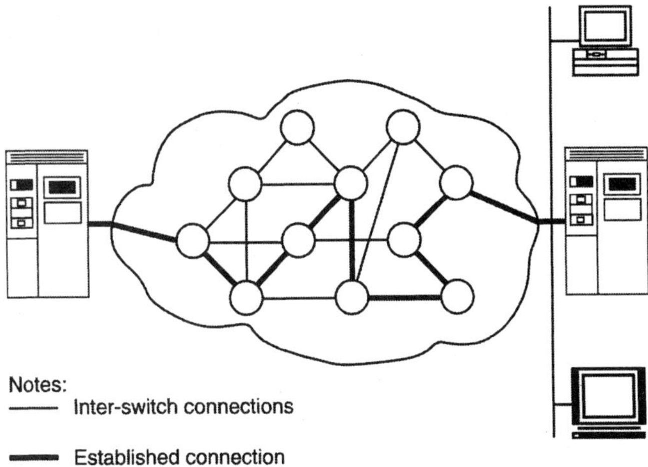

Figure 2-14a. Connection-oriented Network.

In contrast, a connectionless network could be modeled by the postal system. A full source and destination address is attached to the packet (or envelope), and then that information is dropped into the network (or post office box). Each of these packets is routed independently through the network, and through the miracles of packet delivery protocols, delivery to the ultimate destination occurs — we hope — as shown in Figure 2-14b. But, like the postal system, connectionless network delivery is on a "best efforts" basis, which means that they will do all they can to get your information through the network, but there are no guarantees. In other words, there are no guarantees of packet sequentiality, delay, or, for that matter, delivery at all. (At least, no guarantees for the price of a first class stamp. If you want more reliable package delivery, you can call FedEx, but it will cost you considerably more. Similarly, if a connectionless network does not meet your needs, you can always go with a CO system, but the overhead will invariably be higher.) The Internet Protocol (IP) and the User Datagram Protocol (UDP) are examples of connectionless protocols.

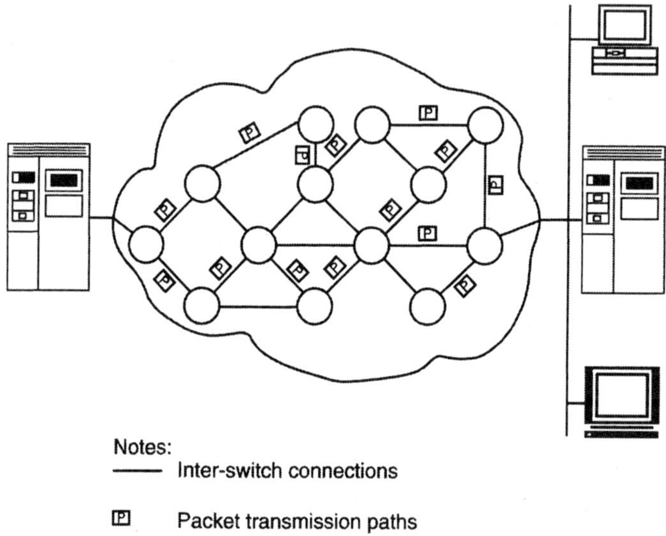

Figure 2-14b. Connectionless Network.

2.3 Internetwork Analysis

As the internetworks increase in their complexity, the network management functions must evolve as well. The migration from host-centric computing to distributed processing that began in the 1970s places a requirement on the individual locations to have their own network management staff. This local staff may handle user-related problems, but refer internetwork failures to a headquarters location that is better equipped to analyze the problem in depth.

The need for high bandwidth channels in support of evolving applications such as video conferencing reduces the economic life of network analysis and management equipment. As new technologies such as fiber optic LANs and high-speed MANs become more popular, internetwork managers must update the diagnostic instruments that they depend on. A decade ago, the typical end-to-end WAN link included analog circuit components in the local loop. If a performance problem or failure was suspected on that link, it would be tested for compliance with analog transmission parameters, such as impulse noise, envelope delay distortion, or phase jitter. Digital circuits that constitute the high-bandwidth pipes require an entirely different analysis approach. Completely different transmission parameters must be analyzed. For example, T1/T3 circuits must be tested for proper data framing; improperly generated signals, such as bipolar violations (BPVs); and special data encoding, such as B8ZS (bipolar with eight zero substitution). If that transmission medium is optical fibers, instead of twisted pairs or coaxial cable, the analyzer must have an optical, instead of electrical, interface.

Multi-protocol internetworks require a greater degree of intelligence within the analyzer itself. Analyzers that were sufficient for one protocol suite, such as IBM's System Network Architecture (SNA), may now be expected to handle several protocol suites, such as SNA, UNIX, and TCP/IP. In addition, support for specific applications, such as database transactions, may also be required.

The protocol analyzer attaches to the internetwork in a passive mode and captures the information transmitted between the various devices. This information can be divided into two categories: data and control.

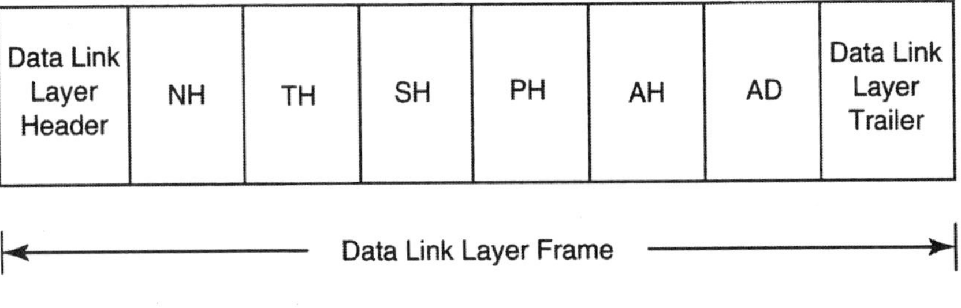

Notes:
NH: Network Layer Header
TH: Transport Layer Header
SH: Session Layer Header
PH: Presentation Layer Header
AH: Application Layer Header
AD: Application Layer Data

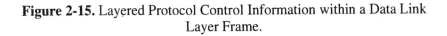

Figure 2-15. Layered Protocol Control Information within a Data Link Layer Frame.

The data is information from the end-user process, such as an electronic mail message. The control information assures that the data transfer obeys the rules of the various protocols that are used. For that reason, this control information is often called protocol control information (PCI). Protocol control information is unique to each protocol that is used within that particular architecture. Thus, if seven different protocols were associated with a particular internetwork function, there would be seven unique PCI elements. Recall from our earlier discussion of open systems architectures that the frame is constructed using a number of protocol elements, or headers (review Figure 2-7). These headers are transmitted along with the data inside the Data Link Layer frame (Figure 2-15). The information to be transmitted originates at the Application Layer, and is therefore called Application Data (AD). The Application Header (AH) is then appended to the data, and that combination (AH + AD) is passed to the next lowest layer (e.g. the Presentation Layer). The Presentation Layer treats this information (AH + AD) as its data, and appends its header (the Presentation Header or PH). The result is PH + AH + AD. This process continues until the entire frame is constructed with the appropriate PCI from each layer.

The protocol analyzer's job can be defined simply: unravel all of the PCI and display it in a user-understandable format. As we will see from the case studies in this text, this process of unraveling is much more complex than it may initially appear, as both LAN and WAN elements are typically involved (Figure 2-16).

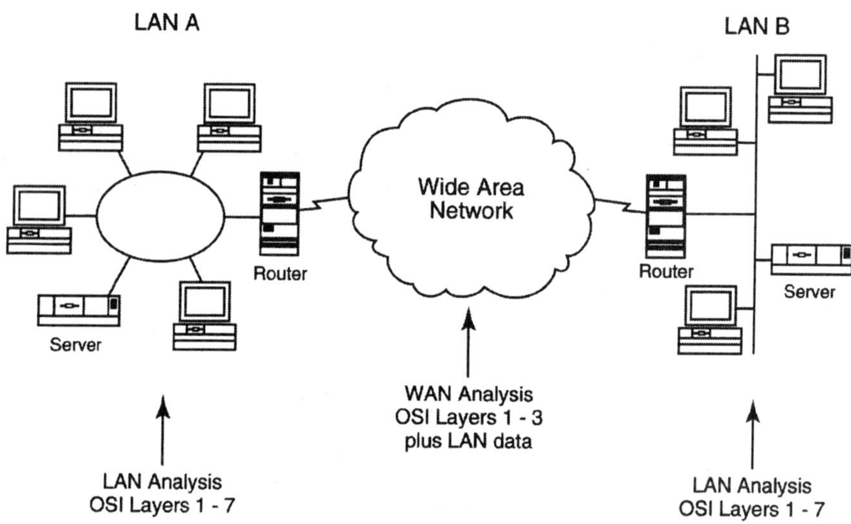

Figure 2-16. LAN and WAN Analysis.

2.4 Analyzing the ARPA Architecture

We will begin our study of TCP/IP analysis by examining problems that can occur at the ARPA architectural model's Network Interface Layer, which makes the connection to local, metropolitan-area, or wide-area networks. The popularity of the Internet protocols has produced RFC (Request for Comments) documents detailing implementations over virtually every type of LAN, MAN, and WAN, including ARCNET, Ethernet, IEEE 802, FDDI, Packet Switched Public Data Networks (PSPDNs) using the ITU-T X.25 protocol, Integrated Services Digital Network (ISDN) connections, frame relay, Asynchronous Transfer Mode (ATM), and many others.

To connect to LANs, the Network Interface Layer must exist in all hosts and routers, although its implementation may change across the internetwork (see Figure 2-17a). Thus, Host A must have a consistent attachment to Router B, but the destination Host Z may be of a different type. In other words, you can start with an Ethernet, traverse a frame relay network, and end with a token ring as long as you maintain consistencies between each transmitter/receiver pair.

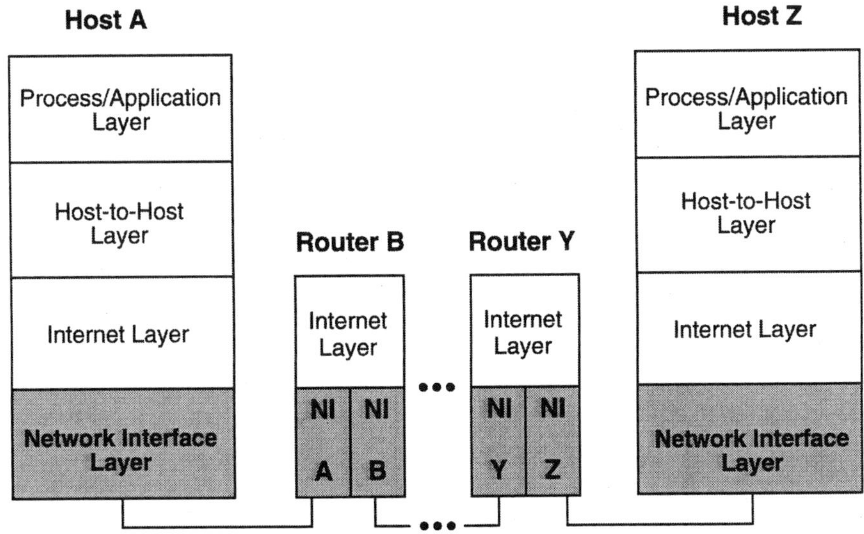

Figure 2-17a. The Network Interface Connection.

Figure 2-17b shows the options for the Network Interface Layer and their supporting RFCs. Recall that the higher layer information (e.g., IP, TCP, and so on) is treated as data inside the transmitted frame (see Figure 2-17c). The headers and trailers are defined by the particular LAN or WAN in use, for example, Ethernet/IEEE 802.3, frame relay, token ring, and so on. Immediately following the Local Network header is the IP header, then the TCP header, and, finally, the higher layer application data, which might be FTP, TELNET, and so forth.

ARPA Layer	Protocol Implementation								OSI Layer
Process / Application	Hypertext Transfer	File Transfer	Electronic Mail	Terminal Emulation	Domain Names	File Transfer	Client / Server	Network Management	Application
	Hypertext Transfer Protocol (HTTP) RFC 2616	File Transfer Protocol (FTP) MIL-STD-1780 RFC 959	Simple Mail Transfer Protocol (SMTP) MIL-STD-1781 RFC 2821	TELNET Protocol MIL-STD-1782 RFC 854	Domain Name System (DNS) RFC 1034, 1035	Trivial File Transfer Protocol (TFTP) RFC 1350	Sun Microsystems Network File System Protocols (NFS) RFC 3530	Simple Network Management Protocol (SNMP) v1: RFC 1157 v2: RFC 1901-10 v3: RFC 3411-18	Presentation Session
Host-to-Host	Transmission Control Protocol (TCP) MIL-STD-1778 RFC 793					User Datagram Protocol (UDP) RFC 768			Transport
Internet	Address Resolution ARP RFC 826 RARP RFC 903			Internet Protocol (IP) MIL-STD-1777 RFC 791			Internet Control Message Protocol (ICMP) RFC 792		Network
Network Interface	Network Interface Cards: Ethernet, Token Ring, MAN and WAN RFC 894, RFC 1042 and others								Data Link
	Transmission Media: Twisted Pair, Coax, Fiber Optics, Wireless Media, etc.								Physical

Figure 2-17b. ARPA Network Interface Layer Protocols.

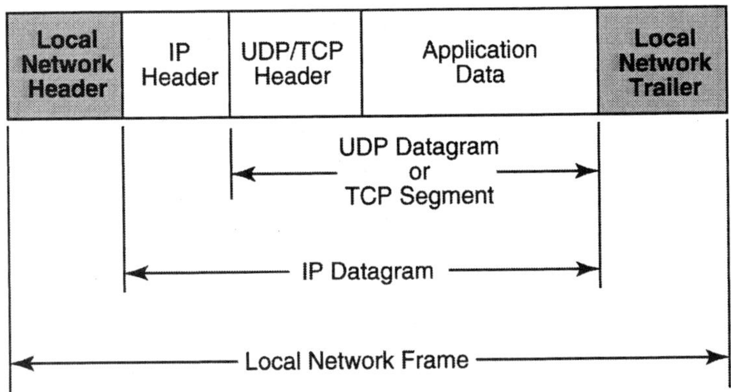

Figure 2-17c. The Internet Transmission Frame.

In summary, the TCP segment comprises the TCP header plus the application data. The IP process treats the TCP segment as data, adds the IP header, and produces the IP datagram. The Local Network process adds the frame header and trailer and transmits that frame on the physical network, such as the Ethernet cable or the X.25 link.

The case studies in the following section will illustrate these protocol interactions and include output from a protocol analyzer for additional details. One case study will consider a LAN issue (Ethernet/IEEE 802.3) and the other will discuss a WAN example (frame relay). Note that, for either case, the protocol analyzer attaches to the network like any other workstation, and captures and then decodes frames of information that were transmitted on the LAN or WAN. When the frames are captured, they are stored in sequential order and then displayed in a number of user-selectable formats. The network manager has the option to select the amount of detail that is included in the decode — showing all layers of the protocol suite, or just focusing in on one layer. The most general display format is called the *summary decode*, which provides information on a frame-by-frame basis. The next level is called the *detail decode*, which shows information on a field-by-field basis within a single frame. The third level is called the *hexadecimal decode*, which displays information on a bit-by-bit basis as it comes across the network. The hexadecimal display will frequently include an ASCII or EBCDIC decode of that information as well. We will use all three of these decodes from time to time in the case studies presented in this text.

A good reference about support for the Network Interface Layer is RFC 1122, "Requirements for Internet Hosts: Communication Layers" [2-6].

2.4.1 Incompatibilities Between Ethernet and IEEE 802.3 Frames

Technical standards ensure that all parties involved with a project or procedure can communicate accurately. This "communication" could be a nut communicating with a bolt (adhering to the same diameter and number of threads per inch) or a terminal communicating with a host computer (adhering to the same character set, such as ASCII). Unfortunately, the Ethernet world has two separate standards that are both loosely termed "Ethernet." The original Ethernet, last published in 1982 by DEC, Intel, and Xerox, is called the Blue Book. The second standard, IEEE 802.3, accommodates elements of the other IEEE LAN standards, such as the IEEE 802.2 Logical Link Control header.

In this case study, a user tries to access some higher layer TCP/IP functions, TCPCON, but can't because the lower layer connection fails due to the confusion of the two LANs on the internet. Let's see what happened.

The internetwork topology consists of several Ethernet segments that connect a number of devices. The NetWare server doubles as an IP router and connects to both local and remote hosts (see Figure 2-18). The network administrator (David, shown in Trace 2.4.1a) wishes to access a network management application to check some SNMP statistics at the remote host. He must first log into his NetWare server. Looking for the nearest file server, he broadcasts a NetWare Service Advertisement Protocol (SAP) packet in Frame 1, then repeats the request every 0.6 seconds. But he receives no response.

Trace 2.4.1b shows the details of the SAP Nearest Service Query and indicates that David's workstation (address H-P 06CA73) was transmitting an Ethernet frame. We know this because the frame header contains an EtherType (8137H) instead of a Length field. Unfortunately, no responses to David's query are returned by the servers that should be operational on the network. A clue is contained in Frame 17, however, which is a keep alive poll transmission from one of the servers. This transmission confirms that at least one server is active and should be responding to David's "Find Nearest File Server" transmissions.

With this clue, David realizes that the server was configured for IEEE 802.3 framing and speculates that the problem might be a frame incompatibility. He reconfigures the workstation by editing the configuration file to specify the IEEE 802.3 frame type.

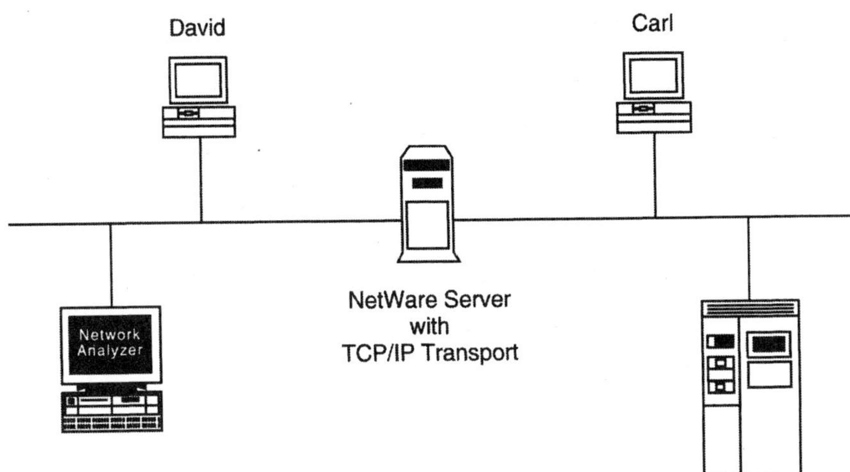

Figure 2-18. Ethernet/IEEE 802.3 Network with Novell NetWare and TCP/IP.

A second attempt (Trace 2.4.1c) proves successful. David's workstation requests the nearest server and receives a response from five servers: NW Svr 2, NW Svr 3, H-P 133A5B, NW Svr 1, and H-P 136A06 (Frames 2 through 6). The NetWare Core Protocol (NCP) algorithm then selects the first responding server (NW Svr 2, shown in Frame 2) and creates a connection to that server in Frame 9. The connection is confirmed and a buffer size accepted in Frames 10 through 12. David is now logged into the server and can finish gathering the SNMP statistics. Trace 2.4.1d examines the NCP Nearest Service Query packet after the workstation reconfiguration. Note that the EtherType field has been replaced with the 802.3 length = 34 octets. All other aspects of the frame are identical. The only difference between the Ethernet and IEEE 802.3 frame formats is the field following the Source Address: Ethernet specifies the Type (the higher layer protocol type, in this case, NetWare) while IEEE 802.3 counts the length of the Data field (in this example, 34 octets). The receiving station cannot tolerate a mistake in the frame format. If it is expecting a length (0022H) and it receives an EtherType (8137H), it rejects the frame because that frame is outside of the range of valid 802.3 length fields (0000–05DCH or 46–1500 decimal). Now that David has successfully logged into the server, he can complete his business with RCONSOLE and TCPCON. The moral of the story: If a newly configured "Ethernet" workstation cannot communicate with its server (but appears to be functioning otherwise), check the frame format. Until the lower layers can communicate, you cannot transmit or receive any TCP/IP-related information.

Trace 2.4.1a. Attempted NetWare Login Summary

Sniffer Network Analyzer data 31-Jan at 4:54:50, file ETHERNET.ENC, Pg 1

SUMMARY	DeltaT	Destination	Source	Summary
M 1		Broadcast	David	SAP C Find nearest file server
2	0.5503	Broadcast	David	SAP C Find nearest file server
3	0.6042	Broadcast	David	SAP C Find nearest file server
4	0.6042	Broadcast	David	SAP C Find nearest file server
5	0.6042	Broadcast	David	SAP C Find nearest file server
6	0.6042	Broadcast	David	SAP C Find nearest file server
7	0.6042	Broadcast	David	SAP C Find nearest file server
8	0.6042	Broadcast	David	SAP C Find nearest file server
9	0.6042	Broadcast	David	SAP C Find nearest file server
10	0.6042	Broadcast	David	SAP C Find nearest file server
11	0.6042	Broadcast	David	SAP C Find nearest file server
12	0.6042	Broadcast	David	SAP C Find nearest file server
13	0.6042	Broadcast	David	SAP C Find nearest file server
14	0.6043	Broadcast	David	SAP C Find nearest file server
15	0.6042	Broadcast	David	SAP C Find nearest file server
16	0.6042	Broadcast	David	SAP C Find nearest file server
17	0.5455	David	NW Svr 2	SAP C Find nearest file server
18	0.0588	Broadcast	David	SAP C Find nearest file server
19	0.6042	Broadcast	David	SAPCFind nearest file server
20	0.6042	Broadcast	David	SAP C Find nearest file server

Trace 2.4.1b. Attempted NetWare Login Details

Sniffer Network Analyzer data 31-Jan at 4:54:50 file ETHERNET.ENC, Pg 1

```
- - - - - - - - - - - - - - - - Frame 1 - - - - - - - - - - - - - - - - -
DLC:    — DLC Header —
DLC:
DLC:    Frame 1 arrived at  14:54:52.9662; frame size is 60 (003C hex) bytes.
DLC:    Destination = BROADCAST FFFFFFFFFFFF, Broadcast
DLC:    Source      = Station H-P   06CA73, David
DLC:    Ethertype  = 8137 (Novell)
DLC:
IPX:    — XNS Header —
IPX:
IPX:    Checksum = FFFF
IPX:    Length = 34
IPX:    Transport control = 00
IPX:              0000 .... = Reserved
```

```
IPX:            .... 0000 = Hop count
IPX:     Packet type = 17 (Novell NetWare)
IPX:
IPX:     Dest   net = 00000000, host = FFFFFFFFFFFF,
         socket = 1106 (NetWare Service Advertising)
IPX:     Source net = 00000000, host = 08000906CA73, socket = 16390 (4006)
IPX:
NSAP:    — NetWare Nearest Service Query —
NSAP:
NSAP:    Server type = 0004 (file server)
```

Trace 2.4.1c. Successful NetWare Login Summary

Sniffer Network Analyzer data 31-Jan at 4:47:46, file IEEE802.ENC, Pg 1

SUMMARY	Delta T	Destination	Source	Summary
M 1		Broadcast	David	SAP C Find nearest file server
2	0.0008	David	NW Svr 2	SAP R ISD
3	0.0003	David	NW Svr 3	SAP R HR
4	0.0004	David	H-P 133A5B	SAP R GL
5	0.0002	David	NW Svr 1	SAP R IC2
6	0.0001	David	H-P 136A06	SAP R ICTEMP
7	0.0018	Broadcast	David	IPX RIP request: find 1 network, 00133ADE
8	0.0005	David	NW Svr 2	IPX RIP response: 1 network, 00133ADE at 1 hop
9	0.0012	NW Svr 2	David	NCP C Create Connection
10	0.0034	David	NW Svr 2	NCP NCP R OK2
11	0.0013	NW Svr 2	David	NCP C Propose buffer size of 1024
12	0.0004	David	NW Svr 2	NCP R OK Accept buffer size of 1024
13	0.0300	David	H-P 11D4BD	SAP R H20
14	0.0016	David	H-P 133A76	SAP R BOOKS
15	0.00272	NW Svr 2	David	NCP C Logout
16	0.0023	David	H-P 133AAA	SAP R OLD
17	0.0008	David	NW Svr 2	NCP R OK
18	0.0012	NW Svr 2	David	NCP R C Get server's clock
19	0.0004	David	N W Svr 2	NCP R OK

Trace 2.4.1d. Successful NetWare Login Details

Sniffer Network Analyzer data 31-Jan at 4:47:46, file IEEE802.ENC, Pg 1

```
- - - - - - - - - - - - - - - - Frame 1 - - - - - - - - - - - - - - - - -
DLC:    — DLC Header —
DLC:
DLC:    Frame 1 arrived at  14:48:13.7824; frame size is 60 (003C hex) bytes.
DLC:    Destination = BROADCAST FFFFFFFFFFFF, Broadcast
DLC:    Source      = Station H-P   06CA73, David
DLC:    802.3 length = 34
DLC:
IPX:    — IPX Header —
IPX:
IPX:    Checksum = FFFF
IPX:    Length = 34
IPX:    Transport control = 00
IPX:              0000 .... = Reserved
IPX:              .... 0000 = Hop count
IPX:    Packet type = 17 (Novell NetWare)
IPX:
IPX:    Dest   net = 00000000, host = FFFFFFFFFFFF,
IPX:    socket = 1106 (NetWare Service Advertising)
IPX:    Source net = 00000000, host = 08000906CA73, socket = 16390 (4006)
IPX:
NSAP:   — NetWare Nearest Service Query —
NSAP:
NSAP:   Server type = 0004 (file server)
```

2.4.2 File Transfers over Frame Relay Networks

As an example of using TCP/IP over wide-area networks, this case study will demonstrate how the File Transfer Protocol operates over a frame relay network. In this case, a user at a branch office needs access to a file that is resident on a host at the headquarters location (see Figure 2-19). A frame relay network connects the various locations, with two permanent virtual connections (PVCs) between the two sites in question. These PVCs operate over a 56 Kbps leased line, transmitting both IPX and IP traffic. The IP traffic is of greater interest to our discussion; this traffic is carried on the PVC identified by Data Link Connection Identifier (DLCI) 140.

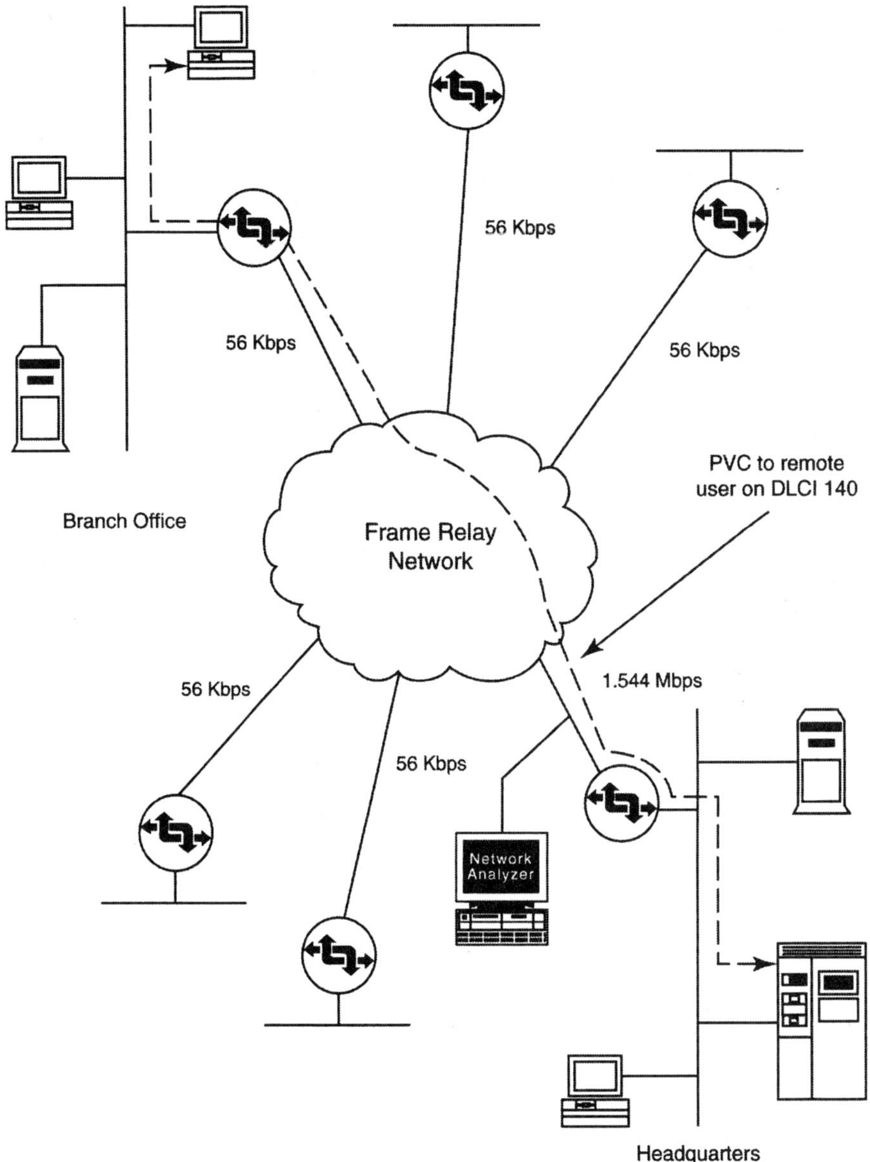

Figure 2-19. TCP/IP over Frame Relay.

Note from Figure 2-19 and Trace 2.4.2a that the data was captured at the host end (or headquarters side) of the connection. As a result, careful attention to the Source/Destination address designations is necessary to avoid confusion. For example, in Frame 1078, the destination specified is the DTE, which is actually the host at headquarters. The source of the data in Frame 1078 is the DCE (the network), which is actually the data coming from the remote user, via the network. (As in previous examples, this trace has been filtered to only show frames relevant to our discussion.)

The remote user initiates the file transfer process in Frame 1078 by entering **TYPE A** at his workstation to indicate to the headquarters host that an ASCII file transfer is being requested. The host confirms the TYPE A transfer in Frame 1079. The ports to be used are defined in Frame 1082 and confirmed in Frame 1083. The user indicates the file to be retrieved in Frame 1088 (Otherlinks.html). The host opens the data connection in Frame 1092, and begins the file transfer in Frame 1093. An acknowledgment from the remote user is seen in Frame 1105, and the host completes the file transfer in Frame 1107 (shown in Trace 2.4.2b).

Trace 2.4.2a. File Transfer Summary

Sniffer Internetwork Analyzer data from 8-Mar at 12:44:38, file FR1.SYC

SUMMARY	Delta T	Destination	Source	Summary
1078	2.5111	DTE	DCE	FTP C PORT=1214 TYPE A<0D0A>
1079	0.0128	DCE	DTE	FTP R PORT=1214 200 Type set to A.<0D0A>
1082	0.0751	DTE	DCE	FTP C PORT=1214 PORT 161,69,133,72,4,254<0D0A>
1083	0.0110	DCE	DTE	FTP R PORT=1214 200 PORT command successful.<0D0A>
1088	0.0491	DTE	DCE	FTP C PORT=1214 RETR Otherlinks.html <0D0A>
1092	0.0552	DCE	DTE	FTP R PORT=1214 150 Opening ASCII mode data connection for Oth...
1107	0.2048	DCE	DTE	FTP R PORT=1214 226 Transfer complete.<0D0A>

The details of the file transfer illustrate how FTP operates over frame relay connections (see Trace 2.4.2b). In Frame 1078, note that the frame relay header (the lines designated FRELAY in Trace 2.4.2b) identifies the DLCI in use (140), the absence of any congestion (either forward or backward), and other details. The analyzer also decodes an Ethertype, a term taken from the original Ethernet standard that specifies the type of data that is contained in this frame. In this case, the Ethertype value (0800H) indicates that an IP header and datagram are contained within this frame relay frame. Next, the IP header identifies the protocol inside the IP datagram (Protocol = 6 (TCP)), along with the IP addresses of the source and destination (the remote workstation and the headquarters host, respectively), indicating that they are on different IP subnetworks. The TCP header defines the host port that will be accessed (Destination Port = 21, the FTP Control port), along with the TCP connection-related information such as sequence and acknowledgment numbers and flags. The FTP layer decode identifies the eight octets of data that were transmitted (TYPE A <0D0A>, where the 0D is a carriage return character, and the 0A is a line feed character).

Readers studying subsequent frames can trace the consistency of the source and destination IP addresses, the orderly progression of the TCP sequence and acknowledgment numbers, and the details of the FTP process at both the remote and host ends of the connection. In the following chapters, we will study these protocol intricacies in greater detail.

Trace 2.4.2b. File Transfer Details

Sniffer Internetwork Analyzer data 8-Mar at 12:44:38, file FR1.SYC

```
- - - - - - - - - - - - - - - - Frame 1078 - - - - - - - - - - - - - - - - -
DLC:      — DLC Header —
DLC:
DLC:      Frame 1078 arrived at 15:07:45.5908; frame size is 52 (0034 hex) bytes.
DLC:      Destination = DTE
DLC:      Source = DCE
DLC:
FRELAY:   — Frame Relay —
FRELAY:
FRELAY:   Address word = 20C1
FRELAY:   0010 00.. 1100 .... = DLCI 140
FRELAY:   .... ..0. .... .... = Response
FRELAY:   .... .... .... 0... = No forward congestion
FRELAY:   .... .... .... .0.. = No backward congestion
FRELAY:   .... .... .... ..0. = Not eligible for discard
FRELAY:   .... .... .... ...1 = Not extended address
FRELAY:
```

```
ETYPE:   Ethertype = 0800 (IP)
ETYPE:
IP:      — IP Header —
IP:
IP:      Version = 4, header length = 20 bytes
IP:      Type of service = 00
IP:            000. .... = routine
IP:            ... 0.... = normal delay
IP:            .... 0... = normal throughput
IP:            .... .0.. = normal reliability
IP:      Total length   = 48 bytes
IP:      Identification = 12828
IP:      Flags          = 4X
IP:            .1.. .... = don't fragment
IP:            ..0. .... = last fragment
IP:      Fragment offset = 0 bytes
IP:      Time to live   = 31 seconds/hops
IP:      Protocol       = 6 (TCP)
IP:      Header checksum = F1B8 (correct)
IP:      Source address      = [XXX.YYY.133.72]
IP:      Destination address = [XXX.YYY.81.1]
IP:      No options
IP:
TCP:     — TCP header —
TCP:
TCP:     Source port           = 1214
TCP:     Destination port      = 21 (FTP)
TCP:     Sequence number       = 3275985
TCP:     Acknowledgment number = 725589307
TCP:     Data offset           = 20 bytes
TCP:     Flags                 = 18
TCP:           ..0. .... = (No urgent pointer)
TCP:           ...1 .... = Acknowledgment
TCP:           .... 1... = Push
TCP:           .... .0.. = (No reset)
TCP:           .... ..0. = (No SYN)
TCP:           .... ...0 = (No FIN)
TCP:     Window                = 8065
TCP:     Checksum              = BC15 (correct)
TCP:     No TCP options
TCP:     [8 byte(s) of data]
TCP:
FTP:     — FTP data —
```

```
FTP:
FTP:     TYPE A<0D0A>
FTP:

- - - - - - - - - - - - - - - - Frame 1079 - - - - - - - - - - - - - - - - -
DLC:     — DLC Header —
DLC:
DLC:     Frame 1079 arrived at 15:07:45.6037; frame size is 64 (0040 hex) bytes.
DLC:     Destination = DCE
DLC:     Source = DTE
DLC:
FRELAY:  — Frame Relay —
FRELAY:
FRELAY:  Address word = 20C1
FRELAY:  0010 00.. 1100 .... = DLCI 140
FRELAY:  .... ..0. .... .... = Response
FRELAY:  .... .... .... 0... = No forward congestion
FRELAY:  .... .... .... .0.. = No backward congestion
FRELAY:  .... .... .... ..0. = Not eligible for discard
FRELAY:  .... .... .... ...1 = Not extended address
FRELAY:
ETYPE:   Ethertype  = 0800 (IP)
ETYPE:
IP:      — IP Header —
IP:
IP:      Version = 4, header length = 20 bytes
IP:      Type of service = 00
IP:            000. .... = routine
IP:            ...0 .... = normal delay
IP:            .... 0... = normal throughput
IP:            .... .0.. = normal reliability
IP:      Total length    = 60 bytes
IP:      Identification  = 7358
IP:      Flags           = 0X
IP:            .0.. .... = may fragment
IP:            ..0. .... = last fragment
IP:      Fragment offset = 0 bytes
IP:      Time to live    = 51 seconds/hops
IP:      Protocol        = 6 (TCP)
IP:      Header checksum = 330B (correct)
IP:      Source address      = [XXX.YYY.81.1]
IP:      Destination address = [XXX.YYY.133.72]
IP:      No options
IP:
```

```
TCP:     — TCP header —
TCP:
TCP:     Source port            = 21 (FTP)
TCP:     Destination port       = 1214
TCP:     Sequence number        = 725589307
TCP:     Acknowledgment number  = 3275993
TCP:     Data offset            = 20 bytes
TCP:     Flags                  = 18
TCP:             ..0. .... = (No urgent pointer)
TCP:             ...1 .... = Acknowledgment
TCP:             .... 1... = Push
TCP:             .... .0.. = (No reset)
TCP:             .... ..0. = (No SYN)
TCP:             .... ...0 = (No FIN)
TCP:     Window                 = 4096
TCP:     Checksum               = 1289 (correct)
TCP:     No TCP options
TCP:     [20 byte(s) of data]
TCP:
FTP:     — FTP data —
FTP:
FTP:     200 Type set to A.<0D0A>
FTP:

- - - - - - - - - - - - - - - - Frame 1082 - - - - - - - - - - - - - - - - -
DLC:     — DLC Header —
DLC:
DLC:     Frame 1082 arrived at 15:07:45.6787; frame size is 70 (0046 hex) bytes.
DLC:     Destination = DTE
DLC:     Source = DCE
DLC:
FRELAY:  — Frame Relay —
FRELAY:
FRELAY:  Address word = 20C1
FRELAY:  0010 00.. 1100 .... = DLCI 140
FRELAY:  .... ..0.  .... .... = Response
FRELAY:  .... ....  .... 0... = No forward congestion
FRELAY:  .... ....  .... .0.. = No backward congestion
FRELAY:  .... ....  .... ..0. = Not eligible for discard
FRELAY:  .... ....  .... ...1 = Not extended address
FRELAY:
ETYPE:   Ethertype  = 0800 (IP)
ETYPE:
```

```
IP:     — IP Header —
IP:
IP:     Version = 4, header length = 20 bytes
IP:     Type of service = 00
IP:             000. .... = routine
IP:             ... 0.... = normal delay
IP:             .... 0... = normal throughput
IP:             .... .0.. = normal reliability
IP:     Total length    = 66 bytes
IP:     Identification  = 13084
IP:     Flags           = 4X
IP:             .1.. .... = don't fragment
IP:             ..0. .... = last fragment
IP:     Fragment offset = 0 bytes
IP:     Time to live    = 31 seconds/hops
IP:     Protocol        = 6 (TCP)
IP:     Header checksum = F0A6 (correct)
IP:     Source address      = [XXX.YYY.133.72]
IP:     Destination address = [XXX.YYY.81.1]
IP:     No options
IP:
TCP:    — TCP header —
TCP:
TCP:    Source port             = 1214
TCP:    Destination port        = 21 (FTP)
TCP:    Sequence number         = 3275993
TCP:    Acknowledgment number   = 725589327
TCP:    Data offset             = 20 bytes
TCP:    Flags                   = 18
TCP:            ..0. .... = (No urgent pointer)
TCP:            ...1 .... = Acknowledgment
TCP:            .... 1... = Push
TCP:            .... .0.. = (No reset)
TCP:            .... ..0. = (No SYN)
TCP:            .... ...0 = (No FIN)
TCP:    Window                  = 8045
TCP:    Checksum                = F946 (correct)
TCP:    No TCP options
TCP:    [26 byte(s) of data]
TCP:
FTP:    — FTP data —
FTP:
FTP:    PORT 161,69,133,72,4,254<0D0A>
FTP:
```

```
- - - - - - - - - - - - - - - - Frame 1083 - - - - - - - - - - - - - - - - -
DLC:      — DLC Header —
DLC:
DLC:      Frame 1083 arrived at 15:07:45.6897; frame size is 74 (004A hex) bytes.
DLC:      Destination = DCE
DLC:      Source = DTE
DLC:
FRELAY:   — Frame Relay —
FRELAY:
FRELAY:   Address word = 20C1
FRELAY:   0010 00.. 1100 .... = DLCI 140
FRELAY:   .... ..0. .... .... = Response
FRELAY:   .... .... .... 0... = No forward congestion
FRELAY:   .... .... .... .0.. = No backward congestion
FRELAY:   .... .... .... ..0. = Not eligible for discard
FRELAY:   .... .... .... ...1 = Not extended address
FRELAY:
ETYPE:    Ethertype = 0800 (IP)
ETYPE:
IP:       — IP Header —
IP:
IP:       Version = 4, header length = 20 bytes
IP:       Type of service = 00
IP:             000. .... = routine
IP:             ...0 .... = normal delay
IP:             .... 0... = normal throughput
IP:             .... .0.. = normal reliability
IP:       Total length   = 70 bytes
IP:       Identification = 7396
IP:       Flags          = 0X
IP:             .0.. .... = may fragment
IP:             ..0. .... = last fragment
IP:       Fragment offset = 0 bytes
IP:       Time to live   = 51 seconds/hops
IP:       Protocol       = 6 (TCP)
IP:       Header checksum = 32DB (correct)
IP:       Source address      = [XXX.YYY.81.1]
IP:       Destination address = [XXX.YYY.133.72]
IP:       No options
IP:
TCP:      — TCP header —
TCP:
TCP:      Source port        = 21 (FTP)
TCP:      Destination port   = 1214
```

```
TCP:     Sequence number        = 725589327
TCP:     Acknowledgment number  = 3276019
TCP:     Data offset            = 20 bytes
TCP:     Flags                  = 18
TCP:           ..0. .... = (No urgent pointer)
TCP:           ...1 .... = Acknowledgment
TCP:           .... 1... = Push
TCP:           .... .0.. = (No reset)
TCP:           .... ..0. = (No SYN)
TCP:           .... ...0 = (No FIN)
TCP:     Window                 = 4096
TCP:     Checksum               = E04C (correct)
TCP:     No TCP options
TCP:     [30 byte(s) of data]
TCP:
FTP:     — FTP data —
FTP:
FTP:     200 PORT command successful.<0D0A>
FTP:

- - - - - - - - - - - - - - - - Frame 1088 - - - - - - - - - - - - - - - - -
DLC:     — DLC Header —
DLC:
DLC:     Frame 1088 arrived at  15:07:45.7388; frame size is 66 (0042 hex) bytes.
DLC:     Destination = DTE
DLC:     Source = DCE
DLC:
FRELAY:  — Frame Relay —
FRELAY:
FRELAY:  Address word = 20C1
FRELAY:  0010 00..  1100 .... = DLCI 140
FRELAY:  .... ..0.  .... .... = Response
FRELAY:  .... ....  .... 0... = No forward congestion
FRELAY:  .... ....  .... .0.. = No backward congestion
FRELAY:  .... ....  .... ..0. = Not eligible for discard
FRELAY:  .... ....  .... ...1 = Not extended address
FRELAY:
ETYPE:   Ethertype  = 0800 (IP)
ETYPE:
IP:      — IP Header —
IP:
IP:      Version = 4, header length = 20 bytes
IP:      Type of service = 00
IP:            000. .... = routine
```

```
IP:          ...0 .... = normal delay
IP:          .... 0... = normal throughput
IP:          .... .0.. = normal reliability
IP:       Total length   = 62 bytes
IP:       Identification = 13340
IP:       Flags          = 4X
IP:          .1.. .... = don't fragment
IP:          ..0. .... = last fragment
IP:       Fragment offset = 0 bytes
IP:       Time to live   = 31 seconds/hops
IP:       Protocol       = 6 (TCP)
IP:       Header checksum = EFAA (correct)
IP:       Source address      = [XXX.YYY.133.72]
IP:       Destination address = [XXX.YYY.81.1]
IP:       No options
IP:
TCP:      — TCP header —
TCP:
TCP:      Source port            = 1214
TCP:      Destination port       = 21 (FTP)
TCP:      Sequence number        = 3276019
TCP:      Acknowledgment number  = 725589357
TCP:      Data offset            = 20 bytes
TCP:      Flags                  = 18
TCP:         ..0. .... = (No urgent pointer)
TCP:         ...1 .... = Acknowledgment
TCP:         .... 1... = Push
TCP:         .... .0.. = (No reset)
TCP:         .... ..0. = (No SYN)
TCP:         .... ...0 = (No FIN)
TCP:      Window                 = 8576
TCP:      Checksum               = B9EE (correct)
TCP:      No TCP options
TCP:      [22 byte(s) of data]
TCP:
FTP:      — FTP data —
FTP:
FTP:      RETR Otherlinks.html<0D0A>
FTP:

- - - - - - - - - - - - - - - - Frame 1092 - - - - - - - - - - - - - - - - -
DLC:      — DLC Header —
DLC:
DLC:      Frame 1092 arrived at 15:07:45.7940; frame size is 118 (0076 hex) bytes.
```

```
DLC:     Destination = DCE
DLC:     Source = DTE
DLC:
FRELAY:  — Frame Relay —
FRELAY:
FRELAY:  Address word = 20C1
FRELAY:  0010 00..  1100 .... = DLCI 140
FRELAY:  .... ..0.  .... .... = Response
FRELAY:  .... ....  .... 0... = No forward congestion
FRELAY:  .... ....  .... .0.. = No backward congestion
FRELAY:  .... ....  .... ..0. = Not eligible for discard
FRELAY:  .... ....  .... ...1 = Not extended address
FRELAY:
ETYPE:   Ethertype  = 0800 (IP)
ETYPE:
IP:      — IP Header —
IP:
IP:      Version = 4, header length = 20 bytes
IP:      Type of service = 00
IP:           000. .... = routine
IP:           ...0 .... = normal delay
IP:           .... 0... = normal throughput
IP:           .... .0.. = normal reliability
IP:      Total length    = 114 bytes
IP:      Identification  = 7442
IP:      Flags           = 0X
IP:           .0.. .... = may fragment
IP:           ..0. .... = last fragment
IP:      Fragment offset = 0 bytes
IP:      Time to live    = 51 seconds/hops
IP:      Protocol        = 6 (TCP)
IP:      Header checksum = 3281 (correct)
IP:      Source address      = [XXX.YYY.81.1]
IP:      Destination address = [XXX.YYY.133.72]
IP:      No options
IP:
TCP:     — TCP header —
TCP:
TCP:     Source port            = 21 (FTP)
TCP:     Destination port       = 1214
TCP:     Sequence number        = 725589357
TCP:     Acknowledgment number  = 3276041
TCP:     Data offset            = 20 bytes
TCP:     Flags                  = 18
```

```
TCP:                ..0. .... = (No urgent pointer)
TCP:                ...1 .... = Acknowledgment
TCP:                .... 1... = Push
TCP:                .... .0.. = (No reset)
TCP:                .... ..0. = (No SYN)
TCP:                .... ...0 = (No FIN)
TCP:        Window              = 4096
TCP:        Checksum            = AB7B (correct)
TCP:        No TCP options
TCP:        [74 byte(s) of data]
TCP:
FTP:        — FTP data —
FTP:
FTP:        150 Opening ASCII mode data connection for Otherlinks.html (1081
            bytes).<0D0A>
FTP:

- - - - - - - - - - - - - - - - Frame 1093 - - - - - - - - - - - - - - - -
DLC:        — DLC Header —
DLC:
DLC:        Frame 1093 arrived at 15:07:45.8005; frame size is 556 (022C hex) bytes.
DLC:        Destination = DCE
DLC:        Source = DTE
DLC:
FRELAY:     — Frame Relay —
FRELAY:
FRELAY:     Address word = 20C1
FRELAY:     0010 00.. 1100 .... = DLCI 140
FRELAY:     .... ..0. .... .... = Response
FRELAY:     .... .... .... 0... = No forward congestion
FRELAY:     .... .... .... .0.. = No backward congestion
FRELAY:     .... .... .... ..0. = Not eligible for discard
FRELAY:     .... .... .... ...1 = Not extended address
FRELAY:
ETYPE:      Ethertype = 0800 (IP)
ETYPE:
IP:         — IP Header —
IP:
IP:         Version = 4, header length = 20 bytes
IP:         Type of service = 00
IP:                000. .... = routine
IP:                ...0 .... = normal delay
IP:                .... 0... = normal throughput
IP:                .... .0.. = normal reliability
```

```
IP:      Total length   = 552 bytes
IP:      Identification  = 7443
IP:      Flags          = 0X
IP:           .0.. .... = may fragment
IP:           ..0. .... = last fragment
IP:      Fragment offset = 0 bytes
IP:      Time to live   = 51 seconds/hops
IP:      Protocol       = 6 (TCP)
IP:      Header checksum = 30CA (correct)
IP:      Source address      = [XXX.YYY.81.1]
IP:      Destination address = [XXX.YYY.133.72]
IP:      No options
IP:
TCP:     — TCP header —
TCP:
TCP:     Source port            = 20 (FTP data)
TCP:     Destination port       = 1278
TCP:     Sequence number        = 1306944001
TCP:     Acknowledgment number  = 4105807
TCP:     Data offset            = 20 bytes
TCP:     Flags                  = 10
TCP:          ..0. .... = (No urgent pointer)
TCP:          ...1 .... = Acknowledgment
TCP:          .... 0... = (No push)
TCP:          .... .0.. = (No reset)
TCP:          .... ..0. = (No SYN)
TCP:          .... ...0 = (No FIN)
TCP:     Window                 = 4096
TCP:     Checksum               = 69D3 (correct)
TCP:     No TCP options
TCP:     [512 byte(s) of data]
TCP:

- - - - - - - - - - - - - - - - Frame 1105 - - - - - - - - - - - - - - - - -
DLC:     — DLC Header —
DLC:
DLC:     Frame 1105 arrived at  15:07:45.9879; frame size is 50 (0032
         hex) bytes.
DLC:     Destination = DTE
DLC:     Source = DCE
DLC:
FRELAY:  — Frame Relay —
FRELAY:
FRELAY:  Address word = 20C1
```

```
FRELAY:  0010 00.. 1100 .... = DLCI 140
FRELAY:  .... ..0. .... .... = Response
FRELAY:  .... .... .... 0... = No forward congestion
FRELAY:  .... .... .... .0.. = No backward congestion
FRELAY:  .... .... .... ..0. = Not eligible for discard
FRELAY:  .... .... .... ...1 = Not extended address
FRELAY:
ETYPE:   Ethertype = 0800 (IP)
ETYPE:
IP:      — IP Header —
IP:
IP:      Version = 4, header length = 20 bytes
IP:      Type of service = 00
IP:            000. .... = routine
IP:            ...0 .... = normal delay
IP:            .... 0... = normal throughput
IP:            .... .0.. = normal reliability
IP:      Total length    = 40 bytes
IP:      Identification  = 14108
IP:      Flags           = 4X
IP:            .1.. .... = don't fragment
IP:            ..0. .... = last fragment
IP:      Fragment offset = 0 bytes
IP:      Time to live    = 31 seconds/hops
IP:      Protocol        = 6 (TCP)
IP:      Header checksum = ECC0 (correct)
IP:      Source address      = [XXX.YYY.133.72]
IP:      Destination address = [XXX.YYY.81.1]
IP:      No options
IP:
TCP:     — TCP header —
TCP:
TCP:     Source port             = 1278
TCP:     Destination port        = 20 (FTP data)
TCP:     Sequence number         = 4105807
TCP:     Acknowledgment number   = 1306944513
TCP:     Data offset             = 20 bytes
TCP:     Flags                   = 10
TCP:           ..0. .... = (No urgent pointer)
TCP:           ...1 .... = Acknowledgment
TCP:           .... 0... = (No push)
TCP:           .... .0.. = (No reset)
TCP:           .... ..0. = (No SYN)
TCP:           .... ...0 = (No FIN)
```

```
TCP:     Window           = 8064
TCP:     Checksum          = FAD9 (correct)
TCP:     No TCP options
TCP:

----------------- Frame 1107 -----------------
DLC:     — DLC Header —
DLC:
DLC:     Frame 1107 arrived at  15:07:45.9987; frame size is 68 (0044 hex) bytes.
DLC:     Destination = DCE
DLC:     Source = DTE
DLC:
FRELAY:  — Frame Relay —
FRELAY:
FRELAY:  Address word = 20C1
FRELAY:  0010 00..  1100 .... = DLCI 140
FRELAY:  .... ..0.  .... .... = Response
FRELAY:  .... ....  .... 0... = No forward congestion
FRELAY:  .... ....  .... .0.. = No backward congestion
FRELAY:  .... ....  .... ..0. = Not eligible for discard
FRELAY:  .... ....  .... ...1 = Not extended address
FRELAY:
ETYPE:   Ethertype  = 0800 (IP)
ETYPE:
IP:      — IP Header —
IP:
IP:      Version = 4, header length = 20 bytes
IP:      Type of service = 00\
IP:            000. .... = routine
IP:            ...0 .... = normal delay
IP:            .... 0... = normal throughput
IP:            .... .0.. = normal reliability
IP:      Total length    = 64 bytes
IP:      Identification  = 7543
IP:      Flags           = 0X
IP:            .0.. .... = may fragment
IP:            ..0. .... = last fragment
IP:      Fragment offset = 0 bytes
IP:      Time to live    = 51 seconds/hops
IP:      Protocol        = 6 (TCP)
IP:      Header checksum = 324E (correct)
```

```
IP:     Source address      = [XXX.YYY.81.1]
IP:     Destination address = [XXX.YYY.133.72]
IP:     No options
IP:
TCP:    — TCP header —
TCP:
TCP:    Source port           = 21 (FTP)
TCP:    Destination port      = 1214
TCP:    Sequence number       = 725589431
TCP:    Acknowledgment number = 3276041
TCP:    Data offset           = 20 bytes
TCP:    Flags                 = 18
TCP:          ..0. .... = (No urgent pointer)
TCP:          ...1 .... = Acknowledgment
TCP:          .... 1... = Push
TCP:          .... .0.. = (No reset)
TCP:          .... ..0. = (No SYN)
TCP:          .... ...0 = (No FIN)
TCP:    Window                = 4096
TCP:    Checksum              = CDC6 (correct)
TCP:    No TCP options
TCP:    [24 byte(s) of data]
TCP:
FTP:    — FTP data —
FTP:
FTP:    226 Transfer complete.<0D0A>
FTP:
```

2.5 Looking Ahead

In this chapter, we studied the concepts of layered network architectures. We considered the lowest layer of the ARPA architecture, the Network Interface Layer, where LAN and WAN protocols operate, and we introduced the concepts of network analysis. In the next chapter, we move up one layer in the ARPA architecture and discuss the Internet Layer, which contains the Internet Protocol and facilitates the routing of datagrams from their source to their intended destination.

2.6 References

[2-1] International Organization for Standardization. *Information Processing Systems — Open Systems Interconnection — Basic Reference Model: The Basic Model,* ISO/IEC 7498-1, 1994.

[2-2] Voelcker, John. "Helping Computers Communicate." *IEEE Spectrum* Vol. 23, No. 3 (March 1986): 61–70 (also published in *Computer Communications: Architectures, Protocols and Standards. 2d ed.*, IEEE Computer Society Press, 1987: 9–18).

[2-3] Leiner, B., et al. "The ARPA Internet Protocol Suite." RS-85-153, included in the *DDN Protocol Handbook,* Volume 2: 2-27–2-49.

[2-4] Cerf, V.G., and R.E. Kahn. "A Protocol for Packet Network Intercommunications." *IEEE Transactions on Communications* (May 1974): 637–648.

[2-5] Padlipsky, M.A. "A Perspective on the ARPANET Reference Model." RFC 871, September 1982.

[2-6] Braden, R. "Requirements for Internet Hosts: Communication Layers." RFC 1122, October 1989.

3

Datagram Addressing and Delivery

In the previous chapter, we discussed the options for the ARPA Network Interface (or Local Network) Layer, ranging from LANs to very high speed WANs. In this chapter, we move up one layer in the ARPA architectural model and explore the Internet Layer. This layer is analogous to the OSI Network Layer and is responsible for delivering a package of data (known as a *datagram*) from its source to its destination via the internetwork. Note that this internetwork may include the Internet, a corporate internet, or any of the LAN and WAN transmission channels we studied in Chapter 2.

To support the data delivery process requires the function of *routing*. Routing gets a data packet, or what is called an Internet Protocol (IP) *datagram*, from one host to another via the internetwork. Several steps occur during the routing process. First, each device must have a Logical address that uniquely identifies it within the internetwork. This internet address is independent of any Hardware or Physical address, such as the 48-bit Data Link Layer address of an Ethernet workstation. A mechanism is also necessary to associate the Logical address within the datagram with the Physical address within the LAN, MAN, or WAN frame. The Address Resolution Protocol (ARP) and the Reverse Address Resolution Protocol (RARP), as well as other variants of that protocol, handle this translation.

Second, you may set up the Logical address to identify individual hosts and/or subnetworks (such as LANs) within a network, and have some mechanism to obtain that Logical address and other necessary configuration information. The Bootstrap Protocol (BOOTP) and the Dynamic Host Configuration Protocol (DHCP) handle those functions. Third, you need a mechanism to associate that Logical address (which machines use) with a host name (which humans use). That process is handled by the Domain Name Service, or DNS. This chapter will cover these addressing-related issues. We will defer a further discussion on how the routing infrastructure is updated and how problems and exception conditions are noted until Chapter 4.

These datagram delivery processes will be the subject of this chapter, beginning with a study of the Internet Protocol and its various functions.

3.1 Internet Protocol

The Internet Protocol (IP) was developed to "provide the functions necessary to deliver a package of bits (an internet datagram) from a source to a destination over an interconnected system of networks" [3-1]. IP is primarily concerned with delivery of the datagram. Equally important, however, are the issues that IP does not address, such as the end-to-end reliable delivery of data or the sequential delivery of data. IP leaves those issues for the Host-to-Host Layer and the implementations of Transmission Control Protocol (TCP), User Datagram Protocol (UDP), or Stream Control Transmission Protocol (SCTP) that reside there. (As mentioned previously, a new version of IP, called IP Version 6 (IPv6) or IP Next Generation (IPng), has been developed and is being implemented. In this text, we will refer to the older version (IP Version 4 or IPv4) as just IP, and the newer version as IPv6.)

The term datagram refers to a package of data transmitted over a connectionless network. *Connectionless* means that no connection between source and destination is established prior to data transmission. Datagram transmission is analogous to mailing a letter. With both a letter and a datagram, you write Source and Destination addresses on the envelope, place the information inside, and drop the package into a mailbox for pickup. But while the post office uses blue or red mailboxes, the Internet uses your network node as the pickup point.

To further illustrate the datagram delivery process, let us suppose that an email application residing on a host in Denver needs to communicate with a host in Washington. Further, let us assume that the Denver LAN is a 10 Mbps IEEE 802.3 network, that a frame relay circuit operating at 1.5 Mbps connects Denver and Washington, and that the Washington LAN is a 100 Mbps IEEE 802.3 network. The Denver host will generate the email message and pass that message down through the various layers of the network architecture until it hits the ARPA Internet layer where an IP datagram will be constructed. This IP datagram will include the email message (or at least a part of that message, depending upon its length), the required upper layer protocol control information (or headers) such as the Transmission Control Protocol (TCP), plus the source and destination IP addresses for that email. The sending host then places this datagram inside an IEEE 802.3 frame and sends it to the router that has a connection to Washington. The router in

Denver will strip off the IEEE 802.3 frame header and trailer, place the datagram inside a frame relay frame (with its own header and trailer), and send that frame over the WAN to a router in Washington at the other end of the link. Once that frame hits the router in Washington, the frame relay header and trailer is stripped off, the destination IP address is examined, the routing table is consulted, and a decision is made to send that datagram to the local IEEE 802.3 interface for delivery. As a result of this process, the IP datagram is encapsulated inside another IEEE 802.3 frame for delivery on that LAN to the destination host in Washington. Note that this end-to-end process involved a number of LAN and WAN connections, each with their own frame formats; however, the IP datagram maintained its integrity from the source and destination. In other words, the user message (the datagram) was placed within a number of envelopes (the various IEEE 802.3 and frame relay frames), but when it reached its ultimate destination (the host in Washington), that message was intact.

Another type of data transmission is a virtual circuit connection, which uses a connection-oriented network. A virtual circuit is analogous to a telephone call, where the Destination address is contacted and a path is defined through the network prior to transmitting data. Note that, with a telephone call, the calling party must signal their request to use network resources (by going off hook and dialing). The network then responds with a progress message (either a ringing or busy signal), the called party answers, and the information exchange begins. When finished, the parties hang up, which releases the network resources to be re-used by another conversation. Because the network is involved in establishing and managing the connection, greater reliability typically results. From this, we use the term *reliable network* to describe a connection-oriented network, and *unreliable network* to describe the connectionless case. For our study, IP is an example of a connectionless protocol, while TCP is an example of a connection-oriented protocol.

In the process of delivering datagrams, IP must deal with *addressing* and *fragmentation*. The address ensures that the datagram arrives at the correct destination, whether it's across town or across the world. Notice from Figure 3-1a that, unlike the Network Interface Layer, the Internet Layer must be implemented with a consistent protocol, such as the Internet Protocol, on an end-to-end basis for addressing consistency. Figure 3-1b illustrates the protocols implemented at the Internet Layer, while Figure 3-1c illustrates the various routing, control, and address resolution protocols that will also be considered in this chapter.

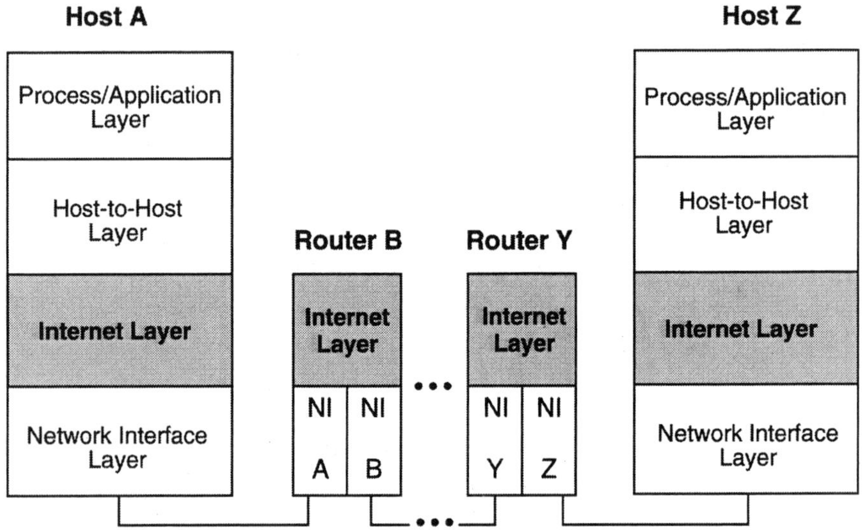

Figure 3-1a. The Internet Connection.

Fragmentation is necessary because the LANs and WANs that any datagram may traverse can have differing frame sizes, and the IP datagram must always fit within the frame, as shown in Figure 3-1d. (We saw this explicitly in the frame structures illustrated in Chapter 2. For example, an Ethernet/IEEE 802.3 frame can accommodate only up to 1,500 octets of data, while FDDI carries up to 4,470 octets.) Specific fields within the IP header handle the addressing and fragmentation functions (see Figure 3-2). Note from the figure that each horizontal group of bits (called a word) is 32 bits wide.

ARPA Layer	Protocol Implementation								OSI Layer
Process / Application	Hypertext Transfer	File Transfer	Electronic Mail	Terminal Emulation	Domain Names	File Transfer	Client / Server	Network Management	Application
	Hypertext Transfer Protocol (HTTP) RFC 2616	File Transfer Protocol (FTP) MIL-STD-1780 RFC 959	Simple Mail Transfer Protocol (SMTP) MIL-STD-1781 RFC 2821	TELNET Protocol MIL-STD-1782 RFC 854	Domain Name System (DNS) RFC 1034, 1035	Trivial File Transfer Protocol (TFTP) RFC 1350	Sun Microsystems Network File System Protocols (NFS) RFC 3530	Simple Network Management Protocol (SNMP) v1: RFC 1157 v2: RFC 1901-10 v3: RFC 3411-18	Presentation Session
Host-to-Host	Transmission Control Protocol (TCP) MIL-STD-1778 RFC 793					User Datagram Protocol (UDP) RFC 768			Transport
Internet	Address Resolution ARP RFC 826 RARP RFC 903			Internet Protocol (IP) MIL-STD-1777 RFC 791			Internet Control Message Protocol (ICMP) RFC 792		Network
Network Interface	Network Interface Cards: Ethernet, Token Ring, MAN and WAN RFC 894, RFC 1042 and others								Data Link
	Transmission Media: Twisted Pair, Coax, Fiber Optics, Wireless Media, etc.								Physical

Figure 3-1b. ARPA Internet Layer Protocols.

BGP RFC 1771
DNS RFCs 1034, 1035
BOOTP RFC 951
DHCP RFC 2131
RIP RFC 1058
RIP-2 RFC 2453
RSVP RFC 2205
TCP RFC 793
UDP RFC 768
ICMP RFC 792
EGP RFC 904
IGMP RFC 1112
IGMPv2 RFC 3376
OSPF RFC 2328
IP RFC 791
ARP RFC 826
RARP RFC 903
InARP RFC 2390
ATMARP InATMARP RFC 2225
Network Interface Layer Protocols

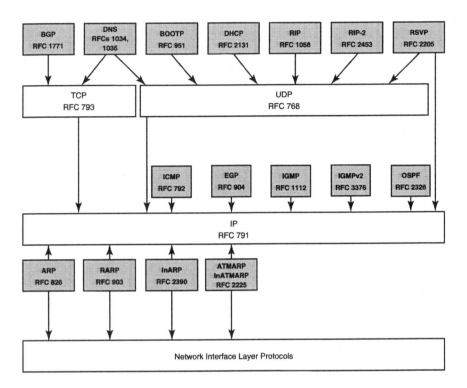

Figure 3-1c. Internet Routing, Control, and Address Resolution Protocols.

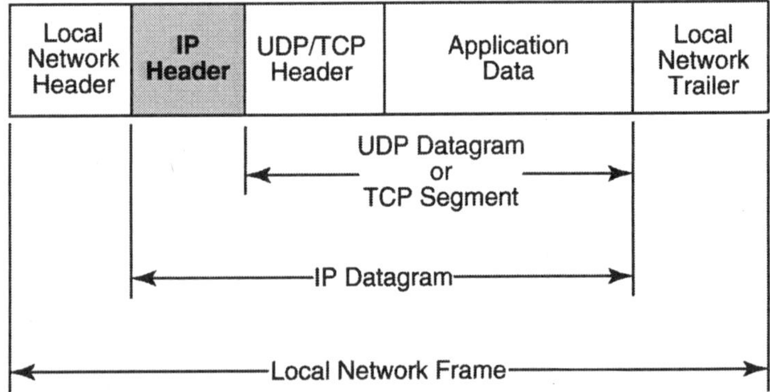

Figure 3-1d. The Internet Transmission Frame and IP Header Position.

The IP header contains a minimum of 20 octets of control information. Version (4 bits) defines the current version of the IP protocol and should equal 4. Internet Header Length (IHL, 4 bits) measures the length of the IP header in 32-bit words. (The minimum value would be five 32-bit words, or 20 octets.) The IHL also provides a measurement (or offset) where the higher layer information, such as the TCP header, begins within that datagram. The Type of Service (8 bits) indicates the quality of service requested for the datagram. Values include:

Bits 0–2: Precedence (or relative importance of this datagram)

111 - Network Control

110 - Internetwork Control

101 - CRITIC/ECP

100 - Flash Override

011 - Flash

010 - Immediate

001 - Priority

000 - Routine

Bit 3: Delay, 0 = Normal Delay, 1 = Low Delay

Bit 4: Throughput, 0 = Normal Throughput, 1 = High Throughput

Bit 5: Reliability, 0 = Normal Reliability, 1 = High Reliability

Bits 6–7: Reserved for future use (set to 0)

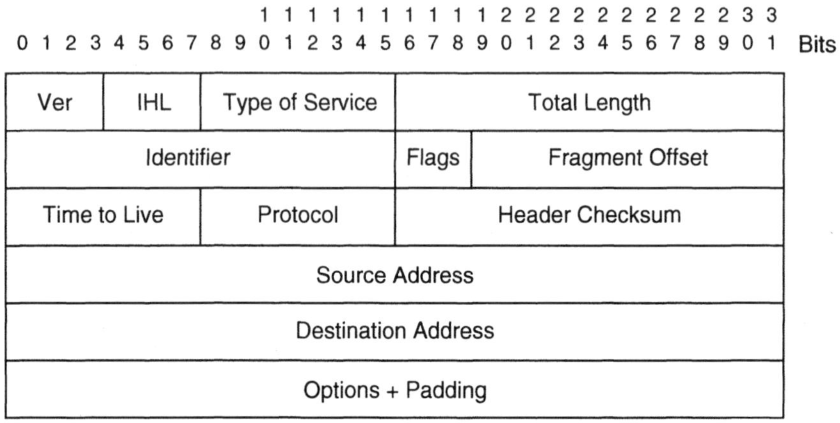

```
                    1 1 1 1 1 1 1 1 1 1 2 2 2 2 2 2 2 2 2 2 3 3
0 1 2 3 4 5 6 7 8 9 0 1 2 3 4 5 6 7 8 9 0 1 2 3 4 5 6 7 8 9 0 1  Bits
```

<table>
<tr><td>Ver</td><td>IHL</td><td colspan="2">Type of Service</td><td colspan="4">Total Length</td></tr>
<tr><td colspan="4">Identifier</td><td>Flags</td><td colspan="3">Fragment Offset</td></tr>
<tr><td colspan="2">Time to Live</td><td colspan="2">Protocol</td><td colspan="4">Header Checksum</td></tr>
<tr><td colspan="8">Source Address</td></tr>
<tr><td colspan="8">Destination Address</td></tr>
<tr><td colspan="8">Options + Padding</td></tr>
</table>

Figure 3-2. Internet Protocol (IPv4) Header Format.

Note that the values and definitions listed above were designed to support government and military applications of years past that may not be applicable in current environments. In contrast, real-time applications, such as voice and video, require some Quality of Service (QoS) differentiation between application data to assure that datagrams containing voice samples (which are very time sensitive) do not get stuck in a queue behind a large file transfer (which is not as time sensitive). This process is known as *differentiated services.* It is specified in RFC 2474 and redefines the Type of Service field in support of these newer applications. We will consider this and other QoS-related issues in Chapter 9. The Total Length field (16 bits) measures the length, in octets, of the IP datagram (IP header plus higher layer information). The 16-bit field allows for a datagram of up to 65,535 octets, although all hosts must be able to handle datagrams of at least 576 octets.

The next 32-bit word contains three fields that deal with datagram fragmentation/reassembly. Recall that the IP datagram may be up to 65,535 octets long. What happens if the endpoint of a WAN that handles such a datagram is attached to an IEEE 802.3 LAN with a maximum data field size of 1,500 octets? IP fragments the large IP datagram into smaller pieces (i.e., fragments) that will fit. The Destination node reassembles all the fragments. The sender assigns the Identification field (16 bits) to help reassemble the fragments into the datagram. Three flags indicate how the fragmentation process will be handled:

Bit 0:	Reserved (set to 0)
Bit 1:	(DF) 0 = May fragment, 1 = Don't fragment
Bit 2:	(MF) 0 = Last fragment, 1 = More fragments

The last field in this word is a 13-bit fragment offset, which indicates where a fragment belongs in the complete message. This offset is measured in 64-bit units. The case study in Section 7.2 will illustrate the fragmentation process.

The next word in the IP header contains a Time-to-Live (TTL) measurement, which is the maximum amount of time the datagram can live within the internet. When TTL = 0, the datagram is destroyed. This field is a failsafe measure that prevents misaddressed datagrams from wandering the internet forever. TTL may be measured in either router hops or seconds. If the measurement is in seconds, the maximum is 255 seconds, or 4.25 minutes (a long time to be lost in today's high-speed internetworks). A default value of TTL = 64 is used with many systems. The TTL field is used with diagnostic utilities, such as PING and TRACEROUTE, which will be discussed in Section 4.3.

The Protocol field (8 bits) following the IP header identifies the higher layer protocol in use. Examples include:

Decimal	Keyword	Description
1	ICMP	Internet Control Message Protocol
6	TCP	Transmission Control Protocol
17	UDP	User Datagram Protocol

The Assigned Numbers site [3-2] lists all current values for this field. A 16-bit header checksum completes the third 32-bit word.

The fourth and fifth words of the IP header contain the Source and Destination addresses, respectively. Recall that we discussed Hardware addresses for the ARPA Network Interface Layer (or OSI Data Link Layer) in Chapter 2. The addresses within the IP header are the Internet Layer (or OSI Network Layer) addresses. The Internet address is a Logical address that gets the IP datagram through the Internet to the correct host and network (LAN, MAN, or WAN).

Options and Padding, if present, may complete the IP header. The Options field, which may be of variable length, carries options within this datagram. Examples of options, as defined in RFC 791, include security, source routing, stream identifier, and timestamp information. Another option, called the Router Alert Option, is defined in RFC 2113 and is also mentioned in conjunction with the Internet Group Management Protocol (IGMP), which we will study in Chapter 4. This option acts as a flag to transit routers, causing them to closely examine the contents of a particular IP datagram while not causing performance penalties for all datagrams going through that router. Finally, the Padding field adds a variable number of zeros to ensure that the IP header ends on a 32-bit boundary.

3.2 Internetwork Addressing

Each 32-bit IP address is divided into Host ID and Network ID sections and may take one of five formats ranging from Class A to Class E, as shown in Figure 3-3. The formats differ in the number of bits they allocate to the Host IDs and Network IDs and are identified by the first four bits.

3.2.1 Address Classes

Class A addresses are designed for very large networks with many hosts, and are frequently referred to as a "/8" prefix (pronounced "slash eight") since the Network ID portion is 8 bits. They are identified by Bit 0 = 0. Bits 1 through 7 identify the network, and Bits 8 through 31 identify the host. With a 7-bit Network ID, only 128 Class A network addresses are available. Of these, network addresses 0 and 127 are reserved [3-3].

The zero (0) address is interpreted as meaning "this," as in "this network." For example, the address [0.0.0.37] could be interpreted as meaning "host number 37 on this network." The Class A network number 127 is assigned the loopback function, that is, a datagram sent by a higher level protocol to a network 127 address should loop back inside the host. Thus, no datagram transmitted to a network 127 address should ever appear on any network anywhere. In addition to the reserved network numbers 0 and 127, there are reserved host numbers that have been defined and are described in Section 3.2.4 below.

The majority of organizations that have distributed processing systems including LANs and hosts use Class B addresses, also called the /16 prefix. Class B addresses are identified with the first 2 bits having a value of

10 (binary). The next 14 identify the network. The remaining 16 bits identify the host. A total of 16,384 Class B network addresses are possible; however, addresses 0 and 16,383 are reserved.

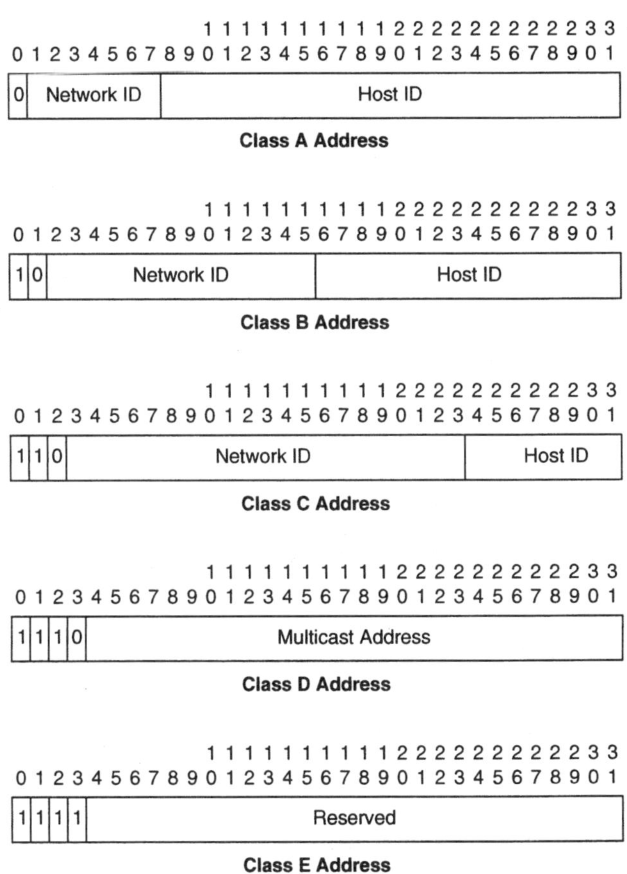

Figure 3-3. IPv4 Address Formats.

Class C addresses are generally used for smaller networks, such as LANs, and are called the /24 prefix. They begin with a binary 110. The next 21 bits identify the network. The remaining 8 bits identify the host. A total of 2,097,152 Class C network addresses are possible, with addresses 0 and 2,097,151 reserved.

Class D addresses begin with a binary 1110 and are intended for multicasting. Class E addresses begin with a binary 1111 and are reserved for future use.

All IP addresses are written in dotted decimal notation, in which each octet is assigned a decimal number from 0 to 255. For example, network [10.55.31.84] is represented in binary as 00001010 00110111 00011111 1010100. The first bit (0) indicates a Class A address, the next 7 bits (0001010) represent the Network ID (decimal 10), and the last 24 bits (00110111 00011111 1010100) represent the Host ID.

Class A addresses begin with 1–127, Class B with 128–191, Class C with 192–223, and Class D with 224–239. Thus, an address of [150.100.200.5] is easily identified as a Class B address. Following is the range of valid addresses, given in dotted decimal notation, for each of these four address classes. Note that the network address 127 is reserved for loopback functions and therefore does not appear in the following list. This loopback address will be discussed further in Section 3.2.4.

Class A: [1.0.0.0] through [126.255.255.255]

Class B: [128.0.0.0] through [191.255.255.255]

Class C: [192.0.0.0] through [223.255.255.255]

Class D: [224.0.0.0] through [239.255.255.255]

3.2.2 Multicast Addresses

Class D addresses are reserved to support multicast applications, and range in value from [224.0.0.0] through [239.255.255.255]. Of these values, the Internet Assigned Numbers Authority (IANA) has assigned specific values for particular multicast functions. For example, the range of addresses between [224.0.0.0] and [224.0.0.255], inclusive, is used for routing protocols and other topology and maintenance functions. Within this range, address [224.0.0.1] identifies "all systems on this subnet" and [224.0.0.2] identifies "all routers on this subnet." Other values have been specified RIP2 routers, [224.0.0.9], DHCP Server/Relay Agents, [224.0.0.12], audio and video transport, and others. Addresses in the range from [239.0.0.0] through [239.255.255.255] are reserved for local applications.

3.2.3 Subnetting

As discussed above, the IP addresses are divided into two fields that identify a network and a host. A central authority assigns the Network ID and the local network administrator assigns the Host ID. Routers send packets to a particular network using the Network ID, and that network completes the delivery to the host. If an organization has two networks, it could request two Network ID assignments from the central authority. But this would cause the routing tables within hosts and routers to expand considerably. The popularity of LANs in the mid-1980s, therefore, inspired the Internet community to revise the IP address structure. The new structure allows for an additional field to identify a subnetwork within an assigned Network ID. Thus the [Network, Host] address format is replaced with the [Network, Subnetwork, Host] format. The space for the Subnetwork field comes from reducing the Host field. The central authority assigns the Network ID, and the individual organization assigns the Subnetwork IDs and the Host IDs on each subnetwork.

The Address Mask differentiates between various subnetworks. A Subnet Mask is a 32-bit number that has ones in the Network ID and Subnetwork ID fields and zeros in the Host ID field:

Address Class	Subnet Mask	Binary
A	[255.0.0.0]	11111111 00000000 00000000 00000000
B	[255.255.0.0]	11111111 11111111 00000000 00000000
C	[255.255.255.0]	11111111 11111111 11111111 00000000

The router or the host implements a mathematical function that performs a logical AND function between the Subnet Mask and a particular IP address to determine whether a datagram can be delivered on the same subnet, or whether it must go through an IP router to another subnet.

A host or router that needs to make a routing decision uses the Subnet Mask for assistance and the logical AND function to arrive at its conclusion. For example, suppose the Destination address is D and my address is M. Further suppose that we are using the Class C default Subnet Mask shown above, [255.255.255.0]. Two calculations are made: Subnet Mask AND Address D, plus Subnet Mask AND Address M. The result of the AND function strips off the host portion of the address, leaving only the network and subnet portions. (Recall that the Host ID field is filled with zeros and that any number AND zero is still zero):

Class C Subnet Mask:	[11111111 11111111 11111111.00000000]
	AND
Class C Address:	[24-bit Net ID .8-bit Host ID]
	Yields
Result:	[24-bit Net ID .00000000]

A similar calculation would be made using the Subnet Mask and Address M. These two calculations would yield two results, each representing a Network ID and a Subnet ID. The two results are then compared and used to make a routing decision. When the two results are identical, Addresses M and D are on the same subnetwork. If Addresses M and D are not equal, the two addresses are on different subnetworks, and the datagram must go to a router for delivery.

Subnet mask values other than the defaults shown above are frequently used. For example, let us consider the Class B address, which allocates 16 bits to the Host ID portion. Since the network administrator has assignment authority over the Host ID portion of the address, it could be divided into 8 bits for Subnet ID and 8 bits for Host ID, or 7 bits for subnet and 9 bits for host, or 9 bits for subnet and 7 bits for host, or any other combination of bits that totaled 16 bits. Following are some further details for the Class B case:

Subnet ID Length	Subnets	Host ID Length	Hosts	Subnet Mask
7 bits	126	9 bits	510	[255.255.254.0]
8 bits	254	8 bits	254	[255.255.255.0]
9 bits	510	7 bits	126	[255.255.255.128]

Thus, a Class B address with a 7 bit Subnet ID could be divided into 126 subnets each with 510 hosts. An address of 10001100 00110101 00001010 110111001 would thus be represented in dotted decimal as [140.53.10.441]. This address would identify Host 441 on Subnet 10 of Network 140.53. Note that the maximum total of subnets and/or hosts is always less than the maximum theoretical number to allow for reserved addresses, as detailed in the following section.

Subnet Masks may be either a fixed length (sometimes called static subnetting), as described above, or a variable length. With the static case, all subnets in that network use the same Subnet Mask. With variable length subnetting, the subnets that comprise that network may use different Subnet Masks. In that way, if a particular subnet only has a small number of hosts, the Subnet Mask can be adjusted to allow fewer bits for the host portion of the mask, thus conserving IP addresses. Likewise, if a particular subnet has a large number of hosts, the host portion can be similarly increased in response. However, some hosts and routers, and some routing protocols, such as RIP version 1, do not support variable length Subnet Masks. RIP version 2 does support variable-length Subnet Masks. Both RIP and RIPv2 will be discussed in Chapter 4.

If the host doesn't know which Subnet Mask to use, it uses the Internet Control Message Protocol (ICMP) to discover the right one. The host broadcasts an ICMP Address Mask Request message and waits for an ICMP Address Mask Reply from a neighboring router. If an Address Mask is used improperly and a datagram is sent to a router erroneously, the router that identifies the misdirected packet returns an ICMP Redirect message. ICMP will be discussed in detail in Chapter 4, and an example of using ICMP to obtain an address mask will be considered in Chapter 7.

RFC 950, "Internet Standard Subnetting Procedure" [3-4], was written to cover subnet-related issues. RFC 1118, "The Hitchhiker's Guide to the Internet" [3-5], discusses UNIX-related subnetting issues.

3.2.4 Reserved Addresses

Several addresses are reserved to identify special purposes [3-6, Sections 4.2.2.11 and 4.2.3.1]:

Address	Interpretation
[Net = 0, Host = 0]	This host on this network
[Net = 0, Host = H]	Specific Host H on this network (Source address only)
[Net = all ones, Host = all ones]	Limited broadcast (within the source's subnetwork)
[Net = N, Host = all ones]	Directed broadcast to network (Destination address only)

[Net = N, Sub = all ones, Host = all ones]	Directed broadcast to all subnets on Network N (Destination address only)
[Net = N, Sub = S, Host = all ones]	Directed broadcast to all hosts on Subnet S, Network N (Destination address only)
[Net = 127, Host = any]	Internal host loopback address

The following two numbers are used for network numbers, but not IP addresses:

[Net = N, Host = 0]	Network N, No Host
[Net = N, Sub = S, Host = 0]:	Subnet S, No Host

Abbreviations:
Net = Network ID
Sub = Subnet ID
Host = Host ID

RFC 1118 [3-5] issues two cautions about IP addresses. First, it notes that BSD 4.2 UNIX systems require additional software for subnetting; BSD 4.3 systems do not. Second, some machines use an IP address of all zeros to specify a broadcast, instead of the more common all ones. BSD 4.3 requires the system administrator to choose the broadcast address. Use caution, since many problems, such as broadcast storms, can result when the broadcast address is not implemented consistently over the network. RFC 1812 [3-6] also discusses cautions in this area.

3.2.5 Network Address Translation

In some circumstances, having a globally unique IP address is not required or desired. For example, stations in a test lab, or those that never communicate to the global Internet, would not need a unique address that is assigned by a central authority. In other cases, perhaps for security reasons, advertising a private network's IP address to the world may not be desired. In either case, assigning a private (or nonglobal) IP address is necessary. This issue is addressed in RFC 1918, "Address Allocation for Private Internets" [3-7]. The IANA has reserved the following blocks of IP addresses for private internets:

[10.0.0.0] – [10.255.255.255]	(A single Class A number)
[172.16.0.0] – [172.31.255.255]	(16 contiguous Class B numbers)
[192.168.0.0] – [192.168.255.255]	(256 contiguous Class C numbers)

Thus, a network administrator could number his network internally using network number 10, and as long as no routing information regarding network number 10 was propagated to the outside world, no confusion would exist. In addition, there could be multiple private networks, all using network number 10. Again, if no routing information was propagated to the outside world, and since many routers administratively prohibit forwarding packets with these addresses outside of the private network, no confusion would result.

However, communication with the outside world requires translation from the internal (private) network address to a global (or universally known) address. This process is called Network Address Translation, or NAT, and is defined in RFC 3022 [3-8]. The NAT process may be performed at a border router between a private internet and some regional router (such as within the Internet), as shown in Figure 3-4. In this example, taken from RFC 3022, two private networks, designated A and B, are both internally numbered with network address 10. Each of these networks is attached to a router with NAT capabilities, which has a globally unique Class C address assigned: [198.76.29.0] for the Network A case, and [198.76.28.0] for the Network B case. These NAT routers, in turn, connect to a regional router.

Suppose that Workstation 1 in Network A wishes to communicate with Workstation 2 in Network B. Workstation 1 would compose an IP datagram using its source address [10.33.96.5] and the globally unique address for Workstation 2 [198.76.28.4]. This packet is then passed to the NAT router, which translates the internal source address of Workstation 1 [10.33.96.5] to a globally unique address [198.76.29.7] that can be forwarded through the regional router. In a similar fashion, the NAT process at Network B translates the globally unique destination address of the datagram [198.76.28.4] to the private destination address [10.81.13.22] used internally within Network B.

In a similar fashion, datagrams from Network B to Network A would also go through address translations.

Thus, NAT can provide many useful functions, including the conservation of global IP addresses, and, when combined with a firewall function, the additional security that comes from keeping private addresses from public disclosure.

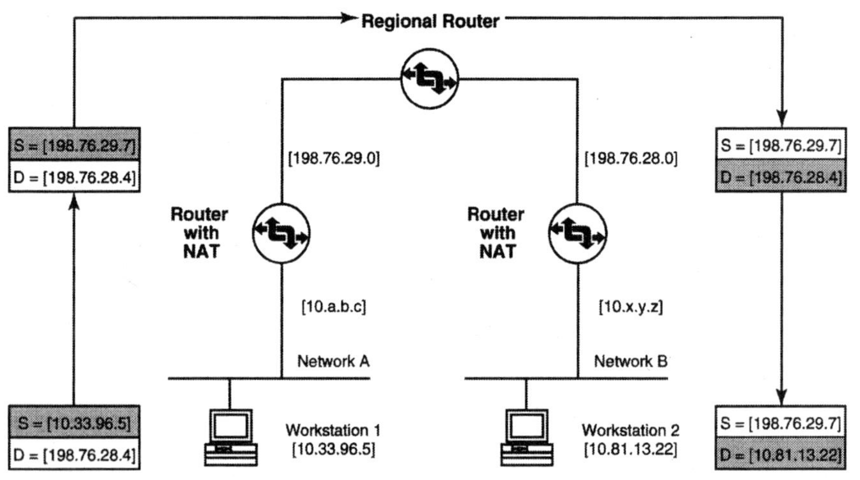

Figure 3-4. Network Address Translation Example.

3.2.6 Classless Inter-Domain Routing

Thus far in our discussion, we have assumed that a device is identified by one of the three classes of IP addresses: A, B, or C. These three address formats have fixed boundaries between the Network ID and the Host ID fields. For example, with a Class A address, there are seven bits in the Network ID field, so there are a maximum of 128 (2^7) theoretical networks. Likewise, there are 24 bits in the Host ID field, so 16,777,216 (2^{24}) hosts can be uniquely identified. (For the purpose of this discussion, we will not consider any reserved addresses that could further reduce the maximum number of theoretical network addresses.) In a similar fashion, Class B addresses can accommodate 16,384 (2^{14}) unique networks, each with 65,536 (2^{16}) unique hosts, and Class C addresses can accommodate 2,097,152 (2^{21}) unique networks, each with 256 (2^8) unique hosts.

But what happens if a network needs addresses for 260 hosts? You immediately exceed the capacity of a Class C address, and must use a Class B. To further complicate matters, what if the Internet community is running out of Class B addresses? Is it reasonable to assign a Class B address to your organization just because you need a few more addresses beyond what a Class C address can supply?

This is the problem that has faced the Internet community in the past few years; the address process called Classless Inter-Domain Routing, or CIDR for short, has been developed to deal with this. The technical and implementation aspects of CIDR are described in a series of RFC documents: RFCs 1517 through 1520 [3-9]. With CIDR, the problem stated above would be solved by assigning two contiguous Class C addresses to the end user's network, thus enabling 512 ($2^9 = 512$) unique host addresses to be assigned. In effect, then, the user has a nine-bit Host ID field.

According to RFC 1519, the objective of CIDR is to allocate contiguous blocks of Class C addresses to organizations or service providers, thus providing for networks that need more than 254 hosts (traditional Class C), but not 65,534 hosts (traditional Class B). The contiguous Class C addresses would accommodate the following number of hosts:

Contiguous Class C Addresses	Maximum Hosts (Theoretical)
1	256
2	512
4	1,024
8	2,048
16	4,096

As before, note that the theoretical maximum number of contiguous host addresses (e.g. 1,024) is reduced by two upon implementation to accommodate the special addressing conditions.

The example given in RFC 1519 and shown in Figure 3-5 illustrates how a block of eight such Class C addresses could be used to provide up to 2,048 contiguous host addresses.

Assume that a block of 2,048 contiguous Class C addresses is allocated to an Internet Service Provider. Further suppose that this block of addresses has a range from [192.24.0.0] through [192.31.255.0]. The first address has value [192.24.0.0], or C0 18 00 00H, or 11000000 00011000 00000000 00000000B. The last address has value [192.31.255.0], or C0 1F FF 00H, or 11000000 00011111 11111111 00000000B.

Further assume that one of the provider's customers needed 2,048 host addresses, the equivalent of eight Class C addresses (256 * 8 = 2,048).

This client would be assigned addresses from [192.24.0.0] or C0 18 00 00H or 11000000 000110000 00000000 00000000B, to [192.24.7.255] or C0 18 07 FFH or 11000000 00011000 00000111 11111111B. An address mask of [255.255.248.0] or FF FF F8 00H or 11111111 11111111 11111000 00000000B would be used to distinguish all of the addresses assigned to this client.

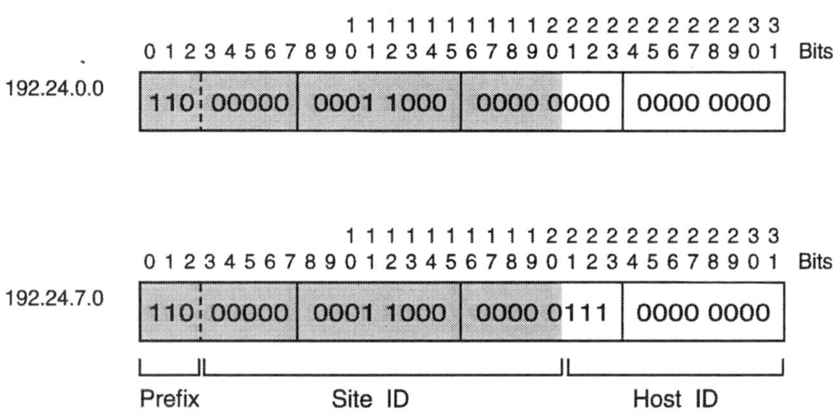

Figure 3-5. CIDR Addressing.

Thus, this block of eight Class C network addresses could be represented as a prefix of 110B (three bits, the Class C network address prefix), followed by a site (or client) identification (18 bits, from 192.24.0 through 192.24.7), followed by the host identification (11 bits, from 000 0000 0000 through 111 1111 1111B). Another way to express the result would be an address prefix of 21 bits (3 plus 18) followed by a host identification of 11 bits. These three sections of the address are illustrated in Figure 3-5.

An example workstation address may be shown as [192.24.7.15]/21, which indicates a 21-bit mask (and the resulting 11-bit host address, as noted above).

Thus, CIDR does not determine the route based on the class of network (A, B, or C) and the resulting Network ID field, but instead on a variable number of high-order bits in the IP address. These bits are referred to as the *address prefix.* When multiplc Class C addresses are assigned to an organization, they are assigned contiguously, which forces them to have the same address prefix. These multiple Class C addresses can then be grouped together into a single routing table entry, which is a process known as *address aggregation*. This route aggregation also minimizes the number of routing table entries that are needed to identify a single CIDR-based route.

3.3 Address Resolution

We have looked at the differences between the Internet address, used by IP, and the Local address, used by the LAN or WAN hardware. We have observed that the IP address is a 32-bit Logical address, but that the Physical address depends on the hardware. For example, an Ethernet/IEEE 802.3 workstation has a 48-bit hardware address, while a frame relay WAN circuit uses a 10-bit (default) Data Link Connection Identifier (DLCI) address.

To illustrate the differences between the hardware and protocol addresses, consider the simple network shown in Figure 3-6a. Each workstation contains a hardware address (HA, such as the 48-bit Ethernet address) and a protocol address (PA, such as a 32-bit IP address). The router has a hardware address for each of its interfaces, plus at least one protocol address. When Workstation 1 needs to communicate with Workstation 2 on the same LAN, it encapsulates the IP datagram inside a frame and sends that frame on its way. Note that the source and destination hardware addresses within both the frame and the packet identify Workstation 1 and Workstation 2 — the respective origin and destination of the message. In this case, Router 3 is not involved in the communication.

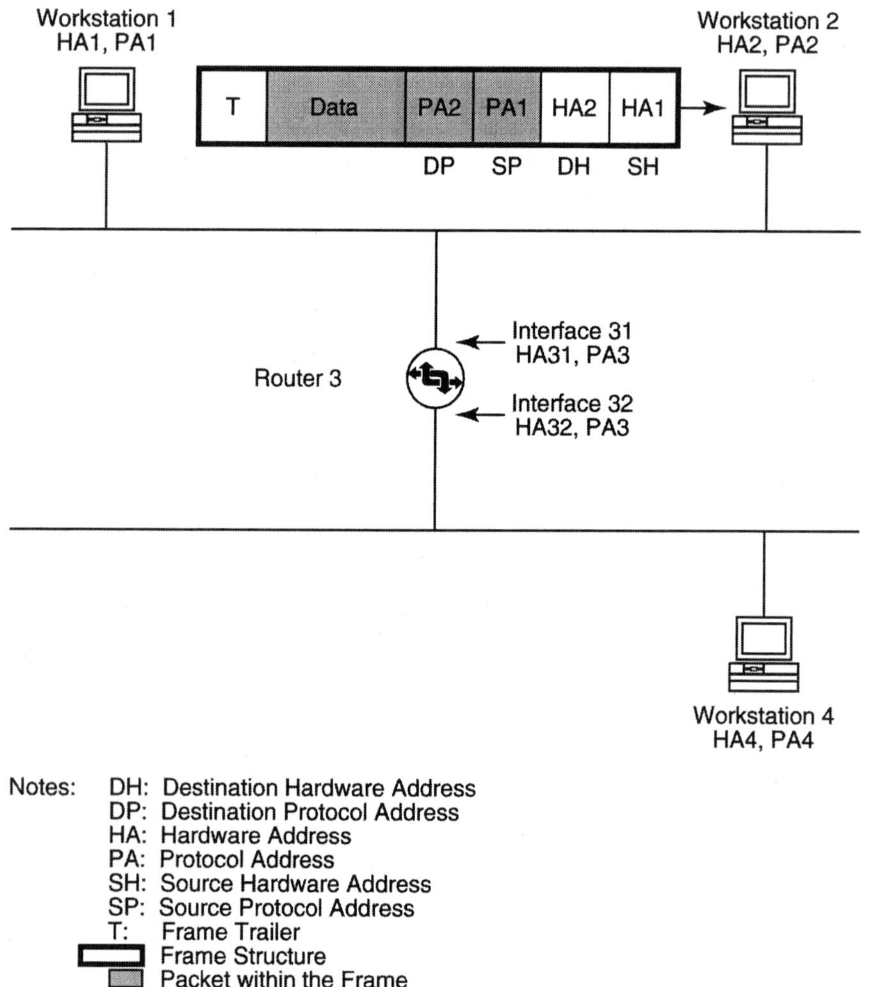

Figure 3-6a. Addressing Functions — Local Delivery.

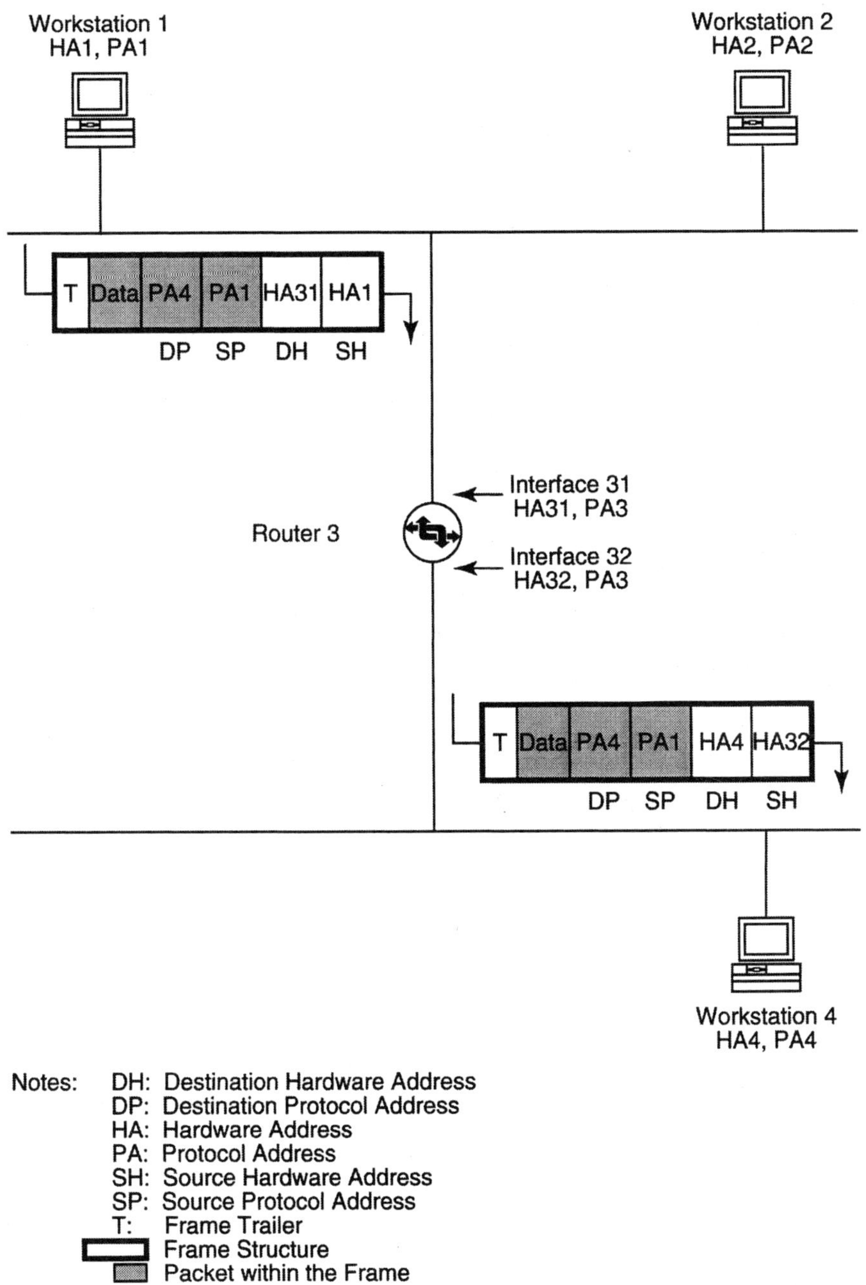

Figure 3-6b. Addressing Functions — Remote Delivery.

When Workstation 1 wants to communicate with Workstation 4, which is not on its LAN, Router 3 must be employed to make the physical connection, as seen in Figure 3-6b. (In other words, the router is your ticket off the local network.) In this case, Workstation 1 encapsulates the IP datagram inside a frame and sends that frame to the incoming interface (called Interface 31 in Figure 3-6b) on Router 3. The router performs a table lookup, determines that the destination address can be reached via its outgoing interface, and then internally sends that frame to Interface 32. As part of this process, Router 3 modifies both the source and destination hardware addresses between the input and output interfaces. The router's second interface (called Interface 32 in Figure 3-6b) is on the same LAN as the destination Workstation 4, which completes the delivery of the frame with its encapsulated IP datagram. Note that the hardware addresses change as the frame moves from Workstation 1 to Router 3 Interface 31 to Router 3 Interface 32 to Workstation 4. The protocol addresses, however, do not change from source to destination, which provides consistency (or an absolute) upon which to base the overall addressing and routing functions.

In many cases, we may know the protocol address of a device (such as the default router IP address that specifies our path off of this LAN), but we may not know the hardware address of that router's interface on our network. If we want to send information to a destination off of this LAN, we must send a frame to the router; to do so, we must know the hardware address of that router's interface. In other situations, we may know the hardware address of a device, but not its protocol address. Thus, since two mapping functions may be necessary (protocol to hardware or hardware to protocol), two mapping protocols have been developed. The Address Resolution Protocol (ARP), described in RFC 826 [3-10], translates from a protocol (IP) address to a hardware address. The Reverse Address Resolution Protocol (RARP), detailed in RFC 903 [3-11], does the opposite, as its name implies.

3.3.1 Address Resolution Protocol

Let's assume that a device, Host X, on an Ethernet wishes to deliver a datagram to another device on the same Ethernet, Host Y. Host X knows Host Y's destination protocol (IP) address, but does not know its hardware (Ethernet) address. It would, therefore, broadcast an ARP packet within an Ethernet frame to determine Host Y's hardware address. The ARP packet is shown in Figure 3-7a. The packet consists of 28 octets, primarily addresses, which are contained within the data field of a LAN frame. The sender broadcasts an ARP packet within a LAN frame requesting the information that it lacks. The device that recognizes its own protocol address responds with the sought-for hardware address. The individual fields of the ARP message show how the protocol operates.

0 1 2 3 4 5 6 7	8 9 1 1 1 1 1 1 (0 1 2 3 4 5)	1 1 1 1 2 2 2 2 (6 7 8 9 0 1 2 3)	2 2 2 2 2 2 3 3 (4 5 6 7 8 9 0 1) Bits
Hardware Type		Protocol Type	
HA Length	PA Length	Operation	
Sender HA (octets 0-3)*			
Sender HA (octets 4-5)		Sender PA (octets 0-1)	
Sender PA (octets 2-3)		Target HA (octets 0-1)	
Target HA (octets 2-5)			
Target PA (octets 0-3)			

* Field lengths assume HA = 6 octets and PA = 4 octets

Figure 3-7a. Address Resolution Protocol (ARP) and Reverse Address Resolution Protocol (RARP) Packet Formats.

The first field, Hardware (2 octets), defines the type of hardware in use. Current values are listed in the IANA online database. Examples include Hardware = 1 (Ethernet), 6 (IEEE 802 Networks), 7 (ARCNET), 11 (LocalTalk), 14 (SMDS), 15 (Frame Relay), and 16 (ATM). The second field, Protocol (2 octets), identifies the protocol address in use. For example, protocol = 0800H would identify IP addresses.

The second word allows the ARP packet to be used with a variety of address structures rather than being restricted to only two, such as IP (32 bits) and IEEE 802 (48 bits). This makes the protocol more adaptive. The Hardware Length (HLEN), 1 octet, and Protocol Length (PLEN), 1 octet, specify the lengths, in octets, of the addresses to be used. Figure 3-7a represents the most common scenario. The Hardware Address (HA) requires 6 octets or 48 bits (HLEN = 6); the Protocol Address (PA) needs 4 octets or 32 bits (PLEN = 4). The Operation field (2 octets) defines an ARP Request = 1 and an ARP Reply = 2.

The next fields contain the addresses themselves. With an ARP Request message, the target hardware address field is unknown and is sent filled with zeros. The ARP Reply packet from the target host inserts the requested address in that field. When it receives the ARP Reply, the originating station records that information in a table, called the ARP cache. The ARP cache reduces internetwork traffic asking to resolve the same address on multiple occasions. Routers' ARP caches have a finite lifetime, which prevents the table from growing too large.

3.3.2 Reverse Address Resolution Protocol

Most hosts on networks are smart enough to remember their own hardware and protocol addresses. But diskless workstations rely on the server for much of their intelligence. Such a workstation would know its hardware address, which is coded into ROM, but the server might assign a protocol address of which the workstation was unaware. The Reverse Address Resolution Protocol (RARP) can discover an unknown protocol address given a known hardware address and a RARP server to supply the answer. Note that the sought-for address is the sender's own IP address, and that the presence of the RARP server is critical to the operation of this protocol.

The process of determining an unknown protocol address is similar to that of finding an unknown hardware address. The same packet structure is used (review Figure 3-7a) with only minor modifications to the field values. The Operation field adds two new values: 3 (RARP Request) and 4 (RARP Reply). When the RARP Request is made, the Sender Hardware address, Sender Protocol address, and Target Hardware address are transmitted. The RARP Reply contains the sought-after Target Protocol address.

3.3.3 Inverse ARP

ARP allows a protocol address, such as an IP address, to be associated with a hardware address, such as Ethernet. In some cases, however, a virtual address identifies the endpoint of the connection, replacing the hardware address. Two examples of this are frame relay and ATM, where a virtual circuit address identifies the connection endpoint. In the case of frame relay, a permanent or switched virtual circuit (PVC or SVC) is identified by a Data Link Connection Identifier (DLCI). When a new PVC or SVC is provisioned (with its associated DLCI), the end user devices know of the PVC's or SVC's existence, but not the IP address associated with that circuit. For these applications, it is necessary to provide a mechanism to discover the Protocol address associated with that particular virtual circuit. The Inverse Address Resolution Protocol (InARP), defined in RFC 2390 [3-12], provides the details of these protocol operations.

Inverse ARP is an extension to ARP, and therefore uses the same packet format (review Figure 3-7a). The hardware address field length may be different from the LAN case; for example, frame relay would use a hardware address length of 2, 3, or 4 octets. Two new operation codes are defined: InARP Request and InARP Reply.

InARP operates in a similar fashion to ARP, with the exception that requests are not broadcast, because the hardware address (such as a DLCI) of the destination station is already known. Instead, the requesting station fills in the Sender and Target Hardware addresses plus the Sender Protocol address, and sends the InARP Request to the distant station, which fills in the missing Target Protocol address.

3.3.4 ATMARP

Another extension to ARP is provided for IP traffic over ATM-based internetworks, and is defined in RFC 2225, "Classical IP and ARP over ATM" [3-13]. The ATMARP protocol is based on both ARP (RFC 826) and Inverse ARP (RFC 2390), discussed above. ATMARP provides a mechanism for associating IP and ATM addresses for communication within a Logical IP Subnetwork. The ATMARP packet format expands on the ARP packet format and is illustrated in Figure 3-7b. The first two fields, Hardware Type (2 octets) and Protocol Type (2 octets), are used in the same way as in the original ARP. The hardware address assigned to the ATM Forum address family is 19 decimal (0013H); however, the Protocol Address field value for IP remains the same (0800H). Four new 1-octet fields define the type and length of the Sender ATM number (SHTL), the Sender ATM subaddress (SSTL), the Target ATM number (THTL), and the Target ATM subaddress (TSTL). The Operation Code (2 octets) defines five operation codes:

Operation Code	Operation
1	ARP Request
2	ARP Reply
8	InARP Request
9	InARP Reply
10	ARP NAK

                    1 1 1 1 1 1 1 1 1 1 2 2 2 2 2 2 2 2 2 2 3 3
0 1 2 3 4 5 6 7 8 9 0 1 2 3 4 5 6 7 8 9 0 1 2 3 4 5 6 7 8 9 0 1 Bits

Bits 0–7	Bits 8–15	Bits 16–23	Bits 24–31
Hardware Type		Protocol Type	
SHTL	SSTL	Operation	
SPA Length	THTL	TSTL	TPA Length
Sender ATM Address (octets 0-3)			
Sender ATM Address (octets 4-7)			
Sender ATM Address (octets 8-11)			
Sender ATM Address (octets 12-15)			
Sender ATM Address (octets 16-19)			
Sender PA			
Target ATM Address (octets 0-3)			
Target ATM Address (octets 4-7)			
Target ATM Address (octets 8-11)			
Target ATM Address (octets 12-15)			
Target ATM Address (octets 16-19)			
Target PA			

* Field lengths assume ATM Address = 20 octets and PA = 4 octets

Figure 3-7b. ATM Address Resolution Protocol (ATMARP) and Inverse ATM Address Resolution Protocol (InATMARP) Packet Formats.

Two 1-octet fields are also included in the third word that specify the Sender Protocol Address (SPA) length and the Target Protocol Address (TPA) length. The four addresses, Sender ATM, Sender Protocol, Target ATM, and Target Protocol, complete the packet. (Note that the format shown in Figure 3-7b assumes ATM address lengths of 20 octets and a Protocol Address length of 4 octets.)

3.3.5 Proxy ARP

When IP is run on a LAN, special attention to the addressing scheme is required. In most cases, it is not desirable for each segment to have a distinct network number; instead, subnetting a single network number is the preferred approach for distinguishing individual LAN segments. The details of this subnetted configuration need not be relevant to those outside the LAN, as long as a packet destined for a host on a particular LAN segment can be delivered correctly. RFC 925, "Multi-LAN Address Resolution" [3-14], deals with these issues; it defines a variation on the ARP process, known as Proxy ARP or promiscuous ARP, to handle these operations.

A proxy is a person or device that acts on behalf of another person or device. Suppose that two networks use the same IP network address, and that we call these two networks A and B. Router R, running ARP, connects these two networks. Router R will become the proxy device. Now suppose that a host on Network A wishes to communicate with another host on Network B, but doesn't know its hardware address. The host on Network A would broadcast an ARP Request seeking the hardware address of the remote host. Router R would intercept the request and reply with HA = R, which would be stored in the host's ARP cache. Subsequent communication between the two hosts would go via R, with Router R using its own lookup table to forward the packet to Network B. In effect, Proxy ARP allows the router to respond to an ARP request that came from one of its attached networks with the information concerning another one of its attached networks. In essence, the router deceives the originating host into thinking that it (i.e. the router) is the correct destination hardware address. In this way, the outside world is insulated from that LAN's address resolution operation.

3.4 Workstation Booting and Configuration

The Bootstrap Protocol (BOOTP), described in RFC 951 [3-15], allows workstation initialization functions that go beyond the address resolution functions of ARP/RARP. The protocol gets it name from the fact that it is meant to be contained within a bootstrap ROM. BOOTP is designed for diskless clients that need information from a server, such as their own IP address, the server's IP address, or the name of a file (i.e., the boot file) to be loaded into memory and then executed. The client broadcasts a Boot Request packet, which is answered by a Boot Reply packet from the server.

One of the significant differences between ARP/RARP and BOOTP is the layer of protocol they address. ARP/RARP packets are contained within local network frames and are transmitted on the local network. BOOTP packets are contained within IP datagrams, contain a UDP header, and are transmitted on the internetwork. The designated server can, therefore, be several router hops away from the client. Two reserved port numbers are used: port 67 for the BOOTP Server and port 68 for the BOOTP Client.

The individual fields of the BOOTP packet are shown in Figure 3-8a. The OpCode (Op, 1 octet) specifies a BOOTREQUEST (Op = 1) or BOOTREPLY (Op = 2). Hardware address Type (Htype, 1 octet) and

Hardware address Length (Hlen, 1 octet) are similar to those fields in the ARP/RARP packet. The Hops field (1 octet) is optional for use in cross-router booting. The Transaction ID (4 octets) correlates the boot requests and responses. The Seconds field (2 octets) allows the client to count the elapsed time since it started the booting sequence. Two unused octets complete the third word of the BOOTP packet.

The next four words designate the various IP addresses. The client states the addresses it knows and the server fills in the rest. These include the Client IP address (filled in by client), Your IP address (filled in by the server if the client does not know its own address), Server IP address, and Gateway router IP address. The Client HA (16 octets), Boot File Name (128 octets), and a Vendor-Specific Area (64 octets), containing information to be sent from the server to the client, complete the packet. RFC 1497 discusses implementation of that vendor-specific field. In summary, BOOTP improves on the ARP concept by allowing the address resolution process to occur across routers. Although it uses IP/UDP, it is small enough to fit within a bootstrap ROM on the client workstation.

RFC 1542 [3-16] provides clarifications and extensions to BOOTP, such as its use with IEEE 802.5 token ring networks. RFC 2132 [3-17] discusses extensions to BOOTP, plus a related protocol, the Dynamic Host Configuration Protocol, which will be discussed in further detail in the next section.

Bits: 0 1 2 3 4 5 6 7 8 9 10 11 12 13 14 15 16 17 18 19 20 21 22 23 24 25 26 27 28 29 30 31

Op	Htype	Hlen	Hops
Transaction ID			
Seconds		Unused	
Client IP Address			
Your IP Address			
Server IP Address			
Gateway IP Address			
Client Hardware Address (16 octets) ⋮			
Server Host Name (64 octets) ⋮			
Boot File Name (128 octets) ⋮			
Vendor-Specific Area (64 octets) ⋮			

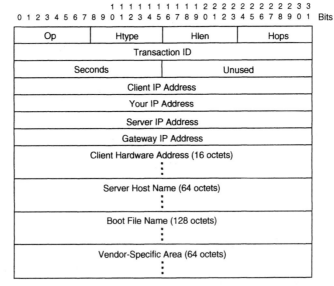

Figure 3-8a. Bootstrap Protocol (BOOTP) Packet Format.

3.4.1 Dynamic Host Configuration Protocol

The Dynamic Host Configuration Protocol (DHCP) provides a superset of BOOTP capabilities to include additional configuration options and automatic allocation of reusable network addresses. DHCP defines a mechanism to transmit configuration parameters to hosts using the ARPA protocol suite, with a message format based on the format of BOOTP. DHCP is defined in RFC 2131 [3-18], and the interoperation between DHCP and BOOTP is discussed in RFC 1534 [3-19].

DHCP is built on a client-server model in which the designated DHCP server allocates network addresses and delivers configuration information to the requesting client (another host). Examples of configuration information include: a default TTL, IP address, subnet mask, maximum transmission unit (MTU), list of default routers, list of static routes, ARP cache timeouts, and others. Two components are therefore included in DHCP: a mechanism for allocating network addresses to clients, and a protocol for delivering these host-specific configuration parameters to a dynamically configured host.

Seven types of DHCP messages are defined in RFC 2131:

1. DHCPDISCOVER: client broadcast to locate available servers.
2. DHCPOFFER: server to client response to the DHCPDISCOVER message, with an offer of configuration parameters.
3. DHCPREQUEST: client broadcast to servers, either:
    a. Requesting offered parameters from one server and implicitly declining offers from all others,
    b. Confirming correctness of previously allocated address after, for example, a system reboot, or
    c. Extending the lease on a particular network address.
4. DHCPACK: server to client with configuration parameters, including committed network address.
5. DHCPNAK: server to client refusing request for configuration parameters.
6. DHCPDECLINE: client to server indicating configuration parameters invalid.
7. DHCPRELEASE: client to server relinquishing network address and canceling remaining lease.

The DHCP message format is based on the BOOTP message format and is illustrated in Figure 3-8b. The first word contains four 1-octet fields: Op, Htype, Hlen, and Hops. The Op Code (Op) defines the message type, with 1 = BOOTREQUEST and 2 = BOOTREPLY. The Hardware Type (Htype) and Hardware Address Length (Hlen) fields define the hardware type and address, respectively, similar to the fields used in the ARP/RARP messages. The Hops field is optionally used by relay-agents when booting via a relay-agent. A client would set the Hops field to zero. The Transaction ID (also called XID) is a 32-bit random number chosen by the client; it is used by the client and the server to associate messages and responses between a client and a server. The Seconds field (2 octets) is filled in by the client, noting the seconds elapsed since the client started to boot. The Flags field (2 octets) contains a one-bit flag, called the broadcast flag, with all other bits set to zero. Details of the use of this flag are discussed in RFC 2131.

Bits 0–31 (0 1 2 3 4 5 6 7 8 9 10 11 12 13 14 15 16 17 18 19 20 21 22 23 24 25 26 27 28 29 30 31)

Bits 0–7	Bits 8–15	Bits 16–23	Bits 24–31
Op	Htype	Hlen	Hops
Transaction ID			
Seconds		Flags	
Client IP Address (ciaddr)			
Your IP Address (yiaddr)			
Server IP Address (siaddr)			
Gateway IP Address (giaddr)			
Client Hardware Address (chaddr – 16 octets)			
Server Name (sname – 64 octets)			
Boot File Name (file – 128 octets)			
Options (variable length)			

Figure 3-8b. Dynamic Host Configuration Protocol (DHCP) Message Format.

The Client IP Address field (4 octets) is set by the client, with its known address or all zeros. The Your IP Address field (4 octets) is set by the server if the Client IP Address field contains all zeros. The Server IP Address field (4 octets) is set by the server, and the Gateway (or router) IP Address is set by the relay-agent. The Client Hardware Address (16 octets) is set by the client. The Server Name is an optional server host name. The Boot File Name (128 octets) is used by the server to return a fully qualified directory path name to the client. Finally, the Options field (variable length) may contain DHCP options as defined in RFC 2132.

3.5 The Domain Name System

The 32-bit IP addresses and the various classes defined for these addresses provide an extremely efficient way to identify devices on an internetwork.Unfortunately, remembering all of those addresses can be overwhelming. To solve that problem, a system of hierarchical naming known as the Domain Name System was developed. DNS is described in RFCs 1034 [3-20] and 1035 [3-21].

DNS is based on several premises. First, it arranges the names hierarchically, like the numbering plan devised for the telephone network. Just as a telephone number is divided into a country code, an area code, an exchange code, and finally a line number, the DNS root is divided into a number of top-level domains, defined in RFC 920. These are:

Domain	**Purpose**
MIL	U.S. Military
GOV	Other U.S. Government
EDU	Educational
COM	Commercial
NET	NICs and NOCs
ORG	Nonprofit Organizations
CON	Two-letter Country Code, e.g. US represents the United States, CA represents Canada, and so on

As with many other addressing-related issues, the rapid growth of the Internet has stressed the number of available domain names. A number of additional top-level domains have been recently added, including .aero, .biz, .coop, .info, .int, .museum, .name, .net, and .pro. Information on the top-level domains can be found on the IANA website [3-22].

Returning to our discussion of top-level domains, specific sites are under each top-level domain. For example, the University of Ferncliff could use the edu domain designation (using the traditional lowercase letters), and it would be shown as ferncliff.edu. The University could then designate names for its departments, such as cs.ferncliff.edu (Computer Science) or ee.ferncliff.edu (Electrical Engineering). A particular host in the Electrical Engineering department could be named voltage.ee.ferncliff.edu. A user with a login on that host could be identified as boomer@voltage.ee.ferncliff.edu. Note that the @ sign separates the user from the remainder of the Host address.

The second DNS premise is that devices are not expected to remember the IP addresses of remote hosts. Rather, Name Servers throughout the internetwork provide this information. The requesting device thus assumes the role of a client, and the Name Server provides the necessary information, known as a resource record, or RR. RRs provide a mapping between domain names and network objects, such as IP addresses. Many different types of RRs are defined in RFCs 1034 and 1035, and types that are used for ISO NSAP addresses are defined in RFC 1706. Examples of RRs include: the A record, which is used to map a host address; the MX record, which provides a mail exchange for the domain and is used with the SMTP; the NS record, which defines the Name Server for a domain; and the PTR record, which is a pointer to another part of the domain name space.

The format for client/server interaction is a DNS message, shown in Figure 3-9. The message header is 12 octets long and describes the type of message. The next four sections provide the details of the query or response.

The first field within the header is an Identifier (16 bits) that correlates the queries and responses. The QR bit identifies the message type as a Query (QR = 0) or a Response (QR = 1). An OPCODE field (4 bits) further defines a Query:

OPCODE	Meaning
0	Standard Query (QUERY)
1	Inverse Query (IQUERY)
2	Server Status request (STATUS)
3–15	Reserved

Bits 0–15				Bits 16–31
ID	QR	OPCODE	Flags	RCODE
QDCOUNT	ANCOUNT			
NSCOUNT	ARCOUNT			

Question Section ⋮
Answer Section ⋮
Authority Section ⋮
Additional Information Section ⋮

Figure 3-9. Domain Name System (DNS) Message Format.

Four Flags are then transmitted to further describe the message:

Bit Number	Meaning
5	Authoritative Answer (AA)
6	Truncation (TC)
7	Recursion Desired (RD)
8	Recursion Available (RA)
9–11	Reserved (set to 0)

A Response Code (RCODE) completes the first word:

Field	Meaning
0	No error
1	Format error
2	Server error
3	Name error
4	Not implemented
5	Refused
6–15	Reserved for future use

The balance of the header contains fields that define the lengths of the remaining four sections:

Field	Meaning
QDCOUNT	Number of Question entries
ANCOUNT	Number of resource records in the Answer section
NSCOUNT	Number of Name Server resource records in the Authority section
ARCOUNT	Number of resource records in the Additional Records section

Following the header is the Question section and the Answer sections (Answer, Authority, or Additional Information).

We will look at an example of a DNS message in Section 7.1.

3.6 Looking Ahead

In this chapter we have covered the basic functions of datagram addressing and delivery, and the associated functions such as address mapping and IP-address-to-hostname lookups. In the next chapter, we will extend this discussion to consider how the routing infrastructure is kept updated with current information, and how errors or exception conditions are reported within the internetwork.

3.7 References

[3-1] Postel, J. "Internet Protocol." RFC 791, September 1981.

[3-2] The Internet Assigned Numbers Authority (IANA) maintains an online database of protocol numbers and parameters at http://www.iana.org/numbers.html. This replaces the former "Assigned Numbers" document, which was last published as RFC 1700 in October 1994.

[3-3] Kirkpatrick, S., et al. "Internet Numbers." RFC 1166, July 1990.

[3-4] Mogul, J., et al. "Internet Standard Subnetting Procedure." RFC 950, August 1995.

3-5] Krol, E. "The Hitchhiker's Guide to the Internet." RFC 1118, September 1989.

[3-6] Baker, F., Editor. "Requirements for IP Version 4 Routers." RFC 1812, June 1995.

[3-7] Rekhter, Y., et al. "Address Allocation for Private Internets." RFC 1918, February 1996.

[3-8] Srisuresh, P., and K. Egevang. "Traditional IP Network Address Translator (Traditional NAT)." RFC 3022, January 2001.

[3-9] Hinden, R. "Applicability Statement for the Implementation of Classless Inter-Domain Routing (CIDR)." RFC 1517, September 1993.

[3-10] Plummer, D. "An Ethernet Address Resolution Protocol, or Converting Network Protocol Addresses to 48-bit Ethernet Addresses for Transmission on Ethernet Hardware." RFC 826, November 1982.

[3-11] Finlayson, R., et al. "A Reverse Address Resolution Protocol." RFC 903, June 1984.

[3-12] Bradley, T., et al. "Inverse Address Resolution Protocol." RFC 2390, September 1998.

[3-13] Laubach, M. "Classical IP and ARP over ATM." RFC 2225, April 1998.

[3-14] Postel, J. "Multi-LAN Address Resolution." RFC 925, October 1984.

[3-15] Gilmore, John. "Bootstrap Protocol (BOOTP)." RFC 951, September 1985.

[3-16] Wimer, W. "Clarifications and Extensions for the Bootstrap Protocol." RFC 1542, October 1993.

[3-17] Alexander, S., and R. Droms. "DHCP Options and BOOTP Vendor Extensions." RFC 2132, March 1997.

[3-18] Droms, R. "Dynamic Host Configuration Protocol." RFC 2131, March 1997.

[3-19] Droms, R. "Interoperation Between DHCP and BOOTP." RFC 1534, October 1993.

[3-20] Mockapetris, P. "Domain Names: Concepts and Facilities." RFC 1034, November 1987.

[3-21] Mockapetris, P. "Domain Names: Implementation and Specification." RFC 1035, November 1987.

[3-22] The Internet Assigned Numbers Authority (IANA) maintains a database of information on top-level domains, including country code designations, at: http://www.iana.org/domain-names.htm.

4

Routing and Intranetwork Communication

So far, we've learned that hosts transmit datagrams and use a 32-bit address to identify the source and destination of a datagram. The host drops the datagram into the internetwork, and the datagram somehow finds its way to its destination. That "somehow" is the work of routers, which examine the destination address, compare that address with internal routing tables, and send the datagram on the correct outgoing communication circuit. However, since the network topology can change, and since circuits can get congested or fail, these routing tables must be kept up to date with current information. Inter-router protocols, such as Open Shortest Path First (OSPF) and the Border Gateway Protocol (BGP), provide this function. In addition, when problems occur within a network, or when connectivity between two endpoints needs to be confirmed, a mechanism to support this testing is required. These functions are handled by the Internet Control Message Protocol (ICMP), which is an adjunct to IP. We will also consider two other protocols: the Internet Group Management Protocol (IGMP), which is closely related to ICMP, and the Resource Reservation Protocol (RSVP), which reserves bandwidth along a path in support of high bandwidth applications, such as voice and video transport. We will begin by looking at the routing protocols.

4.1 Datagram Routing

On the surface, a router has a very simple function: to examine incoming packets and get them on their way to the correct destination as quickly as possible. However, to accomplish this requires several functions. First, the router maintains a routing table that gathers information from other routers about all routes from all routing protocols in operation. This routing information is used to create a forwarding table, which determines the optimum path for each packet and makes the packet forwarding decisions at that router. (In many cases, the router's operating system shields the end user from the distinction between these routing and forwarding functions, and they therefore appear as a single route decision process.) The entries in the forwarding table may be static (i.e., manually built) and fixed for all network conditions, or dynamic (i.e., constructed by the router according to the current topology and conditions). Dynamic routing is considered the better technique because it adapts to changing network conditions. The router uses a metric, or measurement, of the shortest distance between two endpoints to help determine the optimum path. It determines the metric using a number of factors, including the shortest distance, or least cost path, to the destination. The router plugs the metric into one of two algorithms to make a final decision on the correct path. A Distance Vector algorithm makes its choice based on the distance to a remote node. A Link State algorithm also includes information about the status of the various links connecting the nodes and the topology of the network.

The various routers within the network use the Distance Vector or Link State algorithms to inform each other of their current status. Because routers use them for intranetwork communication, the protocols that make use of these algorithms are referred to as Interior Gateway Protocols (IGPs). The Routing Information Protocol (RIP) is an IGP based on a Distance Vector algorithm. The Open Shortest Path First (OSPF) protocol is an IGP based on a Link State algorithm. We'll look at these two algorithms separately in the following sections. If one network wishes to communicate routing information to another network, it uses an Exterior Gateway Protocol (EGP). Two EGPs will also be considered in this chapter: one simply called the Exterior Gateway Protocol, and another called the Border Gateway Protocol (BGP).

4.1.1 Routing Information Protocol

The Routing Information Protocol, described in RFC 1058 [4-1], is used for inter-router (or inter-gateway) communications. RIP is based on a Distance Vector algorithm, sometimes referred to as a Ford-Fulkerson algorithm after its developers L. R. Ford, Jr. and D. R. Fulkerson. A Distance Vector algorithm is one in which the routers periodically exchange information from their routing tables. The routing decision is based on the best path between two devices, which is often the path with the fewest hops or router transversals. The Internet standard warns that many LAN operating systems have their own RIP implementations. Therefore, it is important to adhere to the Internet standard to alleviate interoperability problems. There are, however, enhancements to RIP that are in the Internet Standards track. One example is RFC 2091 [4-2], "Triggered Extensions to RIP to Support Demand Circuits," which discusses the use of RIP over WAN links such as X.25 and ISDN.

RFC 1058 acknowledges several limitations to RIP. RIP allows a path length of 15 hops, which may be insufficient for large internetworks. Routing loops are possible for internetworks containing hundreds of networks because of the time required to transmit updated routing table information. Finally, the metrics used to choose the routing path are fixed and do not allow for dynamic conditions, such as a measured delay or a variable traffic load.

RIP assumes that all devices (hosts and routers) contain a table of RIP-derived routing information. This table contains several entries: the IP address of the destination; the metric, or cost, to get a datagram from the host to the destination; the address of the next router in the path to the destination; a flag indicating whether the routing information has been recently updated; and timers.

Routing information is exchanged via RIP packets, shown in Figure 4-1a, which are transmitted to/from UDP port number 520. The packet begins with a 32-bit header and may contain as many as 25 messages giving details on specific networks. The first field of the header is one octet long and specifies a unique command. Values include:

Command	Meaning
1	Request for routing table information
2	Response containing routing table information
3	Traceon (obsolete)
4	Traceoff (obsolete)
5	Reserved for Sun Microsystems
9	Update Request (from RFC 2091)
10	Update Response (from RFC 2091)
11	Update Acknowledge (from RFC 2091)

The second octet contains a RIP Version Number. Octets 3 and 4 are set equal to zero. The next two octets identify the Address Family being transmitted within that RIP packet; RFC 1058 only defines a value for IP with Address Family ID = 2.

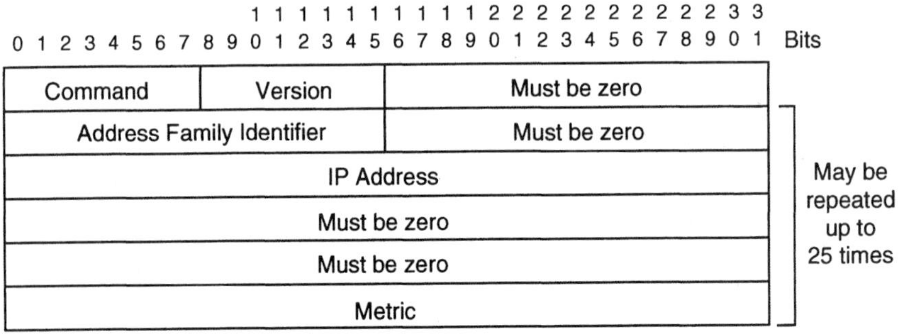

Bits 0–7	Bits 8–15	Bits 16–31
Command	Version	Must be zero
Address Family Identifier		Must be zero
IP Address		
Must be zero		
Must be zero		
Metric		

(Address Family Identifier through Metric: May be repeated up to 25 times)

Figure 4-1a. Routing Information Protocol (RIP) Packet Format.

The balance of the RIP packet contains entries for routing information. Each entry includes the destination IP Address and the Metric to reach that destination. Metric values must be between 1 and 15, inclusive. A metric of 16 indicates that the desired destination is unreachable. Up to 25 of these entries (from the Address Family Identifier through the Metric) may be contained within the datagram.

An extension to the Routing Information Protocol, called RIP version 2, is defined in RFC 2453 [4-3]. This enhancement expands the amount of useful information carried in RIP messages and also adds a measure of security to those messages.

The RIP version 2 packet format is very similar to the original format, containing a four-octet header and up to 25 route entries of 20 octets each in length (see Figure 4-1b). Within the header, the Command, Address Family Identifier, IP Address, and Metric fields are identical to their counterparts used with the original RIP packet format. The Version field specifies version number 2, and the two-octet unused field (filled with all zeros) is ignored. The Route Tag field carries an attribute assigned to a route that must be preserved and readvertised with a route, such as information defining the routing information's origin (either intra- or internetwork). Within each route entry, the Subnet Mask field (four octets) defines the subnet mask associated with a routing entry. The Next Hop field (also four octets) provides the immediate next hop IP address for the packets specified by this routing entry. The spaces now occupied by the Subnet Mask and Next Hop fields were previously filled with all zeros.

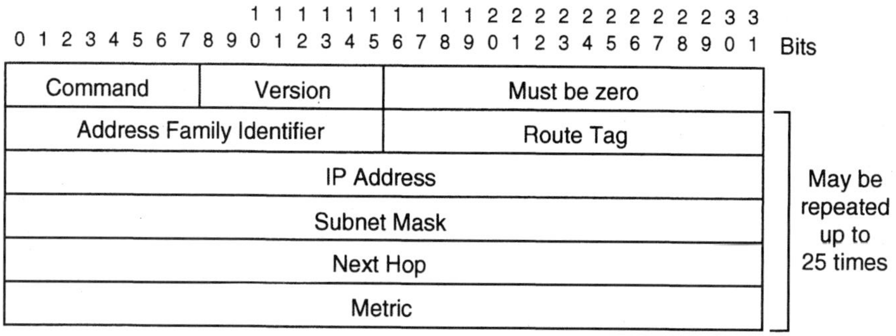

Bits 0–31

Bits 0–7	Bits 8–15	Bits 16–31	
Command	Version	Must be zero	
Address Family Identifier		Route Tag	May be repeated up to 25 times
IP Address			
Subnet Mask			
Next Hop			
Metric			

Figure 4-1b. Routing Information Protocol Version 2 (RIPv2) Packet Format.

The first routing entry in a RIP version 2 packet may contain authentication information to identify the origin of the subsequent data. If that authentication information is included in the packet, the maximum number of routing entries is reduced to 24 (from 25). To use this option, the Address Family Identifier is set for FFFFH, the Route Tag field specifies an Authentication Type, and the remaining 16 octets contain the authentication, such as a password, as shown in Figure 4-1c.

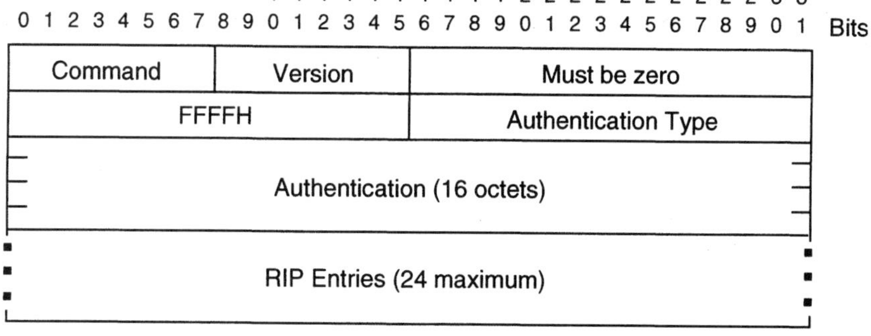

Figure 4-1c. Authenticated RIPv2 Packet Format.

For further details on RIP version 2, consult RFC 2453 or the companion documents RFC 1721 ("RIP version 2 Protocol Analysis"), RFC 1722 ("RIP version 2 Applicability Statement"), or RFC 1724 ("RIP version 2 Management Information Base — MIB").

Many WAN circuits are established based on traffic demand and are disconnected when that traffic subsides or ceases. If routing information was periodically transmitted on that connection (say every 30 seconds), the connection would never be closed, and the WAN circuit costs would increase. With fixed-bandwidth WAN circuits, such as a point-to-point leased line, the periodic transmission of routing updates may significantly impact the amount of end-user data that can be conveyed. As a result of these limitations, RFC 2091 [4-2] was written as a modification to RIP and RIPv2 so that routing information would only be sent on the WAN when a significant change had occurred, such as an update to the routing database or a change in the reachability of a next-hop router. The protocol modification is known as Triggered RIP, in which routing updates are transmitted (or triggered) by one of the following actions:

- When a specific request for a routing update has been received.
- When the routing database is modified by new information from another interface.
- When the circuit manager indicates that a destination has changed from an unreachable to a reachable state.
- When the unit is first powered on (and thus becoming reachable).

To support the Triggered RIP updates, three new packet types are defined: Update Request, Update Response, and Update Acknowledge. The Update Request packet, a request to send all appropriate elements from its routing system, is transmitted when a router is first powered on or a circuit changes from an unreachable to a reachable state. The Update Response is transmitted in response to an Update Request packet, and is sent at periodic intervals until an Update Acknowledge packet is received.

An Update Header is added to the RIP or RIPv2 packet, and it precedes any routing-specific information in that packet, as shown in Figure 4-1d. This header may take on one of two forms. The first is an Update Request Header, which includes a header Version (currently 1), followed by 24 bits of zero, as shown in Figure 4-1e. The second form is used for an Update Response or Update Acknowledge Header, as shown in Figure 4-1f. This header includes the header Version; a Flush field, which is set to 1 in Response headers when the peer is required to start timing out its entries; and a Sequence Number field, which is incremented for every new Update Response packet transmitted.

Command	Version	Must be zero
Update Header		
Address Family Identifier		Zero (RIP) or Route Tag (RIPv2)
RIP Entries		

Figure 4-1d. Update Header Location within the RIP and RIPv2 Packet.

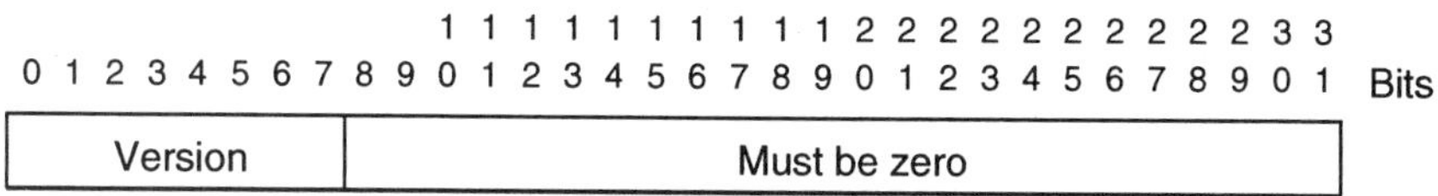

Figure 4-1e. Update Request Header for RIP and RIPv2.

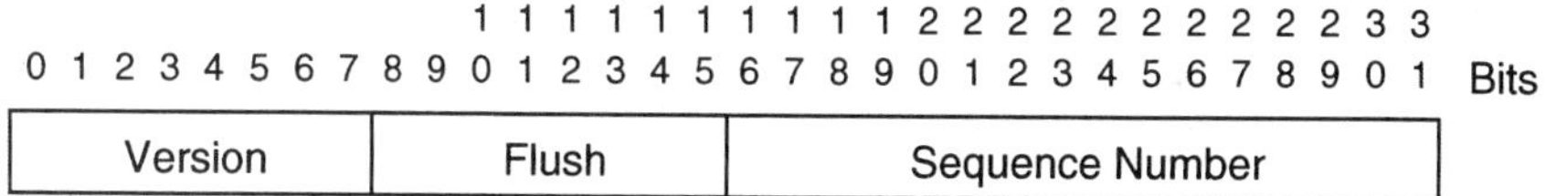

Figure 4-1f. Update Response and Acknowledge Headers for RIP and RIPv2.

4.1.2 Open Shortest Path First Protocol

The Open Shortest Path First protocol is a Link State algorithm that offers several advantages over RIP's Distance Vector algorithm. These advantages include: the ability to configure hierarchical (instead of flat) topologies; to quickly adapt to changes within the internet; to allow for large internetworks; to calculate multiple minimum-cost routes that allow traffic load to be balanced over several paths; to authenticate the exchange of routing table information; and to permit the use of variable-length subnet masks. The protocol uses the IP address and Type of Service field of each datagram (review Figure 3-2) for its operation. An optimum path can be calculated for each Type of Service.

The Internet standard for OSPF version 2 is RFC 2328 [4-4]. OSPF protocol analysis and experience are discussed in RFC 1245 and RFC 1246, respectively. The OSPF version 2 Management Information Base (MIB) is discussed in RFC 1850, and the Applicability Statement for OSPF is found in RFC 1370.

4.1.2.1 Comparing Routing Algorithms

Before considering the improvements of OSPF over RIP, let's review some of the characteristics of Distance Vector algorithms. First, a DVA routes its packets based on the distance, measured in router hops, from the source to the destination. With RIP the maximum hop count is 16, which is a possible limitation for large networks. A DVA-based network is a flat network topology, without a defined hierarchy to subdivide the network into smaller, more manageable pieces. In addition, the hop count measurement does not account for other factors in the communication link, such as the speed of that link or its associated cost. Furthermore, RIP broadcasts its complete routing table to every other router every 30 seconds. As we saw in Figure 4-1a, the RIP packet may contain information for up to 25 routes. If a router's table contains more entries, say 100 routes, then transmitting all of these routes would require a total of four RIP packets. This requires considerable overhead at each router for routing information processing, and it consumes valuable bandwidth on the WAN links in between these routers.

The improvements obtained with a Link State algorithm come in several areas. First, an LSA is based on type of service routing, not hop counts. This allows the network manager to define the least-cost path between two network points based on the actual cost, delay characteristics, reliability factors, and so on. Secondly, OSPF defines a hierarchical, not a flat, network topology. This allows the routing information to be distributed to only a relevant subset of the routers in the internetwork instead of to all of the routers. This hierarchical structure reduces both the router processing time and the bandwidth consumed on the WAN links.

An *autonomous system* (AS), used with an LSA, is defined as a group of routers that exchange routing information via a common routing protocol. The AS is subdivided into areas, which are collections of contiguous routers and hosts that are grouped together, much like the telephone network is divided into area codes. The topology of an area is invisible from outside that area, and routers within a particular area do not know the details of the topology outside of that area. When the AS is partitioned into areas, it is no longer likely (as was the case with a DVA) that all routers in the AS are storing identical topological information in their databases. A router would have a separate topological database for each area it is connected to; however, two routers in the same area would have identical topological databases. A backbone is also defined, which connects the various areas and is used to route a packet between two areas.

Different types of routers are used to connect the various areas. Internal routers operate within a single area, connect to other routers within that area, and maintain information about that area only. An area border router attaches to multiple areas, runs multiple copies of the basic routing algorithm, and condenses the topological information about its attached areas for distribution to the backbone. A backbone router is one that has an interface to the backbone, but it does not have to be an area border router. Lastly, an AS router is one that exchanges information with routers that belong to other Autonomous Systems.

4.1.2.2 OSPF Operation and Packet Formats

The basic routing algorithm for OSPF provides several sequential functions, as defined in RFC 2328, Section 4: discovering a router's neighbors and electing a Designated Router for the network using the OSPF Hello protocol; forming adjacencies between pairs of routers and synchronizing the databases of these adjacent routers; performing calculations of routing tables; and flooding the area with link state advertisements.

These protocol operations are performed using one of five OSPF packets. The OSPF packets are carried within IP datagrams and are designated as IP protocol = 89. If the datagram requires fragmentation, the IP process handles that function. The five OSPF packet types have a common 24-octet header as shown in Figure 4-2a. The first 32-bit word includes fields defining a Version Number (1 octet), an OSPF Packet Type (1 octet), and a Packet Length (2 octets), which measures the length of the OSPF packet including the header. The five packet types defined are:

Type	Packet Name	Protocol Function
1	Hello	Discover/maintain neighbors
2	Database Description	Summarize database contents
3	Link State Request	Database download
4	Link State Update	Database update
5	Link State Acknowledgment	Flooding acknowledgment

The next two fields define the Router ID of the source of that packet (4 octets) and the Area ID (4 octets) that the packet came from. The balance of the OSPF packet header contains a Checksum (2 octets), an Authentication Type (AuType, 2 octets), and an Authentication field (8 octets), used to validate the packet.

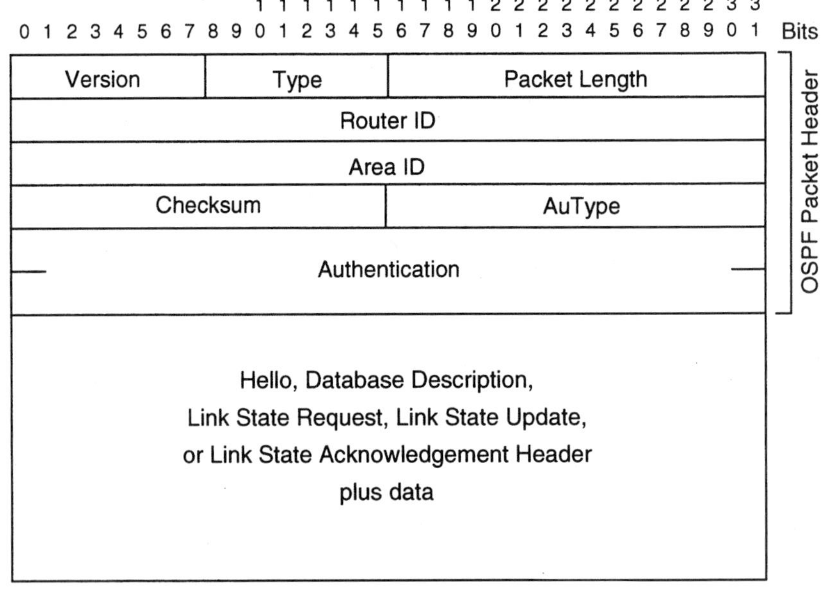

Figure 4-2a. Open Shortest Path First (OSPF) Packet Header Format.

Three types of authentication processes are defined in RFC 2328. Each of these options is configurable on a per-network, or per-network/subnet basis:

AuType	Description
0	Null authentication
1	Simple Password
2	Cryptographic authentication
All others	Reserved for IANA assignment

The null authentication type indicates that the routing exchanges over this network/subnet are not authenticated, and the contents of the 64-bit Authentication field can contain any data, as that field is not examined by the receiver. The Simple Password authentication type uses a 64-bit password, which is configured on a per-network basis. All packets sent on a particular network must have this configured value in the 64-bit Authentication field of their OSPF header. Note that this password is sent as clear text, and could be compromised by an intruder eavesdropping on the network with a protocol analyzer. The third option uses Cryptographic authentication, in which a shared secret key is configured in all routers attached to a common network/subnet. This secret key is used to generate and verify a message digest that is appended to the end of the OSPF packet. The message digest is a function of the OSPF protocol packet and the secret key, and since the key is not sent over the network as clear text, protection is provided against passive attacks. When Cryptographic authentication is used, the Authentication field includes additional parameters, including: a Key ID that identifies the algorithm and secret key in use; the Authentication Data Length, which indicates the length of the message digest appended to the OSPF packet, given in octets; and a Cryptographic Sequence Number, which guards against replay attacks (Figure 4-2b).

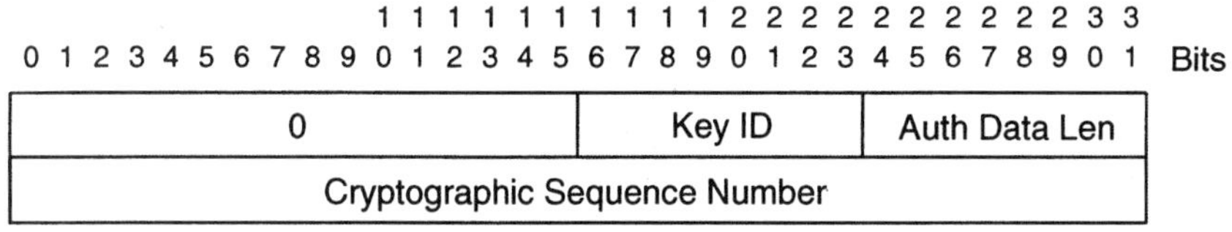

Figure 4-2b. OSPF Packet Header Authentication Field (Cryptographic Authentication).

Following the OSPF packet header is another header specific to the routing information being conveyed. Hello packets (OSPF Packet Type = 1) are periodic transmissions that convey information about neighboring routers (see Figure 4-2c). Fields within the Hello packet include: a Network Mask (4 octets), the network mask associated with this interface; a Hello Interval (2 octets), the number of seconds between this router's Hello packets; Options (1 octet), this router's optional capabilities, as described in RFC 2328, Section A.2; Router Priority (Rtr Pri, 1 octet), this router's router priority, used in Designated (or backup) router election; Router Dead Interval

(4 octets), the number of seconds before declaring a silent router down; Designated Router (4 octets), the identity of the Designated Router for this network; Backup Designated Router (4 octets), the identity of the Backup Designated Router for this network; and Neighbor (4 octets), the Router IDs of each router from whom valid Hello packets have been seen recently (i.e. in the last Router Dead Interval seconds) on the network.

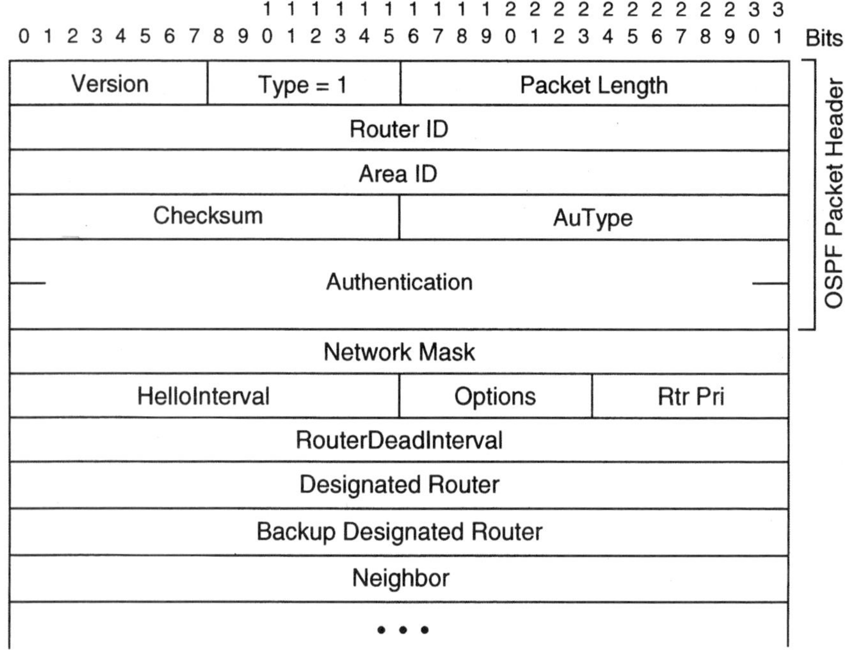

Figure 4-2c. OSPF Hello Packet Format.

Database Description packets (OSPF Packet Type = 2) convey information needed to initialize the topological databases of adjacent devices (see Figure 4-2d). Fields within the Database Description packet include: Interface MTU (2 octets), the size, in octets, of the largest IP datagram that can be sent over the associated interface without fragmentation; Options (1 octet), this router's optional capabilities, as described in RFC 2328, Section A.2; I-bit, the Init bit (when set to one, this packet is the first in the sequence of Database Description packets); M-bit, the More bit (when set to one, it indicates that more Database Description packets are to follow); the MS-bit, the

Master/Slave bit (when set to one, it indicates that the router is the master during the Database Exchange process; otherwise the router is the slave); and Database Description Sequence Number (DD Sequence Number, 4 octets), used to sequence the collection of Database Description Packets. The remainder of the packet consists of a list of the topological database's pieces. Each link state advertisement in the database is described by its link state advertisement header (described below).

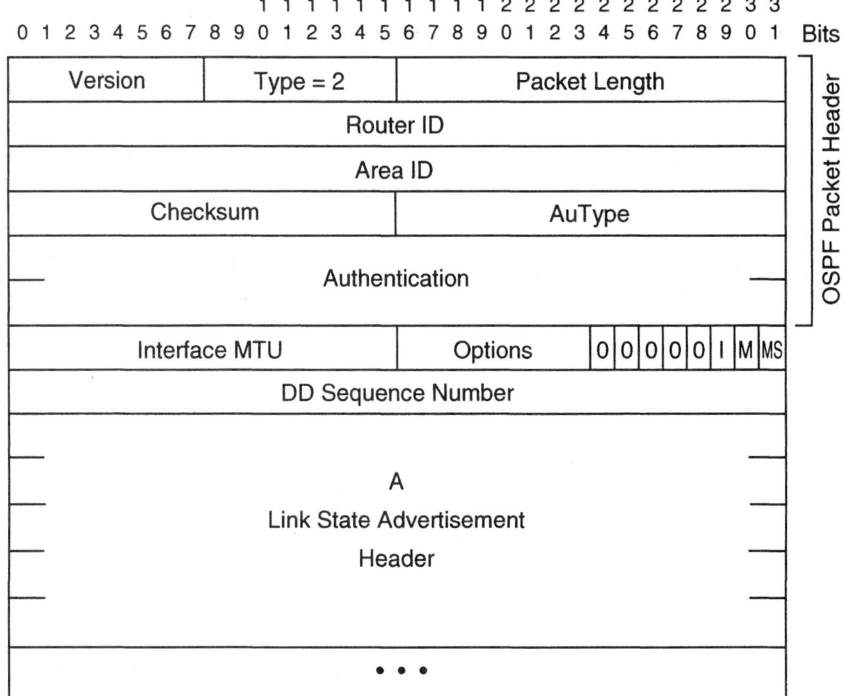

Figure 4-2d. OSPF Database Description Packet Format.

Link State Request packets (OSPF Packet Type = 3) obtain current database information from a neighboring router (see Figure 4-2e). Fields within the Link State Request packet include the Link State type (4 octets), which defines the type of the Link State advertisement and is more fully described below. The Link State ID (4 octets) is a unique identification for the advertisement, and the Advertising Router (4 octets) is the identification of the router that originated the link state advertisement.

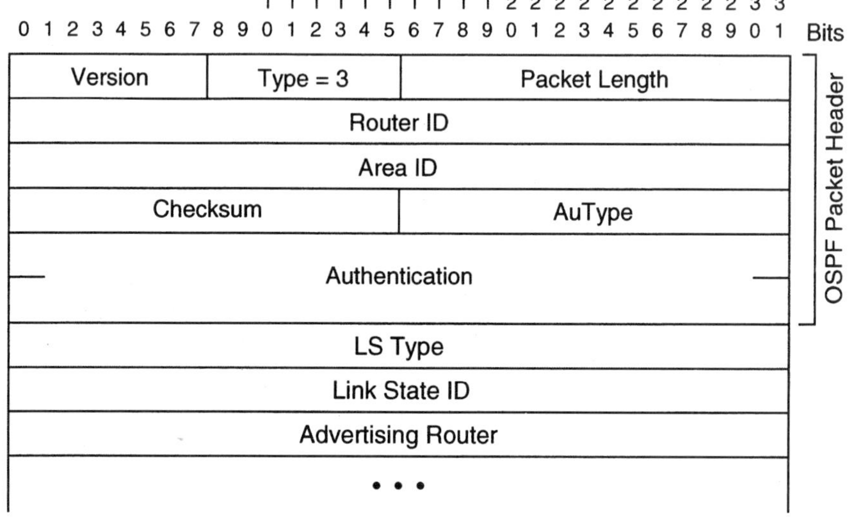

Figure 4-2e. OSPF Link State Request Packet Format.

Link State Update packets (OSPF Packet Type = 4) advertise the status of various links within the internet (see Figure 4-2f). Several link state advertisements may be included in a single packet. The Number advertisements field (# Advertisements, 4 octets) defines the number of link state advertisements included in this update. The body of the Link State Update packet consists of a list of link state advertisements. Each of these advertisements begins with a common 20-octet header, followed by one of five link state advertisements:

LS Type	Advertisement Name	Advertisement Description
1	Router link	Describes the states of the router's interfaces to an area
2	Network link	Lists the routers connected to the network
3	Summary link	Describes a route to a network destination outside the area
4	Summary link	Describes a route to an AS boundary router destination outside the area
5	AS external link	Describes a route to a destination in another Autonomous System

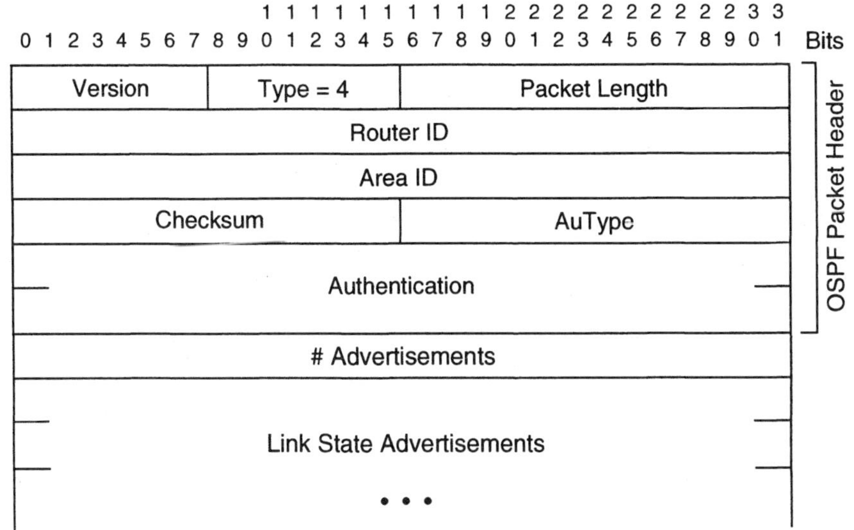

Figure 4-2f. OSPF Link State Update Packet Format.

The Link State header contains enough information to uniquely identify the advertisement (see Figure 4-2g). Fields within the Link State Advertisement header include: Link State Age (LS Age, 2 octets), the time in seconds since the link state advertisement was originated; Options (1 octet), the optional capabilities supported by the described portion of the routing domain and defined in RFC 2328, Section A.2; Link State type (LS Type, 1 octet), the type of link state advertisement, as described in the table above; Link State ID (4 octets), an identifier of the portion of the internet environment that is being described by the advertisement, with the contents of this field dependent on the advertisement's link state type (as above); Advertising Router (4 octets), the Router ID of the router that originated the link state advertisement; Link State Sequence Number (LS Sequence Number, 4 octets), which detects old or duplicate link state advertisements; Link State Checksum (LS Checksum, 2 octets), a checksum for the Link State advertisement; and Length (2 octets), the length, in octets, of the link state advertisement.

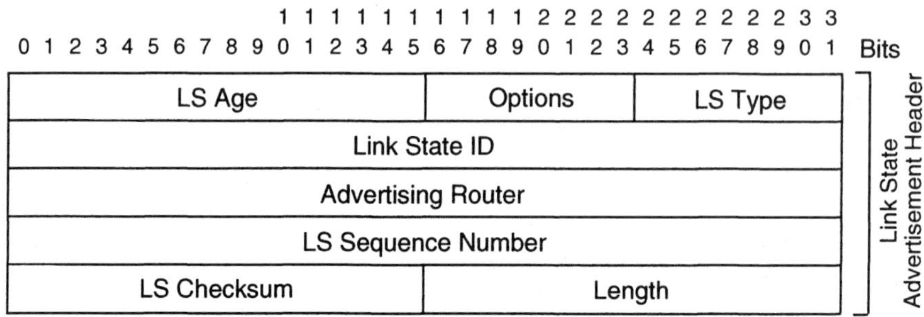

Figure 4-2g. OSPF Link State Advertisement Header Format.

The Router Links Advertisement packet (the Type 1 link state advertisement) describes the state and cost of the router's links (or interfaces) to the area (see Figure 4-2h). This packet begins with the Link State advertisement header (20 octets, described above). Other fields within this packet include: V-bit, which, when set, indicates that the router is an endpoint of an active virtual link that is using the described area; E-bit, which, when set, indicates that the router is an external (or AS boundary) router; B-bit, which, when set, indicates that the router is an area border router; and Number of Links (# Links, 2 octets), the number of router links described by this advertisement.

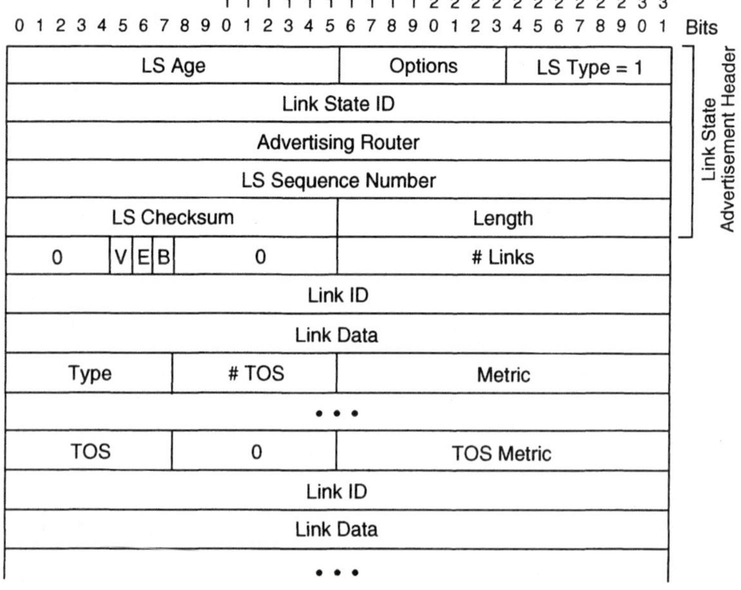

Figure 4-2h. OSPF Router Links Advertisement Packet Format.

Seven fields, consuming 16 octets, are then used to describe each router link. The third descriptive field, Type (1 octet), provides a description of the router link, which may be one of the following:

Type	Description
1	Point-to-point connection to another router
2	Connection to a transit network
3	Connection to a stub network
4	Virtual link

The first descriptive field, Link ID (4 octets), identifies the object that this router link connects to; it depends on the link's Type, described above. The values of the Link ID are:

Type	Link ID
1	Neighboring router's Router ID
2	IP address of Designated Router
3	IP network/subnet number
4	Neighboring router's Router ID

The contents of the Link Data field (4 octets) also depend on the Type field. For example, for connections to stub networks, this field specifies the network's IP address mask. For each link, separate metrics may be specified for each Type of Service (TOS). The Number of TOS field (# TOS, 1 octet) provides the number of different TOS metrics given for this link, not counting the required metric for TOS 0. The Metric field (2 octets) provides the cost for using this router link. The Type of Service field (TOS, 1 octet) indicates the IP Type of Service that this metric refers to, while the TOS Metric field (2 octets) indicates TOS-specific metric information.

The Network Links Advertisement packet (the Type 2 link state advertisement) describes all routers attached to the network, including the Designated Router (see Figure 4-2i). There are two fields in addition to the Link State Advertisement header: Network Mask (4 octets), the IP address mask for the network; and Attached Router (4 octets), the Router IDs of each of the routers attached to the network.

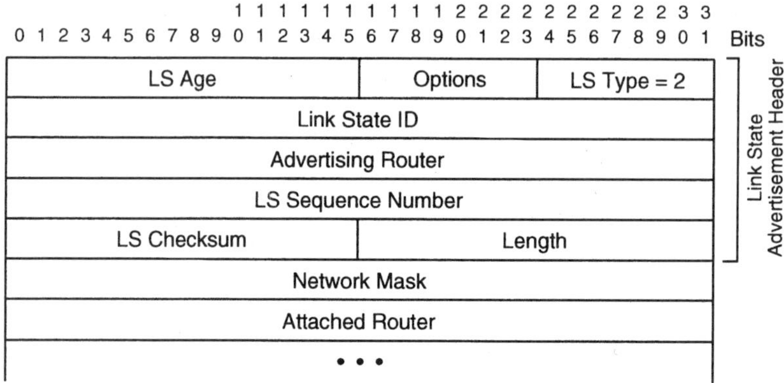

Figure 4-2i. OSPF Network Links Advertisement Packet Format.

The Summary Links Advertisements are originated by area border routers and describe destination links outside of the area. Type 3 link state advertisements are used when the destination is an IP network; when the destination is an AS boundary router, a Type 4 Summary Links Advertisement is used (see Figure 4-2j). There are three fields in addition to the Link State Advertisement header: Network Mask (4 octets) indicates the destination network's IP address mask for Type 3, and is not meaningful and is set to zero for Type 4; Type of Service (TOS, 1 octet) indicates the type of service that the following cost concerns; and Metric (3 octets) indicates the cost of this route, expressed in the same units as the interface costs in the Router Links advertisements.

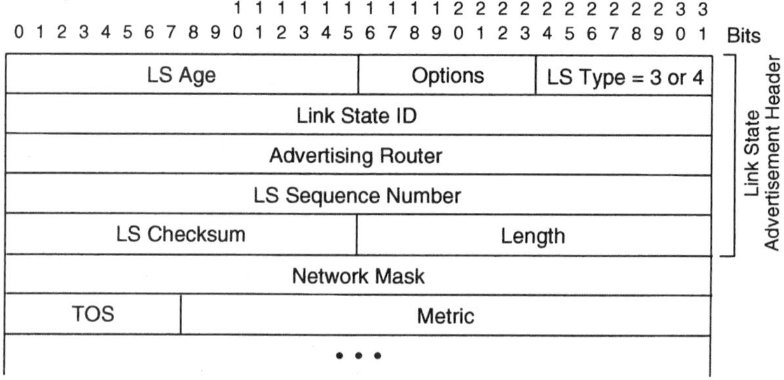

Figure 4-2j. OSPF Summary Links Advertisement Packet Format.

AS External Links Advertisements are originated by AS boundary routers and advertise destinations that are external to the AS (see Figure 4-2k). Fields within the AS External Links Advertisement Packet include: Network Mask (4 octets), the IP address mask for the advertised destination; E-bit, which indicates the type of external metric; Metric (3 octets), the cost of this route; Type of Service (TOS, 7 bits), the type of service that the following cost concerns; TOS Metric (3 octets), which provides TOS-specific metric information; Forwarding Address (4 octets), which specifies where data traffic for the advertised destination will be forwarded to; and External Route Tag (4 octets), a field attached to each external route, which is not used by the OSPF protocol itself but which may be used to communicate information between AS boundary routers.

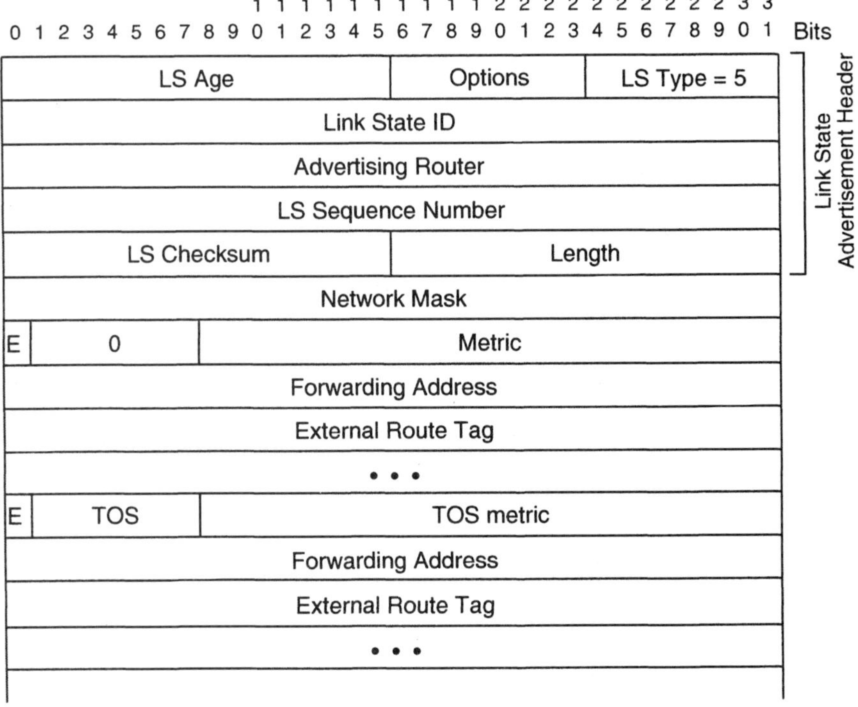

Figure 4-2k. OSPF External Links Advertisement Packet Format.

Finally, Link State Acknowledgment packets (OSPF Type = 5) verify the receipt of database information (see Figure 4-21). The format of this packet is similar to that of the Database Description packet, and contains a list of Link State advertisement headers.

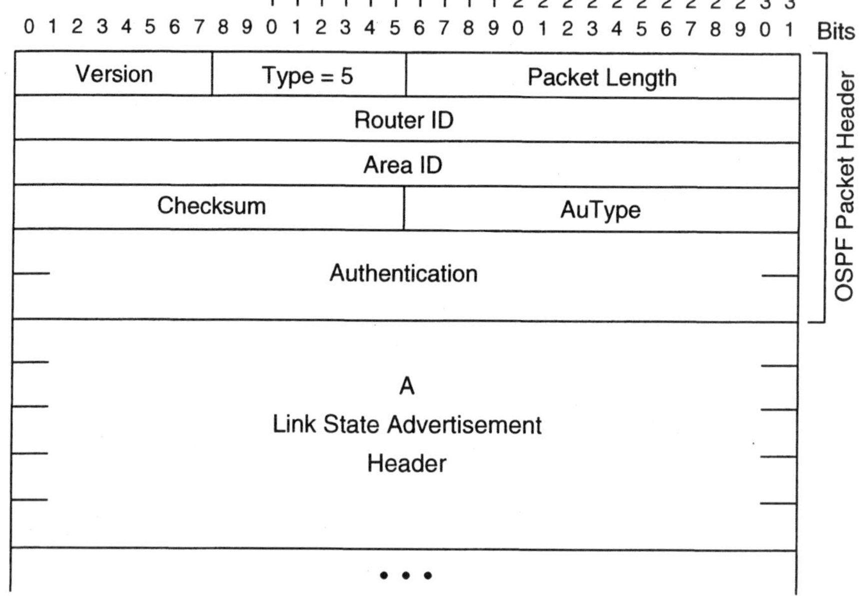

Figure 4-21. OSPF Link State Acknowledgment Packet Format.

4.1.3 Exterior Gateway Protocol (EGP)

Recall from our earlier discussion that there are two general categories of routing protocols: Interior Gateway Protocols and Exterior Gateway Protocols. In short, interior protocols are concerned with routing within a network. Exterior protocols are concerned with routing between Autonomous Systems, which are generally described as groups of routers that all fall within a single administrative domain. In other words, exterior protocols facilitate your communication with networks outside of your router's domain. In this section, we will briefly study the Exterior Gateway Protocol (EGP), defined in RFC 904 [4-5]. In the next section, we will go into greater detail on the Border Gateway Protocol (BGP), defined in RFC 1771 [4-6]. BGP is a higher function replacement for EGP.

EGP is used to convey network reachability information between neighboring gateways (or routers) that are in different Autonomous Systems.

EGP runs over IP and is assigned IP protocol number 8. Three key mechanisms are present in the protocol. The Neighbor Acquisition mechanism allows two neighbors to begin exchanging information using the Acquisition Request and Acquisition Confirm messages. The Neighbor Reachability mechanism maintains real-time information regarding the reachability of its neighbors, using the Hello and I Hear You (I-H-Y) messages. Finally, Update messages are exchanged that carry routing information.

The specific messages, as defined in RFC 904, are:

Message	Function
Request	Request acquisition of neighbor and/or initialize polling variables
Confirm	Confirm acquisition of neighbor and/or initialize polling variables
Refuse	Refuse acquisition of neighbor
Cease	Request deacquisition of neighbor
Cease-ack	Confirm deacquisition of neighbor
Hello	Request neighbor reachability
I-H-U	Confirm neighbor reachability
Poll	Request net-reachability update
Update	Net-reachability update
Error	Error

The general structure of the EGP messages is shown in Figure 4-3. A common message header precedes each message type. The header consists of the EGP Version Number field (1 octet); the Type field (1 octet), which identifies the message type; the Code field (1 octet), which identifies a subtype; the Status field (1 octet), which contains message-specific status information; the Checksum field (2 octets), used for error control; the Autonomous System Number (2 octets), which is an assigned number that identifies the particular autonomous system; and the Sequence Number (2 octets), which maintains state variables.

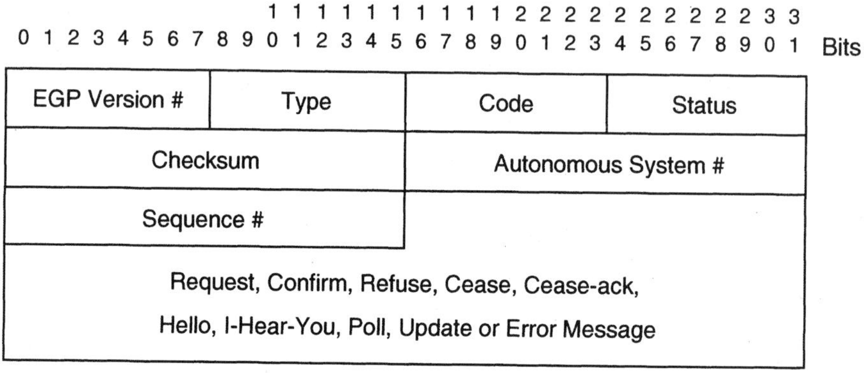

Bits: 0 1 2 3 4 5 6 7 8 9 10 11 12 13 14 15 16 17 18 19 20 21 22 23 24 25 26 27 28 29 30 31

EGP Version #	Type	Code	Status
Checksum		Autonomous System #	
Sequence #			
Request, Confirm, Refuse, Cease, Cease-ack, Hello, I-Hear-You, Poll, Update or Error Message			

Figure 4-3. EGP Message Format.

Further details on the use and specific formats of the various messages can be found in RFC 904.

4.1.4 Border Gateway Protocol

The Border Gateway Protocol (BGP), currently in its fourth version (BGP-4), is defined in RFC 1771 [4-6]. BGP is an inter-Autonomous System (AS) protocol that builds upon and enhances the capabilities of EGP. For example, where EGP runs on IP, BGP runs on TCP, thus ensuring a connection-oriented data flow and greater reliability. BGP is assigned TCP port number 179 (more on TCP port numbers in the next chapter). BGP also supports Classless Inter-Domain Routing (CIDR) and the aggregation of routes.

The system running BGP is called a BGP speaker. Connections between BGP speakers in different Autonomous Systems are called external links, while connections between BGP speakers in the same Autonomous System are called internal links. In a similar fashion, a BGP peer in another AS is referred to as an external peer, while a peer within the same AS is called an internal peer. After the TCP connection has been established in support of BGP, the two BGP systems exchange their entire routing tables. Updates are then sent as those routing tables change. As a result, the BGP speaker will maintain the current version of the routing tables for all of its peers. That routing information is stored within a Routing Information Base, or RIB.

The BGP message consists of a fixed message header that is 19 octets in length, followed by one of four messages: OPEN, UPDATE, NOTIFICATION, or KEEPALIVE. The OPEN, UPDATE, and NOTIFICATION messages add additional information to the BGP message header, while the KEEPALIVE consists of only the annotated message header.

The BGP message header is shown in Figure 4-4a, and consists of three fields plus message-specific information. The Marker field (16 octets) contains a value that the receiver can predict. For example, an OPEN message would use a Marker of all ones. Otherwise the Marker can be incorporated into some authentication mechanism. The Length field (2 octets) indicates the total length of the message, including the header, given in octets. The valid range of the Length field is 19–4,096 octets. The Type field (1 octet) specifies the type of the message as follows:

Type	Message	Function
1	OPEN	The first message sent after transport connection is established
2	UPDATE	Transfers routing information between BGP peers
3	NOTIFICATION	Indicates detection of an error and closure of the connection
4	KEEPALIVE	Periodic confirmation of reachability

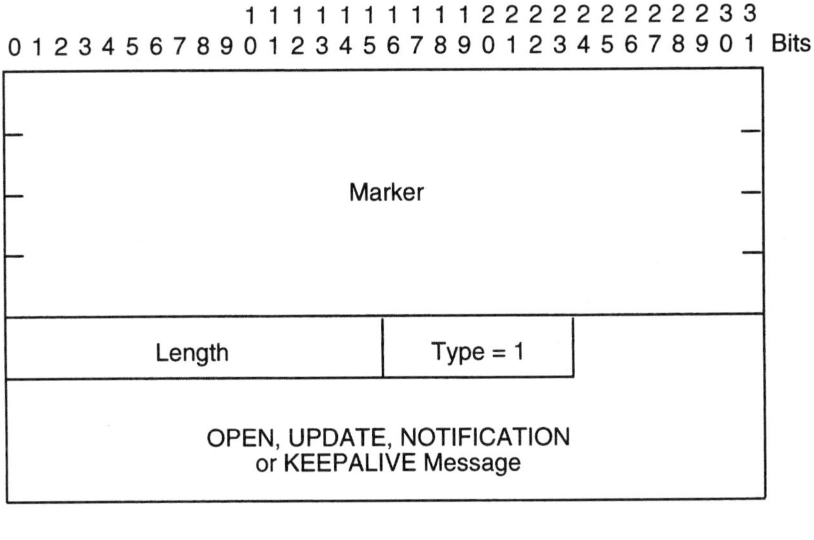

1 1 1 1 1 1 1 1 1 1 2 2 2 2 2 2 2 2 2 2 3 3
0 1 2 3 4 5 6 7 8 9 0 1 2 3 4 5 6 7 8 9 0 1 2 3 4 5 6 7 8 9 0 1 Bits

Marker

Length | Type = 1

OPEN, UPDATE, NOTIFICATION
or KEEPALIVE Message

Figure 4-4a. BGP-4 Message Header.

The OPEN message (message Type = 1) is the first message that is sent by each side following the establishment of the transport (TCP) connection. If the OPEN message is acceptable, then a KEEPALIVE message is returned, confirming the connection. After the OPEN has been confirmed, then UPDATE, KEEPALIVE, and NOTIFICATION messages may be exchanged.

The OPEN message adds five fields, plus optional parameters, to the BGP header (Figure 4-4b). The Version field (1 octet) defines the protocol version of the message (currently 4). The My Autonomous System field (2 octets) indicates the Autonomous System number of the sender. The Hold Time field (2 octets) is the number of seconds that the sender proposes for the value of the Hold Timer. The BGP Identifier field (4 octets) indicates the BGP identifier of the sender. The Optional Parameters Length field (1 octet) indicates the length of any optional parameters, such as Authentication Information, that may be included in this message. If no parameters are present, then the value of this field is zero. The Optional Parameters field may contain a list of optional parameters, encoded as a triplet of <Parameter Type, Parameter Length, Parameter Value>.

The UPDATE message (message Type = 2) is used to transfer routing information between BGP peers. It is used to advertise a single feasible route to a peer, and/or to withdraw multiple feasible routes from service.

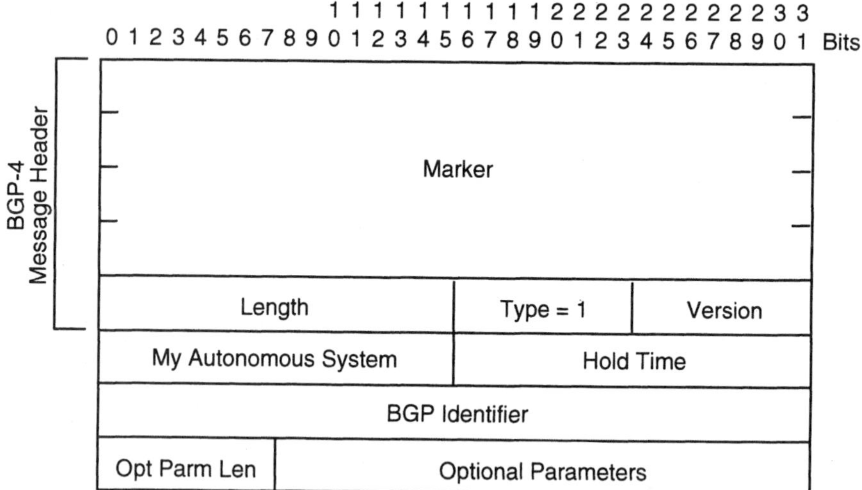

Figure 4-4b. BGP OPEN Message Format.

The UPDATE message adds five fields to the BGP header (Figure 4-4c). The Unfeasible Routes Length field (2 octets) indicates the total length of the Withdrawn Routes field in octets. A value of zero indicates that no Withdrawn Routes are present in this UPDATE message. The Withdrawn Routes field (variable length) contains a list of IP address prefixes for the routes that are being withdrawn from service. The Total Path Attribute Length field (2 octets) indicates the total length of the Path Attributes field in octets. A value of zero indicates that no Network Layer Reachability Information is present in this UPDATE message. The Path Attributes field (variable length) is a sequence of path attributes that are present in every UPDATE message. Each Path Attribute is a variable length triple <Attribute Type, Attribute Length, Attribute Value>. The Network Layer Reachability Information field (variable length) contains a list of IP address prefixes indicating reachability.

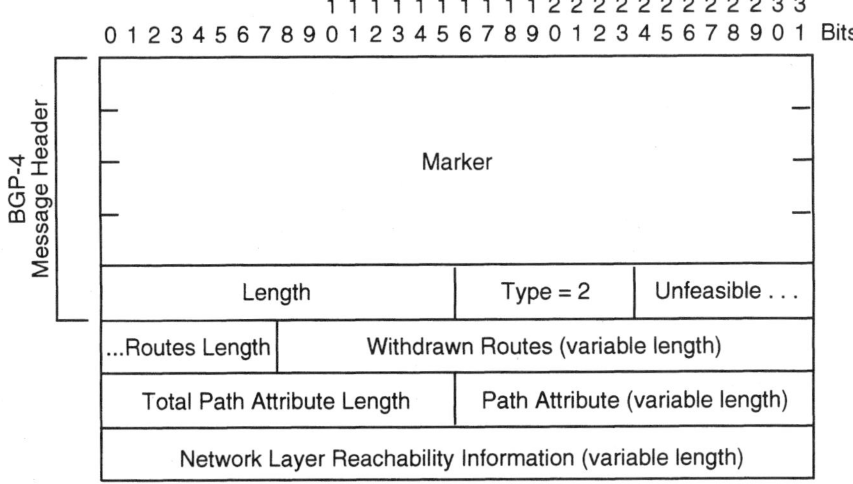

Figure 4-4c. BGP UPDATE Message Format.

The NOTIFICATION message (message Type = 3) is sent when an error condition is detected. The BGP connection is closed immediately after sending the NOTIFICATION message.

The NOTIFICATION message adds three fields to the BGP header (Figure 4-4d). The Error Code field (1 octet) indicates the types of errors that have been defined. The Error Subcode field (1 octet) provides more specific information on the types of errors. The Data field (variable length) is used to diagnose the reason for this NOTIFICATION, and is dependent upon the contents of the Error Code and Error Subcode fields. Specific values for these fields are delineated in RFC 1771.

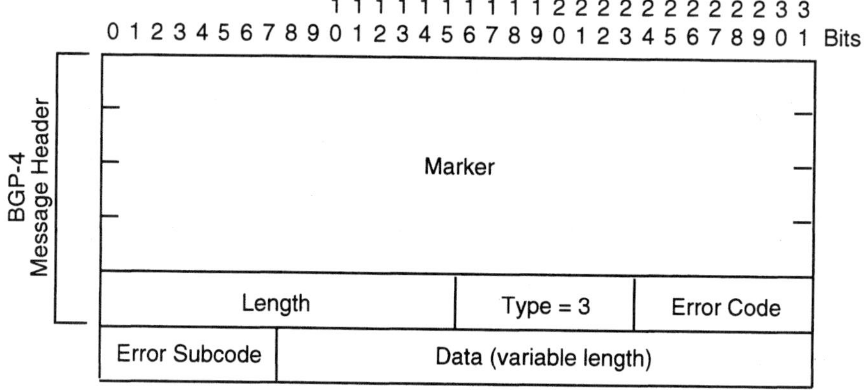

Figure 4-4d. BGP NOTIFICATION Message Format.

The KEEPALIVE message (message Type = 4) is used to determine reachability between peers. KEEPALIVE messages are exchanged often enough so that the Hold Timer does not expire, but they are not sent more frequently than one per second. The KEEPALIVE message (Figure 4-4e) consists of the BGP message header only, without any additional information.

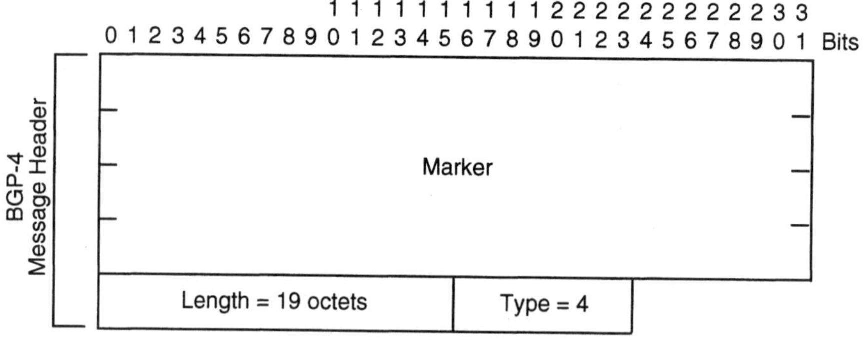

Figure 4-4e. BGP KEEPALIVE Message Format.

Further details on the operation of BGP-4 can be found in RFC 1771.

4.2 Internet Control Message Protocol

If the Internet were flawless, it would always route datagrams to their intended destination without errors, excessive delays, or retransmissions. Unfortunately, this is not the case. As we studied in Chapter 3, IP provides a connectionless service to the attached hosts but requires an additional module, known as the Internet Control Message Protocol (ICMP), to report errors that may have occurred in processing those datagrams. Examples of errors include undeliverable datagrams or incorrect routes. Other uses for the protocol include testing the path to a distant host (known as a PING) or requesting an Address Mask for a particular subnet. ICMP is considered an integral part of IP and must be implemented in IP modules contained in both hosts and routers. The standard for ICMP is RFC 792 [4-7].

ICMP messages are contained within IP datagrams. In other words, ICMP is a user (client) of IP, and the IP header precedes the ICMP message.

Thus, the datagram would include the IP header, the ICMP header, and ICMP data. Protocol = 1 identifies ICMP within the IP header. A Type field within the ICMP header further identifies the purpose and format of the ICMP message. Any data required to complete the ICMP message would then follow the ICMP header.

The standard defines thirteen ICMP message formats, each with a specific ICMP header format. Two of these formats (Information Request/Reply) are considered obsolete and several others share a common message structure. The result is six unique message formats, shown in Figure 4-5. The first three fields are common to all headers. The Type field (1 octet) identifies one of the thirteen unique ICMP messages. These include:

Type Code	ICMP Message
0	Echo Reply
3	Destination Unreachable
4	Source Quench
5	Redirect
8	Echo
11	Time Exceeded
12	Parameter Problem

13	Timestamp
14	Timestamp Reply
15	Information Request (obsolete)
16	Information Reply (obsolete)
17	Address Mask Request
18	Address Mask Reply

The second field is labeled Code (1 octet), and it elaborates on specific message types. For example, the Code field for the Destination Unreachable message indicates whether the network, host, protocol, or port was the unreachable entity. The third field is the Checksum (2 octets) on the ICMP message. Because ICMP messages are of great value in internetwork troubleshooting, we will look at all of the ICMP messages in detail. Internet standards describing ICMP formats and usage include the original standard, RFC 792; implementing congestion control, RFC 896 [4-8]; use of source quench messages, RFC 1016 [4-9]; and Subnet Mask Messages, RFC 950 [4-10].

The Echo message (ICMP Type = 8) tests the communication path from a sender to a receiver via the internet. On many hosts, this function is called PING. The sender transmits an Echo message, which may also contain an Identifier (2 octets) and a Sequence Number (2 octets). Data may be sent with the message. When the destination receives the message, it reverses the Source and Destination addresses, recomputes the Checksum, and returns an Echo Reply (ICMP Type = 0). The contents of the Data field (if any) would also be returned to the sender.

The Destination Unreachable message (ICMP Type = 3) is used when the router or host is unable to deliver the datagram. This message is returned to the Source host of the datagram in question and describes the reason for the delivery problem in the Code field:

Code	Meaning
0	Net Unreachable
1	Host Unreachable
2	Protocol Unreachable
3	Port Unreachable
4	Fragmentation Needed and DF Set
5	Source Route Failed

1 1 1 1 1 1 1 1 1 1 2 2 2 2 2 2 2 2 2 2 3 3
0 1 2 3 4 5 6 7 8 9 0 1 2 3 4 5 6 7 8 9 0 1 2 3 4 5 6 7 8 9 0 1 Bits

Type	Code	Checksum
Identifier		Sequence Number
Data		

Echo and Echo Reply Messages

1 1 1 1 1 1 1 1 1 1 2 2 2 2 2 2 2 2 2 2 3 3
0 1 2 3 4 5 6 7 8 9 0 1 2 3 4 5 6 7 8 9 0 1 2 3 4 5 6 7 8 9 0 1 Bits

Type	Code	Checksum
Unused		
Internet Header + 64 bits of Original Datagram Data		

Destination Unreachable, Source Quench and Time Exceeded Messages

1 1 1 1 1 1 1 1 1 1 2 2 2 2 2 2 2 2 2 2 3 3
0 1 2 3 4 5 6 7 8 9 0 1 2 3 4 5 6 7 8 9 0 1 2 3 4 5 6 7 8 9 0 1 Bits

Type	Code	Checksum
Pointer	Unused	
Internet Header + 64 bits of Original Datagram Data		

Parameter Problem Message

1 1 1 1 1 1 1 1 1 1 2 2 2 2 2 2 2 2 2 2 3 3
0 1 2 3 4 5 6 7 8 9 0 1 2 3 4 5 6 7 8 9 0 1 2 3 4 5 6 7 8 9 0 1 Bits

Type	Code	Checksum
Gateway Internet Address		
Internet Header + 64 bits of Original Datagram Data		

Redirect Message

1 1 1 1 1 1 1 1 1 1 2 2 2 2 2 2 2 2 2 2 3 3
0 1 2 3 4 5 6 7 8 9 0 1 2 3 4 5 6 7 8 9 0 1 2 3 4 5 6 7 8 9 0 1 Bits

Type	Code	Checksum
Identifier		Sequence Number
Originate Timestamp		
Receive Timestamp		
Transmit Timestamp		

Timestamp and Timestamp Reply Messages

1 1 1 1 1 1 1 1 1 1 2 2 2 2 2 2 2 2 2 2 3 3
0 1 2 3 4 5 6 7 8 9 0 1 2 3 4 5 6 7 8 9 0 1 2 3 4 5 6 7 8 9 0 1 Bits

Type	Code	Checksum
Identifier		Sequence Number
Address Mask		

Address Mask Request and Address Mask Reply Messages

Figure 4-5. Internet Control Message Protocol (ICMP) Message Formats.

Routers may use codes 0, 1, 4, or 5. Hosts may use codes 2 or 3. For example, when a datagram arrives at a router, it will do a table lookup to determine the outgoing path to use. If the router determines that the Destination network cannot be reached (i.e., a distance of infinite hops away), a Net Unreachable message will be returned. Similarly, if a host cannot process a datagram because the requested protocol or port is inactive, a Protocol Unreachable or Port Unreachable message, respectively, would be returned. Included in the Destination Unreachable message is the IP header plus the first 64 bits (8 octets) of the datagram in question. This returned data should help the host diagnose the failure in the transmission process.

The advantage of connectionless datagram transmission is its simplicity. The disadvantage is its inability to regulate the traffic on the network. (For an analogy, consider the problem that your local post office faces. To handle the maximum number of letters, it should install enough boxes to handle the holiday rush. However, this might be considered wasteful because many of the boxes are only partially used during the summer.) If a router or host gets congested, it may send a Source Quench message (ICMP Type = 4) to the source of the datagrams asking it to reduce its output. This mechanism is similar to traffic signals that regulate the flow of cars onto a freeway. The Source Quench message does not use the second 32-bit word of the ICMP header, but fills it with zeros. The rest of the message contains the IP header and the first 8 octets of the datagram that triggered the request.

Hosts do not always choose the correct Destination address for a particular datagram and occasionally send one to the wrong router. This is most likely to occur when the host is initialized and its routing tables are incomplete. When such a routing mistake occurs, the router that improperly received the datagram will return a Redirect message to the host identifying a better route. The Code field would contain the following information:

Code	Message
0	Redirect datagrams for the network
1	Redirect datagrams for the host
2	Redirect datagrams for the type of service and network
3	Redirect datagrams for the type of service and host

The Redirect message (ICMP Type = 5) contains the correct router (gateway) address to reach the desired destination. In addition, the IP header plus the first 8 octets of the datagram in question are returned to the source host to aid in the diagnostic process.

Another potential problem of connectionless networks is that datagrams can get lost within the network and can wander for an excessive amount of time. Alternatively, congestion could prevent all fragments of a datagram from being reassembled within the host's required time. Either of these situations can trigger an ICMP Time Exceeded message (ICMP Type = 11). The message defines two codes: Time-to-Live Exceeded in Transmit (code = 0) and Fragment Reassembly Time Exceeded (code = 1). The balance of the message has the same format as the Source Quench message: the second word contains all zeros, and the rest of the message contains the IP header and the first 8 octets of the offending datagram.

When datagrams cannot be processed because of errors, they are discarded and higher layer processes, such as TCP, must recognize the problem and take corrective action. Parameter problems within an IP datagram header (such as an incorrect Type of Service field) may trigger the transmission of an ICMP Parameter Problem message (ICMP Type = 12) to the source of the datagram identifying the location of the problem. The message contains a pointer that identifies the octet with the error. The rest of the message contains the IP datagram header plus the first 8 octets of data, as before.

The Timestamp (ICMP Type = 13) and Timestamp Reply (ICMP Type = 14) messages measure the round-trip transit time between two machines and synchronize their clocks. The first two words of the Timestamp and Timestamp Reply messages are similar to the Echo and Echo Reply messages. The next three fields contain timestamps measured in milliseconds since midnight, Universal Time (UT). The Timestamp Requester fills in the Originate field upon transmission; the recipient fills in the Receive Timestamp when it receives the request. The recipient fills in the Transmit Timestamp when it sends the Timestamp Reply message. The Requester may now estimate the remote processing and round-trip transit times. (Note that these are only estimates because network delay is a highly dynamic and variable measurement.) The remote processing time is the Received Timestamp minus the Transmit Timestamp. The round-trip transit time is the Timestamp Reply message arrival time minus the Originate Timestamp. With these two calculations, the two clocks can be synchronized.

Finally, Address Mask Request (ICMP Type = 17) and Address Mask Reply (ICMP Type = 18) were added to the ICMP message in response to subnetting requirements (RFC 950). It is assumed that the requesting host knows its own Internet address. (If it doesn't, it uses RARP to discover the Internet address.) It then broadcasts the Address Mask Request message to the Destination address [255.255.255.255]. The Address Mask field of the ICMP message would be filled with all zeros. The IP router that knows the correct address mask would respond. As a review of Section 3.2.3, the response for a Class B network (without subnetting) would be [255.255.0.0], while a response for a Class B network using an 8-bit subnet field would be [255.255.255.0].

4.3 Intranetwork Tools: PING and TRACEROUTE

Two software tools, called PING and TRACEROUTE, utilize elements of ICMP to test network connectivity. PING is used to verify network connectivity, while TRACEROUTE discovers the path from a source to a destination.

PING is a simple tool included with most operating systems that support a TCP/IP stack, including Microsoft Windows 9x, 2000, and XP. When invoked, the sending station transmits an ICMP Echo Request message to the desired destination and starts a timer. When the receiver receives that ICMP Echo Request message, it responds with an ICMP Echo Reply message. When that response is received back at the sending station, the timer is stopped and the results are displayed. Below is an example of a PING message sent to the IANA website from a Windows XP workstation:

```
C:\ >ping www.iana.org
Pinging www.iana.org [192.0.34.162] with 32 bytes of data:
Reply from 192.0.34.162: bytes=32 time=347ms TTL=50
Reply from 192.0.34.162: bytes=32 time=310ms TTL=50
Reply from 192.0.34.162: bytes=32 time=320ms TTL=50
Reply from 192.0.34.162: bytes=32 time=321ms TTL=50
Ping statistics for 192.0.34.162:
    Packets: Sent = 4, Received = 4, Lost = 0 (0% loss),
Approximate round trip times in milli-seconds:
    Minimum = 310ms, Maximum = 347ms, Average = 324ms
```

A successful PING response indicates that the target station is connected to the network, is reachable via that network, and is operationally able to respond to the request. If the PING is not successful, further analysis is required to determine if the problem lies with the sending station, the network, or the receiving station, since these details are not returned when the PING fails.

A second software tool, called TRACEROUTE (abbreviated *tracert* in Windows and many other systems), is also based on ICMP, specifically the ICMP Time Exceeded message. This tool allows a station to generate a sequence of User Datagram Protocol (UDP) packets, with the Time to Live (TTL) field within the IP header increasing with each set of packets (TTL = 1, 2, 3, etc., typically transmitted in sets of three packets per TTL value). When those packets reach an intermediate router, but not the final destination, and the TTL expires, an ICMP Time Exceeded message is returned from that intermediate router to the station that initiated the TRACEROUTE. For example, when TTL = 1, the first router in the path will return Time Exceeded messages; when TTL = 2, the second router in the path will return Time Exceeded messages; when TTL = 3, the third router in the path will return Time Exceeded messages, and so on. In this way, the entire path from a source station to its intended destination can be determined. Below is an example of a TRACEROUTE message sent to the IANA website from a Windows XP workstation:

```
C:\ >tracert www.iana.org
Tracing route to www.iana.org [192.0.34.162] over a maximum
of 30 hops:
 1 273 ms 260 ms 260 ms    ts004e01.den-co.concentric.net
                           [207.155.168.6]
 2 270 ms 259 ms 270 ms    rt001e0102.den-co.concentric.net.
                           168.155.207.in-addr.arpa
                           [207.155.168.1]
 3 270 ms 260 ms 260 ms    ge5-0-0.MAR1.Englewood-
                           CO.us.xo.net [207.88.83.17]
 4 270 ms 260 ms 260 ms    p5-1-0-0.RAR1.Denver-CO.us.xo.net
                           [65.106.6.13]
 5 310 ms 310 ms 299 ms    p6-0-0.RAR1.SanJose-CA.us.xo.net
                           [65.106.0.21]
 6 310 ms 309 ms 300 ms    p0-0.IR1.PaloAlto-CA.us.xo.net
                           [65.106.5.194]
 7 320 ms 310 ms 310 ms    [198.32.175.14]
 8 550 ms 320 ms 310 ms    p16-0-
                           10.r20.plalca01.us.bb.verio.net
                           [129.250.3.78]
 9 320 ms 320 ms 310 ms    p64-0-0-
                           0.r20.snjsca04.us.bb.verio.net
                           [129.250.2.71]
10 320 ms 319 ms 320 ms    xe-0-2-
                           0.r21.snjsca04.us.bb.verio.net
                           [129.250.2.73]
11 320 ms 310 ms 320 ms    p16-1-1-
                           1.r21.lsanca01.us.bb.verio.net
                           [129.250.2.186]
```

```
12 320 ms 310 ms 310 ms    ge-1-1-
                           0.a02.lsanca02.us.ra.verio.net
                           [129.250.29.131]
13 310 ms 310 ms 319 ms    ge-1-
                           2.a01.lsanca02.us.ra.verio.net
                           [129.250.46.93]
14 310 ms 310 ms 320 ms    ge-2-3-
                           0.a02.lsanca02.us.ce.verio.net
                           [198.172.117.163]
15 320 ms 310 ms 320 ms    lngw2-isi-1-atm.ln.net
                           [130.152.180.22]
16 320 ms 310 ms 320 ms    [207.151.118.18]
17 320 ms 310 ms 320 ms    www.iana.org [192.0.34.162]
Trace complete.
```

Note from the above results that several Internet Service Provider (ISP) circuits comprised this path (concentric.net, xo.net, verio.net, etc.). The three test messages returned from each hop have approximately the same round trip delay, but this delay increases as the number of hops away from the originating station increases. Also note that the PING and TRACEROUTE response reveals quite a bit of information about the network topology — information that may be considered proprietary. For this and other security reasons, some network managers disable these functions.

Additional details on PING and TRACEROUTE can be found in RFC 2925 [4-11].

4.4 Internet Group Management Protocol

The Internet Group Management Protocol, or IGMP, is an adjunct protocol to the Internet Protocol. IGMP must be implemented on all hosts that wish to receive IP multicasts, which are used for many applications, including voice and video over IP conferencing (review Section 3.2.2 on multicast addressing). IGMP is used by IP hosts to report their group memberships to any immediately neighboring multicast routers. The protocol was originally specified in RFC 1112; IGMP version 2 is documented in RFC 2236, and IGMP version 3 is defined in RFC 3376 [4-12].

The IGMP message is encapsulated within an IP datagram and identified with protocol number = 2. It is sent with TTL = 1 and includes the IP Router Alert option defined in RFC 2113 [4-13]. The message format is very similar to the ICMP message structure, and is shown in Figure 4-6. The message has four fields. The Type field (1 octet) specifies one of five different types of messages. In the case of the Membership Query message, there are two subtype messages, which are distinguished by the group address used. The Type field values and their functions are:

Type Field	Message	Function
11 H	Membership Query	To learn about members and groups. Subtype: General Query message, used to learn which groups have members on an attached network, sent to the All Systems Group, [224.0.0.1]. Subtype: Group-Specific Query message, used to learn if a particular group has any members on an attached network.
22 H	Membership Report (v3)	To indicate that a host has joined a group.

For interoperation with previous versions of IGMP, the following messages are supported:

Type Field	Message	Function
12 H	Membership Report (v1)	For backward compatibility with IGMPv1.
16 H	Membership Report (v2)	For backward compatibility with IGMPv2.
17 H	Leave Group (v2)	To indicate that a host is leaving a group, sent to the All Routers Group, [224.0.0.2].

The Max Response Code field (1 octet) is only meaningful in Membership Query Messages and specifies the maximum time allowed before sending responding Report messages. The unit of measure is in tenths (1/10) of a second. For all other messages, this field is set to 0 and ignored by the receiver. The Checksum field (2 octets) is used for error control purposes. The Group Address field (4 octets) contains an IP address of a type that depends on the message type being transmitted. For a General Query message, this field is set to 0; for a Group-Specific Query, this field is set to the group address of the group being queried; for the Membership Report or Leave Group messages, this field is set to the IP multicast group address of the group being reported or left.

Further details on the state diagrams and other specifics of IGMP operation can be found in RFC 3376.

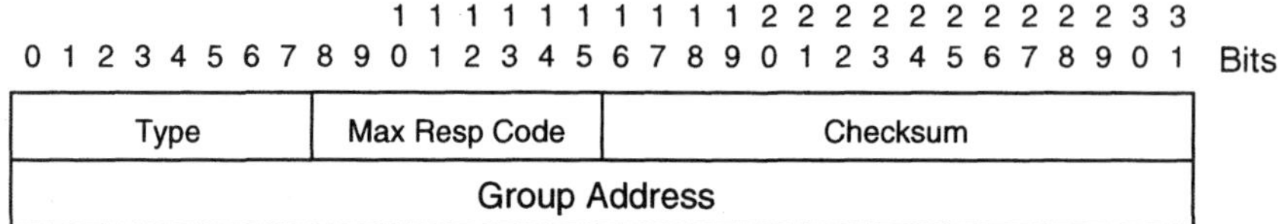

Figure 4-6. Internet Group Management Protocol (IGMP) Message Format.

4.5 Resource Reservation Protocol

As more time-sensitive applications have been developed for the Internet, the need to define a Quality of Service (QoS), and the mechanisms to provide that QoS, have become requirements. The Resource Reservation Protocol (RSVP), defined in RFC 2205 [4-14], is designed to address those requirements. When a host has an application, such as real-time video or multimedia, it may use RSVP to request an appropriate level of service from the network in support of that application. But for RSVP to be effective, every router in that path must support that protocol—something that some router vendors and ISPs are not yet equipped to do. In addition, RSVP is a control protocol, and therefore works in collaboration with—not instead of—traditional routing protocols. In other words, the routing protocol, such as RIP or OSPF, determines which datagrams are forwarded, while RSVP is concerned with the QoS of those datagrams that are forwarded.

RSVP requests that network resources be reserved to support data flowing on a simplex path, and that reservation is initiated and maintained by the receiver of the information. Using this model, it can support both unicast and multicast applications. Reviewing Figure 3-1c, we see that RSVP messages may be sent directly inside IP datagrams (using IP Protocol = 46), or encapsulated inside UDP datagrams (using Ports 1698 and 1699). RSVP defines two basic message types: Reservation Request (or Resv) messages and Path messages. A receiver transmits Resv messages upstream toward the sender. A sender transmits Path messages downstream toward the receivers, following the paths prescribed by the routing protocols that follow the paths of the data. These Path messages store path state information in each node along the way (Figure 4-7).

The RSVP message consists of three sections: a Common Header (8 octets), an Object Header (4 octets), and Object Contents (variable length), as shown in Figure 4-8. The Version field (4 bits) contains the protocol version (currently 1). The Flags field (4 bits) is reserved for future definition. The Message Type field (1 octet) defines one of seven currently defined RSVP messages:

Message Type	Message Name	Function
1	Path	Path message, from sender to receiver, along same path used for the data packets.
2	Resv	Reservation message with reservation requests, carried hop-by-hop from receivers to senders.
3	PathErr	Path error message, reports errors in processing Path messages, and travels upstream toward senders.
4	ResvErr	Reservation error message, reports errors in processing Resv messages, and sent downstream toward receivers.
5	PathTear	Path teardown message, initiated by senders or by a timeout, and sent downstream to all receivers.
6	ResvTear	Reservation teardown message, initiated by receivers or by a timeout, and sent upstream to all matching senders.
7	ResvConf	Reservation confirmation message, acknowledges reservation requests.

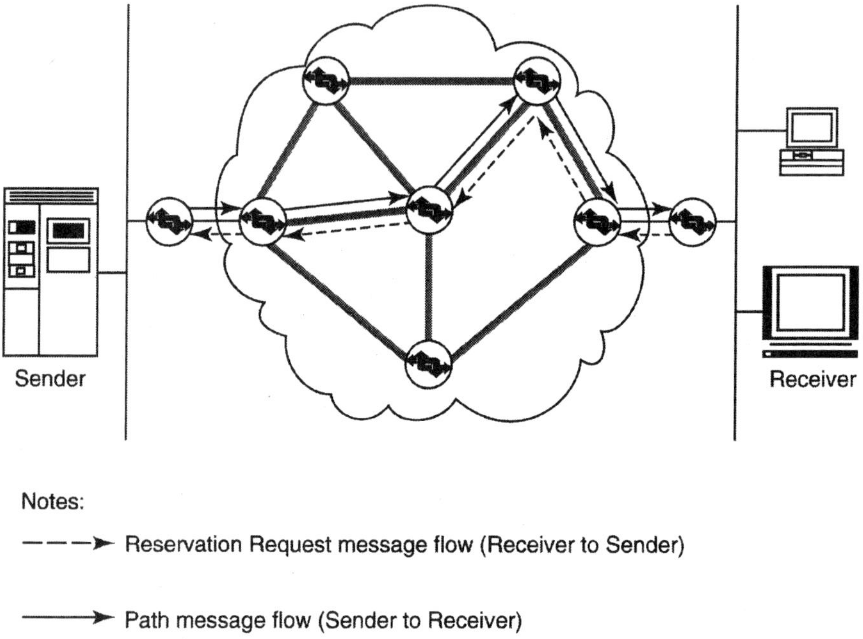

Figure 4-7. RSVP Protocol Operation.

The RSVP Checksum field (2 octets) provides error control. The Send_TTL field (1 octet) is the IP Time to Live field value with which the message was sent. The RSVP Length field (2 octets) is the total length of the RSVP message, given in octets.

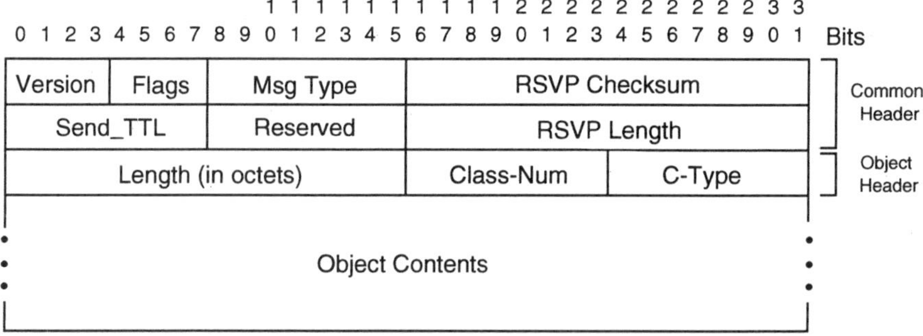

Figure 4-8. RSVP Message Format.

Every object consists of one or more 32-bit words with a 1-octet Object Header. This second header includes a Length field (2 octets), which is the total length of the object (with a minimum of 4 octets, and also a multiple of 4 octets); a Class Number (1 octet), which defines the object class; and a C-Type field (1 octet), which is an object type that is unique with the Class Number. The Object Contents complete the message. Currently defined object classes (which, by convention, are always capitalized), include:

Class-Num	Object Class	Object Function
0	NULL	Null.
1	SESSION	Defines the specific session for other objects that follow, and is required in every RSVP message.
3	RSVP_HOP	The IP address of the RSVP capable sending node.
4	INTEGRITY	Carries cryptographic data to authenticate the originating node and verify contents of the message.
5	TIME_VALUES	The value of the refresh period used by the message creator; it is required in every Path and Resv message.
6	ERROR_SPEC	Specifies an error in a PathErr or ResvErr, or a confirmation in a ResvConf message.
7	SCOPE	Carries a list of sender hosts to which the information in the message is to be forwarded, and may appear in a Resv, ResvErr, or ResvTear message.
8	STYLE	Defines the reservation style, and is required in every Resv message.
9	FLOW_SPEC	Defines a desired QoS, and is used in a Resv message.
10	FILTER_SPEC	Defines a subset of session data packets to receive the desired QoS, and is used in a Resv message.
11	SENDER_TEMPLATE	Contains the sender's IP address, and is required in a Path message.

12	SENDER_TSPEC	Defines the traffic characteristics of a sender's data flow, and is required in a Path message.
13	ADSPEC	Carries One Pass with advertising (OPWA) data, and is used in a Path message.
14	POLICY_DATA	Determines if an associated reservation is administratively permitted, and may be used in Path, Resv, PathErr, or ResvErr messages.
15	RESV_CONFIRM	Carries the IP address of the receiver that requested a confirmation, and may appear in a Resv or ResvConf message.

Details on the operation of RSVP can be found in RFC 2205, with additional details in RFC 2750 [4-15].

4.6 Looking Ahead

In this chapter we have covered a number of protocols that work together to deliver datagrams within the internet from one host to another. Since this transmission is based on the Internet Protocol's connectionless service, guaranteed delivery of those datagrams must be ensured by another process, the Host-to-Host Layer. We will study the two protocols that operate at the Host-to-Host Layer, UDP and TCP, in the next chapter.

4.7 References

[4-1] Hedrick, C. "Routing Information Protocol." RFC 1058, June 1988.

[4-2] Meyer, G., and S. Sherry. "Triggered Extensions to RIP to Support Demand Circuits." RFC 2091, January 1997.

[4-3] Malkin, G. "RIP Version 2 — Carrying Additional Information." RFC 2453, November 1998.

[4-4] Moy, John. "OSPF Version 2." RFC 2328, April 1998.

[4-5] Mills, Dave. "Exterior Gateway Protocol Formal Specification." RFC 904, April 1984.

[4-6] Yehkter, Y., and T. Li. "A Border Gateway Protocol version 4 (BGP-4)." RFC 1771, March 1995.

[4-7] Postel, J. "Internet Control Message Protocol." RFC 792, September 1981.

[4-8] Nagle, John. "Congestion Control in IP/TCP Internetworks." RFC 896, January 1984.

[4-9] Prue, W., et al. "Something a Host Could Do with Source Quench: The Source Quench Introduced Delay (SQuID)." RFC 1016, July 1987.

[4-10] Mogul, J., et al. "Internet Standard Subnetting Procedure." RFC 950, August 1985.

[4-11] White, K. "Definitions of Managed Objects for Remote Ping, Traceroute, and Lookup Operations." RFC 2925, September 2000.

[4-12] Cain, B., et al. "Internet Group Management Protocol, Version 3." RFC 3376, October 2002.

[4-13] Katz, D. "IP Router Alert Option." RFC 2113, February 1997.

[4-14] Braden, R., Editor. "Resource Reservation Protocol (RSVP) Version 1 Functional Specification." RFC 2205, September 1997.

[4-15] Herzog, S. "RSVP Extensions for Policy Control." RFC 2750, January 2000.

5

End-to-End Reliability

Datagrams and virtual circuits both convey information on an end-to-end basis, but each is associated with different benefits and costs. In general terms, datagrams typically require lower overhead and provide lower reliability, while virtual circuits typically offer higher reliability at a cost of additional overhead such as call setup messages. The choice depends on the amount of reliability that is required for the application in question, and the data that is related to that application. For example, an electronic message, which can be easily retransmitted if lost, could be sent as a datagram. File transfers, on the other hand, demand a high degree of reliability and should employ the connection-oriented virtual circuit. The end user, however, is most frequently shielded from these transmission intricacies by either the carrier's service (and associated Service Level Agreement, or SLA), or intelligent applications that hide error processing and retransmissions.

In this chapter, we will study the two protocols that are employed for the reliable transport of end-user information: the User Datagram Protocol (UDP) and the Transmission Control Protocol (TCP). Later in this chapter we will introduce a third protocol, called the Stream Control Transmission Protocol (SCTP), which provides the functions necessary to transport telephony signaling information. UDP, TCP, and SCTP are considered peer protocols, but each has unique functions and characteristics. Other host-related issues are discussed in RFC 1123, "Requirements for Internet Hosts: Application and Support." The first issue to consider is how UDP and TCP fit into the ARPA architectural model.

5.1 The Host-to-Host Connection

In Chapter 2, we learned that the Network Interface Layer handles the physical connection for the LAN or WAN. In Chapters 3 and 4, we explored the Internet Layer, which routes IP datagrams from one device to another on the same network or on a different network via the internetwork. The Internet

Layer also administers the 32-bit Internet addresses (used with IP) and takes care of intranetwork communication (with ICMP).

Figure 5-1a shows the ARPA architectural model; as you can see, the Host-to-Host Layer is the first layer that operates exclusively within the hosts, not in the routers. Assuming that the IP datagram has arrived at the Destination host, what issues must the host (and the architect designing the internetwork) deal with in order to establish and use the communication link between the two endpoints?

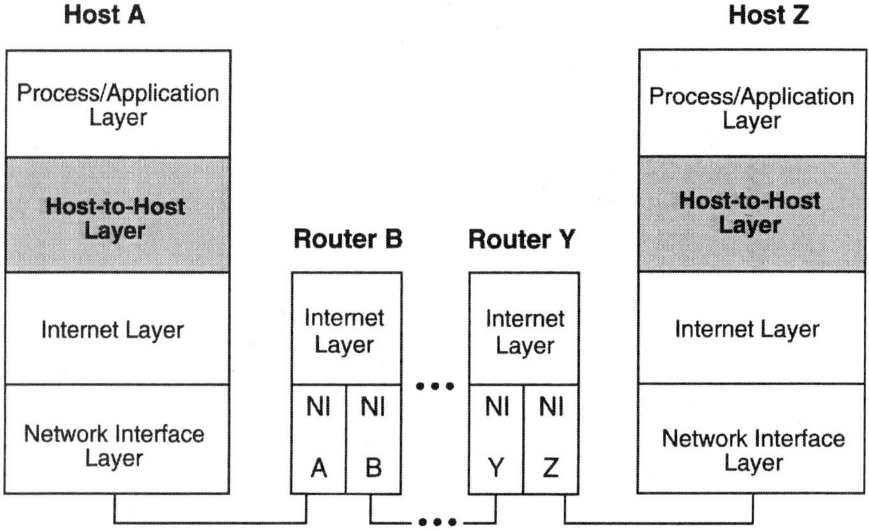

Figure 5-1a. The Host-to-Host Connection.

First, we must assume that the host has more than one process at its disposal. Therefore, the datagram will require additional addressing to identify the host process to which it will apply. This additional address is carried within the UDP or TCP header and is known as a Port address (we'll discuss ports in Section 5.2). Another issue to consider is reliability of the datagram; the hosts will want assurance that the datagram arrived correctly. Both the UDP and TCP headers address this concern. A third consideration is the overhead associated with the host-to-host transmission. The overhead directly relates to the type of connection — TCP or UDP — employed. It is the requirements of the Process/Application Layer that determine the choice between TCP and UDP. As shown in Figure 5-1b, the Hypertext Transfer

ARPA Layer	Protocol Implementation								OSI Layer
Process / Application	Hypertext Transfer	File Transfer	Electronic Mail	Terminal Emulation	Domain Names	File Transfer	Client / Server	Network Management	Application
	Hypertext Transfer Protocol (HTTP) RFC 2616	File Transfer Protocol (FTP) MIL-STD-1780 RFC 959	Simple Mail Transfer Protocol (SMTP) MIL-STD-1781 RFC 2821	TELNET Protocol MIL-STD-1782 RFC 854	Domain Name System (DNS) RFC 1034, 1035	Trivial File Transfer Protocol (TFTP) RFC 1350	Sun Microsystems Network File System Protocols (NFS) RFC 3530	Simple Network Management Protocol (SNMP) v1: RFC 1157 v2: RFC 1901-10 v3: RFC 3411-18	Presentation Session
Host-to-Host	Transmission Control Protocol (TCP) MIL-STD-1778 RFC 793				User Datagram Protocol (UDP) RFC 768				Transport
Internet	Address Resolution ARP RFC 826 RARP RFC 903			Internet Protocol (IP) MIL-STD-1777 RFC 791			Internet Control Message Protocol (ICMP) RFC 792		Network
Network Interface	Network Interface Cards: Ethernet, Token Ring, MAN and WAN RFC 894, RFC 1042 and others								Data Link
	Transmission Media: Twisted Pair, Coax, Fiber Optics, Wireless Media, etc.								Physical

Figure 5-1b. ARPA Host-to-Host Layer Protocols.

Protocol (HTTP), File Transfer Protocol (FTP), Telecommunications Network (TELNET) protocol, and Simple Mail Transfer Protocol (SMTP) run over TCP transport. The Trivial File Transfer Protocol (TFTP), Simple Network Management Protocol (SNMP), and Sun Microsystems Inc.'s Network File System (NFS) protocols work with UDP. The Domain Name System (DNS) may employ either UDP or TCP, depending on the function being performed. We will study these Process/Application Layer protocols in depth in Chapters 8 through 10.

Figure 5-1c shows where the overhead occurs. The Local Network header and trailer delimit the transmission frame and treat the higher layer information as data inside the frame. The IP header is the first component of overhead, followed by either a UDP or TCP header. If UDP is used, it requires eight octets of header overhead. If TCP is used, it requires a minimum of 20 octets of overhead, plus options and padding. As we will see in the following sections, the header overhead relates directly to the rigor of the protocols.

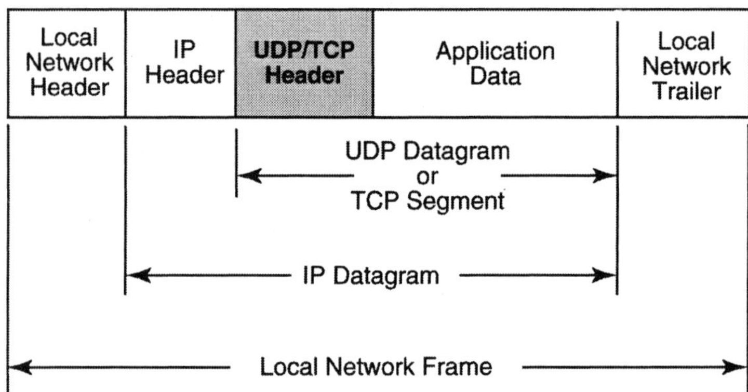

Figure 5-1c. The Internet Transmission Frame and UDP/TCP Header Position.

As noted above, both UDP and TCP use Port addresses to identify the incoming data stream and to multiplex it to the appropriate Host process. We'll look at the functions of the Port addresses next.

5.2 Port Addresses

In our tour of the ARPA architectural model, we've encountered addresses for each layer. Let's review these briefly. The Network Interface (or OSI

Physical and Data Link) Layer requires a Hardware address. The Hardware address is assigned to the network interface card. For Ethernet/IEEE 802.3 LANs, this would be a 48-bit number that resides in a ROM on the board itself. The Hardware address gives a unique identity to each workstation on the LAN or WAN. Because that address resides in hardware, it remains constant as the device is moved from one network to another.

The Internet (or OSI Network) Layer requires a Logical address, which identifies the network to which a host is attached. As we discussed in Section 3.2, this Internet (or IP) address may be further subdivided into Subnetwork and Host IDs. This address is a 32-bit number that the network administrator assigns to the device. If the device were to be moved from one network to another, a different Internet address would be necessary. The ARP and RARP protocols that we studied in Section 3.3 correlate the Internet Layer and Network Interface Layer addresses.

The third layer of addressing is used at the Host-to-Host (or OSI Transport) Layer and completes the ARPA addressing scheme. This address is known as a *Port address*, and it identifies the user process or application within the host. Each host is presumed to have multiple applications available, such as electronic mail, databases, and so on. An identifier, known as a Port number, specifies the process the user wishes to access. These Port numbers are 16 bits long and are divided into three ranges:

- Well-known ports: 0 – 1023
- Registered Ports: 1024 – 49151
- Dynamic and/or Private Ports: 49152 – 65535

The well-known ports are assigned by the IANA and identify common protocol services, such as FTP, TELNET, and SMTP. This service port becomes the incoming point of contact for callers needing access to a commonly used application. In most cases, the same port number is used by both UDP and TCP. The registered port numbers are listed by the IANA and may include vendor-developed protocols or applications that require a unique address, such as Novell NetWare Communication Service Platform (1366), Apple Network License Manager (1381), or Microsoft SQL Server (1433). In many cases, a contact person (name and email address) is listed for these registered port numbers. As their names imply, the dynamic and/or private ports provide other functions, and are neither assigned nor listed by IANA.

A complete listing of the assigned ports is available from the IANA website [3-2]. Some examples are given in Tables 5-1a and 5-1b.

Decimal	Keyword	Description
1	tcpmux	TCP Port Service Multiplexer
5	rje	Remote Job Entry
7	echo	Echo
11	systat	Active Users
13	daytime	Daytime
17	qotd	Quote of the Day
18	msp	Message Send Protocol
19	chargen	Character Generator
20	ftp-data	File Transfer Protocol [Default Data]
21	ftp	File Transfer Protocol [Control]
23	telnet	TELNET
25	smtp	Simple Mail Transfer Protocol
33	dsp	Display Support Protocol
37	time	Time
38	rap	Route Access Protocol
42	nameserver	Host Name Server
43	nicname	Who Is
49	login	Login Host Protocol
53	domain	Domain Name Server
67	bootps	Bootstrap Protocol [Server]
68	bootpc	Bootstrap Protocol [Client]
69	tftp	Trivial File Transfer Protocol
70	gopher	Gopher
79	finger	Finger
80	www-http	World Wide Web HTTP
88	kerberos	Kerberos
92	npp	Network Printing Protocol
93	dcp	Device Control Protocol
101	hostname	NIC Host Name Server
102	iso-tsap	ISO-TSAP
107	rtelnet	Remote TELNET Service
109	pop2	Post Office Protocol — v. 2
110	pop3	Post Office Protocol — v. 3
111	sunrpc	SUN Remote Procedure Call

Table 5-1a. Port Assignments.

Decimal	Keyword	Description
115	sftp	Simple File Transfer Protocol
129	pwdgen	Password Generator Protocol
137	netbios-ns	NetBIOS Name Service
138	netbios-dgm	NetBIOS Datagram Service
139	netbios-ssn	NetBIOS Session Service
143	imap2	Interim Mail Access Protocol v.2
144	news	News
146	iso-tp0	ISO-TP0
147	iso-ip	ISO-IP
152	bftp	Background File Transfer Program
153	sgmp	Simple Gateway Monitoring Protocol
160	sgmp-traps	SGMP-TRAPS
161	snmp	Simple Network Management Protocol
162	snmptrap	SNMPTRAP
163	cmip-manage	CMIP/TCP Manager
164	cmip-agent	CMIP/TCP Agent
165	xns-courier	Xerox
179	bgp	Border Gateway Protocol
190	gacp	Gateway Access Control Protocol
193	srmp	Spider Remote Monitoring Protocol
194	irc	Internet Relay Chat Protocol
199	smux	SNMP Multiplexing
201	at-rmtp	AppleTalk Routing Maintenance Protocol
202	at-nbp	AppleTalk Name Binding Protocol
203	at-3	AppleTalk Unused
204	at-echo	AppleTalk Echo Protocol
205	at-5	AppleTalk Unused
206	at-zis	AppleTalk Zone Information
207	at-7	AppleTalk Unused
208	at-8	AppleTalk Unused
209	tam	Trivial Authenticated Mail Protocol
220	imap3	Interactive Mail Access Protocol — v.3
246	dsp3270	Display Systems Protocol

Table 5-1b. Port Assignments (continued).

In summary, a message sent from one host to another requires three addresses at the source and the destination to complete the communication path. The Port address (16 bits) identifies the user process (or application) and is contained within the UDP or TCP header. The Internet address (32 bits) identifies the network and host that the process is running on and is located inside the IP header. On some hosts, the combination of the Port address and the IP address is referred to as a *socket*. The Hardware address (usually 48 bits) identifies the particular host or workstation on the local network. The Hardware address, which is in the ARPA Network Interface Layer (or OSI Data Link Layer) header, is physically configured on the network interface card and typically burned into a Read-only Memory (ROM) chip.

With that background, let's see how UDP and TCP use the Port address to complete the host-to-host connection.

5.3 User Datagram Protocol

As its name implies, the User Datagram Protocol, described in RFC 768 [5-1], provides a connectionless host-to-host communication path. UDP assumes that IP, which is also connectionless, is the underlying (Internet Layer) protocol. Because this service has minimal overhead, UDP has a relatively small header, as shown in Figure 5-2. The resulting message, consisting of the IP header, UDP header, and user data, is called a UDP datagram.

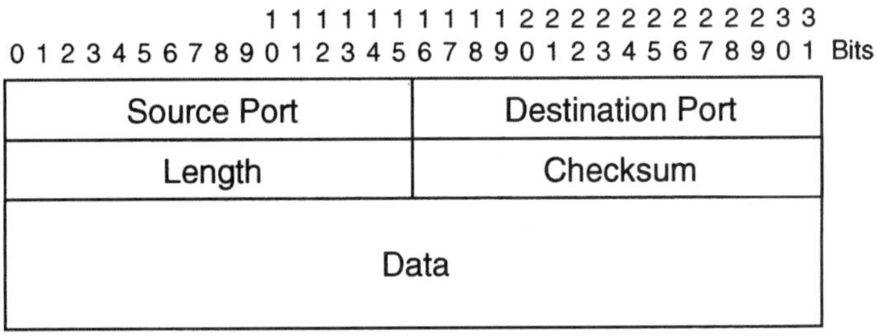

```
                    1 1 1 1 1 1 1 1 1 1 2 2 2 2 2 2 2 2 2 2 3 3
0 1 2 3 4 5 6 7 8 9 0 1 2 3 4 5 6 7 8 9 0 1 2 3 4 5 6 7 8 9 0 1  Bits
```

Bits 0–15	Bits 16–31
Source Port	Destination Port
Length	Checksum
Data	

Figure 5-2. User Datagram Protocol (UDP) Header.

The first two fields are the Source and Destination Port numbers (each 2 octets long), discussed in Section 5.2. The Source Port field is optional, and when not in use is filled with zeros. The Length field (2 octets) is the length of the UDP datagram, which has a minimum value of 8 octets. The Checksum field (2 octets) is also optional and may be filled with zeros if the Upper Layer Protocol (ULP) process does not require a checksum. When the checksum is used, it is calculated from the so-called Pseudo header, which includes the Source and Destination addresses and the Protocol field from the IP header (see Figure 5-3). By including the IP address in its calculations for the checksum, the Pseudo header ensures that the UDP datagram is delivered to the correct Destination network and host. The IP Protocol field within the Pseudo header would equal 17 for UDP. Note that the Pseudo header is used for checksum calculations only, and does not otherwise exist or transmit over the network.

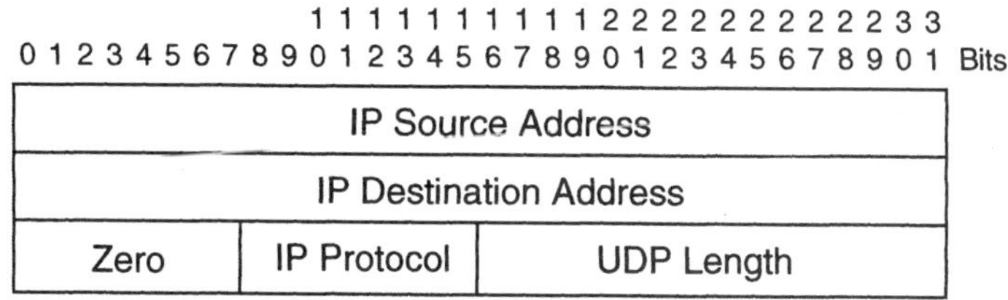

Figure 5-3. UDP Pseudo Header.

In summary, the IP Destination address routes the datagram to the correct host on the specified network, then the UDP Port address routes the datagram to the correct host process. Thus, the UDP adds Port addressing capabilities to IP's datagram service. Examples of host processes that use UDP as the host-to-host protocol include: the Time protocol, Port number 37; the Domain Name Server (DNS), Port number 53; the Bootstrap Protocol (BOOTP) server and client, Port numbers 67 and 68, respectively; the Trivial File Transfer Protocol (TFTP), Port number 69; and the Sun Microsystems Remote Procedure Call (SunRPC), Port number 111. All of these applications assume that if the host-to-host connection were to fail, a higher layer function, such as DNS, would recover. Applications that require more reliability in their end-to-end data transmissions use the more rigorous TCP, which we will discuss next.

5.4 Transmission Control Protocol

In Chapter 1, we learned that the Internet protocols were designed to meet U.S. government and military requirements. These requirements dictated that the data communication system be able to endure battlefield conditions. The Internet protocol designed specifically to meet the rigors of the battlefield was the Transmission Control Protocol, described in RFC 793 [5-2]. Unlike UDP, TCP is a connection-oriented protocol that is responsible for reliable communication between two end processes. The unit of data transferred is called a *stream*, which is simply a sequence of octets. The stream originates at the upper layer protocol process and is subsequently divided into TCP segments, IP datagrams, and Local Network frames. RFC 1180, "A TCP/IP Tutorial" [5-3], offers a useful summary of the way the TCP information fits into the related protocols, such as IP, and the Local Network connection, such as Ethernet. RFC 879, "The TCP Maximum Segment Size and Related Topics" [5-4], describes the relationships between TCP segments and IP datagrams.

TCP handles six functions: basic data transfer, reliability, flow control, multiplexing, connections, and precedence/security. We will discuss these functions in detail in Section 5.5.

The TCP header (see Figure 5-4) has a minimum length of 20 octets. This header contains a number of fields relating to connection management, data flow control, and reliability which UDP did not require. The TCP header starts with two Port addresses (2 octets each) to identify the logical host processes at each end of the connection.

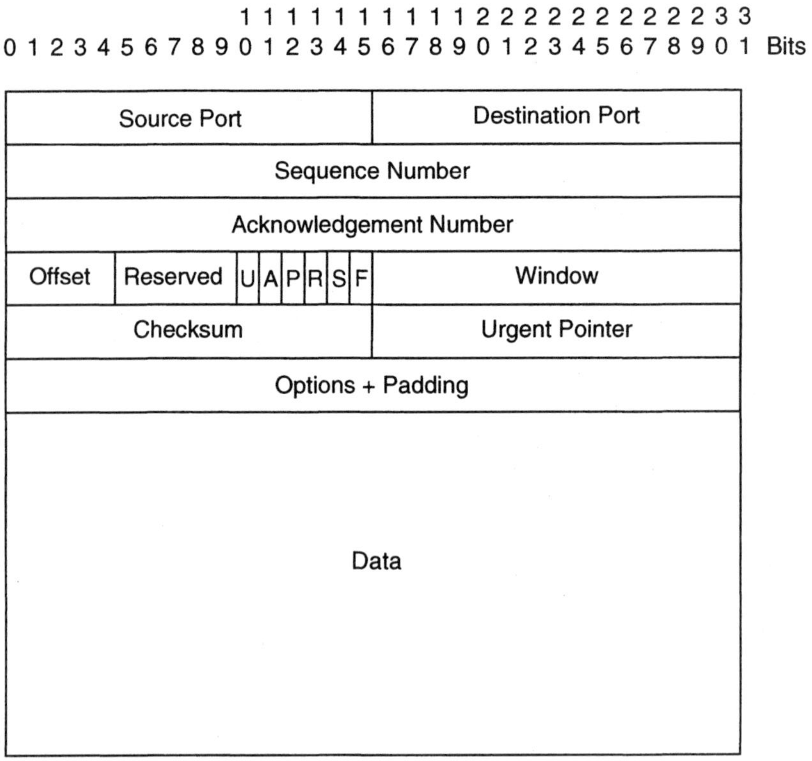

Figure 5-4. Transmission Control Protocol (TCP) Header.

The Sequence Number field (4 octets) is the Sequence number given to the first octet of data. When the SYN flag bit is set, the Sequence number indicates the Initial Sequence Number (ISN) selected. The first data octet sent would then use the next Sequence number [ISN+1]. (For example, if ISN = 100, then the data would begin with SEQ = 101. If the Sequence number was not advanced by one, the process would end up in an endless loop of transmissions and acknowledgments. More details on this will be given in Section 5.5.5.) The Sequence number ensures the correct order of the data stream, which is a fundamental component of reliability.

The Acknowledgment Number field (4 octets) verifies the receipt of data. This protocol process is called Positive Acknowledgment or Retransmission (PAR). The process requires that each unit of data (the octet in the case of TCP) be explicitly acknowledged. If it is not, the sender will time-out and retransmit. The value in the acknowledgment is the next octet (i.e., the next Sequence number) expected from the other end of the connection. When the Acknowledgment field is in use (i.e., during a connection), the ACK flag bit is set.

The next 32-bit word (octets 13–16 in the header) contains a number of fields used for control purposes. The Data Offset field (4 bits) measures the number of 32-bit words in the TCP header. Its value indicates where the TCP header ends and the upper layer protocol (ULP) data begins. The Offset field is necessary because the TCP header has a variable, not fixed, length; therefore, the position of the first octet of ULP data may vary. Since the minimum length of the TCP header is 20 octets, the minimum value of the Data Offset field would be five 32-bit words. The next 6 bits are reserved for future use and are set equal to zero.

Six flags that control the connection and data transfer are transmitted next. Each flag has its own 1-bit field. These flags include:

- URG: Urgent Pointer field significant
- ACK: Acknowledgment field significant
- PSH: Push function
- RST: Reset the connection
- SYN: Synchronize Sequence numbers
- FIN: No more data from sender

We will study the use of these flags in greater detail in Section 5.5.

The Window field (2 octets) provides end-to-end flow control. The number in the Window field indicates the quantity of octets, beginning with the one in the Acknowledgment field, which the sender of the segment can accept. Note that the Window field, like the Acknowledgment field, is bidirectional. Since TCP provides a full-duplex communication path, both ends send control information to their peer process at the other end of the connection. In other words, my host provides both an acknowledgment and a window advertisement to your host, and your host does the same for mine. In this manner, both ends provide control information to their remote partner.

The Checksum field (2 octets) is used for error control. The checksum calculation includes a 12-octet Pseudo header, the TCP header, and ULP data. The TCP Pseudo header (shown in Figure 5-5) is similar to the UDP Pseudo header shown in Figure 5-3. Its purpose is to provide error control on the IP header, the TCP header, and the data. The fields included in the TCP Pseudo header include the Source and Destination addresses, the Protocol, and the TCP Length. The TCP Length field includes the TCP header and ULP data, but not the 12-octet Pseudo header.

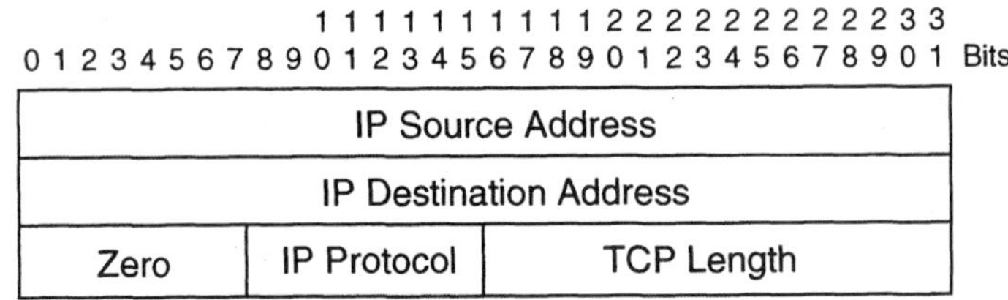

Figure 5-5. TCP Pseudo Header.

The Urgent Pointer field (2 octets) allows the position of urgent data within the TCP segment to be identified. This field is used in conjunction with the Urgent (URG) control flag and points to the Sequence number of the octet that follows the urgent data. In other words, the Urgent pointer indicates the beginning of the routine (nonurgent) data.

Options and Padding fields (both variable in length) complete the TCP header. The Options field is an even multiple of octets in length and specifies options required by the TCP process within the host. One option is the maximum TCP segment size, which mandates the amount of data that the sender of the option is willing to accept. The Padding field contains a variable number of zeros that ensure that the TCP header ends on a 32-bit boundary.

Now that we've explored the fields within the TCP header, we'll see how they provide the six TCP functions: basic data transfer, reliability, flow control, multiplexing, connections, and precedence/security.

5.5 TCP Functions

TCP is a rigorous protocol, rich with the functionality demanded by its government-backed designers. In this section, we'll explore each of the six areas of TCP operation separately, and we'll conclude with a summary of the TCP state diagram. For our network analysis and management, we'll focus on understanding the protocol interactions, not the internal protocol operations.

5.5.1 Basic Data Transfer

A TCP module transfers a series of octets, known as a *segment,* from one host to another. Data flows in both directions, making for a full-duplex connection. The TCP modules at each end determine the length of the segment and indicate this length in the Options field of the TCP header.

Occasionally, a TCP module requires immediate data delivery and can't wait for the segment to fill completely. In that case, an upper layer process would trigger the Push (PSH) flag within the TCP header and tell the TCP module to immediately forward all of the queued data to the receiver.

5.5.2 Reliability

So far, we've assumed that Host B has received all the data sent from Host A. Unfortunately, there's no such network utopia in the real world. The transmitted data could be lost, inadvertently duplicated, delivered out of order, or damaged. If damage occurs, the checksum will fail, alerting the receiver to the problem. The other conditions are more complex, and TCP must have mechanisms to handle them.

The cornerstones of TCP's reliability are its Sequence and Acknowledgment numbers. The Sequence number is logically attached to each outgoing octet. The receiver uses the Sequence numbers to determine whether any octets are missing or have been received out of order. TCP is a Positive Acknowledgment with Retransmission (PAR) protocol. This means that if data is received correctly, the receiving TCP module generates an Acknowledgment (ACK) number. If the transmitting TCP module does not receive an acknowledgment within the specified time, it will retransmit. No Negative Acknowledgments (NAKs) are allowed.

One of the design issues for a TCP/IP-based internetwork is optimizing the waiting time before allowing a retransmission to occur. Internetworks, by definition, contain multiple communication paths. Each of these may have a different propagation delay. In addition, the underlying infrastructure is connectionless (using IP); it is possible that a datagram could merely be delayed via a path with a longer propagation time, but not truly lost. In that case, retransmitting too quickly would consume precious bandwidth and cause confusion at the receiver if two identical messages were (eventually) received. On the other hand, if the datagram *is* truly lost, then delaying the retransmission also delays the receiver's ability to reassemble the entire message. An obvious compromise is in order. RFC 793, page 41, discusses the Retransmission Timer.

Figure 5-6 illustrates the TCP reliability services. (For simplicity, we'll assume that a large data file is being sent from Host A to Host B.) The first segment (SEQ 101–116) is acknowledged without error. The second segment (SEQ 117–132) is not so fortunate and experiences a transmission error. Since that segment never arrives at Host B, no acknowledgment (ACK = 133) is issued by Host B. The Retransmission Timer in Host A expires and retransmits the second segment (SEQ 117–132). When this segment is received, Host B then issues the acknowledgment.

5.5.3 Flow Control

Let's begin by reviewing some terms and concepts; then we'll discuss how they relate to the Flow Control mechanism. Recall from the previous section that the Sequence number is attached by the sending device to the TCP segment and counts each octet transmitted. The Acknowledgment number indicates the next expected Sequence number, that is, the next octet of data expected from the other end of the connection. A third metric, also included within the TCP header, is the Window number. The Window indicates how many more Sequence numbers the sender of the Window is prepared to accept, and is described in RFC 813 [5-5]. The Window controls the flow of data from sender to receiver. Thus, if Host A sends Window = 1024 to Host B, it has much more available buffer space than if Host A sends Window = 10. In the first case, Host A grants Host B permission to transmit 1,024 octets of data past the current Sequence number; in the second case, only 10.

Window = 0 and Window = 1 are two special cases of *sliding window* operations. If a host returns Window = 0, it has shut down communication and will accept no more octets from the other end of the connection. Window = 1 is sometimes known as *stop-and-wait* transmission. The top portion of Figure 5-7 illustrates what happens when Window = 1. (For simplicity, we'll assume that the data is transmitted from Host A to Host B and that acknowledgments go in the opposite direction. In reality, TCP allows full-duplex operation. Therefore, data could also flow from Host B to Host A, with corresponding acknowledgments from Host A to Host B.) If Host A has a large file to send and Host B has set Window = 1, the data transmissions will be limited to one octet before each acknowledgment. Thus, Host A must wait for Host B's acknowledgment before it can transmit each octet of data. Not very efficient, you might say, and few analysts would disagree.

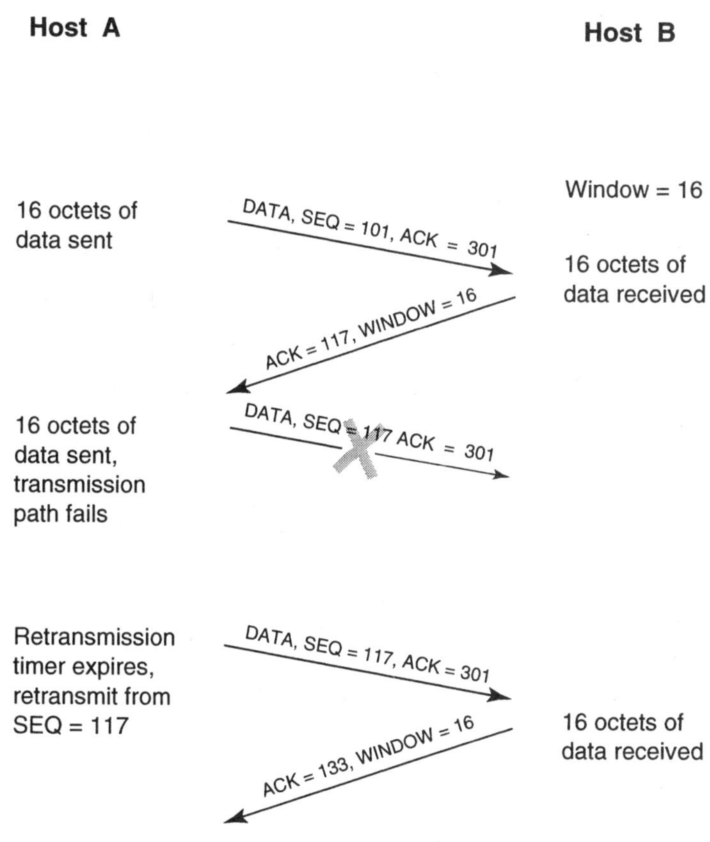

Figure 5-6. TCP Data Retransmissions.

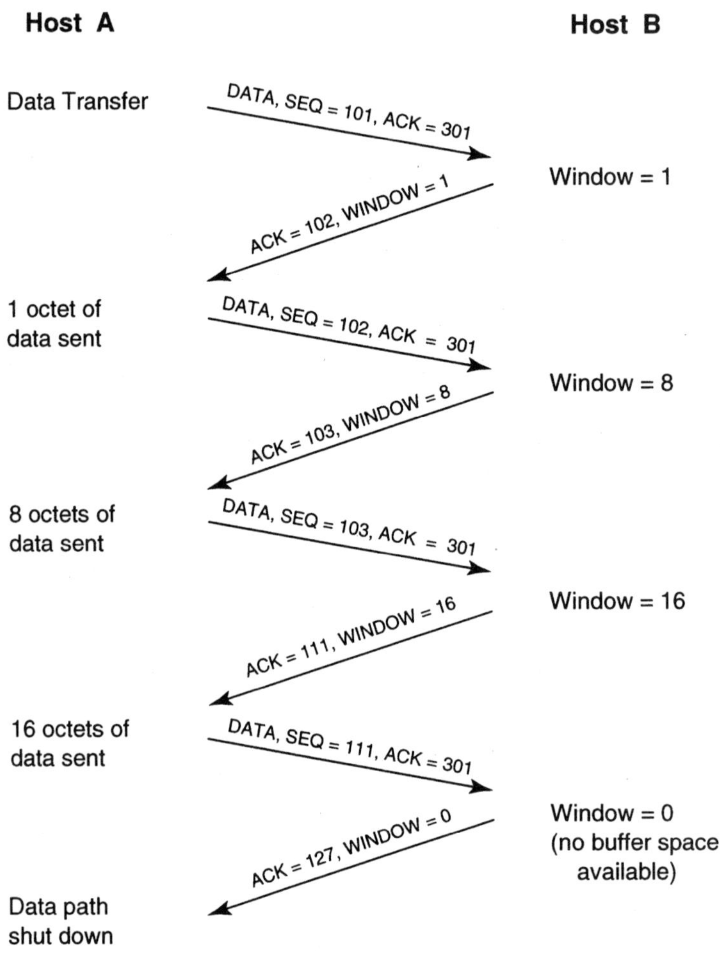

Figure 5-7. TCP Data Transfer (Window Size Varying).

Now suppose that Host B obtains additional buffer space and increases the value to Window = 8. Host A can now send 8 octets of data (SEQ 103–110, inclusive) before requiring an acknowledgment. When the acknowledgment arrives, it indicates the next expected Sequence number (e.g., ACK = 111 would indicate that octets numbered 103–110 arrived correctly and that it is permissible to send 16 more, beginning with SEQ = 111). Similarly, an increase to Window = 32 would allow a corresponding increase in the quantity of data transmitted. We use the term *flow control* to describe the effect of the sliding window operation because you can control the flow of data by adjusting the window value.

The window size is a parameter that requires some optimization. Too many small segments generate needless ACKs, but large segments utilize more buffers. The case study in Section 7.12 provides an example of a window-related problem.

5.5.4 Multiplexing

The TCP module assumes that its associated host may be a multi-function system. Indeed, the host may offer the user an entire menu of applications, and the user may want to use several of them simultaneously. TCP provides a multiplexing function to accommodate these diverse needs. In Section 5.2, we learned that the Port address (16 bits) uniquely identifies an end-user process. The host binds, or associates, a particular port with a process. Some processes, such as the major application protocols FTP, TELNET, and SMTP, have assigned Port numbers. Others may be dynamically assigned as the need arises.

A *socket* is the concatenation of the Internet address and the Port address. A *connection* is the association of a pair of sockets, although a socket is not restricted to one connection. For example, if a host has a port assigned to the TELNET protocol — say, Port 23 — there's nothing to preclude connections from multiple terminals (i.e., remote sockets) to that port. The host can also have other processes, such as the File Transfer Protocol, operating on another port, such as Port 21. Thus, we can say that TCP provides true multiplexing (the term stems from the Greek noun meaning "many paths") of the data connections into and out of a particular host.

5.5.5 Connections

Because TCP is a connection-oriented protocol, a *logical connection* (known as a virtual circuit) must be established between the two end users before any ULP data can be transmitted. The logical connection is a concatenation of many physical connections, such as Host A to LAN A, connecting to Router 1, connecting via a WAN link to Router 2, and so on. Managing this end-to-end system is much easier if you use a single identifier — the logical connection.

The logical connection is established when the ULP process recognizes the need to communicate with a distant peer and passes an OPEN command across the user/TCP interface to the TCP module in the host. The OPEN command is one of a number of primitives described in the TCP standard,

RFC 793. A *primitive* is an event that requests or responds to an action from the other side of the user/TCP interface. The primary TCP primitives include OPEN, SEND, RECEIVE, CLOSE, STATUS, and ABORT. Many parameters traverse the user/TCP interface with the primitive, and therefore more descriptive names such as PASSIVE OPEN or ACTIVE OPEN are often used. Each host operating system may modify these primitives for its particular use.

The OPEN command triggers what is referred to as a *three-way handshake*. The three-way handshake ensures that both ends are prepared to transfer data, reducing the likelihood of data being sent before the distant host can accept it. (If either host becomes confused from an attempted three-way handshake, the confused host sets the Reset (RST) flag, indicating that the segment that arrived was not intended for this connection.) In most cases, one TCP module initiates the connection and another responds. However, it is possible for two TCP modules to request a connection simultaneously; RFC 793 elaborates on this condition.

RFC 793 describes the steps in the three-way handshake as follows:

- Host A to Host B: SYN, my Sequence number is X
- Host B to Host A: ACK, your Sequence number is X
- Host B to Host A: SYN, my Sequence number is Y
- Host A to Host B: ACK, your Sequence number is Y

In other words, each side of the connection sends it own Sequence number to the other end, and receives a confirmation of it in an acknowledgment from the other side. Both sides are then aware of the other's initial Sequence number, and reliable communication can proceed. It is possible (and in most cases preferred) for the second and third steps above (from Host B to Host A) to be combined into a single transmission. This combination then yields a total of three transmissions, illustrating the name "three-way handshake."

In Figure 5-8, the initiating module (Host A) generates a TCP segment with the Synchronize (SYN) flag set (SYN = 1) and an initial Sequence number chosen by the module (e.g., ISN = 100). If the connection request was acceptable, the remote TCP module (Host B) would return a TCP segment containing both an acknowledgment for Host A's ISN [ACK = 101], plus its own ISN [SEQ = 300]. Both the Synchronize and Acknowledgment (ACK) flag bits would be set (SYN = 1, ACK = 1) in that response from Host B. Note that the two TCP modules are not required to have the same

ISN since these numbers are administered locally. An acknowledgment from the initiating module, Host A, is the third step of the three-way handshake. This TCP segment includes the Sequence number [SEQ = 101] plus an acknowledgment for Host B's desired ISN (ACK = 301).

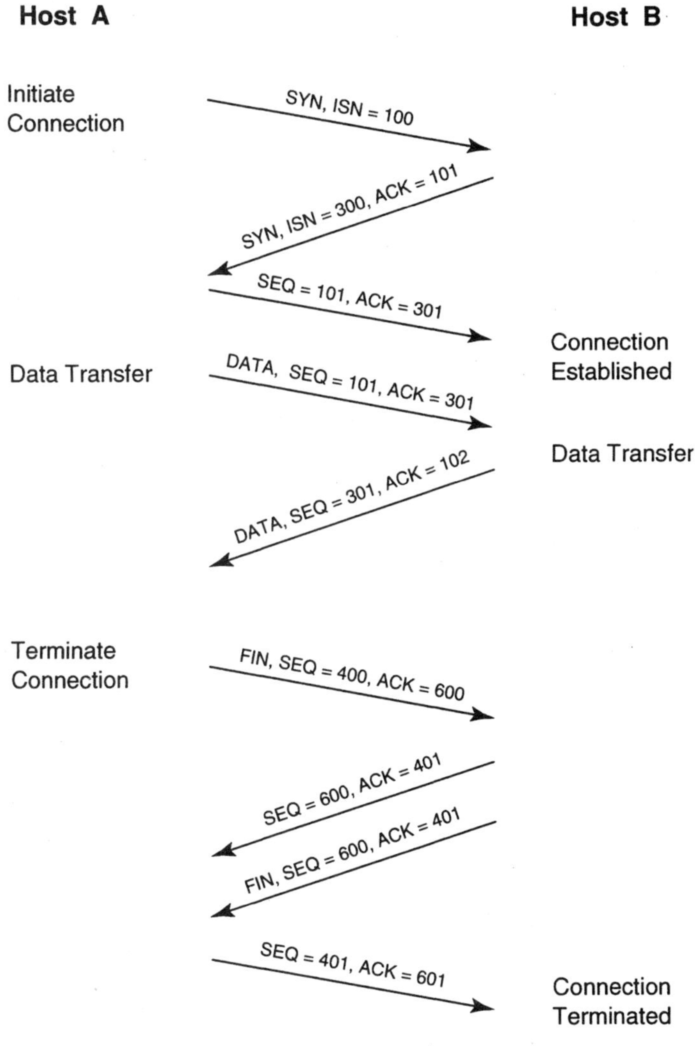

Figure 5-8. TCP Connection Establishment, Data Transfer, and Termination Events.

Note that, by convention, the ACK in the second TCP segment increments (i.e. ACK = 101) even though no data transfer occurred. Similarly, the ACK in the third segment increments (i.e. ACK = 301) even though no data transfer occurred there either. These exceptions to the normal acknowledgment process occur in order to verify receipt of the other host's ISN segment (which may not contain any data). When data transfer does begin (in the fourth and fifth segments), the same Sequence numbers as the previously corresponding ACKs are used (i.e. SEQ = 101 seen in both segments 3 and 4). Another way of saying this is that a segment containing just an ACK (and no data) would not use up a Sequence number; the three-way handshake (with the FIN flag set) *does* use a Sequence number.

To manage connection-related issues, the TCP module maintains a record known as a *Transmission Control Block* (TCB). The TCB stores a number of variables, including the Local and Remote Socket numbers, the security/precedence of the connection, and pointers to identify incoming and outgoing data streams. Two other TCP functions already discussed, reliability and flow control, manage the data transfer process. The Sequence numbers, acknowledgments, and Window numbers are the mechanisms that the reliability, flow control, and data transfer functions use to ensure their proper operation.

Let's assume the data transfer is complete and the connection is no longer required. Because TCP provides a full-duplex connection, either side may initiate the disconnect. The other end, however, must continue to receive data until the remote module has finished sending it. The TCP connection can be terminated in one of three ways: the local host can initiate the termination, the remote host can initiate it, or both can close the connection simultaneously. The shutdown procedure is the same in all three cases. For example, if the local host has completed its business, it can generate a TCP segment with the Finish (FIN) flag set (FIN = 1). The local host will continue to receive data from the remote end until it receives a TCP segment with FIN = 1 from the remote end. When the second Finish segment is acknowledged, the connection is closed.

The bottom of Figure 5-8 illustrates this scenario. Host A has finished transferring its last data segment and terminates the connection by sending [FIN, SEQ = 400, ACK = 600]. Host B acknowledges receipt of Host A's FIN by sending [SEQ = 600, ACK = 401], and then sends its own FIN [FIN, SEQ = 600, ACK = 401]. Host A acknowledges Host B's FIN with [SEQ = 401, ACK = 601], and the connection is closed.

Make a special note of the Sequence and Acknowledgment number progression in these last four segments. The first closing segment [FIN, SEQ = 400, ACK = 600] uses the Sequence and Acknowledgment numbers that were next in line. The second segment is just an acknowledgment for the first [SEQ = 600, ACK = 401], and therefore does not use up a Sequence number. The third segment contains a FIN, and again, by convention, uses the same Sequence number [SEQ = 600] as was found in the previous ACK. The fourth segment (again, operating by convention) increments the Sequence number even though it contains no data ([SEQ = 401], which corresponds with the previous ACK), and also increments the Acknowledgment number [ACK = 601]. This verifies receipt of the FIN contained in the third segment. Once again, the connection termination sequence, also called the modified three-way handshake, causes a special condition for the Sequence and Acknowledgment numbers. This FIN segment (without data) uses a Sequence number, much like the case of the SYN flag illustrated previously.

5.5.6 Precedence/Security

As a military application, TCP required precedence and security. Therefore, these two attributes may be assigned for each connection. The IP Type of Service field and security option may be used to define the requirements associated with that connection. Higher layer protocols, such as TELNET, may specify the attribute required, and the TCP module will then comply on behalf of that ULP.

5.5.7 The TCP Connection State Diagram

Like many computer processes, TCP operation is best summarized with a state diagram, shown in Figure 5-9a. TCP operation progresses from one state to another in response to events, including user calls, incoming TCP segments, and timeouts. The *user calls* are commands that cross the User/TCP interface, that is, the functions that support the communication between user processes. These include OPEN, SEND, RECEIVE, CLOSE, ABORT, and STATUS. When a TCP segment arrives, it must be examined to determine if any of the flags are set, particularly the SYN, ACK, RST, or FIN. The timeouts include the USER TIMEOUT, RETRANSMISSION TIMEOUT, and the TIME-WAIT TIMEOUT. (As is the case with many protocols, TCP will not wait in a particular state for an indefinite period of time, nor will it attempt to recover from an error forever. Many timers, which are not illustrated in Figure 5-9a, are also at work that oversee some of these functions.)

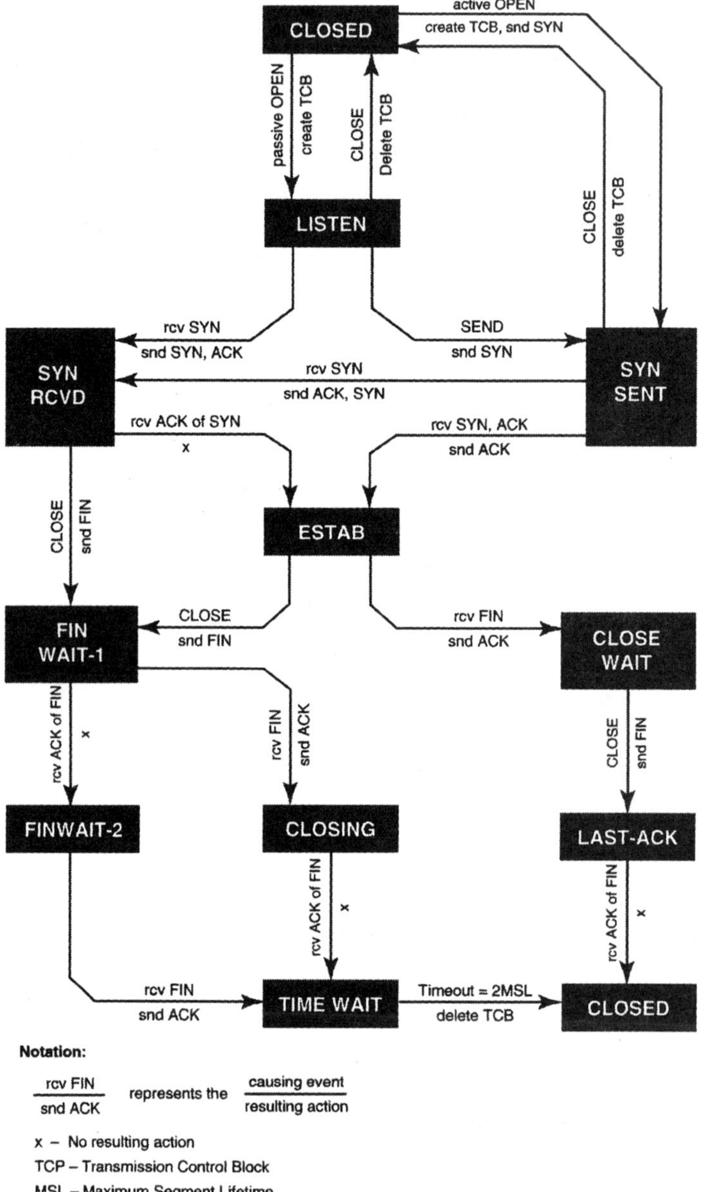

Figure 5-9a. TCP Connection State Diagram.

The first state entered by the TCP process is CLOSED, indicating no connection state at all. A passive OPEN creates a Transmission Control Block (TCB), which moves the process to the LISTEN state. It then waits for a connection from a remote TCP and port. If a SYN is received, a SYN, ACK is sent in response. Should another ACK be received, the connection reaches the ESTABLISHED state. When the data transfer is complete at the near end, a CLOSE is issued and a FIN sent, moving to the state FIN-WAIT-1 and subsequently FIN-WAIT-2 or CLOSING. After waiting to confirm that the remote TCP has received the ACK of its FIN, the TCB is deleted and the connection reaches the CLOSED state. In a similar manner, a FIN received from the other end of the connection moves the process to state CLOSE-WAIT, pending the completion of the local user's requirements for that connection. When the ACK of the FIN has been received, the connection once again reaches the CLOSED state. Further explanations of the various states are given in Figure 5-9b.

The meanings of the states are:

LISTEN – represents waiting for a connection request from any remote TCP and port.

SYN-SENT – represents waiting for a matching connection request after having sent a connection request.

SYN-RECEIVED – represents waiting for a confirming connection request acknowledgment after having both received and sent a connection request.

ESTABLISHED – represents an open connection, data received can be delivered to the user. The normal state for the data transfer phase of the connection.

FIN-WAIT-1 – represents waiting for a connection termination request from the remote TCP, or an acknowledgment of the connection termination request previously sent.

FIN-WAIT-2 – represents waiting for a connection termination request from the remote TCP.

CLOSE-WAIT – represents waiting for a connection termination request from the local user.

CLOSING – represents waiting for a connection termination request acknowledgment from the remote TCP.

LAST-ACK – represents waiting for an acknowledgment of the connection termination request previously sent to the remote TCP (which includes an acknowledgment of its connection termination request).

TIME-WAIT – represents waiting for enough time to pass to be sure the remote TCP received the acknowledgment of its connection termination request.

CLOSED – represents no connection state at all.

Figure 5-9b. TCP Connection States.

5.6 TCP Network Optimization

Many TCP implementations include algorithms that perform specific functions. These are documented in RFC 2581 [5-6] and include:

- **Slow Start:** If too many segments are sent to a distant host, and a slower link, such as a WAN route via an intermediate router, is in the path to that host, congestion at that WAN route may occur. To avoid this, the Slow Start algorithm limits the amount of data that can be transmitted, as controlled by the ACKs received from the distant end.
- **Congestion Avoidance:** The loss of packets is a typical result of congestion between sender and receiver, and this congestion is indicated by the occurrence of a timeout or the receipt of duplicate ACKs. The Congestion Avoidance algorithm, which may operate in conjunction with Slow Start, increases the congestion window variable in linear amounts, thus limiting the data sent into the network.
- **Fast Retransmit:** The receipt of multiple, duplicate ACKs may indicate a lost TCP segment. The sending TCP module will then retransmit the segment that appears to be lost without waiting for a retransmission timer to expire, thus speeding receipt of the data.
- **Fast Recovery:** After the Fast Retransmit algorithm resends an apparently missing segment, Congestion Avoidance, but not Slow Start, is invoked. This process, called the Fast Recovery algorithm, avoids the Slow Start process to keep the sender from reducing the data flow abruptly. The Fast Recovery and Fast Retransmit algorithms may operate in conjunction with each other.

Another topic that is often mentioned in the context of TCP optimization is the bandwidth-delay product, defined in RFC 1323 [5-7]. The bandwidth-delay product measures the amount of data that would be required to "fill the pipe," or, in other words, the amount of buffer space required at both sender and receiver to obtain the maximum throughput on the TCP connection. As its name implies, this parameter is the product of the transmission rate of the communication channel and the round-trip delay of that channel. For example, if the circuit operated at the T1 rate (1,544,000 bits/second) and the round trip delay was 70 milliseconds (0.070 seconds), the bandwidth-delay product would be 108,080 bits, or 13,510 octets, or approximately nine Ethernet/IEEE 802.3 frames. Large bandwidth-delay products, such as those that would result from either a high-speed link (such as DS3 operating at 45 Mbps or OC-3 operating at 155 Mbps) or a very long delay (such as a satellite circuit), introduce new challenges within the window size, packet loss recovery, and round-trip measurement functions. Suggested solutions for these issues are detailed in RFC 1323.

5.7 Stream Control Transport Protocol

When a telephone conversation is established between a calling and a called party, a number of switching centers, called central offices (COs), are involved in establishing the connection. Communication between the CO switches is necessary to establish the circuit paths, measure the duration of the call, render a bill to the appropriate party, disconnect the circuit when the call is over, and so on. This communication is called *signaling*, and it typically employs an ITU-T communication protocol called Signaling System 7 (SS7) to assure interoperability in this international and multivendor environment. When IP networks interface with the Public Switched Telephone Network (PSTN), mechanisms for transporting this signaling information over IP networks must be defined. The architecture for this work is defined in RFC 2719 [5-8], and a new protocol, called the Stream Control Transport Protocol (SCTP), is described in RFC 2960 [5-9] and RFC 3286 [5-10].

The Stream Control Transport Protocol is a peer of both UDP and TCP, and it provides reliable transport services to IP-based networks. RFC 2960 defines several key services provided by SCTP that are required to support signaling transport, but may also benefit other applications:

- acknowledged error-free nonduplicated transfer of user data
- data fragmentation to conform to discovered path MTU size
- sequenced delivery of user messages within multiple streams, with an option for order-of-arrival delivery of individual user messages
- optional bundling of multiple user messages into a single SCTP packet
- network-level fault tolerance through supporting of multi-homing at either or both ends of an association

The architecture that supports these functions is based on a peer relationship between SCTP endpoints, which is called an SCTP association (Figure 5-10). Note that the SCTP transport services support the user applications (such as signaling) for transport over an IP-based network.

The SCTP transport service includes a number of functions (Figure 5-11). The association startup is initiated by a request from the SCTP user and provides for a four-way handshake for enhanced security. For association takedown, a graceful close of that association can be initiated by request from the end user, or an ungraceful close (called an ABORT) can occur upon request from the end user, or when error conditions result.

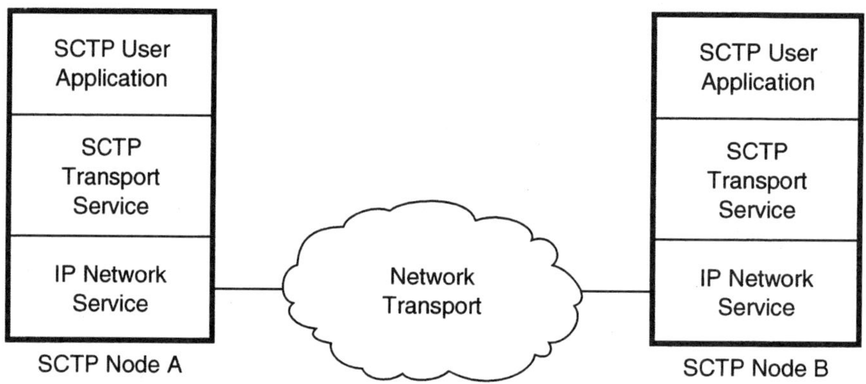

Figure 5-10. An SCTP Association.

Within SCTP, the stream is a sequence of user messages that are delivered in order to the upper layer protocol. (In contrast, TCP considers a stream to be a sequence of octets, not user messages.) When the association is initiated, the number of streams to be supported by the association is negotiated between the two endpoints. A stream sequence number is then assigned to each message, and those messages are delivered in sequence at the receiving side.

SCTP supports data fragmentation by assuring that the SCTP packet that is passed to the lower layer (e.g. IP) conforms to the maximum transmission unit (MTU) size of the path in use.

The acknowledgment and congestion avoidance function is responsible for the retransmission of a packet when an acknowledgment has not been received in a timely fashion. The acknowledgment function is supported by a Transmission Sequence Number (TSN) that is assigned to each element of user data.

The SCTP packet consists of a common header followed by one or more "chunks" of information, where the chunks could either be user data or control information (Figure 5-12). Multiple user messages can be bundled into a single SCTP packet, which requires a chunk bundling (packet assembly) function at the transmitter, and a corresponding disassembly function at the receiver.

The SCTP common header (Figure 5-12) includes a Verification Tag and a 32-bit Checksum field in support of packet validation. The Verification Tag value is chosen by each endpoint during the association startup. The checksum protects against corrupted data packets within the network. Packets that arrive without the expected Verification Tag, or with an invalid checksum, are discarded.

The path management function monitors the reachability of an intended destination, and also manages a set of destination transport addresses that are communicated to the far end of the association.

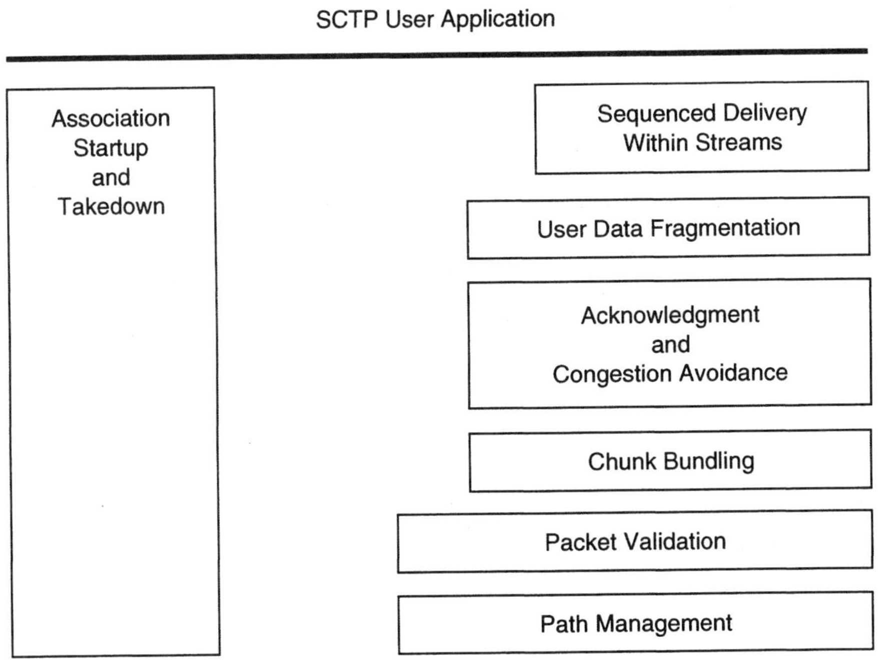

SCTP User Application

Association Startup and Takedown

Sequenced Delivery Within Streams

User Data Fragmentation

Acknowledgment and Congestion Avoidance

Chunk Bundling

Packet Validation

Path Management

Figure 5-11. Functional View of the SCTP Transport Service.

The SCTP packet consists of a common header followed by a variable number of chunks containing user data or control information (review Figure 5-12). Multiple chunks may be bundled into one SCTP packet (up to the MTU size), except for the INIT, INIT ACK, and SHUTDOWN COMPLETE chunks, which must not be bundled. Long data messages that do not fit into one SCTP packet can be fragmented by the data fragmentation function discussed above. The header consists of four fields: Source Port Number, Destination Port Number, Verification Tag, and the Checksum.

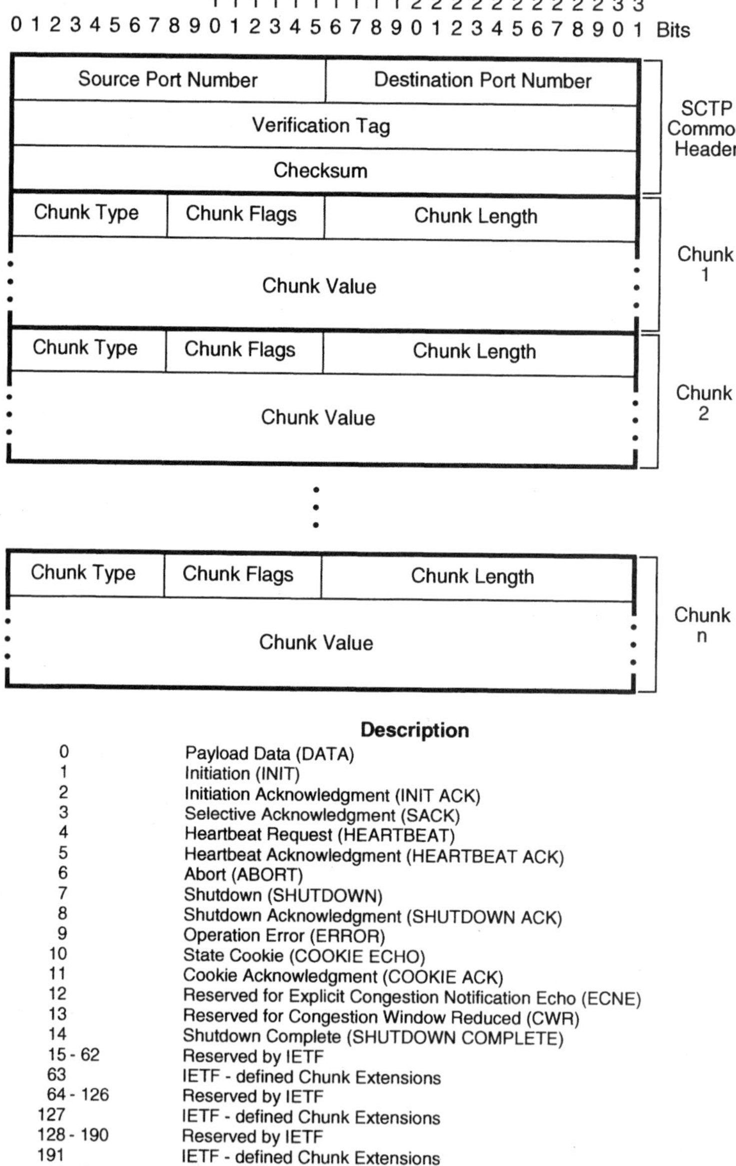

	Description
0	Payload Data (DATA)
1	Initiation (INIT)
2	Initiation Acknowledgment (INIT ACK)
3	Selective Acknowledgment (SACK)
4	Heartbeat Request (HEARTBEAT)
5	Heartbeat Acknowledgment (HEARTBEAT ACK)
6	Abort (ABORT)
7	Shutdown (SHUTDOWN)
8	Shutdown Acknowledgment (SHUTDOWN ACK)
9	Operation Error (ERROR)
10	State Cookie (COOKIE ECHO)
11	Cookie Acknowledgment (COOKIE ACK)
12	Reserved for Explicit Congestion Notification Echo (ECNE)
13	Reserved for Congestion Window Reduced (CWR)
14	Shutdown Complete (SHUTDOWN COMPLETE)
15 - 62	Reserved by IETF
63	IETF - defined Chunk Extensions
64 - 126	Reserved by IETF
127	IETF - defined Chunk Extensions
128 - 190	Reserved by IETF
191	IETF - defined Chunk Extensions
192 - 254	Reserved by IETF
255	IETF - defined Chunk Extension

Figure 5-12. SCTP Packet Format and Chunk Values.

Each chunk consists of three common fields, plus a variable length value. The Chunk Type field defines the type of data or control information contained in that chunk. The Chunk Flags are bits that are used with some of the chunk types for further processing. The Chunk Length represents the size of the chunk, given in octets, including the Chunk Type, Chunk Flags, Chunk Length, and Chunk Value fields. If the Chunk Value is zero-length, the Chunk Length field would be set to 4 (octets, or 32 bits).

The DATA chunk provides an example of a chunk format (Figure 5-13). This chunk begins with a Type field (0, indicating a DATA chunk) and a Reserved field (which is set to all zeros). Three one-bit flags specify chunk sequencing and fragmentation functions: the Unordered (U) bit, which, when set to 1, indicates that this is an unordered DATA chunk without an assigned Stream Sequence Number; the Beginning (B) fragment bit, which, when set to 1, indicates the first fragment of a user message; and the Ending (E) fragment bit, which, when set to 1, indicates the last fragment of a user message. The Length field indicates the length of the DATA chunk, given in octets, measuring from the beginning of the Type field to the end of the User Data field. The Transmission Sequence Number (TSN) is a 32-bit sequence number that is used internally by SCTP and attached to each chunk containing user data to acknowledge receipt and detect duplicate deliveries. The Stream Identifier S is a 16-bit number that identifies the stream to which this user data belongs. The Stream Sequence Number n is a 16-bit number that represents the stream sequence number of the user data within Stream S. The Payload Protocol Identifier is a 32-bit number that identifies the type of upper layer information contained in this DATA chunk. The User Data field is variable in length and contains the payload user data.

A second example is that of the INIT chunk, which is used to initiate associations between two SCTP endpoints (Figure 5-14). The INIT chunk begins with the Type (set to 1 for the INIT chunk), Chunk Flags (which are set to 0 and ignored by the receiver), and Chunk Length. Next come two sections of parameters: a set of mandatory fixed parameters, followed by a set of optional variable parameters. The fixed parameters include a 32-bit Initiate Tag that the receiver of the INIT will record and use in its Verification Tag field; a 32-bit Advertised Receiver Window Credit (a_rwnd) that indicates the amount of dedicated buffer space (given in octets) that has been reserved for this association; a 16-bit Number of Outbound Streams that defines the number of outbound streams that the sender of this INIT chunk wishes to create in this association; a 16-bit Number of Inbound Streams that defines the maximum number of streams the sender of this INIT chunk allows the peer end to create in this association; and a 32-bit Initial Transmission Sequence Number (TSN). The optional parameters support IPv4 and IPv6 addresses, the life-span of the State Cookie, communication of the Host Name to the SCTP peer at the other end of the connection, and a list of address types that are supported.

```
                    1 1 1 1 1 1 1 1 1 1 2 2 2 2 2 2 2 2 2 2 3 3
0 1 2 3 4 5 6 7 8 9 0 1 2 3 4 5 6 7 8 9 0 1 2 3 4 5 6 7 8 9 0 1 Bits
+---------------+-----------+-+-+-+-------------------------------+
|   Type = 0    | Reserved  |U|B|E|            Length             |
+---------------+-----------+-+-+-+-------------------------------+
|                              TSN                                |
+---------------------------------+-------------------------------+
|      Stream Identifier S        |  Stream Sequence Number n     |
+---------------------------------+-------------------------------+
|                  Payload Protocol Identifier                    |
+-----------------------------------------------------------------+
:                 User Data (seq n of Stream S)                   :
+-----------------------------------------------------------------+
```

Figure 5-13. SCTP DATA Chunk Format.

```
                    1 1 1 1 1 1 1 1 1 1 2 2 2 2 2 2 2 2 2 2 3 3
0 1 2 3 4 5 6 7 8 9 0 1 2 3 4 5 6 7 8 9 0 1 2 3 4 5 6 7 8 9 0 1 Bits
+---------------+-----------------+-------------------------------+
|   Type = 1    |   Chunk Flags   |         Chunk Length          |
+---------------+-----------------+-------------------------------+
|                          Initiate Tag                           |
+-----------------------------------------------------------------+
|          Advertised Receiver Window Credit (a_rwnd)             |
+---------------------------------+-------------------------------+
|  Number of Outbound Streams     |   Number of Inbound Streams   |
+---------------------------------+-------------------------------+
|                          Initial TSN                            |
+-----------------------------------------------------------------+
:            Optional / Variable - Length Parameters              :
+-----------------------------------------------------------------+
```

Figure 5-14. SCTP INIT Chunk Format.

Reviewing Figure 5-12, it is obvious that many other chunk formats have been defined, which are all described in detail in RFC 2960.

5.8 Looking Ahead

This chapter marks the three-quarter point in our journey from the bottom to the top of the ARPA protocol stack. We have now made the LAN, MAN, or WAN hardware connection at the Network Interface Layer, transmitted the datagrams at the Internet Layer, and ensured the reliability of those datagrams at the Host-to-Host Layer. In the next chapter, we will consider thc next generation of the Internet Protocol, called IPv6.

5.9 References

[5-1] Postel, J. "User Datagram Protocol." RFC 768, August 1980.

[5-2] Postel, J., Editor. "Transmission Control Protocol." RFC 793, September 1981.

[5-3] Socolofsky, T., et al. "A TCP/IP Tutorial." RFC 1180, January 1991.

[5-4] Postel, J. "The TCP Maximum Segment Size and Related Topics." RFC 879, November 1983.

[5-5] Clark, David D. "Window and Acknowledgement Strategy in TCP." RFC 813, July 1982.

[5-6] Stevens, W. "TCP Congestion Control." RFC 2581, April 1999.

[5-7] Jacobson, V., et al. "TCP Extensions for High Performance." RFC 1323, May 1992.

[5-8] Ong, L., et al. "Framework Architecture for Signaling Transport." RFC 2719, October 1999.

[5-9] Stewart, R., et al. "Stream Control Transmission Protocol." RFC 2960, October 2000.

[5-10] Ong, L., and J. Yoakum. "An Introduction to the Stream Control Transmission Protocol." RFC 3286, May 2002.

6

The Next Generation IPv6

Like many emerging systems, the Internet has suffered growing pains. In the 1970s, when many of these protocols were first developed, the majority of users were limited to educational and government environments, not the general public, so there were plenty of addresses to go around. The term "electronic commerce" had not yet been coined, so the need for enhanced network security to support credit card transactions was not an issue. Furthermore, real-time communication over an IP network, such as voice and video, had not yet been developed, so the requirements to support these functions in a connectionless environment had yet to be defined.

What a difference a few years make! Those of us who have been associated with networking for several decades look back and marvel at the speed at which this industry has evolved. And, like many previous systems, such as the 8-bit processor, improvements are required in order to keep up with current end-user requirements. The Internet Protocol is no exception to this rule — updates are required in order to maintain optimum performance and effectiveness. For IP, those enhancements began in the early 1990s and were dubbed Internet Protocol — Next Generation, or IPng. We now know that work by its formal protocol name, IP version 6, or IPv6, which is the subject of this chapter. We will begin with a brief history of the protocol's development.

6.1 IPng Development

In December 1993, RFC 1550 [6-1] was distributed, titled "IP: Next Generation (IPng) White Paper Solicitation." This RFC invited any interested party to submit comments regarding any specific requirements for the IPng or any key factors that should be considered during the IPng selection process. Twenty-one responses were submitted that addressed a variety of topics, including: security (RFC 1675), a large corporate user's view (RFC 1687), a cellular industry view (RFC 1674), and a cable television industry view (RFC

1686). Other interested organizations, such as the Institute of Electrical and Electronics Engineers (IEEE), also supported the IPng enhancement efforts.

The IPng Area commissioned RFC 1726, "Technical Criteria for Choosing IP: The Next Generation (IPng)" [6-2], to define a set of criteria that would be used in the IPng evaluation process. Seventeen criteria were noted:

- Scale — The IPng Protocol must scale to allow the identification and addressing of at least 10^{12} end systems and 10^9 individual networks.
- Topological Flexibility — The routing architecture and protocols of IPng must allow for many different network topologies.
- Performance — A state-of-the-art, commercial-grade router must be able to process and forward IPng traffic at speeds capable of fully utilizing common, commercially available, high-speed media at the time. Hosts must be able to achieve data transfer rates with IPng that are comparable to the rates achieved with IPv4, using similar levels of host resources.
- Robust Service — The network service and its associated routing and control protocols must be robust.
- Transition — The protocol must have a straightforward transition plan from the current IPv4.
- Media Independence — The protocol must work across an internetwork of many different LAN, MAN, and WAN media, with individual link speeds ranging from ones-of-bits per second to hundreds of gigabits per second.
- Unreliable Datagram Service — The protocol must support an unreliable datagram delivery service.
- Configuration, Administration, and Operation — The protocol must permit easy and largely distributed configuration and operation. The automatic configuration of hosts and routers is required.
- Secure Operation — IPng must provide a secure network layer.
- Unique Naming — IPng must assign all IP-Layer objects in global, ubiquitous, Internet unique names.
- Access and Documentation — The protocols that define IPng, its associated protocols, and the routing protocols must be published in the standards track RFCs, be freely available, and require no licensing fees for implementation.

- Multicast — The protocol must support both unicast and multicast packet transmission.
- Extensibility — The protocol must be extensible; it must be able to evolve to meet the future service needs of the Internet. In addition, as IPng evolves, it must allow different versions to coexist on the same network.
- Network Servicc — The protocol must allow the network to associate packets with particular service classes and provide them with the services specified by those classes.
- Mobility — The protocol must support mobile hosts, networks, and internetworks.
- Control Protocol — The protocol must include elementary support for testing and debugging networks.
- Private Networks — IPng must allow users to build private internetworks on top of the basic Internet infrastructure, supporting both IP-based and non–IP-based internetworks.

6.1.1 Functional Capabilities of IPv6

RFC 1752 [6-3] describes the important features of IPng, formally designated IPv6. These capabilities of IPv6 include:

- "Expanded addressing and routing — Increasing the IP address field from 32 to 128 bits in length, which allows for a much greater number of addressable nodes, more levels of addressing hierarchy, defining new types of addresses, and so on."
- Simplified header format — Eliminating or making optional some of the IPv4 header fields to reduce the packet handling overhead, thus providing some compensation for the larger addresses. Even with the addresses, which are four times as long, the IPv6 header is only 40 octets in length, compared with 20 octets for IPv4.
- Extension headers and options — IPv6 options are placed in separate headers, located after the core IPv6 header information, such that processing at every intermediate stop between source and destination may not be required.
- Authentication and privacy — Required support in all implementations of IPv6 to authenticate the sender of a packet and encrypt the contents of that packet, as required.

- Autoreconfiguration — Support from node address assignments up to the use of the Dynamic Host Reconfiguration Protocol (DHCP).
- Source routes — Support for a header that supports the Source Demand Routing Protocol (SDRP), such that a source-selected route may complement the route determined by the existing routing protocols.
- Simple and flexible transition — A transition plan with four basic requirements:
    1. Incremental upgrade — Existing IPv4 hosts can be upgraded at any time without a dependency on other hosts or routers being upgraded.
    2. Incremental deployment — New IPv6 hosts and routers can be installed at any time without any pre-requisites.
    3. Easy addressing — When existing installed IPv4 hosts or routers are upgraded to IPv6, they may continue to use their existing address without needing a new assigned address.
    4. Low start-up costs — Little or no preparation work is needed in order to upgrade existing IPv4 systems to IPv6 or to deploy new IPv6 systems.
- Quality of service capabilities — A new capability is added to enable the labeling of packets belonging to particular traffic "flows" for which the sender has requested special handling, such as nondefault quality of service or "real-time" service.

Thus the foreseeable exhaustion of the available IPv4 address space gave the Internet community the impetus to consider revisions to this widely deployed protocol. But the other factors listed above, such as support for real-time applications, enhanced security, and other design criteria, may be equally important, or possibly more important, to individual network managers. The ongoing development of IPv6 is the responsibility of the IETF IPv6 Working Group, which is documented on their website [6-4].

6.1.2 IPv6 Terminology

RFC 2460, section 2 [6-5], defines some terms that will be relevant to our discussion of the IPv6 architecture. These definitions are paraphrased below:

- Node — A device that implements IPv6.
- Router — A node that forwards IPv6 packets not explicitly addressed to itself [see note below].
- Host — Any node that is not a router [see note below].
- Upper layer — A protocol layer immediately above IPv6, such as ICMP, OSPF, TCP, UDP, and so on.
- Link — A communication facility or medium over which nodes can communicate at the Data Link Layer, such as Ethernet, token ring, frame relay, PPP, and so on.
- Neighbors — Nodes attached to the same link.
- Interface — A node's attachment to a link.
- Address — An IPv6-layer identifier for an interface or a set of interfaces.
- Packet — An IPv6 header plus payload.
- Link MTU — The maximum transmission unit, or packet size, given in octets, that can be conveyed in one piece over a link.
- Path MTU — The minimum link MTU of all the links in a path between a source node and a destination node.

Note: It is possible, though unusual, for a device with multiple interfaces to be configured to forward non-self-destined packets arriving from some set (fewer than all) of its interfaces, and to discard non-self-destined packets arriving from its other interfaces. Such a device must obey the protocol requirements for routers when receiving packets from, and interacting with neighbors over, the former (forwarding) interfaces. It must obey the protocol requirements for hosts when receiving packets from, and interacting with neighbors over, the latter (nonforwarding) interfaces.

From the definitions above, notice that an IPv6 node may deploy router, host, or both router and host functions. Secondly, the term upper layer has a slightly different meaning in the context of IPv6. When used here, it means the next highest protocol layer. Examples cited in the specification (such as the Internet Control Message Protocol — ICMP — or the Open Shortest Path First — OSPF protocol) are both routing (OSI Network Layer) protocols. In more traditional networking parlance, the term upper layer might be construed to mean an Application Layer Protocol, such as the File Transfer Protocol (FTP) or the Simple Network Management Protocol (SNMP). Finally, the distinction between the link MTU and the path MTU implies that there are mechanisms that keep track of these values. As we will discover in subsequent chapters, routers can assign a link MTU to a workstation on that link, and hosts can discover the path MTU prior to sending a packet to the desired destination.

6.2 The IPv6 Header

The IPv6 packet is carried within a local network frame much like in IPv4; however, the IPv6 header consists of two parts. These are the IPv6 base header, plus optional extension headers (Figure 6-1). With or without any optional extension headers, a fixed size constraint on the local network frame must be respected. For example, the maximum amount of data that can be carried in an Ethernet/IEEE 802.3 frame is 1,500 octets. If extension headers are added to the IPv6 packet, less application data can be sent. The host and/or its operating system should have a mechanism to manage this.

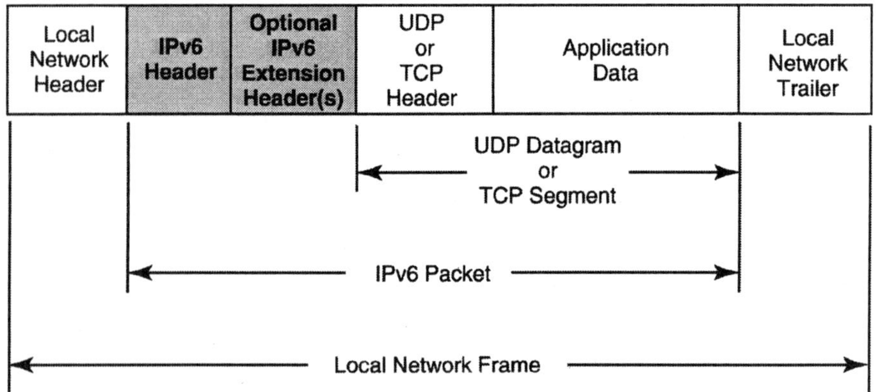

Figure 6-1. Internet Transmission Frame with IPv6.

The IPv6 header is 40 octets long, with eight fields and two addresses (Figure 6-2). Compare this with the IPv4 header, which is 20 octets long, has ten fields, two addresses, and options.

1 1 1 1 1 1 1 1 1 1 2 2 2 2 2 2 2 2 2 2 3 3
0 1 2 3 4 5 6 7 8 9 0 1 2 3 4 5 6 7 8 9 0 1 2 3 4 5 6 7 8 9 0 1 Bits

Version	Traffic Class	Flow Label
Payload Length	Next Header	Hop Limit
Source Address		
Destination Address		

Figure 6-2. IPv6 Header Format.

6.2.1 Version Field

The Version field is four bits long and identifies the version of the protocol (Figure 6-3). For IPv6, Version = 6. Note that this is the only field with a function and position that is consistent between IPv4 and IPv6. Everything else is different in some fashion. Having this field at the beginning of the packet allows for quick identification of the IP version, and for passing of that packet to the appropriate protocol process: IPv4 or IPv6.

1 1 1 1 1 1 1 1 1 1 2 2 2 2 2 2 2 2 2 2 3 3
0 1 2 3 4 5 6 7 8 9 0 1 2 3 4 5 6 7 8 9 0 1 2 3 4 5 6 7 8 9 0 1 Bits

Version	Traffic Class	Flow Label
Payload Length	Next Header	Hop Limit
Source Address		
Destination Address		

Figure 6-3. Version Field.

6.2.2 Traffic Class Field

The Traffic Class field (Figure 6-4a) is eight bits long and is intended for originating nodes and/or forwarding routers to identify and distinguish between different classes or priorities of IPv6 packets. (In the first publication of the IPv6 Specification, RFC 1883, this field was called the Priority field, reflecting this function. Enhancements to this work renamed it the Class field, with a length of four bits. Further work at the IPng Meeting at the

Munich IETF plenary in August 1997 expanded this field to eight bits and reduced the Flow Label field to 20 bits (from the previous 24) [6-6]. The new term Traffic Class, defined in RFC 2460, further identifies the specific purpose and function of this field.)

This field replaces the functions that were provided by the IPv4 Type of Service field, allowing differentiation between categories of packet transfer service. This function is often referred to as "differentiated services." At the time of this writing, many experiments are being conducted in this area of technology, especially in support of time-dependent signal transport, such as voice or video over IP, and differentiated services is one of several techniques currently under industry review.

Three general requirements for the Traffic Class field are stated in RFC 2460:

- For packets that are originated within a node by any upper layer protocol, that upper layer protocol would specify the value of the Traffic Class field bits. The default value is zero.
- Nodes that support a particular function that uses the Traffic Class bits may change the values of those bits in packets that they originate, forward, or receive. If a node does not support that particular function, it should not change any of the Traffic Class bits.
- Upper layer protocols should not assume that the values of the Traffic Class bits in a received packet are the same values that were originally transmitted. In other words, an intermediate node may be permitted to change the Traffic Class bits in transit.

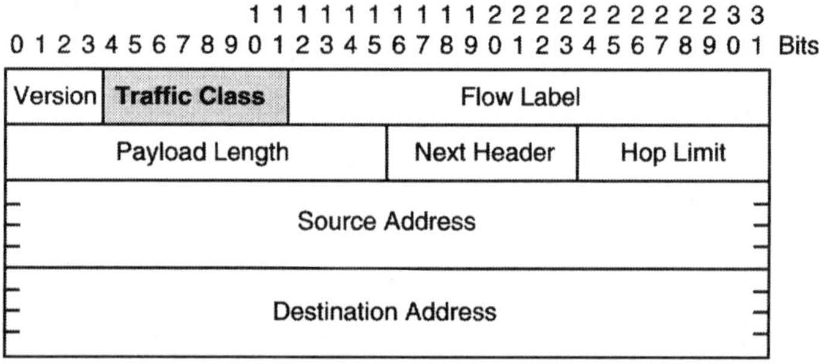

```
                    1 1 1 1 1 1 1 1 1 1 2 2 2 2 2 2 2 2 2 2 3 3
0 1 2 3 4 5 6 7 8 9 0 1 2 3 4 5 6 7 8 9 0 1 2 3 4 5 6 7 8 9 0 1  Bits
```

Version	**Traffic Class**	Flow Label	
Payload Length		Next Header	Hop Limit
Source Address			
Destination Address			

Figure 6-4a. Traffic Class Field.

Two other documents, RFC 2474 [6-7] and RFC 2475 [6-8], discuss the concepts and implementation of differentiated services, which are intended to discriminate between various types of service, requiring per-flow state and signaling at every hop. RFC 2474 defines a Differentiated Services (called "Diffserv") field, which is intended to replace the IPv4 Type of Service field, as shown in Figure 3-2, and the IPv6 Traffic Class field, as shown in Figure 6-4a. RFC 2475 is more general in nature and describes an architecture for differentiated services and the functions to be provided.

This architecture is described in two components: one deals with the forwarding of the packets themselves, and the other deals with the policies that determine the parameters used in the forwarding path. An analogy is drawn from the difference between packet forwarding and packet routing. Forwarding is the per-packet process that determines (from a routing table) which interface a packet should be sent from. (In other words, if the packet header identifies the subnet in Kansas City, then send this packet out on interface number 5.) Routing is a more complex process that determines the entries in that routing table, and (possibly more importantly) the policies that determine how that table is constructed. (For example, if the link to Kansas City goes down, then send the packet via Chicago instead of sending it via Denver.) As discussed in RFC 2474, forwarding path behaviors are better understood than the policies that configure the parameters that affect that forwarding path.

RFC 2474 concentrates on the forwarding path component that determines the per-hop behavior (PHB) of packets, rather than the policy and parameter configuration component. The PHBs would include specific treatment that an individual packet receives, along with queue management issues that are required to effect that special treatment. A sufficiently defined PHB should allow the construction of predictable services.

RFC 2474 further defines the format for the Differentiated Services (DS) field, which contains two subfields (Figure 6-4b). The Differentiated Services Codepoint (DSCP) subfield selects the PHB that a packet experiences at each node. The Explicit Congestion Notification (ECN) field was initially called the Currently Unused (CU) field in RFC 2474. Further work in congestion control methods that was documented in RFC 3168 [6-9] defined these bits for congestion notification purposes.

Reviewing Section 3.1, recall that the IPv4 Type of Service field consists of three parts: a three-bit Precedence field, three bits that specify flags (Delay, Throughput, and Reliability, or DTR), and two bits that are reserved. RFC 2474 defines a group of codepoints, known as the Class Selector codepoints, that preserve some degree of backward compatibility with the IPv4 Precedence bits. For these Class Selector codepoints, the bit pattern for the DSCP subfield would be XXX000 (in binary, where the X bits could take on the value of either zero or one). Note that the three "X" bits correspond with the same bit positions of the Precedence bits. The three "0" bits are in the same bit position as the DTR bits; however, RFC 2474 states that no attempt is made to maintain backward compatibility with these bit flags. In addition, the codepoint with value 000000 is assigned as the default PHB, which is defined as the "common, best-effort" forwarding behavior. (Noting the comparison with the Precedence field, this would correspond with the value for "Routine" precedence.) Other codepoint values have been grouped in pools, with one pool reserved for standards-based assignments, and others for experimental and local use purposes. RFC 2474 describes these assignments in greater detail. RFC 3260 [6-10] is also a useful reference, as it clarifies many Diffserv implementation issues.

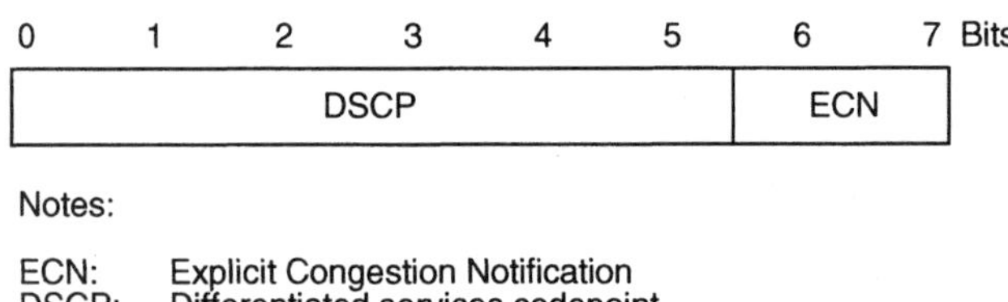

Figure 6-4b. Differentiated Services Field.

6.2.3 Flow Label Field

The Flow Label field (Figure 6-5) is 20 bits long, and may be used by a host to request special handling for certain packets, such as those with a nondefault quality of service or real-time quality of service. In the first version of the IPv6 Specification, RFC 1883, this field was 24 bits long, but four of these bits have now been allocated to the Traffic Class field, as discussed in the previous section.

A flow is a sequence of packets sent either to a unicast or a multicast destination that need special handling by the intervening IPv6 routers. All packets belonging to the same flow must be sent with the same source address, destination address, and flow label. An example of a flow would be packets supporting a real-time service, such as audio or video. The Flow Label field is used by that source to label those packets that require special handling by the IPv6 node. If a host or router does not support Flow Label field functions, the field is set to zero on origination and ignored on reception.

0–3	4–11	12–31
Version	Traffic Class	**Flow Label**

0–15	16–23	24–31
Payload Length	Next Header	Hop Limit
Source Address		
Destination Address		

Figure 6-5. Flow Label Field.

Multiple data flows may exist between a source and a destination, as well as data traffic that is not associated with a particular flow. A unique flow is identified by the combination of a source address and a nonzero flow label. The flow label is a pseudo-random number chosen from the range of 1 to FFFFFH (where H denotes hexadecimal notation). That label is used as a hash key by routers to look up the state associated with that flow.

RFC 1809, "Using the Flow Label Field in IPv6" [6-11], describes some of the earlier research on the subject. Like the Class field, the Flow Label field is the subject of current research and may change as industry experience matures.

6.2.4 Payload Length Field

The Payload Length field (Figure 6-6) is a 16-bit unsigned integer that measures the length, given in octets, of the payload (i.e., the balance of the IPv6 packet that follows the base IPv6 header). Note that optional extension headers are considered part of the payload, along with any upper layer protocols, such as TCP, FTP, and so on.

0–3	4–11	12–31
Version	Traffic Class	Flow Label

0–15	16–23	24–31
Payload Length	Next Header	Hop Limit
Source Address		
Destination Address		

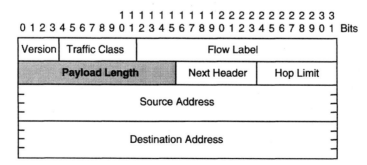

Figure 6-6. Payload Length Field.

The Payload Length field is similar to the IPv4 Total Length field, except that the two measurements operate on different fields. The Payload Length (IPv6) measures the data after the header, while the Total Length (IPv4) measures the data and the header.

Payloads greater than 65,535 are allowed and are called jumbo payloads. To indicate a jumbo payload, the value of the Payload Length is set to zero and the actual payload length is specified within an option that is carried in a Hop-by-Hop extension header. We will consider the format of this option in Section 6.3.3.

6.2.5 Next Header Field

The Next Header field (Figure 6-7) is eight bits long and identifies the header immediately following the IPv6 header. This field uses the same values as the IPv4 Protocol field. Examples are:

Value	Header
0	Hop-by-Hop Options
1	ICMPv4
4	IP in IP (encapsulation)
6	TCP
17	UDP
43	Routing
44	Fragment
50	Encapsulating Security Payload
51	Authentication
58	ICMPv6
59	None (no next header)
60	Destination Options

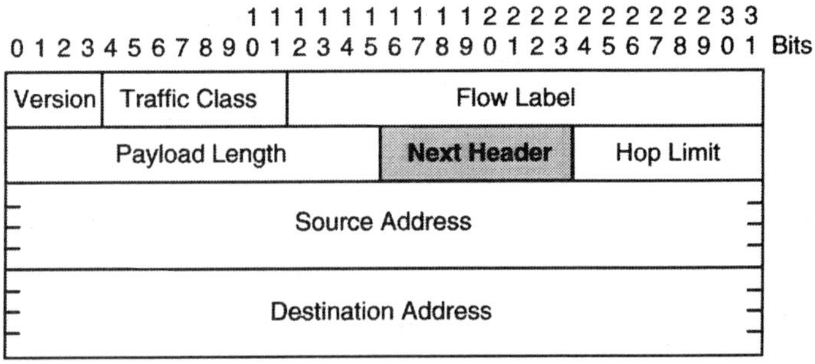

```
                    1 1 1 1 1 1 1 1 1 1 2 2 2 2 2 2 2 2 2 2 3 3
0 1 2 3 4 5 6 7 8 9 0 1 2 3 4 5 6 7 8 9 0 1 2 3 4 5 6 7 8 9 0 1 Bits
Version | Traffic Class | Flow Label
Payload Length | Next Header | Hop Limit
Source Address
Destination Address
```

Figure 6-7. Next Header Field.

An IPv6 packet, which consists of an IPv6 packet header plus its payload, may consist of zero, one, or more extension headers, as shown in Figure 6-8. Many of the extension headers also employ a Next Header field. Note the values of the Next Header fields in each example shown in the figure. In the first case, no extension headers are required, the Next Header = TCP, and the TCP header and any upper layer protocol data follows. In the second case, a Routing header is required. Therefore, the IPv6 Next Header = Routing; in the Routing header, Next Header = TCP, and the TCP header and any upper layer protocol data follows. In the third case, both the Routing and Fragment headers are required, with the Next Header fields identified accordingly.

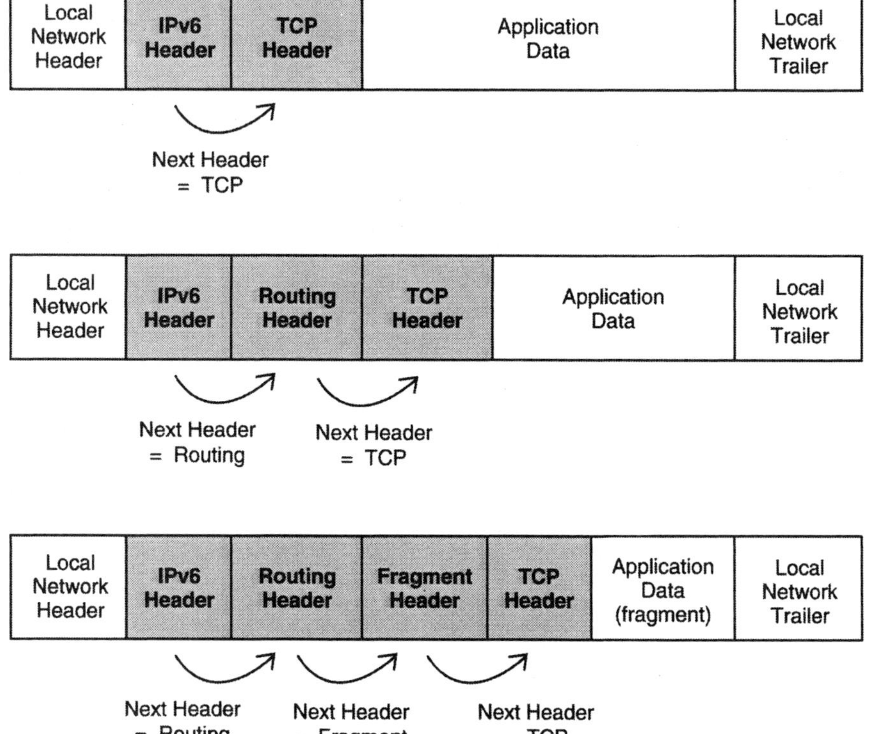

Figure 6-8. Next Header Field Operation.

6.2.6 Hop Limit Field

The Hop Limit field (Figure 6-9) is eight bits long and is decremented by one by each node that forwards the packet. When the Hop Limit equals zero, the packet is discarded and an error message is returned. This field is similar to the Time to Live (TTL) field found in IPv4, with one key exception. The Hop Limit field (IPv6) measures the maximum number of hops that can occur as the packet is forwarded by various nodes. The TTL field (IPv4) can be measured in either hops or seconds. Note that with the Hop Limit used in IPv6, the time basis is no longer available.

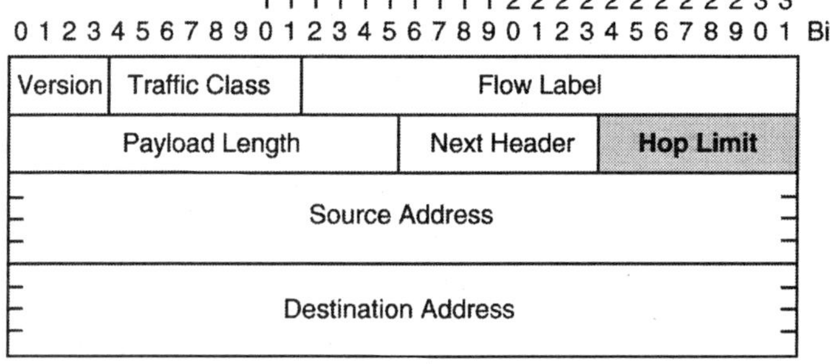

Figure 6-9. Hop Limit Field.

6.2.7 Source Address Field

The Source Address field (Figure 6-10) is a 128-bit field that identifies the originator of the packet. The format of this field is further defined in RFC 3513, "IP Version 6 Addressing Architecture" [6-12].

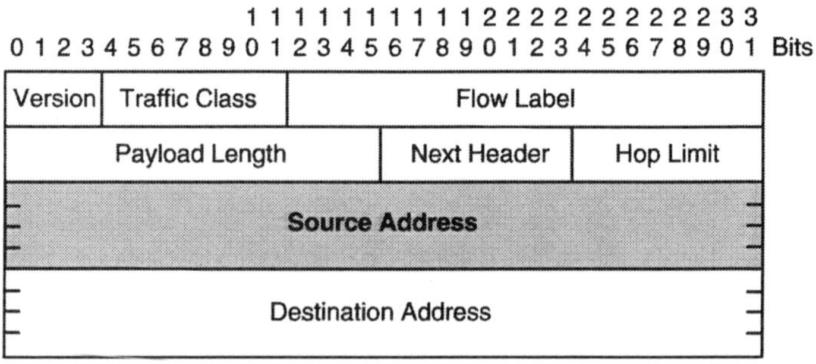

1 1 1 1 1 1 1 1 1 1 2 2 2 2 2 2 2 2 2 2 3 3
0 1 2 3 4 5 6 7 8 9 0 1 2 3 4 5 6 7 8 9 0 1 2 3 4 5 6 7 8 9 0 1 Bits

Version | Traffic Class | Flow Label
Payload Length | Next Header | Hop Limit
Source Address
Destination Address

Figure 6-10. Source Address Field.

6.2.8 Destination Address Field

The Destination Address field (Figure 6-11) is a 128-bit field that identifies the intended recipient of the packet. An important distinction is that the intended recipient may not be the ultimate recipient, as a Routing header may be employed to specify the path that the packet takes from its source, through intermediate destination(s), and on to its final destination.

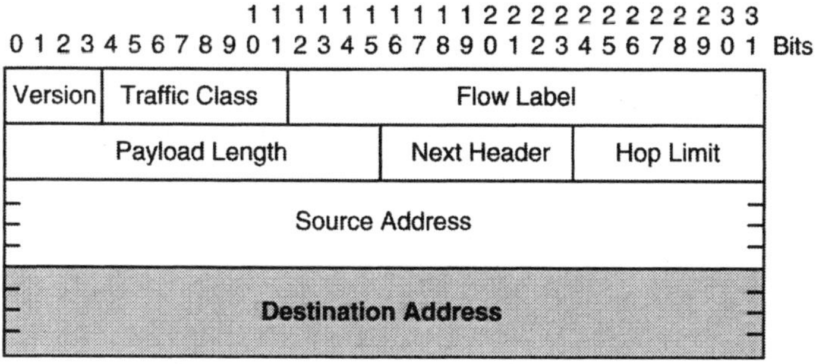

```
                    1 1 1 1 1 1 1 1 1 1 2 2 2 2 2 2 2 2 2 2 3 3
0 1 2 3 4 5 6 7 8 9 0 1 2 3 4 5 6 7 8 9 0 1 2 3 4 5 6 7 8 9 0 1 Bits
| Version | Traffic Class |            Flow Label                |
|      Payload Length          | Next Header |    Hop Limit      |
|                        Source Address                          |
|                     Destination Address                        |
```

Figure 6-11. Destination Address Field.

6.3 Extension Headers

The IPv6 design simplified the existing IPv4 header by placing many of the existing fields in optional headers. In this way, the processing of ordinary packets is not complicated by undue overhead, while the more complex conditions are still provided for. As we have seen, an IPv6 packet, which consists of an IPv6 packet plus its payload, may consist of zero, one, or more extension headers. Each extension header is an integral multiple of eight octets in length to retain the eight-octet alignment for subsequent headers. For optimum protocol performance, these extension headers are placed in a specific order.

6.3.1 Extension Header Order

RFC 2460 recommends that the extension headers be placed in the IPv6 packet in a particular order:

- IPv6 header
- Hop-by-Hop Options header
- Destination Options header (for options to be processed by the first destination that appears in the IPv6 Destination Address field, plus subsequent destinations listed in the Routing header)
- Routing header
- Fragment header

- Authentication header (as detailed in RFC 2402)
- Encapsulating Security Payload header (as detailed in RFC 2406)
- Destination Options header (for options to be processed by the final destination only)
- Upper Layer Protocol header (TCP and so on)

Figure 6-12 illustrates the IPv6 and optional headers, with their suggested order.

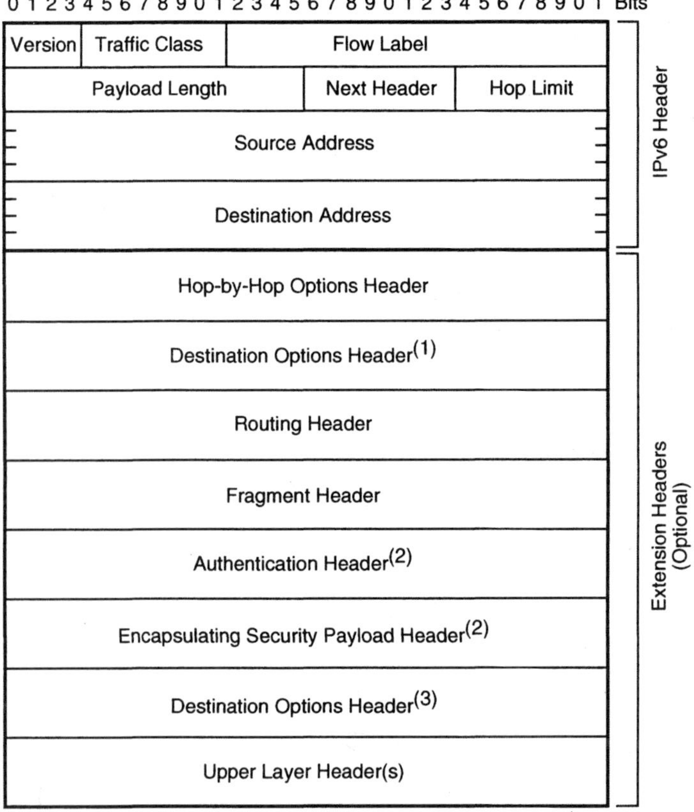

Notes:

(1) For options to be processed by the first destination that appears in the IPv6 Destination Address field plus subsequent destinations listed in the Routing header.

(2) Additional recommendations regarding the relative order of the Authentication and Encapsulating Security Payload headers are given in RFC 2406.

(3) For options to be processed only by the final destination of the packet.

Figure 6-12. IPv6 Packet Format with Optional Extension Headers.

6.3.2 Extension Header Options

Two of the extension headers, the Hop-by-Hop Options header and the Destination Options header, may carry one or more options that further identify parameters or network operation. These options are encoded using the Type-Length-Value (TLV) format that is specified by the Abstract Syntax Notation One (ASN.1) message description language. (TLV is widely used within communication protocols, including the Simple Network Management Protocol, SNMP.) The option format (Figure 6-13) includes an eight-bit Option Type field that identifies the option in question; an eight-bit Opt Data Len field that specifies the length of the Option Data field, given in octets; and a variable length Option Data field.

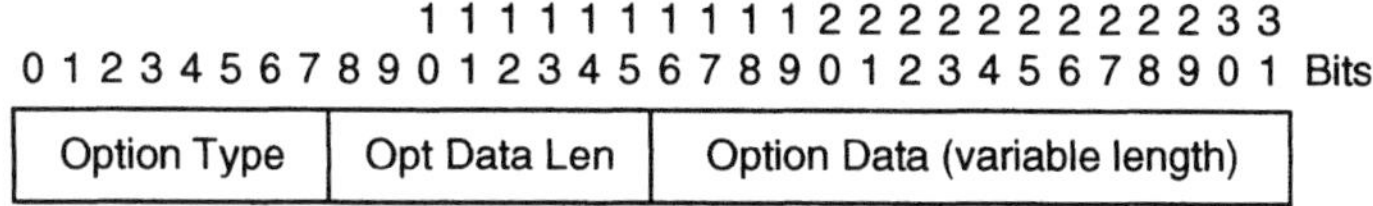

Figure 6-13. TLV Encoded Options Format.

The two highest-order bits of the Option Type field specify how to handle options that are unrecognizable at the processing IPv6 node:

Value	Action
00	Skip over the option and continue processing the header.
01	Discard the packet.
10	Discard the packet and send an ICMP Parameter Problem message (unrecognized Option Type) to the source.
11	Discard the packet and send an ICMP Parameter Problem message (unrecognized Option Type) to the source (only if destination was not multicast).

The third highest-order bit of the Option Type field specifies whether or not the Option Data of that option can change en route to the packet's final destination.

Value	Action
0	Option data does not change en route.
1	Option data may change en route.

In addition, there are two options that are used, as necessary, to pad the options such that the extension header contains a multiple of eight octets. The Pad1 Option (Figure 6-14) is used to insert one octet of padding into the Options area of a header. Note that this option is a special case (noted by Type = 0) that does not have Opt Data Len or Option Data fields.

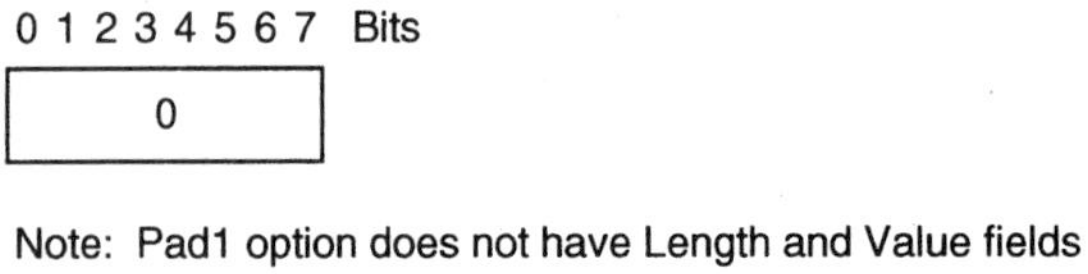

0 1 2 3 4 5 6 7 Bits
0

Note: Pad1 option does not have Length and Value fields

Figure 6-14. Pad1 Option Format.

The PadN Option (Figure 6-15) is used to insert two or more octets of padding into the Options area of a header. Note that this option has a Type field = 1. If the padding desired was n octets, the Opt Data Len field would contain the value of n - 2, and the Option Data field would contain n - 2 zero-valued octets.

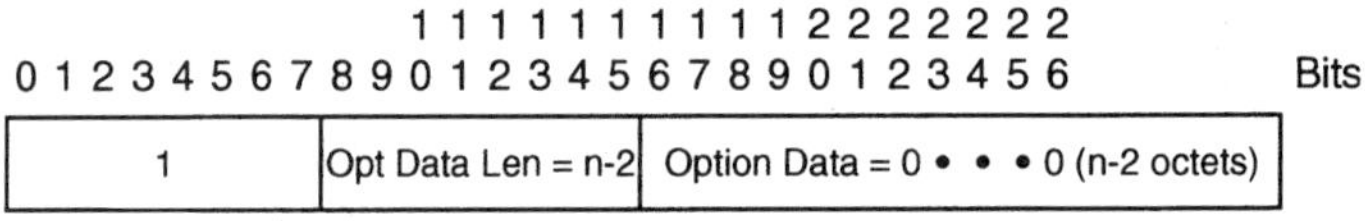

0 1 2 3 4 5 6 7	8 9 1 1 1 1 1 1 0 1 2 3 4 5	1 1 1 1 2 2 2 2 2 2 2 6 7 8 9 0 1 2 3 4 5 6 Bits
1	Opt Data Len = n-2	Option Data = 0 • • • 0 (n-2 octets)

Figure 6-15. PadN Option Format.

Guidelines for designing new options are presented in Appendix B of RFC 2460.

6.3.3 Hop-by-Hop Options Header

The Hop-by-Hop Options header carries optional information that must be examined by every node along a packet's delivery path (see Figure 6-16). As a result, the Hop-by-Hop Options header, when present, must immediately follow the IPv6 header. (The other extension headers are not examined or processed by any node along a packet's delivery path until the packet reaches its intended destination or destinations.) The presence of the Hop-by-Hop Options header is identified by a value of 0 in the Next Header field of the IPv6 header. This header contains two fields, plus options.

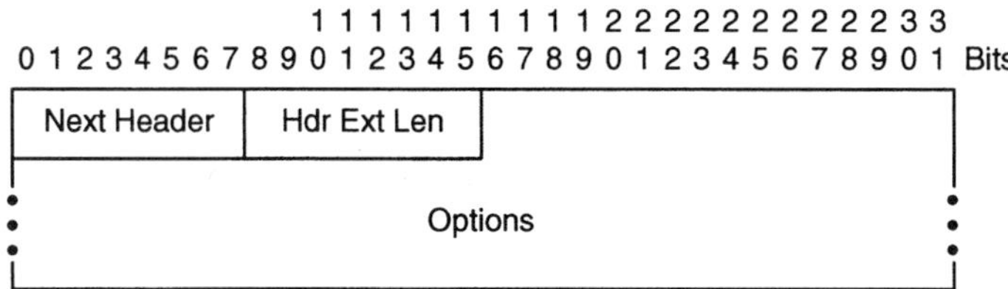

Figure 6-16. Hop-by-Hop Options Header Format.

The Next Header field is eight bits long and identifies the header immediately following the Hop-by-Hop Options header. This field uses the same protocol values as those assigned for use with the IPv4 Protocol field, but is extended with additional values (such as those defining Hop-by-Hop, Routing, Fragment, etc. headers) that are not found with IPv4-only applications.

The Header Extension Length (Hdr Ext Len) field is eight bits long and measures the length of the Hop-by-Hop Options header in eight-octet units, not counting the first eight octets.

The Options field is variable in length, as long as the complete Hop-by-Hop Options header is an integer and a multiple of eight octets in length. One option is currently defined, the Jumbo Payload option [6-13], which is used to send IPv6 packets that are between 65,536 and 4,294,967,295 octets in length (Figure 6-17). This option is defined by Option Type = 194 (or C2H), Opt Data Len = 4 (octets), and a four-octet field that carries the length of the jumbo packet in octets (excluding the IPv6 base header, but including the Hop-by-Hop Options header and any other headers). The Jumbo Payload Length must be greater than 65,535. In addition, the Payload Length field = 0 (to indicate a special condition) when the Jumbo Payload is used. Further, this option must not be used in a packet that carries a fragment header, as it would make little sense to send a very large packet (for efficiency) only to require that it be broken up into smaller pieces (for transmission).

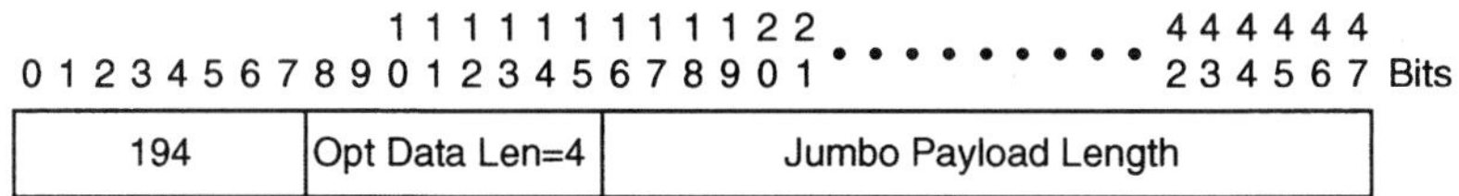

Figure 6-17. Jumbo Payload Option Format.

6.3.4 Destination Options Header

The Destination Options header carries optional information that needs to be examined only by a packet's destination node(s), as shown in Figure 6-18. The presence of the Destination Options header is identified by a value of 60 in the preceding header's Next Header field. This header contains two fields plus options.

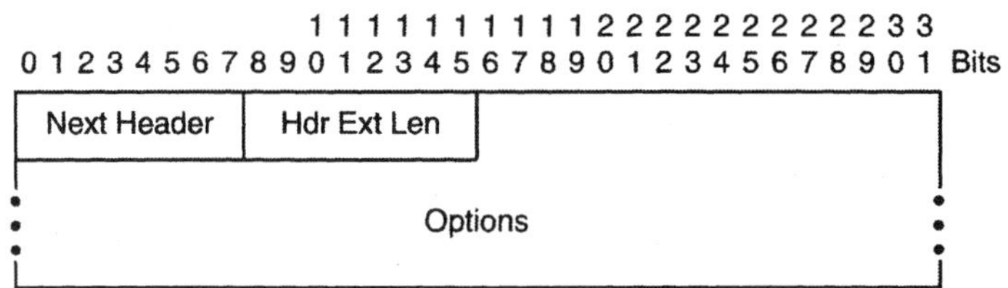

Figure 6-18. Destination Options Header Format.

The Next Header field is eight bits long and identifies the header immediately following the Destination Options header. This field uses the same values as the IPv4 Protocol field, as discussed above.

The Header Extension Length (Hdr Ext Len) field is eight bits long and measures the length of the Destination Options header in eight-octet units, not counting the first eight octets.

The Options field is variable in length, such that the complete Destination Options header is an integer multiple of eight octets in length. Only two options are defined in RFC 2460 — the Pad1 option, used to insert one octet of padding into the Options area of a header; and PadN, used to insert two or more octets of padding into the Options area of a header.

6.3.5 Routing Header

The Routing header lists one or more intermediate nodes that are "visited" on the path from the source to the destination (see Figure 6-19a). The presence of the Routing header is identified by a value of 43 in the preceding header's Next Header field. This header contains four fields, plus type-specific data.

The Next Header field is eight bits long and identifies the header immediately following the Routing header. This field uses the same values as the IPv4 Protocol field, as discussed above.

The Header Extension Length (Hdr Ext Len) field is eight bits long and measures the length of the Routing header in eight-octet units, not counting the first eight octets.

The Routing Type field is eight bits long and identifies a particular Routing header variant. (RFC 2460 defines one variant, Routing Type 0, which is described below.)

The Segments Left field is eight bits long and indicates the number of route segments remaining, or, in other words, the number of explicitly listed intermediate nodes still to be visited before reaching the final destination.

The Type-Specific field is variable in length, with a format defined by the particular Routing Type variant.

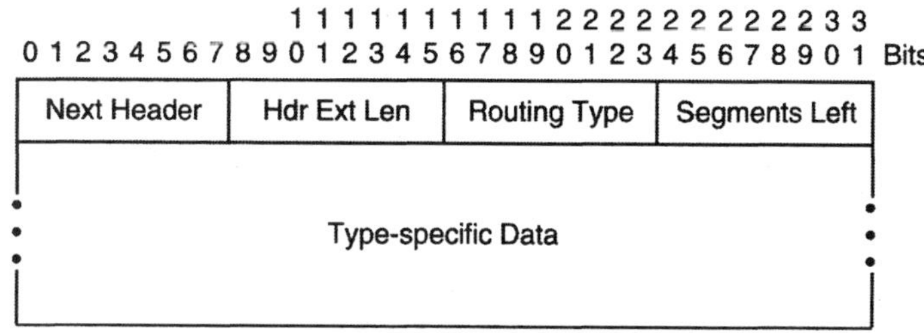

Figure 6-19a. Routing Header Format.

RFC 2460 defines a single variant, the Routing Type 0 header, which contains an ordered list of addresses that are to be visited along that packet's path (Figure 6-19b). For this header, the Next Header field is defined as above; however, the Hdr Ext Len field contains a number equal to two times the number of addresses in the header. For example, if there were n addresses in the header, the Hdr Ext Len field would contain the value 2n. The Routing Type field would indicate Type = 0. The Segments Left field would be as above, and the Reserved field would be set to zero for transmission, and ignored on reception. A list of 128-bit addresses, numbered 1 to n, would complete the Routing Type 0 Header.

1 1 1 1 1 1 1 1 1 1 2 2 2 2 2 2 2 2 2 2 3 3
0 1 2 3 4 5 6 7 8 9 0 1 2 3 4 5 6 7 8 9 0 1 2 3 4 5 6 7 8 9 0 1 Bits

Next Header | Hdr Ext Len = 2n | Routing Type = 0 | Segments Left

Reserved

Address [1]

Address [2]

Address [n]

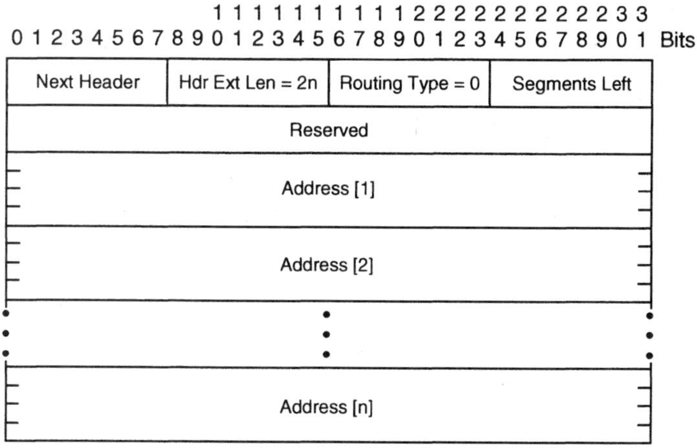

Figure 6-19b. Routing Header Format (Type 0).

RFC 2460 gives an example of the usage of the Routing header (Figure 6-20). In this example, note that three intermediate nodes (and four segments) separate the Source node (S) from the Destination node (D).

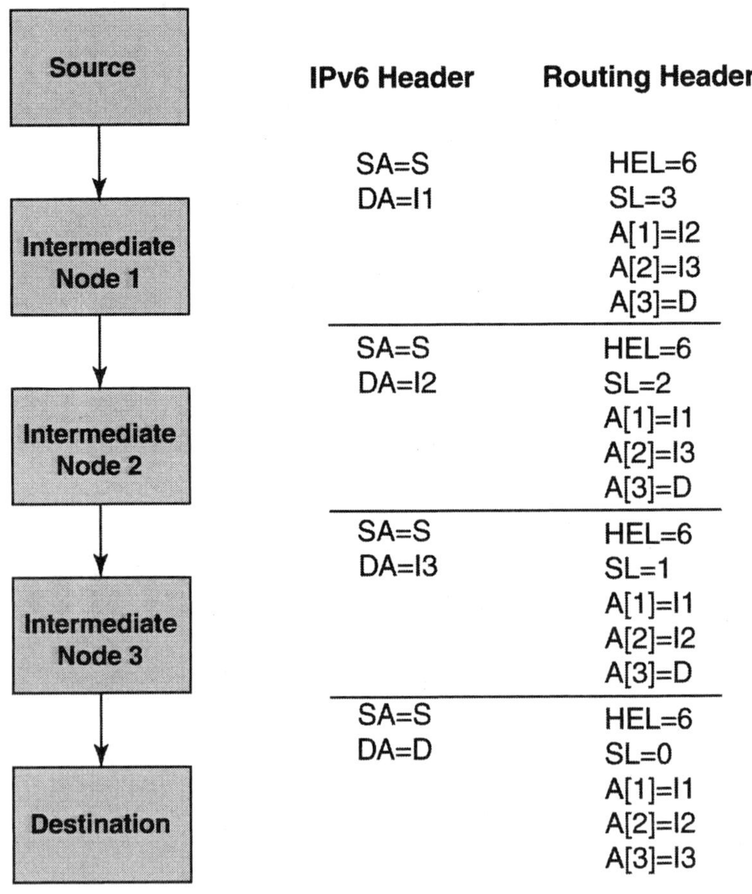

Notes:

A[n]: IPv6 address
DA: Destination address
HEL: Header extension length
In: Intermediate Node n
SA: Source address
SL: Segments left

Figure 6-20. Routing Header Usage.

To travel from the Source node to Intermediate Node 1, the IPv6 base header uses Source Address (SA) = S, and Destination Address (DA) = Intermediate Node 1 (I1). The Routing header specifies a Header Extension Length = 6, Segments Left = 3, and the addresses of the three remaining nodes along the path: Intermediate Node 2 (I2), Intermediate Node 3 (I3), and the Destination node (D).

To travel from I1 to I2, the routing algorithm exchanges the IPv6 Destination Address with the first address in the address list (I2). Note that theSource Address (SA = S) will be consistent for all segments.

To travel from I2 to I3, the routing algorithm exchanges the IPv6 Destination Address with the second address in the address list (I3).

To travel from I3 to the final destination (D), the routing algorithm exchanges the IPv6 Destination Address with the third address in the address list (D). Note that the final state of the IPv6 header addresses is now SA = S and DA = D, and the Routing header addresses list the intermediate nodes, I1, I2, and I3, in the order that they were visited.

6.3.6 Fragment Header

The Fragment header (see Figure 6-21) is used by an IPv6 source to send packets that are larger than would fit in the path maximum transmission unit (MTU) to their destinations. The presence of the Fragment header is identified by a value of 44 in the preceding header's Next Header field. Note that fragmentation for IPv6 is only done at the source node, not at intermediate routers along the packet's delivery path; this is a procedural change from IPv4.

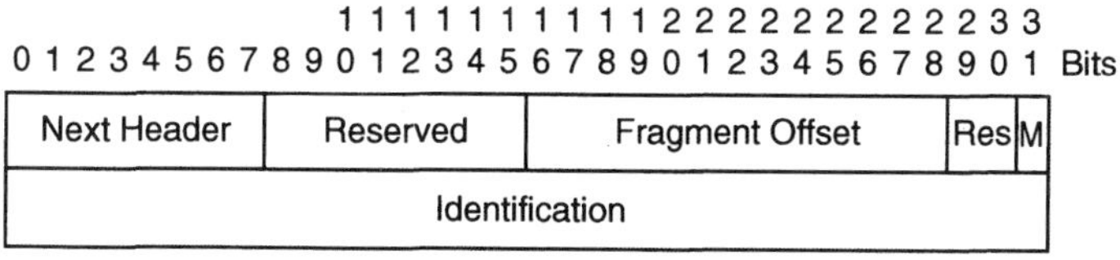

```
                    1 1 1 1 1 1 1 1 1 1 2 2 2 2 2 2 2 2 2 2 3 3
0 1 2 3 4 5 6 7 8 9 0 1 2 3 4 5 6 7 8 9 0 1 2 3 4 5 6 7 8 9 0 1  Bits
```

Next Header	Reserved	Fragment Offset	Res	M
Identification				

Figure 6-21. Fragment Header Format.

The Fragment header contains six fields. The Next Header field is eight bits long and identifies the header immediately following the Fragment header. This field uses the same values as the IPv4 Protocol field, as discussed above.

The Reserved field is eight bits long and is reserved for future use. This field is initialized to zero for transmission and is ignored on reception.

The Fragment Offset field is a 13-bit unsigned integer that measures the offset, in eight-octet units, of the data following this header, relative to the start of the fragmentable part of the original packet.

The Reserved field is two bits long and is reserved for future use. This field is initialized to zero for transmission and is ignored on reception.

The M flag is one bit long and determines whether more fragments are coming (M = 1) or whether this is the last fragment (M = 0).

The Identification field is 32 bits long and uniquely identifies the fragmented packet(s) during the reassembly process. This field is generated by the source node.

A packet requiring fragmentation is considered to consist of two parts — an unfragmentable part and a fragmentable part (Figure 6-22a). The unfragmentable part includes the IPv6 header, plus any extension headers that must be processed en route to the destination; these may include a Hop-by-Hop header and a Routing header. The fragmentable part is the balance of the packet, which may include any extension headers that are processed at the final destination node(s), the upper layer headers, and application data.

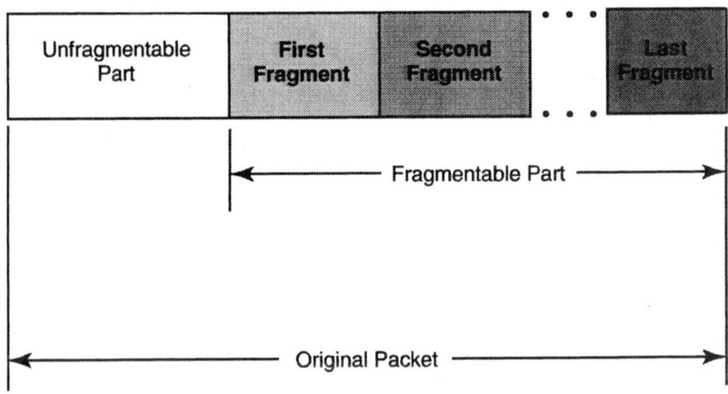

Figure 6-22a. Original Packet Requiring Fragmentation.

The fragmentable part of the original packet is divided into fragments that are an integral multiple of eight octets (except perhaps for the last fragment, which may not be an integral multiple of eight octets), as shown in Figure 6-22b. Each fragment packet consists of three parts: the unfragmentable part of the original packet, a Fragment header, and a data fragment. The unfragmentable part of each fragment contains a revised Payload Length field (within the IPv6 header portion) that matches the length of this fragment, and a Next Header field = 44 (indicating that a Fragment header comes next). The lengths of the fragments are chosen such that the resulting fragment packets fit within the MTU of the path to the packets' destination(s). (A related process, called Path MTU Discovery, will be explored later in this chapter.) At the destination node(s), a process called reassembly is used to reconstruct the original packet from the fragment packets. The reassembly process is also described in RFC 2460.

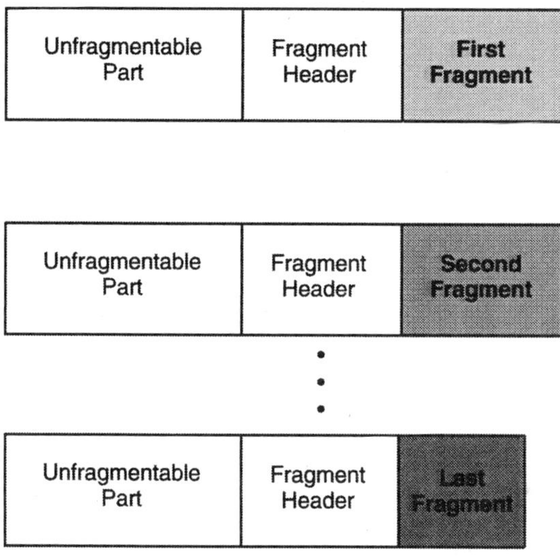

Figure 6-22b. Fragment Packets.

6.3.7 Authentication Header

Ensuring secure data transmissions has become an increasingly important issue for network managers. The Internet community has addressed these issues in RFC 2401, "Security Architecture for the Internet Protocol" [6-14].

Two headers are discussed in RFC 2401 that provide the IP security mechanisms. The Authentication header is defined in RFC 2402 [6-15]. The IP Encapsulating Security Payload (ESP) is defined in RFC 2406 [6-16]. These two mechanisms may be used separately, or jointly, as security needs dictate.

The Authentication header (Figure 6-23) provides connectionless integrity and data origin authentication for IP datagrams, plus optional protection against replays. The presence of the Authentication header is identified by a value of 51 in the preceding header's Next Header field. This header contains six fields.

The Next Header field is eight bits long and identifies the header immediately following the Authentication header. This field uses the same values as the IPv4 Protocol field, as discussed above.

The Payload Len field is eight bits long and provides the length of the Authentication field in 32-bit words, minus two (i.e. the first eight octets of the Authentication header are not counted). The minimum value is one, which consists of the 96-bit authentication value (three 32-bit words), less the value two (3 - 2 = 1). This minimum is only used in the case of a "null" authentication algorithm, employed for debugging purposes.

Bits 0–7	Bits 8–15	Bits 16–31
Next Header	Payload Len	Reserved
Security Parameters Index (SPI)		
Sequence Number		
Authentication Data (variable length)		

Figure 6-23. Authentication Header Format.

The Reserved field is sixteen bits long and is reserved for future use. This field is initialized to zero for transmission. It is included in the Authentication Data calculation, but is otherwise ignored on reception.

The Security Parameters Index (SPI) field is an arbitrary 32-bit value that identifies the security association (SA) for this datagram, relative to the destination IP address contained in the IP header with which this security header is associated, and relative to the security protocol employed. The security association, as defined in RFC 2401, is a simple, logical connection that is created for security purposes. All traffic that traverses an SA has the same security processing. The SA may comprise many parameters, including the Authentication algorithm, authentication algorithm keys, the Encryption algorithm, encryption algorithm keys, and others. According to RFC 2402, the value of SPI = 0 may be used for local, implementation-specific purposes. Other values, in the range of 1–255, are reserved for future use by the Internet Assigned Numbers Authority (IANA).

The Sequence Number field contains a 32-bit number that monotonically increases. Both the sender's counter and the receiver's counter are initialized to zero when a security association is established.

The Authentication Data is a variable-length field that contains the Integrity Check Value (ICV) for this packet. This field must be an integral multiple of 32 bits in length.

6.3.8 Encapsulating Security Payload Header

The proposed Encapsulating Security Payload (ESP) header (Figure 6-24) is designed to provide confidentiality, data origin authentication, connectionless integrity, an anti-replay service, and limited traffic flow confidentiality. The service or services provided depend on the security association and its implementation. The presence of the ESP header is identified by a value of 50 in the preceding header's Next Header field. This header contains seven fields, some of which are mandatory, and some of which are optional, depending on the security association.

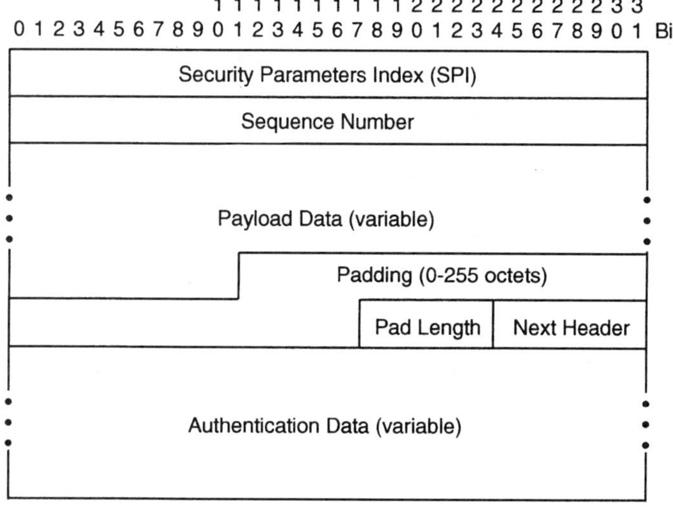

Figure 6-24. Encapsulating Security Payload Header Format.

The Security Parameters Index (SPI) field is an arbitrary 32-bit value that identifies the security association for this datagram, relative to the destination IP address contained in the IP header with which this security header is associated, and relative to the security protocol employed. The SPI field is mandatory.

The Sequence Number field contains a 32-bit number that monotonically increases. The sender's counter and receiver's counter are initialized to zero when a security association is established. The Sequence Number field is mandatory.

The Payload Data field is a variable length field containing data described by the Next Header field. The Payload Data field is mandatory.

The Padding field may optionally contain 0–255 octets of pad information, as required by the security implementation. The Pad Length field indicates the number of pad octets (0–255) that are immediately preceding. The Pad Length field is mandatory. The Next Header field is eight bits long and identifies the header immediately following the ESP header. This field uses the same values as the IPv4 Protocol field, as discussed above. The Next Header field is mandatory.

The Authentication Data is a variable-length field containing an Integrity Check Value (ICV). The length of this field depends on the authentication function that is selected. The Authentication Data field is optional, and is included only if that security association has selected authentication service.

6.3.9 No Next Header

The value of 59 in the Next Header field of an IPv6 packet or any of the extension headers indicates that nothing follows that header (Figure 6-25). As such, this is called a "No Next Header."

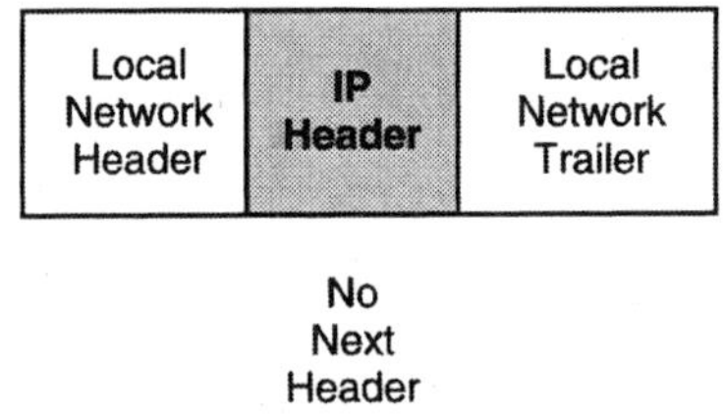

Figure 6-25. No Next Header Format.

6.4 IPv6 Addressing

The IPv6 address structure finds its roots in the CIDR structure, which includes an Address Prefix, a Site ID, and a Host ID. For IPv6, however, there will be multiple address prefixes, and each of these may have multiple structures similar to the Site ID and Host ID. As a basis, the IPv6 addressing architecture document, RFC 3513 [6-2], defines three different types of IPv6 addresses:

- Unicast — An identifier to a single interface. A packet sent to a unicast address is delivered to the interface identified by that address, as shown in Figure 6-26. RFC 1887 [6-17] describes IPv6 unicast addresses in greater detail.
- Anycast — An identifier for a set of interfaces (typically belonging to different nodes). A packet sent to an anycast address is delivered to one of the interfaces identified by that address (the nearest one, according to the routing protocol's measure of distance), as shown in Figure 6-27. RFC 1546 [6-18] describes the concepts of anycast service in greater detail.
- Multicast — An identifier for a set of interfaces (typically belonging to different nodes). A packet sent to a multicast address is delivered to all interfaces identified by that address, as shown in Figure 6-28.

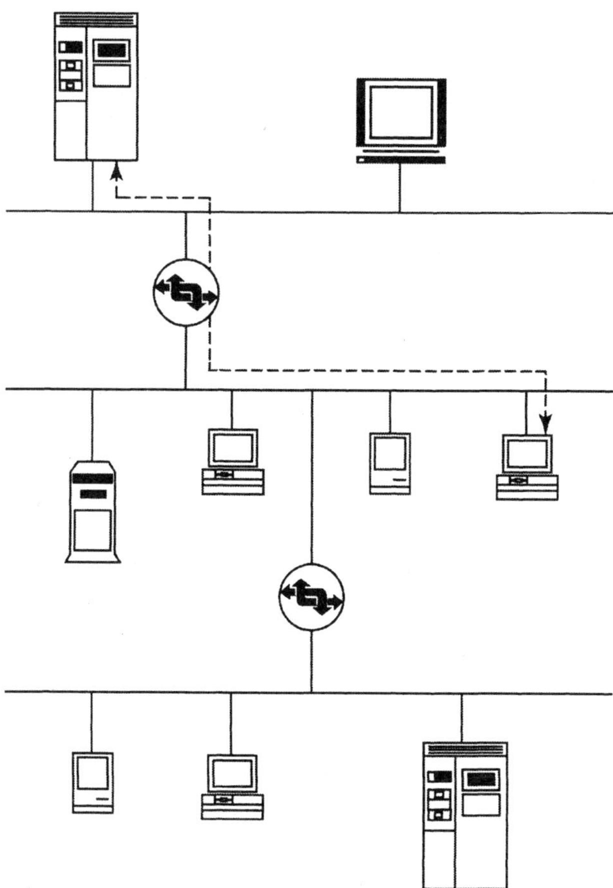

Figure 6-26. Unicast Addressing.

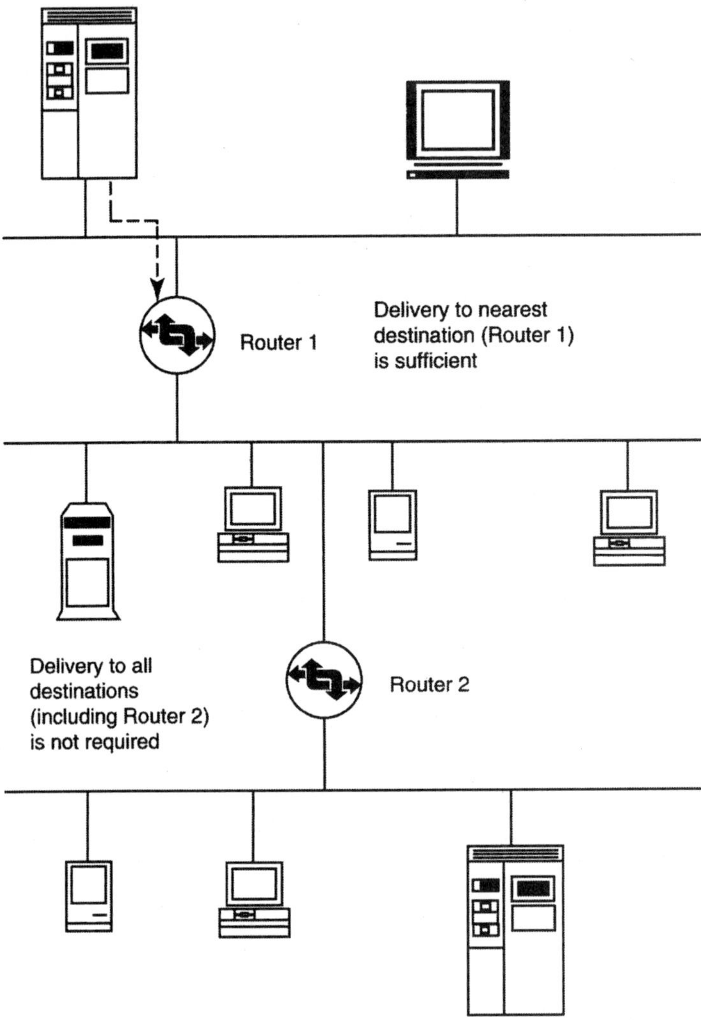

Figure 6-27. Anycast Addressing.

Note that the term broadcast does not appear, because the broadcast function is replaced by the multicast definition. Also note that IPv6 addresses of all types are assigned to interfaces, not nodes; one node (such as a router) may have multiple interfaces, and therefore multiple unicast addresses. In addition, a single interface may be assigned multiple addresses.

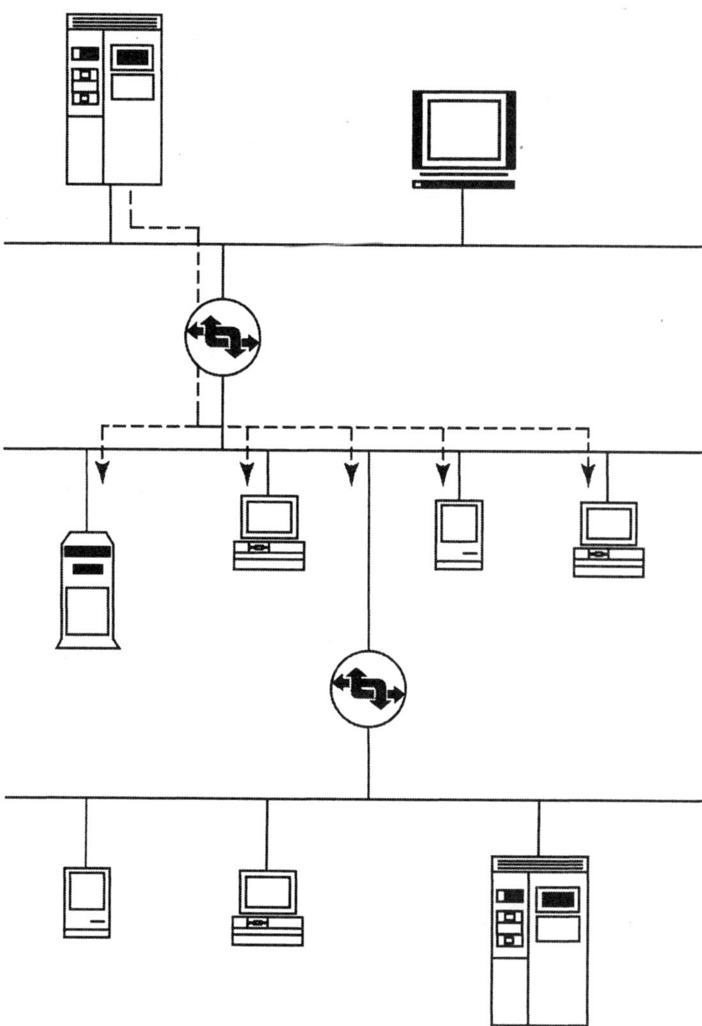

Figure 6-28. Multicast Addressing.

6.4.1 Address Representation

IPv4 addresses are typically represented in dotted decimal notation. As such, a 32-bit address is divided into four eight-bit sections, and each section is represented by a decimal number between 0 and 255:

128.138.213.13

Since IPv6 addresses are 128 bits long, a different method of representation is required. As specified in RFC 3513, the preferred representation is:

x:x:x:x:x:x:x:x

where each x represents 16 bits, and each of those 16-bit sections is defined in hexadecimal. For example, an IPv6 address could be of the form:

FEDC:BA98:7654:3210:FEDC:BA98:7654:3210

Note that each of the 16-bit sections is separated by colons, and that four hexadecimal numbers are used to represent each 16-bit section. Should any one of the 16-bit sections contain leading zeros, those zeros are not required. For example:

1080:0000:0000:0000:0008:0800:200C:417A

may be simplified to:

1080:0:0:0:8:800:200C:417A

If long strings of zeros appear in an address, a double colon (::) may be used to indicate one or more groups of 16 bits of zeros, which further simplifies the example shown above:

1080::8:800:200C:417A

The use of the double colon is restricted to appearing only once in an address, although it may be used to compress either the leading or trailing zeros in an address. For example, a loopback address of:

0:0:0:0:0:0:0:1

could be simplified as:

::1

When IPv6 addresses are expressed in text, it is common to delineate them by address and prefix length:

ipv6-address/prefix-length

where the IPv6 address is expressed in one of the notations listed above, and the prefix length is a decimal value that specifies the number of the leftmost bits of the address comprising the prefix. For example:

12AB:0000:0000:CD30:0000:0000:0000:0000/60

indicates that the 60-bit prefix (in hexadecimal) is:

12AB00000000CD3

6.4.2 Addressing Architecture

The 128-bit IPv6 address may be divided into a number of subfields to provide maximum flexibility for both current and future address representations. The leading bits, called the Format Prefix, define the specific type of IPv6 address. RFC 3513 defines a number of these prefixes, as shown in Figure 6-29. Additional work that documents specifics of particular addressing configurations is defined in References [6-19] through [6-29].

Note that address space has been allocated for NSAP, global unicast, multicast, and other types of addresses. At the time of this writing, fifteen percent of the address space has been allocated, and the remaining 85% has been reserved for future use.

A multicast address begins with the binary value 11111111; any other prefix identifies a unicast address. Anycast addresses are part of the allocation for unicast addresses and are not given a unique identifier.

Reviewing Figure 6-29, note that RFC 3513 defines two additional constraints regarding the IPv6 addressing architecture. First, several types of special addresses are assigned out of the 0000 0000 Format Prefix. These are the Unspecified address, the Loopback address, and the Compatibility addresses that contain embedded IPv4 addresses. All of these addresses will be discussed in subsequent sections. Second, some IPv6 addresses contain embedded interface identifiers. Format Prefixes 001 through 111, except for the Multicast addresses (Format Prefix 1111 1111), are all required to have 64-bit interface identifiers specified in the IEEE EUI-64 format. The EUI-64 format will also be discussed in detail later in this chapter.

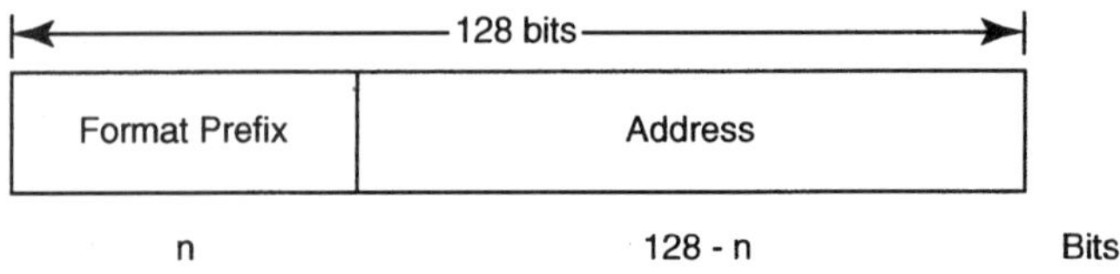

Prefix (binary)	Allocation
0000 0000	Unassigned*
0000 0001	Unassigned
0000 001	Reserved for NSAP Allocation
0000 01	Unassigned
0000 1	Unassigned
0001	Unassigned
001	Global Unicast
010	Unassigned
011	Unassigned
100	Unassigned
101	Unassigned
110	Unassigned
1110	Unassigned
1111 0	Unassigned
1111 10	Unassigned
1111 110	Unassigned
1111 1110 0	Unassigned
1111 1110 10	Link Local Unicast Addresses
1111 1110 11	Site Local Unicast Addresses
1111 1111	Multicast Addresses

*Note:

The Unspecified Address, the Loopback Address, and the IPv6 Addresses with Embedded IPv4 Addresses are assigned out of the 0000 0000 binary prefix space.

Figure 6-29. IPv6 Addressing Architecture.

6.4.2.1 Unicast Addresses

A number of forms for unicast addresses have been defined for IPv6, some with more complex structures that provide for hierarchical address assignments. The simplest form is a unicast address with no internal structure, in other words, with no address-defined hierarchy (Figure 6-30).

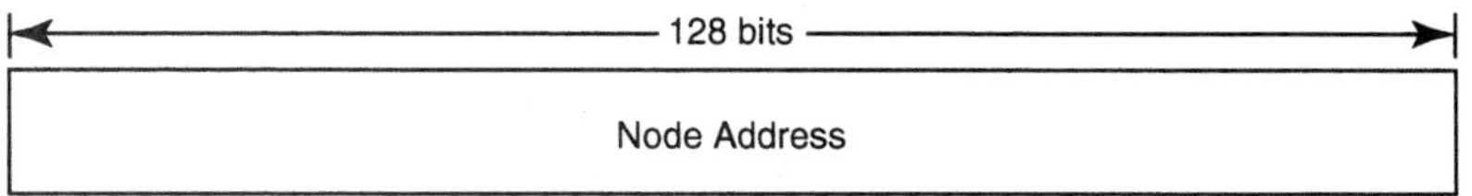

Figure 6-30. Unicast Address without Internal Structure.

The next possibility would be to specify a Subnet Prefix within the 128-bit address, thus dividing the address into a Subnet Prefix (with n bits) and an Interface ID (with 128 - n bits), as shown in Figure 6-31.

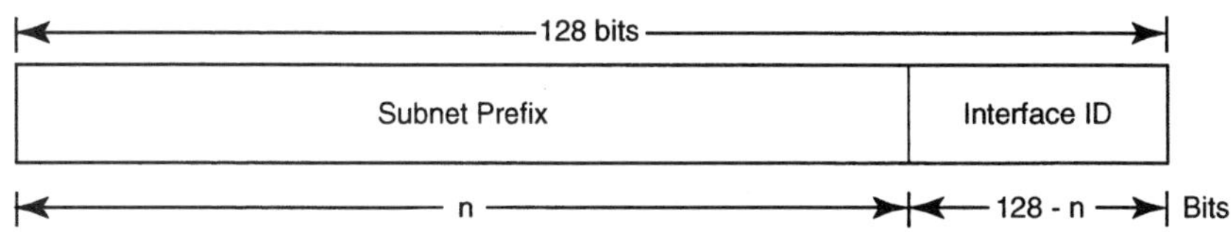

Figure 6-31. Unicast Address with Subnet.

6.4.2.2 Interface Identifiers

The Institute of Electrical and Electronics Engineers (IEEE) administers the addressing scheme for all local networks that adhere to the IEEE Project 802 series of standards. These include Carrier Sense Multiple Access with Collision Detection (CSMA/CD) networks, such as IEEE 802.3 10BASE-T or IEEE 802.5 token ring.

An IEEE 802 address consists of two parts: a Company ID and an Extension ID. The IEEE assigns Company IDs (sometimes called manufacturers' IDs) to organizations manufacturing network interface hardware. The company, in turn, assigns the Extension ID (sometimes called the Board ID). Taken together, the Company ID and Extension ID become a unique identifier (or address) for this network hardware; it is typically embedded in a read only memory (ROM) on that network board.

In the past, the IEEE has allocated 24 bits each to the Company ID and the Extension ID, yielding a 48-bit address. The IEEE has recently enhanced this addressing scheme to expand the Extension ID field to 40 bits, thus accommodating more hardware interfaces (approximately 1 trillion — $10^{12)}$ per manufacturer. This new scheme, which provides for addresses that are 64 bits long, is called EUI-64 and is shown in Figure 6-32 [6-30].

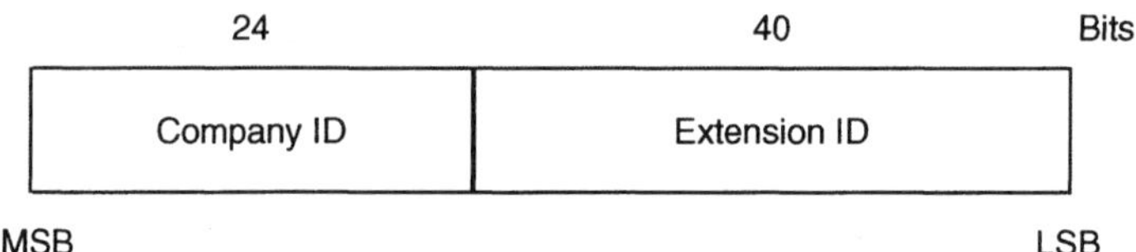

Figure 6-32. IEEE EUI-64 Address.

Within the IEEE addressing scheme (both 48-bit and EUI-64), there are two flag bits: the Individual/Group (I/G) bit and the Universal/Local (U/L) bit. These flag bits identify whether the address is an Individual Address (I/G = 0), a Group Address (I/G = 1), Universally (or globally) administered (U/L = 0), or Locally administered (U/L = 1).

Many of the IPv6 addresses incorporate an Interface ID field, which is defined using a modified version of the EUI-64 format, and is detailed in Appendix A of RFC 3513. The EUI-64 address with universal (or global) significance is illustrated in Figure 6-33 (note the seventh bit of the first octet: U/L = 0). The EUI-64 address with a local significance (used for interface identifiers on links or nodes with IPv6) is illustrated in Figure 6-34. Note that, in this case, the seventh bit of the first octet has been inverted (U/L = 1).

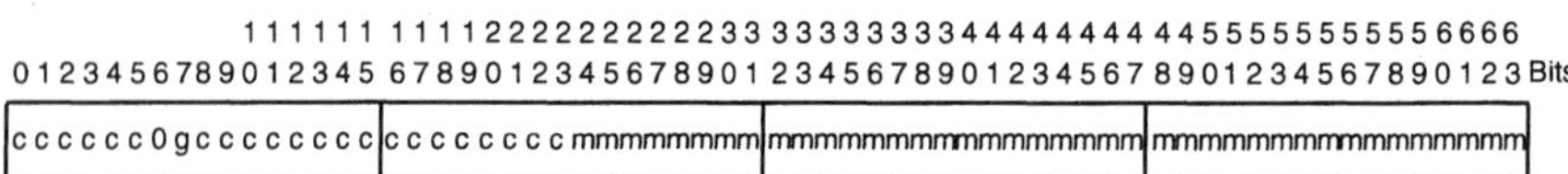

Notes:
c: Assigned Company ID bits
g: Individual/Group bit
m: Manufacturer-selected Extension Identifier bits
0: Universal/Local bit (U/L=0)

Figure 6-33. IEEE EUI-64 Address (Universal).

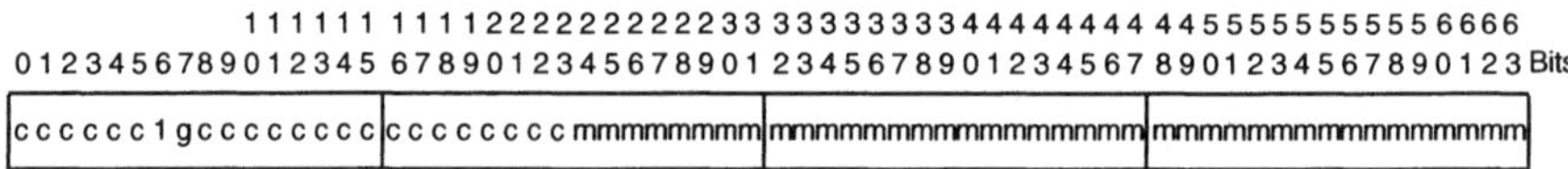

Notes:
c: Assigned Company ID bits
g: Individual/Group bit
m: Manufacturer-selected Extension Identifier bits
1: Universal/Local bit (U/L=1)

Figure 6-34. IEEE EUI-64 Address (Local).

To convert from the 48-bit address to the EUI-64 address, 16 bits are inserted between the Company ID and the Extension ID, as shown in Figure 6-35. These 16 bits are represented by the hexadecimal characters FFFE, or 1111 1111 1111 1110 in binary, which are shown in the lower portion of Figure 6-35 between the Company ID (c bits) and the manufacturer-defined Extension ID (m bits).

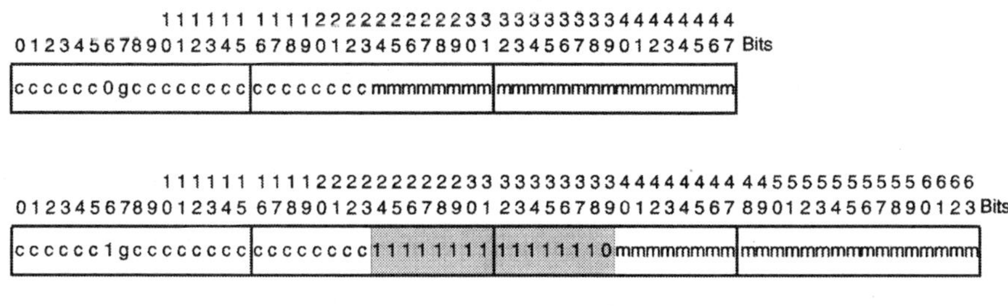

Notes:
c: Assigned Company ID bits
g: Individual/Group bit
m: Manufacturer-selected Extension Identifier bits
0/1: Universal/Local bit

Figure 6-35. IEEE EUI-64 Address (48- to 64-bit Compatibility).

Some interfaces, such as AppleTalk and ARCnet, do not adhere to the IEEE addressing scheme. To create an EUI-64 formatted identifier for these types of links, the node identifier is placed in the right-most bits, and all zeros are filled in to the left. In this case, U/L = 0, indicating a local scope address. Figure 6-36 shows the example of an AppleTalk node identifier of 4FH carried in the EUI-64 format.

```
                   111111 1111222222222233 3333333333444444 4455555555556666
0123456789012345 6789012345678901 2345678901234567 8901234567890123 Bits
0000000000000000 0000000000000000 0000000000000000 0000000001001111
```

Figure 6-36. IEEE EUI-64 Address (With Nonglobal Identifiers).

6.4.2.3 Unspecified and Loopback Addresses

Some special addresses are also defined in RFC 3513.

The address 0:0:0:0:0:0:0:0 (also represented as 0::0, or simply ::) is defined as the unspecified address, which indicates the absence of an address (Figure 6-37). This address might be used on startup when a node has not yet had an address assigned. The unspecified address may never be assigned to any node.

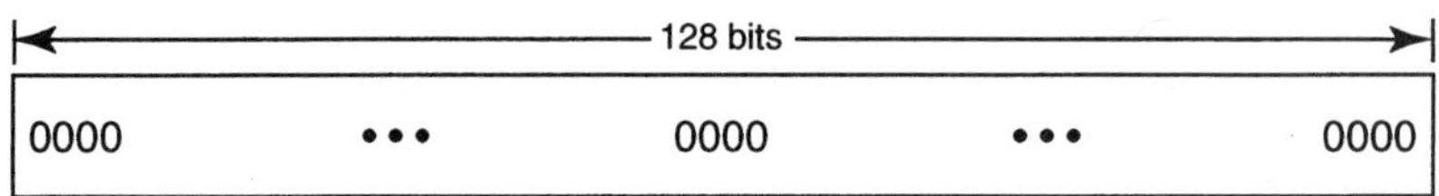

Figure 6-37. Unspecified Address.

The address 0:0:0:0:0:0:0:1 (also represented as 0::1, or simply ::1) is defined as the loopback address (Figure 6-38). This address is used by a node to send a packet to itself. The loopback address may never be assigned to any interface. An IPv6 packet with the destination address of the loopback address must never be sent outside a single node, and must never be forwarded by an IPv6 router.

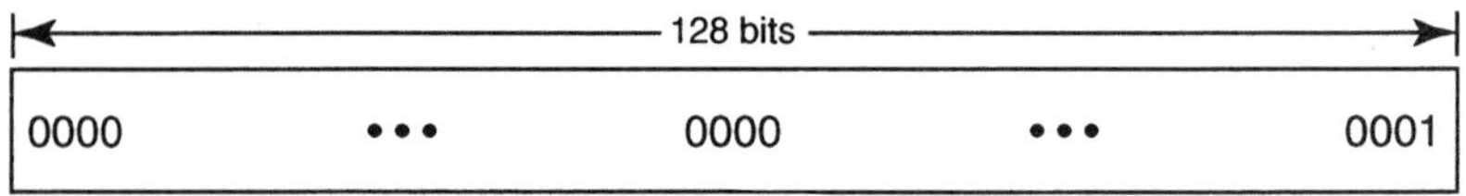

Figure 6-38. Loopback Address.

6.4.2.4 Global Unicast Addresses

Many communication networks, such as the global telephone network, are based on a hierarchical addressing scheme. A hierarchy facilitates easier scaling and routing. For example, calls within North America require the North American Zone Code (1), the Area Code (e.g. 303), the Central Office Code (e.g. 555), and the Line Number (1212). Calls from North America to London require the International Access Code (011), the Country Code (44 for the United Kingdom), a City Code (71 for London), and then the local telephone number. If the network grows, you can add another Area Code or Country Code, which facilitates the scaling challenge. If you are calling within the country, there is no need to dial the International Access Code or the Country Code, which facilitates the routing challenge.

For IPv6, the hierarchy for the global unicast address, also known as the aggregatable global unicast address, is organized into three levels: a public topology, a site topology, and an interface identifier, as documented in RFC 2374 [6-19]. The public topology is the collection of providers and exchanges that provide public Internet transit service. This public topology is identified using the Global Routing Prefix portion of the Global Unicast Address, is typically hierarchically structured, and is assigned to a site (Figure 6-39). The second level of addressing identifies a specific link within that site and is called the Subnet ID. The third level of addressing identifies a specific interface on that subnet and is called the Interface ID. Note that the Global Routing Prefix and Subnet ID can be a variable number of bits (designated n and m, respectively). Global unicast addresses that start with a binary 000 have no constraint on the size or structure of the Interface ID field. Global unicast addresses that start with other than binary 000 have a 64-bit Interface ID field, yielding a Global Routing Prefix and Subnet ID (m + n) sum of 64 bits.

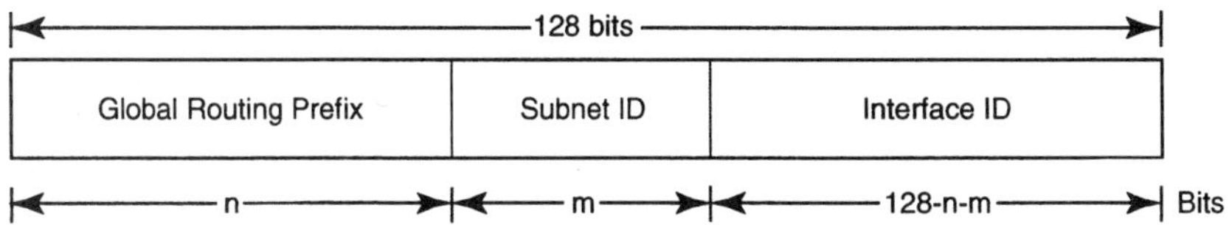

Figure 6-39. Aggregatable Global Unicast Address.

Examples of global unicast addresses that begin with a binary 000 are the IPv6 addresses that contain embedded IPv4 addresses, which are described in detail in the next section.

6.4.2.5 Compatibility Addresses

Two transition addresses have been defined for IPv4/IPv6 transition networks.

The first such address is called an IPv4-compatible IPv6 address (Figure 6-40). It is used when two IPv6 devices (such as hosts or routers) need to communicate via an IPv4 routing infrastructure. The devices at the edge of the IPv4 would use this special unicast address, which carries an IPv4 address in the low-order 32 bits. This process is called automatic tunneling. Note that the prefix is 96 bits of all zeros.

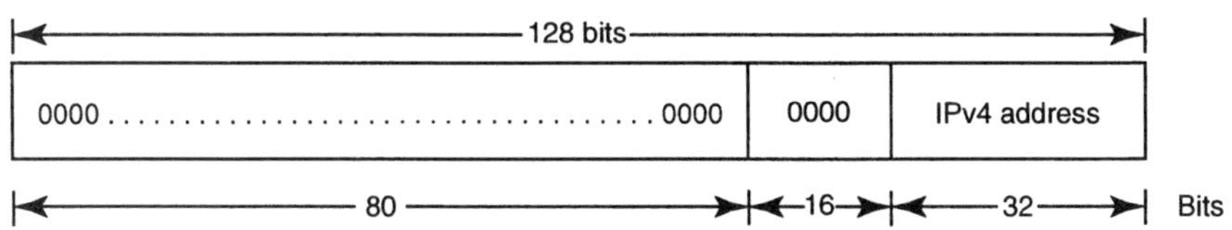

Figure 6-40. IPv4-Compatible IPv6 Address.

The second type of transition address is called an IPv4-mapped IPv6 address (Figure 6-41). This address is used by IPv4-only nodes that do not support IPv6. For example, an IPv6 host would use an IPv4-mapped IPv6 address to communicate with another host that only supports IPv4. Note that the prefix is 80 bits of zeros, followed by 16 bits of ones.

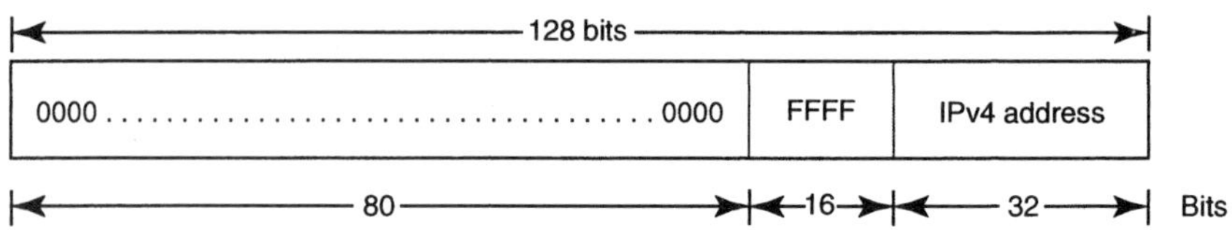

Figure 6-41. IPv4-Mapped IPv6 Address.

6.4.2.6 Local Use Addresses

Two addresses are defined for local use only. The Link-Local address is used for a single link and is intended for auto-address configuration, neighbor discovery, or when no routers are present. The Link-Local address (Figure 6-42) begins with the Format Prefix 1111111010 and includes a 64-bit Interface ID field. Routers never forward packets with Link-Local source or destination addresses to other links.

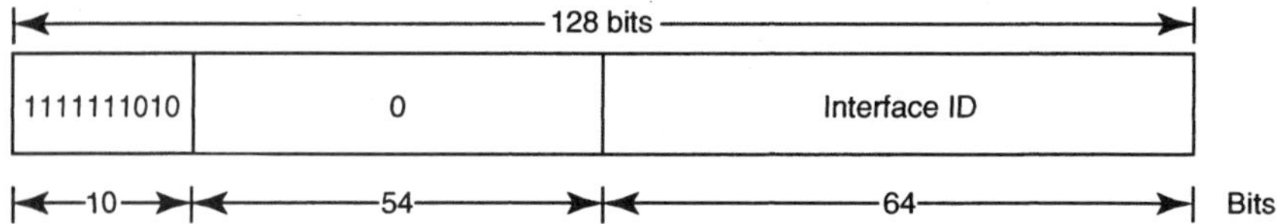

Figure 6-42. Unicast Link-Local Address.

The Site-Local address is used by organizations that have not yet connected to the Internet. Instead of fabricating an IPv6 address, they may use the Site-Local address. Routers never forward packets with Site-Local source addresses outside of that site. This address (Figure 6-43) begins with the Format Prefix 1111111011 and includes both a 16-bit Subnet ID field and a 64-bit Interface ID field.

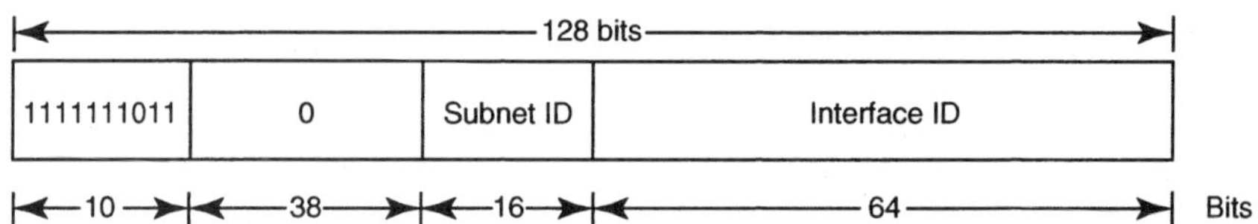

Figure 6-43. Unicast Site-Local Address.

6.4.2.7 Testing Addresses

A special allocation of addresses has been proposed for the purpose of testing IPv6 software and is described in RFC 2471 [6-23]. These addresses are only to be used for IPv6 testing and are not routable in the Internet. The testing address format is based on the Global Unicast Address, with the Global Routing Prefix subdivided into three fields: the Format Prefix (FP), the Top Level Aggregation Identifier (TLA ID), and the Next-Level Aggregation Identifier (NLA ID). The Subnet ID field is called the Site-Level Aggregation Identifier (SLA ID). These fields are assigned as follows (Figure 6-44):

FP: 001 in binary (Assigned to Global Unicast Addresses)
TLA ID: 1FFE in hexadecimal (assigned for 6bone testing)
NLA ID: Assigned by the TLA administrator
SLA ID: Assigned by the individual organization
Interface ID: An interface identifier for that link (Ethernet, token ring, etc.)

Thus, if the FP and TLA ID fields are combined, the testing address will begin with a 3FFE H, which is easily recognized as a testing address prefix.

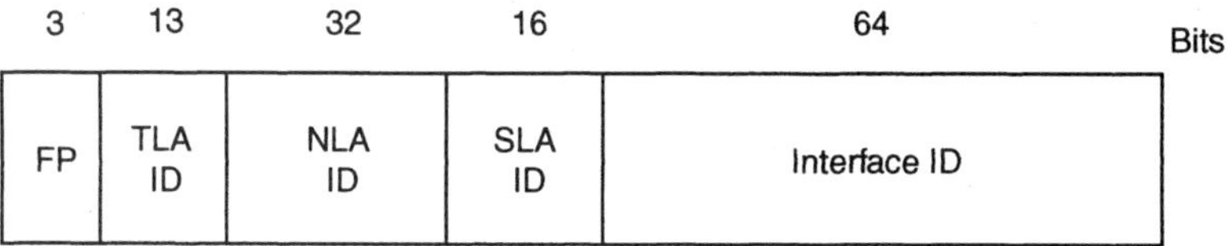

Notes:

FP: Format Prefix (001)
TLA ID: Top-Level Aggregation Identifier (1FFEH)
(assigned by the TLA administrator)
NLA ID: Next-Level Aggregation Identifier
(create an addressing hierarchy and identify sites)
SLA ID: Site-Level Aggregation Identifier
(assigned by the local administrator)
Interface ID: Interface Identifiers

Figure 6-44. Testing Address.

6.4.2.8 Anycast Addresses

An anycast address is one that is assigned to multiple interfaces, typically on different nodes. A packet with an anycast destination address is routed to the nearest interface having that address, as measured by the routing protocol's definition of distance. The concept of anycasting within IP-based internetworks was first proposed in RFC 1546.

RFC 3513 notes several possible uses for the anycast address:

- Identifying a set of routers belonging to an Internet Service Provider (ISP).
- Identifying the set of routers attached to a particular subnet.
- Identifying the set of routers that provide entry to a particular routing domain.

Two restrictions are placed on anycast addresses. First, they must not be used as a source address for an IPv6 packet. Second, an anycast address may only be assigned to routers, not hosts.

One anycast address is predefined and required: the Subnet-Router anycast address (Figure 6-45). This address begins with a variable-length subnet prefix and concludes with zeros for filler. All routers on that subnet must support this anycast address. It is intended to be used for applications where a node needs to communicate with one member of a group of routers on a remote subnet.

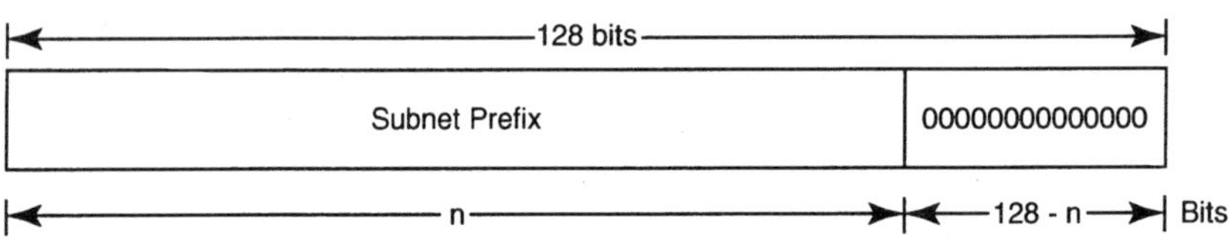

Figure 6-45. Subnet-Router Anycast Address (Required).

Additional work has been proposed that defines a set of reserved anycast addresses within each subnet prefix, as described in Reference [6-24]. Additional anycast address assignments are expected to be defined in the future.

6.4.2.9 Multicast Addresses

The multicast address identifies a group of nodes, and each of these nodes may belong to multiple multicast groups. Multicast addresses are defined in RFC 3513 and documented in further detail in RFC 2375 [6-25].

The multicast address (Figure 6-46) begins with the Format Prefix 11111111 and includes three additional fields. The Flags field contains four one-bit flags. The three most significant flag bits are reserved for future use and are initialized to zero. The fourth flag is called the T, or transient, bit. When T = 0, the multicast address is a permanently assigned (or well-known) multicast address, assigned by the global Internet numbering authority. When T = 1, a transient (or nonpermanently assigned) multicast address is indicated.

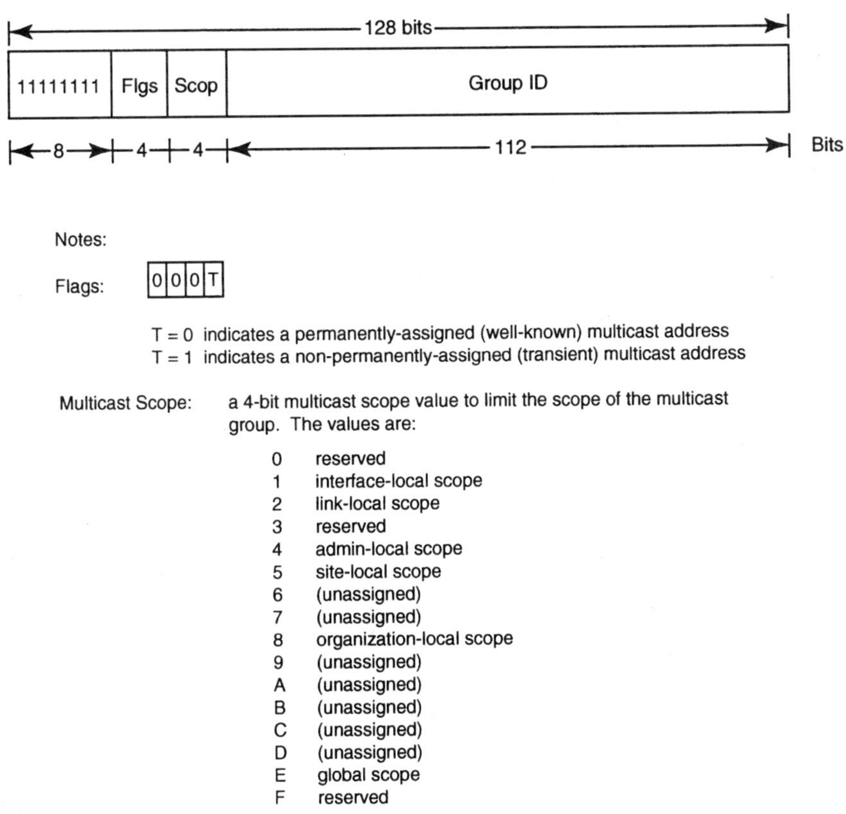

Figure 6-46. Multicast Address.

The Scop field is a four-bit field that is used to limit the scope of the multicast group, such as interface-local, link-local, or site-local. The Group ID field identifies the multicast group, either permanent or transient, within the given scope. Multicast addresses may not be used as source addresses in IPv6 datagrams or appear in any routing header.

RFC 3513 documents a number of predefined multicast addresses. For example, FF01:0:0:0:0:0:0:1 identifies the group of all IPv6 nodes within scope 1 (interface-local), while the address FF02:0:0:0:0:0:0:1 (or FF02::1) identifies all nodes within scope 2 (link-local). The second address would have the meaning "all nodes on this link." In a similar fashion, three all routers addresses have been designated: FF01:0:0:0:0:0:0:2, FF02:0:0:0:0:0:0:2, and FF05:0:0:0:0:0:0:2. These identify the group of all IPv6 routers within scope 1 (interface-local), 2 (link-local), or 5 (site-local). For example, the address FF02:0:0:0:0:0:0:2 (or FF02::2) has the meaning "all routers on this link." A number of other multicast addresses have also been predefined, such as the solicited-node address, which is a function of a node's unicast and anycast addresses. These other addresses are documented for reference in RFC 3513.

6.4.2.10 Required Addresses for Nodes

RFC 3513 summarizes the following requirements for nodes and their addresses:

A host is required to recognize the following addresses as identifying itself:

- Its Link-Local Address for each interface
- Assigned Unicast Addresses
- Loopback Address
- All-Nodes Multicast Address
- Solicited-Node Multicast Address for each of its assigned unicast and anycast addresses
- Multicast Addresses of all other groups to which the host belongs

A router is required to recognize the following addresses as identifying itself:

- The Subnet-Router Anycast Addresses for the interfaces it is configured to act as a router on
- All other Anycast Addresses with which the router has been configured
- All-Routers Multicast Address
- Solicited-Node Multicast Address for each of its assigned unicast and anycast addresses

- Multicast Addresses of all other groups to which the router belongs

The only address prefixes that should be predefined in an implementation are the following:

- Unspecified Address
- Loopback Address
- Multicast Prefix (FFH)
- Local-Use Prefixes (Link-Local and Site-Local)
- Predefined Multicast Addresses
- IPv4-Compatible Prefixes

Implementations should assume all other addresses are unicast unless specifically configured (e.g., anycast addresses).

6.5 Intranetwork Communication

ICMPv6 is a new version of the Internet Control Message Protocol and is documented in RFC 2463 [6-31]. ICMPv6 includes functions from the Internet Group Membership Protocol version 3 (IGMPv3), specified in RFC 3376 [6-32], plus extensions to the IGMP work, known as Multicast Listener Discovery (MLD) and documented in RFC 2710, that supports IPv6 [6-33]. ICMPv6 is considered an integral part of IPv6 and must be implemented by every IPv6 node.

ICMPv6 is used to report packet processing errors, intranetwork communication path diagnosis such as the PING (ICMPv6 Echo Request), and multicast membership reporting. In addition, there are extensions to ICMPv6 that define additional message types defined by other protocols or processes. The Neighbor Discovery process allows nodes on the same link to discover each other's presence, to determine other nodes' respective addresses, and to find routers for paths to other networks. Neighbor Discovery is documented in RFC 2461 [6-34]. Another extension, Path MTU Discovery, defined in RFC 1981 [6-35], allows a host to determine the maximum transmission unit (MTU) size along the path to a destination, thus optimizing that communication path.

All of these processes provide communication within the IPv6 internetwork. We will begin by looking at the fundamental messages defined for ICMPv6.

6.5.1 ICMPv6 Error Messages

With respect to IPv6, an ICMPv6 message would be considered an upper layer protocol. As such, the ICMPv6 message is preceded by an IPv6 header and zero or more IPv6 extension headers (review Figure 6-12). The header preceding the ICMPv6 header will have a Next Header field value of 58.

ICMPv6 messages have three fields that are common to all messages, plus a variable length message body whose contents depend on the type of message being transmitted (Figure 6-47). The common fields are:

- Type, eight bits long, which indicates the type of message.
- Code, eight bits long, which creates an additional level of message granularity and depends on the message type being sent.
- Checksum, 16 bits long, which is used to detect data corruption in the ICMPv6 message and parts of the IPv6 header.

ICMPv6 messages are grouped into two categories: error messages and informational messages. (Note from Figure 6-47 that the IGMP and Neighbor Discovery messages are also considered information messages.) The two categories are readily identified by the high-order bit of the message Type field.

IPv6 nodes that discard a packet may send an ICMPv6 error message to the original destination. There are four cases in which this may occur, and each of these circumstances produces a different error message — the destination was not reachable, the packet was too big, the packet exceeded its allowable time to live, or a parameter problem occurred. Each error message format is similar, with an eight-octet section that contains the Type, Code, Checksum, and a parameter field, plus a copy (or partial copy) of the originally transmitted packet.

Error messages have a zero in the high-order bit, and therefore have message types from 0–127:

Type	**Message**
1	Destination Unreachable
2	Packet Too Big
3	Time Exceeded
4	Parameter Problem

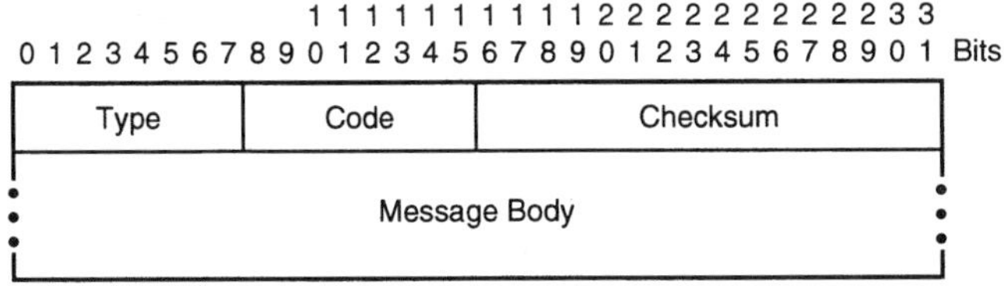

Type Field Value	Meaning
0-127	ICMPv6 error messages
1	Destination unreachable
2	Packet too big
3	Time exceeded
4	Parameter problem
128-255	ICMPv6 informational messages
128	Echo request
129	Echo reply
130	Multicast Listener query (1)
131	Multicast Listener report (1)
132	Multicast Listener done (1)
133	Router solicitation (2)
134	Router advertisement (2)
135	Neighbor solicitation (2)
136	Neighbor advertisement (2)
137	Redirect (2)

Notes:
(1) Defined as part of the Multicast Listener Discovery protocol
(2) Defined as part of the Neighbor Discovery protocol

Figure 6-47. ICMPv6, Multicast Listener Discovery, and Neighbor Discovery Messages.

6.5.2 ICMPv6 Informational Messages

Informational messages have a one in the high-order bit, and therefore have message types 128–255:

Type	Message
128	Echo Request
129	Echo Reply
130	Multicast Listener Query
131	Multicast Listener Report
132	Multicast Listener Done
133	Router Solicitation
134	Router Advertisement
135	Neighbor Solicitation
136	Neighbor Advertisement
137	Redirect

Message types 128–129 are the familiar Echo Request/Reply messages that are similar to their counterparts in IPv4. Message types 130–132 are defined by the Multicast Listener Discovery Protocol, and message types 133–137 are defined by the Neighbor Discovery Protocol. Both of these are examples of extensions to the ICMPv6 message set.

Elements from the Internet Group Management Protocol version 3 (IGMPv3), specified in RFC 3376, were merged into a new protocol called Multicast Listener Discovery (MLD), which is documented in RFC 2710 [6-33]. MLD is considered to be a subprotocol of ICMPv6. The purpose of MLD is to enable IPv6 routers to discover nodes that wish to receive multicast packets on its directly attached links, and to discover which multicast addresses are of interest to those neighboring nodes. The discovered information is then provided to the multicast routing protocol that is being used by that router so that it can deliver the multicast information to all of its links on which there are nodes that wish to listen. These nodes are called Multicast Listeners.

There are three types of Multicast Listener Discovery Messages: Query, Report, and Done. The Multicast Listener Query is used to learn if a particular multicast address or addresses has listener(s) on a specific link. There are two subtypes of Query messages: the General Query, which is used to learn which multicast addresses have listeners on a specific link; and the Multicast Address-Specific Query, which is used to learn if a particular multicast address has any listeners on an attached link. General Queries are sent to the Link-Local all-nodes multicast address (FF02::1), as discussed in Section 6.4.2.9 above. Nodes respond to a Query by generating a Multicast Listener Report message. The address being reported is carried in both the IPv6 Destination Address field and the MLD Multicast Address field. When a node ceases to listen to a multicast address on an interface, it should send a Multicast Listener Done message to the link-local all-routers multicast address (FF02::2), with the address that it is ceasing to listen to in its Multicast Address field.

The Neighbor Discovery Protocol, defined in RFC 2461 [6-34], combines the functions of three IPv4 mechanisms: the Address Resolution Protocol (ARP), from RFC 826; ICMP Router Discovery messages, from RFC 1256; and the ICMP Redirect message, from RFC 792. Nodes (hosts or routers) use Neighbor Discovery to determine the Data Link Layer addresses for their neighbors on attached links, and to purge cached values that are no longer valid. Nodes also use this protocol to actively track which neighbors are reachable and which are not, and also to detect Data Link Layer addresses that have changed. Hosts use Neighbor Discovery to find neighboring routers that are willing to forward packets on their behalf. According to RFC 2461, the Neighbor Discovery protocol defines mechanisms for solving the following problems:

- Router Discovery: How hosts locate routers that reside on an attached link.
- Prefix Discovery: How hosts discover the set of address prefixes that define which destinations are on-link for an attached link. (Nodes use prefixes to distinguish destinations that reside on-link from those only reachable through a router.)
- Parameter Discovery: How a node learns such link parameters as the link MTU or such Internet parameters as the hop limit value to place in outgoing packets.
- Address Autoconfiguration: How nodes automatically configure an address for an interface.
- Address Resolution: How nodes determine the Link Layer address of an on-link destination (such as a neighbor) given only the destination's IP address.
- Next-Hop Determination: The algorithm for mapping an IP destination address into the IP address of the neighbor to which traffic for the destination should be sent. The Next-Hop can be a router or the destination itself.
- Neighbor Unreachability Detection: How nodes determine that a neighbor is no longer reachable. For neighbors used as routers, alternate default routers can be tried. For both routers and hosts, address resolution can be performed again.
- Duplicate Address Detection: How a node determines that an address it wishes to use is not already in use by another node.
- Redirect: How a router informs a host of a better first-hop node to reach a particular destination.

Combining the IPv4 ARP, Router Discovery, and Redirect functions into the IPv6 Neighbor Discovery process allows for an efficient and consistent way of disseminating information. Five ICMPv6 messages facilitate the Neighbor Discovery information: Router Solicitation, Router Advertisement, Neighbor Solicitation, Neighbor Advertisement, and Redirect.

The Router Solicitation Message is transmitted by a host to prompt routers to generate Router Advertisement messages quickly. Routers transmit Router Advertisement messages on a periodic basis, or in response to a host's Router Solicitation message.

Neighbor Solicitation messages are sent by nodes to request the Data Link Layer address of a target node, while also providing their own Data Link Layer address to that target. Neighbor Solicitations are multicast when the node needs to resolve an address, and unicast when the node seeks to verify the reachability of a neighbor. Neighbor Advertisement messages are sent by nodes in response to Neighbor Solicitation messages, or are sent unsolicited to propagate new information quickly.

Finally, Redirect messages are sent by routers to inform a host of a better first-hop node on the path to a destination.

6.5.3 Path MTU Discovery Process

The Path MTU Discovery Process is documented in RFC 1981 [6-35]. Recall that a link's MTU is the maximum transmission unit size, given in octets, that can be conveyed in one piece over that link. For example, the MTU of an Ethernet/IEEE 802.3 LAN is 1,500 octets. Also recall that fragmentation is a host (not a router) responsibility, which implies that the host must have some knowledge of the network topology. (Otherwise, it would not know if fragmentation were required or not.) Path MTU Discovery answers these topology questions for the host.

The Path MTU Process is illustrated in Figure 6-48; it begins with the Source Node assuming that the Path MTU (PMTU) is also the MTU of the first hop (a known value). For example, if you know that the first hop is on an FDDI ring, with an MTU of 4,352 octets, then you will assume that the entire path has an MTU of 4,352 octets. In other words, you assume that the first hop and the Destination Node are on the same type of network topology.

The Source Node then transmits a packet and checks to see whether an ICMPv6 Packet Too Big message is received. If no message is received, then the entire path has an MTU of 4,352 octets and the Source Node can continue transmitting. If a Packet Too Big Message is received, then the PMTU must be reduced by the MTU that is returned in that message. A packet is transmitted again, this time with a smaller MTU, and a similar test is made. This process continues until no Packet Too Big Messages are returned, at which time the PMTU has been discovered and transmission may proceed. On a periodic basis, the PMTU is retested to see if its value can be increased.

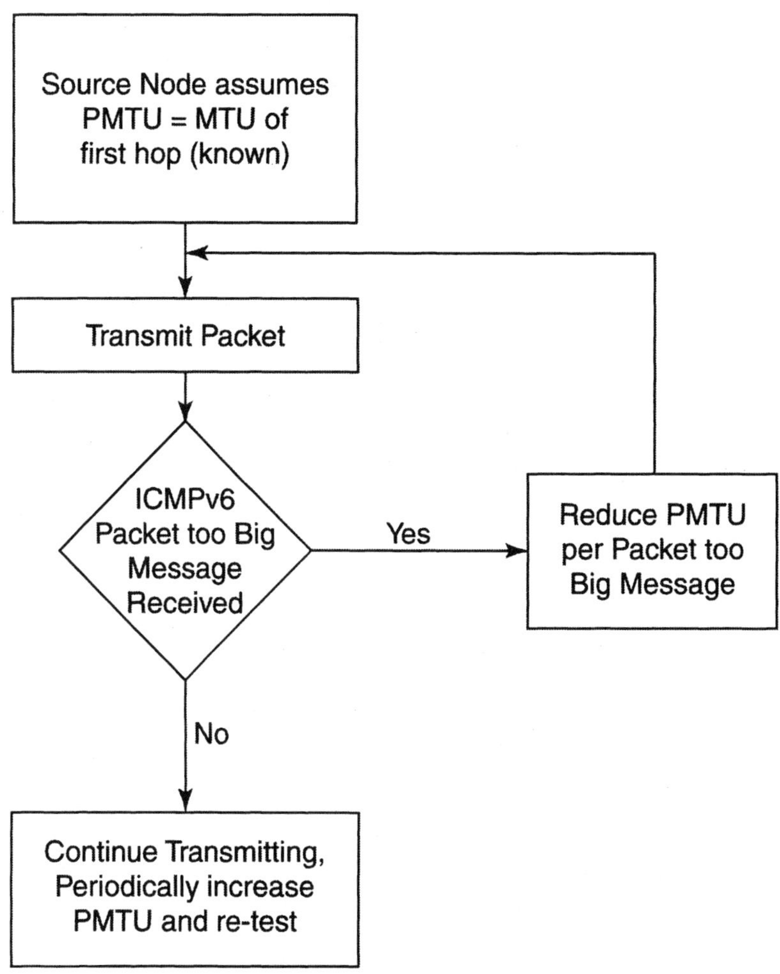

Figure 6-48. Path MTU Discovery.

6.6 Station Configuration

The word autoconfiguration is best described by its two roots: auto, meaning self, and configuration, meaning a functional arrangement. According to RFC 2462 [6-36], the autoconfiguration process includes creating a Link-Local address and verifying its uniqueness on the link, as well as determining what information should be autoconfigured (addresses, other information, or both). Note that the autoconfiguration process specified in RFC 2462 applies to hosts only; it is assumed that routers are configured by some other means.

There are three methods for obtaining addresses: a stateless mechanism, a stateful mechanism, or both. Both stateless and stateful autoconfiguration may be used simultaneously. Which type of autoconfiguration is in use is specified by control flags sent in the Router Advertisement messages.

In a stateful autoconfiguration model, hosts obtain addresses, configuration information, parameters, and so on, from a server. That server maintains a database containing the necessary information and keeps tight control over the address assignments. The stateful autoconfiguration model for IPv6 is defined by the Dynamic Host Configuration Protocol for IPv6 (DHCPv6).

In contrast, stateless autoconfiguration requires no manual configuration of hosts, minimal (or no) configuration of routers, and no additional servers. The stateless approach is used when a site is not concerned about the specific addresses that are used, as long as they are unique and routable.

6.6.1 Stateless Autoconfiguration

With stateless autoconfiguration, a host generates its own address using two elements of information: locally available information (i.e. available from the host itself), plus information advertised by routers. The host part is called an interface identifier, which identifies an interface on a subnet. The router part comes from an address prefix that identifies the subnet associated with a link. The derived address is a combination of these two elements. If a router does not exist on a subnet, the host can still generate a special type of address called the Link-Local address (review Figure 6-42). The Link-Local address may only be used for communication between nodes attached to the same link.

Note that the stateless autoconfiguration process, as defined in RFC 2462, applies to hosts only, not routers. (Because hosts obtain some of their address information from routers, those routers must be configured using some other means.) The only exception to this rule is that routers can generate their own Link-Local addresses, and can verify the uniqueness of these addresses on the link, when they are booted or rebooted.

IPv6 addresses are "leased" to an interface for a particular period of time, which may be infinite. Associated with the address is a lifetime indicating how long it can be bound to that interface. Upon expiration of the lifetime, both the binding and the address become invalid, and the address may be reassigned to another interface in the Internet. In support of these bindings, the assigned address may have two phases: preferred, meaning that use of that address is unrestricted; and deprecated, indicating that further use of the address is discouraged in anticipation of an invalid binding.

The Stateless Autoconfiguration process is illustrated in Figures 6-49a and 6-49b. This process begins with the generation of a Link-Local address for that interface.

The Link-Local address is generated by combining the Link-Local address prefix (1111 1110 10, as shown in Figure 6-42) with a 64-bit interface identifier. The interface identifier is specific to the LAN or WAN topology in use. In most cases, it is derived from the hardware address that resides in a ROM on the network interface card.

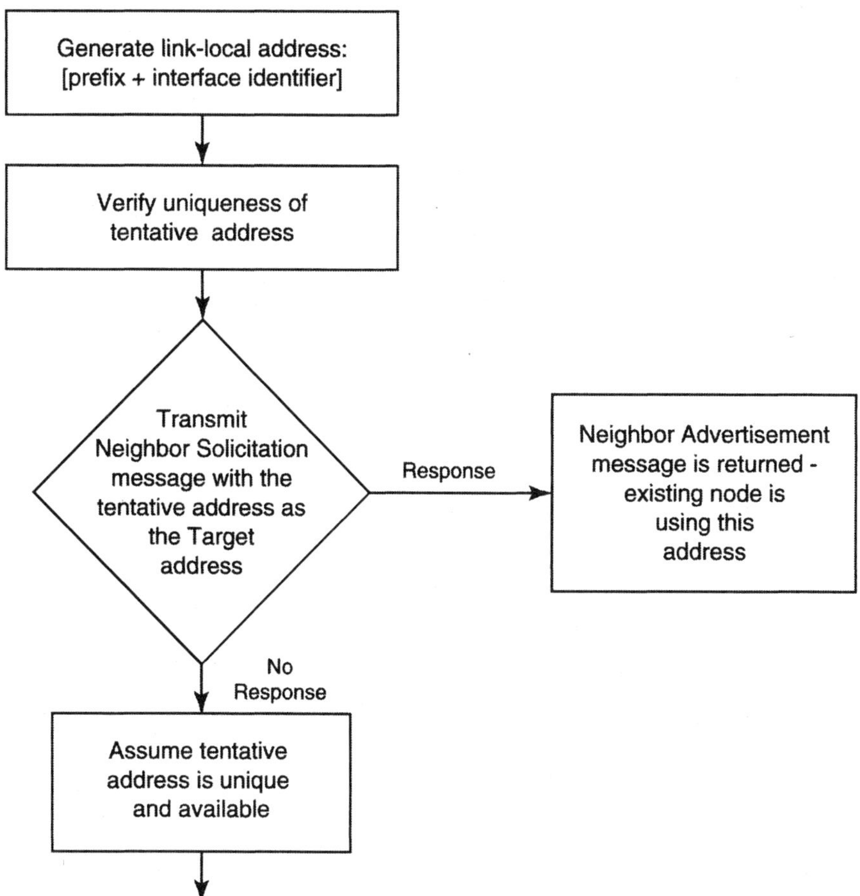

Figure 6-49a. The Stateless Autoconfiguration Process.

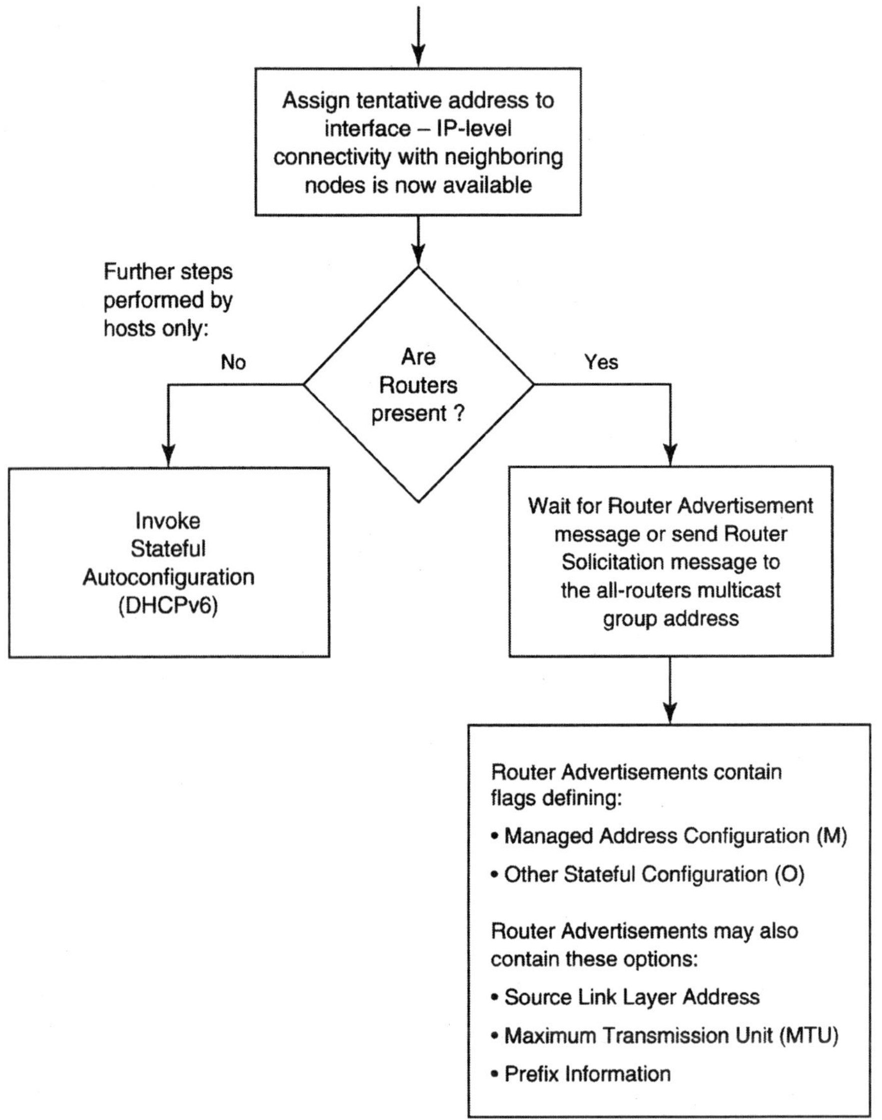

Figure 6-49b. The Stateless Autoconfiguration Process, continued.

The next step determines the uniqueness of the tentative address that has been derived by combining the Link-Local prefix and the interface identifier. In this step, a Neighbor Solicitation message is transmitted with the tentative address as the target address. If another node is using this address, a Neighbor Advertisement message is returned. In this event, autoconfiguration stops and some manual intervention is required. If no Neighbor Advertisement responses are returned, the tentative address is considered unique and IP-level connectivity with the neighboring nodes is now possible. Note that both hosts and routers can generate Link-Local addresses using this part of the autoconfiguration process.

The next phase is performed by hosts only; it involves listening for the Router Advertisement messages that routers periodically transmit, or forcing an immediate Router Advertisement message by transmitting a Router Solicitation message (Figure 6-49b). If no Router Advertisements are received, meaning that no routers are present, a stateful method, such as DHCPv6, should be used to complete the configuration process.

6.6.2 Dynamic Host Configuration Protocol

In some cases, such as when a duplicate address exists or routers are not present, a stateful autoconfiguration process must be used. The Dynamic Host Configuration Protocol version 6 (DHCPv6) provides these configuration parameters to Internet nodes; it is documented in Reference [6-37]. DHCPv6 consists of two elements: a protocol that delivers node-specific configuration information from a DHCPv6 server to a client, and a mechanism for allocating network addresses and other parameters to IPv6 nodes.

DHCPv6 is built on a client/server model, which relies on a number of Request and Reply messages for communication of these parameter details. Several types of functional DHCPv6 nodes are defined:

- DHCPv6 Client: A node that initiates requests on a link to obtain configuration parameters.
- DHCPv6 Server: A node that responds to requests from clients to provide configuration parameters. The server may or may not be on the same link as the client.
- DHCPv6 Relay Agent: A node that acts as an intermediary to deliver DHCPv6 messages between clients and servers, and is on the same link as the client.

Communication between DHCPv6 nodes uses the following well-known multicast addresses:

FF02:0:0:0:0:0:1:2 Link-Local All-DHCP-Agents-and-Servers multicast group
FF05:0:0:0:0:0:1:3 Site-Local All-DHCP-Servers multicast group

All of the DHCPv6 messages have a similar format, which begins with a Message Type (Msg-type) field indicating the specific function. Configuration parameters, which are called extensions, are included in the DHCPv6 messages and documented in Reference [6-37]. Extensions have been defined to specify an IP address, Timezones, Domain Name Server, Directory Agent, Network Time Protocol Server, Network Information Server, Transmission Control Protocol (TCP) parameters, Client-Server Authentication, and many other parameters.

The DHCPv6 messages are:

- SOLICIT: A client sends a Solicit message to locate servers.
- ADVERTISE: A server sends an Advertise message to indicate that it is available for DHCP service, in response to a Solicit message received from a client.
- REQUEST: A client sends a Request message to request configuration parameters, including IP addresses, from a specific server.
- CONFIRM: A client sends a Confirm message to any available server to determine whether the addresses it was assigned are still appropriate to the link.
- RENEW: A client sends a Renew message to the server that originally provided the client's addresses and configuration parameters to extend the lifetimes on the addresses assigned to the client and to update other configuration parameters.
- REBIND: A client sends a Rebind message to any available server to extend the lifetimes on the addresses assigned to the client and to update other configuration parameters.
- REPLY: A server sends a Reply message containing assigned addresses and configuration parameters in response to a Solicit, Request, Renew, or Rebind message received from a client.
- RELEASE: A client sends a Release message to the server that assigned addresses to the client to indicate that the client will no longer use one or more of the assigned addresses.

- DECLINE: A client sends a Decline message to a server to indicate that the client has determined that one or more addresses assigned by the server are already in use on the link.
- RECONFIGURE: A server sends a Reconfigure message to a client to inform the client that the server has new or updated configuration parameters.
- INFORMATION-REQUEST: A client sends an Information-request message to a server to request configuration parameters without the assignment of any IP addresses to the client.
- RELAY-FORW: A relay agent sends a Relay-forward message to relay messages to servers, either directly or through another relay agent.
- RELAY-REPL: A server sends a Relay-reply message to a relay agent containing a message that the relay agent delivers to a client.

Details regarding the formats and usage of these messages can be found in Reference [6-37].

6.6.3 Network-specific Support for IPv6

Support for a number of IPv6-based network infrastructures, both LAN and WAN, has been developed. This includes Ethernet, documented in RFC 2464; Fiber Distributed Data Interface (FDDI), documented in RFC 2467; Token Ring, documented in RFC 2470; Point-to-Point Protocol WANs, documented in RFC 2472; ATM networks, documented in RFC 2492; ARCNET, documented in RFC 2497; frame relay WANs, documented in RFC 2590; and IEEE 1394 networks, documented in RFC 3146. In this section, we will look at three representative examples: Ethernet, PPP, and frame relay.

6.6.3.1 Ethernet over IPv6

Ethernet, originally developed by Digital Equipment Corporation (DEC) — now part of Hewlett-Packard Company — Intel Corporation, and Xerox Corporation, has been traditionally popular with TCP/IP-based internetworks. Support for IPv6 over Ethernet networks is documented in RFC 2464 [6-38]. (Note that RFC 2464 documents support for Ethernet networks, not IEEE 802.3 networks, which have similar, but not identical, frame formats.)

The Ethernet frame may carry as much as 1,500 octets of data in the information field; therefore, we would say the maximum transmission unit (MTU) for Ethernet is 1500 octets. This size may be reduced by a Router Advertisement packet specifying a smaller MTU, as detailed in RFC 2461. The Ethernet Type (Ethertype) field contains the value 86DDH to specify IPv6 (Figure 6-50).

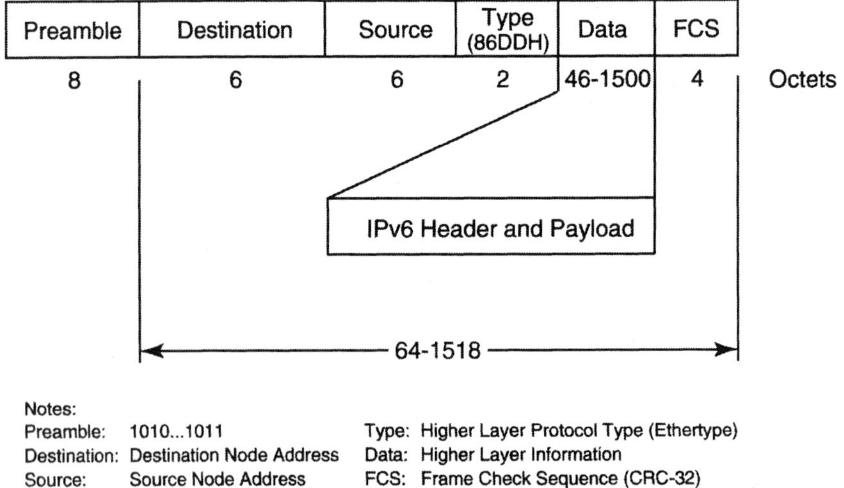

Figure 6-50. Ethernet Frame with IPv6 Packet.

The Link-Local address is formed by prepending the Link-Local prefix (FE80::0) to the interface identifier. For Ethernet networks, the interface identifier is the 48-bit Ethernet address, expanded in the center with the hexadecimal characters FFFE (review Figure 6-35) to create an EUI-64 compatible 64-bit address (Figure 6-51).

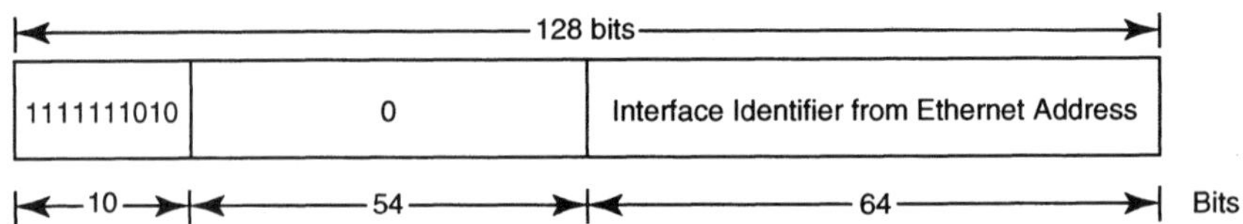

Figure 6-51. Link-Local Address for Ethernet.

For multicast addresses, an IPv6 address with a multicast destination address (e.g. DST) is transmitted to the Ethernet multicast address that begins with the value 3333H and ends with the last four octets of the DST address. (Note from Figure 6-52 that the value 3333H occupies the first two octets of the Ethernet multicast address, and the last four octets of the 16-octet IPv6 address (designated DST13, DST14, DST15, and DST16) occupy the last four octets of the Ethernet address.)

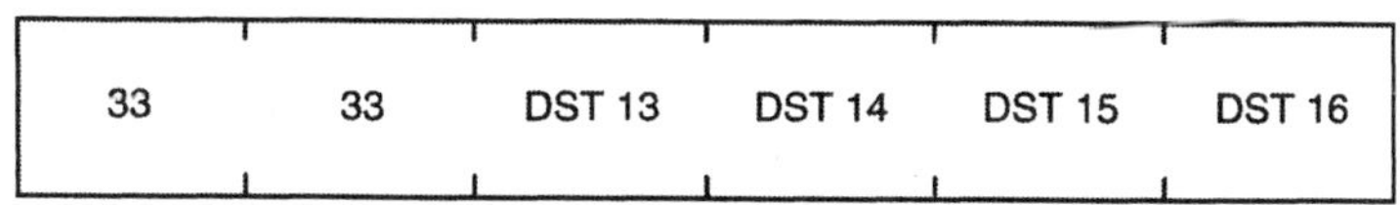

Figure 6-52. Ethernet Multicast Address Mapping.

Further details on Ethernet support are found in RFC 2464 [6-38].

6.6.3.2 IPv6 over PPP

The Point to Point Protocol (PPP) is used extensively for transmission of TCP/IP traffic over WAN links. Support for IPv6 over PPP is documented in RFC 2472 [6-39]. PPP consists of three elements: an encapsulation (or framing) format for serial links; a Link Control Protocol (LCP) for establishing, configuring, and testing the link connection; and a family of Network Control Protocols (NCPs) for establishing and configuring different Network Layer protocols. For example, the NCP for establishing and configuring IPv6 over PPP is called the IPv6 Control Protocol, or IPV6CP.

The PPP frame is shown in Figure 6-53. Note that either one IPv6 packet or one IPV6CP packet would fit inside the Information field of that PPP frame. The Protocol field defines the type of packet that is carried: 0057H indicates IPv6, while 8057H indicates IPV6CP.

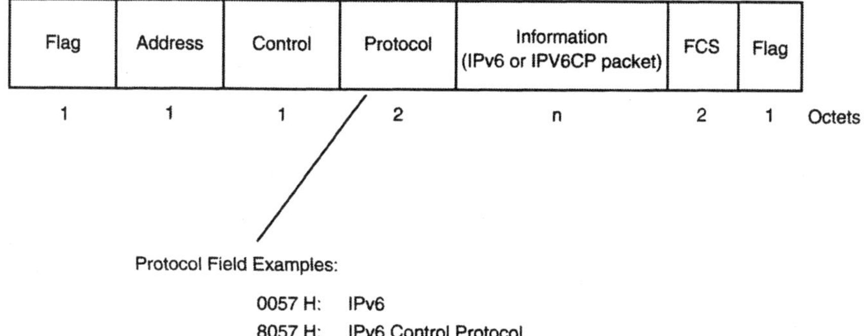

Figure 6-53. PPP Frame with IPv6 or IPV6CP Packet.

For LANs such as Ethernet or token ring, the interface identifier used with the Stateless Address Autoconfiguration process is based on the hardware address, which is typically resident in a ROM on the network interface card. For PPP links, the interface identifier may be selected using one of the following methods (listed in the order of preference):

1. If an IEEE global identifier (either EUI-48 or EUI-64) is available anywhere on the node, then that address should be used.
2. If an IEEE global identifier is not available, then a different source of uniqueness, such as a machine serial number, should be used.
3. If a good source of uniqueness cannot be found, a random number should be generated.

The resulting Link-Local address for PPP is shown in Figure 6-54.

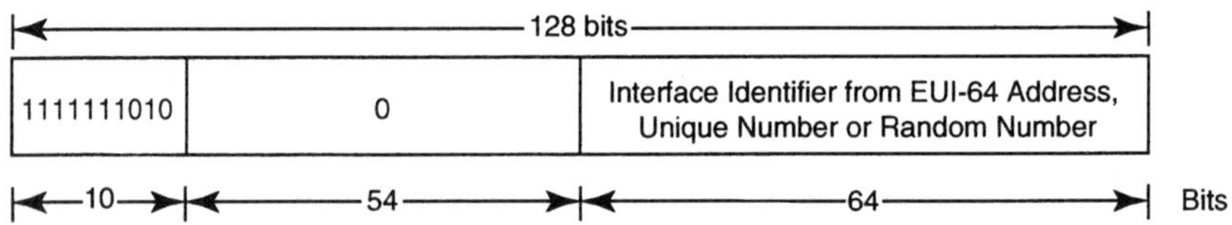

Figure 6-54. Link-Local Address for PPP.

The IPV6CP allows for IPv6 parameters to be negotiated during link startup. Two options, the Interface Identifier and the IPv6 Compression Protocol, have been defined. The Interface Identifier option facilitates the negotiation of a unique 64-bit interface identifier for that link, using one of the three alternatives listed above. The IPv6-Compression-Protocol option provides a way to negotiate the use of a specific IPv6 packet compression protocol. The current values for the IPv6-Compression-Protocol field are found in the most recent "Assigned Numbers" document (review Reference [3-2]).

6.6.3.3 IPv6 over Frame Relay

Support for IPv6 over frame relay networks is defined in RFC 2590 [6-40], which draws heavily upon RFC 2427, "Multiprotocol Interconnect over Frame Relay" [6-41]. Frame relay networks are used to connect endpoints, such as hosts and routers, across a wide area network. The connections are made by virtual circuits, which are identified by a Data Link Connection Identifier, or DLCI. The DLCI may be 10, 17, or 23 bits in length. When multiple endpoints exist, the virtual circuits may form a fully connected mesh network (where all endpoints are connected to all other endpoints), or a partially connected mesh network (where each endpoint may not be connected to all other endpoints). Each virtual circuit has its own transmission characteristics, such as data throughput, frame size, and so on. Two different types of virtual circuits are defined: permanent virtual circuits (PVCs), which are established by administrative assignments, and switched virtual circuits (SVCs), which are established dynamically.

The frame relay frame format is derived from the ISO High Level Data Link (HDLC) frame format, but it combines the HDLC Address and Control fields into a single Address field (Figure 6-55a). The default frame relay frame size is 1,600 octets, which allows 1,592 octets as the default MTU size for IPv6 (not counting the Flag characters that begin and end the frame).

The frame relay Address field format is defined in ITU-T Recommendation Q.922 and illustrated in Figure 6-55b. Note that there are two possible lengths for this Address field: two and four octets. The two-octet Address field is used for 10-bit DLCIs, and the four-octet Address field is used for 17- or 23-bit DLCIs.

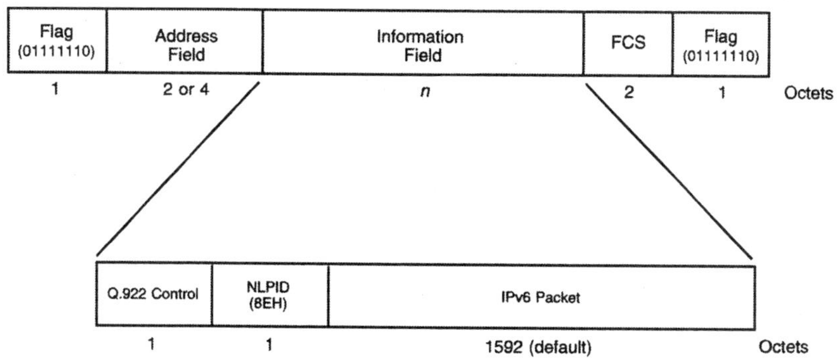

Figure 6-55a. Frame Relay Frame with IPv6 Packet.

RFC 2590 defines the method for constructing the frame relay interface identifier, using three fields: the "EUI bits" field, the "Mid" field, and the "DLCI" field. In addition, the "Mid" field can be constructed in several different ways. Readers needing those technical details are referred to RFC 2590. The resulting Link-Local address is illustrated in Figure 6-56.

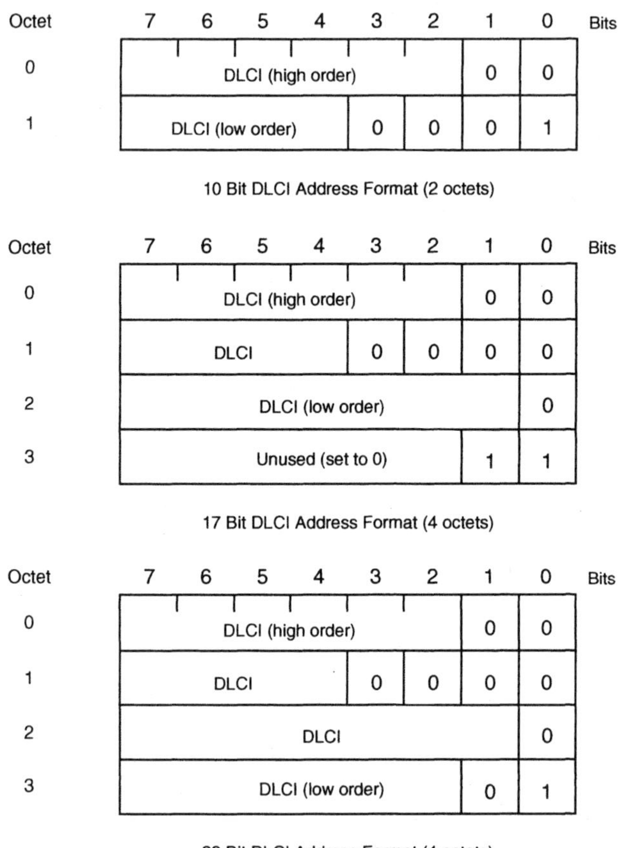

Figure 6-55b. Frame Relay Address Field Formats.

Extensions to the Neighbor Discovery protocol in support of frame relay networks have been proposed in RFC 3122 [6-42]. These extensions are known as Inverse Neighbor Discovery (IND). They allow a frame relay node to discover the equivalent of link-layer addresses that identify the local node at its location, and also identify the local node from a remote location (i.e. the other end of the virtual connection). The IND protocol operates similarly to Neighbor Discovery, where a node requesting a target IP address transmits a solicitation message, and the target node responds with an advertisement that contains the information requested. Processing details for these messages are also discussed in RFC 3122.

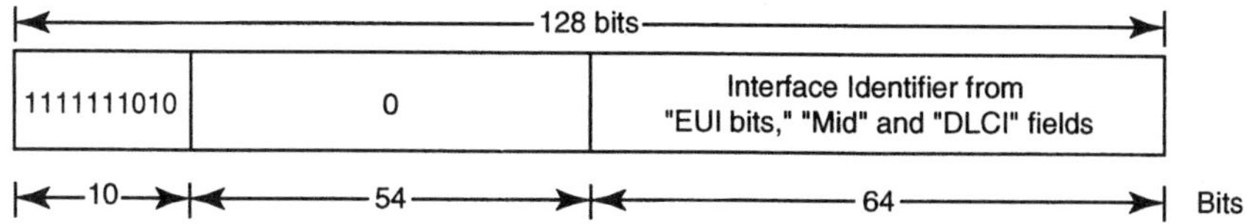

Figure 6-56. Link-Local Address for Frame Relay.

6.7 Routing Enhancements

As we have seen, the most dramatic difference between IPv4 and IPv6 is in the new addressing structure, which expands the number of addressable nodes by many orders of magnitude. This new structure will affect any device that deals with IPv6 addresses, including routers and hosts, plus their associated operating systems and protocols. In this section, we will examine enhancements to three routing protocols that will be required to support IPv6, and we will consider updates to RIP, OSPF, and BGP. In the next section, we will consider host-related enhancements.

6.7.1 Routing Information Protocol — Next Generation

The Routing Information Protocol is one of the most widely used Interior Gateway Protocols; it was originally defined in 1988 and documented in RFC 1058. Support for RIP with IPv6 is called RIPng and is documented in RFC 2080 [6-43].

RIP is a Distance Vector Algorithm–based protocol, with a history that dates back to the early days of the ARPAnet. RIP is designed for networks of moderate size, with a few limitations:

- The protocol is limited to networks whose longest path (or network diameter) is 15 hops.
- The protocol depends on a process called "counting to infinity" to resolve certain situations, such as routing loops. This process may consume a large amount of network bandwidth before resolution.
- The protocol depends on fixed metrics to compare alternative routes, without regard for real-time parameters such as delay, reliability, or load.

RIPng is the protocol that allows routers to exchange information for computing routes through an IPv6-based internetwork. Each router that implements RIPng is assumed to have a routing table that has an entry for each reachable IPv6 destination. Each entry contains the following:

- The IPv6 prefix of the destination.
- A metric that indicates the total cost of getting a datagram from the router to that destination.
- The IPv6 address of the next router along the path to the destination, called the next hop.
- A Route Change Flag that indicates whether the information about that route has recently changed.
- Various timers, such as a 30-second timer that triggers the transmission of routing table information to neighboring routers.

RIPng is a User Datagram Protocol (UDP)–based protocol that sends and receives packets on UDP port number 521. The RIPng packet (Figure 6-57) includes three fields: Command (Request or Response), Version (1), and a Route Table Entry (RTE). Each Route Table Entry (Figure 6-58) includes the IPv6 Prefix, the Route Tag (to separate internal from external routes), a Prefix Length field (to determine the number of significant bits in the Prefix), and the Metric (to define the current metric for the destination).

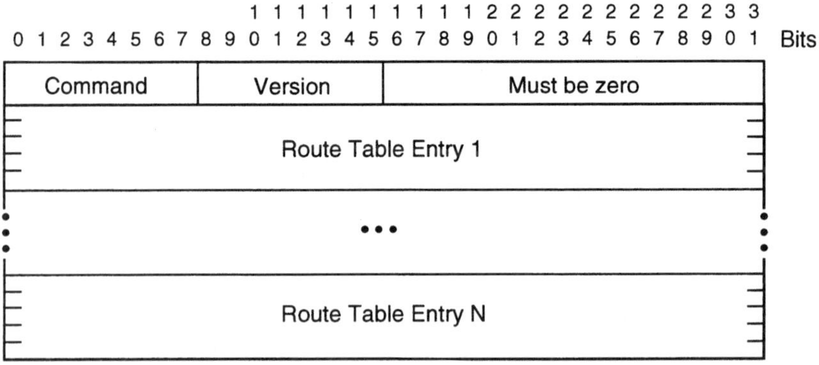

Figure 6-57. RIPng Packet Format.

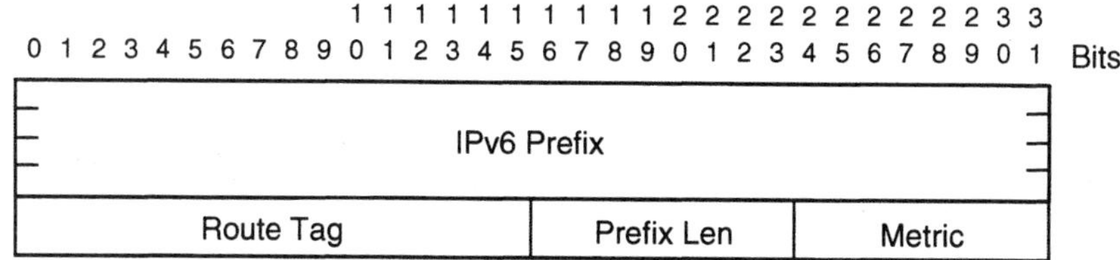

Figure 6-58. Route Table Entry Format.

RIPng also provides the ability to specify the immediate next hop IPv6 address for packets. This next hop is specified by a special RTE, The Next Hop Route Table Entry (Figure 6-59). The Next Hop RTE is identified by the value of FFH in the Metric field. The Prefix field specifies the IPv6 address of the next hop; the Route Tag and the Prefix Length are set to zero on transmission and ignored on reception.

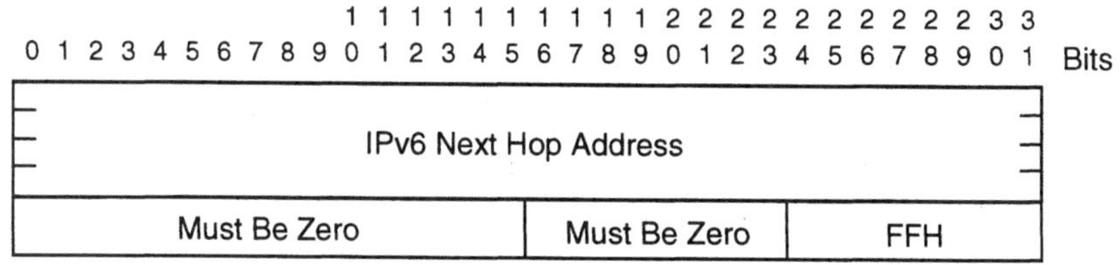

Figure 6-59. Next Hop RTE Format.

Version 1 of RIPng supports two commands: Request and Response. A Request is used to ask for all or part of a routing table. In most cases, Requests are sent as multicasts from the RIPng port (port 521). If information for only one router is needed, that request would be sent directly to that router from a port other than the RIPng port. There are three types of Responses: a response to a specific query; a regular update, which is an unsolicited response sent every 30 seconds to every neighboring router; and a triggered update caused by a route change. Specific details regarding the processing of the Request and Response packets are given in RFC 2080.

6.7.2 Open Shortest Path First Protocol for IPv6

The Open Shortest Path First (OSPF) protocol for IPv6 operates using a Link State Algorithm (LSA) and is defined in RFC 2740 [6-44]. An LSA offers several advantages over a Distance Vector Algorithm, such as that used with RIP. These include the ability to do the following: configure hierarchical (instead of flat) topologies; more quickly adapt to internetwork changes; allow for larger internetworks; calculate multiple minimum-cost routes that allow the traffic load to be balanced over several paths; and permit the use of variable-length subnet masks.

There are five packet types defined for OSPF: Hello, Database Description, Link State Request, Link State Update, and Link State Acknowledgment. These packet types begin with a consistent packet header (Figure 6-60). The most notable difference between this proposed header and the one currently defined for OSPF for IPv4 is the absence of the Authentication field (review Figure 4-2a). Since IPv6 has its own Authentication header available, that function is removed from the OSPF for IPv6 header to avoid redundancy.

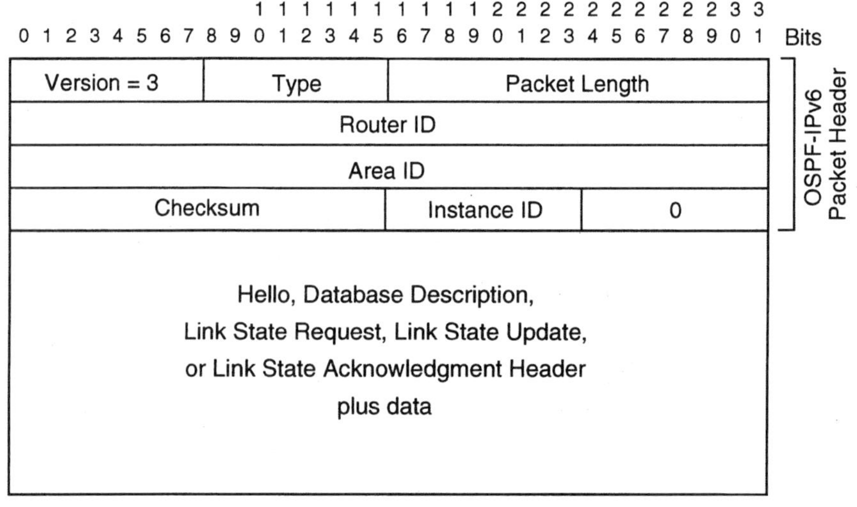

Figure 6-60. OSPF for IPv6 Header.

Some additional changes to OSPF have been proposed to provide support for IPv6, including:

- Protocol processing on a per-link basis, not on a per-subnet basis, since multiple subnets can be assigned to a single link with IPv6. (Recall that RFC 2460 defines an IPv6 link as a "communication facility or medium over which nodes can communicate at the link layer.")

- Removal of addressing semantics from the OSPF packet headers, leaving a network protocol–independent core. For example, IPv6 addresses are not carried in OSPF packets, with the exception of the Link State Advertisement payloads carried in the Link State Update packets.
- New Flooding Scope for Link State Advertisements (Link-local scope).
- Support to run multiple OSPF protocol instances per link.
- Use of Link-Local addresses, which are not forwarded by routers.
- Removing authentication from the OSPF packet header, as this function is now covered with the IPv6 Authentication and Encapsulating Security Payload headers.
- Packet format changes: new version number (3), removal of the Authentication field, and so on.
- New Link State Advertisement (LSA) formats to distribute IPv6 address resolution and next hop resolution, plus new processes to handle unknown LSA types.
- Updated support for stub areas.
- Consistent identification of all neighboring routers on a given link by their OSPF Router ID.
- Removal of Type of Service (TOS) semantics, with the provision that the IPv6 Flow Label field may be used for this function in the future.

Details on these proposed changes are provided in RFC 2740.

6.7.3 Border Gateway Protocol

An autonomous system exchanges routing information with another autonomous system using an Exterior Gateway Protocol, or EGP. The most prevalent is the Border Gateway Protocol (BGP), which has gone through several iterations. These include BGP-1 (RFC 1105), BGP-2 (RFC 1163), BGP-3 (RFC 1267), and BGP-4 (RFCs 1771, 1772, 1773, and 1774). The primary function of a BGP system is to exchange network reachability information with other BGP systems. This information includes data on the list of ASs that the reachability information traverses, which allows the construction of a connectivity graph.

BGP-4 uses the Transmission Control Protocol (TCP) for greater reliability of communication between ASs. The BGP-4 message header is defined in RFC 1771 [6-45], Section 4.1. This header is 19 octets long and supports one of four message types (Figure 6-61):

- OPEN: initiates the BGP connection.
- UPDATE: used to transfer routing information between BGP peers.
- KEEPALIVE: exchanged on a periodic basis to determine peer reachability.
- NOTIFICATION: sent when an error condition is detected; causes the BGP connection to be closed.

After a TCP connection is established, the first message sent is an OPEN message. If the OPEN is acceptable to the other end of the connection, a KEEPALIVE message is returned in confirmation. Once the OPEN has been confirmed, UPDATE, KEEPALIVE, and NOTIFICATION messages may be exchanged.

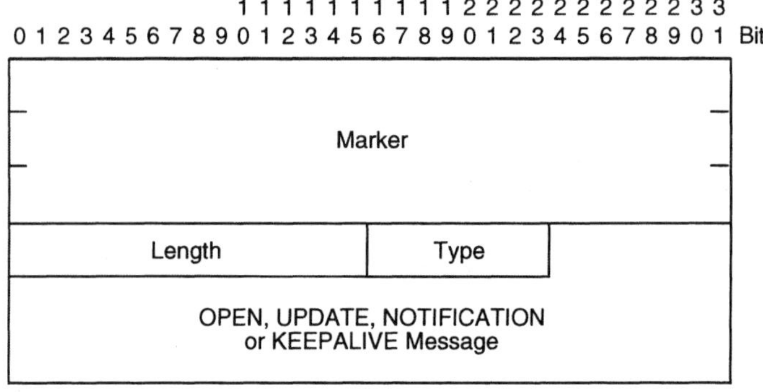

Figure 6-61. BGP-4 Message Header.

Multiprotocol extensions to BGP-4 have been defined in RFC 2858 [6-46] that allow it to carry multiple Network Layer protocols, including IPv6. As with other changes to routing protocols, these BGP-4 extensions must include support for the IPv6 addressing structure, the ability to distinguish between various Network Layer protocols, and so on. Details regarding the use of these protocol extensions are documented in RFC 2545 [6-47].

6.8 Upper Layer Protocol and Host Issues

Since IP is at the core of the ARPA architecture, changes at the lower layer affect the operation of the upper layers. In this section we will consider upper layer protocol issues, as well as enhancements that effect hosts, including updates to DNS and application programming interfaces.

6.8.1. Upper Layer Protocol Issues

Four design and implementation issues concerning upper layers are considered in section 8 of the IPv6 specification, RFC 2460 [6-5]: upper layer checksums, maximum packet lifetimes, maximum upper layer payload size, and responding to packets with routing headers. These issues will be discussed individually.

6.8.1.1 Upper Layer Checksums

Both the TCP and UDP headers include a 16-bit checksum field, which is used to verify the reliability of the data delivered to the upper layer process. But, in addition to the data itself being reliable, that data must be delivered to the correct destination. To incorporate this destination element into the overall checksum calculation, a special header, called a Pseudo header, was devised. This Pseudo header includes the source and destination addresses for that packet. The checksum calculation then includes the Pseudo header, the UDP or TCP header, and the upper layer protocol headers and data in its algorithm (Figure 6-62a). Since the Pseudo header includes the addresses, the destination element is included in the reliability function. In addition, ICMPv6 uses this same Pseudo header in its calculation, unlike ICMP for IPv4.

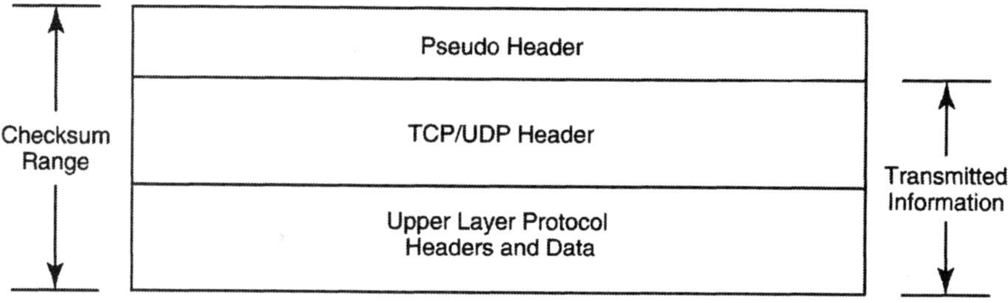

Figure 6-62a. Pseudo Header Position and Checksum Range.

For IPv6, which uses larger addresses, the Pseudo header format is revised (Figure 6-62b). The IPv6 Pseudo header includes fields for the 128-bit Source and Destination addresses, the Upper Layer Packet Length, and a

Next Header field. If the IPv6 packet contains a Routing header, the Destination address field in the Pseudo header contains the final address for that packet. The Upper Layer Packet Length field is the length of the upper layer header (e.g. TCP) plus data. The Next Header field identifies the upper layer protocol, such as TCP (Next Header = 6), UDP (Next Header = 17), or ICMP (Next Header = 58). Note that the contents of this field may not have the same value as the IPv6 header if extension headers are included in the IPv6 packet.

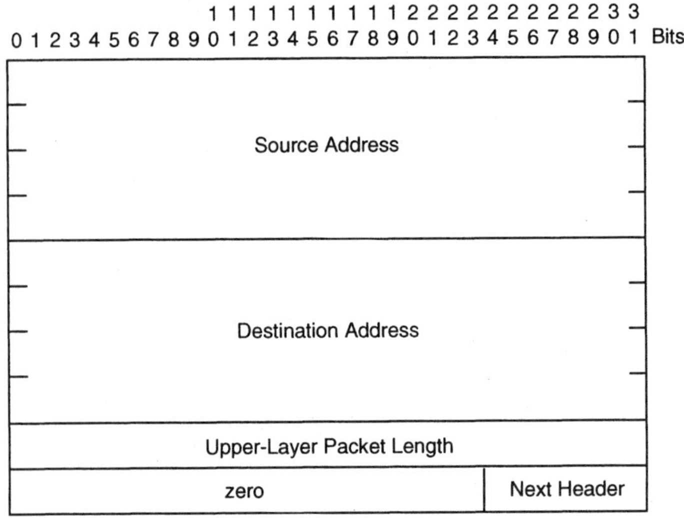

Figure 6-62b. Pseudo Header for Use with IPv6.

6.8.1.2 Maximum Packet Lifetimes

The IPv4 header included a Time-to-Live field that could be measured in seconds or hops. In IPv6, this function is performed by the Hop Limit field, which is measured only in hops. For IPv4 applications, most implementations measure the Time-to-Live in hops (not seconds); however, any upper layer protocols that need a time basis for measurement will need to be revised, such that some other mechanism is provided to detect and discard obsolete packets.

6.8.1.3 Maximum Upper Layer Payload Size

The maximum segment size (MSS) is the maximum payload size available for the upper layer data. In some cases, this number may be constrained by lower layer functions, such as the availability of fragmentation (at the ARPA

Internet Layer or OSI Network Layer) or the type of local network being used (at the ARPA Network Interface Layer or OSI Data Link Layer).

For the sake of illustration, let's assume that fragmentation is not in use, and that each upper layer segment is carried in a single Ethernet frame. Recall that each Ethernet frame can handle up to 1,500 octets of data. For a TCP/IPv4 segment, 20 octets would be used by the TCP header and 20 octets would be used by the IPv4 header, leaving 1,460 octets for the upper layer information (FTP, TELNET, etc.). Since IPv6 has a minimum header length of 40 octets (in contrast with the IPv4 length of 20 octets), a TCP/IPv6 segment would have an MSS of 1,440 octets.

6.8.1.4 Responding to Packets with Routing Headers

The Routing header is used to specify the route that a packet takes from source to destination (review Section 6.3.5 and Figures 6-19 and 6-20). When a packet containing a Routing header is received by a host, the upper layer protocols in that host must not automatically assume that a reversal of the specified route is correct and/or appropriate. Depending on the circumstances, there are three possible kinds of response packets that can be sent:

- A response packet that does not contain a Routing header.
- A response packet that contains a Routing header that was not derived by reversing the original route, such as a Routing header that was derived by the responding host itself.
- A response packet that contains a Routing header that was derived by reversing the original route, if the integrity and authenticity of the received Source address and Routing header have been verified.

Additional details on these upper layer issues are found in the IPv6 specification, RFC 2460 [6-5].

6.8.2. Domain Name System Extensions

The Domain Name System, or DNS, is a distributed database that provides a naming hierarchy for machines that are attached to the Internet. DNS has two key functions: a specification for the names themselves, and a method of mapping host names to IP addresses. In specifying machine names, a hierarchy exists that defines a number of top-level domains, such as .com (commercial), .edu (educational), .gov (government), .net (networks), and so on.

While applications may identify process destinations with machine names, communication protocols require numeric addresses. Therefore, the second key function of DNS is to provide a machine name lookup service similar to the telephone white pages. You supply a host name, and a DNS server returns an IP address.

But heretofore, all DNS information has been based on the assumption that IPv4 addresses, called Type A Resource Records, were being stored and retrieved. To support IPv6, enhancements to DNS, which are documented in RFC 1886 [6-48], have been developed. These are:

- A new Resource Record Type (AAAA), which stores a single 128-bit IPv6 address.
- A new special domain, IP6.INT, which is used to look up a host name given its IPv6 address. (These queries are sometimes referred to as Pointer Queries.)
- New procedures for existing query types that perform additional section processing. (DNS messages include a field called "Additional Information" which may contain other types of records, such as name server, mail exchange, or mailbox information.) If these additional records are queried, the name server must add any relevant IPv4 addresses and any relevant IPv6 addresses to the response.

Further details on these extensions can be found in RFC 1886.

6.8.3 Application Programming Interfaces

IPv4 is implemented in the Berkeley Software Distribution (BSD) version 4.x of the UNIX operating system, which defines an application programming interface (API) through which TCP/IP-based applications can communicate. This is informally known as the socket interface. Many other operating systems, including Linux, Mac OS, and Windows XP, also have APIs that are used by TCP/IP and IPv6 applications.

In order for the upper layer applications to access and therefore take advantage of the new IPv6 features, the APIs must be enhanced. These API extensions have been defined in two categories: basic and advanced. The Basic Socket Extensions are documented in RFC 3493 [6-49] and include:

- New address data structures that can accommodate the 16-octet IPv6 address (enhanced from the 4-octet IPv4 address), plus support for special addresses such as the IPv4-mapped IPv6 address and IPv6 Loopback address.

- Name-to-address translation functions, including support for IPv6 address lookup functions within the Domain Name System (DNS). Both forward lookup (nodename-to-address translation) and reverse lookup (address-to-nodename translation) are supported.
- Address conversion functions, which convert IPv6 addresses between binary and text formats.
- New options that are defined in support of IPv6-specific functions: unicast hop limit, sending and receiving multicast packets, joining and leaving multicast groups, plus support for the IPv6 Traffic Class and Flow Label fields.

The Advanced Socket Extensions are documented in RFC 3542 [6-50] and include:

- Definitions for IPv6 raw sockets (those that bypass the Transport Layer (UDP or TCP)) that are used for ICMPv6, and are also used for datagrams with a Next Header field that the operating system kernel does not process.
- Packet information that is transferred between the application and the outgoing/received packet, including the source/destination addresses, the outgoing/received interface, the outgoing/received hop limit, and the outgoing next hop address.
- Support for extension header processing, including the Hop-by-Hop, Destination Options, and Routing headers.
- Other advanced functions, such as sending with the minimum MTU, sending without fragmentation, Path MTU Discovery and UDP, and determining the current Path MTU.

At the time of this writing, IPv6 host implementations have been announced by over thirty router and host manufacturers. For a current listing of IPv6 host implementations, see the IPv6 industry website listed in Reference [6-51].

6.9 IPv4 to IPv6 Transition Strategies

The benefits derived from a new protocol must be balanced with the costs associated with making a transition from the existing systems. These logistical and technical issues have been addressed in a document entitled "Transition Mechanisms for IPv6 Hosts and Routers," RFC 2893 [6-52]. Transition issues specific to routing infrastructures are addressed in another document, "Routing Aspects of IPv6 Transition," RFC 2185 [6-53]. These efforts were the result of the work of the IETF Next Generation Transition (ngtrans) Working Group, which is documented in Reference [6-54].

The developers of IPv6 recognized that not all systems would upgrade from IPv4 to IPv6 in the immediate future, and that some might not upgrade for years. Some of that delay could be attributed to the supply and demand for available addresses. In regions such as the United States where IPv4 addresses have historically been more readily available, migration to IPv6 may not occur as quickly as in other regions where IPv4 addresses are in shorter supply. To complicate matters, most internetworks are heterogeneous systems, with various routers, hosts, and so on, manufactured by different vendors. If such a multivendor system were to be upgraded at one time, IPv6 capabilities would be required on all of the individual elements before the larger project could be attempted. Another (much larger) issue becomes the worldwide Internet, which operates across 24 different time zones. Upgrading this system in a single process would be even more difficult.

Given the above constraints, it therefore becomes necessary to develop strategies for IPv4 and IPv6 to coexist, until such time as IPv6 becomes the preferred option. At the time of this writing, two mechanisms for this coexistence have been proposed: a dual IP layer and IPv6 over IPv4 tunneling. These two complementary technologies will be discussed in the following sections.

6.9.1 Transition Definitions

The following terms, which relate to nodes and addresses for use in the transition architectures, are defined in RFC 2893:

- IPv4-only node: A host or router that implements only IPv4 and does not understand IPv6. The installed base of IPv4 hosts and routers existing before the transition begins are IPv4-only nodes.
- IPv6/IPv4 node: A host or router that implements both IPv4 and IPv6.
- IPv6-only node: A host or router that implements IPv6 and does not implement IPv4.

- IPv6 node: Any host or router that implements IPv6. IPv6/IPv4 and IPv6-only nodes are both IPv6 nodes.
- IPv4 node: Any host or router that implements IPv4. IPv6/IPv4 and IPv4-only nodes are both IPv4 nodes.
- IPv4-compatible IPv6 address: An IPv6 address, assigned to an IPv6/IPv4 node, which bears the high-order 96-bit prefix 0:0:0:0:0:0 and an IPv4 address in the low-order 32-bits. IPv4-compatible addresses are used by the automatic tunneling mechanism.
- IPv6-native address: The remainder of the IPv6 address space. An IPv6 address that bears a prefix other than 0:0:0:0:0:0.

To address the dual nodes further, those devices have three different modes of operation:

- IPv6-only operation: An IPv6/IPv4 node with its IPv6 stack enabled and its IPv4 stack disabled.
- IPv4-only operation: An IPv6/IPv4 node with its IPv4 stack enabled and its IPv6 stack disabled.
- Pv6/IPv4 operation: An IPv6/IPv4 node with both stacks enabled.

6.9.2 Transition Mechanisms

The transition mechanisms provide the ways and means of implementing a transition strategy. RFC 2893 considers two mechanisms: a dual IP layer and IPv6 over IPv4 tunneling. For the tunneling mechanism, there are two alternatives: configured tunneling and automatic tunneling. With configured tunneling, the tunnel endpoint is determined by configuration information that exists at the tunnel entry point (or encapsulating node). For automatic tunneling, a special address, the IPv4-compatible IPv6 address, is used to derive the tunnel endpoint. In the sections that follow, we will see examples of these two alternatives, plus other methods that are defined in other Internet documents.

6.9.2.1 Dual IP Stacks

The simplest mechanism for IPv4 and IPv6 coexistence is for both of the protocol stacks to be implemented on the same device. That device, which could be either a host or a router, is then referred to as an IPv6/IPv4 node. The IPv6/IPv4 node has the capability to send and receive both IPv4 and IPv6 packets, and can therefore interoperate with an IPv4 device using IPv4 packets and with an IPv6 device using IPv6 packets (Figure 6-63). The dual IP stacks may also work in conjunction with the two tunneling techniques described above.

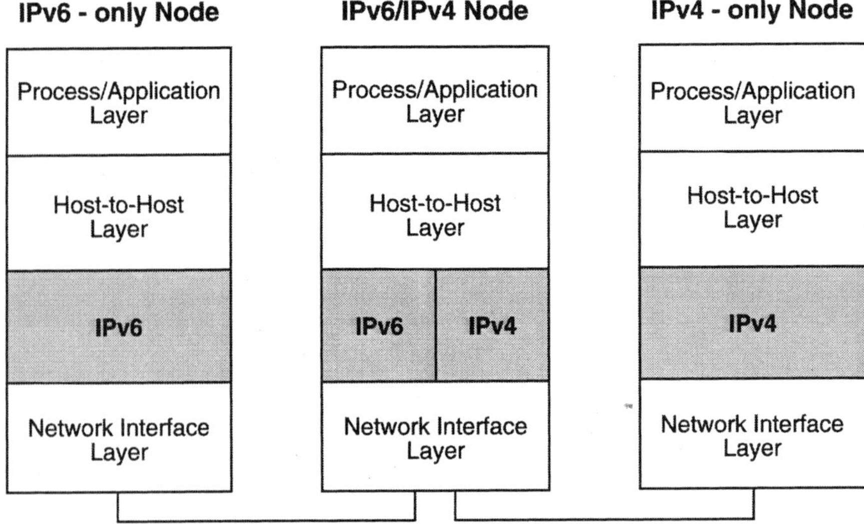

Figure 6-63. Dual IP Stack Architecture.

The IPv6/IPv4 node would be configured with addresses that support both protocols, and those addresses may or may not be related to each other. Other address-related functions, such as the Dynamic Host Configuration Protocol (DHCP) or the Bootstrap Protocol (BOOTP), may also be involved in this process. In addition, an IPv6/IPv4 node must support both the A and AAAA resource record types within the Domain Name System (DNS).

6.9.2.2 Tunneling

Tunneling is a process whereby information from one protocol is encapsulated inside the frame or packet of another architecture, thus enabling the original data to be carried over that second architecture. The tunneling scenarios for IPv6/IPv4 are designed to enable an existing IPv4 infrastructure to carry IPv6 packets by encapsulating the IPv6 information inside IPv4 datagrams. Note that the major emphasis of the transition mechanisms is on tunneling IPv6 packets over an existing IPv4 infrastructure, since the majority of today's networks are IPv4 systems. As the transition process matures, however, it is expected that some IPv4-only systems may then tunnel packets over newly installed IPv6 infrastructures. But for most of us, this second scenario is a few years off.

The encapsulation process is illustrated in Figure 6-64a. Note that the resulting IPv4 datagram contains both an IPv4 header and an IPv6 header, plus all of the upper layer information, such as the TCP header, application data, and so on. The reverse process, decapsulation, is illustrated in Figure 6-64b. In this case, the IPv4 header is removed, leaving only the IPv6 packet. Within the IPv4 header, the Protocol field would have a value of 41, identifying an IPv6 payload.

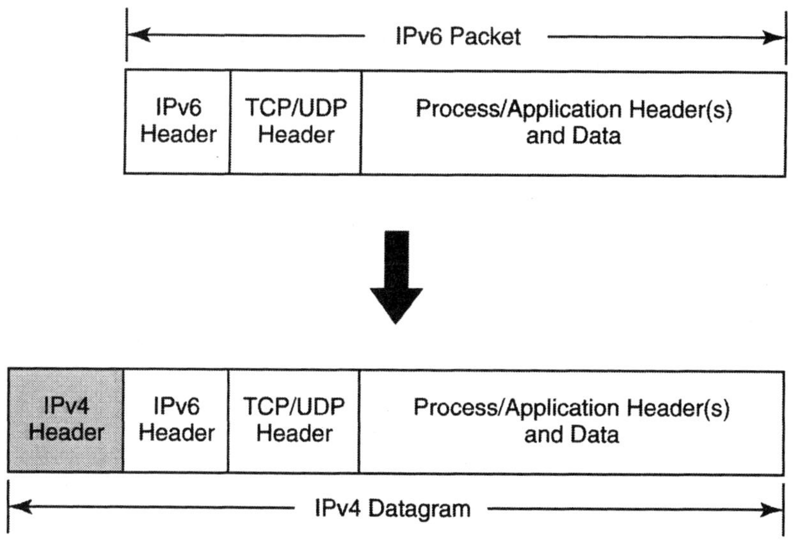

Figure 6-64a. Encapsulating IPv6 in IPv4.

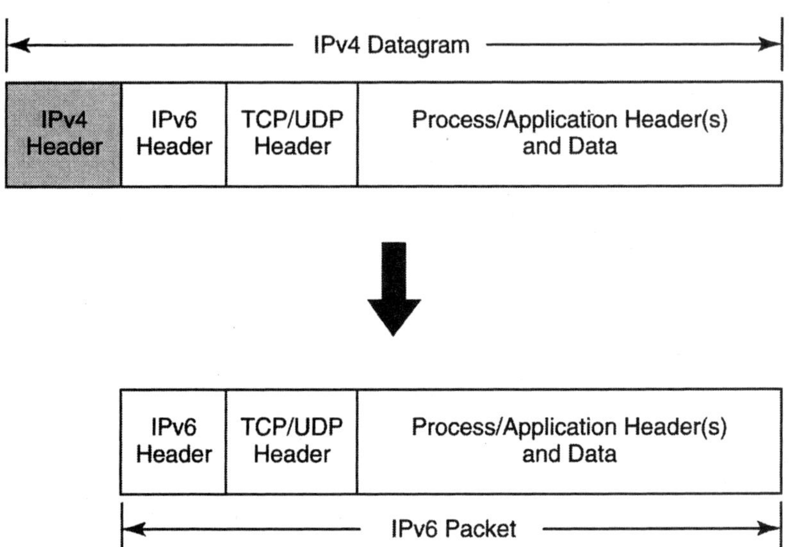

Figure 6-64b. Decapsulating IPv6 from IPv4.

The tunneling process involves three distinct steps: encapsulation, decapsulation, and tunnel management. At the encapsulating node (or tunnel entry point), the IPv4 header is created and the encapsulated packet is transmitted. At the decapsulating node (or tunnel exit point), the IPv4 header is removed and the IPv6 packet is processed. In addition, the encapsulating node may maintain configuration information regarding the tunnels that are established, such as the maximum transfer unit (MTU) size that is supported in that tunnel.

RFC 2893 defines four possible tunnel configurations that could be established between routers and hosts:

- Router-to-Router: IPv6/IPv4 routers that are separated by an IPv4 infrastructure tunnel IPv6 packets between themselves. In this case, the tunnel would span one segment of the packet's end-to-end path.
- Host-to-Router: An IPv6/IPv4 host tunnels IPv6 packets to an IPv6/IPv4 router that is reachable via an IPv4 infrastructure. In this case, the tunnel would span the first segment of the packet's end-to-end path.

- Host-to-Host: IPv6/IPv4 hosts that are interconnected by an IPv4 infrastructure can tunnel IPv6 packets across the IPv4 infrastructure. In this case, the tunnel spans the packet's entire end-to-end path.
- Router-to-Host: IPv6/IPv4 routers can tunnel IPv6 packets to an IPv6/IPv4 host that is the final destination. In this case, the tunnel would span only the final segment of the packet's end-to-end path.

For a tunnel to operate, addresses of both the tunnel endpoint and the packet's destination must be known, and these two addresses are not necessarily the same. The manner in which the tunnel endpoint address is determined defines one of three types of tunnels: a configured tunnel, an automatic tunnel, or a combined tunnel. These alternatives will be explored in the following sections.

6.9.2.2.1 Configured Tunneling

Configured tunneling is defined in RFC 2983 as IPv6-over-IPv4 tunneling where the IPv4 tunnel endpoint address is determined by configuration information on the encapsulating node. The tunnels can be either unidirectional or bidirectional. Bidirectional configured tunnels behave as virtual point-to-point links.

From the four tunneling scenarios that were discussed previously, the router-to-router and host-to-router terminate on a router, which then decapsulates the information and forwards the IPv6 packet to its final destination. Note that the tunnel endpoint address is different from the final destination endpoint address. (Recall that, by definition, routers forward packets, in contrast to hosts, which run applications. The final destination is therefore assumed to be a host, not a router.) Because the tunnel endpoint and the destination endpoint addresses differ, the node performing the tunneling determines the tunnel endpoint (i.e. a router) from some configuration information. For this reason, this type of tunneling is called configured tunneling.

Let's look at the various scenarios individually. The first case occurs when two IPv6 hosts are separated by an IPv4 infrastructure (Figure 6-65). The source host, H_1, is an IPv6 node configured for either IPv6-only or IPv6/IPv4. H_1 generates an IPv6 packet and sends that packet to the first router, R_1, a dual node. A configured tunnel exists across the IPv4 infrastructure between R_1 and R_2. When the entry router sees the IPv6 destination address, it encapsulates the IPv6 packet inside an IPv4 datagram and sends it across the IPv4 internetwork. At the tunnel endpoint, the exit router decapsulates the IPv6 packet and sends it to the final destination, H_2.

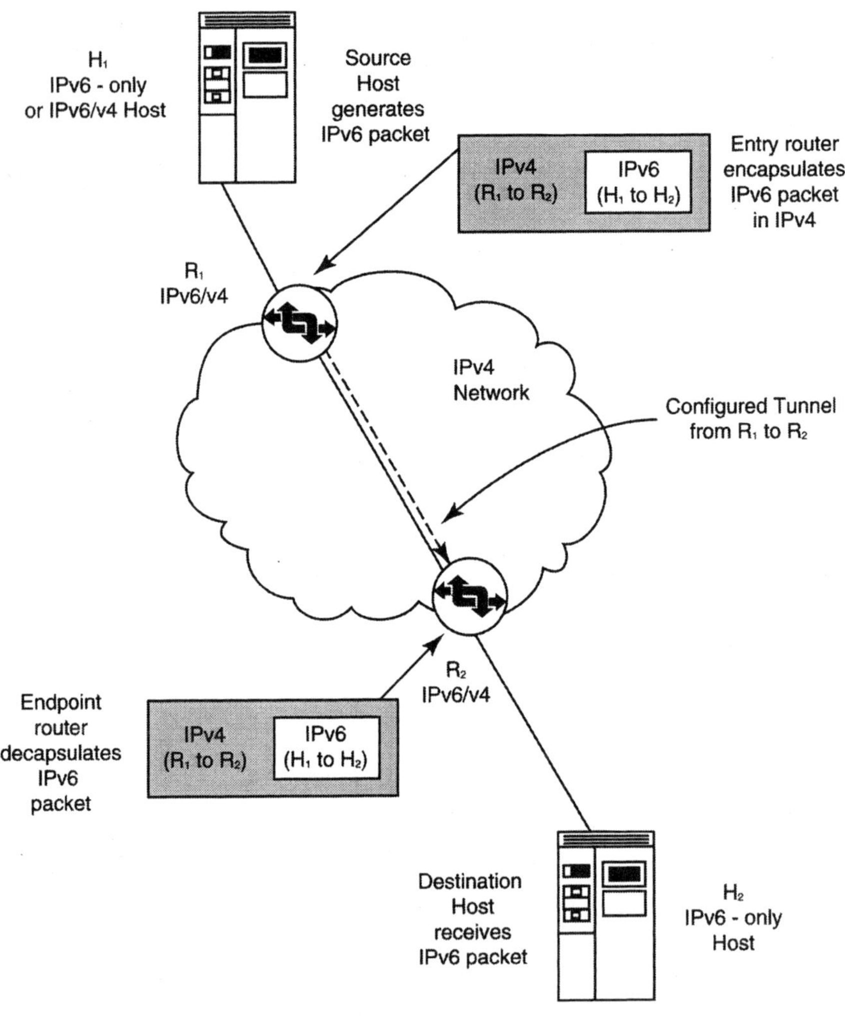

Figure 6-65. Router-to-Router Configured Tunnel.

The internetwork could also be configured such that the tunnel entry point is at the originating host (Figure 6-66). In this case, the source host, H_1, must be a dual host, and the tunnel is configured between H_1 and R_2. Encapsulation occurs at the source host, H_1, and decapsulation occurs at the tunnel endpoint, R_2. (In this case, the destination host, H_2, does not have an IPv4-compatible address available.)

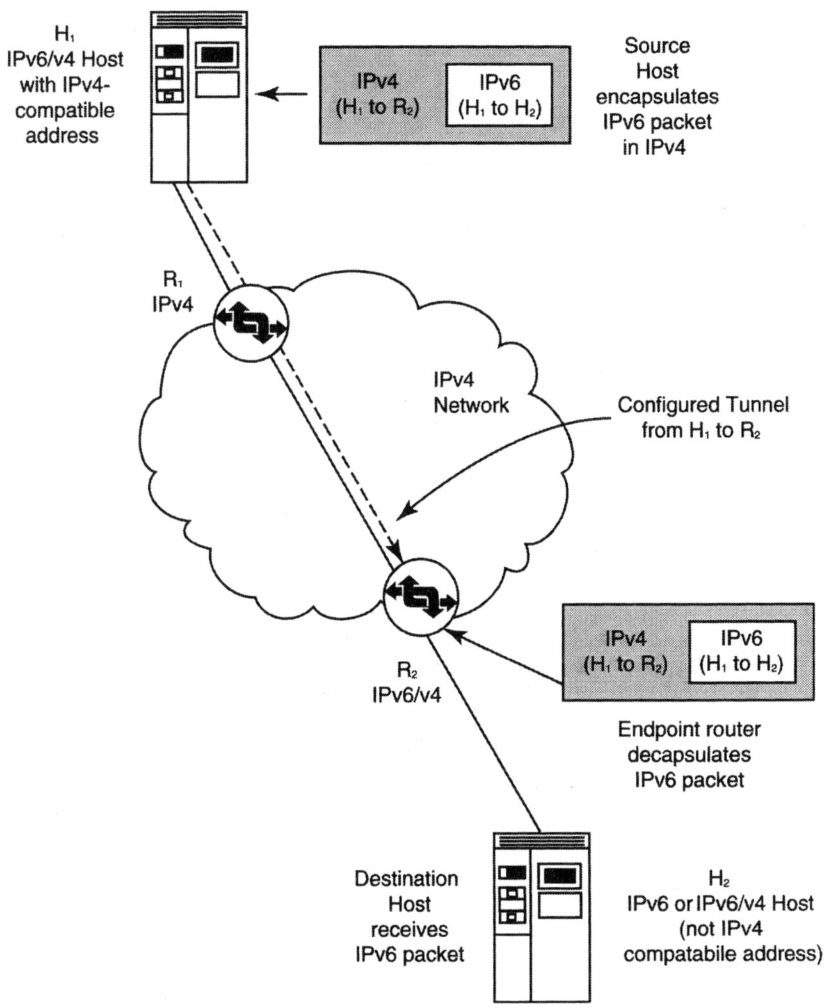

Figure 6-66. Host-to-Router Configured Tunnel.

6.9.2.2.2 Automatic Tunneling

Automatic tunneling is defined in RFC 2893 as IPv6-over-IPv4 tunneling where the IPv4 tunnel endpoint address is determined from the IPv4 address embedded in the IPv4-compatible destination address of the IPv6 packet being tunneled.

From the four tunneling scenarios that were discussed previously, the host-to-host and router-to-host tunnels terminate on a host. Note that the tunnel endpoint address and the IPv6 packet endpoint address both identify the same device (the host). But note that the tunnel endpoint is an IPv4 address, while the host address is an IPv6 address. If the IPv4 and IPv6 addresses can be correlated, the tunneling process is considerably simplified.

This is the purpose of the IPv4-compatible IPv6 address. Recall that the 32-bit IPv4 address occupies the lower 32 bits of the IPv6 address, and the balance of the address is filled with all zeros (Figure 6-67). Since the IPv4 destination address can be easily derived from the IPv4-compatible IPv6 address, this type of tunneling is called automatic tunneling.

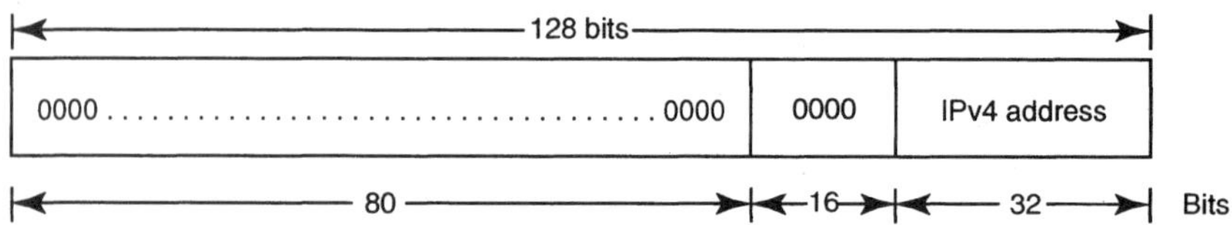

Figure 6-67. IPv4-compatible IPv6 Address.

If two hosts have IPv4-compatible IPv6 addresses, they can communicate across an IPv4 infrastructure using automatic tunneling (Figure 6-68). The source host, H_1, generates an IPv6 packet, and also encapsulates that packet inside an IPv4 datagram. The IPv4 addresses for the source (H_1) and the destination (H_2) can be easily derived from the IPv4-compatible addresses known to H_1 by simply masking off the high-order 96 bits of zero (review Figure 6-67). The address derivation process automatically yields the tunnel entry and exit points, hence the term automatic tunneling. Host-to-host automatic tunneling allows IPv6 to be deployed on the hosts with no changes required to the routing infrastructure.

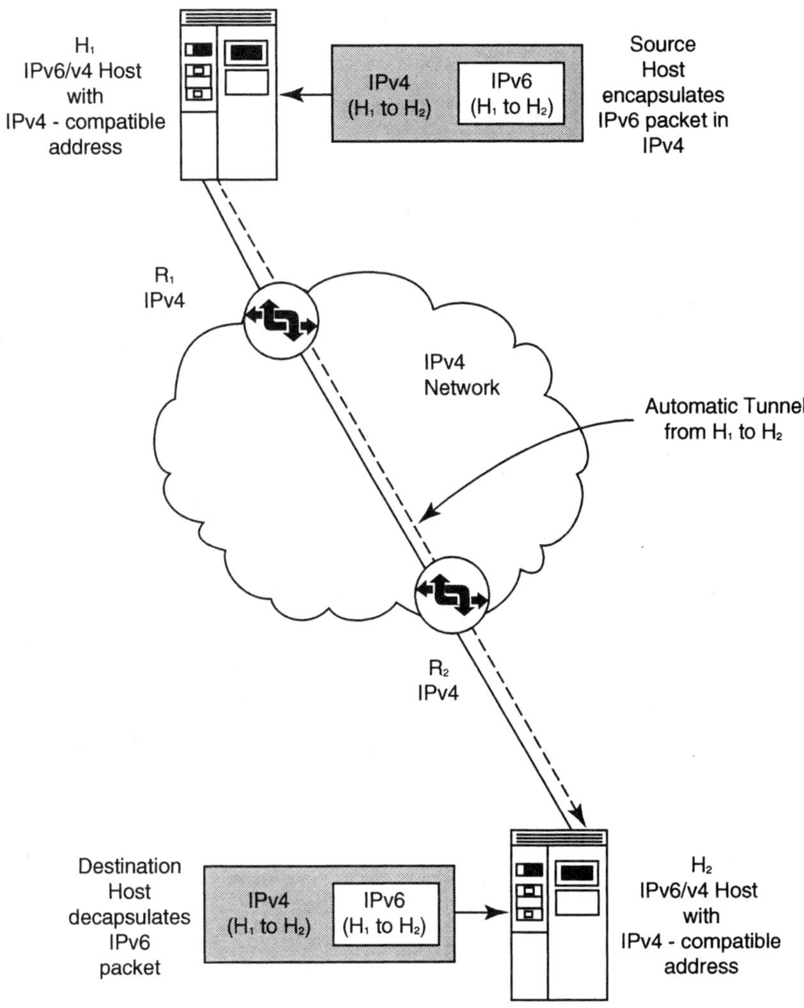

Figure 6-68. Host-to-Host Automatic Tunnel.

In a similar manner, a dual router, upon receiving an IPv6 packet destined for a host with an IPv4-compatible address, can automatically tunnel that packet to its endpoint (Figure 6-69).

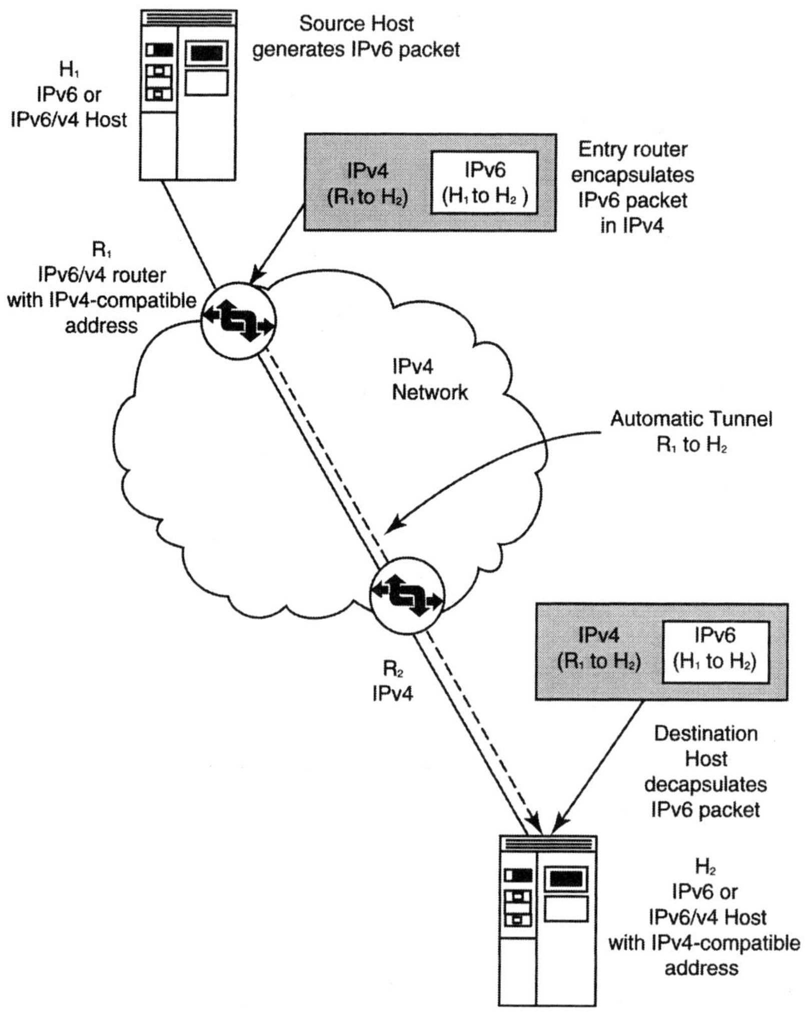

Figure 6-69. Router-to-Host Automatic Tunnel.

6.9.2.2.3 Combined Tunnels

RFC 2185 [6-53] also discusses tunneling, but from the perspective of a tunnel's interaction with neighboring routers. In addition, that document discusses various combinations of automatic tunnels so that communication can be achieved in both directions. (Recall from Figures 6-65, 6-66, 6-68, and 6-69 that our focus was on one-way communication.) RFC 2185 summarizes the various communication paths that are possible between two hosts, A and B (Figure 6-70).

Host A	Host B	Communication Paths
IPv4-compatible addr. no local IPv6 router	IPv4-compatible addr. no local IPv6 router	Host-to-Host tunneling in both directions
IPv4-compatible addr. no local IPv6 router	IPv4-compatible addr. local IPv6 router	A->B: Host-to-Host tunneling B->A: IPv6 forwarding plus Router-to-Host tunnel
IPv4-compatible addr. no local IPv6 router	Incompatible addr. local IPv6 router	A->B: Host-to-Router tunnel plus IPv6 forwarding B->A: IPv6 forwarding plus Router-to-Host tunnel
IPv4-compatible addr. local IPv6 router	IPv4-compatible addr. local IPv6 router	End-to-End native IPv6 in both directions
IPv4-compatible addr. local IPv6 router	Incompatible addr. local IPv6 router	End-to-End native IPv6 in both directions
Incompatible addr. local IPv6 router	Incompatible addr. local IPv6 router	End-to-End native IPv6 in both directions

Figure 6-70. Automatic Tunneling Combinations.

Two key assumptions are made for this figure: when possible, native IPv6 communication is preferred over tunneling, and, if tunneling is needed, host-to-host tunneling (going end-to-end) is preferred over host-to-router tunneling.

Referring to Figure 6-70, if two hosts have IPv4-compatible addresses, and no local IPv6 routers are available, then host-to-host tunneling should be used in both directions (review Figure 6-68). If one of these hosts has a local IPv6 router, then a host-to-host tunnel can be used in one direction (Figure 6-68), and a router-to-host tunnel in the other (Figure 6-69). If one host has an IPv4-compatible address and the other does not, then a host-to-router tunnel may be used in one direction (Figure 6-66), and a router-to-host tunnel in the other direction (Figure 6-69). Anytime that an IPv6 router is available, regardless of the type of IPv6 address used (IPv4-compatible or not), native IPv6 packets are sent in both directions.

The next section will discuss the routing conditions in greater detail.

6.9.3 Transition Routing Terms

A number of routing issues affect IPv4 to IPv6 transitions; these are described in RFC 2185. The following terms, which relate to transition routing architectures, are defined in that document:

- Border router: A router that forwards packets across routing domain boundaries.
- Routing domain: A collection of routers that coordinate routing knowledge using a single routing protocol.
- Routing region (or region): Collection of routers, interconnected by a single Internet protocol (e.g., IPv6), that coordinate their routing knowledge using routing protocols from a single Internet protocol stack. A routing region may be a superset of a routing domain.
- Tunneling: Encapsulation of protocol A within protocol B, such that A treats B as though it were a Data Link layer.
- Reachability information (or reachability): Information describing the set of reachable destinations that can be used for packet forwarding decisions.
- Address prefix: The high-order bits in an address.
- Routing prefix: Address prefix that expresses destinations that have addresses with the matching address prefixes. The routing prefix is used by routers to advertise what systems they are capable of reaching.
- Route leaking: Advertisement of network layer reachability information across routing region boundaries.

Some of the issues also described in RFC 2185 are: the routing for IPv4 datagrams (using IPv4 addresses), the routing of IPv6 packets (using IPv6-native or IPv4-compatible addresses), the operation of configured tunnels, and the operation of automatic tunnels. It is assumed that dual routers may independently support IPv4 and IPv6 (i.e. two instances of RIP or two instances of OSPF, with each instance supporting one version of the protocol). It would be conceivable for a single routing protocol to support both IPv4 and IPv6, although, to date, none of these have been developed. When required, one protocol may inject routing information into another protocol's region to facilitate tunnel configurations or other routing decisions.

6.9.4 Transition Routing Scenarios

RFC 2185 provides several examples that illustrate routing scenarios for packet transmission between two regions: A, which contains IPv6/IPv4 routers, and B, which contains IPv4-only routers (Figure 6-71a). Some of the hosts in each region are dual hosts; others support IPv4 only.

In the first example, H_1, an IPv4-only host, sends a message to H_8, a dual host (Figure 6-71b). Since H_1 only supports IPv4, it transmits an IPv4 datagram. That datagram can traverse Region A (the IPv6/IPv4 routing region) and Region B (the IPv4-only routing region) with no change, using normal IPv4 routing methods.

A return message, from H_8 to H_1, would also be transmitted as an IPv4 datagram and would also use the IPv4 routing infrastructure (Figure 6-71c).

Now assume a message from H_3 to H_8 (Figure 6-71d). Since H_8 is a dual host, but resident in an IPv4-only domain, it should have an IPv4-compatible address assigned. Since both H_3 and H_8 have IPv6 capabilities, H_3 can send an IPv6 packet as far as a boundary router (R_2 or R_4). At that point, it is encapsulated inside an IPv4 packet and sent via a Router-to-Host tunnel to the final destination, H_8.

The return path from H_8 to H_3 may use one of two possible tunnels, depending on the type of address that is assigned to H_3. If H_3 has an IPv4-compatible address, and we continue to assume that H_8 also has an IPv4-compatible address (since it is in an IPv4-only region), then Host-to-Host automatic tunneling may be used (Figure 6-71e).

If H_3 has an IPv6-only address (i.e. not IPv4-compatible), then the Host-to-Host automatic tunnel will not work, since the IPv4 destination address cannot be derived from the IPv6 address. A Host-to-Router configured tunnel from H_8 to a Region A border router (R_2 or R_4) will be used instead (Figure 6-71f). Source Host H_8 will generate an IPv6 packet, encapsulate it in an IPv4 datagram, and send it via a configured tunnel to Region A (R_2 or R_4). One of those dual routers will decapsulate the IPv6 packet, discard the IPv4 header, and send it to its final destination, H_3.

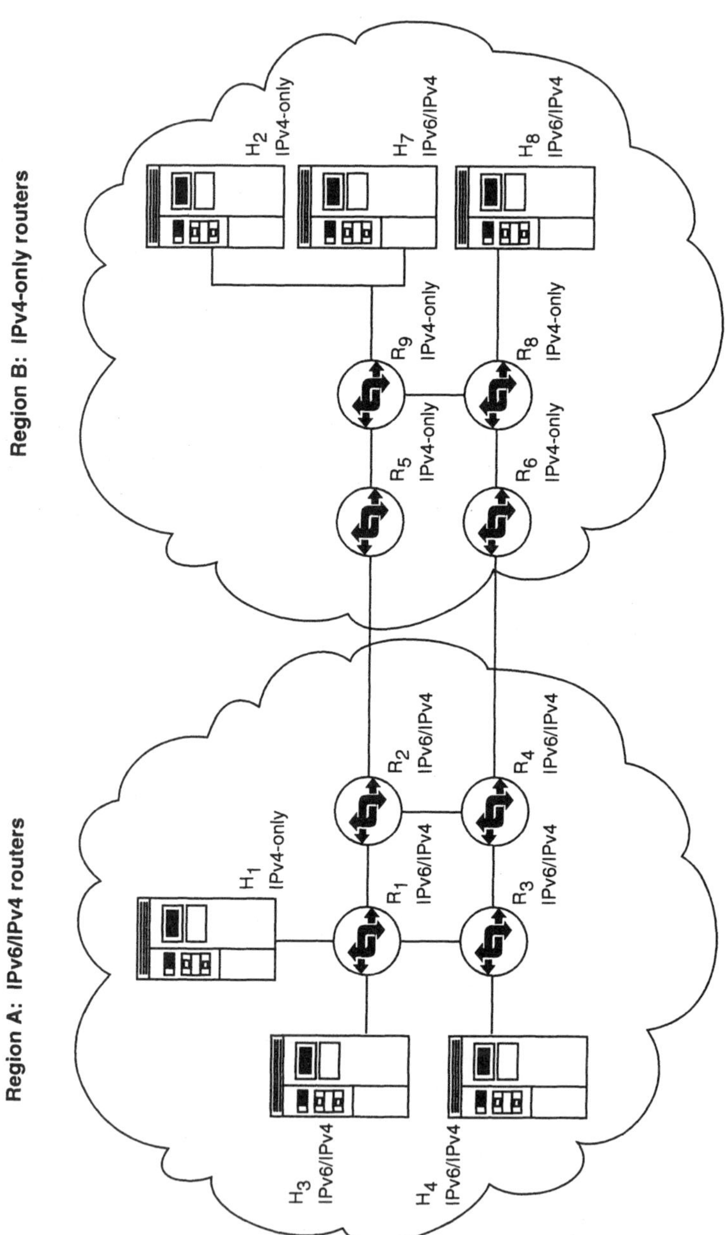

Figure 6-71a. Routing Example.

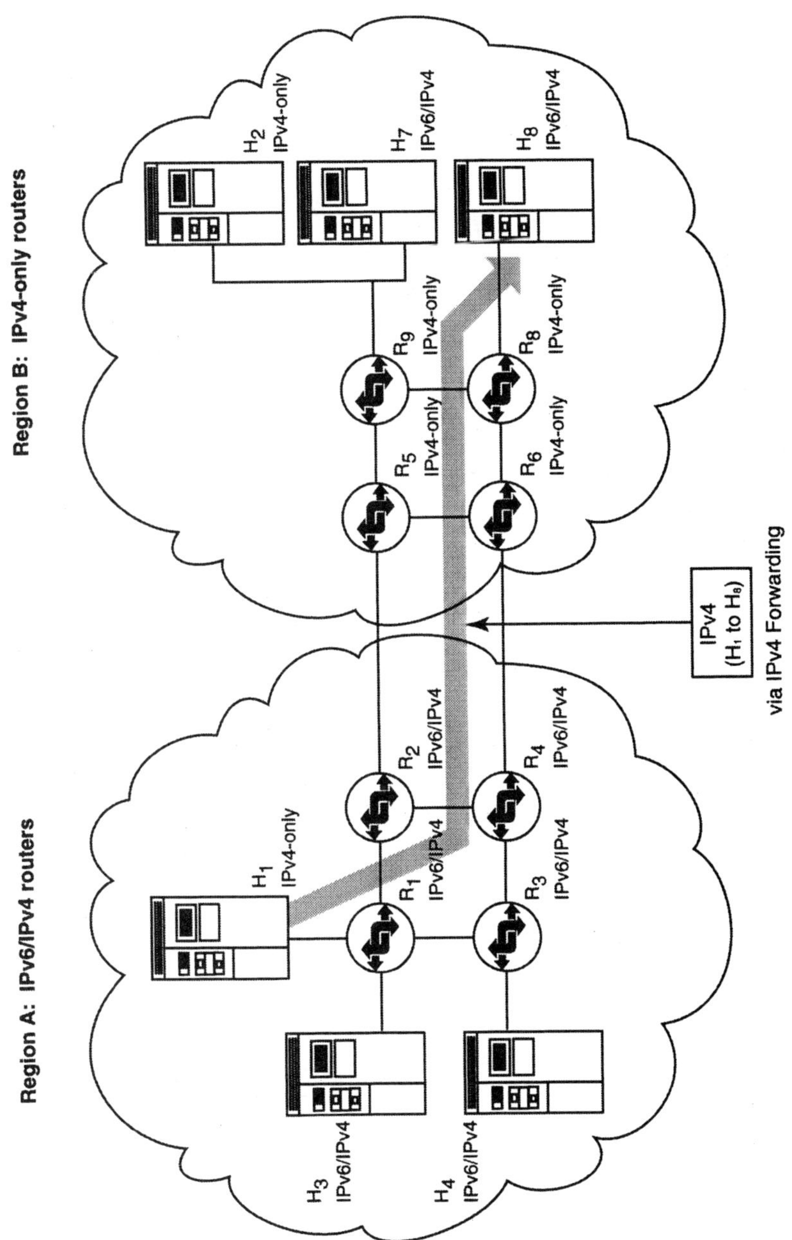

Figure 6-71b. Routing Example: IPv4 Datagram from H_1 to H_8 (via IPv4 Forwarding).

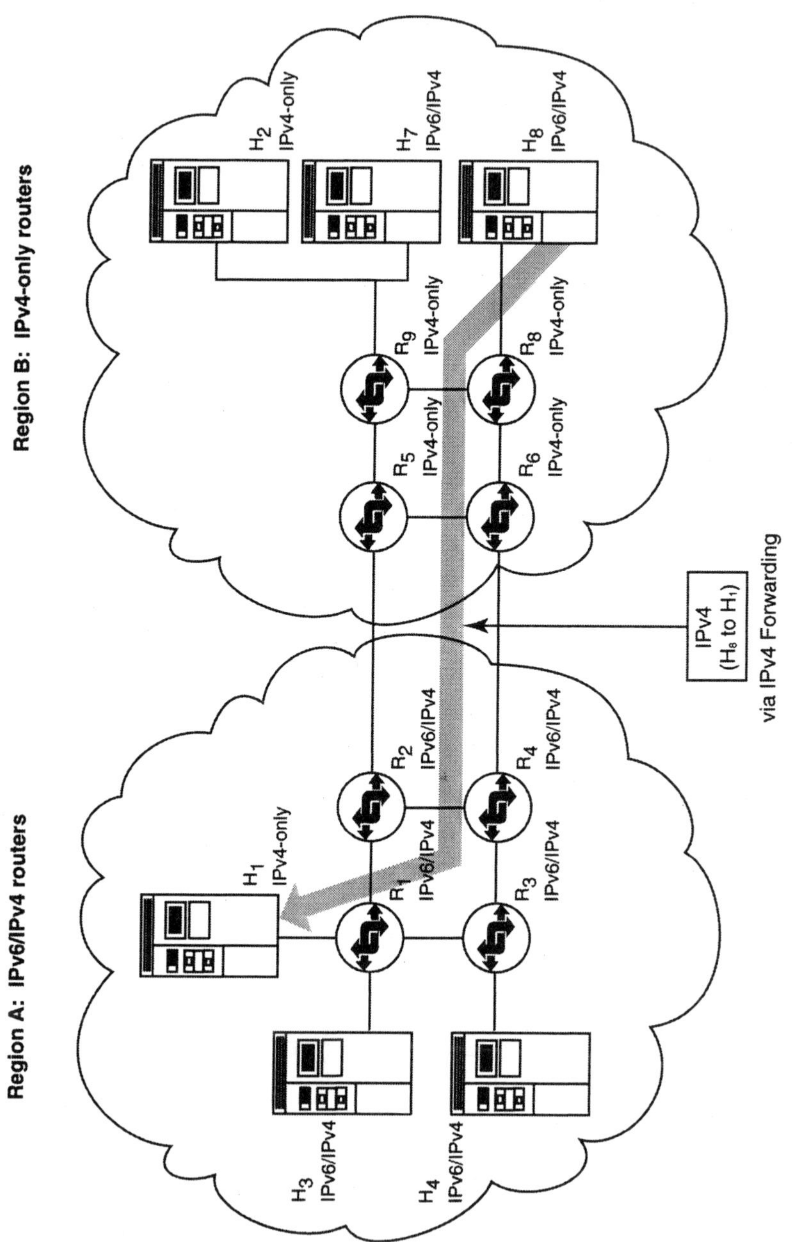

Figure 6-71c. Routing Example: IPv4 Datagram from H_8 to H_1 (via IPv4 Forwarding).

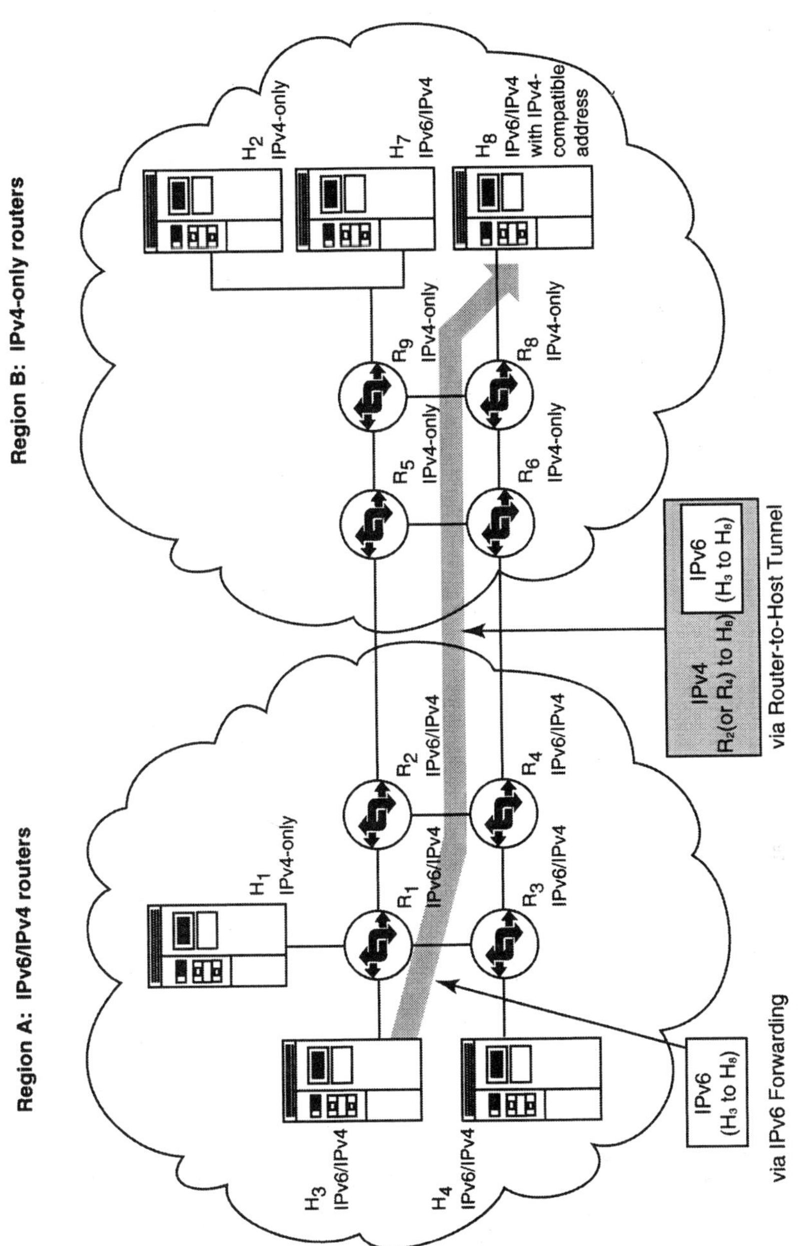

Figure 6-71d. Routing Example: IPv6 Packet from H_3 to H_8 (via Router-to-Host Tunnel).

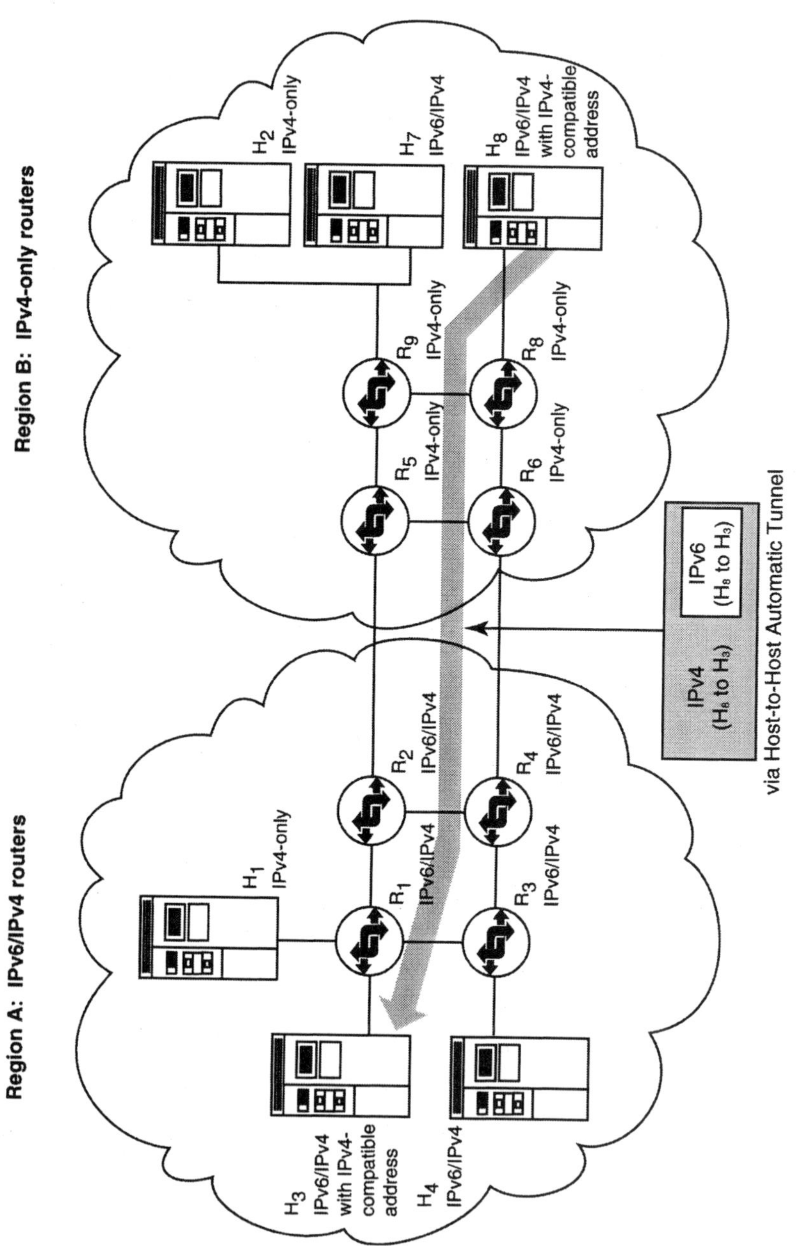

Figure 6-71e. Routing Example: IPv6 Packet from H_8 to H_3 (via Host-to-Host Automatic Tunnel).

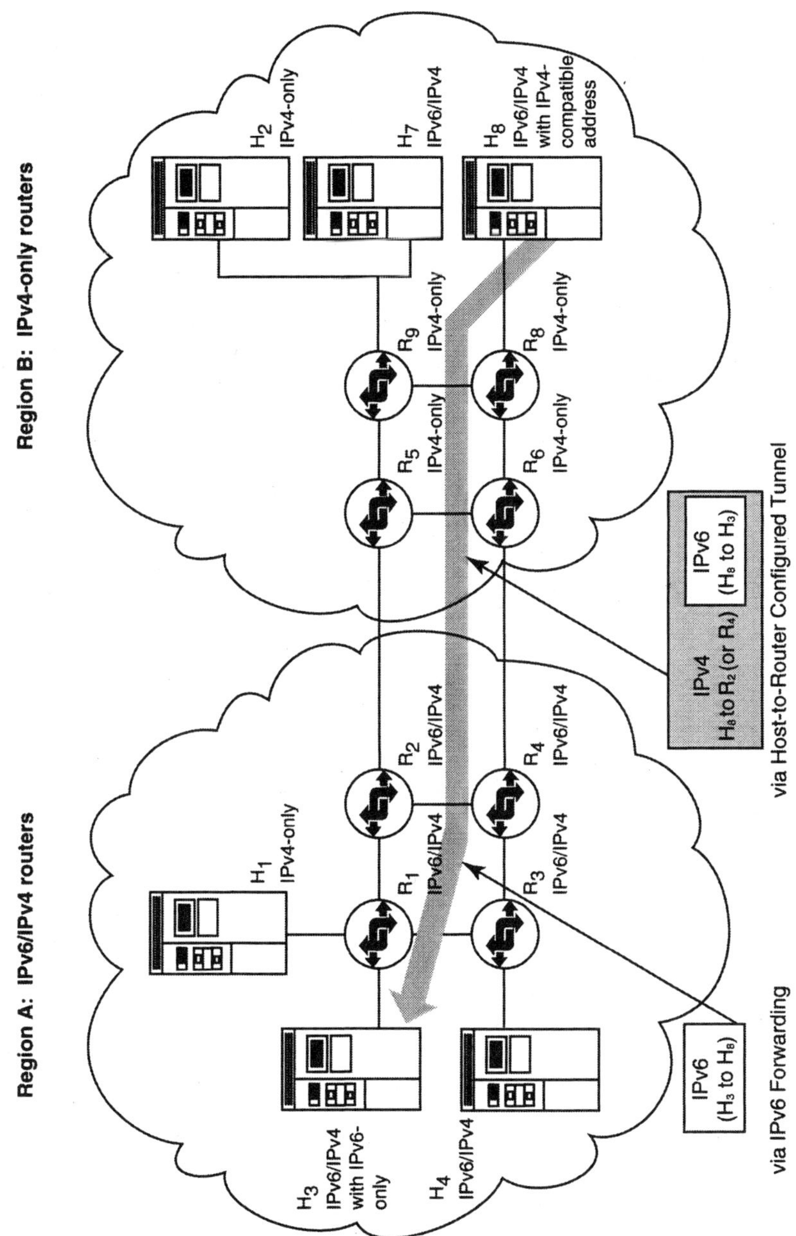

Figure 6-71f. Routing Example: IPv6 Packet from H_8 to H_3 (via Host-to-Router Configured Tunnel).

Many of the IPv4-to-IPv6 transition schemes are first tested on the 6Bone Network, a worldwide testing environment for IPv6 protocol development and research. The 6Bone website [6-55] contains a wealth of information that would benefit anyone considering any IPv4-to-IPv6 upgrade plans.

6.10 Looking Ahead

In the last four chapters, we have considered the technical and operational characteristics of IP and its related protocols. In the next chapter, we will examine these processes further with the aid of case studies and output from a protocol analyzer.

6.11 References

[6-1] Bradner, S., and A. Mankin. “IP: Next Generation (IPng) White Paper Solicitation.” RFC 1550, December 1993.

[6-2] Partridge, C., and F. Kastenholz. “Technical Criteria for Choosing IP: The Next Generation (IPng).” RFC 1726, December 1994.

[6-3] Bradner, S., and A. Mankin. “The Recommendation for the IP Next Generation Protocol.” RFC 1752, January 1995.

[6-4] The IETF IPv6 Working Group maintains a website with current information regarding IPv6 development and documentation activities at http://www.ietf.org/html.charters/ipv6-charter.html.

[6-5] Deering, S., and R. Hinden. “Internet Protocol, Version 6 (IPv6) Specification.” RFC 2460, December 1998.

[6-6] Minutes of the IPng Working Group Meetings are available at: http://playground.sun.com/pub/ipng/html/#MINUTES.

[6-7] Nichols, K., et al. “Definition of the Differentiated Services Field (DS Field) in the IPv4 and IPv6 Headers.” RFC 2474, December 1998.

[6-8] Blake, S., et al. “An Architecture for Differentiated Services.” RFC 2475, December 1998.

[6-9] Ramakrishnan, K., et al. "The Addition of Explicit Congestion Notification (ECN) to IP." RFC 3168, September 2001.

[6-10] Grossman, D. "New Terminology and Clarifications for Diffserv." RFC 3260, April 2002.

[6-11] Partridge, C. "Using the Flow Label Field in IPv6." RFC 1809, June 1995.

[6-12] Hinden, R., and S. Deering. "IP Version 6 Addressing Architecture." RFC 3513, April 2003.

[6-13] Borman, D., et al. "IPv6 Jumbograms." RFC 2675, August 1999.

[6-14] Kent, S., and R. Atkinson. "Security Architecture for the Internet Protocol." RFC 2401, November 1998.

[6-15] Kent, S., and R. Atkinson. "IP Authentication Header." RFC 2402, November 1998.

[6-16] Kent, S., and R. Atkinson. "IP Encapsulating Security Payload." RFC 2406, November 1998.

[6-17] Rekhter, Y., and T. Li. "An Architecture for IPv6 Unicast Address Allocation." RFC 1887, December 1995.

[6-18] Partridge, C., T. Mendez, and W. Milliken. "Host Anycasting Service." RFC 1546, November 1993.

[6-19] Hinden, R., S. Deering, and M. O'Dell. "An IPv6 Aggregatable Global Unicast Address Format." RFC 2374, July 1998.

[6-20] Bound, J., B. Carpenter, D. Harrington, J. Houldsworth, A. Lloyd. "OSI NSAPs and IPv6." RFC 1888, August 1996.

[6-21] Hinden, R., and M. O'Dell. "Proposed TLA and NLA Assignment Rules." RFC 2450, December 1998.

[6-22] Hinden, R., et al. "Initial IPv6 Sub-TLA ID Assignments." RFC 2928, September 2000.

[6-23] Hinden, R., R. Fink, and J. Postel. "IPv6 Testing Address Allocation." RFC 2471, December 1998.

[6-24] Johnson, David B., and Steven E. Deering. "Reserved IPv6 Subnet Anycast Addresses." RFC 2526, March 1999.

[6-25] Hinden, R., and S. Deering. "IPv6 Multicast Address Assignments." RFC 2375, July 1998.

[6-26] Hinden, R., et al. "Format for Literal IPv6 Addresses in URLs." RFC 2732, December 1999.

[6-27] Haberman, B., and D. Thaler. "Unicast-Prefix-based IPv6 Multicast Addresses." RFC 3306, August 2002.

[6-28] Draves, R. "Default Address Selection for Internet Protocol version 6 (IPv6)." RFC 3484, February 2003.

[6-29] Blanchet, M. "A Flexible Method for Managing the Assignment of Bits of an IPv6 Address Block." RFC 3531, April 2003.

[6-30] IEEE, Guidelines for 64-bit Global Identifier (EUI-64™) Registration Authority, http://standards.ieee.org/regauth/oui/tutorials/EUI64.html, April 2003.

[6-31] Conta, A., and S. Deering. "Internet Control Message Protocol (ICMPv6) for the Internet Protocol Version 6 (IPv6) Specification." RFC 2463, December 1998.

[6-32] Cain, B., et al. "Internet Group Management Protocol, Version 3." RFC 3376, October 2002.

[6-33] Deering, S., et al. "Multicast Listener Discovery (MLD) for IPv6." RFC 2710, October 1999.

[6-34] Narten, Thomas, et al. "Neighbor Discovery for IP Version 6 (IPv6)." RFC 2461, December 1998.

[6-35] McCann, J., et al. "Path MTU Discovery for IP version 6." RFC 1981, August 1996.

[6-36] Thomson, Susan, and Thomas Narten. "IPv6 Stateless Address Autoconfiguration." RFC 2462, December 1998.

[6-37] Droms, R., et al. "Dynamic Host Configuration Protocol for IPv6 (DHCPv6)." Work in Progress, November 2002.

[6-38] Crawford, Matt. "Transmission of IPv6 Packets over Ethernet." RFC 2464, December 1998.

[6-39] Haskin, D., and E. Allen. "IP Version 6 over PPP." RFC 2472, December 1998.

[6-40] Conta, A., et al. "Transmission of IPv6 Packets over Frame Relay Networks Specification." RFC 2590, May 1999.

[6-41] Brown, C., and A. Malis. "Multiprotocol Interconnect over Frame Relay." RFC 2427, September 1998.

[6-42] Conta, A. "Extensions to IPv6 Neighbor Discovery for Inverse Discovery Specification." RFC 3122, June 2001.

[6-43] Malkin, G., and R. Minnear. "RIPng for IPv6." RFC 2080, January 1997.

[6-44] Coltun, R., et al. "OSPF for IPv6." RFC 2740, December 1999.

[6-45] Rekhter, Y., and T. Li. "A Border Gateway Protocol 4 (BGP-4)." RFC 1771, March 1995.

[6-46] Bates, Tony, et al. "Multiprotocol Extensions for BGP-4." RFC 2858, June 2000.

[6-47] Marques, P., and F. Dupont. "Use of BGP-4 Multiprotocol Extensions for IPv6 Inter-Domain Routing." RFC 2545, March 1999.

[6-48] Thomson, S., and C. Huitema. "DNS Extensions to Support IP Version 6." RFC 1886, December 1995.

[6-49] Gilligan, R., et al. "Basic Socket Interface Extensions for IPv6." RFC 3493, February 2003.

[6-50] Stevens, W., et al. "Advanced Sockets Application Program Interface (API) for IPv6." RFC 3542, May 2003.

[6-51] Information on current IPv6 host and router implementations is available at the IPv6 Industry website: http://playground.sun.com/pub/ipng/html/ipng-implementations.html.

[6-52] Gilligan, R., and E. Nordmark. "Transition Mechanisms for IPv6 Hosts and Routers." RFC 2893, August 2000.

[6-53] Callon, R., and D. Haskin. "Routing Aspects of IPv6 Transition." RFC 2185, September 1997.

[6-54] The work of the IETF Next Generation Transition (ngtrans) Working Group concluded in February 2003. However, the results of their work are archived at http://www.ietf.org/html.charters/OLD/ngtrans-charter.html.

[6-55] Activities and results for the 6Bone Network, a testbed for IPv6 research and development activities, can be found at http://www.6bone.net.

7

Case Studies in Packet Transport

It has been said that a picture is worth a thousand words. For network managers, that "picture" of the network is best portrayed by using a protocol analyzer to illustrate the communication activities and interactions between the various hosts, routers, and other devices on the network. In this chapter, we will look at case studies that illustrate packet transport functions for both IP and IPv6, which we studied in Chapters 3–6. As with the case studies in Chapter 2, we will use the Sniffer protocol analyzer from Network Associates, Inc., as our window through which to view the clear picture of our network that we are seeking.

7.1 Login to a Remote Host

In our first example, we'll look at the processes necessary to log into a remote host using the TELNET utility (see Trace 7.1a). The network administrator, Paul, resides on an Ethernet segment (Network 132) in one location, and the remote host that he needs to access resides on a different segment (Network 129) in another part of the country (see Figure 7-1). Thus, this internetwork consists of two LANs plus the communication circuits and routers that connect these two locations.

When Paul launches the TELNET application, it invokes the other processes necessary to complete the communication. First, the Domain Name System (DNS) Server must be found. To do so, the workstation broadcasts an ARP Request packet within an Ethernet frame identifying the Protocol Address, PA = [132.XXX.128.1], for the Internet Protocol (IP), but with an unknown Hardware Address (HA = 0), as shown in Frame 1 of Trace 7.1b. The Name Server responds in Frame 2, identifying its HA = 08002006B501H. Paul now knows how to reach the Name Server. (Expert analysts will note a subtle difference between the address representations in Frame 2 of Trace 7.1a (Summary) and Frame 2 of Trace 7.1b (Details). In the Summary trace, the entire hardware address is represented: HA = 08002006B501. In the Detail trace, the Sniffer decodes the Manufacturer ID portion of the Sender's hardware address (080020) and substitutes the manufacturer's name (Sun). The balance of the address (06B501) remains the same. The substituted name (Name Svr) comes from the Sniffer's manually entered database of node names.)

Paul's workstation next locates the address for the remote host fs1.cam.nist.gov that he wishes to log into. The DNS request is transmitted in Frame 3, with a response given in Frame 4 of Trace 7.1c indicating that fs1.cam.nist.gov is located at [129.XXX.80.33]. Recall that Paul is located on Network 132 and the Name Server says that his desired destination is located on another network (Network 129). A router must now spring into action. Paul's workstation broadcasts an ARP Request packet looking for the router (address [132.XXX.1.1] in Frame 5 of Trace 7.1a). The router responds with its hardware address in Frame 6 (HA = 000093E0807BH).

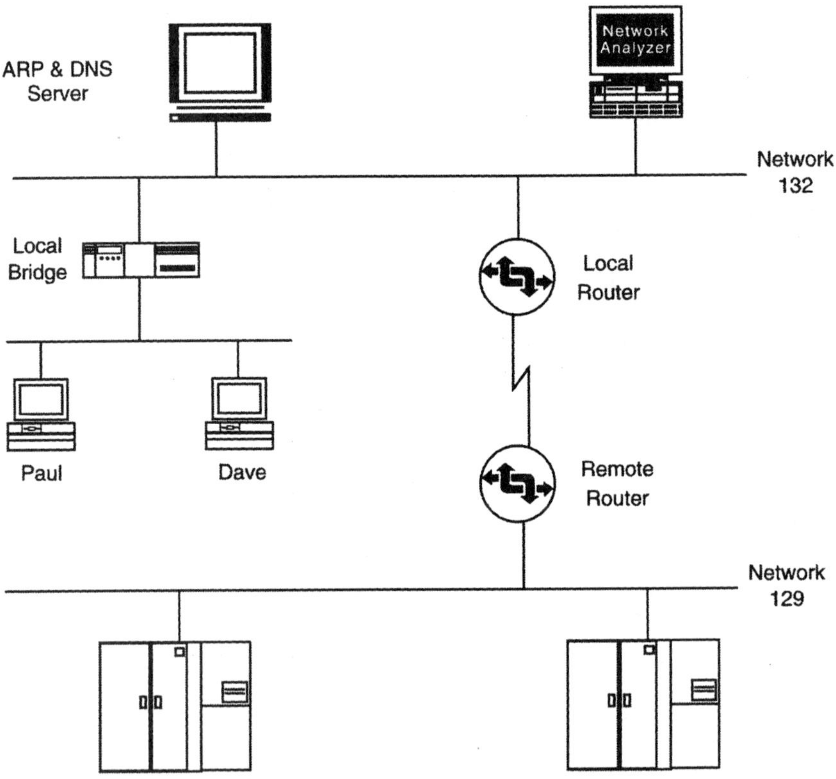

Figure 7-1. Login to Remote Host.

Paul now has all the information he needs. In Frame 7 he initiates a TCP connection (via a three-way handshake, which we studied in Chapter 5) with the remote host. Note that the destination port is 23 (D = 23 in Frame 7 in the summary Trace 7.1a), indicating the well-known port address for the TELNET application. The details of the Connection Establishment message verify that Paul knows where to find his desired host (see Trace 7.1d). In the Data Link Control (DLC, or frame) header, Paul addresses this message to the Proteon router (address ProteonE0807B). We know that this frame is transmitted on a DIX Ethernet (not an IEEE 802.3) LAN because of the EtherType field identified IP = 0800 H. (Had it been an IEEE 802.3 frame, a Length field would have been shown instead of the EtherType field.) The IP header identifies Paul's workstation as the source of the datagram [132.XXX.129.15], with the destination address of the remote host on a different network [129.XXX.80.33]. The TCP header specifies that the Destination Port = 23 (TELNET). The TCP flags indicate that Paul wants to establish a connection, since the Synchronize flag is set (SYN = 1). Paul also specifies that he can accept a maximum segment size of 512 octets.

In the subsequent frames (8 through 20 of Trace 7.1a), Paul's workstation and the remote host negotiate various TELNET parameters, such as the Terminal Type (Frames 10 through 12), Echo (Frames 15 and 16), and Login (Frames 17 through 20). Once Paul is logged in, he can complete his business with the remote host.

To summarize, Paul's workstation undertakes the following steps:

1. Identifies DNS Name Server (ARP).
2. Identifies remote host address (DNS).
3. Identifies required router (ARP).
4. Establishes connection with remote host (TCP).
5. Negotiates parameters with remote host (TELNET).
6. Initiates remote terminal session (TELNET).

In the next case study, we will examine how IP fragments long messages.

Trace 7.1a. Login to Remote Host Summary

Sniffer Network Analyzer data 21-Jun at 13:37:00, LOGIN.ENC, Pg 1

SUMMARY	Delta T	Destination	Source	Summary
M 1		Broadcast	Paul	ARP C PA=[132.XXX.128.1] PRO=IP
2	0.0006	Paul	Name Svr	ARP R PA=[132.XXX.128.1] HA=08002006B501 PRO=IP
3	0.0016	Name Svr	Paul	DNS C ID=3 OP=QUERY NAME=fs1.cam.nist.gov
4	0.0026	Paul	Name Svr	DNS R ID=3 STAT=OK NAME=fs1.cam.nist.gov
5	0.0088	Broadcast	Paul	ARP C PA=[132.XXX.1.1] PRO=IP
6	0.0010	Paul	Router	ARP R PA=[132.XXX.1.1] HA=000093E0807B PRO=IP
7	0.0016	Router	Paul	TCP D=23 S=3133 SYN SEQ=9947492 LEN=0 WIN=1024
8	0.1081	Paul	Router	TCP D=3133 S=23 SYN ACK=9947493 SEQ=80448000 LEN=0 WIN=4096
9	0.0032	Router	Paul	TCP D=23 S=3133 ACK=80448001 WIN=1024
10	2.1818	Paul	Router	Telnet R PORT=3133 IAC Do Terminal type
11	0.0029	Router	Paul	TCP D=23 S=3133 ACK=80448004 WIN=1021
12	0.0027	Router	Paul	Telnet C PORT=3133 IAC Will Terminal type
13	0.1134	Paul	Router	Telnet R PORT=3133 IAC SB...
14	0.0070	Router	Paul	Telnet C PORT=3133 IA SB ...
15	0.1912	Paul	Router	Telnet R PORT=3133 IAC Will Echo
16	0.0053	Router	Paul	Telnet C PORT=3133 IAC Do Echo
17	0.2566	Paul	Router	Telnet R PORT=3133 login:
18	0.0033	Router	Paul	Telnet C PORT=3133 IAC Do Suppress go-ahead
19	0.1113	Paul	Router	Telnet R PORT=3133 IAC Don't Echo

```
  20    0.2591   Router     Paul        TCP D=23 S=3133
                                          ACK=80448057 WIN=968
```

Trace 7.1b. Login to Remote Host ARP/RARP Details
Sniffer Network Analyzer data 21-Jun at 13:37:00, file LOGIN.ENC, Pg 1

```
--------------- Frame 1 ----------------
DLC:     — DLC Header —
DLC:
DLC:     Frame 1 arrived at 13:37:06.3280; frame size is 60 (003C hex) bytes.
DLC:     Destination = BROADCAST FFFFFFFFFFFF, Broadcast
DLC:     Source = Station 3Com 1AB9BE, Paul
DLC:     Ethertype = 0806 (ARP)
DLC:
ARP:     — ARP/RARP Frame —
ARP:
ARP:     Hardware type = 1 (10Mb Ethernet)
ARP:     Protocol type = 0800 (IP)
ARP:     Length of hardware address = 6 bytes
ARP:     Length of protocol address = 4 bytes
ARP:     Opcode 1 (ARP request)
ARP:     Sender's hardware address = 3Com 1AB9BE, Paul
ARP:     Sender's protocol address = [132.XXX.129.15]
ARP:     Target hardware address = 000000000000, 000000000000
ARP:     Target protocol address = [132.XXX.128.1]
ARP:

--------------- Frame 2 ----------------
DLC:     — DLC Header —
DLC:
DLC:     Frame 2 arrived at 13:37:06.3287; frame size is 60 (003C hex) bytes.
DLC:     Destination = Station 3Com 1AB9BE, Paul
DLC:     Source      = Station Sun 06B501, Name Svr
DLC:     Ethertype  = 0806 (ARP)
DLC:
ARP:     — ARP/RARP Frame —
ARP:
ARP:     Hardware type = 1 (10Mb Ethernet)
ARP:     Protocol type = 0800 (IP)
ARP:     Length of hardware address = 6 bytes
ARP:     Length of protocol address = 4 bytes
ARP:     Opcode 2 (ARP reply)
ARP:     Sender's hardware address = Sun 06B501, Name Svr
ARP:     Sender's protocol address = [132.XXX.128.1]
```

```
ARP:        Target hardware address = 3Com 1AB9BE, Paul
ARP:        Target protocol address = [132.XXX.129.15]
ARP:
```

Trace 7.1c. Login to Remote Host DNS Details
Sniffer Network Analyzer data 21-Jun at 13:37:00, LOGIN.ENC, Pg 1

```
- - - - - - - - - - - - - - - - Frame 3 - - - - - - - - - - - - - - - - -
DLC:        — DLC Header —
DLC:
DLC:        Frame 3 arrived at 13:37:06.3304; frame size is 76 (004C hex) bytes.
DLC:        Destination = Station Sun 06B501, Name Svr
DLC:        Source      = Station 3Com 1AB9BE, Paul
DLC:        Ethertype   = 0800 (IP)
DLC:
IP:         — IP Header —
IP:
IP:         Version = 4, header length = 20 bytes
IP:         Type of service = 20
IP:                   001. .... = priority
IP:                   ...0 .... = normal delay
IP:                   .... 0... = normal throughput
IP:                   .... .0.. = normal reliability
IP:         Total length = 62 bytes
IP:         Identification = 30
IP:         Flags = 0X
IP:         .0.. .... = may fragment
IP:         ..0. .... = last fragment
IP:         Fragment offset = 0 bytes
IP:         Time to live = 64 seconds/hops
IP:         Protocol = 17 (UDP)
IP:         Header checksum = 701A (correct)
IP:         Source address = [132.XXX.129.15]
IP:         Destination address = [132.XXX.128.1]
IP:         No options
IP:
UDP:        — UDP Header —
UDP:
UDP:        Source port = 1116 (Domain)
UDP:        Destination port = 53
UDP:        Length = 42
UDP:        Checksum = 54EF (correct)
UDP:
DNS:        — Internet Domain Name Service Header —
```

```
DNS:
DNS:        ID = 3
DNS:        Flags = 01
DNS:        0... .... = Command
DNS:        .000 0... = Query
DNS:        .... ..0. = Not truncated
DNS:        .... ...1 = Recursion desired
DNS:        Flags = 0X
DNS:        ...0 .... = Unicast packet
DNS:        Question count = 1, Answer count = 0
DNS:        Authority count = 0, Additional record count = 0
DNS:
DNS:        Question section:
DNS:                Name = fs1.cam.nist.gov
DNS:                Type = Host address (A,1)
DNS:                Class = Internet (IN,1)
DNS:
DNS:        [Normal end of "Internet Domain Name Service Header".]
DNS:

- - - - - - - - - - - - - - - - Frame 4 - - - - - - - - - - - - - - - - -
DLC:        — DLC Header —
DLC:
DLC:        Frame 4 arrived at 13:37:06.3330; frame size is 92 (005C hex) bytes.
DLC:        Destination = Station 3Com  1AB9BE, Paul
DLC:        Source      = Station Sun   06B501, Name Svr
DLC:        Ethertype  = 0800 (IP)
DLC:
IP:         — IP Header —
IP:
IP:         Version = 4, header length = 20 bytes
IP:         Type of service = 00
IP:                 000. .... = routine
IP:                 ...0 .... = normal delay
IP:                 .... 0... = normal throughput
IP:                 .... .0.. = normal reliability
IP:         Total length = 78 bytes
IP:         Identification = 60111
IP:         Flags = 0X
IP:         .0.. .... = may fragment
IP:         ..0. .... = last fragment
IP:         Fragment offset = 0 bytes
IP:         Time to live = 30 seconds/hops
IP:         Protocol = 17 (UDP)
```

```
IP:         Header checksum = A778 (correct)
IP:         Source address = [132.XXX.128.1]
IP:         Destination address = [132.XXX.129.15]
IP:         No options
IP:
UDP:        — UDP Header —
UDP:
UDP:        Source port = 53 (Domain)
UDP:        Destination port = 1116
UDP:        Length = 58
UDP:        No checksum
UDP:
DNS:        — Internet Domain Name Service Header —
DNS:
DNS:        ID = 3
DNS:        Flags = 85
DNS:        1... .... = Response
DNS:        .... .1.. = Authoritative answer
DNS:        .000 0... = Query
DNS:        .... ..0. = Not truncated
DNS:        Flags = 8X
DNS:        ...0 .... = Unicast packet
DNS:        1... .... = Recursion available
DNS:        Response code = OK (0)
DNS:        Question count = 1, Answer count = 1
DNS:        Authority count = 0, Additional record count = 0
DNS:
DNS:        Question section:
DNS:                Name = fs1.cam.nist.gov
DNS:                Type = Host address (A,1)
DNS:                Class = Internet (IN,1)
DNS:        Answer section:
DNS:                Name = fs1.cam.nist.gov
DNS:                Type = Host address (A,1)
DNS:                Class = Internet (IN,1)
DNS:                Time-to-live = 2589119 (seconds)
DNS:                Address = [129.XXX.80.33]
DNS:
DNS:        [Normal end of "Internet Domain Name Service Header".]
DNS:
```

Trace 7.1d. Login to Remote Host TCP Details

Sniffer Network Analyzer data 21-Jun at 13:37:00, LOGIN.ENC, Pg 1

```
- - - - - - - - - - - - - - - - Frame 7 - - - - - - - - - - - - - - - - -
DLC:      — DLC Header —
DLC:
DLC:      Frame 7 arrived at 13:37:06.3445; frame size is 60 (003C hex) bytes.
DLC:      Destination = Station PrteonE0807B, Router
DLC:      Source      = Station 3Com  1AB9BE, Paul
DLC:      Ethertype   = 0800 (IP)
DLC:
IP:       — IP Header —
IP:
IP:       Version = 4, header length = 20 bytes
IP:       Type of service = 00
IP:              000. .... = routine
IP:              ...0 .... = normal delay
IP:              .... 0... = normal throughput
IP:              .... .0.. = normal reliability
IP:       Total length = 44 bytes
IP:       Identification = 31
IP:       Flags = 0X
IP:       .0.. .... = may fragment
IP:       ..0. .... = last fragment
IP:       Fragment offset = 0 bytes
IP:       Time to live = 64 seconds/hops
IP:       Protocol = 6 (TCP)
IP:       Header checksum = A3D3 (correct)
IP:       Source address = [132.XXX.129.15]
IP:       Destination address = [129.XXX.80.33]
IP:       No options
IP:
TCP:      — TCP Header —
TCP:
TCP:      Source port = 3133
TCP:      Destination port = 23 (Telnet)
TCP:      Initial sequence number = 9947492
TCP:      Data offset = 24 bytes
TCP:      Flags = 02
TCP:      ..0. .... = (No urgent pointer)
TCP:      ...0 .... = (No acknowledgment)
TCP:      .... 0... = (No push)
TCP:      .... .0.. = (No reset)
TCP:      .... ..1. = SYN
```

```
TCP:        .... ...0 = (No FIN)
TCP:        Window = 1024
TCP:        Checksum = EAAF (correct)
TCP:
TCP:        Options follow
TCP:        Maximum segment size = 512
TCP:
```

7.2 Fragmenting Long Messages

In this case study, we will investigate how a host fragments a long message into multiple IP datagrams for transmission on a TCP/IP-based internetwork. In this example, the source is a station (DEC 029487) that wishes to send a large SUN RPC file to another station (Intrln 00027C0) on the same LAN. Three frames are needed to transmit the message (see Trace 7.2 and Figure 7-2). The message requires a User Datagram Protocol (UDP) header to identify the source and destination ports at the communicating hosts.

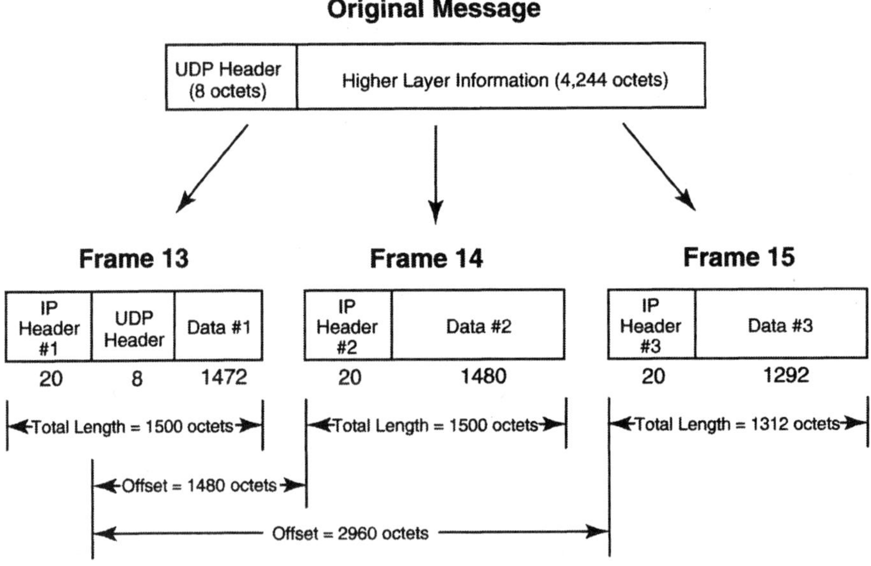

Figure 7-2. IPv4 Fragments.

The message contains 4,244 octets of data plus an 8-octet UDP header, for a total of 4,252 octets. The message is divided into three frames, with 1,472, 1,480, and 1,292 octets of data each (Frames 13, 14, and 15, respectively). Frame 13 has a total length of 1,514 octets (matching the maximum Ethernet frame size, including Source Address, Destination Address, Type, and Information fields) and indicates that more fragments, associated with message ID = 59738, are on the way. The next higher layer protocol within the first fragment is UDP (Protocol = 17, UDP). The Source and Destination addresses [XXX.YYY.0.33] and [XXX.YYY.0.10] identify the network. The UDP header indicates that the total message length is 4,252 octets. This includes the UDP header (8 octets) plus the data (4,244 octets). Thus, the 1,500 octets of Frame 13 consist of the IP header (20 octets), the UDP header (8 octets), and the first data fragment (1,472 octets).

The second frame also contains 1,514 octets of information (14 octets of Ethernet frame header plus 1,500 octets of data). The IP header indicates that this fragment is 1,500 octets in length and that more fragments associated with message ID = 59738 are coming. The fragment offset is 1,480, indicating that the data within Frame 14 starts 1,480 octets after the beginning of the original datagram. Note that the second IP header comprises 20 octets of Frame 14's data, but it does not count in the offset calculation because the IP header is not part of the original message.

The third fragment (Frame 15) contains 1,312 octets of information, consisting of a 20-octet IP header and 1,292 octets of data. This third IP header uses the same message ID = 59738 but indicates that no more fragments are coming (the More Fragments bit = 0). This fragment fits 2,960 octets after the beginning of the original message.

Now let's see if the IP module in the Source host was working properly. The UDP header in Frame 13 says that the total length of this message is 4,252 octets. This is divided into the UDP header (8 octets), Data Fragment 1 (1,472 octets), Data Fragment 2 (1,480 octets), and Data Fragment 3 (1,292 octets). Each fragment requires a 20-octet IP header, which does not count in the total. Each frame requires 14 octets of Ethernet header, which also does not count in the total. Reviewing Figure 7-2, we can see that the fragment offsets are also correct: 1,480 (8 + 1,472) for Fragment 2, and 2,960 (8 + 1,472 + 1,480) for Fragment 3. We can therefore conclude that the IP module within station DEC 029487 was functioning properly.

Trace 7.2. IP Fragments

Sniffer Network Analyzer data 10-Dec at 11:20:38, file TCPIP.ENC, Pg 1

```
- - - - - - - - - - - - - - - - Frame 13 - - - - - - - - - - - - - - - - -
SUMMARY  Delta T  Destination   Source      Summary
13       0.0152   Intrln0027C0  DEC029487   DLC Ethertype=0800,
                                              size=1514 bytes
                                            IP D=[XXX.YYY.0.10]
                                              S=[XXX.YYY.0.33]
                                              LEN=1480 ID=59738
                                            UDP D=2049 S=1026
                                              LEN=4252
DLC:     — DLC Header —
DLC:
DLC:     Frame 13 arrived at 11:20:35.9514; frame size is 1514 (05EA hex) bytes.
DLC:     Destination = Station Intrln0027C0
DLC:     Source      = Station DEC 029487
DLC:     Ethertype   = 0800 (IP)
DLC:
IP:      — IP Header —
IP:
IP:      Version = 4, header length = 20 bytes
IP:      Type of service = 00
IP:               000. .... = routine
IP:               ...0 .... = normal delay
IP:               .... 0... = normal throughput
IP:               .... .0.. = normal reliability
IP:      Total length = 1500 bytes
IP:      Identification = 59738
IP:      Flags = 2X
IP:      .0.. .... = may fragment
IP:      ..1. .... = more fragments
IP:      Fragment offset = 0 bytes
IP:      Time to live = 255 seconds/hops
IP:      Protocol = 17 (UDP)
IP:      Header checksum = 6421 (correct)
IP:      Source address = [XXX.YYY.0.33]
IP:      Destination address = [XXX.YYY.0.10]
IP:      No options
IP:
UDP:     — UDP Header —
UDP:
UDP:     Source port = 1026
UDP:     Destination port = 2049 (Sun RPC)
```

```
UDP:        Length = 4252 (not all data contained in this fragment)
UDP:        No checksum
UDP:
UDP:        [1472 byte(s) of data]
UDP:

- - - - - - - - - - - - - - - - Frame 14 - - - - - - - - - - - - - - - -
SUMMARY  Delta T  Destination   Source     Summary
14       0.0022   Intrln0027C0  DEC029487  DLC Ethertype=0800,
                                             size=1514 bytes
                                             IP D=[XXX.YYY.0.10]
                                             S=[XXX.YYY.0.33]
                                             LEN=1480 ID=59738
                                             UDP continuation ID=59738
DLC:        — DLC Header —
DLC:
DLC:        Frame 14 arrived at 11:20:35.9536; frame size is 1514 (05EA hex) bytes.
DLC:        Destination = Station Intrln0027C0
DLC:        Source      = Station DEC 029487
DLC:        Ethertype   = 0800 (IP)
DLC:
IP:         — IP Header —
IP:
IP:         Version = 4, header length = 20 bytes
IP:         Type of service = 00
IP:                 000. .... = routine
IP:                 ...0 .... = normal delay
IP:                 .... 0... = normal throughput
IP:                 .... .0.. = normal reliability
IP:         Total length = 1500 bytes
IP:         Identification = 59738
IP:         Flags = 2X
IP:         .0.. .... = may fragment
IP:         ..1. .... = more fragments
IP:         Fragment offset = 1480 bytes
IP:         Time to live = 255 seconds/hops
IP:         Protocol = 17 (UDP)
IP:         Header checksum = 6368 (correct)
IP:         Source address = [XXX.YYY.0.33]
IP:         Destination address = [XXX.YYY.0.10]
IP:         No options
IP:
UDP:        [1480 byte(s) of data, continuation of IP ident=59738]
```

```
- - - - - - - - - - - - - - - - Frame 15 - - - - - - - - - - - - - - - - -
SUMMARY  Delta T  Destination  Source     Summary
15       0.0019   Intrln0027C0 DEC029487  DLC Ethertype=0800,
                                            size=1326 bytes
                                            IP D=[XXX.YYY.0.10]
                                            S=[XXX.YYY.0.33]
                                            LEN=1292 ID=59738
                                            UDP continuation ID=59738
DLC:     — DLC Header —
DLC:
DLC:     Frame 15 arrived at 11:20:35.9556; frame size is 1326 (052E hex) bytes.
DLC:     Destination = Station Intrln0027C0
DLC:     Source      = Station DEC 029487
DLC:     Ethertype   = 0800 (IP)
DLC:
IP:      — IP Header —
IP:
IP:      Version = 4, header length = 20 bytes
IP:      Type of service = 00
IP:                000. .... = routine
IP:                ...0 .... = normal delay
IP:                .... 0... = normal throughput
IP:                .... .0.. = normal reliability
IP:      Total length = 1312 bytes
IP:      Identification = 59738
IP:      Flags = 0X
IP:      .0.. .... = may fragment
IP:      ..0. .... = last fragment
IP:      Fragment offset = 2960 bytes
IP:      Time to live = 255 seconds/hops
IP:      Protocol = 17 (UDP)
IP:      Header checksum = 836B (correct)
IP:      Source address = [XXX.YYY.0.33]
IP:      Destination address = [XXX.YYY.0.10]
IP:      No options
IP:
UDP:     [1292 byte(s) of data, continuation of IP ident=59738]
```

7.3 Duplicate IP Addresses

In Chapter 3, we looked at the differences between the Physical (Hardware) address and the Logical (Internet Protocol) address on any internet node. Recall that a ROM on the network interface card (e.g. Ethernet) typically contains the Physical address, while the Logical address is assigned by the network administrator manually or by an automated process such as DHCP (review Section 3.4.1). Let's see what happens when human error affects address assignments.

In this scenario, two engineers, Wayne and Benoit, wish to establish TELNET sessions with a router. The router's TELNET capabilities allow administrators to access its configuration files for network management. Wayne establishes his session first (see Trace 7.3a). The TCP connection is established in Frames 1 through 3, and the TELNET session initiates beginning in Frame 5. Wayne's session appears to be proceeding normally until Benoit starts to transmit in Frame 43. Benoit sends an ARP broadcast looking for the same router (Frame 43), and the router responds in Frame 44. Benoit then establishes his TCP connection in Frames 45 through 48, and, like Wayne, initiates a TELNET session beginning in Frame 49. Benoit doesn't realize, however, that his presence on the internetwork has caused Wayne's connection to fail. Let's see why.

Details of Wayne's TCP connection message (Frame 1) are shown in Trace 7.3b. Note that Wayne is communicating to the router using IP source address [131.195.116.250] on the same Class B network. Wayne is accessing the TELNET port on the router, and is using Sequence Number = 265153482. The router acknowledges the use of this sequence number in its response, shown in Frame 2.

In Trace 7.3c, Benoit's TCP connection message looks similar (Frame 45). Benoit claims that his Source Address = [131.195.116.250] (the same as Wayne's) and that the Destination Address = [131.195.116.42]. The TCP header also identifies the same Destination Port (23, for TELNET), but uses a different Sequence Number (73138176).

We can now see why Wayne's TELNET connection failed. When Benoit established a connection with the router using the same IP source address as Wayne's, confusion resulted. The router was examining the IP source address, not the Hardware (Data Link Layer) address. As a result, it was unable to differentiate between the duplicate IP addresses.

We traced the problem to a duplicate entry on the network manager's address database. Unknowingly, he had given both Wayne and Benoit the same IP address for their workstations. After discovering this mistake, Wayne changed his workstation configuration file to incorporate a unique IP address and no further problems occurred.

Trace 7.3a. Duplicate IP Address Summary
Sniffer Network Analyzer data 11-Oct at 10:49:04, IPDUPLIC.ENC, Pg 1

SUMMARY		Delta T	Destination	Source	Summary
M	1		Router	Wayne	TCP D=23 S=2588 SYN SEQ=265153482 LEN=0 WIN=1024
	2	0.0014	Wayne	Router	TCP D=2588 S=23 SYN ACK=265153483 SEQ=331344504 LEN=0 WIN=0
	3	0.0016	Router	Wayne	TCP D=23 S=2588 ACK=331344505 WIN=1024
	4	0.0019	Wayne	Router	TCP D=2588 S=23 ACK=265153483 WIN=2144
	5	0.0048	Wayne	Router	Telnet R PORT=2588 IAC Will Echo
	6	0.0304	Wayne	Router	Telnet R PORT=2588 <0D><0A>
	7	0.0896	Router	Wayne	Telnet C PORT=2588 IAC Do Echo
	8	0.3010	Wayne	Router	TCP D=2588 S=23 ACK=265153486 WIN=2141
	9	0.0312	Router	Wayne	Telnet C PORT=2588 IAC Do Suppress go-ahead
	10	0.3005	Wayne	Router	TCP D=-2588 S=23 ACK=265153489 WIN=2138
	11	0.4320	Router	Wayne	Telnet C PORT=2588 c
	12	0.3000	Wayne	Router	TCP D=2588 S=23 ACK=265153490 WIN=2137
	13	0.0016	Router	Wayne	Telnet C PORT=2588 d ACK=265153491 WIN=2136
	14	0.2984	Wayne	Router	TCP D=2588 S=23
	15	0.0016	Router	Wayne	Telnet C PORT=2588 2
	16	0.2985	Wayne	Router	TCP D=2588 S=23 ACK=265153492 WIN=2135
	17	0.0016	Router	Wayne	Telnet C PORT=2588 <0D><0A>
	18	0.0024	Wayne	Router	Telnet R PORT=2588

				<0D><0A>CD_BAS1_2>
19	0.1206	Router	Wayne	TCP D=23 S=2588 ACK=331344857 WIN=1012
20	0.7757	Router	Wayne	Telnet C PORT=2588 s
21	0.0025	Wayne	Router	Telnet R PORT=2588 s
22	0.1552	Router	Wayne	TCP D=23 S=2588 ACK=331344858 WIN=1023
23	0.1939	Router	Wayne	Telnet C PORT=2588 h
24	0.0030	Wayne	Router	Telnet R PORT=2588 h
25	0.1326	Router	Wayne	TCP D=23 S=2588 ACK=331344859 WIN=1023
26	0.0158	Router	Wayne	Telnet C PORT=2588
27	0.0030	Wayne	Router	Telnet R PORT=2588
28	0.1458	Router	Wayne	TCP D=23 S=2588 ACK=331344860 WIN=1023
29	0.0410	Router	Wayne	Telnet C PORT=2588 i
30	0.0026	Wayne	Router	Telnet R PORT=2588 i
31	0.1212	Router	Wayne	TCP D=23 S=2588 ACK=331344861 WIN=1023
32	0.1134	Router	Wayne	Telnet C PORT=2588 n
33	0.0034	Wayne	Router	Telnet R PORT=2588 n
34	0.1575	Router	Wayne	TCP D=23 S=2588 ACK=331344862 WIN=1023
35	0.2039	Router	Wayne	Telnet C PORT=2588 t
36	0.0027	Wayne	Router	Telnet R PORT=2588 t
37	0.1229	Router	Wayne	TCP D=23 S=2588 ACK=331344863 WIN=1023
38	0.1116	Router	Wayne	Telnet C PORT=2588 <0D><0A>
39	0.0332	Wayne	Router	Telnet R PORT=2588 <0D><0A><0D><0A> Ethernet 0 line protocol is...
40	0.0026	Router	Wayne	TCP D=23 S=2588 ACK=331345399 WIN=1024
41	0.0165	Wayne	Router	Telnet R PORT=2588 ute output rate 9325 bits/sec, 2 packets/sec<0D>...
42	0.2203	Router	Wayne	TCP D=23 S=2588 ACK=331345750 WIN=649
43	41.4005	Broadcast	Benoit	ARP C PA=[131.195.116.42] PRO=IP
44	0.0007	Benoit	Router	ARP R PA=[131.195.116.42]

				HA=00000C00A145 PRO=IP
45	0.0013	Router	Benoit	TCP D=23 S=15165 SYN SEQ=73138176 LEN=0 WIN=2048
46	0.0015	Benoit	Router	TCP D=15165 S=23 SYN ACK=73138177 SEQ=331390708 LEN=0 WIN=0
47	0.0031	Router	Benoit	TCP D=23 S=15165 ACK=331390709 WIN=2048
48	0.0018	Benoit	Router	TCP D=15165 S=23 ACK=73138177 WIN=2144
49	0.0068	Benoit	Router	Telnet R PORT=15165 IAC Will Echo
50	0.0297	Benoit	Router	Telnet R PORT=15165 <0D><0A>
51	0.2341	Router	Benoit	Telnet C PORT=15165 IAC Do Echo
52	0.2997	Benoit	Router	TCP D=15165 S=23 ACK=73138180 WIN=2141
53	0.0323	Router	Benoit	Telnet C PORT=15165 IAC Do Suppress go-ahead
54	0.2995	Benoit	Router	TCP D=15165 S=23 ACK=73138183 WIN=2138

Trace 7.3b. Duplicate IP Address Details (Original station)
Sniffer Network Analyzer data 11-Oct at 10:49:04, IPDUPLIC.ENC, Pg 1

- - - - - - - - - - - - - - - Frame 1 - - - - - - - - - - - - - - - - -	
DLC:	— DLC Header —
DLC:	
DLC:	Frame 1 arrived at 10:49:08.4044; frame size is 60 (003C hex) bytes.
DLC:	Destination = Station Cisco 00A145, Router
DLC:	Source = Station Intrln06C202, Wayne
DLC:	Ethertype = 0800 (IP)
DLC:	
IP:	— IP Header —
IP:	
IP:	Version = 4, header length = 20 bytes
IP:	Type of service = 00
IP:	000. .... = routine
IP:	...0 .... = normal delay
IP:	.... 0... = normal throughput

```
IP:         .... .0.. = normal reliability
IP:         Total length = 44 bytes
IP:         Identification = 13
IP:         Flags = 0X
IP:         .0.. .... = may fragment
IP:         ..0. .... = last fragment
IP:         Fragment offset = 0 bytes
IP:         Time to live = 64 seconds/hops
IP:         Protocol = 6 (TCP)
IP:         Header checksum = 8A14 (correct)
IP:         Source address = [131.195.116.250]
IP:         Destination address = [131.195.116.42]
IP:         No options
IP:
TCP:        — TCP Header —
TCP:
TCP:        Source port = 2588
TCP:        Destination port = 23 (Telnet)
TCP:        Initial sequence number = 265153482
TCP:        Data offset = 24 bytes
TCP:        Flags = 02
TCP:        ..0. .... = (No urgent pointer)
TCP:        ...0 .... = (No acknowledgment)
TCP:        .... 0... = (No push)
TCP:        .... .0.. = (No reset)
TCP:        .... ..1. = SYN
TCP:        .... ...0 = (No FIN)
TCP:        Window = 1024
TCP:        Checksum = 9DAF (correct)
TCP:
TCP:        Options follow
TCP:        Maximum segment size = 1460
TCP:

- - - - - - - - - - - - - - - - Frame 2 - - - - - - - - - - - - - - - -
DLC:        — DLC Header —
DLC:
DLC:        Frame 2 arrived at 10:49:08.4059; frame size is 60 (003C hex) bytes.
DLC:        Destination = Station Intrln06C202, Wayne
DLC:        Source      = Station Cisco 00A145, Router
DLC:        Ethertype  = 0800 (IP)
DLC:
IP:         — IP Header —
IP:
```

```
IP:     Version = 4, header length = 20 bytes
IP:     Type of service = 00
IP:     000. .... = routine
IP:     ...0 .... = normal delay
IP:     .... 0... = normal throughput
IP:     .... .0.. = normal reliability
IP:     Total length = 44 bytes
IP:     Identification = 0
IP:     Flags = 0X
IP:     .0.. .... = may fragment
IP:     ..0. .... = last fragment
IP:     Fragment offset = 0 bytes
IP:     Time to live = 255 seconds/hops
IP:     Protocol = 6 (TCP)
IP:     Header checksum = CB20 (correct)
IP:     Source address = [131.195.116.42]
IP:     Destination address = [131.195.116.250]
IP:     No options
IP:
TCP:    — TCP Header —
TCP:
TCP:    Source port = 23 (Telnet)
TCP:    Destination port = 2588
TCP:    Initial sequence number = 331344504
TCP:    Acknowledgment number = 265153483
TCP:    Data offset = 24 bytes
TCP:    Flags = 12
TCP:    ..0. .... = (No urgent pointer)
TCP:    ...1 .... = Acknowledgment
TCP:    .... 0... = (No push)
TCP:    .... .0.. = (No reset)
TCP:    .... ..1. = SYN
TCP:    .... ...0 = (No FIN)
TCP:    Window = 0
TCP:    Checksum = A367 (correct)
TCP:
TCP:    Options follow
TCP:    Maximum segment size = 1460
TCP:
```

Trace 7.3c. Duplicate IP Address Details (Duplicate station)
Sniffer Network Analyzer data 11-Oct at 10:49:04, IPDUPLIC.ENC, Pg 1

```
- - - - - - - - - - - - - - - - Frame 45 - - - - - - - - - - - - - - - - -
DLC:    — DLC Header —
DLC:
DLC:    Frame 45 arrived at 10:49:54.6082; frame size is 60 (003C hex) bytes.
DLC:    Destination = Station Cisco 00A145, Router
DLC:    Source      = Station Intrln05E253, Benoit
DLC:    Ethertype   = 0800 (IP)
DLC:
IP:     — IP Header —
IP:
IP:     Version = 4, header length = 20 bytes
IP:     Type of service = 10
IP:             000. .... = routine
IP:             ...1 .... = low delay
IP:             .... 0... = normal throughput
IP:             .... .0.. = normal reliability
IP:     Total length = 44 bytes
IP:     Identification = 1
IP:     Flags = 0X
IP:     .0.. .... = may fragment
IP:     ..0. .... = last fragment
IP:     Fragment offset = 0 bytes
IP:     Time to live = 64 seconds/hops
IP:     Protocol = 6 (TCP)
IP:     Header checksum = 8A10 (correct)
IP:     Source address = [131.195.116.250]
IP:     Destination address = [131.195.116.42]
IP:     No options
IP:
TCP:    — TCP Header —
TCP:
TCP:    Source port = 15165
TCP:    Destination port = 23 (Telnet)
TCP:    Initial sequence number = 73138176
TCP:    Data offset = 24 bytes
TCP:    Flags = 02
TCP:    ..0. .... = (No urgent pointer)
TCP:    ...0 .... = (No acknowledgment)
TCP:    .... 0... = (No push)
TCP:    .... .0.. = (No reset)
TCP:    .... ..1. = SYN
```

```
TCP:        .... ...0 = (No FIN)
TCP:        Window = 2048
TCP:        Checksum = 5FCA (correct)
TCP:
TCP:        Options follow
TCP:        Maximum segment size = 1460
TCP:

- - - - - - - - - - - - - - - - Frame 46 - - - - - - - - - - - - - - - - -
DLC:        — DLC Header —
DLC:
DLC:        Frame 46 arrived at 10:49:54.6097; frame size is 60 (003C hex) bytes.
DLC:        Destination = Station Intrln05E253, Benoit
DLC:        Source      = Station Cisco 00A145, Router
DLC:        Ethertype  = 0800 (IP)
DLC:
IP:         — IP Header —
IP:
IP:         Version = 4, header length = 20 bytes
IP:         Type of service = 00
IP:                 000. .... = routine
IP:                 ..0 .... = normal delay
IP:                 .... 0... = normal throughput
IP:                 .... .0.. = normal reliability
IP:         Total length = 44 bytes
IP:         Identification = 0
IP:         Flags = 0X
IP:         .0.. .... = may fragment
IP:         ..0. .... = last fragment
IP:         Fragment offset = 0 bytes
IP:         Time to live = 255 seconds/hops
IP:         Protocol = 6 (TCP)
IP:         Header checksum = CB20 (correct)
IP:         Source address = [131.195.116.42]
IP:         Destination address = [131.195.116.250]
IP:         No options
IP:
TCP:        — TCP Header —
TCP:
TCP:        Source port = 23 (Telnet)
TCP:        Destination port = 15165
TCP:        Initial sequence number = 331390708
TCP:        Acknowledgment number = 73138177
TCP:        Data offset = 24 bytes
```

```
TCP:        Flags = 12
TCP:        ..0. .... = (No urgent pointer)
TCP:        ...1 .... = Acknowledgment
TCP:        .... 0... = (No push)
TCP:        .... .0.. = (No reset)
TCP:        .... ..1. = SYN
TCP:        .... ...0 = (No FIN)
TCP:        Window = 0
TCP:        Checksum = B505 (correct)
TCP:
TCP:        Options follow
TCP:        Maximum segment size = 1460
TCP:
```

7.4 Incorrect Address Mask

As discussed in Section 3.2.3, system implementers can use subnetworks to form a hierarchical routing structure within an internetwork. Subnet addressing works as follows: A 32-bit IP address is comprised of a Network ID plus a Host ID. For example, a Class B network uses 16 bits for the network portion and 16 bits for the host portion. If an internet has multiple physical networks (e.g., LANs), the 16-bit host portion of the address can be further divided into a subnetwork address (representing the particular physical network, such as an Ethernet) and a host address (representing a particular device on that Ethernet). Class B addresses commonly use 8-bit subnetting. This would give a Network ID of 16 bits, a Subnetwork ID of 8 bits, and a Host ID of 8 bits. Thus, it could uniquely identify up to 254 subnetworks (i.e., LANs), each having up to 254 hosts (i.e., workstations, servers, and so on). Recall that the all zeros and all ones addresses are not allowed, thus reducing the theoretical limit of 256 subnetworks and 256 hosts to 254 of each (256 - 2 = 254). The first 16 bits (the Network ID) would deliver the datagram to the access point for the network (the router). The router would then decide which of the 254 subnetworks this datagram was destined for. The router uses a Subnet Mask to make that decision. Subnet Masks are stored within the host's configuration files, and can be entered manually or obtained using an automatic process such as the ICMP Address Mask Request/Reply messages. This case study will show what happens when the host uses an incorrect Subnet Mask.

In this example, a network manager, Paul, wishes to check the status of a host on a different segment but connected via a bridge (see Figure 7-1). Paul's workstation software stores a number of parameters, including the

Subnet Mask. The network in question is a Class B network, without subnetting. The Subnet Mask should be set for [255.255.0.0], corresponding with a Network ID of 16 bits and a Host ID of 16 bits. Since subnetting is not used, all datagrams for the local network should be delivered directly, but datagrams destined for another network should go through the router. Paul uses the ICMP Echo (PING) command to check the status of his host (a Sun 4 Server), but there's a delay in the ICMP Echo Reply. Let's see what happened.

Paul's workstation first broadcasts an ARP message looking for a router (Frames 1– 2 in Trace 7.4a). This step is unexpected, since a router is not required for this transaction. Next, it attempts an ICMP Echo message (Frame 3). If all systems are functioning properly, an ICMP Echo Reply should follow the ICMP Echo immediately. In Paul's case, however, the ICMP Redirect message in Frame 4 follows the ICMP Echo. The Redirect message (see Trace 7.4b) indicates that it is redirecting datagrams for the host (ICMP Code = 1), and that the correct router (gateway) for this operation is [132.XXX.132.12]. Paul's workstation then sends another ARP Request, looking for address [132.XXX.132.12] in Frame 5. The Sun Server responds with HA = 0800200C1DC3H (Frame 6). The ICMP Echo and Echo Reply then proceed as expected in Frames 7 and 8. The question remains: Why did Paul's workstation access the router, causing the ICMP Redirect Message? The second ICMP Echo (Frame 7) sent from Paul's workstation [132.XXX.129.15] to the Sun Server [132.XXX.132.12] provides a clue. Note that both devices are on the same network [132.XXX]. Since this is a Class B network without subnetting, the Subnet Mask should be [255.255.0.0]. When Paul examined the Subnet Mask in his workstation parameters, he found that it had been set for a Class B network with an 8-bit subnet address field, which is [255.255.255.0]. When his workstation calculated the Subnet Mask, it came up with:

```
Subnet Mask            11111111  11111111  11111111  00000000
    AND
Source Address         10000100  XXXXXXXX  10000001  00001111
[132.XXX.129.15]
Result #1              10000100  XXXXXXXX  10000001  00000000


Subnet Mask            11111111  11111111  11111111  00000000
    AND
Destination Address 1000100     XXXXXXXX  10000100  00001100
[132.XXX.132.12]
Result #2              1000100   XXXXXXXX  10000100  00000000
```

Because the workstation found that Result #1 and Result #2 were different, it incorrectly concluded that the two devices were on different subnetworks, requiring the assistance of the router. When Paul reconfigured his workstation's Subnet Mask to [255.255.0.0], the ICMP Echo and Echo Replies proceeded without router intervention.

Trace 7.4a. Subnet Mask Misconfiguration Summary
Sniffer Network Analyzer data 13-Feb at 14:50:26, SUBNETC.ENC, Pg 1

SUMMARY	Delta T	Destination	Source	Summary
M 1		Broadcast	Paul	ARP C PA=[132.XXX.1.1] PRO=IP
2	0.0011	Paul	Router	ARP R PA=[132.XXX.1.1] HA=000093E0A0BF PRO=IP
3	0.0022	Router	Paul	ICMP Echo
4	0.0021	Paul	Router	ICMP Redirect (Redirect datagrams for the host)
5	11.7626	Broadcast	Paul	ARP C PA=[132.XXX.132.12] PRO=IP
6	0.0008	Paul	SunSvr	ARP R PA=[132.XXX.132.12] HA=0800200C1DC3 PRO=IP
7	0.0023	SunSvr	Paul	ICMP Echo
8	0.0016	Paul	SunSvr	ICMP Echo reply

Trace 7.4b. Subnet Mask Misconfiguration ICMP Details
Sniffer Network Analyzer data 13-Feb at 14:50:26, SUBNETC.ENC, Pg 1

```
- - - - - - - - - - - - - - - - Frame 3 - - - - - - - - - - - - - - - -
ICMP:        — ICMP Header —
ICMP:
ICMP:        Type = 8 (Echo)
ICMP:        Code = 0
ICMP:        Checksum = 719E (correct)
ICMP:        Identifier = 1315
ICMP:        Sequence number = 1
ICMP:        [256 bytes of data]
ICMP:
ICMP:        [Normal end of "ICMP Header".]
ICMP:
```

```
- - - - - - - - - - - - - - - - Frame 4 - - - - - - - - - - - - - - - - -
ICMP:          — ICMP Header —
ICMP:
ICMP:          Type = 5 (Redirect)
ICMP:          Code = 1 (Redirect datagrams for the host)
ICMP:          Checksum = 738C (correct)
ICMP:          Gateway internet address = [132.XXX.132.12]
ICMP:          IP header of originating message (description follows)
ICMP:
IP:            — IP Header —
IP:
IP:            Version = 4, header length = 20 bytes
IP:            Type of service = 00
IP:                      000. .... = routine
IP:                      ...0 .... = normal delay
IP:                      .... 0... = normal throughput
IP:            .... .0.. = normal reliability
IP:            Total len gth = 284 bytes
IP:            Identification = 2
IP:            Flags = 0X
IP:            .0.. .... = may fragment
IP:            ..0. .... = last fragment
IP:            Fragment offset = 0 bytes
IP:            Time to live = 63 seconds/hops
IP:            Protocol = 1 (ICMP)
IP:            Header checksum = 6C7D (correct)
IP:            Source address = [132.XXX.129.15]
IP:            Destination address = [132.XXX.132.12]
IP:            No options
ICMP:
ICMP:          [First 8 byte(s) of data of originating message]
ICMP:
ICMP:          [Normal end of "ICMP Header".]
ICMP:

- - - - - - - - - - - - - - - - Frame 7 - - - - - - - - - - - - - - - - -
ICMP:          — ICMP Header —
ICMP:
ICMP:          Type = 8 (Echo)
ICMP:          Code = 0
ICMP:          Checksum = 4070 (correct)
ICMP:          Identifier = 13905
ICMP:          Sequence number = 1
ICMP:          [256 bytes of data]
```

```
ICMP:
ICMP:         [Normal end of "ICMP Header".]
ICMP:

- - - - - - - - - - - - - - - - Frame 8 - - - - - - - - - - - - - - - -
ICMP:         — ICMP Header —
ICMP:
ICMP:         Type = 0 (Echo reply)
ICMP:         Code = 0
ICMP:         Checksum = 4870 (correct)
ICMP:         Identifier = 13905
ICMP:         Sequence number = 1
ICMP:         [256 bytes of data]
ICMP:
ICMP:         [Normal end of "ICMP Header".]
ICMP:
```

7.5 Using ICMP Echo Messages

ICMP messages can answer many questions about the health of the network. The ICMP Echo/Echo Reply messages, commonly known as the PING, are probably the most frequently used. You invoke PING from your host operating system to test the path to a particular host. If all is well, a message will return verifying the existence of the path to the host or network. One caution is in order, however: unpredictable results can occur if you PING an improper destination address. For example, PINGing address [255.255.255.255] (limited broadcast within this subnet) may cause excessive internetwork traffic. Let's look at an example.

One lonely weekend, a network administrator decides to test the paths to some of the hosts on his internet. He has two ways to accomplish this. He could send an ICMP Echo message to each host separately, or he could send a directed broadcast to all hosts on his network and subnetwork. He decides to enter the Destination address [129.XXX.23.255] that PINGs all of the hosts on Class B network 129.XXX, subnet 23. Trace 7.5a shows the result. Frame 683 is the ICMP Echo (PING); the network administrator receives 27 frames containing ICMP Echo Reply messages. Note that all of the reply messages are directed to the originator of the ICMP Echo message, workstation 020C5D. Also note that the Nortel Networks (formerly Bay Networks) routers on this network did not respond to the PING because their design protects against such a transmission. Other hosts or routers may be designed with such a safeguard in place as well.

Details of the ICMP Echo (Frame 683) and the first ICMP Echo Reply (Frame 684) are given in Trace 7.5b. The Echo's Destination address is set for broadcast, and it indicates that the Echo Replies come back to the originating station. As the trace shows, the message originator [129.XXX.23.146] has specified that all hosts on subnet 23 should respond by setting the Destination address to [129.99.23.255]. The first response (Frame 684) is from device [129.XXX.23.17], followed by responses from 26 other hosts (Frames 685 through 712). Also note that the ICMP header contains an Identifier (18501) that correlates Echo and Echo Reply messages in case PINGs to/from different hosts occur simultaneously. The 56 octets of data transmitted with the Echo message are returned with the Echo Reply.

We can draw one clear conclusion from this exercise: The ICMP Echo message can be a very valuable troubleshooting tool, but make sure of your destination before you initiate the command. A PING to a broadcast address could have a great impact on the internetwork traffic.

Trace 7.5a. ICMP Echo to IP Address [X.X.X.255] Summary
Sniffer Network Analyzer data 17-Feb at 15:31:08, PINGTST4.ENC, Pg 1

SUMMARY	Delta T	Destination	Source	Summary
683		Broadcast	SilGrf 020C5D	ICMP Echo
684	0.0004	SilGrf 020C5D	SilGrf 060C44	ICMP Echo reply
685	0.0004	SilGrf 020C5D	SilGrf 020FF9	ICMP Echo reply
686	0.0003	SilGrf 020C5D	Sun 106604	ICMP Echo reply
687	0.0003	SilGrf 020C5D	Sun 0F5DC2	ICMP Echo reply
688	0.0002	SilGrf 020C5D	Sun 08FE6B	ICMP Echo reply
689	0.0003	SilGrf 020C5D	SilGrf 021193	ICMP Echo reply
690	0.0001	SilGrf 020C5D	SilGrf 02137B	ICMP Echo reply
691	0.0004	SilGrf 020C5D	Sun 094668	ICMP Echo reply
692	0.0003	SilGrf 020C5D	Prteon1064D6	ICMP Echo reply
693	0.0002	SilGrf 020C5D	Sun 062A16	ICMP Echo reply
694	0.0002	SilGrf 020C5D	Sun 00DAFF	ICMP Echo reply
695	0.0002	SilGrf 020C5D	CMC A00666	ICMP Echo reply
696	0.0002	SilGrf 020C5D	Sun 00E849	ICMP Echo reply
697	0.0003	SilGrf 020C5D	Sun 005513	ICMP Echo reply
698	0.0001	SilGrf 020C5D	NSC 010212	ICMP Echo reply
699	0.0002	SilGrf 020C5D	00802D0020F8	ICMP Echo reply
700	0.0001	SilGrf 020C5D	Exceln 231835	ICMP Echo reply
701	0.0002	SilGrf 020C5D	00802D002202	ICMP Echo reply
702	0.0001	SilGrf 020C5D	DEC 0A136F	ICMP Echo reply
703	0.0001	SilGrf 020C5D	DECnet00F760	ICMP Echo reply
704	0.0001	SilGrf 020C5D	Sun 0045AD	ICMP Echo reply

```
705    0.0001    SilGrf 020C5D   Intel 0361A8     ICMP Echo reply
706    0.0001    SilGrf 020C5D   Sun 0A73A2       ICMP Echo reply
707    0.0001    SilGrf 020C5D   DEC 0D0B05       ICMP Echo reply
708    0.0001    SilGrf 020C5D   Sun 00F725       ICMP Echo reply
709    0.0001    SilGrf 020C5D   SilGrf 020C38    ICMP Echo reply
712    0.0211    SilGrf 020C5D   Intrln 008248    ICMP Echo reply
```

Trace 7.5b. ICMP Echo to IP Address [X.X.X.255] Details
Sniffer Network Analyzer data 17-Feb at 15:31:08, PINGTST4.ENC, Pg 1

```
- - - - - - - - - - - - - - - - Frame 683 - - - - - - - - - - - - - - - - -
DLC:     — DLC Header —
DLC:
DLC:     Frame 683 arrived at 15:31:50.3581; frame size is 102 (0066 hex) bytes.
DLC:     Destination = BROADCAST FFFFFFFFFFFF, Broadcast
DLC:     Source      = Station SilGrf020C5D
DLC:     Ethertype  = 0800 (IP)
DLC:
IP:      — IP Header —
IP:
IP:      Version = 4, header length = 20 bytes
IP:      Type of service = 00
IP:            000. .... = routine
IP:            ...0 .... = normal delay
IP:            .... 0... = normal throughput
IP:            .... .0.. = normal reliability
IP:      Total length = 84 bytes
IP:      Identification = 16898
IP:      Flags = 0X
IP:      .0.. .... = may fragment
IP:      ..0. .... = last fragment
IP:      Fragment offset = 0 bytes
IP:      Time to live = 255 seconds/hops
IP:      Protocol = 1 (ICMP)
IP:      Header checksum = 474F (correct)
IP:      Source address = [129.XXX.23.146]
IP:      Destination address = [129.XXX.23.255]
IP:      No options
IP:
ICMP:    — ICMP Header —
ICMP:
ICMP:    Type = 8 (Echo)
ICMP:    Code = 0
ICMP:    Checksum = 851E (correct)
```

```
ICMP:      Identifier = 18501
ICMP:      Sequence number = 0
ICMP:      [56 bytes of data]
ICMP:
ICMP:      [Normal end of "ICMP Header".]
ICMP:

---------------- Frame 684 ----------------
DLC:       — DLC Header —
DLC:
DLC:       Frame 684 arrived at 15:31:50.3585; frame size is 98 (0062 hex) bytes.
DLC:       Destination = Station SilGrf020C5D
DLC:       Source      = Station SilGrf060C44
DLC:       Ethertype  = 0800 (IP)
DLC:
IP:        — IP Header —
IP:
IP:        Version = 4, header length = 20 bytes
IP:        Type of service = 00
IP:                 000. .... = routine
IP:                 ...0 .... = normal delay
IP:                 .... 0... = normal throughput
IP:                 .... .0.. = normal reliability
IP:        Total length = 84 bytes
IP:        Identification = 985
IP:        Flags = 0X
IP:        .0.. .... = may fragment
IP:        ..0. .... = last fragment
IP:        Fragment offset = 0 bytes
IP:        Time to live = 255 seconds/hops
IP:        Protocol = 1 (ICMP)
IP:        Header checksum = 8666 (correct)
IP:        Source address = [129.XXX.23.17]
IP:        Destination address = [129.XXX.23.146]
IP:        No options
IP:
ICMP:      — ICMP Header —
ICMP:
ICMP:      Type = 0 (Echo reply)
ICMP:      Code = 0
ICMP:      Checksum = 8D1E (correct)
ICMP:      Identifier = 18501
ICMP:      Sequence number = 0
ICMP:      [56 bytes of data]
```

ICMP:
ICMP: [Normal end of "ICMP Header".]
ICMP:

7.6 Misdirected Datagrams

In Chapter 4, we studied the operation of the Internet Control Message Protocol (ICMP) and saw how the ICMP Redirect message (ICMP Type = 5) corrects routing problems. A unique code within the ICMP message specifies the datagrams to be redirected (review Figure 4-5).

In this example, a workstation wishes to communicate with a remote host on another subnetwork (see Figure 7-3). Unfortunately, the workstation's configuration file is incorrect, and its initial attempt to communicate with the remote host fails (see Trace 7.6a). From out of the internet comes an intelligent router to the rescue! Router 234 (the workstation's default router) recognizes the error and issues an ICMP Redirect in Frame 4. The workstation obliges, switches its transmissions to the correct path (Router 235) in Frame 5, and successfully establishes a TELNET session with the remote host (Frames 6 through 20). Let's look inside the ICMP messages and see what happened (Trace 7.6b).

The workstation's initial transmission (Frame 3) is directed to Router 234, with the datagram destined for the remote Host address [XXX.YYY.0.154]. We know that a router will be involved in the communication because the Source address (the workstation address) [XXX.YYY.0.152] and the Host address [XXX.YYY.0.154] are on different subnetworks. Router 234 responds to this initial transmission by issuing an ICMP Redirect in Frame 4. We can now identify the IP address of Router 234 (the Source address [XXX.YYY.0.234] of the ICMP Redirect). The ICMP header indicates a Redirect for the host (ICMP Code = 1) and the correct path, which is via Router 235, address [XXX.YYY.0.235].

The workstation sees the error of its ways and changes the Destination hardware address to be Router 235 (Frame 5). Note that the Source and Destination IP addresses did not change between Frame 3 and Frame 5. In other words, the workstation knew its desired destination, but just didn't know how to get there. Frame 6 shows the confirmation of the redirected path; that datagram is a response from the remote host (IP Source address = [XXX.YYY.0.154]).

Upon further study, the network manager discovered that the workstation's default gateway had been set incorrectly. When this parameter was changed to reflect Router 235 (instead of Router 234) the problem did not reoccur.

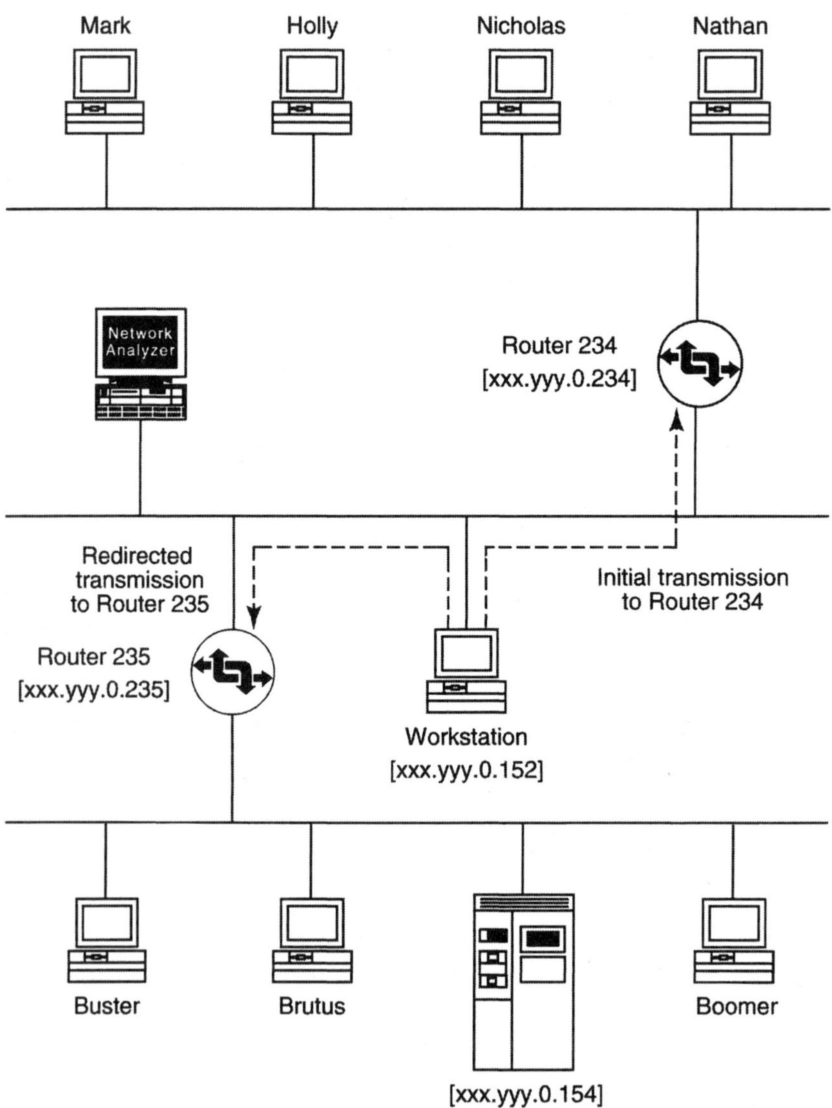

Figure 7-3. Misdirected Datagram Topology.

Trace 7.6a. Misdirected Datagram Summary

Sniffer Network Analyzer data 2-Mar at 10:17:44, RICMP.ENC, Pg 1

SUMMARY	Delta T	Destination	Source	Summary
3	0.0162	Router 234	Workstation	Telnet R PORT=2909 <0D><0A>
4	0.0029	Workstation	Router 234	ICMP Redirect (Redirect datagrams for the host)
5	0.0118	Router 235	Workstation	Telnet R PORT=2909 $
6	0.0941	Workstation	Router 235	TCP D=23 S=2909 ACK=813932917 WIN=4096
7	3.7848	Workstation	Router 235	Telnet C PORT=2909 p
8	0.0088	Router 235	Workstation	Telnet R PORT=2909 p
9	0.0065	Workstation	Router 235	TCP D=23 S=2909 ACK=813932918 WIN=4096
10	0.3042	Workstation	Router 235	Telnet C PORT=2909 s
11	0.0082	Router 235	Workstation	Telnet R PORT=2909 s
12	0.0881	Workstation	Router 235	TCP ACK=813932919 WIN=4096
13	0.0053	Workstation	Router 235	Telnet C PORT=2909
14	0.0080	Router 235	Workstation	Telnet R PORT=2909
15	0.1858	Workstation	Router 235	TCP D=23 S=2909 ACK=813932920 WIN=4096
16	0.5343	Workstation	Router 235	Telnet C PORT=2909 -
17	0.0078	Router 235	Workstation	Telnet R PORT=2909 -
18	0.0577	Workstation	Router 235	TCP D=23 S=2909 ACK=813932921 WIN=4096
19	0.1744	Workstation	Router 235	Telnet C PORT=2909 a
20	0.0078	Router 235	Workstation	Telnet R PORT=2909 a

Trace 7.6b. Misdirected Datagram Details

Sniffer Network Analyzer data 2-Mar at 10:17:44, RICMP.ENC, Pg 1

```
- - - - - - - - - - - - - - - - Frame 3 - - - - - - - - - - - - - - - - -
DLC:      — DLC Header —
DLC:
DLC:      Frame 3 arrived at 10:21:01.5697; frame size is 74 (004A hex) bytes.
DLC:      Destination = Station XXXXXX 002461, Router 234
DLC:      Source      = Station XXXXXX 01BB41, Workstation
```

```
DLC:        Ethertype  = 0800 (IP)
DLC:
IP:         — IP Header —
IP:
IP:         Version = 4, header length = 20 bytes
IP:         Type of service = 00
IP:                   000. .... = routine
IP:                   ...0 .... = normal delay
IP:                   .... 0... = normal throughput
IP:                   .... .0.. = normal reliability
IP:         Total length = 42 bytes
IP:         Identification = 2931
IP:         Flags = 0X
IP:         .0.. .... = may fragment
IP:         ..0. .... = last fragment
IP:         Fragment offset = 0 bytes
IP:         Time to live = 30 seconds/hops
IP:         Protocol = 6 (TCP)
IP:         Header checksum = 31A8 (correct)
IP:         Source address = [XXX.YYY.0.152]
IP:         Destination address = [XXX.YYY.0.154]
IP:         No options
IP:
TCP:        — TCP Header —
TCP:
TCP:        Source port = 23 (Telnet)
TCP:        Destination port = 2909
TCP:        Sequence number = 813932913
TCP:        Acknowledgment number = 63520058
TCP:        Data offset = 20 bytes
TCP:        Flags = 18
TCP:        ..0. .... = (No urgent pointer)
TCP:        ...1 .... = Acknowledgment
TCP:        .... 1... = Push
TCP:        .... .0.. = (No reset)
TCP:        .... ..0. = (No SYN)
TCP:        .... ...0 = (No FIN)
TCP:        Window = 9116
TCP:        Checksum = 0105 (correct)
TCP:        No TCP options
TCP:        [2 byte(s) of data]
TCP:
Telnet::    — Telnet Data —
Telnet::
```

```
Telnet::        <0D><0A>
Telnet::

- - - - - - - - - - - - - - - - Frame 4 - - - - - - - - - - - - - - - - -
DLC:            — DLC Header —
DLC:
DLC:            Frame 4 arrived at 10:21:01.5726; frame size is 70 (0046 hex) bytes.
DLC:            Destination = Station XXXXXX 01BB41, Workstation
DLC:            Source      = Station XXXXXX 002461, Router 234
DLC:            Ethertype  = 0800 (IP)
DLC:
IP:             — IP Header —
IP:
IP:             Version = 4, header length = 20 bytes
IP:             Type of service = 00
IP:                         000. .... = routine
IP:                         ...0 .... = normal delay
IP:                         .... 0... = normal throughput
IP:                         .... .0.. = normal reliability
IP:             Total length = 56 bytes
IP:             Identification = 0
IP:             Flags = 0X
IP:             .0.. .... = may fragment
IP:             ..0. .... = last fragment
IP:             Fragment offset = 0 bytes
IP:             Time to live = 255 seconds/hops
IP:             Protocol = 1 (ICMP)
IP:             Header checksum = 5CA3 (correct)
IP:             Source address = [XXX.YYY.0.234]
IP:             Destination address = [XXX.YYY.0.152]
IP:             No options
IP:
ICMP:           — ICMP Header —
ICMP:
ICMP:           Type = 5 (Redirect)
ICMP:           Code = 1 (Redirect datagrams for the host)
ICMP:           Checksum = EE0C (correct)
ICMP:           Gateway internet address = [XXX.YYY.0.235]
ICMP:           IP header of originating message (description follows)
ICMP:
IP:             — IP Header —
IP:
IP:             Version = 4, header length = 20 bytesIP:
IP:             Type of service = 00
```

```
IP:                 000. .... = routine
IP:                 ...0 .... = normal delay
IP:                 .... 0... = normal throughput
IP:                 .... .0.. = normal reliability
IP:         Total length = 42 bytes
IP:         Identification = 2931
IP:         Flags = 0X
IP:         .0.. .... = may fragment
IP:         ..0. .... = last fragment
IP:         Fragment offset = 0 bytes
IP:         Time to live = 29 seconds/hops
IP:         Protocol = 6 (TCP)
IP:         Header checksum = 32A8 (correct)
IP:         Source address = [XXX.YYY.0.152]
IP:         Destination address = [XXX.YYY.0.154]
IP:         No options
ICMP:
ICMP:       [First 8 byte(s) of data of originating message]
ICMP:
ICMP:       [Normal end of "ICMP Header".]
ICMP:

- - - - - - - - - - - - - - - - Frame 5 - - - - - - - - - - - - - - - - -
DLC:        — DLC Header —
DLC:
DLC:        Frame 5 arrived at 10:21:01.5845; frame size is 74 (004A hex) bytes.
DLC:        Destination = Station XXXXXX 00244F, Router 235
DLC:        Source      = Station XXXXXX 01BB41, Workstation
DLC:        Ethertype  = 0800 (IP)
DLC:
IP:         — IP Header —
IP:
IP:         Version = 4, header length = 20 bytes
IP:         Type of service = 00
IP:                 000. .... = routine
IP:                 ...0 .... = normal delay
IP:                 .... 0... = normal throughput
IP:                 .... .0.. = normal reliability
IP:         Total length = 42 bytes
IP:         Identification = 2932
IP:         Flags = 0X
IP:         .0.. .... = may fragment
IP:         ..0. .... = last fragment
IP:         Fragment offset = 0 bytes
```

```
IP:         Time to live = 30 seconds/hops
IP:         Protocol = 6 (TCP)
IP:         Header checksum = 31A7 (correct)
IP:         Source address = [XXX.YYY.0.152]
IP:         Destination address = [XXX.YYY.0.154]
IP:         No options
IP:
TCP:        — TCP Header —
TCP:
TCP:        Source port = 23 (Telnet)
TCP:        Destination port = 2909
TCP:        Sequence number = 813932915
TCP:        Acknowledgment number = 63520058TCP: Data offset = 20 bytes
TCP:        Flags = 18
TCP:        ..0. .... = (No urgent pointer)
TCP:        ...1 .... = Acknowledgment
TCP:        .... 1... = Push
TCP:        .... .0.. = (No reset)
TCP:        .... ..0. = (No SYN)
TCP:        ... ...0 = (No FIN)
TCP:        Window = 9116
TCP:        Checksum = E9EC (correct)
TCP:        No TCP options
TCP:        [2 byte(s) of data]
TCP:
Telnet::    — Telnet data —
Telnet::
Telnet::    $
Telnet::

- - - - - - - - - - - - - - - - Frame 6 - - - - - - - - - - - - - - - - -
DLC:        — DLC Header —
DLC:
DLC:        Frame 6 arrived at 10:21:01.6787; frame size is 60 (003C hex) bytes.
DLC:        Destination = Station XXXXXX 01BB41, Workstation
DLC:        Source      = Station XXXXXX 00244F, Router 235
DLC:        Ethertype  = 0800 (IP)
DLC:
IP:         — IP Header —
IP:
IP:         Version = 4, header length = 20 bytes
IP:         Type of service = 00
IP:                   000. .... = routine
IP:                   ...0 .... = normal delay
```

```
IP:                 .... 0... = normal throughput
IP:                 .... .0.. = normal reliability
IP:          Total length = 40 bytes
IP:          Identification = 37695
IP:          Flags = 0X
IP:          .0.. .... = may fragment
IP:          ..0. .... = last fragment
IP:          Fragment offset = 0 bytes
IP:          Time to live = 58 seconds/hops
IP:          Protocol = 6 (TCP)
IP:          Header checksum = 8DDD (correct)
IP:          Source address = [XXX.YYY.0.154]
IP:          Destination address = [XXX.YYY.0.152]
IP:          No options
IP:
TCP:         — TCP Header —
TCP:
TCP:         Source port = 2909
TCP:         Destination port = 23 (Telnet)
TCP:         Sequence number = 63520058
TCP:         Acknowledgment number = 813932917
TCP:         Data offset = 20 bytes
TCP:         Flags = 10
TCP:         ..0. .... = (No urgent pointer)
TCP:         ...1 .... = Acknowledgment
TCP:         .... 0... = (No push)
TCP:         .... .0.. = (No reset)
TCP:         .... ..0. = (No SYN)
TCP:         .... ...0 = (No FIN)
TCP:         Window = 4096
TCP:         Checksum = 21B1 (correct)
TCP:         No TCP options
TCP:
```

7.7 Using OSPF and BGP

In this case study, we will look at how OSPF and BGP are used within an internetwork to communicate reachability information among the various routers. The internetwork in question is based on an Ethernet backbone [140.XXX.6.Y], with three routers extending from that backbone that connect to various Ethernet and WAN segments (see Figure 7-4). Three routers participate in this communication: the Broadway Lab router [140.XXX.6.253], which runs BGP and OSPF; the Aeronautics Lab router [140.XXX.6.248], which runs OSPF; and the Science Lab router [140.XXX.6.250], which runs BGP. For the purposes of BGP communication, the Broadway Lab router connects to UUnet (part of MCI, with autonomous system number 701), and the Science Lab connects to the Science Network (autonomous system number 297), plus other autonomous systems beyond the Science Network (which are not illustrated in Figure 7-4).

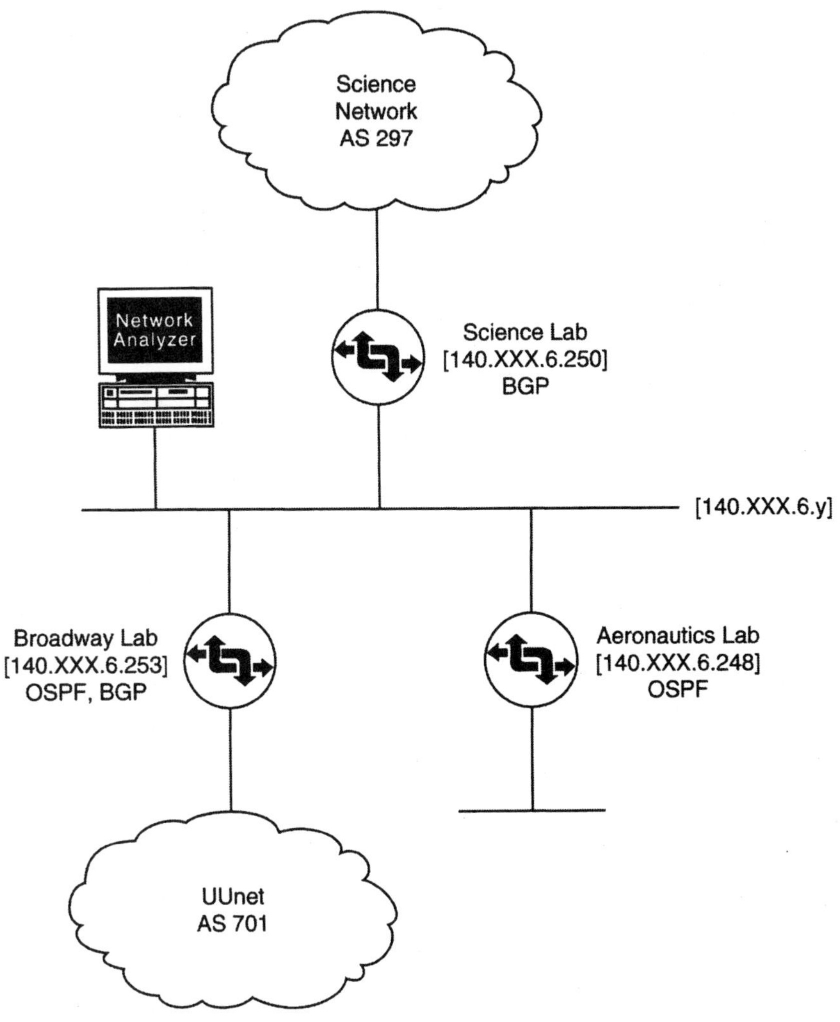

Figure 7-4. Router Communication Using OSPF and BGP.

A summary of the router communication, filtered to show only the BGP and OSPF traffic, is shown in Trace 7.7a. Note that four frames carry OSPF information, and one frame carries BGP information. The Destination address for the OSPF messages uses the All Routers multicast address: [224.0.0.5]. The Destination address for the BGP messages uses the BGP peer of interest.

Trace 7.7a. OSPF and BGP Message Summary

Sniffer Network Analyzer data from 23-Apr at 10:23:56, file NIS5.ENC, Pg 1

SUMMARY	Delta T	Destination	Source	Summary
2693	0.8539	[224.0.0.5]	Broadway	OSPF Link State Update ID=[140.XXX.1.1]
3652	1.6463	[224.0.0.5]	Aeronautics	OSPF Link State Acknowledgment ID=[140.XXX.241.100]
3862	0.4159	[224.0.0.5]	Aeronautics	OSPF Hello ID=[140.XXX.241.100]
4608	2.1827	[224.0.0.5]	Broadway	OSPF Hello ID=[140.XXX.1.1]
5841	3.2997	Broadway	Science	BGP Update

The details of an OSPF Link State Update message from Frame 2693 are shown in Trace 7.7b. Note that the Destination address at the Data Link layer is also a multicast address (01005E000005), which is then converted into the multicast IP address of [224.0.0.5], according to the procedure outlined in RFC 1112, section 6.4. The EtherType specifies IP as the next highest layer, and the IP header follows. Within the IP header, note that the next highest protocol specified is OSPF (Protocol = 89), the Source IP address is the address of the Broadway router [140.XXX.6.253], and the Destination IP address is the multicast address discussed above [224.0.0.5].

The first portion of the OSPF message is the OSPF header that indicates a Link State Update message (review Figure 4-2a). The Link State Update header follows next, beginning with the Number of Advertisements field (review Figure 4-2f). The Link State Advertisement Header follows next, beginning with the Link State Age field (review Figure 4-2g). This particular advertisement is an AS External Link Advertisement (Link State = 5, as shown in Figure 4-2k), which advertises destinations that are external to that autonomous system. Finally, there are seven link state advertisements, each noting a different destination network's IP address:

Link State Advertisement	Link State ID
1	[140.XXX.172.0]
2	[140.XXX.173.0]
3	[140.XXX.174.0]
4	[140.XXX.187.0]
5	[140.XXX.253.0]
6	[140.XXX.254.0]
7	[140.XXX.148.0]

Trace 7.7b. OSPF Link State Update Message Details
Sniffer Network Analyzer data from 23-Apr at 10:23:56, file NIS5.ENC, Pg 1

```
---------------- Frame 2693 ----------------
DLC:     — DLC Header —
DLC:
DLC:     Frame 2693 arrived at 10:24:00.9119; frame size is 314 (013A hex) bytes.
DLC:     Destination = Multicast 01005E000005
DLC:     Source      = Station 00E0F92F20A0
DLC:     Ethertype   = 0800 (IP)
DLC:
IP:      — IP Header —
IP:
IP:      Version = 4, header length = 20 bytes
IP:      Type of service = C0
IP:            110. .... = internetwork control
IP:            ...0 .... = normal delay
IP:            .... 0... = normal throughput
IP:            .... .0.. = normal reliability
IP:      Total length    = 300 bytes
IP:      Identification  = 12462
IP:      Flags           = 0X
IP:            .0.. .... = may fragment
IP:            ..0. .... = last fragment
IP:      Fragment offset = 0 bytes
IP:      Time to live    = 1 seconds/hops
IP:      Protocol        = 89 (OSPFIGP)
IP:      Header checksum = 135D (correct)
IP:      Source address      = [140.XXX.6.253], Broadway
IP:      Destination address = [224.0.0.5]
IP:      No options
IP:
OSPF:    — OSPF Header —
OSPF:
```

```
OSPF:    Version = 2,  Type = 4 (Link State Update),  Length = 280
OSPF:    Router ID      = [140.XXX.1.1]
OSPF:    Area ID        = [0.0.0.0]
OSPF:    Header checksum = 1D65 (correct)
OSPF:    Authentication: Type = 0 (No Authentication),  Value = 00 00 00 00
         00 00 00 00
OSPF:
OSPF:    Number of Advertisements = 7
OSPF:    Link State Advertisement # 1
OSPF:    Link state age       = 4 (seconds)
OSPF:    Optional capabilities = 00
OSPF:              .0.. .... = Opaque-LSAs not forwarded
OSPF:              ..0. .... = Demand Circuit bit
OSPF:              ...0 .... = External Attributes bit
OSPF:              .... 0... = no NSSA capability
OSPF:              .... .0.. = no multicast capability
OSPF:              .... ..0. = no external routing capability
OSPF:              .... ...0 = no Type of Service routing capability
OSPF:    Link state type      = 5 (AS external link)
OSPF:    Link state ID        = [140.XXX.172.0]
OSPF:    Advertising Router   = [140.XXX.172.254]
OSPF:    Sequence number      = 2147484307,  Checksum = 3FB3
OSPF:    Length               = 36
OSPF:    Network mask         = [255.255.255.0]
OSPF:    Type of service   = 80
OSPF:              1... .... = Type 2 external metric
OSPF:              .... .000 = routine
OSPF:    Metric            = 130
OSPF:    Forwarding Address = [0.0.0.0]
OSPF:    External Route Tag = 0x00000000
OSPF:    Local Info      = 0x00000000
OSPF:
OSPF:    Link State Advertisement # 2
OSPF:    Link state age       = 4 (seconds)
OSPF:    Optional capabilities = 00
OSPF:              .0.. .... = Opaque-LSAs not forwarded
OSPF:              ..0. .... = Demand Circuit bit
OSPF:              ...0 .... = External Attributes bit
OSPF:              .... 0... = no NSSA capability
OSPF:              .... .0.. = no multicast capability
OSPF:              .... ..0. = no external routing capability
OSPF:              .... ...0 = no Type of Service routing capability
OSPF:    Link state type      = 5 (AS external link)
OSPF:    Link state ID        = [140.XXX.173.0]
```

```
OSPF:     Advertising Router   = [140.XXX.172.254]
OSPF:     Sequence number      = 2147484307,   Checksum = 34BD
OSPF:     Length               = 36
OSPF:     Network mask         = [255.255.255.0]
OSPF:     Type of service      = 80
OSPF:               1... .... = Type 2 external metric
OSPF:               .... .000 = routine
OSPF:     Metric               = 130
OSPF:     Forwarding Address = [0.0.0.0]
OSPF:     External Route Tag = 0x00000000
OSPF:     Local Info           = 0x00000000
OSPF:
OSPF:     Link State Advertisement # 3
OSPF:     Link state age       = 4 (seconds)
OSPF:     Optional capabilities = 00
OSPF:               .0.. .... = Opaque-LSAs not forwarded
OSPF:               ..0. .... = Demand Circuit bit
OSPF:               ...0 .... = External Attributes bit
OSPF:               .... 0... = no NSSA capability
OSPF:               .... .0.. = no multicast capability
OSPF:               .... ..0. = no external routing capability
OSPF:               .... ...0 = no Type of Service routing capability
OSPF:     Link state type      = 5 (AS external link)
OSPF:     Link state ID        = [140.XXX.174.0]
OSPF:     Advertising Router   = [140.XXX.172.254]
OSPF:     Sequence number      = 2147484307,   Checksum = 29C7
OSPF:     Length               = 36
OSPF:     Network mask         = [255.255.255.0]
OSPF:     Type of service      = 80
OSPF:               1... .... = Type 2 external metric
OSPF:               .... .000 = routine
OSPF:     Metric               = 130
OSPF:     Forwarding Address = [0.0.0.0]
OSPF:     External Route Tag = 0x00000000
OSPF:     Local Info           = 0x00000000
OSPF:
OSPF:     Link State Advertisement # 4
OSPF:     Link state age       = 4 (seconds)
OSPF:     Optional capabilities = 00
OSPF:               .0.. .... = Opaque-LSAs not forwarded
OSPF:               ..0. .... = Demand Circuit bit
OSPF:               ...0 .... = External Attributes bit
OSPF:               .... 0... = no NSSA capability
OSPF:               .... .0.. = no multicast capability
```

```
OSPF:                    .... ..0. = no external routing capability
OSPF:                    .... ...0 = no Type of Service routing capability
OSPF:         Link state type      = 5 (AS external link)
OSPF:         Link state ID        = [140.XXX.187.0]
OSPF:         Advertising Router   = [140.XXX.172.254]
OSPF:         Sequence number      = 2147484307,   Checksum = 994A
OSPF:         Length               = 36
OSPF:         Network mask         = [255.255.255.0]
OSPF:         Type of service      = 80
OSPF:                    1... .... = Type 2 external metric
OSPF:                    .... .000 = routine
OSPF:         Metric               = 130
OSPF:         Forwarding Address = [0.0.0.0]
OSPF:         External Route Tag = 0x00000000
OSPF:         Local Info           = 0x00000000
OSPF:
OSPF:         Link State Advertisement # 5
OSPF:         Link state age       = 4 (seconds)
OSPF:         Optional capabilities = 00
OSPF:                    .0.. .... = Opaque-LSAs not forwarded
OSPF:                    ..0. .... = Demand Circuit bit
OSPF:                    ...0 .... = External Attributes bit
OSPF:                    .... 0... = no NSSA capability
OSPF:                    .... .0.. = no multicast capability
OSPF:                    .... ..0. = no external routing capability
OSPF:                    .... ...0 = no Type of Service routing capability
OSPF:         Link state type      = 5 (AS external link)
OSPF:         Link state ID        = [140.XXX.253.0]
OSPF:         Advertising Router   = [140.XXX.172.254]
OSPF:         Sequence number      = 2147484317,   Checksum = ACEA
OSPF:         Length               = 36
OSPF:         Network mask         = [255.255.255.0]
OSPF:         Type of service      = 80
OSPF:                    1... .... = Type 2 external metric
OSPF:                    .... .000 = routine
OSPF:         Metric               = 130
OSPF:         Forwarding Address = [0.0.0.0]
OSPF:         External Route Tag = 0x00000000
OSPF:         Local Info           = 0x00000000
OSPF:
OSPF:         Link State Advertisement # 6
OSPF:         Link state age       = 4 (seconds)
OSPF:         Optional capabilities = 00
OSPF:                    .0.. .... = Opaque-LSAs not forwarded
```

```
OSPF:                    ..0. .... = Demand Circuit bit
OSPF:                    ...0 .... = External Attributes bit
OSPF:                    .... 0... = no NSSA capability
OSPF:                    .... .0.. = no multicast capability
OSPF:                    .... ..0. = no external routing capability
OSPF:                    .... ...0 = no Type of Service routing capability
OSPF:          Link state type      = 5 (AS external link)
OSPF:          Link state ID        = [140.XXX.254.0]
OSPF:          Advertising Router   = [140.XXX.172.254]
OSPF:          Sequence number      = 2147484307,   Checksum = B5EA
OSPF:          Length               = 36
OSPF:          Network mask         = [255.255.255.0]
OSPF:          Type of service      = 80
OSPF:                    1... .... = Type 2 external metric
OSPF:                    .... .000 = routine
OSPF:          Metric               = 130
OSPF:          Forwarding Address = [0.0.0.0]
OSPF:          External Route Tag = 0x00000000
OSPF:          Local Info       = 0x00000000
OSPF:
OSPF:          Link State Advertisement # 7
OSPF:          Link state age       = 4 (seconds)
OSPF:          Optional capabilities = 00
OSPF:                    .0.. .... = Opaque-LSAs not forwarded
OSPF:                    ..0. .... = Demand Circuit bit
OSPF:                    ...0 .... = External Attributes bit
OSPF:                    .... 0... = no NSSA capability
OSPF:                    .... .0.. = no multicast capability
OSPF:                    .... ..0. = no external routing capability
OSPF:                    .... ...0 = no Type of Service routing capability
OSPF:          Link state type      = 5 (AS external link)
OSPF:          Link state ID        = [192.XXX.148.0]
OSPF:          Advertising Router   = [140.XXX.172.254]
OSPF:          Sequence number      = 2147484307,   Checksum = B637
OSPF:          Length               = 36
OSPF:          Network mask         = [255.255.255.0]
OSPF:          Type of service      = 80
OSPF:                    1... .... = Type 2 external metric
OSPF:                    .... .000 = routine
OSPF:          Metric               = 130
OSPF:          Forwarding Address = [0.0.0.0]
OSPF:          External Route Tag = 0x00000000
OSPF:          Local Info       = 0x00000000
OSPF:
```

The details of an OSPF Link State Acknowledgment message from Frame 3652 are shown in Trace 7.7c. This packet is multicast from a different router, the Aeronautics Lab router, and acknowledges some previously received Link State Acknowledgments. In Frame 3652, we see the OSPF header, which specifies the Link State Acknowledgment (Type = 5). A Link State Advertisement header (review Figure 4-2g) is next, followed by a Router Links Advertisement message, with Link State Type = 1 (review Figure 4-2h). A total of eight Link State Advertisements are included in this message. Advertisements 2–8 in this Link State Acknowledgment correspond with Advertisements 1–7 in Frame 2693, above.

Link State Advertisement	Link State Type	Link State ID
1	Router links	[140.XXX.172.254]
2	AS external links	[140.XXX.172.0]
3	AS external links	[140.XXX.173.0]
4	AS external links	[140.XXX.174.0]
5	AS external links	[140.XXX.187.0]
6	AS external links	[140.XXX.253.0]
7	AS external links	[140.XXX.254.0]
8	AS external links	[140.XXX.148.0]

Trace 7.7c. OSPF Link State Acknowledgment Message Details
Sniffer Network Analyzer data from 23-Apr at 10:23:56, file NIS5.ENC, Pg 1

```
- - - - - - - - - - - - - - - - Frame 3652 - - - - - - - - - - - - - - - - -
DLC:        — DLC Header —
DLC:
DLC:        Frame 3652 arrived at 10:24:02.5582; frame size is 218 (00DA hex) bytes.
DLC:        Destination = Multicast 01005E000005
DLC:        Source      = Station Cisco 599718
DLC:        Ethertype   = 0800 (IP)
DLC:
IP:         — IP Header —
IP:
IP:         Version = 4, header length = 20 bytes
IP:         Type of service = C0
IP:                 110. .... = internetwork control
IP:                 ...0 .... = normal delay
IP:                 .... 0... = normal throughput
IP:                 .... .0.. = normal reliability
IP:         Total length    = 204 bytes
```

```
IP:       Identification  = 33912
IP:       Flags           = 0X
IP:                 .0.. .... = may fragment
IP:                 ..0. .... = last fragment
IP:       Fragment offset = 0 bytes
IP:       Time to live    = 1 seconds/hops
IP:       Protocol        = 89 (OSPFIGP)
IP:       Header checksum = BFF7 (correct)
IP:       Source address      = [140.XXX.6.248], Aeronautics
IP:       Destination address = [224.0.0.5]
IP:       No options
IP:
OSPF:     — OSPF Header —
OSPF:
OSPF:     Version = 2,  Type = 5 (Link State Acknowledgment),  Length = 184
OSPF:     Router ID       = [140.XXX.241.100]
OSPF:     Area ID         = [0.0.0.0]
OSPF:     Header checksum = 176E (correct)
OSPF:     Authentication: Type = 0 (No Authentication),  Value = 00 00 00 00
          00 00 00 00
OSPF:
OSPF:     Link State Advertisement Header # 1
OSPF:     Link state age        = 4 (seconds)
OSPF:     Optional capabilities = 02
OSPF:               .0.. .... = Opaque-LSAs not forwarded
OSPF:               ..0. .... = Demand Circuit bit
OSPF:               ...0 .... = External Attributes bit
OSPF:               .... 0... = no NSSA capability
OSPF:               .... .0.. = no multicast capability
OSPF:               .... ..1. = external routing capability
OSPF:               .... ...0 = no Type of Service routing capability
OSPF:     Link state type       = 1 (Router links)
OSPF:     Link state ID         = [140.XXX.172.254]
OSPF:     Advertising Router    = [140.XXX.172.254]
OSPF:     Sequence number       = 2147484346,  Checksum = 9A4B
OSPF:     Length                = 48
OSPF:
OSPF:     Link State Advertisement Header # 2
OSPF:     Link state age        = 4 (seconds)
OSPF:     Optional capabilities = 00
OSPF:               .0.. .... = Opaque-LSAs not forwarded
OSPF:               ..0. .... = Demand Circuit bit
OSPF:               ...0 .... = External Attributes bit
OSPF:               .... 0... = no NSSA capability
```

```
OSPF:                     .... .0.. = no multicast capability
OSPF:                     .... ..0. = no external routing capability
OSPF:                     .... ...0 = no Type of Service routing capability
OSPF:        Link state type      = 5 (AS external link)
OSPF:        Link state ID        = [140.XXX.172.0]
OSPF:        Advertising Router   = [140.XXX.172.254]
OSPF:        Sequence number      = 2147484307,   Checksum = 3FB3
OSPF:        Length               = 36
OSPF:
OSPF:        Link State Advertisement Header # 3
OSPF:        Link state age       = 4 (seconds)
OSPF:        Optional capabilities = 00
OSPF:                     .0.. .... = Opaque-LSAs not forwarded
OSPF:                     ..0. .... = Demand Circuit bit
OSPF:                     ...0 .... = External Attributes bit
OSPF:                     .... 0... = no NSSA capability
OSPF:                     .... .0.. = no multicast capability
OSPF:                     .... ..0. = no external routing capability
OSPF:                     .... ...0 = no Type of Service routing capability
OSPF:        Link state type      = 5 (AS external link)
OSPF:        Link state ID        = [140.XXX.173.0]
OSPF:        Advertising Router   = [140.XXX.172.254]
OSPF:        Sequence number      = 2147484307,   Checksum = 34BD
OSPF:        Length               = 36
OSPF:
OSPF:        Link State Advertisement Header # 4
OSPF:        Link state age       = 4 (seconds)
OSPF:        Optional capabilities = 00
OSPF:                     .0.. .... = Opaque-LSAs not forwarded
OSPF:                     ..0. .... = Demand Circuit bit
OSPF:                     ...0 .... = External Attributes bit
OSPF:                     .... 0... = no NSSA capability
OSPF:                     .... .0.. = no multicast capability
OSPF:                     .... ..0. = no external routing capability
OSPF:                     .... ...0 = no Type of Service routing capability
OSPF:        Link state type      = 5 (AS external link)
OSPF:        Link state ID        = [140.XXX.174.0]
OSPF:        Advertising Router   = [140.XXX.172.254]
OSPF:        Sequence number      = 2147484307,   Checksum = 29C7
OSPF:        Length               = 36
OSPF:
OSPF:        Link State Advertisement Header # 5
OSPF:        Link state age       = 4 (seconds)
OSPF:        Optional capabilities = 00
```

```
OSPF:              .0.. .... = Opaque-LSAs not forwarded
OSPF:              ..0. .... = Demand Circuit bit
OSPF:              ...0 .... = External Attributes bit
OSPF:              .... 0... = no NSSA capability
OSPF:              .... .0.. = no multicast capability
OSPF:              .... ..0. = no external routing capability
OSPF:              .... ...0 = no Type of Service routing capability
OSPF:       Link state type      = 5 (AS external link)
OSPF:       Link state ID        = [140.XXX.187.0]
OSPF:       Advertising Router   = [140.XXX.172.254]
OSPF:       Sequence number      = 2147484307,   Checksum = 994A
OSPF:       Length               = 36
OSPF:
OSPF:       Link State Advertisement Header # 6
OSPF:       Link state age       = 4 (seconds)
OSPF:       Optional capabilities = 00
OSPF:              .0.. .... = Opaque-LSAs not forwarded
OSPF:              ..0. .... = Demand Circuit bit
OSPF:              ...0 .... = External Attributes bit
OSPF:              .... 0... = no NSSA capability
OSPF:              .... .0.. = no multicast capability
OSPF:              .... ..0. = no external routing capability
OSPF:              .... ...0 = no Type of Service routing capability
OSPF:       Link state type      = 5 (AS external link)
OSPF:       Link state ID        = [140.XXX.253.0]
OSPF:       Advertising Router   = [140.XXX.172.254]
OSPF:       Sequence number      = 2147484317,   Checksum = ACEA
OSPF:       Length               = 36
OSPF:
OSPF:       Link State Advertisement Header # 7
OSPF:       Link state age       = 4 (seconds)
OSPF:       Optional capabilities = 00
OSPF:              .0.. .... = Opaque-LSAs not forwarded
OSPF:              ..0. .... = Demand Circuit bit
OSPF:              ...0 .... = External Attributes bit
OSPF:              .... 0... = no NSSA capability
OSPF:              .... .0.. = no multicast capability
OSPF:              .... ..0. = no external routing capability
OSPF:              .... ...0 = no Type of Service routing capability
OSPF:       Link state type      = 5 (AS external link)
OSPF:       Link state ID        = [140.XXX.254.0]
OSPF:       Advertising Router   = [140.XXX.172.254]
OSPF:       Sequence number      = 2147484307,   Checksum = B5EA
OSPF:       Length               = 36
```

```
OSPF:
OSPF:     Link State Advertisement Header # 8
OSPF:     Link state age       = 4 (seconds)
OSPF:     Optional capabilities = 00
OSPF:               .0.. .... = Opaque-LSAs not forwarded
OSPF:               ..0. .... = Demand Circuit bit
OSPF:               ...0 .... = External Attributes bit
OSPF:               .... 0... = no NSSA capability
OSPF:               .... .0.. = no multicast capability
OSPF:               .... ..0. = no external routing capability
OSPF:               .... ...0 = no Type of Service routing capability
OSPF:     Link state type      = 5 (AS external link)
OSPF:     Link state ID        = [192.149.148.0]
OSPF:     Advertising Router   = [140.XXX.172.254]
OSPF:     Sequence number      = 2147484307,  Checksum = B637
OSPF:     Length               = 36
OSPF:
```

Frames 3862 and 4608 in Trace 7.7d contain OSPF Hello messages, sent from the Aeronautics and Broadway routers, respectively. Within the OSPF header, the Type = 1 field indicates a Hello message and the Router ID indicates that this message came from network [140.XXX.241.100]. That network's Designated router is [140.XXX.6.253], the Broadway router, and its Backup Designated router is [140.XXX.6.248], the Aeronautics Lab router. In Frame 4608, the Router ID indicates that it came from network [140.XXX.1.1]. That network's Designated router is [140.XXX.6.253], the Broadway router, and its Backup Designated router is [140.XXX.6.248], the Aeronautics Lab router.

Trace 7.7d. OSPF Hello Message Details

Sniffer Network Analyzer data from 23-Apr at 10:23:56, file NIS5.ENC, Pg 1

```
- - - - - - - - - - - - - - - - Frame 3862 - - - - - - - - - - - - - - - - -
DLC:      — DLC Header —
DLC:
DLC:      Frame 3862 arrived at 10:24:02.9741; frame size is 82 (0052 hex) bytes.
DLC:      Destination = Multicast 01005E000005
DLC:      Source      = Station Cisco 599718
DLC:      Ethertype   = 0800 (IP)
DLC:
IP:       — IP Header —
IP:
IP:       Version = 4, header length = 20 bytes
```

```
IP:      Type of service = C0
IP:            110. .... = internetwork control
IP:            ...0 .... = normal delay
IP:            .... 0... = normal throughput
IP:            .... .0.. = normal reliability
IP:      Total length    = 68 bytes
IP:      Identification  = 33913
IP:      Flags           = 0X
IP:            .0.. .... = may fragment
IP:            ..0. .... = last fragment
IP:      Fragment offset = 0 bytes
IP:      Time to live    = 1 seconds/hops
IP:      Protocol        = 89 (OSPFIGP)
IP:      Header checksum = C07E (correct)
IP:      Source address      = [140.XXX.6.248], Aeronautics
IP:      Destination address = [224.0.0.5]
IP:      No options
IP:
OSPF:    — OSPF Header —
OSPF:
OSPF:    Version = 2,  Type = 1 (Hello),  Length = 48
OSPF:    Router ID       = [140.XXX.241.100]
OSPF:    Area ID         = [0.0.0.0]
OSPF:    Header checksum = C98D (correct)
OSPF:    Authentication: Type = 0 (No Authentication), Value = 00 00 00 00 00 00 00 00
OSPF:
OSPF:    Network mask            = [255.255.255.0]
OSPF:    Hello interval          = 10 (seconds)
OSPF:    Optional capabilities   = 02
OSPF:          .0.. .... = Opaque-LSAs not forwarded
OSPF:          ..0. .... = Demand Circuit bit
OSPF:          ...0 .... = External Attributes bit
OSPF:          .... 0... = no NSSA capability
OSPF:          .... .0.. = no multicast capability
OSPF:          .... ..1. = external routing capability
OSPF:          .... ...0 = no Type of Service routing capability
OSPF:    Router priority         = 1
OSPF:    Router dead interval    = 40 (seconds)
OSPF:    Designated router       = [140.XXX.6.253]
OSPF:    Backup designated router = [140.XXX.6.248]
OSPF:    Neighbor (1)            = [140.XXX.1.1]
OSPF:
```

```
---------------- Frame 4608 ----------------
DLC:    — DLC Header —
DLC:
DLC:    Frame 4608 arrived at 10:24:05.1568; frame size is 82 (0052 hex) bytes.
DLC:    Destination = Multicast 01005E000005
DLC:    Source      = Station 00E0F92F20A0
DLC:    Ethertype   = 0800 (IP)
DLC:
IP:     — IP Header —
IP:
IP:     Version = 4, header length = 20 bytes
IP:     Type of service = C0
IP:             110. .... = internetwork control
IP:             ...0 .... = normal delay
IP:             .... 0... = normal throughput
IP:             .... .0.. = normal reliability
IP:     Total length    = 68 bytes
IP:     Identification  = 12492
IP:     Flags           = 0X
IP:             .0.. .... = may fragment
IP:             ..0. .... = last fragment
IP:     Fragment offset = 0 bytes
IP:     Time to live    = 1 seconds/hops
IP:     Protocol        = 89 (OSPFIGP)
IP:     Header checksum = 1427 (correct)
IP:     Source address      = [140.XXX.6.253], Broadway
IP:     Destination address = [224.0.0.5]
IP:     No options
IP:
OSPF:   — OSPF Header —
OSPF:
OSPF:   Version = 2,  Type = 1 (Hello),  Length = 48
OSPF:   Router ID     = [140.XXX.1.1]
OSPF:   Area ID       = [0.0.0.0]
OSPF:   Header checksum = C98D (correct)
OSPF:   Authentication: Type = 0 (No Authentication),
        Value = 00 00 00 00 00 00 00 00
OSPF:
OSPF:   Network mask          = [255.255.255.0]
OSPF:   Hello interval        = 10 (seconds)
OSPF:   Optional capabilities   = 02
OSPF:           .0.. .... = Opaque-LSAs not forwarded
OSPF:           ..0. .... = Demand Circuit bit
OSPF:           ...0 .... = External Attributes bit
```

```
OSPF:                 .... 0... = no NSSA capability
OSPF:                 .... .0.. = no multicast capability
OSPF:                 .... ..1. = external routing capability
OSPF:                 .... ...0 = no Type of Service routing capability
OSPF:      Router priority       = 1
OSPF:      Router dead interval  = 40 (seconds)
OSPF:      Designated router     = [140.XXX.6.253]
OSPF:      Backup dcsignated router = [140.XXX.6.248]
OSPF:      Neighbor (1)          = [140.XXX.241.100]
OSPF:
```

Our last example, Trace 7.7e, illustrates a BGP Update message sent from the Science Lab router to the Broadway router. Note that BGP messages are sent over TCP, using TCP Port 179. Within the BGP header (review Figure 4-4a), note that the Message Type field indicates an Update message (Type = 2), as shown in Figure 4-4c. The Unfeasible Routes length is zero, indicating that no routes are being withdrawn from service. Three Path Attributes follow: an Origin, indicating that the Network Layer Reachability information is interior to the originating AS; an AS path, which is followed by a sequence of AS path segments (513 and 297); and a Next Hop, which defines the IP address of the border router that should be used as the next hop to the Network Layer Reachability field. Finally, the Network Layer Reachability field contains the IP address of interest: [204.XXX.66.0].

Trace 7.7e. BGP Update Message Details

```
---------------- Frame 5841 ----------------
DLC:       — DLC Header —
DLC:
DLC:       Frame 5841 arrived at 10:24:08.4565; frame size is 99 (0063 hex) bytes.
DLC:       Destination = Station 00E0F92F20A0
DLC:       Source      = Station Cisco 389872
DLC:       Ethertype   = 0800 (IP)
DLC:
IP:        — IP Header —
IP:
IP:        Version = 4, header length = 20 bytes
IP:        Type of service = C0
IP:                  110. .... = internetwork control
IP:                  ...0 .... = normal delay
IP:                  .... 0... = normal throughput
IP:                  .... .0.. = normal reliability
```

```
IP:          Total length    = 85 bytes
IP:          Identification  = 1738
IP:          Flags           = 0X
IP:                  .0.. .... = may fragment
IP:                  ..0. .... = last fragment
IP:          Fragment offset = 0 bytes
IP:          Time to live    = 1 seconds/hops
IP:          Protocol        = 6 (TCP)
IP:          Header checksum = 8ACA (correct)
IP:          Source address      = [140.XXX.6.250], Science
IP:          Destination address = [140.XXX.6.253], Broadway
IP:          No options
IP:
TCP:         — TCP Header —
TCP:
TCP:         Source port            = 179 (BGP)
TCP:         Destination port       = 37590
TCP:         Sequence number        = 3540261264
TCP:         Acknowledgment number  = 641647944
TCP:         Data offset            = 20 bytes
TCP:         Flags                  = 18
TCP:                 ..0. .... = (No urgent pointer)
TCP:                 ...1 .... = Acknowledgment
TCP:                 .... 1... = Push
TCP:                 .... .0.. = (No reset)
TCP:                 .... ..0. = (No SYN)
TCP:                 .... ...0 = (No FIN)
TCP:         Window                 = 15852
TCP:         Checksum               = CAEE (correct)
TCP:         No TCP options
TCP:         [45 byte(s) of data]
TCP
BGP:         — BGP Message —
BGP:
BGP:         16 byte Marker  (all 1's)
BGP:         Length      = 45
BGP:         BGP type    = 2 (Update)
BGP:
BGP:         Unfeasible Routes Length   = 0
BGP:         No Withdrawn Routes in this Update
BGP:         Path Attribute Length   = 18 bytes
BGP:         Attribute Flags = 4X
BGP:                 0... .... = Well-known
BGP:                 .1.. .... = Transitive
```

```
BGP:                  ..0. .... = Complete
BGP:                  ...0 .... = 1 byte Length
BGP:        Attribute type code     = 1 (Origin)
BGP:        Attribute Data Length   = 1
BGP:        Origin type             = 0 (IGP)
BGP:        Attribute Flags = 4X
BGP:                  0... .... = Well-known
BGP:                  .1.. .... = Transitive
BGP:                  ..0. .... = Complete
BGP:                  ...0 .... = 1 byte Length
BGP:        Attribute type code     = 2 (AS Path)
BGP:        Attribute Data Length   = 4
BGP:        AS Identifier           = 513
BGP:        AS Identifier           = 297
BGP:        Attribute Flags = 4X
BGP:                  0... .... = Well-known
BGP:                  .1.. .... = Transitive
BGP:                  ..0. .... = Complete
BGP:                  ...0 .... = 1 byte Length
BGP:        Attribute type code     = 3 (Next Hop)
BGP:        Attribute Data Length   = 4
BGP:        Next Hop                = [140.XXX.6.250], Science
BGP:
BGP:        Network Layer Reachability Information:
BGP:        IP Prefix Length = 24 bits, IP subnet mask [255.255.255.0]
BGP:        IP address [204.XXX.66.0]
DLC:        — Frame too short —
```

7.8 Examining BOOTP with UDP Transport

The Bootstrap Protocol (BOOTP), described in RFC 951 [7-1] and RFC 2132 [7-2], allows a client device to obtain its bootup parameters from a server. BOOTP runs over the UDP transport and uses two defined ports: Port 67 (BOOTP Server) and Port 68 (BOOTP Client). Because it uses the connectionless UDP as the transport, the connection depends on the end-user processes for reliability. The BOOTP message has a standard format (review Figure 3-8a) for both client requests and server replies, which adds to the protocol's simplicity.

Once the client locates its server and boot file, it uses the Trivial File Transfer Protocol (TFTP) to obtain the actual file, as described in RFC 906 [7-3]. Let's look at what happens when the client can't find its server.

In this example, a Retix bridge is configured to obtain its SNMP parameters from a Sun workstation located on another segment (see Figure

7-5). Upon power-up, the bridge broadcasts a BOOTP request, looking for the BOOTP server (see Trace 7.8a). Unfortunately, the administrator of the Sun workstation has not loaded the BOOTP daemon on the Sun workstation, so the BOOTP requests go unanswered. Note that the BOOTP request packets are staggered in time and transmitted at relatively long intervals (2 to 17 seconds apart).

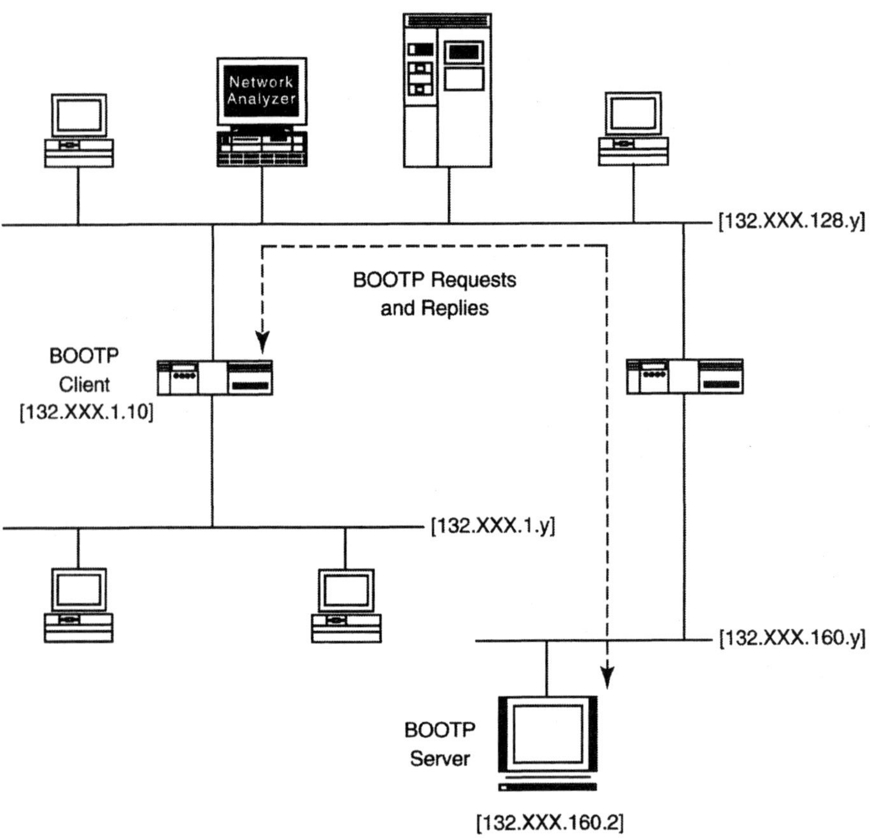

Figure 7-5. Booting Remote Bridge Using BOOTP.

When the bridge administrator realizes that his device is not operating properly, he studies the repeated BOOTP requests and theorizes that the BOOTP server is not listening and therefore not responding. His assumption is correct. After the BOOTP daemon is loaded on the Sun workstation, the bridge receives a response (Frame 4 of Trace 7.8b). The bridge then requests its boot file from the Sun workstation and receives its information in Frame 6. An acknowledgment from the bridge (Frame 7) completes the transaction.

Details of the scenario (Trace 7.8c) show how the BOOTP and TFTP protocols work together. The BOOTP request (Frame 3) contains the boot file name (90034CF1) that the bridge requires. The BOOTP reply (Frame 4) contains a server-assigned IP address for the client [132.XXX.1.10] and the address of the BOOTP server [132.XXX.160.2]. The location of the boot file name (/tftpboot/90034CF1) is also included. Next, the bridge sends a TFTP Read request in Frame 5. The BOOTP server responds with a data packet containing the configuration parameters the bridge requires. A TFTP ACK from the bridge completes the transaction. With the BOOTP daemon properly installed on the Sun workstation, the bridge can receive its configuration parameters and begin initialization.

Trace 7.8a. Bridge BOOTP Unanswered Request Summary
Sniffer Network Analyzer data 5-Sep at 14:02:52, RETXBOOT.ENC, Pg 1

SUMMARY	Delta T	Destination	Source	Summary
M 1		090077000001	Retix 034CF1	DSAP=80, I frame
2	10.8868	Broadcast	Retix 034CF1	BOOTP Request
3	2.2928	Broadcast	Retix 034CF1	BOOTP Request
4	9.8320	Broadcast	Retix 034CF1	BOOTP Request
5	2.0745	Broadcast	Retix 034CF1	BOOTP Request
6	11.0334	Broadcast	Retix 034CF1	BOOTP Request
7	2.1841	Broadcast	Retix 034CF1	BOOTP Request
8	6.1166	CMC 614107	Retix 034CF1	ARP R PA=[0.0.0.0] HA=080090034CF1 PRO=IP
9	7.2104	Broadcast	Retix 034CF	BOOTP Request
10	2.0746	Broadcast	Retix 034CF1	BOOTP Request
11	8.1893	090077000001	Retix 034CF1	DSAP=80, I frame
12	7.2135	Broadcast	Retix 034CF1	BOOTP Request
13	2.2931	Broadcast	Retix 034CF1	BOOTP Request
14	15.4028	Broadcast	Retix 034CF1	BOOTP Request
15	2.0747	Broadcast	Retix 034CF1	BOOTP Request
16	13.2180	Broadcast	Retix 034CF1	BOOTP Request
17	2.2931	Broadcast	Retix 034CF1	BOOTP Request
18	8.7395	Broadcast	Retix 034CF1	BOOTP Request
19	2.0746	Broadcast	Retix 034CF1	BOOTP Request
20	17.6970	Broadcast	Retix 034CF1	BOOTP Request
21	2.1838	Broadcast	Retix 034CF1	BOOTP Request
22	7.7562	Broadcast	Retix 034CF1	BOOTP Request

```
  23      2.0747  Broadcast      Retix 034CF1    BOOTP Request
  24     12.1260  Broadcast      Retix 034CF1    BOOTP Request
  25      2.1836  Broadcast      Retix 034CF1    BOOTP Request
  26      5.4624  Broadcast      Retix 034CF1    BOOTP Request
  27      2.0746  Broadcast      Retix 034CF1    BOOTP Request
  28      5.4625  Broadcast      Retix 034CF1    BOOTP Request
  29      2.2931  Broadcast      Retix 034CF1    BOOTP Request
  30     16.4951  Broadcast      Retix 034CF1    BOOTP Request
  31      2.0746  Broadcast      Retix 034CF1    BOOTP Request
```

Trace 7.8b. Bridge BOOTP Request/Reply Summary
Sniffer Network Analyzer data 16-Mar at 09:48:26, BOOTP.ENC, Pg 1

```
SUMMARY  Delta T   Destination    Source          Summary
M  1               090077000001   Retix 034CF1    DSAP=80, I frame
   2      6.6956   090077000001   Retix 034CF1    DSAP=80, I frame
   3      4.1908   Broadcast      Retix 034CF1    BOOTP Request
   4      0.5307   Retix 034CF1   Sun 0AB646      BOOTP Reply
   5      0.0053   Sun 0AB646     Retix 034CF1    TFTP Read request
                                                  File=/tftpboot/
                                                  90034CF1
   6      0.2037   Retix 034CF1   Sun   0AB646    TFTP Data packet
                                                  NS=1 (Last)
   7      0.0033   Sun 0AB646     Retix 034CF1    TFTP Ack NR=1
   8     50.2652   090077000001   Retix 034CF1    DSAP=80, I frame
```

Trace 7.8c. Bridge BOOTP Request/Reply Details
Sniffer Network Analyzer data 16-Mar at 09:48:26, BOOTP.ENC, Pg 1

```
- - - - - - - - - - - - - - - Frame 3 - - - - - - - - - - - - - - - -
BOOTP:     — BOOTP Header —
BOOTP:
BOOTP:     Boot record type       = 1 (Request)
BOOTP:     Hardware address type  = 1 10Mb Ethernet
BOOTP:     Hardware address length = 6 bytes
BOOTP:
BOOTP:     Hops = 0
BOOTP:     Transaction id = 0000063F
BOOTP:     Elapsed boot time = 0 seconds
BOOTP:
BOOTP:     Client self-assigned IP address = [0.0.0.0] (Unknown)
BOOTP:     Client hardware address = Retix 034CF1
BOOTP:
BOOTP:     Host name = "
```

```
BOOTP:      Boot file name = "90034CF1"
BOOTP:
BOOTP:      [Vendor specific information]
BOOTP:

-------------- Frame 4 ----------------
BOOTP:      — BOOTP Header —
BOOTP:
BOOTP:      Boot record type        = 2 (Reply)
BOOTP:      Hardware address type   = 1 10Mb Ethernet
BOOTP:      Hardware address length = 6 bytes
BOOTP:
BOOTP:      Hops = 0
BOOTP:      Transaction id = 0000063F
BOOTP:      Elapsed boot time = 0 seconds
BOOTP:
BOOTP:      Client self-assigned IP address = [0.0.0.0] (Unknown)
BOOTP:      Client server-assigned IP address = [132.XXX.1.10]
BOOTP:      Server IP address = [132.XXX.160.2]
BOOTP:      Gateway IP address = [132.XXX.160.2]
BOOTP:      Client hardware address = Retix 034CF1
BOOTP:
BOOTP:      Host name = "
BOOTP:      Boot file name = "/tftpboot/90034CF1"
BOOTP:
BOOTP:      [Vendor specific information]
BOOTP:

--------------- Frame 5 ----------------
TFTP:       — Trivial File Transfer —
TFTP:
TFTP:       Opcode = 1 (Read request)
TFTP:       File name = "/tftpboot/90034CF1"
TFTP:       Mode = "octet"
TFTP:
TFTP:       [Normal end of "Trivial File Transfer".]
TFTP:

--------------- Frame 6 ----------------
TFTP:       — Trivial File Transfer —
TFTP:
TFTP:       Opcode = 3 (Data packet)
TFTP:       Block number = 1
TFTP:       [160 bytes of data] (Last frame)
```

```
TFTP:
TFTP:      [Normal end of "Trivial File Transfer".]
TFTP:

--------------- Frame 7 -----------------
TFTP:      — Trivial File Transfer —
TFTP:
TFTP:      Opcode = 4 (Ack)
TFTP:      Block number = 1
TFTP:
TFTP:      [Normal end of "Trivial File Transfer".]
TFTP:
```

7.9 Establishing and Terminating TCP Connections

In Chapter 5, we learned that the TCP connection must be established with a three-way handshake prior to any transmission of data. Once the data has been transferred, a modified three-way handshake terminates the connection. Let's study how these operations function, using a PC and its server as examples.

In this case study, a PC wishes to establish several logical connections to a Sun server via a single Ethernet physical connection (Figure 7-6). Different Port numbers identify the different logical connections (created using TCP's multiplexing capabilities). Let's follow the steps of one of these logical connections from establishment to data transfer to termination (see Trace 7.9a).

A TCP segment identifies the connection establishment with the Synchronize bit set (SYN = 1) and no data in the segment (LEN = 0). We can see this in Frame 706, along with an initial Sequence number, ISN = 1988352000. Also note that the Sun server is initiating this connection and that it is advertising a rather large window size (WIN = 24576). The Source port (S = 20) identifies the File Transfer Protocol (FTP), and the Destination port (D = 1227) is assigned by the Sun server. The PC responds with a similar synchronize segment in Frame 707; the receipt of the server's segment is acknowledged (ACK = 1988352001), the PC's ISN is sent (SEQ = 201331252), but a smaller window size is allowed (WIN = 1024). Note that the Source and Destination Port numbers are now reversed, since the original source (the Sun server) has now become the PC's destination. Frame 708 completes the connection establishment, with the server acknowledging the PC's ISN (ACK = 201331253).

The details of the connection establishment are shown in Trace 7.9b. Note that one TCP option (the maximum segment size) is used. Maximum segment size = 1460 suggests transmission over an Ethernet/IEEE 802.3 LAN. The Ethernet/IEEE 802.3 can accommodate 1,500 octets of data within the frame. Of these, 20 octets are used for the TCP header and another 20 octets for the IP header, leaving 1,460 octets for the TCP segment. Data transfer may now proceed.

Data transfer begins in Frame 710 with a TCP segment of 1,024 octets (LEN = 1024). (Note that Frame 709 belongs to a previously established FTP connection destined for another port (D = 1219) on the same PC. While Frame 709 illustrates TCP's port multiplexing capabilities, it has nothing to do with the current discussion since it belongs to another logical connection.) The PC acknowledges receipt of the data in Frame 713 by sending ACK = 1988353025 (1988352001 + 1024 = 1988353025). The PC will not permit any more data at this time, and it indicates this by shutting its window (WIN = 0). In Frame 715, the PC's processor has caught up with its backlog (note that it was servicing Port 1219 in Frame 714) and restores data flow by sending WIN = 1024. The server responds with another 1,024 octets in Frame 716. This process proceeds normally until the server completes its business.

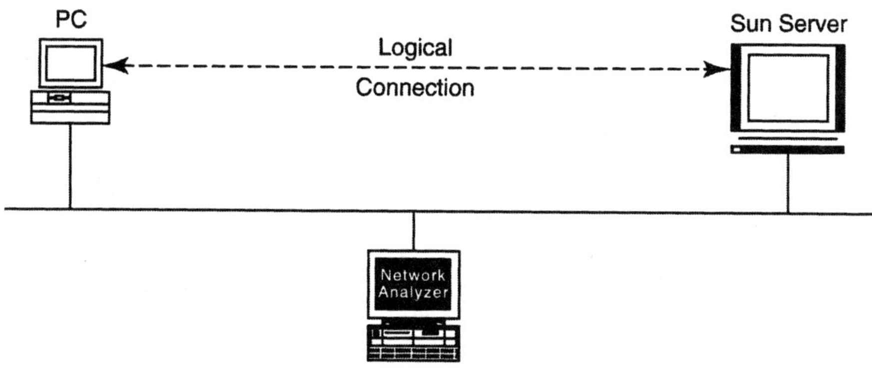

Figure 7-6. TCP Connection Establishment/Termination.

In Frame 779, the server finishes its business and sends the Finish flag (FIN = 1) along with its last 114 octets of data. The PC responds with an acknowledgment in Frame 780 (ACK = 1988384884), then sends a second segment (Frame 781) containing both an ACK and a FIN. Note that the details in Trace 7.9c indicate that the PC does not shut down the connection, but permits up to 910 more octets of data from the other end (Window = 910). The server has said its piece, however; it acknowledges the FIN and closes its end of the connection in Frame 782. Both ends of the logical connection are now closed. In the following example, we'll see what happens when both ends are unable to maintain the connection and the Reset (RST) flag is required.

Trace 7.9a. TCP Connection Establishment and Termination Summary
Sniffer Network Analyzer data 27-Mar at 09:04:54, TCPMEDLY.ENC, Pg 1

SUMMARY	Delta T	Destination	Source	Summary
706	0.0027	PC	Sun Server	TCP D=1227 S=20 SYN SEQ=1988352000 LEN=0 WIN=24576
707	0.0139	Sun Server	PC	TCP D=20 S=1227 SYN ACK=1988352001 SEQ=201331252 LEN=0 WIN=1024
708	0.0006	PC	Sun Server	TCP D=1227 S=20 ACK=201331253 WIN=24576
709	0.0007	PC	Sun Server	FTP R PORT=1219 150 ASCII data connection
710	0.0993	PC	Sun Server	TCP D=1227 S=20 ACK=201331253 SEQ=1988352001 LEN=1024 WIN=24576
711	0.0733	Sun Server	PC	TCP D=21 S=1219 ACK=1972865105 WIN=951
712	0.0008	PC	Sun Server	FTP R PORT=1219 226 ASCII Transfer complete <0D><0A>
713	0.0529	Sun Server	PC	TCP D=20 S=1227 ACK=1988353025 WIN=0

714	0.0535	Sun Server	PC	TCP D=21 S=1219 ACK=1972865135 WIN=956
715	0.0790	Sun Server	PC	TCP D=20 S=1227 ACK=1988353025 WIN=1024
716	0.0015	PC	Sun Server	TCP D=1227 S=20 ACK=201331253 SEQ=1988353025 LEN=1024 WIN=24576
717	0.1379	Sun Server	PC	TCP D=20 S=1227 ACK=1988354049 WIN=0
718	0.0644	Sun Server	PC	TCP D=20 S=1227 ACK=1988354049 WIN=1024
719	0.0015	PC	Sun Server	TCP D=1227 S=20 ACK=201331253 SEQ=1988354049 LEN=1024 WIN=24576
720	0.0553	Sun Server	PC	TCP D=20 S=1227 ACK=1988355073 WIN=1024
721	0.0015	PC	Sun Server	TCP D=1227 S=20 ACK=201331253 SEQ=1988355073 LEN=1024 WIN=24576
722	0.0575	Sun Server	PC	TCP D=20 S=1227 ACK=1988356097 WIN=1024
.				
.				
.				
779	0.0008	PC	Sun Server	TCP D=1227 S=20 FIN ACK=201331253 SEQ=1988384769 LEN=114 WIN=24576
780	0.0204	Sun Server	PC	TCP D=20 S=1227 ACK=1988384884 WIN=910
781	0.0616	Sun Server	PC	TCP D=20 S=1227 FIN ACK=1988384884 SEQ=201331253 LEN=0 WIN=910

```
    782     0.0006   PC              Sun Server    TCP D=1227 S=20
                                                     ACK=201331254
                                                     WIN=24576
```

Trace 7.9b. TCP Connection Synchronization (SYN) Details
Sniffer Network Analyzer data 27-Mar at 09:04:54, TCPMEDLY.ENC, Pg 1

```
- - - - - - - - - - - - - - - Frame 706 - - - - - - - - - - - - - - - - -
TCP:        — TCP Header —
TCP:
TCP:        Source port = 20 (FTP data)
TCP:        Destination port = 1227
TCP:        Initial sequence number = 1988352000
TCP:        Data offset = 24 bytes
TCP:        Flags = 02
TCP:        ..0. .... = (No urgent pointer)
TCP:        ...0 .... = (No acknowledgment)
TCP:        .... 0... = (No push)
TCP:        .... .0.. = (No reset)
TCP:        .... ..1. = SYN
TCP:        .... ...0 = (No FIN)
TCP:        Window = 24576
TCP:        Checksum = D02F (correct)
TCP:
TCP:        Options follow
TCP:        Maximum segment size = 1460
TCP:

- - - - - - - - - - - - - - - Frame 707 - - - - - - - - - - - - - - - - -
TCP:        — TCP Header —
TCP:
TCP:        Source port = 1227
TCP:        Destination port = 20 (FTP data)
TCP:        Initial sequence number = 201331252
TCP:        Acknowledgment number = 1988352001
TCP:        Data offset = 24 bytes
TCP:        Flags = 12
TCP:        ..0. .... = (No urgent pointer)
TCP:        ...1 .... = Acknowledgment
TCP:        .... 0... = (No push)
TCP:        .... .0.. = (No reset)
TCP:        .... ..1. = SYN
TCP:        .... ...0 = (No FIN)
TCP:        Window = 1024
```

```
TCP:        Checksum = 0DEB (correct)
TCP:
TCP:        Options follow
TCP:        Maximum segment size = 1460
TCP:

- - - - - - - - - - - - - - - Frame 708 - - - - - - - - - - - - - - - - -
TCP:        — TCP Header —
TCP:
TCP:        Source port = 20 (FTP data)
TCP:        Destination port = 1227
TCP:        Sequence number = 1988352001
TCP:        Acknowledgment number = 201331253
TCP:        Data offset = 20 bytes
TCP:        Flags = 10
TCP:        ..0. .... = (No urgent pointer)
TCP:        ...1 .... = Acknowledgment
TCP:        .... 0... = (No push)
TCP:        .... .0.. = (No reset)
TCP:        .... ..0. = (No SYN)
TCP:        .... ...0 = (No FIN)
TCP:        Window = 24576
TCP:        Checksum = C9A7 (correct)
TCP:        No TCP options
TCP:
```

Trace 7.9c. TCP Connection Termination (FIN) Details
Sniffer Network Analyzer data 27-Mar at 09:04:54, TCPMEDLY.ENC, Pg 1

```
- - - - - - - - - - - - - - - Frame 779 - - - - - - - - - - - - - - - - -
TCP:        —TCP Header —
TCP:
TCP:        Source port = 20 (FTP data)
TCP:        Destination port = 1227
TCP:        Sequence number = 1988384769
TCP:        Acknowledgment number = 201331253
TCP:        Data offset = 20 bytes
TCP:        Flags = 19
TCP:        ..0. .... = (No urgent pointer)
TCP:        ...1 .... = Acknowledgment
TCP:        .... 1... = Push
TCP:        .... .0.. = (No reset)
TCP:        .... ..0. = (No SYN)
TCP:        .... ...1 = FIN
```

```
TCP:        Window = 24576
TCP:        Checksum = 492C (correct)
TCP:        No TCP options
TCP:        [114 byte(s) of data]
TCP:

--------------- Frame 780 ----------------
TCP:        — TCP Header —
TCP:
TCP:        Source port = 1227
TCP:        Destination port = 20 (FTP data)
TCP:        Sequence number = 201331253
TCP:        Acknowledgment number = 1988384884
TCP:        Data offset = 20 bytes
TCP:        Flags = 10
TCP:        ..0. .... = (No urgent pointer)
TCP:        ...1 .... = Acknowledgment
TCP:        .... 0... = (No push)
TCP:        .... .0.. = (No reset)
TCP:        .... ..0. = (No SYN)
TCP:        .... ...0 = (No FIN)
TCP:        Window = 910
TCP:        Checksum = A5A6 (correct)
TCP:        No TCP options
TCP:

--------------- Frame 781 ----------------
TCP:        — TCP Header —
TCP:
TCP:        Source port = 1227
TCP:        Destination port = 20 (FTP data)
TCP:        Sequence number = 201331253
TCP:        Acknowledgment number = 1988384884
TCP:        Data offset = 20 bytes
TCP:        Flags = 11
TCP:        ..0. .... = (No urgent pointer)
TCP:        ...1 .... = Acknowledgment
TCP:        .... 0... = (No push)
TCP:        .... .0.. = (No reset)
TCP:        .... ..0. = (No SYN)
TCP:        .... ...1 = FIN
TCP:        Window = 910
TCP:        Checksum = A5A5 (correct)
TCP:        No TCP options
TCP:
```

```
--------------- Frame 782 ----------------
TCP:        — TCP Header —
TCP:
TCP:        Source port = 20 (FTP data)
TCP:        Destination port = 1227
TCP:        Sequence number = 1988384884
TCP:        Acknowledgment number = 201331254
TCP:        Data offset = 20 bytes
TCP:        Flags = 10
TCP:        ..0. .... = (No urgent pointer)
TCP:        ...1 .... = Acknowledgment
TCP:        .... 0... = (No push)
TCP:        .... .0.. = (No reset)
TCP:        .... ..0. = (No SYN)
TCP:        .... ...0 = (No FIN)
TCP:        Window = 24576
TCP:        Checksum = 4933 (correct)
TCP:        No TCP options
TCP:
```

7.10 Reset TCP Connection

The TCP header contains six flags that manage the virtual circuit. In Chapter 5, we saw how the Acknowledgment (ACK), Synchronize (SYN), and Finish (FIN) flags are used for connection management and data transfer under normal conditions. One of the remaining flags, known as the Reset (RST) flag, is used when a TCP segment arrives that is not intended for the current connection. The RST flag is also used if a TCP module detects a fatal error or if the application process unilaterally decides to close the connection. One possible scenario is when one TCP module crashes during a session. The term used to describe this condition is a half-open connection because one end is maintaining its Sequence numbers and other transmission-related parameters while the other is not. When the crashed host returns to life, its Sequence numbers are unlikely to be the same as those it was using prior to the crash. As a result, it sends unexpected (and thus unacknowledged) Sequence numbers to the host at the other end of the link.

When the crashed host realizes that it has caused this confusion, it sends a Reset to its distant partner, which triggers a three-way handshake to re-establish the connection. Let's see how the Reset flag is used in a failure scenario.

In this case, a Sun workstation is communicating with a remote host located in another part of the country. A TELNET session is in progress over the internet, with the workstation emulating a host terminal. Two routers connect the local and remote Ethernet networks (see Figure 7-7). Without warning, the user loses his response from the host and starts hitting the Return key, hoping for a miracle (sound familiar?). We can see his frustration in Trace 7.10a. In Frame 3 all appears to be fine, but after almost two minutes of waiting (119.8658 seconds) the Sun sends a carriage return. (Note the <0D> <00> output pattern, indicating that an ASCII Carriage Return <0D> has been transmitted.) The Sun sends these characters (<0D> <00> <0D> <00>...) at increments of 2 seconds apart, then 4 seconds, 8, 16, and so on, attempting to wake up the remote host.

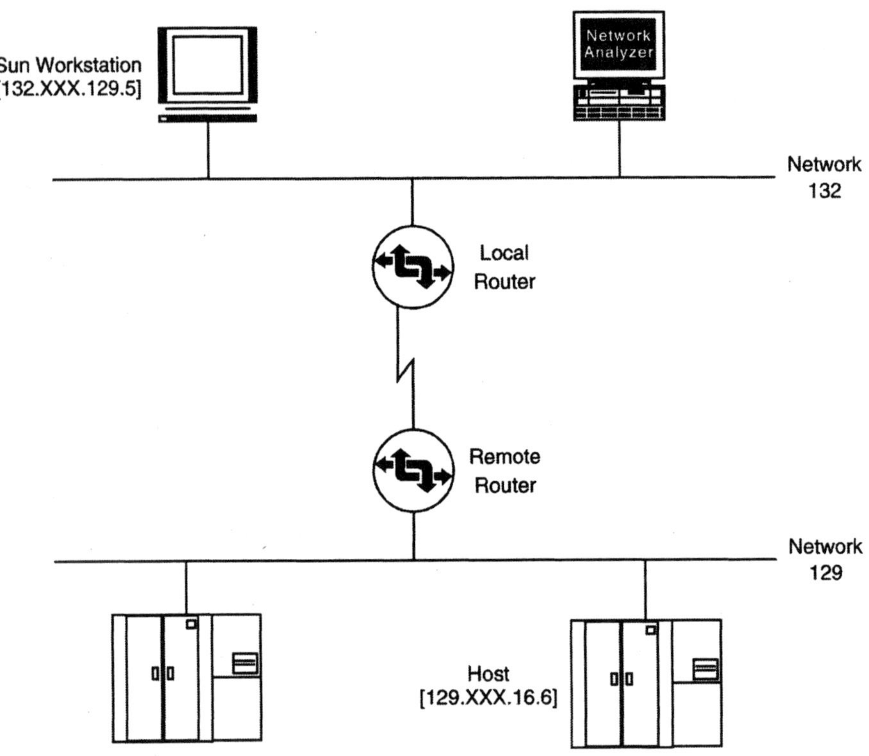

Figure 7-7. TCP Connection Reset.

Unfortunately, the efforts are in vain. The remote host gets confused, sends a TCP Reset (RST = 1) in Frame 13, and disables future receptions by setting Window = 0. The Sun resets the connection, but also acknowledges the last data octet that it received in Frame 3 (ACK = 20881447). Clearly, the Sun does not want to end the conversation since it keeps Window = 4096.

The problem is traced to a bad Ethernet card connecting the remote host to its local network. The card's operation is intermittent, working part of the time—such as when the TELNET session was initialized—then failing for no apparent reason. When the failure occurred, it caused a half-open TCP connection, and ultimately the mysterious TCP Reset. When the faulty Ethernet card was replaced, no further problems occurred.

Trace 7.10a. TCP Connection Reset (RST) Summary
Sniffer Network Analyzer data 14-Sep at 11:43:30, TCPRST.ENC, Pg 1

SUMMARY	Delta T	Destination	Source	Summary
M 1		Router	Sun 01DF5E	TCP D=23 S=1169 ACK=20881319 WIN=4096
2	5.0808	Sun 01DF5E	Router	Telnet R PORT=1169 FILE: ROUT (NO CHANGES)
3	0.1189	Router	Sun 01DF5E	TCP D=23 S=1169 ACK=20881447 WIN=4096
4	119.8658	Router	Sun 01DF5E	Telnet C PORT=1169 <0D><00>...
5	0.8820	Router	Sun 01DF5E	Telnet C PORT=1169 <0D><00>...
6	2.0002	Router	Sun 01DF5E	Telnet C PORT=1169 <0D><00>...
7	4.0001	Router	Sun 01DF5E	Telnet C PORT=1169 <0D><00>...
8	8.0006	Router	Sun 01DF5E	Telnet C PORT=1169 <0D><00>...
9	16.0010	Router	Sun 01DF5E	Telnet C PORT=1169 <0D><00>...
10	32.0020	Router	Sun 01DF5E	Telnet C PORT=1169 <0D><00>...
11	64.0039	Router	Sun 01DF5E	Telnet C PORT=1169 <0D><00>...
12	64.0040	Router	Sun 01DF5E	Telnet C PORT=1169 <0D><00>...

13	0.2722	Sun 01DF5E	Router	TCP D=1169 S=23 RST WIN=0
14	0.0021	Router	Sun 01DF5E	TCP D=23 S=1169 RST ACK=20881447 WIN=4096

Trace 7.10b. TCP Connection Reset (RST) Details
Sniffer Network Analyzer data 14-Sep at 11:43:30, TCPRST.ENC, Pg 1

```
- - - - - - - - - - - - - - - Frame 13 - - - - - - - - - - - - - - - - -
DLC:        — DLC Header —
DLC:
DLC:        Frame 13 arrived at 12:24:26.1927; frame size is 60 (003C hex) bytes.
DLC:        Destination = Station Sun   01DF5E
DLC:        Source      = Station PrteonE0807B, Router
DLC:        Ethertype  = 0800 (IP)
DLC:
IP:         — IP Header —
IP:
IP:         Version = 4, header length = 20 bytes
IP:         Type of service = 00
IP:                  000. .... = routine
IP:                  ...0 .... = normal delay
IP:                  .... 0... = normal throughput
IP:                  .... .0.. = normal reliability
IP:         Total length = 40 bytes
IP:         Identification = 25254
IP:         Flags = 0X
IP:         .0.. .... = may fragment
IP:          ..0. .... = last fragment
IP:         Fragment offset = 0 bytes
IP:         Time to live = 19 seconds/hops
IP:         Protocol = 6 (TCP)
IP:         Header checksum = AE75 (correct)
IP:         Source address = [129.XXX.16.6]
IP:         Destination address = [132.XXX.129.5]
IP:         No options
IP:
TCP:        — TCP Header —
TCP:
TCP:        Source port = 23 (Telnet)
TCP:        Destination port = 1169
TCP:        Sequence number = 20881447
TCP:        Data offset = 20 bytes
```

```
TCP:        Flags = 04
TCP:        ..0. .... = (No urgent pointer)
TCP:        ...0 .... = (No acknowledgment)
TCP:        .... 0... = (No push)
TCP:        .... .1.. = Reset
TCP:        .... ..0. = (No SYN)
TCP:        .... ...0 = (No FIN)
TCP:        Window = 0
TCP:        Checksum = 25EE (correct)
TCP:        No TCP options
TCP:

- - - - - - - - - - - - - - - Frame 14 - - - - - - - - - - - - - - - - -
DLC:        — DLC Header —
DLC:
DLC:        Frame 14 arrived at 12:24:26.1948; frame size is 60 (003C hex) bytes.
DLC:        Destination = Station PrteonE0807B, Router
DLC:        Source      = Station Sun   01DF5E
DLC:        Ethertype  = 0800 (IP)
DLC:
IP:         — IP Header —
IP:
IP:         Version = 4, header length = 20 bytes
IP:         Type of service = 00
IP:                   000. .... = routine
IP:                   ...0 .... = normal delay
IP:                   .... 0... = normal throughput
IP:                   .... .0.. = normal reliability
IP:         Total length = 40 bytes
IP:         Identification = 25255
IP:         Flags = 0X
IP:         .0.. .... = may fragment
IP:         ..0. .... = last fragment
IP:         Fragment offset = 0 bytes
IP:         Time to live = 30 seconds/hops
IP:         Protocol = 6 (TCP)
IP:         Header checksum = A374 (correct)
IP:         Source address = [132.XXX.129.5]
IP:         Destination address = [129.XXX.16.6]
IP:         No options
IP:
TCP         — TCP Header —
TCP:
TCP:        Source port = 1169
```

```
TCP:        Destination port = 23 (Telnet)
TCP:        Sequence number = 398210404
TCP:        Acknowledgment number = 20881447
TCP:        Data offset = 20 bytes
TCP:        Flags = 14
TCP:        ..0. .... = (No urgent pointer)
TCP:        ...1 .... = Acknowledgment
TCP:        .... 0... = (No push)
TCP:        .... .1.. = Reset
TCP:        .... ..0. = (No SYN)
TCP:        .... ...0 = (No FIN)
TCP:        Window = 4096
TCP:        Checksum = 15EE (correct)
TCP:        No TCP options
TCP:
```

7.11 Using the Finger User Information Protocol

In Chapter 4, we discovered that the ICMP Echo (PING) command can test the transmission path between two devices on an internetwork. Another utility, known as the Finger User Information Protocol (Finger), described in RFC 1288 [7-4], also provides some end-to-end testing. Finger provides an interface to a database of users attached to a particular host, called the Remote User Information Program (RUIP).

Finger consists of a query/response interaction based on TCP transport (Figure 7-8). To initiate Finger, a TCP connection is established with Port 79 (the Finger port) on the remote host. Then the local host's Finger utility sends a query to RUIP at the remote host. The remote host responds with the information requested. When used in an internet environment, the Finger utility not only checks the end-to-end communication path (like the ICMP Echo), but it also verifies that the remote host knows of the remote user's existence. Let's find out how to use the Finger protocol.

The network manager wishes to check on one of his users, Kevin Anderson. From his PC, he establishes a TCP connection with the Sun server to which Kevin is attached (see Trace 7.11a). Note that the Destination port requested in the initial TCP connection segment, Frame 801, is the well-known Finger port, 79 (D = 79). The three-way handshake is completed in Frame 803, and the network manager's RUIP sends a query to the server requesting information for the user. The details of the query, shown in Trace

7.11b, indicate that the user name or user id (kpa) is sent with the query. Note that the Push (PSH) flag is used to send data on its way immediately.

The Sun server responds with an acknowledgment (Frame 805) followed by the RUIP response (Frame 806). The response contains 174 octets of data that pertain to user kpa. We learn the following about the user:

```
Login name: kpa   In real life: Kevin P. Anderson
Directory: /home/h0008/kpa    Shell: /bin/csh
Last login Wed Mar 27 09:22 on ttyp0 from h0009z
No unread mail
No plan
```

With the answer transmitted, the server's RUIP closes the connection in Frame 807 by setting the Finish (FIN) flag. The PC acknowledges the server's FIN (Frame 808) and sends a FIN of its own (Frame 809). The server acknowledges the final transaction in Frame 810.

In all, it took only 10 frames to learn about user kpa, his directory, shell, last login, and so on. Some of this information may be considered sensitive for security reasons, and network administrators are advised to read the security issues detailed in the Finger standard, RFC 1288. However, if you can surmount these concerns, the Finger protocol can be a valuable addition to your bag of troubleshooting techniques.

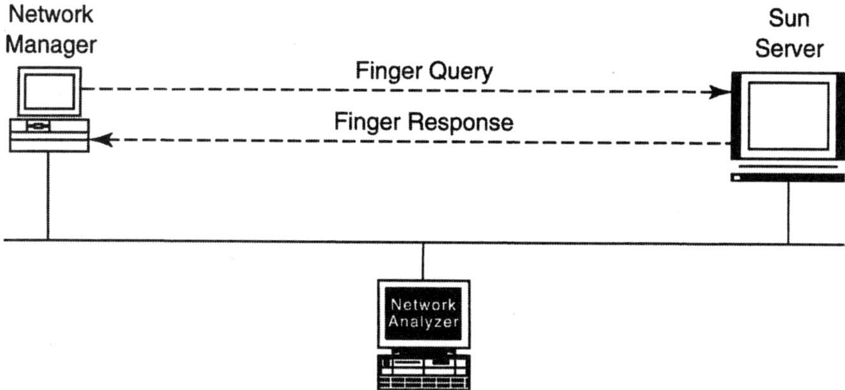

Figure 7-8. Finger User Information Protocol Operation.

Trace 7.11a. Finger User Information Summary
Sniffer Network Analyzer data 27-Mar at 09:04:54, TCPMEDLY.ENC, Pg 1

SUMMARY	Delta T	Destination	Source	Summary
801	57.9373	Sun Server	PC	TCP D=79 S=1228 SYN SEQ=134222388 LEN=0 WIN=1024
802	0.0009	PC	Sun Server	TCP D=1228 S=79 SYN ACK=134222389 SEQ=2009792000 LEN=0 WIN=4096
803	0.0130	Sun Server	PC	TCP D=79 S=1228 ACK=2009792001 WIN=1024
804	0.0441	Sun Server	PC	TCP D=79 S=1228 ACK=2009792001 SEQ=134222389 LEN=5 WIN=1024
805	0.1348	PC	Sun Server	TCP D=1228 S=79 ACK=134222394 WIN=4096
806	0.0433	PC	Sun Server	TCP D=1228 S=79 ACK=134222394 SEQ=2009792001 LEN=174 WIN=4096
807	0.0002	PC	Sun Server	TCP D=1228 S=79 FIN ACK=134222394 SEQ=2009792175 LEN=0 WIN=4096
808	0.0225	Sun Server	PC	TCP D=79 S=1228 ACK=2009792176 WIN=850
809	0.4897	Sun Server	PC	TCP D=79 S=1228 FIN ACK=2009792176 SEQ=134222394 LEN=0 WIN=850
810	0.0006	PC	Sun Server	TCP D=1228 S=79 ACK=134222395 WIN=4096

Trace 7.11b. Finger User Information Details
Sniffer Network Analyzer data 27-Mar at 09:04:54, TCPMEDLY.ENC, Pg 1

```
--------------- Frame 804 ----------------
TCP:        — TCP Header —
TCP:
TCP:        Source port = 1228
TCP:        Destination port = 79 (Finger)
TCP:        Sequence number = 134222389
TCP:        Acknowledgment number = 2009792001
TCP:        Data offset = 20 bytes
TCP:        Flags = 18
TCP:        ..0. .... = (No urgent pointer)
TCP:        ...1 .... = Acknowledgment
TCP:        .... 1... = Push
TCP:        .... .0.. = (No reset)
TCP:        .... ..0. = (No SYN)
TCP:        .... ...0 = (No FIN)
TCP:        Window = 1024
TCP:        Checksum = 2B9A (correct)
TCP:        No TCP options
TCP:        [5 byte(s) of data]
TCP:

ADDR  HEX                                                ASCII
0000  08 00 20 09 42 A4 00 00  C0 3C 55 17 08 00 45 00   .. .B....<U...E.
0010  00 2D 02 C3 00 00 40 06  63 75 8B B1 FE 5F 8B B1   .-....@.cu..._..
0020  FE D0 04 CC 00 4F 08 00  12 35 77 CA FE 01 50 18   .....O...5w...P.
0030  04 00 2B 9A 00 00 6B 70  61 0D 0A 00               ..+...kpa...

--------------- Frame 805 ----------------
TCP:        — TCP Header —
TCP:
TCP:        Source port = 79 (Finger)
TCP:        Destination port = 1228
TCP:        Sequence number = 2009792001
TCP:        Acknowledgment number = 134222394
TCP:        Data offset = 20 bytes
TCP:        Flags = 10
TCP:        ..0. .... = (No urgent pointer)
TCP:        ...1 .... = Acknowledgment
TCP:        .... 0... = (No push)
TCP:        .... .0.. = (No reset)
TCP:        .... ..0. = (No SYN)
```

```
TCP:        .... ...0 = (No FIN)
TCP:        Window = 4096
TCP:        Checksum = F61F (correct)
TCP:        No TCP options
TCP:
```

```
ADDR  HEX                                                ASCII
0000  00 00 C0 3C 55 17 08 00  20 09 42 A4 08 00 45 00  ...<U....B...E.
0010  00 28 D6 EE 00 00 3C 06  93 4E 8B B1 FE D0 8B B1  .(....<..N.....
0020  FE 5F 00 4F 04 CC 77 CA  FE 01 08 00 12 3A 50 10  ._.O..w.....:P.
0030  10 00 F6 1F 00 00 64 2E  62 79 6E 61              ..... d.byna
```

```
- - - - - - - - - - - - - - - - Frame 806 - - - - - - - - - - - - - - - - -
TCP:        — TCP Header —
TCP:
TCP:        Source port = 79 (Finger)
TCP:        Destination port = 1228
TCP:        Sequence number = 2009792001
TCP:        Acknowledgment number = 134222394
TCP:        Data offset = 20 bytes
TCP:        Flags = 18
TCP:        ..0. .... = (No urgent pointer)
TCP:        ...1 .... = Acknowledgment
TCP:        .... 1... = Push
TCP:        .... .0.. = (No reset)
TCP:        .... ..0. = (No SYN)
TCP:        .... ...0 = (No FIN)
TCP:        Window = 4096
TCP:        Checksum = CAA6 (correct)
TCP:        No TCP options
TCP:        [174 byte(s) of data]
TCP:
```

```
ADDR  HEX                                                ASCII
0000  00 00 C0 3C 55 17 08 00  20 09 42 A4 08 00 45 00  ...<U....B...E.
0010  00 D6 D6 F3 00 00 3C 06  92 9B 8B B1 FE D0 8B B1  ......<.........
0020  FE 5F 00 4F 04 CC 77 CA  FE 01 08 00 12 3A 50 18  ._.O..w......:P.
0030  10 00 CA A6 00 00 4C 6F  67 69 6E 20 6E 61 6D 65  ....Login name
0040  3A 20 6B 70 61 20 20 20  20 20 20 20 09 09 09 49  : kpa    ...I
0050  6E 20 72 65 61 6C 20 6C  69 66 65 3A 20 48 42 4F  n real life:....
0060  0D 0A 44 69 72 65 63 74  6F 72 79 3A 20 2F 68 6F  ..Directory: /ho
0070  6D 65 2F 68 30 30 30 38  2F 6B 70 61 20 20 20 20  me/h0008/kpa
0080  20 20 20 20 20 20 09 53  68 65 6C 6C 3A 20 2F 62  .Shell: /b
0090  69 6E 2F 63 73 68 0D 0A  4C 61 73 74 20 6C 6F 67  in/csh..Last log
00A0  69 6E 20 57 65 64 20 4D  61 72 20 32 37 20 30 39  in Wed Mar 27 09
```

```
00B0  3A 32 32 20 6F 6E 20 74  74 79 70 30 20 66 72 6F  :22 on ttyp0 fro
00C0  6D 20 68 30 30 30 39 7A  0D 0A 4E 6F 20 75 6E 72  m h0009z..No unr
00D0  65 61 64 20 6D 61 69 6C  0D 0A 4E 6F 20 50 6C 61  ead mail.No Pla
00E0  6E 2E 0D 0A                                        n...
```

7.12 Optimizing the TCP Window Size

When a distant host sends a value of Window = 0, the transmission path to the host effectively shuts down. If you receive a large window (say, Window = 8192), you can transmit a reasonable amount of data. Depending on the application, there should be an optimum window size that won't cause the transmitter to wait for unnecessary acknowledgments (i.e., the window is too small), yet won't overwhelm the receiver with more data than it can handle (i.e., the window is too large). As in "Goldilocks and the Three Bears," you need to find the window size that is "just right."

Two documents provide useful information about optimizing the window size: RFC 879, "The TCP Maximum Segment Size and Related Topics" [7-5], and RFC 813, "Window and Acknowledgement Strategy in TCP" [7-6]. Let's study an internetwork that needs some help in this area.

In the following case study, the internetwork consists of a headquarters location connected to three remote Ethernet segments (see Figure 7-9). Each remote segment contains a UNIX host plus a number of workstations. The remote segments are connected via bridges and 64 Kbps leased lines. The remote hosts contact these peers on a periodic basis to transfer files using a File Transfer Protocol (FTP) facility. In addition, the PC users use TELNET for remote host access. The problem is that while two hosts on the same segment can communicate without a problem, the hosts can't contact other hosts at a remote location without excessive delays and interrupts of the FTP session. Let's see why these problems occur.

To study the problem, the network engineer captures the data between one host on the headquarters segment (Hdqtr Host) and another host on a remote segment. He places the analyzer on the headquarters segment to capture the information coming to and from the remote host via the bridge (see Trace 7.12a). The initial analysis does not reveal any frames with errors. The remote host initiates a TCP connection in Frames 2 through 4, then starts an FTP session. The user identifies himself (Buster, in Frame 9), a password is transmitted (Frame 11), and an account is sent (Frame 15). The login completes in Frame 17, and the user can proceed. The only indication that a problem exists is the delays between frames, shown in Frames 9, 15, 17, and 19. These delays range from 6 to 18 seconds and indicate a transmission problem between the two communicating hosts.

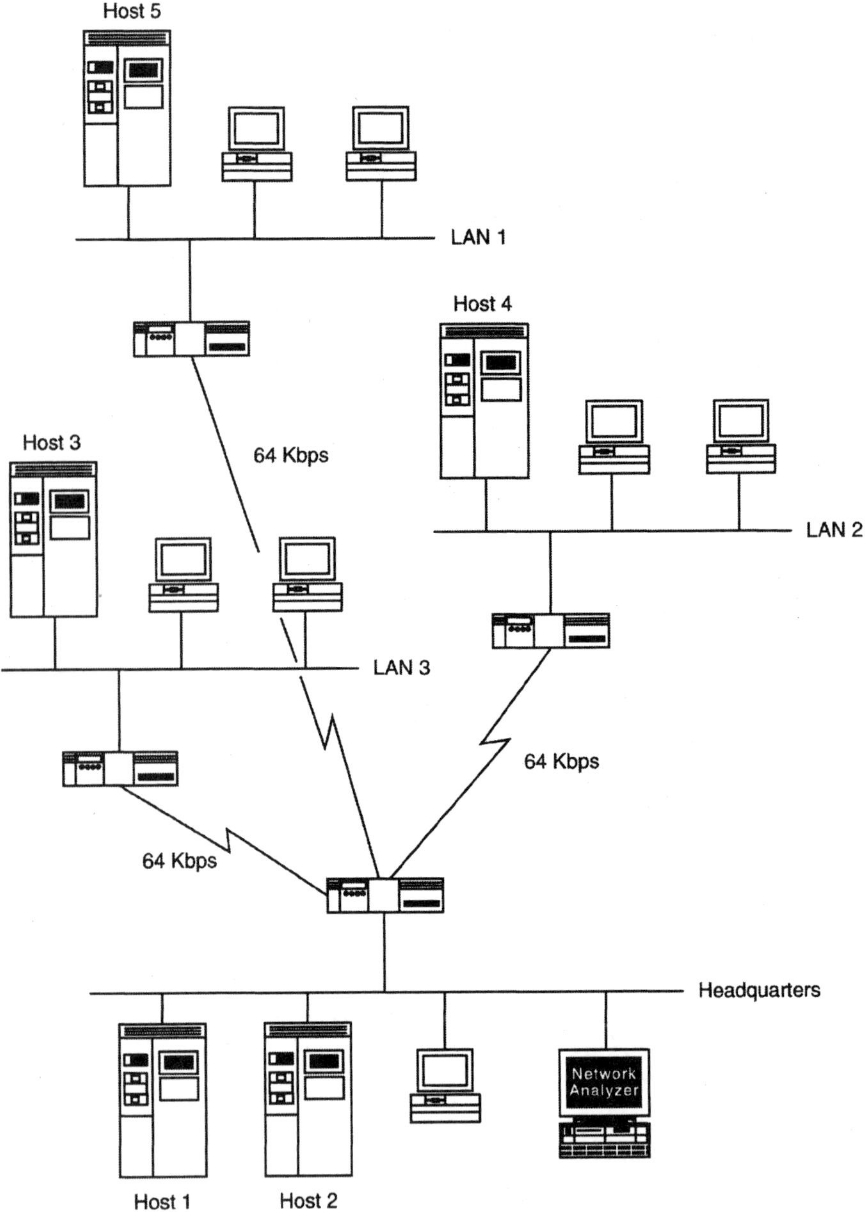

Figure 7-9. Host-to-Host Connections via WAN Bridges.

The headquarters host initiates a second TCP data connection beginning in Frame 26. Note that the Destination Ports are different. The first connection (Frames 2 through 4) uses Destination Port = 21 (FTP Control) and Source Port = 4979. The second connection uses Source Port = 20 (FTP data) and Destination Port = 4980. By all appearances, the two hosts should have established a full-duplex connection. Note also that both hosts return Window = 0 in Frames 26 and 27, but subsequently revise that number in Frames 28, 29, and 30.

Data transfer begins in Frame 31 with 512 octets of information. Successive frames convey 512 and 256 octets of data, respectively. An unexpected association of Sequence and Acknowledgment numbers occurs, however, in Trace 7.12b:

Frame	Data (octets)	Beginning Sequence	Ending Sequence	Acknowledgment
31	512	55278459	55278970	53214721
32	512	55278971	5279482	53214721
33	256	55279483	5279738	53214721
34	0	53214721	53214721	55278971

Frames 31 through 33 contain data from the headquarters host to the remote host. During this time no data is received in the other direction, since the Acknowledgment number remains constant (ACK = 53214721). Frame 34 contains an acknowledgment from the remote host, but note that it is the acknowledgment for the third previous segment (Frame 31). In other words, Frames 32 and 33 are still en route to the remote host, somewhere in the link between the two bridges.

The data transfer continues for a number of segments until the headquarters host abruptly terminates the connection in Frame 149 via the RST flag (see Trace 7.12c). The FTP connection is then closed and no additional data can be transferred. The immediately preceding frames provide the answer:

Frame	Sequence	Acknowledgment	Outstanding Data (octets)
144	55310931		
145		55306995	3,936
146	55311373		
147		55306995	4,378

The headquarters host transmits 442 octets of data in Frame 144; the remote host acknowledges a lesser amount (ACK = 55306995), leaving 3,936 octets of data unacknowledged (55310931 - 55306995 = 3,936). The problem worsens in Frames 146 and 147, with 4,378 unacknowledged octets (55311373 - 55306995 = 4,378).

Somewhere between the processing of Frames 144 and 146 a threshold within the host that counts the maximum allowable outstanding octets is crossed, causing the headquarters host to reset the connection. This threshold is half the allowable window size. The host has Window = 8,192, half of which is 4,096. In Frame 145, 3,936 octets are outstanding; in Frame 146 that number increases to 4,378, exceeding the threshold of 4,096.

The reason for the large amount of outstanding data is traced to inadequate I/O buffers on the WAN side of the bridges. Adding I/O buffers improves transmission. A second parameter, the discard threshold, is also increased inside the bridges. The discard threshold determines the maximum number of MAC-level frames that can be queued within the bridge. Setting this parameter to a higher value also improves the internetwork response. With these two bridge parameters adjusted, no further problems occur.

Trace 7.12a. TCP Window Management Summary

Sniffer Network Analyzer data 23-Jan at 03:10:38, TRACE.ENC, Pg 1

SUMMARY	Delta T	Destination	Source	Summary
M 1		Bridge	Hdqtr Host	ARP R PA=[143.130.1.5] HA=08000B001180 PRO=IP
2	0.0248	Hdqtr Host	Bridge	TCP D=21 S=4979 SYN SEQ=44830720 LEN=0 WIN=4096
3	0.5002	Bridge	Hdqtr Host	TCP D=4979 S=21 SYN ACK=44830721 SEQ=55214045 LEN=0 WIN=8192
4	0.0258	Hdqtr Host	Bridge	TCP D=21 S=4979 ACK=55214046 WIN=4096
5	0.2294	Bridge	Hdqtr Host	TCP D=4979 S=21 ACK=44830721 WIN=8192
6	1.1449	Bridge	Hdqtr Host	TCP D=4979 S=21 ACK=44830721 WIN=8192

7	2.7069	Bridge	Hdqtr Host	FTP R PORT=4979 220 1100JD1100 Service ready for new user <0D><0A>
8	0.0635	Hdqtr Host	Bridge	TCP D=21 S=4979 ACK=55214090 WIN=4096
9	14.7769	Hdqtr Host	Bridge	FTP C PORT=4979 USER Buster<0D><0A>
10	0.1605	Bridge	Hdqtr Host	FTP R PORT=4979 331 User name okay, need password.<0D><0A>
11	0.0465	Hdqtr Host	Bridge	FTP C PORT=4979 PASS pass<0D><0A>
12	0.2198	Bridge	Hdqtr Host	TCP D=4979 S=21 ACK=44830745 WIN=8192
13	0.7356	Bridge	Hdqtr Host	FTP R PORT=4979 332 Need account for login. <0D><0A>
14	0.0597	Hdqtr Host	Bridge	TCP D=21 S=4979 ACK=55214155 WIN=4096
15	6.7912	Hdqtr Host	Bridge	FTP C PORT=4979 ACCT Rufus<0D><0A>
16	0.2171	Bridge	Hdqtr Host	TCP D=4979 S=21 ACK=44830761 WIN=8192
17	14.4493	Bridge	Hdqtr Host	FTP R PORT=4979 230 User logged in, proceed <0D><0A>
18	0.1431	Hdqtr Host	Bridge	TCP D=21 S=4979 ACK=55214186 WIN=4096
19	18.7942	Hdqtr Host	Bridge	FTP C PORT=4979 PORT 143,130,2,1,19,116 <0D><0A>
20	0.2157	Bridge	Hdqtr Host	TCP D=4979 S=21 ACK=44830786 WIN=8192
21	0.2449	Bridge	Hdqtr Host	FTP R PORT=4979 200 Command okay. <0D><0A>

22	0.0551	Hdqtr Host	Bridge	FTP C PORT=4979 RETR faithful.jog <0D><0A>
23	0.2195	Bridge	Hdqtr Host	TCP D=4979 S=21 ACK=44830809 WIN=8192
24	2.0346	Bridge	Hdqtr Host	FTP R PORT=4979 150 File status okay; about to open data conn...
25	0.0371	Hdqtr Host	Bridge	TCP D=21 S=4979 ACK=55214259 WIN=4096
26	1.0401	Bridge	Hdqtr Host	TCP D=4980 S=20 SYN SEQ=55278458 LEN=0 WIN=0
27	0.0275	Hdqtr Host	Bridge	TCP D=20 S=4980 SYN ACK=55278459 SEQ=53214720 LEN=0 WIN=0
28	0.2321	Bridge	Hdqtr Host	TCP D=4980 S=20 ACK=53214721 WIN=8192
29	0.0279	Hdqtr Host	Bridge	TCP D=20 S=4980 ACK=55278459 WIN=4096
30	0.0099	Hdqtr Host	Bridge	TCP D=20 S=4980 ACK=55278459 WIN=8192
31	0.9281	Bridge	Hdqtr Host	TCP D=4980 S=20 ACK=53214721 SEQ=55278459 LEN=512 WIN=8192
32	0.0070	Bridge	Hdqtr Host	TCP D=4980 S=20 ACK=53214721 SEQ=55278971 LEN=512 WIN=8192
33	0.0025	Bridge	Hdqtr Host	TCP D=4980 S=20 ACK=53214721 SEQ=55279483 LEN=256 WIN=8192
34	0.3243	Hdqtr Host	Bridge	TCP D=20 S=4980 ACK=55278971 WIN=8192
35	0.0792	Bridge	Hdqtr Host	TCP D=4980 S=20 ACK=53214721

				SEQ=55279739 LEN=512 WIN=8192
36	0.0033	Bridge	Hdqtr Host	TCP D=4980 S=20 ACK=53214721 SEQ=55280251 LEN=204 WIN=8192
37	0.0873	Hdqtr Host	Bridge	TCP D=20 S=4980 ACK=55280251 WIN=8192
38	0.0514	Bridge	Hdqtr Host	TCP D=4980 S=20ACK=53214721 SEQ=55280455 LEN=512 WIN=8192
39	0.0040	Bridge	Hdqtr Host	TCP D=4980 S=20 ACK=53214721 LEN=512 WIN=8192
40	0.0042	Bridge	Hdqtr Host	TCP D=4980 S=20 ACK=53214721 SEQ=55281479 LEN=432 WIN=8192
41	0.1563	Hdqtr Host	Bridge	TCP D=20 S=4980 ACK=55281479 WIN=8192
42	0.0448	Bridge	Hdqtr Host	TCP D=4980 S=20 ACK=53214721 SEQ=55281911 LEN=512 WIN=8192
43	0.0042	Bridge	Hdqtr Host	TCP D=4980 S=20 ACK=53214721 SEQ=55282423 LEN=512 WIN=8192
44	0.0045	Bridge	Hdqtr Host	TCP D=4980 S=20 ACK=53214721 SEQ=55282935 LEN=361 WIN=8192
45	0.1611	Hdqtr Host	Bridge	TCP D=20 S=4980 ACK=55281911 WIN=8192
.				
.				
.				
128	0.0018	Bridge	Hdqtr Host	TCP D=4980 S=20 ACK=53214721 SEQ=55306733 LEN=37 WIN=8192

129	0.2577	Hdqtr Host	Bridge	TCP D=20 S=4980 ACK=55303791 WIN=8192
130	0.0573	Bridge	Hdqtr Host	TCP D=4980 S=20 ACK=53214721 SEQ=55306770 LEN=225 WIN=8192
131	0.0051	Bridge	Hdqtr Host	TCP D=4980 S=20 ACK=53214721 SEQ=55306995 LEN=512 WIN=8192
132	0.0037	Bridge	Hdqtr Host	TCP D=4980 S=20 ACK=53214721 SEQ=55307507 LEN=352 WIN=8192
133	0.0416	Hdqtr Host	Bridge	TCP D=20 S=4980 ACK=55304243 WIN=8192
134	0.0378	Bridge	Hdqtr Host	TCP D=4980 S=20 ACK=53214721 SEQ=55307859 LEN=512 WIN=8192
135	0.0628	Hdqtr Host	Bridge	TCP D=20 S=4980 ACK=55305267 WIN=8192
136	0.0429	Bridge	Hdqtr Host	TCP D=4980 S=20 ACK=53214721 SEQ=55308371 LEN=512 WIN=8192
137	0.0044	Bridge	Hdqtr Host	TCP D=4980 S=20 ACK=53214721 SEQ=55308883 LEN=512 WIN=8192
138	0.0521	Hdqtr Host	Bridge	TCP D=20 S=4980 ACK=55305709 WIN=8192
139	0.0285	Bridge	Hdqtr Host	TCP D=4980 S=20 ACK=53214721 SEQ=55309395 LEN=512 WIN=8192
140	0.1706	Hdqtr Host	Bridge	TCP D=20 S=4980 ACK=55306221 WIN=8192

141	0.0368	Bridge	Hdqtr Host	TCP D=4980 S=20 ACK=53214721 SEQ=55309907 LEN=512 WIN=8192
142	0.1640	Hdqtr Host	Bridge	TCP D=20 S=4980 ACK=55306995 WIN=8192
143	0.0393	Bridge	Hdqtr Host	TCP D=4980 S=20 ACK=53214721 SEQ=55310419 LEN=512 WIN=8192
144	0.0048	Bridge	Hdqtr Host	TCP D=4980 S=20 ACK=53214721 SEQ=55310931 LEN=442 WIN=8192
145	0.1554	Hdqtr Host	Bridge	TCP D=20 S=4980 ACK=55306995 WIN=8192
146	0.0211	Bridge	Hdqtr Host	TCP D=4980 S=20 ACK=53214721 SEQ=55311373 LEN=59 WIN=8192
147	0.1795	Hdqtr Host	Bridge	TCP D=20 S=4980 ACK=55306995 WIN=8192
148	0.1987	Hdqtr Host	Bridge	TCP D=20 S=4980 ACK=55306995 WIN=8192
149	1.4094	Bridge	Hdqtr Host	TCP D=4980 S=20 RST WIN=0
150	0.5261	Bridge	Hdqtr Host	FTP R PORT=4979 426 Connection is closed: A DDP ABORT WAS EX ECUTED...
151	0.0648	Hdqtr Host	Bridge	TCP D=21 S=4979 ACK=55214312 WIN=4096

Trace 7.12b. TCP Window Management Acknowledgments
Sniffer Network Analyzer data 23-Jan at 03:10:38, TRACE.ENC, Pg 1

```
- - - - - - - - - - - - - - - Frame 31 - - - - - - - - - - - - - - - -
TCP:        — TCP Header —
TCP:
TCP:        Source port = 20 (FTP data)
TCP:        Destination port = 4980
TCP:        Sequence number = 55278459
TCP:        Acknowledgment number = 53214721
TCP:        Data offset = 20 bytes
TCP:        Flags = 10
TCP:        ..0. .... = (No urgent pointer)
TCP:        ...1 .... = Acknowledgment
TCP:        .... 0... = (No push)
TCP:        .... .0.. = (No reset)
TCP:        .... ..0. = (No SYN)
TCP:        .... ...0 = (No FIN)
TCP:        Window = 8192
TCP:        Checksum = 27AC (correct)
TCP:        No TCP options
TCP:        [512 byte(s) of data]
TCP:

- - - - - - - - - - - - - - - Frame 32 - - - - - - - - - - - - - - - -
TCP:        — TCP Header —
TCP:
TCP:        Source port = 20 (FTP data)
TCP:        Destination port = 4980
TCP:        Sequence number = 55278971
TCP:        Acknowledgment number = 53214721
TCP:        Data offset = 20 bytes
TCP:        Flags = 10
TCP:        ..0. .... = (No urgent pointer)
TCP:        ...1 .... = Acknowledgment
TCP:        .... 0... = (No push)
TCP:        .... .0.. = (No reset)
TCP:        .... ..0. = (No SYN)
TCP:        .... ...0 = (No FIN)
TCP:        Window = 8192
TCP:        Checksum = 16E8 (correct)
TCP:        No TCP options
TCP:        [512 byte(s) of data]
TCP:
```

```
--------------- Frame 33 -----------------
TCP:        — TCP Header —
TCP:
TCP:        Source port = 20 (FTP data)
TCP:        Destination port = 4980
TCP:        Sequence number = 55279483
TCP:        Acknowledgment number = 53214721
TCP:        Data offset = 20 bytes
TCP:        Flags = 10
TCP:        ..0. .... = (No urgent pointer)
TCP:        ...1 .... = Acknowledgment
TCP:        .... 0... = (No push)
TCP:        .... .0.. = (No reset)
TCP:        .... ..0. = (No SYN)
TCP:        .... ...0 = (No FIN)
TCP:        Window = 8192
TCP:        Checksum = 0C2D (correct)
TCP:        No TCP options
TCP:        [256 byte(s) of data]
TCP:

--------------- Frame 34 -----------------
TCP:        — TCP Header —
TCP:
TCP:        Source port = 4980
TCP:        Destination port = 20 (FTP data)
TCP:        Sequence number = 53214721
TCP:        Acknowledgment number = 55278971
TCP:        Data offset = 20 bytes
TCP:        Flags = 10
TCP:        ..0. .... = (No urgent pointer)
TCP:        ...1 .... = Acknowledgment
TCP:        .... 0... = (No push)
TCP:        .... .0.. = (No reset)
TCP:        .... ..0. = (No SYN)
TCP:        .... ...0 = (No FIN)
TCP:        Window = 8192
TCP:        Checksum = D84E (correct)
TCP:        No TCP options
TCP:
```

Trace 7.12c. TCP Window Management Reset Condition
Sniffer Network Analyzer data 23-Jan at 03:10:38, TRACE.ENC, Pg 1

```
- - - - - - - - - - - - - - - Frame 144 - - - - - - - - - - - - - - - - -
TCP:        — TCP Header —
TCP:
TCP:        Source port = 20 (FTP data)
TCP:        Destination port = 4980
TCP:        Sequence number = 55310931
TCP:        Acknowledgment number = 53214721
TCP:        Data offset = 20 bytes
TCP:        Flags = 10
TCP:        ..0. .... = (No urgent pointer)
TCP:        ...1 .... = Acknowledgment
TCP:        .... 0... = (No push)
TCP:        .... .0.. = (No reset)
TCP:        .... ..0. = (No SYN)
TCP:        .... ...0 = (No FIN)
TCP:        Window = 8192
TCP:        Checksum = 1AD5 (correct)
TCP:        No TCP options
TCP:        [442 byte(s) of data]
TCP:

- - - - - - - - - - - - - - - Frame 145 - - - - - - - - - - - - - - - - -
TCP:        — TCP Header —
TCP:
TCP:        Source port = 4980
TCP:        Destination port = 20 (FTP data)
TCP:        Sequence number = 53214721
TCP:        Acknowledgment number = 55306995
TCP:        Data offset = 20 bytes
TCP:        Flags = 10
TCP:        ..0. .... = (No urgent pointer)
TCP:        ...1 .... = Acknowledgment
TCP:        .... 0... = (No push)
TCP:        .... .0.. = (No reset)
TCP:        .... ..0. = (No SYN)
TCP:        .... ...0 = (No FIN)
TCP:        Window = 8192
TCP:        Checksum = 6AD6 (correct)
TCP:        No TCP options
TCP:
```

```
--------------- Frame 146 -----------------
TCP:       — TCP Header —
TCP:
TCP:       Source port = 20 (FTP data)
TCP:       Destination port = 4980
TCP:       Sequence number = 55311373
TCP:       Acknowledgment number = 53214721
TCP:       Data offset = 20 bytes
TCP:       Flags = 10
TCP:       ..0. .... = (No urgent pointer)
TCP:       ...1 .... = Acknowledgment
TCP:       .... 0... = (No push)
TCP:       .... .0.. = (No reset)
TCP:       .... ..0. = (No SYN)
TCP:       .... ...0 = (No FIN)
TCP:       Window = 8192
TCP:       Checksum = E735 (correct)
TCP:       No TCP options
TCP:       [59 byte(s) of data]
TCP:

--------------- Frame 147 -----------------
TCP:       — TCP Header —
TCP:
TCP:       Source port = 4980
TCP:       Destination port = 20 (FTP data)
TCP:       Sequence number = 53214721
TCP:       Acknowledgment number = 55306995
TCP:       Data offset = 20 bytes
TCP:       Flags = 10
TCP:       ..0. .... = (No urgent pointer)
TCP:       ...1 .... = Acknowledgment
TCP:       .... 0... = (No push)
TCP:       .... .0.. = (No reset)
TCP:       .... ..0. = (No SYN)
TCP:       .... ...0 = (No FIN)
TCP:       Window = 8192
TCP:       Checksum = 6AD6 (correct)
TCP:       No TCP options
TCP:
```

```
--------------- Frame 148 -----------------
TCP:        — TCP Header —
TCP:
TCP:        Source port = 4980
TCP:        Destination port = 20 (FTP data)
TCP:        Sequence number = 53214721
TCP:        Acknowledgment number = 55306995
TCP:        Data offset = 20 bytes
TCP:        Flags = 10
TCP:        ..0. .... = (No urgent pointer)
TCP:        ...1 .... = Acknowledgment
TCP:        .... 0... = (No push)
TCP:        .... .0.. = (No reset)
TCP:        .... ..0. = (No SYN)
TCP:        .... ...0 = (No FIN)
TCP:        Window = 8192
TCP:        Checksum = 6AD6 (correct)
TCP:        No TCP options
TCP:

--------------- Frame 149 -----------------
TCP:        — TCP Header —
TCP:
TCP:        Source port = 20 (FTP data)
TCP:        Destination port = 4980
TCP:        Sequence number = 55311432
TCP:        Data offset = 20 bytes
TCP:        Flags = 04
TCP:        ..0. .... = (No urgent pointer)
TCP:        ...0 .... = (No acknowledgment)
TCP:        .... 0... = (No push)
TCP:        .... .1.. = Reset
TCP:        .... ..0. = (No SYN)
TCP:        .... ...0 = (No FIN)
TCP:        Window = 0
TCP:        Checksum = 7ABA (correct)
TCP:        No TCP options
TCP:

--------------- Frame 150 -----------------
FTP:        — FTP Data —
FTP:
FTP:        426 Connection is closed: A DDP ABORT WAS EXECUTED...
FTP:
```

```
---------------- Frame 151 ----------------
TCP:        — TCP Header —
TCP:
TCP:        Source port = 4979
TCP:        Destination port = 21 (FTP)
TCP:        Sequence number = 44830809
TCP:        Acknowledgment number = 55214312
TCP:        Data offset = 20 bytes
TCP:        Flags = 10
TCP:        ..0. .... = (No urgent pointer)
TCP:        ...1 .... = Acknowledgment
TCP:        .... 0... = (No push)
TCP:        .... .0.. = (No reset)
TCP:        .... ..0. = (No SYN)
TCP:        .... ...0 = (No FIN)
TCP:        Window = 4096
TCP:        Checksum = D30A (correct)
TCP:        No TCP options
TCP:
```

7.13 Data Transport Using IPv6

In Chapter 6 we considered the enhancements that have been provided with IPv6, including a larger address field size, flow labels, optional extension headers, and so on. In this example, two workstations are communicating across an Ethernet segment using IPv6. In the following trace files, we will see how standard Internet applications are encapsulated within IPv6 for transport to another IPv6-based workstation, and we will also observe how some of the IPv6-native processes, such as Neighbor Solicitation and Advertising, operate.

To start the communication between the two workstations, a TCP three-way handshake is required (Trace 7.13a). Note from Frame 321 that the TCP header immediately follows the IPv6 header, which is indicated in the IPv6 Next Header field (TCP). By reviewing Figure 6-44, we can also observe that the IPv6 workstations are identified with IPv6 Testing Addresses, which begin with a hexadecimal 3FFE (a portion of these addresses are disguised to protect the source). Within the TCP header, we see that the Destination Port indicates the FTP Control channel (Port 21), and the Synchronize flag (SYN) is set indicating the three-way handshake. In addition, the Maximum Segment Size option is set for 1,440 octets, indicating that no fragmentation will be used by the source station (the 3Com station) for this TCP session. (We can derive this information by noting that the physical infrastructure is an Ethernet LAN with an MTU of 1,500 octets. If we allow 40 octets for the IPv6 header, and 20 octets for the TCP header, we have 1,440 octets left over for the upper layer protocol data, which becomes the Maximum Segment Size.) Frame 322 is the second leg of the three-way handshake (the second SYN) and Frame 323 is the third leg of the process and the final acknowledgment. Note that a different Maximum Segment Size is transmitted from the Xircom station (33,160 octets), indicating that it is capable of handling message fragmentation.

Trace 7.13a. TCP over IPv6 3-way Handshake Details
Sniffer Network Analyzer data 08-Jul at 11:01:35, IPV6TEST.ENC, Pg 1

```
- - - - - - - - - - - - - - - - Frame 321 - - - - - - - - - - - - - - -
DLC:        — DLC Header —
DLC:
DLC:        Frame 321 arrived at 14:32:00.1457; frame size is 78 (004E hex) bytes.
DLC:        Destination = Station XircomF35AC8
DLC:        Source      = Station 3Com  B81E32
DLC:        Ethertype   = 86DD (IPv6)
DLC:
IPv6:       — IPv6 Header —
IPv6:
IPv6:       Version           = 6
IPv6:       Traffic Octet     = 0
IPv6:       Differentiated Services Field : 0x00 (DS Codepoint - Default PHB)
IPv6:       Currently Unused Field        : 0x00
IPv6:       Flow Label        = 0x00000
IPv6:       Payload Length    = 24
IPv6:       Next Header       = 6 (TCP)
IPv6:       Hop Limit         = 64
IPv6:       Source address      = 3ffe:U:V:W:X:Y:Z:cad5
IPv6:       Destination address = 3ffe:U:V:W:X:Y:Z:4d0
IPv6:
TCP:        — TCP Header —
TCP:
TCP:        Source port            =  1035
TCP:        Destination port       =   21 (FTP-ctrl)
TCP:        Initial sequence number = 2829780403
TCP:        Next expected Seq number= 2829780404
TCP:        Data offset            = 24 bytes
TCP:        Reserved Bits: Reserved for Future Use
TCP:        Flags                  = 02
TCP:        ..0. .... = (No urgent pointer)
```

```
TCP:        ...0 .... = (No acknowledgment)
TCP:        .... 0... = (No push)
TCP:        .... .0.. = (No reset)
TCP:        .... ..1. = SYN
TCP:        .... ...0 = (No FIN)
TCP:        Window              = 16384
TCP:        Checksum            = 095E (correct)
TCP:        Urgent pointer      = 256
TCP:
TCP:        Options follow
TCP:        Maximum segment size = 1440
TCP:

- - - - - - - - - - - - - - - - - - - - - Frame 322 - - - - - - - - - - - - - - - - - - - -
DLC:        — DLC Header —
DLC:
DLC:        Frame 322 arrived at  14:32:00.7104; frame size is 78 (004E hex) bytes.
DLC:        Destination = Station 3Com  B81E32
DLC:        Source      = Station XircomF35AC8
DLC:        Ethertype   = 86DD (IPv6)
DLC:
IPv6:       — IPv6 Header —
IPv6:
IPv6:       Version          = 6
IPv6:       Traffic Octet    = 0
IPv6:       Differentiated Services Field : 0x00 (DS Codepoint - Default PHB)
IPv6:       Currently Unused Field        : 0x00
IPv6:       Flow Label       = 0x00000
IPv6:       Payload Length   = 24
IPv6:       Next Header      = 6 (TCP)
IPv6:       Hop Limit        = 53
IPv6:       Source address   = 3ffe:U:V:W:X:Y:Z:4d0
IPv6:       Destination address = 3ffe:U:V:W:X:Y:Z:cad5
IPv6:
TCP:        — TCP Header —
TCP:H
TCP:        Source port         =  21 (FTP-ctrl)
TCP:        Destination port    =  1035
TCP:        Initial sequence number = 2665481639
TCP:        Next expected Seq number= 2665481640
TCP:        Acknowledgment number   = 2829780404
TCP:        Data offset         = 24 bytes
TCP:        Reserved Bits: Reserved for Future Use
TCP:        Flags               = 12
TCP:        ..0. .... = (No urgent pointer)
```

```
TCP:        ...1 .... = Acknowledgment
TCP:        .... 0... = (No push)
TCP:        .... .0.. = (No reset)
TCP:        .... ..1. = SYN
TCP:        .... ...0 = (No FIN)
TCP:        Window             = 16384
TCP:        Checksum           = E9DC (correct)
TCP:        Urgent pointer     = 0
TCP:
TCP:        Options follow
TCP:        Maximum segment size = 33160
TCP:

- - - - - - - - - - - - - - - - - - - - Frame 323 - - - - - - - - - - - - - - - - - - - -
DLC:        — DLC Header —
DLC:
DLC:        Frame 323 arrived at  14:32:00.7107; frame size is 74 (004A hex) bytes.
DLC:        Destination = Station XircomF35AC8
DLC:        Source      = Station 3Com  B81E32
DLC:        Ethertype   = 86DD (IPv6)
DLC:
IPv6:       — IPv6 Header —
IPv6:
IPv6:       Version            = 6
IPv6:       Traffic Octet      = 0
IPv6:       Differentiated Services Field : 0x00 (DS Codepoint - Default PHB)
IPv6:       Currently Unused Field        : 0x00
IPv6:       Flow Label         = 0x00000
IPv6:       Payload Length     = 20
IPv6:       Next Header        = 6 (TCP)
IPv6:       Hop Limit          = 64
IPv6:       Source address     = 3ffe:U:V:W:X:Y:Z:cad5
IPv6:       Destination address = 3ffe:U:V:W:X:Y:Z:4d0
IPv6:
TCP:        — TCP Header —
TCP:
TCP:        Source port            =  1035
TCP:        Destination port       =    21 (FTP-ctrl)
TCP:        Sequence number        = 2829780404
TCP:        Next expected Seq number= 2829780404
TCP:        Acknowledgment number  = 2665481640
TCP:        Data offset            = 20 bytes
TCP:        Reserved Bits: Reserved for Future Use
TCP:        Flags                  = 10
```

```
TCP:        ..0. .... = (No urgent pointer)
TCP:        ...1 .... = Acknowledgment
TCP:        .... 0... = (No push)
TCP:        .... .0.. = (No reset)
TCP:        .... ..0. = (No SYN)
TCP:        .... ...0 = (No FIN)
TCP:        Window           = 17280
TCP:        Checksum          = 44EE (correct)
TCP:        Urgent pointer     = 13568
TCP:        No TCP options
TCP:
```

After the TCP session has been established, the application information (FTP in this case) is transported across the network independent of the underlying network infrastructure, IPv4 or IPv6, as shown in Trace 7.13b. (For brevity, all protocol layers are shown in Frames 324 and 325, and, thereafter, only the upper layers (TCP and FTP) are illustrated.) The FTP functions include the host response (Frame 326), user login (Frames 327–330), directory change and acknowledgments (Frames 331–333), and data transfer (Frame 334). Since the TCP functions are assuring reliable application delivery, the FTP process does not know (nor does it care) that the underlying infrastructure is IPv6 instead of IPv4.

Trace 7.13b. FTP via IPv6 Network

```
-------------------- Frame 324 --------------------
DLC:        — DLC Header —
DLC:
DLC:        Frame 324 arrived at  14:32:09.5760; frame size is 80 (0050 hex)
bytes.
DLC:        Destination = Station 3Com  B81E32
DLC:        Source     = Station XircomF35AC8
DLC:        Ethertype  = 86DD (IPv6)
DLC:
IPv6:       — IPv6 Header —
IPv6:
IPv6:       Version          = 6
IPv6:       Traffic Octet     = 0
IPv6:       Differentiated Services Field : 0x00 (DS Codepoint - Default PHB)
IPv6:       Currently Unused Field      : 0x00
IPv6:       Flow Label        = 0x78EBB
IPv6:       Payload Length     = 26
```

```
IPv6:     Next Header      = 6 (TCP)
IPv6:     Hop Limit        = 53
IPv6:     Source address   = 3ffe:U:V:W:X:Y:Z:4d0
IPv6:     Destination address = 3ffe:U:V:W:X:Y:Z:cad5
IPv6:
TCP:      — TCP Header —
TCP:
TCP:      Source port          =  21 (FTP-ctrl)
TCP:      Destination port     =  1034
TCP:      Sequence number      = 2334476307
TCP:      Next expected Seq number= 2334476313
TCP:      Acknowledgment number  = 916850913
TCP:      Data offset          = 20 bytes
TCP:      Reserved Bits: Reserved for Future Use (Not shown in the Hex Dump)
TCP:      Flags                = 18
TCP:                ..0. .... = (No urgent pointer)
TCP:                ...1 .... = Acknowledgment
TCP:                .... 1... = Push
TCP:                .... .0.. = (No reset)
TCP:                .... ..0. = (No SYN)
TCP:                .... ...0 = (No FIN)
TCP:      Window               = 16384
TCP:      Checksum             = 4E20 (correct)
TCP:      Urgent pointer       = 0
TCP:      No TCP options
TCP:      [6 Bytes of data]
TCP:
FTP:      — File Transfer Data Protocol —
FTP:
FTP:      Line  1:  220-
FTP:

------------------- Frame 325 -------------------
DLC:      — DLC Header —
DLC:
DLC:      Frame 325 arrived at  14:32:09.6813; frame size is 74 (004A hex) bytes.
DLC:      Destination = Station XircomF35AC8
DLC:      Source      = Station 3Com  B81E32
DLC:      Ethertype   = 86DD (IPv6)
DLC:
IPv6:     — IPv6 Header —
IPv6:
IPv6:     Version          = 6
IPv6:     Traffic Octet    = 0
```

```
IPv6:     Differentiated Services Field : 0x00 (DS Codepoint - Default PHB)
IPv6:     Currently Unused Field      : 0x00
IPv6:     Flow Label          = 0x00000
IPv6:     Payload Length      = 20
IPv6:     Next Header         = 6 (TCP)
IPv6:     Hop Limit           = 64
IPv6:     Source address      = 3ffe:U:V:W:X:Y:Z:cad5
IPv6:     Destination address = 3ffe:U:V:W:X:Y:Z:4d0
IPv6:
TCP:      — TCP Header —
TCP:
TCP:      Source port           =  1034
TCP:      Destination port      =    21 (FTP-ctrl)
TCP:      Sequence number       = 916850913
TCP:      Next expected Seq number= 916850913
TCP:      Acknowledgment number   = 2334476313
TCP:      Data offset           = 20 bytes
TCP:      Reserved Bits: Reserved for Future Use (Not shown in the Hex Dump)
TCP:      Flags                 = 10
TCP:                ..0. .... = (No urgent pointer)
TCP:                ...1 .... = Acknowledgment
TCP:                .... 0... = (No push)
TCP:                .... .0.. = (No reset)
TCP:                .... ..0. = (No SYN)
TCP:                .... ...0 = (No FIN)
TCP:      Window                = 17274
TCP:      Checksum              = 8517 (correct)
TCP:      Urgent pointer        = 13568
TCP:      No TCP options
TCP:

- - - - - - - - - - - - - - - - - - - - Frame 326 - - - - - - - - - - - - - - - - - - - -
TCP:      — TCP Header —
TCP:
TCP:      Source port           =    21 (FTP-ctrl)
TCP:      Destination port      =  1034
TCP:      Sequence number       = 2334476313
TCP:      Next expected Seq number= 2334476374
TCP:      Acknowledgment number   = 916850913
TCP:      Data offset           = 20 bytes
TCP:      Reserved Bits: Reserved for Future Use (Not shown in the Hex Dump)
TCP:      Flags                 = 18
TCP:                ..0. .... = (No urgent pointer)
TCP:                ...1 .... = Acknowledgment
```

```
TCP:                  .... 1... = Push
TCP:                  .... .0.. = (No reset)
TCP:                  .... ..0. = (No SYN)
TCP:                  .... ...0 = (No FIN)
TCP:        Window              = 16384
TCP:        Checksum            = D8E0 (correct)
TCP:        Urgent pointer      = 0
TCP:        No TCP options
TCP:        [61 Bytes of data]
TCP:
FTP:        — File Transfer Data Protocol —
FTP:
FTP:        Line 1: 220 apple.kame.net FTP server (NetBSD-ftpd 20020615) ready.
FTP:

- - - - - - - - - - - - - - - - - - - - Frame 327 - - - - - - - - - - - - - - - - - - - -
TCP:        — TCP Header —
TCP:
TCP:        Source port         = 1034
TCP:        Destination port    =   21 (FTP-ctrl)
TCP:        Sequence number     = 916850913
TCP:        Next expected Seq number= 916850929
TCP:        Acknowledgment number  = 2334476374
TCP:        Data offset         = 20 bytes
TCP:        Reserved Bits: Reserved for Future Use (Not shown in the Hex Dump)
TCP:        Flags               = 18
TCP:                  ..0. .... = (No urgent pointer)
TCP:                  ...1 .... = Acknowledgment
TCP:                  .... 1... = Push
TCP:                  .... .0.. = (No reset)
TCP:                  .... ..0. = (No SYN)
TCP:                  .... ...0 = (No FIN)
TCP:        Window              = 17213
TCP:        Checksum            = 3223 (correct)
TCP:        Urgent pointer      = 0
TCP:        No TCP options
TCP:        [16 Bytes of data]
TCP:
FTP:        — File Transfer Data Protocol —
FTP:
FTP:        Line 1: USER anonymous
FTP:
```

```
-------------------- Frame 328 --------------------
TCP:      — TCP Header —
TCP:
TCP:      Source port          =  21 (FTP-ctrl)
TCP:      Destination port     =  1034
TCP:      Sequence number      = 2334476374
TCP:      Next expected Seq number= 2334476423
TCP:      Acknowledgment number  = 916850929
TCP:      Data offset          = 20 bytes
TCP:      Reserved Bits: Reserved for Future Use (Not shown in the Hex Dump)
TCP:      Flags                = 18
TCP:                ..0. .... = (No urgent pointer)
TCP:                ...1 .... = Acknowledgment
TCP:                .... 1... = Push
TCP:                .... .0.. = (No reset)
TCP:                .... ..0. = (No SYN)
TCP:                .... ...0 = (No FIN)
TCP:      Window               = 16384
TCP:      Checksum             = FE3D (correct)
TCP:      Urgent pointer       = 0
TCP:      No TCP options
TCP:      [49 Bytes of data]
TCP:
FTP:      — File Transfer Data Protocol —
FTP:
FTP:      Line  1:  331 Guest login ok, type your name as password.
FTP:

-------------------- Frame 329 --------------------
TCP:      — TCP Header —
TCP:
TCP:      Source port          =  1034
TCP:      Destination port     =   21 (FTP-ctrl)
TCP:      Sequence number      = 916850929
TCP:      Next expected Seq number= 916850943
TCP:      Acknowledgment number  = 2334476423
TCP:      Data offset          = 20 bytes
TCP:      Reserved Bits: Reserved for Future Use (Not shown in the Hex Dump)
TCP:      Flags                = 18
TCP:                ..0. .... = (No urgent pointer)
TCP:                ...1 .... = Acknowledgment
TCP:                .... 1... = Push
TCP:                .... .0.. = (No reset)
TCP:                .... ..0. = (No SYN)
```

```
TCP:          .... ...0 = (No FIN)
TCP:      Window             = 17164
TCP:      Checksum           = BE0E (correct)
TCP:      Urgent pointer     = 0
TCP:      No TCP options
TCP:      [14 Bytes of data]
TCP:
FTP:      — File Transfer Data Protocol —
FTP:
FTP:      Line  1:  PASS boomer
FTP:

- - - - - - - - - - - - - - - - - - - - Frame 330 - - - - - - - - - - - - - - - - - - - -
TCP:      — TCP Header —
TCP:
TCP:      Source port        =  21 (FTP-ctrl)
TCP:      Destination port   =  1034
TCP:      Sequence number    = 2334476423
TCP:      Next expected Seq number= 2334476471
TCP:      Acknowledgment number  = 916850943
TCP:      Data offset        = 20 bytes
TCP:      Reserved Bits: Reserved for Future Use (Not shown in the Hex Dump)
TCP:      Flags              = 18
TCP:          ..0. .... = (No urgent pointer)
TCP:          ...1 .... = Acknowledgment
TCP:          .... 1... = Push
TCP:          .... .0.. = (No reset)
TCP:          .... ..0. = (No SYN)
TCP:          .... ...0 = (No FIN)
TCP:      Window             = 16384
TCP:      Checksum           = 10E7 (correct)
TCP:      Urgent pointer     = 0
TCP:      No TCP options
TCP:      [48 Bytes of data]
TCP:
FTP:      — File Transfer Data Protocol —
FTP:
FTP:      Line  1:  230 Guest login ok, access restrictions apply.
FTP:

- - - - - - - - - - - - - - - - - - - - Frame 331 - - - - - - - - - - - - - - - - - - - -
TCP:      — TCP Header —
TCP:
TCP:      Source port        =  1034
```

```
TCP:        Destination port      =  21 (FTP-ctrl)
TCP:        Sequence number       = 916850943
TCP:        Next expected Seq number= 916850964
TCP:        Acknowledgment number  = 2334476471
TCP:        Data offset           = 20 bytes
TCP:        Reserved Bits: Reserved for Future Use (Not shown in the Hex Dump)
TCP:        Flags                 = 18
TCP:                ..0. .... = (No urgent pointer)
TCP:                ...1 .... = Acknowledgment
TCP:                .... 1... = Push
TCP:                .... .0.. = (No reset)
TCP:                .... ..0. = (No SYN)
TCP:                .... ...0 = (No FIN)
TCP:        Window                = 17116
TCP:        Checksum              = 7DB0 (correct)
TCP:        Urgent pointer        = 0
TCP:        No TCP options
TCP:        [21 Bytes of data]
TCP:
FTP:        — File Transfer Data Protocol —
FTP:
FTP:        Line 1: CWD /pub/kame/misc/
FTP:

- - - - - - - - - - - - - - - - - - - - Frame 332 - - - - - - - - - - - - - - - - - - - -
TCP:        — TCP Header —
TCP:
TCP:        Source port           =  21 (FTP-ctrl)
TCP:        Destination port      =  1034
TCP:        Sequence number       = 2334476471
TCP:        Next expected Seq number= 2334476477
TCP:        Acknowledgment number  = 916850964
TCP:        Data offset           = 20 bytes
TCP:        Reserved Bits: Reserved for Future Use (Not shown in the Hex Dump)
TCP:        Flags                 = 18
TCP:                ..0. .... = (No urgent pointer)
TCP:                ...1 .... = Acknowledgment
TCP:                .... 1... = Push
TCP:                .... .0.. = (No reset)
TCP:                .... ..0. = (No SYN)
TCP:                .... ...0 = (No FIN)
TCP:        Window                = 16384
TCP:        Checksum              = 4D46 (correct)
TCP:        Urgent pointer        = 0
```

```
TCP:     No TCP options
TCP:     [6 Bytes of data]
TCP:
FTP:     — File Transfer Data Protocol —
FTP:
FTP:     Line  1:  250-
FTP:

-------------------- Frame 333 --------------------
TCP:     — TCP Header —
TCP:
TCP:     Source port            =  1034
TCP:     Destination port       =    21 (FTP-ctrl)
TCP:     Sequence number        = 916850964
TCP:     Next expected Seq number= 916850964
TCP:     Acknowledgment number  = 2334476477
TCP:     Data offset            = 20 bytes
TCP:     Reserved Bits: Reserved for Future Use (Not shown in the Hex Dump)
TCP:     Flags                  = 10
TCP:              ..0. .... = (No urgent pointer)
TCP:              ...1 .... = Acknowledgment
TCP:              .... 0... = (No push)
TCP:              .... .0.. = (No reset)
TCP:              .... ..0. = (No SYN)
TCP:              .... ...0 = (No FIN)
TCP:     Window                 = 17110
TCP:     Checksum               = 84E4 (correct)
TCP:     Urgent pointer         = 13568
TCP:     No TCP options
TCP:

-------------------- Frame 334 --------------------
TCP:     — TCP Header —
TCP:
TCP:     Source port            =    21 (FTP-ctrl)
TCP:     Destination port       =  1034
TCP:     Sequence number        = 2334476477
TCP:     Next expected Seq number= 2334476906
TCP:     Acknowledgment number  = 916850964
TCP:     Data offset            = 20 bytes
TCP:     Reserved Bits: Reserved for Future Use (Not shown in the Hex Dump)
TCP:     Flags                  = 18
TCP:              ..0. .... = (No urgent pointer)
TCP:              ...1 .... = Acknowledgment
```

```
TCP:                  .... 1... = Push
TCP:                  .... .0.. = (No reset)
TCP:                  .... ..0. = (No SYN)
TCP:                  .... ...0 = (No FIN)
TCP:        Window            = 16384
TCP:        Checksum          = DD14 (correct)
TCP:        Urgent pointer    = 0
TCP:        No TCP options
TCP:        [429 Bytes of data]
TCP:
FTP:        — File Transfer Data Protocol —
FTP:
FTP:        Line 1: racoon-*.tgz
FTP:        Line 2: <09>Racoon IKE daemon. This is included in KAME kit, but
FTP:        provided here
FTP:        Line 3: <09>separately for people who tries to use this on other IP
FTP:        v6/IPsec stacks.
FTP:        Line 4: <09>The development of racoon is ongoing.
FTP:        Line 5:
FTP:        Line 6: socks64-v10r3-*.tgz
FTP:        Line 7: <09>socks64, modified socks5 for IPv4/v6 support.
FTP:        Line 8:
FTP:        Line 9: icecast-11[34]-*
FTP:        Line 10: <09>we drop access permission because these versions have
FTP:        security hole.
FTP:        Line 11: 250 CWD command successful.
FTP:
```

The third function involves communication between the two IPv6 workstations, using the ICMPv6 Echo Request and Response messages to check connectivity on their data link. Note in Trace 7.13c that the IPv6 Next Header field indicates that an ICMPv6 header will follow, and that the IPv6 addresses begin with 3FFE H, specifying the Testing Address as in the previous example. The ICMPv6 Echo Request message is identified by a Type field value of 128, and transmits 32 octets of test information in Frame 649. This transmission is confirmed in Frame 650 by an ICMPv6 Echo Reply message, with a Type Field value of 129.

Trace 7.13c. ICMPv6 Echo Request/Response Messages

```
-------------------- Frame 649 --------------------
DLC:      — DLC Header —
DLC:
DLC:      Frame 649 arrived at  14:34:22.5526; frame size is 94 (005E hex) bytes.
DLC:      Destination = Station XircomF35AC8
DLC:      Source      = Station 3Com  B81E32
DLC:      Ethertype   = 86DD (IPv6)
DLC:
IPv6:     — IPv6 Header —
IPv6:
IPv6:     Version          = 6
IPv6:     Traffic Octet    = 0
IPv6:     Differentiated Services Field : 0x00 (DS Codepoint - Default PHB)
IPv6:     Currently Unused Field        : 0x00
IPv6:     Flow Label       = 0x00000
IPv6:     Payload Length   = 40
IPv6:     Next Header      = 58 (ICMPv6)
IPv6:     Hop Limit        = 64
IPv6:     Source address   = 3ffe:U:V:W:X:Y:Z:cad5
IPv6:     Destination address = 3ffe:U:V:W:X:Y:Z:4d0
IPv6:
ICMPv6:   — ICMPv6 Header —
ICMPv6:
ICMPv6:   Type             = 128 (Echo Request Message)
ICMPv6:   Code             = 0
ICMPv6:   Checksum         = 0x3997
ICMPv6:   Identifier       = 0
ICMPv6:   Sequence Number  = 4
ICMPv6:   [32 Bytes of data]
ICMPv6:

-------------------- Frame 650 --------------------
DLC:      — DLC Header —
DLC:
DLC:      Frame 650 arrived at  14:34:23.1173; frame size is 94 (005E hex) bytes.
DLC:      Destination = Station 3Com  B81E32
DLC:      Source      = Station XircomF35AC8
DLC:      Ethertype   = 86DD (IPv6)
DLC:
IPv6:     — IPv6 Header —
IPv6:
IPv6:     Version          = 6
```

```
IPv6:      Traffic Octet      = 0
IPv6:      Differentiated Services Field : 0x00 (DS Codepoint - Default PHB)
IPv6:      Currently Unused Field     : 0x00
IPv6:      Flow Label       = 0x00000
IPv6:      Payload Length     = 40
IPv6:      Next Header      = 58 (ICMPv6)
IPv6:      Hop Limit        = 53
IPv6:      Source address     = 3ffe:U:V:W:X:Y:Z:4d0
IPv6:      Destination address = 3ffe:U:V:W:X:Y:Z:cad5
IPv6:
ICMPv6:    — ICMPv6 Header —
ICMPv6:
ICMPv6:    Type            = 129 (Echo Reply Message)
ICMPv6:    Code            = 0
ICMPv6:    Checksum          = 0x3897
ICMPv6:    Identifier        = 0
ICMPv6:    Sequence Number     = 4
ICMPv6:    [32 Bytes of data]
ICMPv6:
```

The final example illustrates the ICMPv6 Neighbor Solicitation and Advertisement messages, which provide a mechanism for workstations to become aware of neighbors on the same IPv6 segment (Trace 7.13d). The Neighbor Solicitation message (the query, asking what other stations are my neighbors) is sent in Frame 651, and the Neighbor Advertisement message (the response, indicating the presence of a neighbor on that segment) is returned in Frame 652.

Note that the source and destination IPv6 addresses begin with the Link Local address prefix of FE8H (or 1111 1111 10 in binary) that was shown in Figure 6-42. From your review of Figure 6-42 see that the second half of the Link Local Address contains a 64-bit Interface Identifier, which uniquely identifies the workstation in question. Present Interface IDs assigned by the IEEE are 48-bit quantities, which are converted to the required 64-bit number by inserting the hexadecimal sequence FF FE; changing the Individual/Group and Universal/Local bits is also required (review Figure 6-35). We see this sequence in the source and destination IPv6 addresses that are carried within the Neighbor Solicitation and Advertisement messages. For example, the IPv6 Link Local address for the Xircom station begins with FE 80H, includes the FF FEH in the middle, and ends with F3 5A C8H, the last portion of the Xircom hardware address.

Also observe that the Neighbor Advertisement includes three flags that provide details regarding this station: we know that it has routing capabilities; that this message was sent in response to the Neighbor Solicitation message, not just a random advertisement; but that the information does not update the link-layer address that has already been cached.

Trace 7.13d. IPv6 Neighbor Solicitation/Advertisement Messages

```
-------------------- Frame 651 --------------------
DLC:      — DLC Header —
DLC:
DLC:      Frame 651 arrived at  14:34:24.0646; frame size is 86 (0056 hex) bytes.
DLC:      Destination = Station XircomF35AC8
DLC:      Source      = Station 3Com  B81E32
DLC:      Ethertype   = 86DD (IPv6)
DLC:
IPv6:     — IPv6 Header —
IPv6:
IPv6:     Version           = 6
IPv6:     Traffic Octet     = 0
IPv6:     Differentiated Services Field : 0x00 (DS Codepoint - Default PHB)
IPv6:     Currently Unused Field        : 0x00
IPv6:     Flow Label        = 0x00000
IPv6:     Payload Length    = 32
IPv6:     Next Header       = 58 (ICMPv6)
IPv6:     Hop Limit         = 255
IPv6:     Source address    = fe80::206:5bff:feb8:1e32
IPv6:     Destination address = fe80::210:a4ff:fef3:5ac8
IPv6:
ICMPv6:   — ICMPv6 Header —
ICMPv6:
ICMPv6:   Type              = 135 (Neighbor Solicitation)
ICMPv6:   Code              = 0
ICMPv6:   Checksum          = 0x2224
ICMPv6:   Reserved          = 0x00000000
ICMPv6:   Target Address    = fe80::210:a4ff:fef3:5ac8
ICMPv6:
ICMPv6:   Options follow
ICMPv6:   Type              = 1 (Source Link-Layer Address)
ICMPv6:   Length            = 1 (units of 8 octets)
ICMPv6:   Link Layer Address  = Station 3Com  B81E32
ICMPv6:
```

```
-------------------- Frame 652 --------------------
DLC:      — DLC Header —
DLC:
DLC:      Frame 652 arrived at 14:34:24.0655; frame size is 78 (004E hex) bytes.
DLC:      Destination = Station 3Com  B81E32
DLC:      Source      = Station XircomF35AC8
DLC:      Ethertype   = 86DD (IPv6)
DLC:
IPv6:     — IPv6 Header —
IPv6:
IPv6:     Version          = 6
IPv6:     Traffic Octet    = 0
IPv6:     Differentiated Services Field : 0x00 (DS Codepoint - Default PHB)
IPv6:     Currently Unused Field        : 0x00
IPv6:     Flow Label       = 0x00000
IPv6:     Payload Length   = 24
IPv6:     Next Header      = 58 (ICMPv6)
IPv6:     Hop Limit        = 255
IPv6:     Source address   = fe80::210:a4ff:fef3:5ac8
IPv6:     Destination address = fe80::206:5bff:feb8:1e32
IPv6:
ICMPv6:   — ICMPv6 Header —
ICMPv6:
ICMPv6:   Type             = 136 (Neighbor Advertisement)
ICMPv6:   Code             = 0
ICMPv6:   Checksum         = 0x5416
ICMPv6:   R/S/O Flags      = CX
ICMPv6:             1... .... = sender is a router
ICMPv6:             .1.. .... = sent in response to Neighbor Solicitation
ICMPv6:             ..0. .... = does not update cached link-layer address
ICMPv6:   Reserved         = 0x00000000
ICMPv6:   Target Address   = fe80::210:a4ff:fef3:5ac8
ICMPv6:   No Neighbor Discovery options
ICMPv6:
```

7.14 Looking Ahead

In this chapter we have examined a number of protocols that work together to deliver datagrams within the internet from one host to another. Since this transmission is based on the Internet Protocol's connectionless service, guaranteed delivery of those datagrams must be ensured by another process, the Host-to-Host Layer. We also examined the two protocols that operate at the Host-to-Host Layer, UDP and TCP. Now we need some application data to send! We'll discuss these topics in the next section by studying protocols for file transfer, electronic mail, remote host access, and other end-user applications.

7.15 References

[7-1] Croft, W., et al. "Bootstrap Protocol (BOOTP)." RFC 951, September 1985.

[7-2] Alexander, S., and R. Droms. "DHCP Options and BOOTP Vendor Extensions." RFC 2132, March 1997.

[7-3] Finlayson, Ross. "Bootstrap Loading Using TFTP." RFC 906, June 1984.

[7-4] Zimmerman, D. "The Finger User Information Protocol." RFC 1288, December 1991.

[7-5] Postel, J. "The TCP Maximum Segment Size and Related Topics." RFC 879, November 1983.

[7-6] Clark, David D. "Window and Acknowledgement Strategy in TCP." RFC 813, July 1982.

8

Data Transport

The previous chapters have dealt with the Internet communications infrastructure that transports packets from a sender to a receiver using IP, TCP, UDP, and other protocols. Chapters 8 through 10 will consider the end-user applications that are communicated over that infrastructure. This chapter will look at applications supporting data transport, Chapter 9 will examine applications supporting multimedia transport, such as voice and video, and Chapter 10 will consider case studies that illustrate the operation of these end-user applications.

8.1 The Process/Application Connection

The Process/Application Layer sits at the very top of the ARPA architectural model. Unlike the Host-to-Host, Internet, and Network Interface Layers, which are transparent to end users, the Process/Application Layer is accessed by users directly via the host's operating system. End users use this layer's functions to perform computer operations such as file transfer, electronic mail, and so on (see Figure 8-1a).

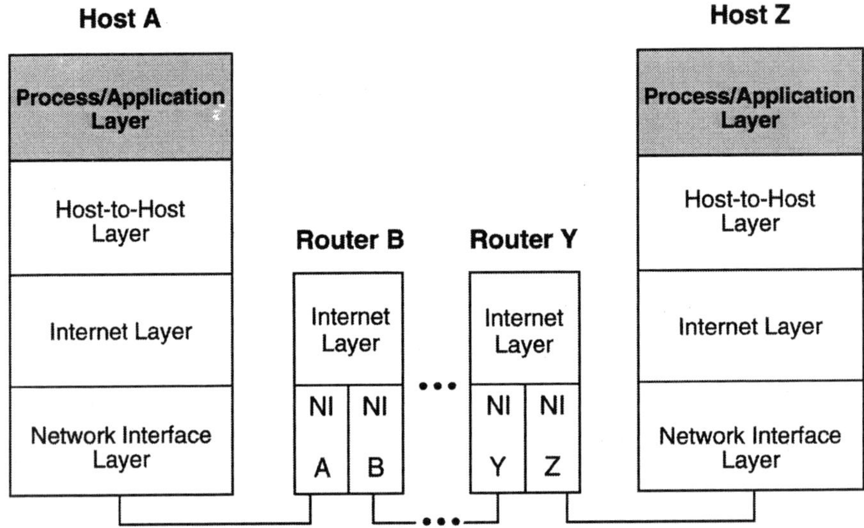

Figure 8-1a. The Process/Application Connection.

Figure 8-1b shows examples of the significant Process/Application protocols and the standards that describe them. Since these protocols provide functions that relate to the OSI Session, Presentation, and Application Layers, they are sometimes referred to as Upper Layer Protocols (ULPs). Figure 8-1c shows the position of the Application (ULP) data within the transmission frame. The lower layer headers and trailers all serve to reliably transfer the ULP information from one host to another via the internetwork.

ARPA Layer	Protocol Implementation								OSI Layer
Process / Application	Hypertext Transfer	File Transfer	Electronic Mail	Terminal Emulation	Domain Names	File Transfer	Client / Server	Network Management	Application
	Hypertext Transfer Protocol (HTTP) RFC 2616	File Transfer Protocol (FTP) MIL-STD-1780 RFC 959	Simple Mail Transfer Protocol (SMTP) MIL-STD-1781 RFC 2821	TELNET Protocol MIL-STD-1782 RFC 854	Domain Name System (DNS) RFC 1034, 1035	Trivial File Transfer Protocol (TFTP) RFC 1350	Sun Microsystems Network File System Protocols (NFS) RFC 3530	Simple Network Management Protocol (SNMP) v1: RFC 1157 v2: RFC 1901-10 v3: RFC 3411-18	Presentation Session
Host-to-Host	Transmission Control Protocol (TCP) MIL-STD-1778 RFC 793					User Datagram Protocol (UDP) RFC 768			Transport
Internet	Address Resolution ARP RFC 826 RARP RFC 903			Internet Protocol (IP) MIL-STD-1777 RFC 791			Internet Control Message Protocol (ICMP) RFC 792		Network
Network Interface	Network Interface Cards: Ethernet, Token Ring, MAN and WAN RFC 894, RFC 1042 and others								Data Link
	Transmission Media: Twisted Pair, Coax, Fiber Optics, Wireless Media, etc.								Physical

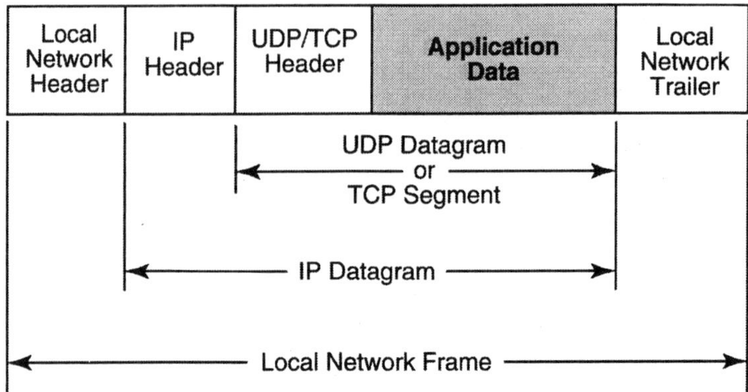

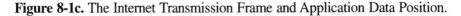

Figure 8-1c. The Internet Transmission Frame and Application Data Position.

To complete our tour of the ARPA architectural model, we will study the significant Process/Application protocols, beginning with TFTP. Several references provide details on the protocols discussed in this chapter. RFC 1123, "Requirements for Internet Hosts: Application and Support" [8-1], details requirements that hosts must provide to properly support file transfer, remote host access, and electronic mail. Specific details and parameters for these protocols are provided in the IANA online documents, Reference [8-2].

8.2 File Transfers using TFTP

The Trivial File Transfer Protocol (TFTP), described in RFC 1350 [8-3], reads and writes files or mail messages from one host to another. As an example, TFTP can be used to transfer files that provide system initialization information, such as that required for the startup of IP telephones. TFTP offers no functions other than file transfer — its strength is its simplicity — and it transfers the 512-octet blocks of data without excessive overhead. Because it is implemented on UDP transport, TFTP is one of the easiest ULPs to implement, but it does not guarantee data reliability.

TFTP defines five packet types, which are distinguished by an Opcode (operation code) field (see Figure 8-2):

Opcode	Operation
1	Read Request (RRQ)
2	Write Request (WRQ)
3	Data (DATA)
4	Acknowledgment (ACK)
5	Error (ERROR)

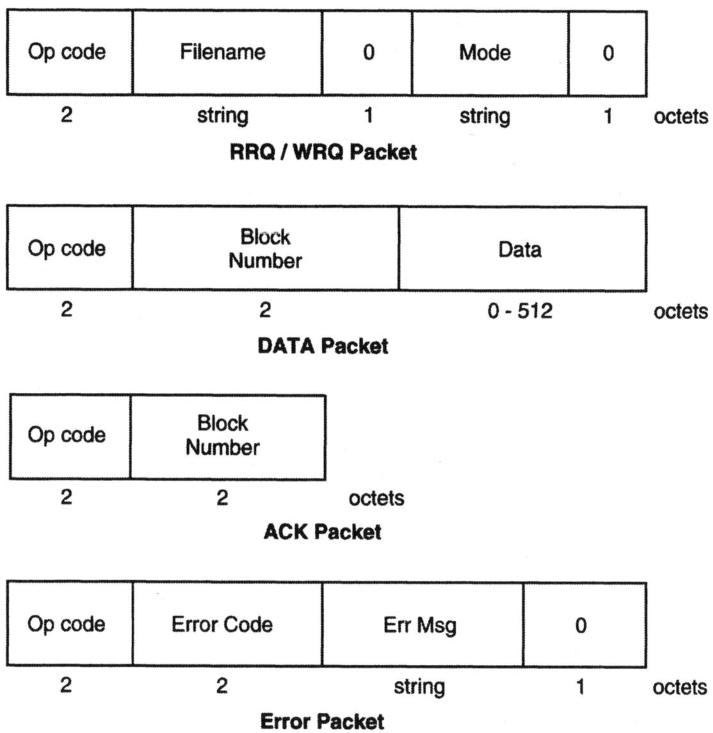

Figure 8-2. Trivial File Transfer Protocol (TFTP) Packet Formats.

The Read Request (RRQ, Opcode = 1) and Write Request (WRQ, Opcode = 2) packets have the same structure. Following the Opcode (2 octets), a string of netascii characters specifies the filename. (The netascii code is an 8-bit code defined by the ANSI standard X3.4-1968.) An octet containing zero terminates the filename. The Mode field (also a string) specifies which of three data transfer modes are to be used. The choices are netascii, octet (raw 8-bit bytes), and mail. The mail mode is defined as netascii characters destined for a user instead of a host. The Mode field is terminated by an octet containing zero.

The Data packet (Opcode = 3) transfers information. A Block number (2 octets) follows the Opcode and identifies the particular 512-octet block of data being sent. The Data field (0–512 octets in length) carries the actual information. Blocks less than 512 octets in length (i.e., 0–511 octets) indicate the end of an atomic unit of transmission.

The ACK packet (Opcode = 4) is used for acknowledgment. It contains a Block Number field (2 octets) that corresponds to the similar number in the Data packet being acknowledged. For simplicity, TFTP incorporates the lock-step acknowledgment, which requires each data packet to be acknowledged prior to the transmission of another. In other words, because TFTP operates over UDP, not TCP, it has no provisions for a window mechanism, and all packets (except ERROR packets) must be acknowledged. ACK or ERROR packets acknowledge DATA and WRQ packets; DATA or ERROR packets acknowledge RRQ or ACK packets. ERROR packets require no acknowledgment.

The ERROR packet (Opcode = 5) may be used to acknowledge any of the other four packet types. It contains a 2-octet error code that describes the problem:

Value	Meaning
0	Not defined, see error message (if any)
1	File not found
2	Access violation
3	Disk full or allocation exceeded
4	Illegal TFTP operation
5	Unknown transfer ID
6	File already exists
7	No such user

An error message (ErrMsg) consisting of a netascii string followed by a zero completes the packet. When errors occur, an ERROR packet is transmitted and the connection is terminated. Hosts generate ERROR packets for three types of events: when the host cannot satisfy a request such as locating a file; when the host receives a delayed or duplicate packet; or when the host loses access to a resource such as a disk during the transfer.

Figure 8-3 shows TFTP's operation. Host A issues an RRQ or WRQ and receives a response of either DATA (for RRQ) or ACK (for WRQ). Each host initiating a connection chooses a random number between 0 and 65,535 for use as a Transfer Identifier (TID). The TID passes to UDP, which uses it as a Port address. When the RRQ or WRQ is initially transmitted from Host A, it selects a TID to identify its end of the connection and designates Destination = 69, the TFTP Port number. If the connection is accepted, the remote host, Host B, returns its TID B subscript as the source with the TID A subscript as the destination. If a WRQ was the initial transmission, an ACK with Block number = 0 is returned. If the transmission was an RRQ, a DATA packet with Block number = 1 is returned. Data transfer then proceeds in 512-octet blocks, with each host identifying the appropriate Source and Destination TIDs with each DATA or ACK packet. The receipt of a DATA packet with less than 512 octets signals the termination of the connection. If errors occur during transmission, they generate an ERROR packet containing the appropriate Error Code. An example of this is provided in Section 10.1.

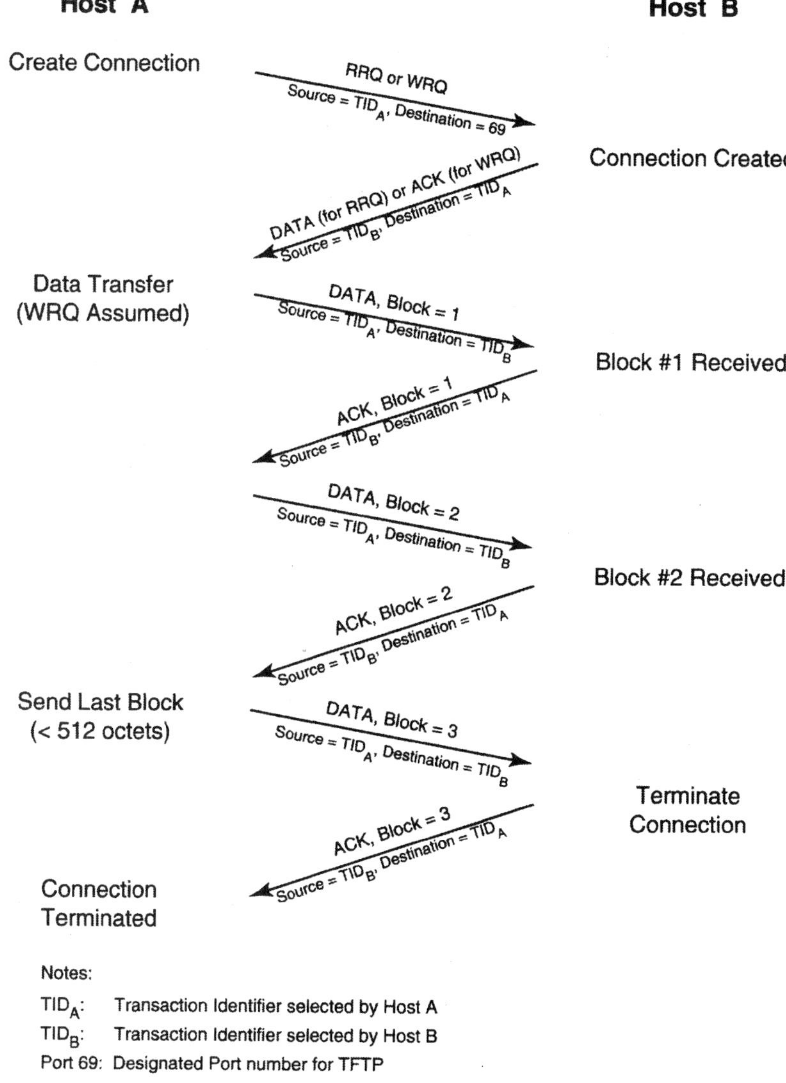

Figure 8-3. TFTP Connection, Data Transfer, and Termination.

In addition to RFC 1350, several other documents provide details on TFTP operation. These include: RFC 1785, "TFTP Option Negotiation Analysis"; RFC 2347, "TFTP Option Extension"; RFC 2348, "TFTP Blocksize Option"; and RFC 2349, "TFTP Timeout Interval and Transfer Size Options."

8.3 File Transfers using FTP

One of the most popular Process/Application protocols is the File Transfer Protocol (FTP), described in RFC 959 [8-4]. RFC 2228 discusses FTP Security Extensions and updates RFC 959. As its name implies, FTP allows local or remote client and server machines to share files and data using TCP's reliable transport.

The complete FTP service includes a User-FTP and a Server-FTP, as shown in Figure 8-4. The User-FTP includes a User Interface (UI), a User Protocol Interpreter (PI), and a User Data Transfer Process (DTP). The Server-FTP includes a Server-PI and a Server-DTP, but excludes the user interface. The User-PI initiates the logical control connection, which uses the TELNET protocol. The user (or client) uses an internally assigned Port number to connect to Server Port number 21, designated for FTP control. The data to be transferred passes from another self-assigned port on the User-DTP to Port number 20 (designated FTP data) on the Server-DTP.

Thus, two Port numbers are used for the two logical communication paths: control and data.

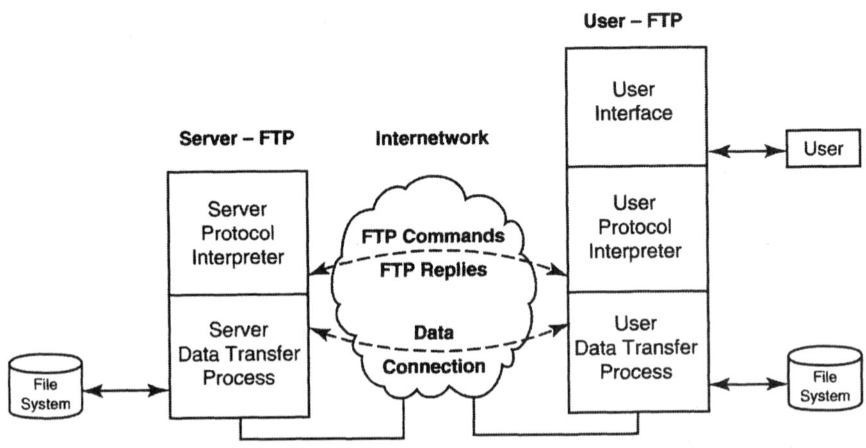

Figure 8-4. File Transfer Protocol (FTP) Model.

8.3.1 Data Representation, Data Structures, and Transmission Modes

Three parameters are required to completely specify the data or file to be transferred: the Data Representation, the Data Structure, and the Transmission Mode. The data may be represented in one of four ways: ASCII (the default is 8-bit characters); EBCDIC (8-bit characters); Image (8-bit contiguous bytes), which transfers binary data; and Local, which uses a byte size defined by the local host. Similarly, there are three types of data structures: the File Structure (the default), which is a continuous sequence of data bytes; the Record Structure, which consists of continuous records; and the Page Structure, which is comprised of independent, indexed pages. Three transmission modes are available, including Stream Mode (a transmitted stream of bytes), Block Mode (a header plus a series of data blocks), and Compressed Mode (used for maximizing bandwidth). The proper operation of the data transfer between the User and Server Data Transfer Processes depends on the control commands that go between the User and Server Protocol Interpreters. We'll look at these commands next.

8.3.2 FTP Commands

FTP commands and replies are communicated between the Server and User Protocol Interpreters (review Figure 8-4). These commands are defined in three categories: Access Control, Transfer Parameter, and Service. In many cases, the actual FTP commands are hidden from the end user by the graphical user interface of the host operating system, and may only be seen in the forms shown below when a command line interface is used.

8.3.2.1 Access Control Commands

The Access Control commands determine which users may access a particular file; these commands are invoked by the Server-FTP.

Command Code and Argument	Usage
USER <SP> <username> <CRLF>	Identifies the user
PASS <SP> <password> <CRLF>	User's password
ACCT <SP> <account-information> <CRLF>	User's account
CWD <SP> <pathname> <CRLF>	Change working directory
CDUP <CRLF>	Change to Parent directory
SMNT <SP> <pathname> <CRLF>	Structure Mount
REIN <CRLF>	Reinitialize, terminating the user
QUIT <CRLF>	Logout
<SP>	Space character
<CRLF>	Carriage Return, Line Feed characters

8.3.2.2 Transfer Parameter Commands

The Transfer Parameter commands alter the default parameters used to transfer data on an FTP connection:

Command Code and Argument	Usage
PORT <SP> <host-port> <CRLF>	Specifies the data port to be used
PASV <CRLF>	Requests the Server DTP to listen on a data port
TYPE <SP> <type-code> <CRLF>	Representation type: ASCII, EBCDIC, Image, or Local
STRU <SP> <structure-code> <CRLF>	File structure: File, Record, or Page
MODE <SP> <mode-code> <CRLF>	Transmission mode: Stream, Block, or Compressed

8.3.2.3 Service Commands

Service commands define the file operation requested by the user. The convention for the pathname argument is determined by the FTP Server's local conventions:

Command Code and Argument	Usage
RETR <SP> <pathname> <CRLF>	Retrieve a copy of the file from the other end
STOR <SP> <pathname> <CRLF>	Store data at the Server
STOU <CRLF>	Store Unique
APPE <SP> <pathname> <CRLF>	Append
ALLO <SP> <decimal-integer> [<SP> R <SP> <decimal-integer>] <CRLF>	Allocate storage
REST <SP> <marker> <CRLF>	Restart transfer at checkpoint
RNFR <SP> <pathname> <CRLF>	Rename from
RNTO <SP> <pathname> <CRLF>	Rename to
ABOR <CRLF>	Abort previous service command
DELE <SP> <pathname> <CRLF>	Delete file at Server
RMD <SP> <pathname> <CRLF>	Remove directory
MKD <SP> <pathname> <CRLF>	Make directory
PWD <CRLF>	Print working directory
LIST [<SP> <pathname>] <CRLF>	List files or text
NLST [<SP> <pathname>] <CRLF>	Name list
SITE <SP> <string> <CRLF>	Site parameters
SYST <CRLF>	Determine operating system
STAT [<SP> <pathname>] <CRLF>	Status
HELP [<SP> <string>] <CRLF>	Help information
NOOP <CRLF>	No operation

8.3.3 FTP Replies

An FTP reply consists of a three-digit number and a space, and is followed by one line of text. Each digit of the reply is significant. The first digit (value 1–5) determines whether the response is good, bad, or incomplete. The second and third digits are encoded to provide additional details regarding the reply. The values for the first digit are:

1yz	Positive Preliminary reply
2yz	Positive Completion reply
3yz	Positive Intermediate reply
4yz	Transient Negative Completion reply
5yz	Permanent Negative Completion reply

The values for the second digit are:

x0z	Syntax
x1z	Information
x2z	Connections
x3z	Authentication and accounting
x4z	Unspecified as yet
x5z	File system

The third digit gives a finer definition for each function category specified by the second digit. An example would be:

211:	System status
212:	Directory status
213:	File status
214:	Help message

The FTP specification, RFC 959, elaborates in great detail on the states and conditions that trigger these reply messages.

8.3.4 FTP Operation

A typical scenario in which FTP is used to retrieve a file from a remote host begins when the user initiates a connection to the remote host by entering FTP [host address] from a command line interface. For example, to retrieve an RFC, the user would enter *ftp.rfc-editor.org* to access the RFC Editor's server that contains Internet documentation such as RFCs. The host responds by asking for the username and password. If the desired file is not in the root directory, the user must change to the proper subdirectory. For example, the user would enter *cd in-notes* to change to the subdirectory that contains the RFCs. (Note that some host systems abbreviate certain commands. For example, the *cwd* command to change the current directory becomes *cd*. Most systems support the *help* command, which lists the commands or abbreviations accepted by the system.) The third step is for the user to tell the host the action required, such as file transfer. For example, to retrieve an RFC, the command would be *get rfcnnnn.txt*, where *nnnn* represents the number of the desired RFC. If the file transfer requires a different mode (such as binary), the user must specify that mode before invoking the *get* command. The file transfer would then begin, and the user would terminate the FTP connection (using the FTP *quit* command) when all business was completed. In some cases, a web browser may be used to access the ftp site, and then the typical point and click operation is used to identify the desired subdirectory and file to be retrieved. Another example of an FTP session is given in the case study in Section 10.2.

8.4 Terminal Emulation

TELNET, which stands for Telecommunications Network, is a protocol that allows a user (or client) at a terminal to access a remote host (or server). TELNET operates with TCP transport using Port number 23 and allows the terminal and host to exchange 8-bit characters of information in a half-duplex manner. The primary standard for TELNET, RFC 854 [8-5], discusses the three objectives of the protocol; RFC 855 [8-6] considers the various TELNET options.

TELNET's first objective is to define the Network Virtual Terminal (NVT). The NVT is a hypothetical representation of a network-standard terminal, also called a canonical (or standard) form. When both ends of the connection convert their data representations into the canonical form, they can communicate regardless of whether one end is, say, a DEC VT-100 and the other an SNA Host. The defined NVT format is the 7-bit USASCII code, transmitted in 8-bit octets. Figure 8-5 illustrates the conversion process from the local terminal/host format to the NVT format. The conversion typically occurs inside the devices, although an ancillary device such as a terminal server may perform the conversion for a number of similar devices.

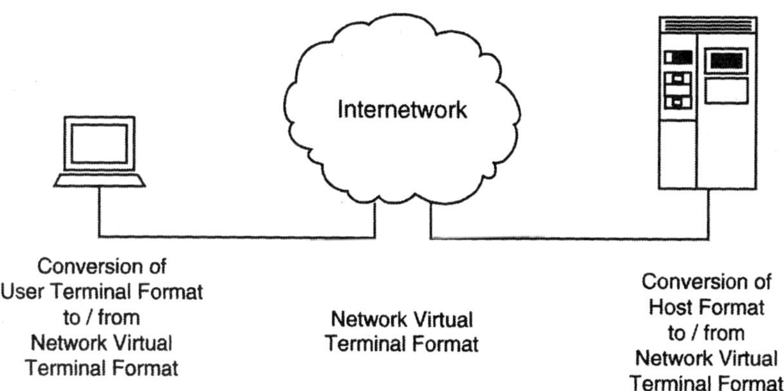

Figure 8-5. TELNET Network Virtual Terminal Operation.

The second objective of TELNET is to allow clients and servers to negotiate various options. This feature allows you to use a number of different terminals (some intelligent and some not so intelligent) with equal facility. To begin the negotiation, either of the two ends of the connection states their desire to negotiate a particular option. The other end of the connection can accept or reject the proposal. Telnet defines four negotiation options: WILL, WON'T, DO, and DON'T. WILL XXX indicates that party's desire (or offer) to begin performing option XXX. DO XXX or DON'T XXX are the returned positive or negative acknowledgments, respectively. DO XXX indicates that party's desire (or request) that the other party begin performing option XXX. WILL XXX and WON'T XXX are the returned positive and negative acknowledgments, respectively. For example, suppose the terminal wanted to use binary transmission. It would send a DO Binary Transmission to the remote host. The host could then respond with either a WILL Binary Transmission (a positive acknowledgment) or a WON'T Binary Transmission (a negative acknowledgment). If that terminal does not want its characters echoed across the TELNET connection it would send WON'T Echo; if the remote host agrees that no characters will be echoed it would return DON'T Echo.

The third concept of TELNET is one of symmetry in the negotiation syntax. This symmetry allows either the client or the server end of the connection to request a particular option as required, thus optimizing the service provided by the other party.

TELNET commands consist of a mandatory two-octet sequence and an optional third octet. The first octet is always the Interpret as Command (IAC) character, Code 255. The second octet is the code for one of the commands listed below. The third octet is used when options are to be negotiated, and contains the option number of interest. For example, the command "IAC DO Terminal Type" would be represented by FF FD 18 (hexadecimal), corresponding to the Codes 255 253 24 (which are represented in decimal). RFC 854 defines the following TELNET commands:

Command	Code	Meaning
SE	240	End of subnegotiation parameters
NOP	241	No operation
Data Mark	242	The data stream portion of a Synch. This should always be accompanied by a TCP Urgent notification.
Break	243	NVT character BRK
Interrupt Process	244	The function IP
Abort output	245	The function AO
Are You There	246	The function AYT
Erase character	247	The function EC
Erase Line	248	The function EL
Go ahead	249	The GA signal
SB	250	Indicates that what follows is subnegotiation of the indicated option.
WILL (option code)	251	Indicates the desire to begin performing, or confirmation that you are now performing, the indicated option.
WON'T (option code)	252	Indicates the refusal to perform, or to continue performing, the indicated option.
DO (option code)	253	Indicates the request that the other party perform, or confirmation that you are expecting the other party to perform, the indicated option.
DON'T (option code)	254	Indicates the demand that the other party stop performing, or confirmation that you are no longer expecting the other party to perform, the indicated option.
IAC	255	Data Byte 255

The IANA assigned numbers website lists the following TELNET options, along with any known reference documents that provide further information:

Option	Name	Reference
0	Binary Transmission	RFC 856
1	Echo	RFC 856
2	Reconnection	NIC 50005
3	Suppress Go Ahead	RFC 858
4	Approx Message Size Negotiation	DIX Ethernet standard
5	Status	RFC 859
6	Timing Mark	RFC 860
7	Remote Controlled Trans and Echo	RFC 726
8	Output Line Width	NIC 50005
9	Output Page Size	NIC 50005
10	Output Carriage-Return Disposition	RFC 652
11	Output Horizontal Tab Stops	RFC 653
12	Output Horizontal Tab Disposition	RFC 654
13	Output Formfeed Disposition	RFC 655
14	Output Vertical Tabstops	RFC 656
15	Output Vertical Tab Disposition	RFC 657
16	Output Linefeed Disposition	RFC 658
17	Extended ASCII	RFC 698
18	Logout	RFC 727
19	Byte Macro	RFC 735
20	Data Entry Terminal	RFC 1043
21	SUPDUP	RFC 736
22	SUPDUP Output	RFC 749
23	Send Location	RFC 779
24	Terminal Type	RFC 1091
25	End of Record	RFC 885
26	TACACS User Identification	RFC 927
27	Output Marking	RFC 933
28	Terminal Location Number	RFC 946
29	Telnet 3270 Regime	RFC 1041
30	X.3 PAD	RFC 1053
31	Negotiate About Window Size	RFC 1073
32	Terminal Speed	RFC 1079
33	Remote Flow Control	RFC 1372

34	Linemode	RFC 1184
35	X Display Location	RFC 1096
36	Environment	RFC 1408
37	Authentication	RFC 2941
38	Encryption	RFC 2946
39	New Environment	RFC 1572
40	TN3270E	RFC 1647
41	XAUTH	
42	CHARSET	RFC 2066
43	Telnet Remote Serial Port	
44	Com Port Control Option	RFC 2217
45	Telnet Suppress Local Echo	
46	Telnet Start TLS	
47	KERMIT	RFC 2840
48	SEND-URL	
49	FORWARD_X	
50	Unassigned	
.		
.		
.		
137	Unassigned	
138	TELOPT PRAGMA Logon	
139	TELOT SSPI Logon	
140	TELOPT PRAGMA Heartbeat	
255	Extended-Options-List	RFC 861

Note the large number of documents available to describe individual options, which are further described in Reference [8-2]. The case studies in Sections 10.3, 10.4, and 10.5 will consider TELNET operation.

8.5 Electronic Mail

The Simple Mail Transfer Protocol (SMTP) is based on a straightforward model of client/server computing. The model includes a sender and a receiver, both of which have access to a file system for message storage. Some type of communication channel links sender and receiver, completing the model. SMTP is intended to be a dependable message delivery system, although it does not provide absolute end-to-end reliability. It is based on TCP transport, however, which increases its effectiveness. The standard governing the SMTP system is specified in RFC 2821 [8-7], while the message format is contained in "Standard for the Format of ARPA Internet Text Messages," RFC 2822 [8-8]. In addition, RFC 2157 discusses mapping between X.400 and RFC 822/MIME message bodies.

8.5.1 Message Transfer

The transfer of an electronic message can be divided into several distinct stages, all supported by the SMTP model (see Figure 8-6). First, the user provides input to an interface system, known as the user agent, which facilitates the entry of the mail message. Then the message is sent to the Sender-SMTP, which assigns an arbitrary Port number to the process and establishes a TCP connection with Port number 25 on its peer (Receiver-SMTP). While establishing that connection, the receiver identifies itself to the sender. Next, the mail message is transferred using the RFC 2822 format (discussed in the following section). Finally, the sender signals its desire to terminate the connection, which is acknowledged by the receiver. After that acknowledgment, the TCP connection is closed.

As was the case with FTP, many of these functions and commands are hidden from the end user by the graphical user interface within the workstation's operating system.

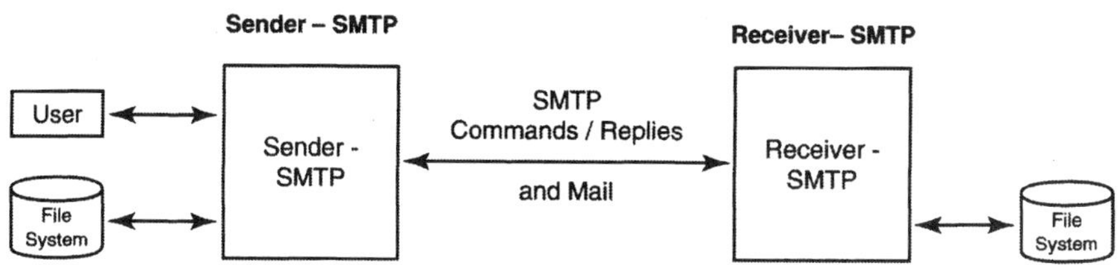

Figure 8-6. Simple Mail Transfer Protocol (SMTP) Model.

8.5.2 Message Formats

RFC 2822 [8-8] (which is based on the previous standard, RFC 822) defines the message formats used with SMTP. A message consists of a header, which contains a number of fields, and the message text. A blank line separates the header from the text. Many of the header fields are optional and depend on local implementation (yielding the potential for interoperability problems between mail systems); however, some variation of the examples below will be present in most systems. The following examples of mail formats are taken from Appendix A of RFC 2822.

8.5.2.1 A Message from One Person to Another

This example is called a canonical message, with a single author (John Doe), a single recipient (Mary Smith), a subject, date, message identifier, and a textual message in the body:

```
From: John Doe <jdoe@machine.example>
To: Mary Smith <mary@example.net>
Subject: Saying Hello
Date: Fri, 21 Nov 1997 09:55:06 -0600
Message-ID: <1234@local.machine.example>

This is a message just to say hello.
So, "Hello".
```

If John's secretary Michael actually sent the message, though John was the author and wants replies to this message to go back to him, the sender field would be used:

```
From: John Doe <jdoe@machine.example>
Sender: Michael Jones <mjones@machine.example>
To: Mary Smith <mary@example.net>
Subject: Saying Hello
Date: Fri, 21 Nov 1997 09:55:06 -0600
Message-ID: <1234@local.machine.example>

This is a message just to say hello.
So, "Hello".
```

8.5.2.2 Using Different Types of Mailboxes

This message includes multiple addresses in the destination fields and also uses several different forms of addresses.

```
From: "Joe Q. Public" <john.q.public@example.com>
To: Mary Smith <mary@x.test>, jdoe@example.org, Who?
<one@y.test>
Cc: <boss@nil.test>, "Giant; \"Big\" Box"
<sysservices@example.net>
Date: Tue, 1 Jul 2003 10:52:37 +0200
Message-ID: <5678.21-Nov-1997@example.com>

Hi everyone.
```

This example shows several different addressing constructs. First, the display names for *Joe Q. Public* and *Giant; "Big" Box* need to be enclosed in double-quotes because the former contains a period and the latter contains both semicolon and double-quote characters (the double-quote characters appear as a quoted-pair construct). Conversely, the display name for *Who?* could appear without them because the question mark is legal within that self-contained or indivisible part (called an atom). In addition, the *jdoe@example.org* and *boss@nil.test* have no display names associated with them at all, and *jdoe@example.org* uses the simpler address form without the angle brackets.

8.5.2.3 Using Group Addresses

For certain applications, the message should be delivered to a group of recipients instead of a single recipient:

```
From: Pete <pete@silly.example>
To: A Group:Chris Jones <c@a.test>,joe@where.test,John
<jdoe@one.test>;
Cc: Undisclosed recipients:;
Date: Thu, 13 Feb 1969 23:32:54 -0330
Message-ID: <testabcd.1234@silly.example>

Testing.
```

In this message, the "To:" field has a single group recipient named *A Group*, which contains three addresses, and a "Cc:" field with an empty group recipient named *Undisclosed recipients*.

8.5.2.4 Reply Messages

The following is a series of three messages that make up a conversation thread between John and Mary. John first sends a message to Mary, Mary then replies to John's message, and then John replies to Mary's reply message. Of special interest are the "Message-ID:", "References:", and "In-Reply-To:" fields in each message, which illustrate the thread binding these messages.

```
From: John Doe <jdoe@machine.example>
To: Mary Smith <mary@example.net>
Subject: Saying Hello
Date: Fri, 21 Nov 1997 09:55:06 -0600
Message-ID: <1234@local.machine.example>

This is a message just to say hello.
So, "Hello".
```

When sending replies, the Subject field is often retained, though prepended with "Re:".

```
From: Mary Smith <mary@example.net>
To: John Doe <jdoe@machine.example>
Reply-To: "Mary Smith: Personal Account"
<smith@home.example>
Subject: Re: Saying Hello
Date: Fri, 21 Nov 1997 10:01:10 -0600
Message-ID: <3456@example.net>
In-Reply-To: <1234@local.machine.example>
References: <1234@local.machine.example>

This is a reply to your hello.
```

Note the "Reply-To:" field in the above message. When John replies to Mary's message above, the reply should go to the address in the "Reply-To:" field instead of the address in the "From:" field.

```
To: "Mary Smith: Personal Account"
<smith@home.example>
From: John Doe <jdoe@machine.example>
Subject: Re: Saying Hello
Date: Fri, 21 Nov 1997 11:00:00 -0600
Message-ID: <abcd.1234@local.machine.tld>
In-Reply-To: <3456@example.net>
References: <1234@local.machine.example>
<3456@example.net>

This is a reply to your reply.
```

8.5.2.5 Resent Messages

We'll use the message from the above example as a starting point:

```
From: John Doe <jdoe@machine.example>
To: Mary Smith <mary@example.net>
Subject: Saying Hello
Date: Fri, 21 Nov 1997 09:55:06 -0600
Message-ID: <1234@local.machine.example>
```

```
This is a message just to say hello.
So, "Hello".
```

Suppose that Mary, upon receiving this message, wishes to send a copy of the message to Jane such that the message would appear to have come straight from John. If Jane replies to the message, the reply should go back to John, and all of the original information, like the date the message was originally sent to Mary, the message identifier, and the original addressee, would be preserved. In this case, resent fields are prepended to the message:

```
Resent-From: Mary Smith <mary@example.net>
Resent-To: Jane Brown <j-brown@other.example>
Resent-Date: Mon, 24 Nov 1997 14:22:01 -0800
Resent-Message-ID: <78910@example.net>
From: John Doe <jdoe@machine.example>
To: Mary Smith <mary@example.net>
Subject: Saying Hello
Date: Fri, 21 Nov 1997 09:55:06 -0600
Message-ID: <1234@local.machine.example>

This is a message just to say hello.
So, "Hello".
```

If Jane, in turn, wished to resend this message to another person, she would prepend her own set of resent header fields to the above and send that.

8.5.2.6 Messages with Trace Fields

As messages are sent through the transport system as described in RFC 2821, trace fields are prepended to the message. These fields may be used by anti-spam software to try and locate the true sender of the message. The following is an example of what those trace fields might look like. Note that there is some folding white space in the first message since these lines can be long.

```
Received: from x.y.test
   by example.net
   via TCP
   with ESMTP
   id ABC12345
   for <mary@example.net>;  21 Nov 1997 10:05:43 -
```

```
0600
Received: from machine.example by x.y.test; 21 Nov
1997 10:01:22 -0600
From: John Doe <jdoe@machine.example>
To: Mary Smith <mary@example.net>
Subject: Saying Hello
Date: Fri, 21 Nov 1997 09:55:06 -0600
Message-ID: <1234@local.machine.example>

This is a message just to say hello.
So, "Hello".
```

Additional message examples are presented in RFC 2822.

8.5.3 SMTP Commands

SMTP commands are comprised of a command code followed by an argument, and are defined in Section 4.1 of RFC 2821. The command codes are four alphabetic characters in either upper or lower case. The command code is separated from the argument by one or more space characters. The argument field carries addresses, parameters, and so on. The SMTP commands are:

Command Code	Usage
EHLO	Used to start the SMTP session
HELO	Identifies Sender-SMTP to Receiver-SMTP
MAIL	Deliver mail data to mailbox(es)
RCPT	Identify mail data recipient
DATA	The mail data
RSET	Abort current mail transaction
VRFY	Verify that the argument identifies a user
EXPN	Verify that the argument identifies a mailing list
HELP	Send information
NOOP	No operation
QUIT	Send OK reply, then close channel
SEND	Deliver mail to terminal(s), obsoleted in RFC 2821
SOML	Send or Mail, obsoleted in RFC 2821
SAML	Send and Mail, obsoleted in RFC 2821
TURN	Exchange Sender/Receiver roles, deprecated in RFC 2821

8.5.4 SMTP Replies

The SMTP Reply messages are also defined in RFC 2821. They are three digits long, and each digit has special significance. These replies are similar to the FTP codes in that the first digit is more general, whereas the third digit is more specific. The values for the first digit are:

1yz	Positive Preliminary reply
2yz	Positive Completion reply
3yz	Positive Intermediate reply
4yz	Transient Negative Completion reply
5yz	Permanent Negative Completion reply

The values for the second digit are:

x0z	Syntax
x1z	Information
x2z	Connection
x3z	Unspecified as yet
x4z	Unspecified as yet
x5z	Mail system

The third digit gives a finer definition for each of the function categories specified by the second digit. A complete listing of the reply codes from RFC 2821, in numerical order, is:

Code	Meaning
211	System status, or system help reply
214	Help message
220	<domain> Service ready
221	<domain> Service closing transmission channel
250	Requested mail action okay, completed
251	User not local; will forward to <forward-path>
252	Cannot VRFY user, but will accept message and attempt delivery
354	Start mail input; end with <CRLF>.<CRLF>
421	<domain> Service not available, closing transmission channel
450	Requested mail action not taken: mailbox unavailable
451	Requested action aborted: local error in processing
452	Requested action not taken: insufficient system storage
500	Syntax error, command unrecognized

501 Syntax error in parameters or arguments
502 Command not implemented
503 Bad sequence of commands
504 Command parameter not implemented
550 Requested action not taken: mailbox unavailable
551 User not local; please try <forward-path>
552 Requested mail action aborted: exceeded storage allocation
553 Requested action not taken: mailbox name not allowed
554 Transaction failed

If you require more information, turn to RFC 2821 for specifics on command usage, state diagrams that detail the command implementation, and command/reply sequences. We will study an example of SMTP in Section 10.6.

8.5.5 Multipurpose Internet Mail Extensions

SMTP defined the format for the message headers; however, it made the assumption that the contents of that message would be ASCII text. In many cases, however, email users wish to transmit other types of messages, such as a graphic or an audio or video clip. The Multipurpose Internet Mail Extensions, or MIME, provides for encoding techniques to allow these other types of messages to be transmitted after the RFC 2822 (or SMTP) headers. For example, MIME encoding may be used to support applications associated with Voice over IP systems. Technical details for MIME are specified in five parts, RFCs 2045–2049 (References [8-9] through [8-13]). Part One (RFC 2045) discusses the format of Internet message bodies; Part Two (RFC 2046) discusses the general structure of the MIME media typing system; Part Three (RFC 2047) describes extensions to RFC 822 to allow non-US-ASCII text data in mail fields; Part Four (RFC 2048) discusses IANA registration procedures for MIME-related facilities; and Part Five (RFC 2049) provides MIME conformance criteria, examples, and a bibliography.

RFC 2045 defines additional headers that further specify the contents of the electronic message. These headers include:

- MIME-Version header: Specifies the MIME version for this message.
- Content-Type header: Specifies the type and subtype of data in the body of the message. Seven data types have been defined:

    - Text: Textual information in a number of different character sets.
    - Image: For transmitting still image (or picture) data; includes subtypes for .jpeg and .gif files.
    - Audio: For transmitting audio or voice data, which requires an output device such as a speaker or a telephone.
    - Video: For transmitting video or moving image data; includes a subtype for .mpeg files.
    - Application: Used to transmit application or binary data; includes subtypes for octet-stream and PostScript files.
    - Multipart: Used to combine different types of data into a single message.
    - Message: Used for encapsulating another mail message.
- Content-Transfer Encoding header: Used to specify an auxiliary encoding that was applied to the data in order to allow it to pass through mail transport mechanisms.
- Content ID and Content Description headers: Further describe the data in the message body.

Below is an example, taken from Appendix A of RFC 2049, which illustrates the use of the above headers. What follows is the outline of a complex multipart message. This message has five parts to be displayed serially: two introductory plain text parts, an embedded multipart message, a text/enriched object, and a closing encapsulated text message in a non-ASCII character set. The embedded multipart message has two parts to be displayed in parallel, a picture and an audio fragment.

```
MIME-Version: 1.0
From: Nathaniel Borenstein <nsb@nsb.fv.com>
To: Ned Freed <ned@innosoft.com>
Date:  Fri,07 Oct 1994 16:15:05 00700 (PDT)
Subject: A multipart example
Content-Type: multipart/mixed;
              boundary=unique-boundary-1
```

```
This is the preamble area of a multipart message.
Mail readers that understand multipart format
should ignore this preamble.
If you are reading this text, you might want to
consider changing to a mail reader that understands
how to properly display multipart messages.
-unique-boundary-1

... Some text appears here...
[Note that the blank between the boundary and the start
of the text in this part means no header fields were
given and this is text in the US-ASCII character set.
It could have been done with explicit typing as in the
next part.]

-unique-boundary-1
Content-type: text/plain; charset=US-ASCII
This could have been part of the previous part,
but illustrates explicit versus implicit typing of body
parts.

-unique-boundary-1
      Content-Type: multipart/parallel;
      boundary=unique-boundary-2

-unique-boundary-2
Content-Type: audio/basic
Content-Transfer-Encoding: base64

... base64-encoded 8000 Hz single-channel
   mu-law-format audio data goes here....

-unique-boundary-2
Content-Type: image/jpeg
Content-Transfer-Encoding: base64

... base64-encoded image data goes here....

-unique-boundary-2
```

```
-unique-boundary-1
Content-type: text/enriched

This is <bold><italic>enriched.</italic></bold>
<smaller>as defined in RFC 1896</smaller>
<nl><nl>Isn't it
<bigger><bigger>cool?</bigger></bigger>

-unique-boundary-1
Content-Type: message/rfc822

From: (mailbox in US-ASCII)
To: (address in US-ASCII)
Subject: (subject in US-ASCII)
Content-Type: Text/plain; charset=ISO-8859-1
Content-Transfer-Encoding: Quoted-printable

... Additional text in ISO-8859-1 goes here ...

-unique-boundary-1
```

As the above example illustrates, MIME provides for a great deal of flexibility for message content specification. Readers needing further technical details should consult RFCs 2045–2049.

8.6 NetBIOS

NetBIOS, the Network Basic Input Output System, was developed by IBM and Sytek, Inc. (later known as Hughes LAN Systems, Inc.) for use with the IBM PC Network program. Just as ROMBIOS enables a PC's operating system and application programs to access its local I/O devices, NetBIOS provides applications access to network devices. NetBIOS is considered an OSI Session Layer interface and has become a de facto standard. The expansion of LANs into internetworks and the popularity of TCP/IP as the primary internetworking protocol suite have created a need for the NetBIOS interface to operate over the Internet protocols. From an architectural perspective, combining NetBIOS support at the OSI Session Layer with the TCP or UDP protocols at the Transport Layer is a natural way to support the numerous existing LAN applications in a distributed internetwork environment.

Just as different vendors have written ROMBIOS routines specific to their PCs, LAN operating system vendors have also come up with their own implementations of NetBIOS. RFC 1001 [8-14] and RFC 1002 [8-15] define a standard for NetBIOS support within the Internet community; hopefully, the rigorous detail contained in these RFCs will eliminate the multiple-implementation (semi-proprietary) issue that has crept into LAN environments.

Before delving into the specifics of NetBIOS use within the context of the Internet protocols, you need some background in the operation of NetBIOS. NetBIOS provides four types of primitives: Name Service, Session Service, Datagram Service, and Miscellaneous functions. Application programs use these services to locate network resources, establish and terminate connections, and transfer data.

The Name Service permits you to refer to an application, representing a resource, by a name on the internetwork. The name consists of 16 alphanumeric characters, which may be either exclusive (unique to an application) or shared (used by a group). The application registers the name to ensure that no other applications raise any objections to its use. The Name Service primitives are Add Name, Add Group Name, and Delete Name. Figure 8-7a illustrates the NetBIOS Name Service header.

The Session Service is used for the reliable exchange of data between two NetBIOS applications. Each data message may be from 0 to 131,071 octets in length. The Session Service primitives are somewhat analogous to the TCP primitives discussed in Chapter 5 and include: Call, Listen, Hang Up, Send, Receive, and Session Status. Figure 8-7b illustrates the NetBIOS Session Service header.

The Datagram Service provides unreliable, nonsequenced, and connectionless data transfer. The data may be transferred in two ways. In one technique, the datagram sender registers a name under which the data will be sent and specifies the name to which it will be sent. The second technique broadcasts the datagram. The Datagram Service primitives are Send Datagram, Send Broadcast Datagram, Receive Datagram, and Receive Broadcast Datagram. Figure 8-7c illustrates the NetBIOS Datagram Service header.

Miscellaneous functions generally control the operation of the network interface and are implementation-dependent. These functions include Reset, Cancel, Adapter Status, Unlink, and Remote Program Load. IBM's token ring implementation added the Find Name primitive, which determines whether a particular name is registered on the network. (Four of the NetBIOS primitives listed above – Reset, Session Status, Unlink, and Remote Program

Load — are considered local implementation issues that do not impact interoperability, and are therefore considered outside the scope of the Internet specification.)

RFC 1001 defines three types of NetBIOS end nodes and two NetBIOS support servers. The end nodes include the Broadcast (B), Point-to-Point (P), and Mixed Mode (M) types. Each of these types is specified by the operations it is allowed to perform. The NetBIOS support servers include the NetBIOS Name Server nodes (NBNS) and the NetBIOS Datagram Distribution nodes (NBDD). The NBNS manages and validates names used within the internetwork. NBNS formats the NetBIOS names to be valid Domain Name System (DNS) names and allows the NBNS to function in a fashion similar to the DNS Query service. The NBDD extends the NetBIOS datagram functions to the internet, which does not support multicast or broadcast transmissions. All of the nodes and servers are combined in various topologies of local and interconnected networks; RFC 1001 defines these details.

8.6.1 NetBIOS Name Service

NetBIOS's operation over UDP or TCP transport begins with the NetBIOS Name Service registering the name of the application to be used. The Name Query is the process by which the IP address associated with a NetBIOS name is discovered. Depending on the type of node in use (B, P, or M), the queries are either broadcast or directed to the NBNS. The exact procedures are described in RFC 1001.

NetBIOS Name Service messages use Port number 137 and are compatible with the DNS header format shown in Figure 8-7a. The structure includes a header followed by four entries — the Question section, Answer section, Authority section, and Additional Information section. A typical Name Query message would include the NetBIOS name in the Question section, followed by a Response message that provides details about the name, including its IP address. When an IP address has been found for the target name, either the Session Service (using TCP) or the Datagram Service (using UDP) may be implemented.

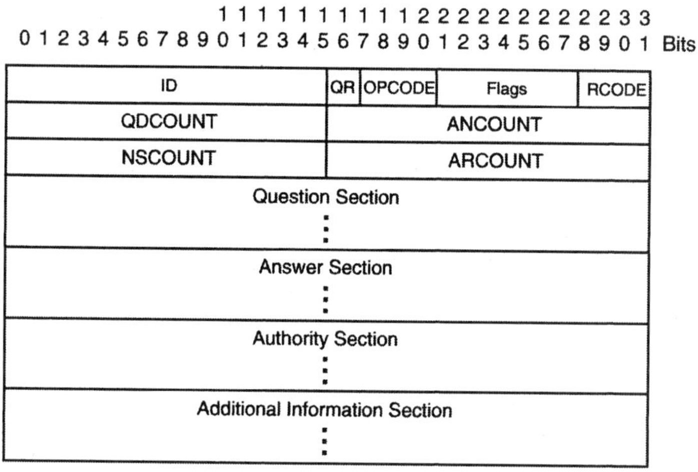

```
                    1 1 1 1 1 1 1 1 1 1 2 2 2 2 2 2 2 2 2 2 3 3
0 1 2 3 4 5 6 7 8 9 0 1 2 3 4 5 6 7 8 9 0 1 2 3 4 5 6 7 8 9 0 1 Bits
```

ID	QR	OPCODE	Flags	RCODE
QDCOUNT	ANCOUNT			
NSCOUNT	ARCOUNT			
Question Section ⋮				
Answer Section ⋮				
Authority Section ⋮				
Additional Information Section ⋮				

Figure 8-7a. NetBIOS Name Service Header.

8.6.2 NetBIOS Session Service

The NetBIOS Session Service uses Port number 139 and is implemented in three phases: session establishment, steady state, and session close. Session establishment determines the IP address and the TCP port of the called name and establishes a TCP connection with the remote device. Steady state provides data transfer and keep-alive functions. Session close terminates the session and triggers the close of the TCP session. Figure 8-8 illustrates these conditions; note the differences between the TCP and NetBIOS functions.

Figure 8-7b shows the format of the NetBIOS Session Service header. The Session Service header consists of a 4-octet header and a trailer that depends on the type of packet being transmitted. Three fields comprise the header: a Session Type field (1 octet), a Flags field (1 octet), and a Length field (2 octets). Values for the Session Type field are:

Value (hexadecimal)	Packet Type
00	Session message
81	Session request
82	Positive Session response
83	Negative Session response
84	Retarget Session response
85	Session Keep Alive

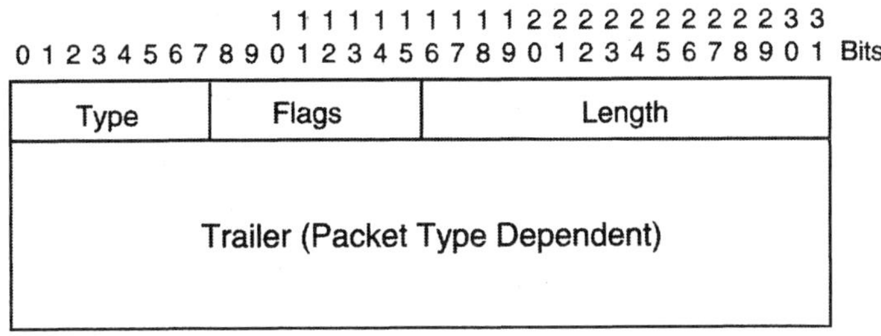

Figure 8-7b. NetBIOS Session Service Header.

The Flags field (1 octet) uses only Bit 7; all others are set to zero. Bit 7 is used as an extension to the Length field, which specifies the number of octets contained in the trailer (i.e., nonheader) fields. When used in combination with the Flag Extension Bit, the cumulative length of the Trailer field(s) has a maximum value of 128K octets.

8.6.3 NetBIOS Datagram Service

NetBIOS datagrams use UDP transport with Port number 138. The complete NetBIOS datagram includes the IP header (20 octets), UDP header (8 octets), NetBIOS datagram header (14 octets), and the NetBIOS data. The NetBIOS data consists of the Source and Destination NetBIOS names (255 octets each) and up to 512 octets of NetBIOS user data. The complete NetBIOS datagram can be up to 1,064 octets in length, but it may need fragmentation if the maximum IP datagram length is 576 octets.

NetBIOS datagrams require a Name Query operation to determine the IP address of the destination name. The NetBIOS datagram can then be transmitted within a UDP datagram or multiple UDP datagrams, as required. Three transmission modes are available: unicast, which transmits to a unique NetBIOS name; multicast, which transmits to a group NetBIOS name; and broadcast, which uses the Send Broadcast Datagram primitive.

Figure 8-7c illustrates the NetBIOS Datagram header. The Msg Type field (1 octet) defines the datagram function:

Value (hexadecimal)	Msg_Type
10	Direct_Unique_Datagram
11	Direct_Group Datagram
12	Broadcast datagram
13	Datagram error
14	Datagram Query request
15	Datagram Positive Query response
16	Datagram Negative Query response

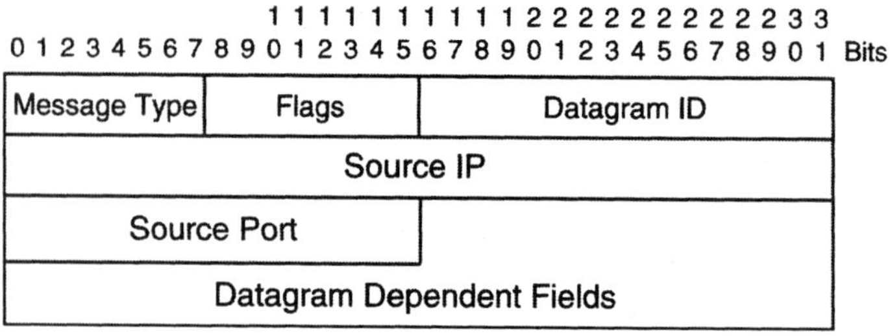

Bits 0–7	Bits 8–15	Bits 16–31
Message Type	Flags	Datagram ID
Source IP		
Source Port		
Datagram Dependent Fields		

Figure 8-7c. NetBIOS Datagram Header.

The Flags (1 octet) define the first datagram fragment, whether more fragments will follow, and the type of source node (B, P, or M). The remaining fields support datagram service, with a Datagram ID (2 octets), the Source IP address (4 octets), Source Port (2 octets), plus datagram-specific user data fields, such as a Datagram Length (2 octets), a Packet Offset (2 octets), and others, depending upon the message requirements. We'll see an example of NetBIOS packet operation in Section 10.7.

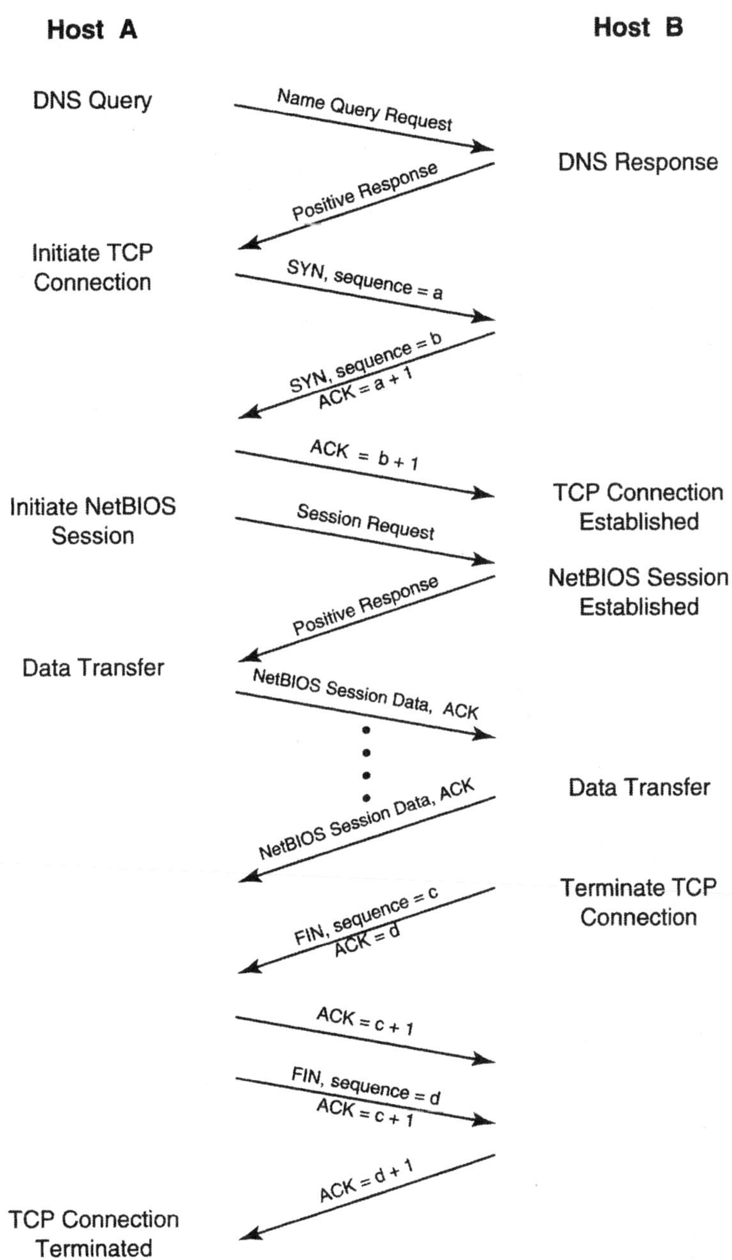

Figure 8-8. TCP and NetBIOS Connection Establishment/Disconnect Events.

8.7 Hypertext Transfer Protocol

The Hypertext Transfer Protocol (HTTP), defined in RFC 2616 [8-16], with security implementations described in RFC 2817 [8-17] and RFC 2818 [8-18], is a protocol used for communication with multimedia systems and is the key mechanism used for communication within the World Wide Web (WWW). HTTP is built on a client/server paradigm and uses TCP for transport. According to RFC 2616, HTTP can also be used for communication between user agents and proxies or gateways into other Internet-based protocol systems, including SMTP, the Network News Transfer Protocol (NNTP, defined in RFC 977), FTP, the Internet Gopher Protocol (defined in RFC 1436), and the Wide Area Information Service (WAIS, defined in RFC 1625).

The client/server interaction can be implemented using one of three topologies (Figures 8-9a, 8-9b, and 8-9c). In the simplest case, the User Agent sends a request chain for a particular resource to the Origin Server (Figure 8-9a). The Origin Server replies with a response chain containing the resource of interest. Note that a single connection links the User Agent and the Origin Server.

If a single link is not sufficient, various intermediaries may be inserted in the connection (Figure 8-9b). These intermediaries may be: a proxy, which may reformat the request; a gateway, which may translate the request to a protocol understandable by the Server; or a tunnel, which is a relay point between two connections, such as a firewall.

In some cases, communication to the distant Origin Server may not be necessary, because an intermediate point may have cached the requested information as the result of a previous (and possibly unrelated) request chain (Figure 8-9c). For example, an intermediary in the path to a popular web server (such as www.usatoday.com) may cache some of the more frequently requested web pages, such that the request chain need not travel the entire length of the connection to complete the transaction.

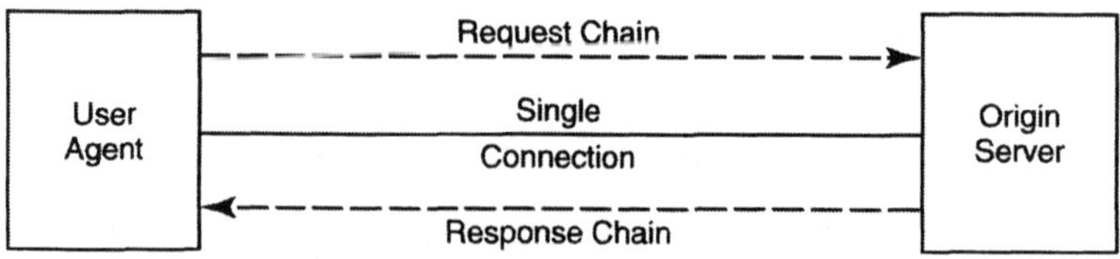

Figure 8-9a. HTTP Communication with Single Connection.

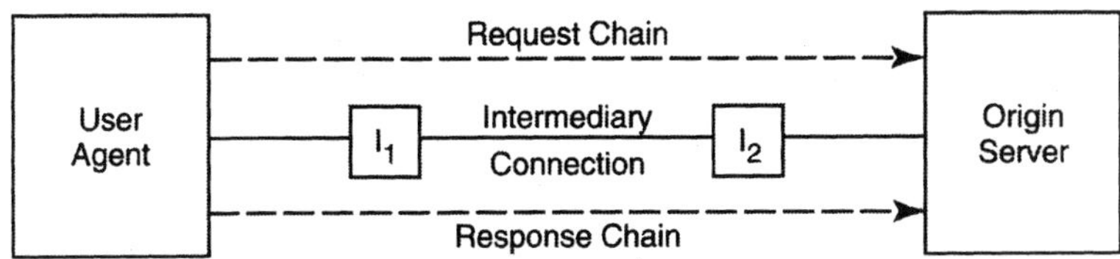

Figure 8-9b. HTTP Communication with Intermediary Connection.

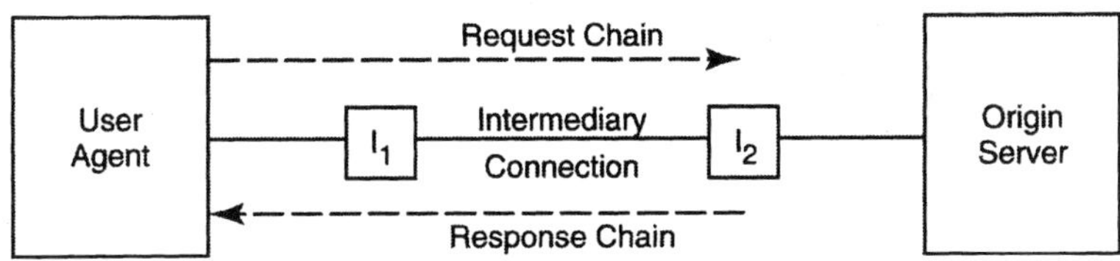

Figure 8-9c. HTTP Communication with Shortened Request/Response Chain (Cache at I2).

8.7.1 Uniform Resource Identifiers

One element of the request chain identifies the information that is being requested. The client identifies that resource using a Uniform Resource Identifier (URI), defined in RFC 1630 [8-19]. The URI may be represented as a Uniform Resource Name (URN), defined in RFC 1737 [8-20], or a Uniform Resource Locator (URL), defined in RFC 1738 [8-21]. URLs are written as follows:

```
<scheme>:<scheme-specific-part>
```

For example:

```
ftp://ftp.rfc-editor.org/in-notes/rfc1738.txt
```

In this case, *ftp* specifies the protocol in use (the File Transfer Protocol), *ftp.rfc-editor.org* specifies the host name, *in-notes* specifies the path within that host, and *rfc1738.txt* specifies the filename of interest.

A second example:

```
http://www.ietf.org/maillists.html
```

In this case, *http* specifies the Hypertext Transfer Protocol, *www.ietf.org* specifies the host name, and *maillists.html* specifies an HTML filename of interest.

8.7.2 HTTP Request Messages

HTTP is a request/response protocol. The client sends a Request message to the server, and the server then replies with a Response message. Request messages have the following format:

```
Request Line
Headers (general-headers, request-headers, entity-
   headers)
CRLF
Message Body
```

The Request-Line includes a Method Token, a Request-URI, and a protocol version, and ends with a carriage return and line feed. The Method Token indicates the operation to be performed on the resource, and could include: OPTIONS, GET, HEAD, POST, PUT, DELETE, TRACE, or CONNECT. The Request-URI identifies the resource upon which to apply the request. RFC 2616 gives the following example of a Request-Line:

```
GET http://www.w3.org/pub/WWW TheProject.html HTTP/1.1
```

The Headers may include General Headers, Request Headers, and Entity Headers. General Header fields are described in RFC 2616, section 4.5, and are applicable to both Request and Response messages:

- Cache-Control: Specifies directives that must be obeyed.
- Connection: Allows sender to specify options for that connection.
- Date: Date and time message was originated.
- Pragma: Used to include implementation-specific directives.

- Transfer-Encoding: Indicates what transformation has been applied to the message body.
- Trailer: Indicates that the given set of header fields is present in the trailer of a message encoded with chunked transfer-coding.
- Upgrade: Specifies additional communication protocols supported.
- Via: Used by gateways and proxies to indicate intermediate protocols and recipients.
- Warning: Carries additional information about the status or transformation of a message that might not be reflected in the message.

Request Header fields are described in RFC 2616, section 5.3, and allow the client to pass additional information about the request, and about the client itself, to the server:

- Accept: Specifies media types acceptable for the response.
- Accept-Charset: Indicates character sets acceptable for the response.
- Accept-Encoding: Restricts the content-encoding values.
- Accept-Language: Restricts the set of natural languages.
- Authorization: Contains credentials for the user agent.
- Expect: Indicates that particular server behaviors are required by the client.
- From: The email address of the human controlling the requesting user agent.
- Host: The Internet host and Port number of the requested resource.
- If-Match: Used with a method to make it conditional.
- If-Modified-Since: Used with the GET method to make it conditional.
- If-None-Match: Used with a method to make it conditional.
- If-Range: Used to request all or part of an entity.
- If-Unmodified-Since: Used with a method to make it conditional.
- Max-Forwards: Limits the number of proxies or gateways that can forward the request.

- Proxy-Authorization: Allows the client to identify itself to a proxy.
- Range: Specifies a portion of a resource.
- Referer: The address URI from which the Request-RI was obtained.
- TE: Indicates what extension transfer-codings it is willing to accept in the response and whether or not it is willing to accept trailer fields in a chunked transfer-coding.
- User-Agent: Contains information about the user agent originating the request.

Entity Header fields are described in RFC 2616, section 7.1, and define optional metainformation about the entity-body, or, if no body is present, about the resource identified by the request:

- Allow: Lists the methods supported by the resource.
- Content-Encoding: Modifies the media-type.
- Content-Language: Describes the natural language(s).
- Content-Location: Supplies the resource location for the entity.
- Content-MD5: Provides an end-to-end message integrity check.
- Content-Range: Specifies where the partial body should be inserted.
- Content-Type: Indicates the media type of the entity-body.
- Expires: Gives the date/time after which the response is stale.
- Last-Modified: The date/time the origin was last modified.

The CRLF includes a carriage return and line feed, which inserts a blank line in the Request Header.

The Message Body (if included in the HTTP message) is used to carry the entity-body that is associated with the request or response. (The message-body and the entity-body differ only when a transfer encoding has been applied, as discussed in RFC 2616, sections 4.3 and 14.41.) Information that is contained in the Message Body may be formatted using the Hypertext Markup Language (HTML), defined in RFC 2854 [8-22].

8.7.3 HTTP Response Messages

HTTP servers issue a Response message in reply to a Request message. Response messages have the following format:

```
Status Line
Headers (general-headers, response-headers,
   entity-headers)
CRLF
Message Body
```

The Status-Line consists of a protocol version followed by a numeric Status Code and its associated textual phrase. The Status Code is a three-digit integer result code that indicates the disposition of the request. RFC 2616, section 6.1, describes the Status Codes. The first digit of the Status Code defines the class of response. The last two digits do not have any categorization role. There are five values for the first digit:

- 1xx: Informational — Request received, continuing process.
- 2xx: Success — The action was successfully received, understood, and accepted.
- 3xx: Redirection — Further action must be taken in order to complete the request.
- 4xx: Client Error — The request contains bad syntax or cannot be fulfilled.
- 5xx: Server Error — The server failed to fulfill an appar ently valid request.

The individual values of the numeric Status Codes defined for HTTP/1.1, with their associated meanings, are:

Status Code	Meaning
100	Continue
101	Switching Protocols
200	OK
201	Created
202	Accepted
203	Nonauthoritative Information
204	No Content
205	Reset Content

206	Partial Content
300	Multiple Choices
301	Moved Permanently
302	Found
303	See Other
304	Not Modified
305	Use Proxy
307	Temporary Redirect
400	Bad Request
401	Unauthorized
402	Payment Required
403	Forbidden
404	Not Found
405	Method Not Allowed
406	Not Acceptable
407	Proxy Authentication Required
408	Request Time-out
409	Conflict
410	Gone
411	Length Required
412	Precondition Failed
413	Request Entity Too Large
414	Request-URI Too Large
415	Unsupported Media Type
416	Requested Range not Satisfiable
417	Expectation Failed
500	Internal Server Error
501	Not Implemented
502	Bad Gateway
503	Service Unavailable
504	Gateway Time-out
505	HTTP Version not supported

The Response Header fields are described in RFC 2616, section 6.2, and allow the server to pass additional information to the client that cannot be placed in the Status-Line:

- Accept-Ranges: Allows the server to indicate its acceptance of range requests for a resource.
- Age: Time since response was generated at the server.
- ETag: Provides the current value of the entity tag for the requested variant.
- Location: Redirects the recipient to another URI location.
- Proxy-Authenticate: Included with Proxy Authentication Required response.
- Public: Lists the methods supported by the server.
- Retry-After: Indicates how long the resource will be unavailable.
- Server: Identifies the software at the origin server.
- Vary: Indicates that the server has selected a representation of the response.
- WWW-Authenticate: Contains the challenge indicating the authentication scheme and parameters.

The other fields in the HTTP Response message are similar to their counterparts in the Request message as described above.

8.7.4 HTTP Over Transport Layer Security

The popularity of electronic commerce applications over the Internet has identified the need for secure web-based connections. The overall process is described as Transport Layer Security (TLS), formerly known as Secure Socket Layer (SSL), and it provides a mechanism for private end-to-end connections. These formats are defined in RFC 2817 [8-17] and RFC 2818 [8-18], and provide for authentication (identifying the source of the information), encryption (keeping that information from unauthorized disclosure), and data compression. Conceptually, HTTP/TLS operates much like HTTP over TCP, with the exception that the default port number is 443 instead of port 80, which is used with HTTP. In addition, the URI format is modified with a protocol identifier of *https* (instead of http) to indicate that security is involved in this connection. Complete TLS specifications can be found in RFC 2246 [8-23].

8.8 Looking Ahead

This chapter has examined application protocols that facilitate the transport of data, such as end-user files and emails. In Chapter 9, we will consider multimedia applications that transport voice, fax, and video information.

8.9 References

[8-1] Braden, R. "Requirements for Internet Hosts: Application and Support." RFC 1123, October 1989.

[8-2] The Internet Assigned Numbers Authority (IANA) maintains an online database of protocol numbers and parameters at http://www.iana.org/numbers.html. This replaces the former "Assigned Numbers" document, which was last published as RFC 1700 in October 1994.

[8-3] Sollins, K. R. "The TFTP Protocol (Revision 2)." RFC 1350, July 1992.

[8-4] Postel, J. "File Transfer Protocol." RFC 959, October 1985.

[8-5] Postel, J., and J. Reynolds. "Telnet Protocol Specification." RFC 854, May 1983.

[8-6] Postel, J., and J. Reynolds. "Telnet Option Specifications." RFC 855, May 1983.

[8-7] Klensin, J., editor. "Simple Mail Transfer Protocol." RFC 2821, April 2001.

[8-8] Resnick, P., editor. "Standard for the Format of ARPA Internet Text Messages." RFC 2822, April 2001.

[8-9] Freed, N., and N. Borenstein. "Multipurpose Internet Mail Extensions (MIME) Part One: Format of Internet Message Bodies." RFC 2045, November 1996.

[8-10] Freed, N., and N. Borenstein. "Multipurpose Internet Mail Extensions (MIME) Part Two: Media Types." RFC 2046, November 1996.

[8-11] Moore, K. "MIME (Multipurpose Internet Mail Extensions) Part Three: Message Header Extensions for Non-ASCII Text." RFC 2047, November 1996.

[8-12] Freed, N., et al. "Multipurpose Internet Mail Extensions (MIME) Part Four: Registration Procedures." RFC 2048, November 1996.

[8-13] Freed, N., and N. Borenstein. "Multipurpose Internet Mail Extensions (MIME) Part Five: Conformance Criteria and Examples." RFC 2049, November 1996.

[8-14] NetBIOS Working Group. "Protocol Standard for a NetBIOS Service on TCP/UDP Transport: Concepts and Methods." RFC 1001, March 1987.

[8-15] NetBIOS Working Group. "Protocol Standard for a NetBIOS Service on a TCP/UDP Transport: Detailed Specifications." RFC 1002, March 1987.

[8-16] Fielding, R., et al. "Hypertext Transfer Protocol — HTTP/ 1.1." RFC 2616, June 1999.

[8-17] Khare, R. et al. "Upgrading to TLS within HTTP/1.1." RFC 2817, May 2000.

[8-18] Rescorla, E. "HTTP Over TLS." RFC 2818, May 2000.

[8-19] Berners-Lee, T., et al. "Uniform Resource Identifiers in WWW, a Unifying Syntax for the Expression of Names and Addresses of Objects on the Network as used in the World-Wide Web." RFC 1630, June 1994.

[8-20] Sollins, K., and L. Masinter. "Functional Requirements for Uniform Resource Names." RFC 1737, December 1994.

[8-21] Berners-Lee, T., et al. "Uniform Resource Locators (URL)." RFC 1738, December 1994.

[8-22] Connolly, D., and L. Masinter. "The text/html Media Type." RFC 2854, June 2000.

[8-23] Dierks, T., and C. Allen. "The TLS Protocol." RFC 2246, January 1999.

9

Converged Networks and Multimedia Transport

The TCP/IP protocol suite, in support of the ARPANET, was originally designed in the early 1970s to support data applications, such as file transfers, email, and remote terminal access, that were required for government and military communication. Since that time, however, a number of real-time applications, such as voice and video, have been developed that could use TCP/IP and the Internet infrastructure for transport. These applications are often associated with the term *converged networks.*

In the 1980s, the term *convergence* was rarely used. Instead, telephone providers and subscribers used the term *integrated* to describe their vision of using a single telephone line to their home or business, and then being able to use that multifunction line for a variety of applications. These included voice communications (one or possibly two conversations over that single line), data applications for remote host access (stay-at-home shopping, remote electric meter reading, and so on), and even video applications (à la the famed Picturephone developed by Bell Telephone Laboratories in the 1950s). At that time, ISDN proponents envisioned that these new technologies would displace the currently deployed analog Public Switched Telephone Network (PSTN) within a few years.

Unfortunately, the implementation of ISDN was more of a challenge than the architects of the technology envisioned. When the telephone companies began to deploy ISDN, it was discovered that a large number of the local loops, or, in other words, the physical circuits from the telephone company central offices to the end-users' premises (homes or small businesses), would not support this high-speed data transmission. In particular, some of the loops were too long, and some included systems to optimize the network for voice transmission, which would not support the transmission of high-speed data.

As a result, ISDN deployments were primarily focused on the requirements for new construction. For example, if a new housing development was under construction, and new telephone service was required, then making that new service compatible with ISDN technology made sense. On the other hand, retrofitting existing outside plant systems to support ISDN was much more expensive and was often not undertaken. As an end result, there are still parts of the United States that do not have ISDN service available. So, in that respect, we could say that the promise of converged networks based on ISDN technologies met with limited success.

But before dismissing the concept of converged voice and data networks as just one more vendor architecture or carrier offering, we need to look at another significant trend coming from the data side of telecommunications — the ubiquity of the Internet Protocol. At the present time, it would be very difficult to find any type of computing platform, from the smallest cellphone or personal digital assistant (PDA) to the largest mainframe computer, which does not support IP. Thus, combining these two trends — the need for a combined voice/data network and the ubiquity of IP — we have a generalization of what a converged network might look like.

In this chapter, we will consider converged network applications and the additional protocols that are necessary to support them, beginning with a discussion on the characteristics that distinguish voice and data networks [9-1].

9.1 Voice and Data Network Characteristics

Most enterprise networks are really a "network of networks," with distinct infrastructures that address specific requirements. For example, there may be a separate network for circuit-switched voice and fax, a private tie-line network for intra-company voice and fax, a centralized processing data network such as one based on IBM's System Network Architecture (SNA), a distributed processing data network supporting client/server applications, a private internet or intranet for communication between employees, a public internet or extranet for communication with customers and/or suppliers, and systems and networks supporting other remote access configurations.

In general, voice networking infrastructures are connection-oriented, while data networking infrastructures are connectionless. But in order to determine which network type will be most prevalent in the years to come, let's consider some specifics of each technology.

9.1.1 Voice Network Characteristics

Traditional voice networks, such as the PSTN, are connection-oriented. To initiate a call, the end user takes their telephone off-hook, which signals the telephone company Central Office (C.O.) that service is requested (Figure 9-1). The end user then enters the destination telephone number via the numeric keypad (or rotary dial, if you are into antiques!). The destination telephone number becomes input to the signaling system between the various

C.O. switches to set up the call along the path from the source to the destination. The last C.O. in the chain signals the destination user's telephone that an incoming call is in process by sending ringing current through the line, thus creating an audible signal. When the end user (or that person's answering machine) takes the phone off-hook, the circuit is completed and communication can proceed.

One of the key elements that makes the PSTN so effective is its ubiquity, or universal service — the premise that states that basic telephone service should be an affordable commodity to anyone who wants it. As a result, reaching (almost) any end user via the PSTN should be possible.

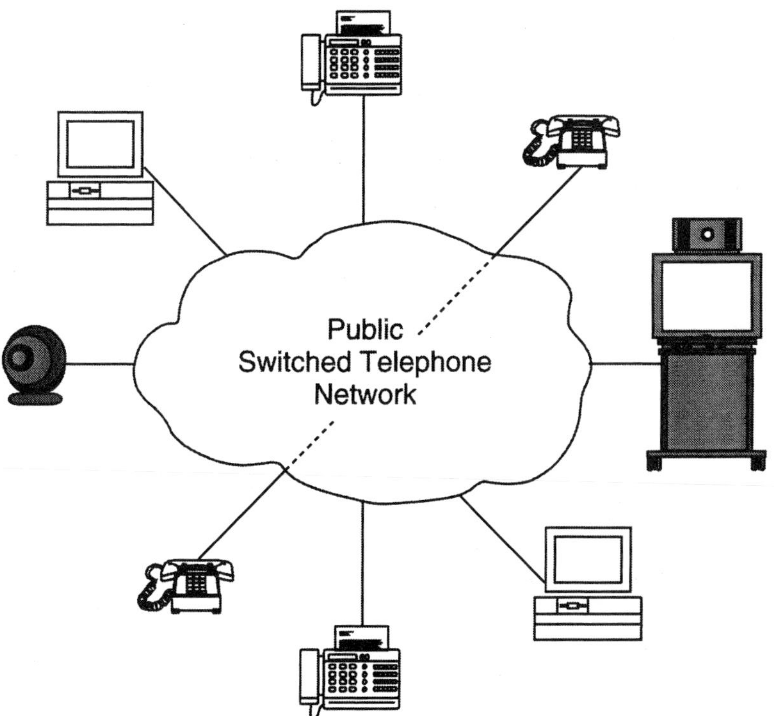

Figure 9-1. Public Switched Telephone Network.

The current PSTN creates internal circuits that provide a bandwidth of 64 Kbps, an element that is known as a Digital Signal Level 0 (or DS0) channel. (However, various transmission impairments such as signal attenuation and noise may reduce the actual bandwidth that the end subscribers realize on their local loop to less than 64 Kbps.) For now, suffice it to say that the 64 Kbps of bandwidth that is provided with a DS0 (which dates back several decades) is more than is required with current fax or voice compression technologies. For example, a fax transmission uses only 9.6 or 14.4 Kbps of bandwidth, while the balance goes unused. We will see how this "leftover" bandwidth can be used for other conversations with voice compression technologies, thus providing some of the economies of scale that help justify the investment in VoIP and FoIP equipment.

In some cases, switched network connections, which provide 64 Kbps per channel, are not adequate for the amount of information that needs to traverse the connection. For example, two PBXs may have a required number of simultaneous connections between two locations. For those applications, a private voice network can be configured, with trunk circuits provisioned according to the bandwidth requirements (Figure 9-2). As an example, if a dozen or so simultaneous conversations were expected between two locations, a Digital Signal Level 1 (or DS1) circuit would likely be employed, which provides 24 of the DS0 (64 Kbps) channels. As one might expect, the bandwidth of the trunk circuit is primarily limited by the depth of your checkbook — the carrier is quite content to provide more bandwidth as long as you are willing to pay for it. For example, the next generally available step in the digital multiplexing hierarchy is known as DS3, which operates at 44.736 Mbps and provides the equivalent of 672 DS0 channels between two locations. Not too many organizations need this amount of bandwidth just for voice, but when other media, such as data, fax, and video, are considered, then that amount of bandwidth is much easier to justify.

Legacy voice networks are also likely to incorporate a Private Branch Exchange (PBX) network, with connections to both private and public (PSTN) wide area network links. Adjunct processors, such as voice mail systems, are likely to be incorporated into that network architecture. In most cases, vendor-proprietary signaling protocols are used for communication between the user stations and the PBX switch, and are also used between PBXs and adjunct processors such as voice mail systems. Industry standard protocols, such as the ITU-T Signaling System 7 (SS7), are used between central offices within the PSTN, thus assuring end-to-end interoperability. Telephone network-based addressing, defined in ITU-T recommendations such as E.164 *Numbering Plan for the ISDN Era,* is typically used.

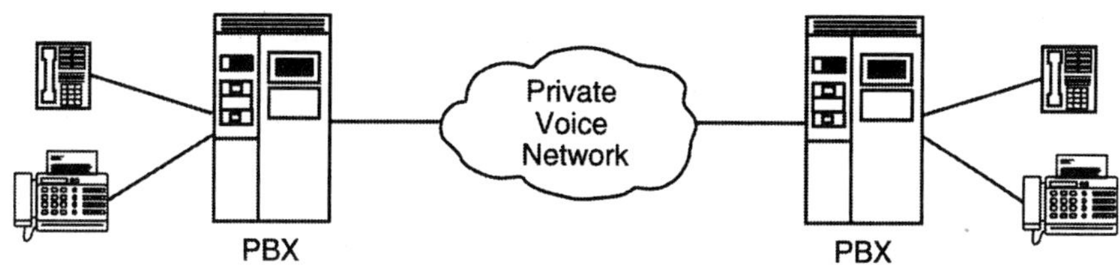

Figure 9-2. Private Voice Network.

For the duration of a particular call, however, resources along the path of the circuit have been reserved on behalf of the communicating parties, and, as a result, the parties will pay for the network services for the time that these resources have been reserved. The reservation of resources provides some predictable characteristics of that connection, including a constant delay for that unique path and sequential delivery of the information (the first word that I speak is the first word that you hear, and so on).

But if we were to summarize the characteristics of the telephone network, the term reliable would have to be used. Since the PSTN is considered a national resource, regulatory bodies such as the Federal Communications Commission (FCC) and state Public Utility Commissions (PUCs) monitor a number of PSTN operational characteristics. These include: the number and duration of service outages, the time it takes to repair and restore service in the event of an outage, prices the carriers charge for their services, and so on. The term "five nines of reliability" comes from an industry standard for central office switches, which are designed for two hours of downtime in a period of forty years of service. If all goes according to design, the network downtime would be calculated as:

```
2 hours of downtime/40 years * 365 days/year *
24 hours/day = 2/350,400 or 0.00000571
```

If this number is converted to availability (or uptime) by subtracting it from 1, you have:

```
Network availability = 1 - 0.00000571 = 0.99999429
```

Converting this to a percentage yields:

```
99.999429 percent availability (or reliability)
```

Hence, the term "five nines of reliability."

The ultra-reliable performance of the PSTN can be both a blessing and a curse. It is a blessing in that it rarely fails. (For example, when was the last time that your home telephone failed? In contrast, when was the last time that your PC locked up? It has been said that the good thing about conventional telephone service is that you don't have to reboot the end-user equipment!) It is a curse in that the end users know how reliable the telephone network is, and will expect that same level of service from a converged voice/data network.

9.1.2 Data Network Characteristics

In contrast to voice networks, which are generally connection-oriented, data network infrastructures are typically connectionless. Note that a distinction needs to be made between the networking infrastructure and the applications that run over that infrastructure. The applications may require a connection orientation, which would be typically handled by the ARPA Host-to-Host Layer (or OSI Transport Layer). For example, the File Transfer Protocol requires some degree of reliability and synchronization between the sending and receiving file transfer processes. Those functions are handled by TCP, which runs at the ARPA Host-to-Host Layer. IP, which is a connectionless protocol and runs at the (next lower) ARPA Internet Layer, would be considered the defining infrastructure protocol.

The origins of data networking are frequently traced to the development of packet switching and the X.25 protocol in the late 1960s. Packet switching technologies were developed, in part, in response to U.S. Government requirements for secure and survivable military communication. The premise was to divide a large message into a number of smaller elements (called packets), and to then send these packets via different routes. It was much harder for an adversary to eavesdrop on the message when that message was broken up into smaller pieces and transmitted via different paths (thus yielding some degree of network security). In addition, routing around network failures, thus making the network more robust or survivable, was also possible. Other benefits were also derived from this concept, such as load balancing between these different routes.

Early data networks relied heavily on leased line and packet-switched (using X.25) connections, migrating to Internet connections as that transmission medium became more prevalent (Figure 9-3). From the host-to-host applications that were prevalent in the 1960s and 1970s, local area networks

(LANs) became the data network of choice in the 1980s. Private data networks that comprised both the local network infrastructure (such as Ethernet LANs) and the wide area connections became the next step in the evolutionary process (Figure 9-4).

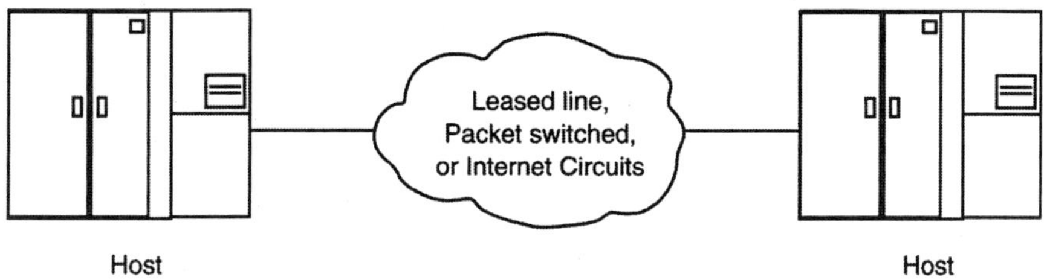

Figure 9-3. Public Data Network.

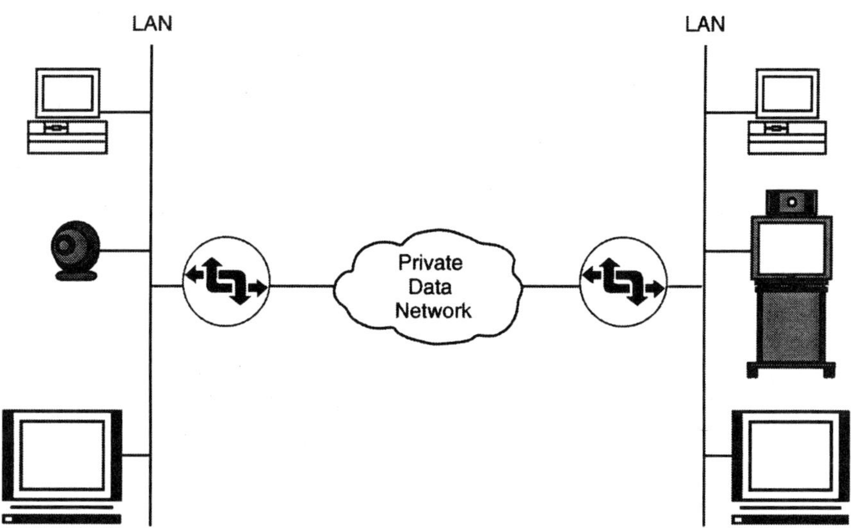

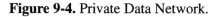

Figure 9-4. Private Data Network.

In contrast to the ultra-reliable, ubiquitous PSTN, data networks offer what is called "best-efforts" service. In other words, if your transmission on the Ethernet collides with mine, both are destroyed, and then the retransmission mechanism built into Ethernet must be employed for subsequent transmission attempts. But that's the way it goes — the Ethernet undertook its "best efforts" to get the transmission through. At best, both transmissions will be delayed. Packet switching networks are another example of best-efforts service, as packets that arrive in error or too late may be discarded, requiring the upper layer protocol to request a retransmission. These multiple attempts yield overall network reliability that is quite a bit less than the 99.999 percent reliability provided on the PSTN. But when the transmission is successful, you have access to a huge amount of network bandwidth — perhaps megabits per second of bandwidth — and are not limited to the 64 Kbps channels that are typically provisioned in voice networks.

As a similar contrast, data networks are typically priced based on consumption. If you migrate from 10 Mbps Ethernet to Fast Ethernet to Gigabit Ethernet, you will pay more per port. In a similar fashion, if your WAN connections migrate from a DS1 circuit to a DS3, you also pay more. In contrast, PSTN pricing tends to be relatively flat, because the bandwidth consumed is a consistent 64 Kbps. For example, your PSTN rate may be $0.05 per minute anytime of the day, to any destination in the United States.

Legacy data networks are also likely to be based on the Internet Protocol (IP) and to incorporate IP-based addressing schemes. Most host systems, from handheld Personal Digital Assistants (PDAs) to the largest mainframe processors, have IP-based connectivity options. As a result, these legacy IP networks are likely to contain a wide variety of processing and storage capabilities. In many cases, the legacy data network is also a distributed system, incorporating WAN connectivity such as frame relay circuits between sites. Because of the packet-switched nature of this network, protocols such as the Open Shortest Path First (OSPF) or Border Gateway Protocol (BGP) are deployed to keep the routing information up to date. Note that signaling protocols are not typically required. With the connectionless environment, call setups are not required; furthermore, any WAN links are likely to be permanent connections.

Finally, there is a big difference in the speed at which innovation is required within data networks. These technologies move very quickly. The sales, marketing, engineering, customer support, and other departments within data networking organizations must be prepared to move quickly as well, as the end users of this service demand higher and higher data transmission rates, amounts of memory and information storage, and so on. Contrast this with the innovation that is required in the PSTN. The process of making a telephone call today, and one several decades ago, is about the same. The PSTN infrastructure, such as the internal signaling systems, has markedly improved. However, since many of these improvements are not readily visible to the end user, the service presented to those end users may appear to be quiescent.

That brings us back to the subject of network convergence, and the implementation of an integrated voice/data network, such as that shown in Figure 9-5. Note that elements from both public and private voice and data networks will be present. Our objective, however, will be to take the most favorable characteristics of each network type and design the new system to optimally handle both voice and data transport with equal facility. Our network infrastructure of choice will be based on IP.

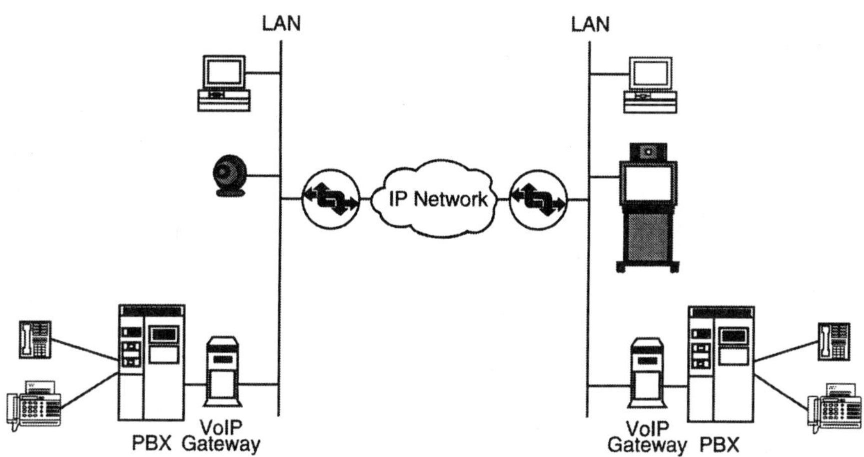

Figure 9-5. Integrated Voice/Data Network.

A final observation regarding converged networks can be seen. In most organizations, the *growth* of voice traffic has been relatively flat in the past few years, while the *growth* of data traffic has increased much more aggressively, as shown in Figure 9-6. Note that at this point in our discussion we are considering growth patterns, not actual traffic volumes. You will also notice that the exact point of crossover on the time axis has been omitted, as this would be a network-specific characteristic that varies from enterprise to enterprise.

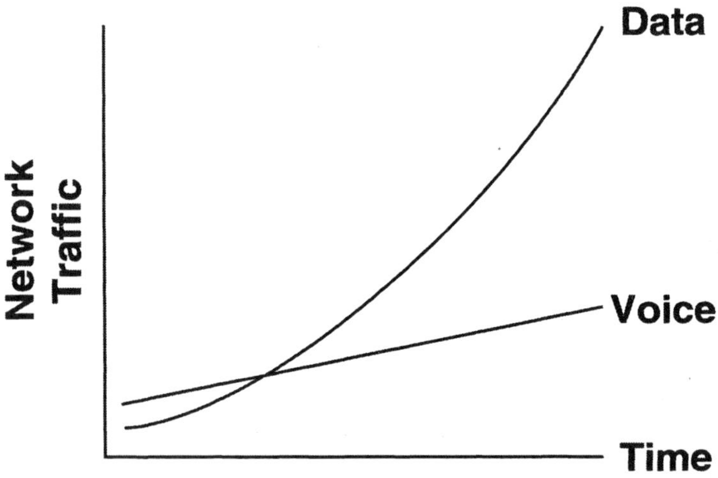

Figure 9-6. Typical Voice and Data Network Growth Patterns.

For example, a high technology research center with an ATM switched infrastructure and fiber optic trunk connections may have seen its growth in data traffic exceed that of voice traffic some time ago. In contrast, a call center or service operations center may continue to have more voice than data, and may not have an equivalent amount of voice and data traffic growth (the crossover point) for some time to come. And when you consider that most voice traffic today is still 64 Kbps traffic (not compressed), then the amount of traffic (measured in bits per second) that leaves your office may be greater in the voice case (today) than in the data case. In any event, however, consider that your data traffic growth should exceed that of your voice network at some time in the future.

To evaluate this premise, consider the following generalizations. The aggregate voice network (considering the PSTN, private networks, and so on) supports one primary application (voice) and a few data applications (such as modem traffic) that consume relatively small amounts of bandwidth (typically 64 Kbps per channel or less). The aggregate data network (including the Internet, private internetworks, virtual private networks, and application-specific networks such as SNA backbones) supports many data applications. Much of this is heavily influenced by the large growth in the number of users that are attached to the Internet. The most prominent of this Internet traffic is World Wide Web information, which is clearly growing at exponential rates fueled by electronic commerce and other interactive applications. It would not be an exaggeration to depict the generalized growth in voice traffic as being relatively flat (or linear), while the growth in data traffic has some positive slope (and possibly an exponential slope), as depicted in Figure 9-6.

9.2 Converged Network Architecture and Applications

Thus far, we have seen that the concept of an integrated voice/data network has been around since the development of ISDN, that the well-understood Internet Protocol can be used as an enabling transport system, and that the growth in voice traffic has been relatively flat in the past few years while the growth in data traffic has increased much more rapidly. But prior to looking at the architecture of a converged network, let's first consider its advantages.

9.2.1 Realizing the Advantages of the Converged Network

The advantages of a converged network are many: a reduction in overall complexity, since there would be fewer protocols, operating systems, and so on to manage; greater utilization efficiency of WAN circuits, since separate voice and data connections could be combined into a larger, more cost-effective pipe; one integrated management system instead of separate systems to support the voice and data networks; plus new applications that leverage the integration of voice, data, video, and other media.

To realize these advantages will require a number of elements — some from traditional voice networks, such as telephones, fax machines, and PBXs; some from traditional data networks, such as terminals, hosts, and servers; and some from new systems, such as gateways and gatekeepers, that are designed as the glue to hold all of these disparate systems together. Figure 9-7 illustrates some of these various elements, which could incorporate many different subsystems, including the PSTN, dedicated WAN connections, IP routers, wiring hubs and end stations, and many others [9-2].

A call over an IP network could start with a local PSTN connection and the conversion of the analog voice signal to a digital pulse stream at the client end station, PBX, or gateway. That pulse stream will likely undergo some signal processing, including silence suppression and echo cancellation. The gateway may consult with a gatekeeper and/or Domain Name Service (DNS) server to obtain information about the desired destination, such as its transmission capabilities and IP address. The voice signal is then converted into packets, transmitted over the IP-based network to a destination gateway, and the digital pulse stream is reconstructed. As a final step, the digital pulse stream is then converted into an analog signal and delivered to the desired destination.

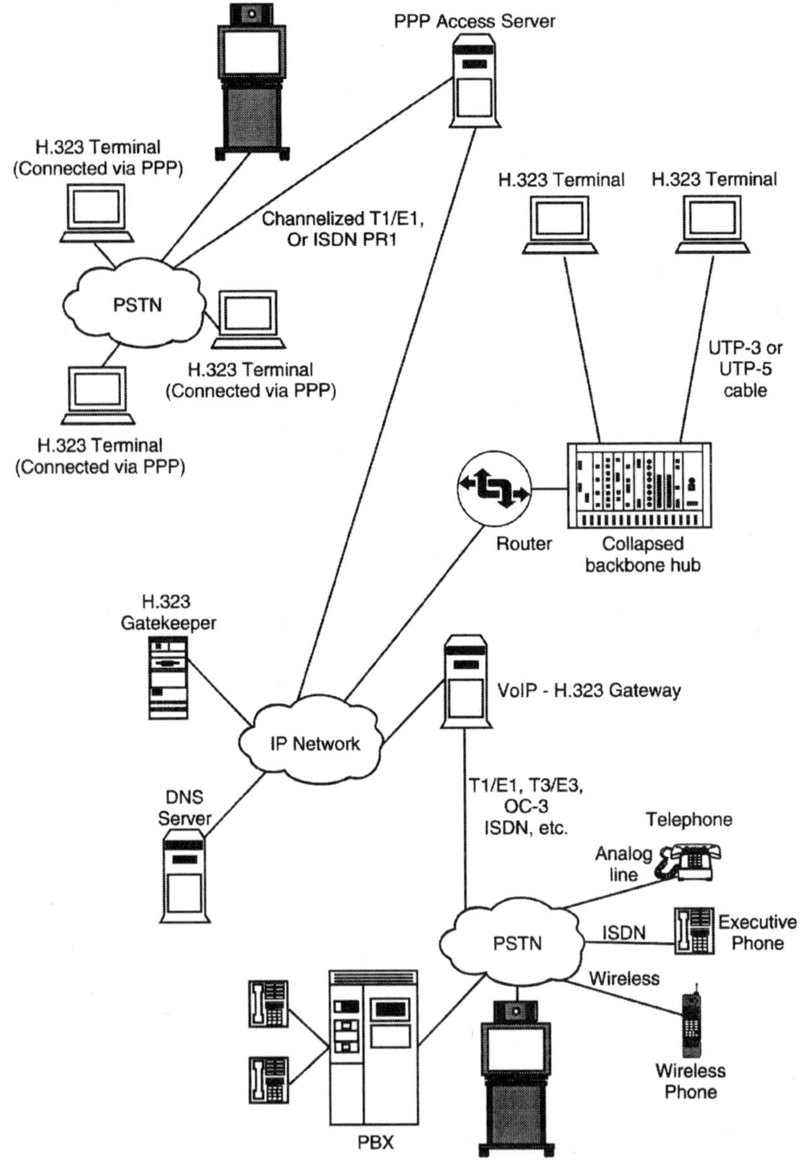

Figure 9-7. Voice over IP Network Elements (Possible Configuration, Not Mandatory). *Source: Voice over IP Forum.*

9.2.2 Converged Network Infrastructure Challenges

As we have seen, the converged network brings many advantages, but also a key challenge:

> *How do you take a connection-oriented application (such as voice or video) and transmit it over a connectionless network (such as an IP network, or the Internet), and do so reliably?*

This challenge should be considered from three perspectives: network infrastructure, continuing network management, and end-user quality of service.

The network infrastructure has likely been designed in support of voice communications or data communications, but not both. In other words, let us suppose that you are considering adding IP voice services or video conferencing to your existing internetwork. Does the network have the capacity for these additional, bandwidth-intensive applications?

For the network manager, one of the key challenges may be education. Most separate voice and data networks also had separate management staffs and systems. Thus, if you had responsibility for the voice WAN, you were only peripherally involved in issues concerning the local data networks. As a result, you were not likely to develop expertise in LAN technologies. If you are now responsible for both voice and data elements, you must develop network management and analysis expertise in both these areas.

But perhaps the greatest challenge is keeping the end users satisfied with the service provided by your internetwork. Even though your end users may not be technologists who understand the interworkings of computer communication systems, they know how to operate the telephone, and they know from experience how reliable the telephone system is. They have been conditioned to expect a high quality of service. Meeting anything less than their existing expectations is likely to be unacceptable.

9.2.3 Converged Network Standards and Components

Perhaps more important than any other issue surrounding converged networks is the one of reliability, and for such a system to be successful requires adherence to published standards. Many organizations have been hard at work developing standards and protocol suites that address this challenge. At the forefront of these efforts are the European Telecommunications Standards Institute (ETSI), the Internet Engineering Task Force (IETF), and the ITU-T. Here are a few examples of the work that has been produced:

- ETSI: Telecommunications and Internet Protocol Harmonization Over Networks (TIPHON) [9-3].
- ITU-T: H.225.0, H.323, and H.245 standards for multimedia communications [9-4].
- IETF: Resource Reservation Protocol (RSVP), Real Time Protocol (RTP), and the Session Initiation Protocol (SIP) for multimedia communication over IP-based networks.

The converged network incorporates elements from both the voice and data environments, but requires additional components and protocol operations to enable the internetwork communication. For example, voice networks are connection oriented and require call signaling protocols to establish a call. In contrast, data networks are connectionless (the end stations simply drop packets into the network), but they require routing protocols to update the path information and increase the likelihood of successful packet delivery. In addition, stations on the voice network are addressed using a telephone number entered from a touch-tone keypad. Stations on the data network are addressed according to the protocol deployed, such as a 32-bit IP address. These address and protocol conversions occur at several enabling systems (which may go by different names than those listed below, depending on the standard they adhere to):

- Gateways, which convert packet-based user information to/from circuit-based (or streaming) information, and also handle any signaling protocol issues, such as call setups and disconnects.
- Gatekeepers, which provide network management functions, such as address translations and bandwidth oversight for the network as a whole. Thus, a gatekeeper could prevent a video conference from consuming all available bandwidth and leaving network users without basic connectivity.
- Domain Name System Servers, which work with the gatekeepers to provide address translation and management functions.
- Terminals, which provide the multimedia interface and signal processing functions such as converting the analog voice to a digital bit stream. Much of this signal processing is done in a codec chip, short for coder/decoder.

The protocols that operate in these enabling systems have been derived from two different organizations: the International Telecommunication Union — Telecommunication Standards Sector (ITU-T) and the Internet Engineering Task Force (IETF). We will look at the IETF protocols that support ARPA-based networks in the next section.

9.3 ARPA Protocols Supporting Converged Networks

Thus far, we have looked at the underlying infrastructure and protocols that are used to route and reliably deliver IP datagrams through an internetwork. These protocols, such as IP, RIP, OSPF, UDP, TCP, and the others we studied, were developed with data applications in mind: email, file transfers, remote terminal access, and so on. In this section, we will briefly consider some of the additional protocols, primarily at the higher layers, that are required to support real-time and multimedia applications. A summary of these protocols, plus their dependencies, is shown in Figure 9-8.

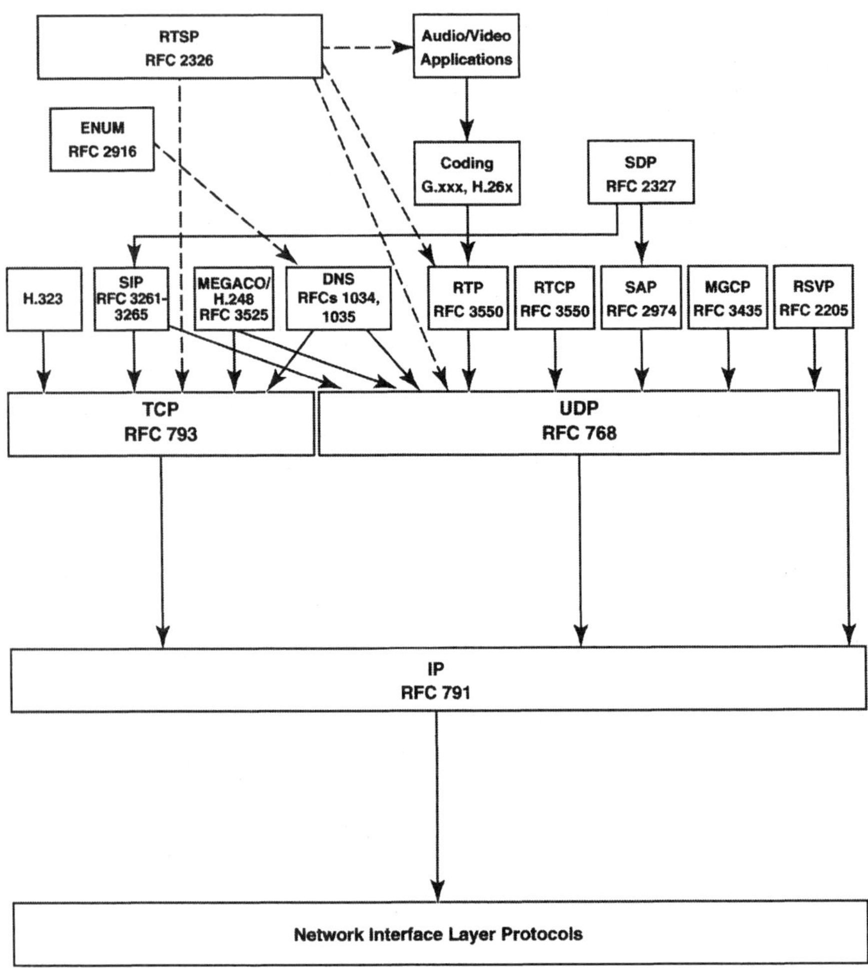

Figure 9-8. Voice over IP Protocols.

9.3.1 Multicast IP

The term *multi* is derived from a Latin term that means much or many. In the case of IP datagrams, the objective is to send one packet and have it be received at many destinations. This one to many packet transport mechanism can be used in a number of applications, including: audio and video conferencing, news broadcasts, stock quotation distribution, and distance learning.

When datagrams are multicast, instead of individual copies of the message being sent to each recipient, the bandwidth of the communications infrastructure is conserved. A single stream of information from one source is delivered to the destination end stations, instead of multiple streams from that one source being delivered to those end stations (Figure 9-9).

The concepts of multicast transmission over IP-based internetworks were first described in RFC 1112, "Host Extensions for IP Multicasting," published in 1989 [9-5]. To support multicast applications, the end stations must be able to support the Internet Group Management Protocol (IGMP, also described in RFC 1112 and further defined in RFC 2236 [9-6]), which enables multicast routers to determine which end stations on their attached subnets are members of multicast groups. These groups are identified by special multicast addresses. IPv4 Class D addresses are reserved to support multicast applications, and range in value from 224.0.0.0 through 239.255.255.255.

From these values, the Internet Assigned Numbers Authority (IANA) has assigned specific values for particular multicast functions. For example, the range of addresses between 224.0.0.0 and 224.0.0.255, inclusive, is used for routing protocols and other topology and maintenance functions. Within this range, address 224.0.0.1 identifies "all systems on this subnet" and 224.0.0.2 identifies "all routers on this subnet." Other values have been specified RIP2 routers (224.0.0.9), DHCP Server/Relay Agents (224.0.0.12), audio and video transport, and others. Addresses in the range from 239.0.0.0 through 239.255.255.255 are reserved for local applications. A complete listing of the assigned multicast addresses is available in Reference [9-7].

In order to propagate these datagrams to multiple destinations, the routers within the network infrastructure operate with modified routing protocols. Two examples of these multicast routing protocols are the Distance Vector Multicast Routing Protocol (DVMRP), described in RFC 1075 [9-8], and the Multicast Extensions to OSPF (MOSPF), described in RFC 1584 [9-9]. These multicast routing protocols construct logical spanning trees, which describe how the multicast traffic flows to the end stations.

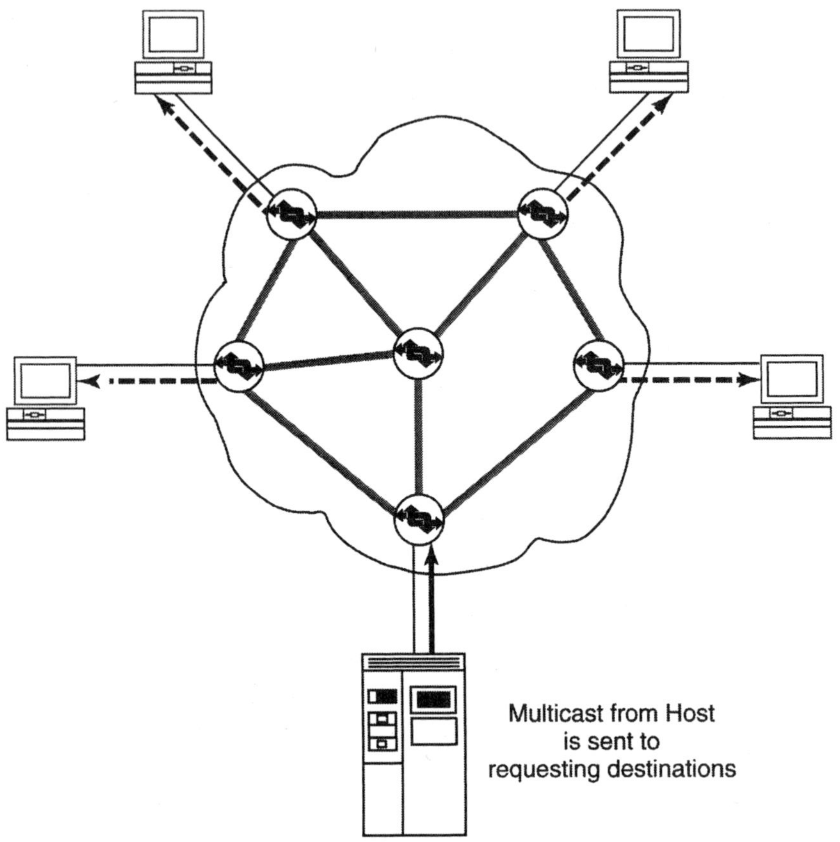

Figure 9-9. IP Multicast Operation.

9.3.2 Real-Time Transport Protocol

Recall from our earlier discussions that IP networks provide connectionless (CON) data transport services. One of the main characteristics of CON networks is their absence of delay characteristics. In other words, you drop a packet into the network, and if all works as planned, it reaches the intended destination. However, you cannot make any statements regarding the timeframe that will be required for that packet to travel from the source to the destination. Perhaps it will travel slowly, or perhaps it will travel quickly (or perhaps it will get clobbered and not get there at all, but we optimists will dismiss that alternative for the moment and assume successful packet delivery).

So the next question becomes, how do we take data from an application, such as voice or video, which is real-time, and send it through a network that cannot (by definition and design) guarantee reliable delivery? The answer is found in the use of the Real-time Transport Protocol (RTP), which is defined in RFC 3550 [9-10]. A companion protocol, the RTP Control Protocol (RTCP), monitors the quality of service and conveys information about the participants in the communication session. RTCP will be the discussed in the following section.

RTP provides end-to-end delivery services for data that requires real-time support, such as interactive audio and video. According to RFC 3550, the services provided by RTP include payload type identification, sequence numbering, timestamping, and delivery monitoring. Applications typically run RTP on top of UDP to make use of UDP's multiplexing and checksum services, and as such both RTP and UDP contribute parts of the transport protocol functionality. Provisions are defined, however, to use RTP with other underlying network or transport protocols.

It is also important to note the functions that RTP does not provide. For example, RTP itself does not provide any mechanism to ensure timely delivery or provide other quality-of-service guarantees, but relies on the lower layer services for these functions. It does not guarantee packet delivery or prevent out-of-order packet delivery, nor does it assume that the underlying network is reliable and deliver packets in sequence. The sequence numbers included in RTP allow the receiver to reconstruct the sender's packet sequence, but sequence numbers might also be used to determine the proper location of a packet, for example in video decoding, without necessarily decoding packets in sequence.

There are two parts of RTP defined in RFC 3550: the real-time transport protocol (RTP), which carries data that has real-time properties; and the RTP control protocol (RTCP), which monitors the quality of service and conveys information about the participants in an ongoing session. In addition to the protocol specification given in RFC 3550, a companion document, RFC 3551 [9-11], provides a profile specification that defines a set of payload type codes and their mapping to payload formats, such as various media encodings. For example, RFC 3551 defines a profile specification with minimal session control. In other words, this profile is designed for sessions where no negotiation or membership control is used. Examples of the use of RTP and its profiles are given in both RFC 3550 and RFC 3551.

Before discussing the RTP packet format, a few definitions, taken from RFC 3550, are in order:

- RTP payload: The data transported by RTP in a packet, for

example audio samples or compressed video data.

- RTP packet: A data packet consisting of the fixed RTP header, a possibly empty list of contributing sources (as defined below), and the payload data.
- RTCP packet: A control packet consisting of a fixed header part similar to that of RTP data packets, followed by structured elements that vary depending on the RTCP packet type. The formats are defined in RFC 3550, Section 6.
- Port: The addressing mechanism that uniquely identifies different applications within a host. For example, the File Transfer Protocol (FTP) and the Simple Network Management Protocol (SNMP) have different port numbers. RTP depends on the lower layer protocol to provide some mechanism such as ports to multiplex the RTP and RTCP packets of a session.
- Transport address: The combination of a network address and port that identifies a transport-level endpoint, for example an IP address and a UDP port. Packets are transmitted from a source transport address to a destination transport address.
- RTP session: The association among a set of participants communicating with RTP. For each participant, the session is defined by a particular pair of destination transport addresses (one network address plus a port pair for RTP and RTCP). The destination transport address pair may be common for all participants, as in the case of IP multicast, or may be different for each, as in the case of individual unicast network addresses plus a common port pair. In a multimedia session, each medium is carried in a separate RTP session with its own RTCP packets. The multiple RTP sessions are distinguished by different port number pairs and/or different multicast addresses.
- Synchronization source (SSRC): The source of a stream of RTP packets, identified by a 32-bit numeric SSRC identifier carried in the RTP header so as not to be dependent upon the network address. All packets from a synchronization source form part of the same timing and sequence number space, so a receiver groups packets by synchronization source for play back. Examples of synchronization sources include the sender of a stream of packets derived from a signal source such as a microphone or a camera, or an RTP mixer (as defined be-

low). A synchronization source may change its data format, for example its audio encoding mechanism, over time.

- Contributing source (CSRC): A source of a stream of RTP packets that has contributed to the combined stream produced by an RTP mixer (as defined below). The mixer inserts a list of the SSRC identifiers of the sources that contributed to the generation of a particular packet into the RTP header of that packet. This list is called the CSRC list. An example application is audio conferencing, where a mixer indicates all the talkers whose speech was combined to produce the outgoing packet, allowing the receiver to indicate the current talker even though all the audio packets contain the same SSRC identifier (that of the mixer).
- End system: An application that generates the content to be sent in RTP packets and/or consumes the content of received RTP packets. An end system can act as one or more synchronization sources in a particular RTP session, but typically acts only as one.
- Mixer: An intermediate system that receives RTP packets from one or more sources, possibly changes the data format, combines the packets in some manner, and then forwards a new RTP packet. Since the timing among multiple input sources will not generally be synchronized, the mixer will make timing adjustments among the streams and generate its own timing for the combined stream. Thus, all data packets originating from a mixer will be identified as having the mixer as their synchronization source.
- Translator: An intermediate system that forwards RTP packets with their synchronization source identifier intact. Examples of translators include devices that convert encodings without mixing, replicators from multicast to unicast, and application-level filters in firewalls.
- Monitor: An application that receives RTCP packets sent by participants in an RTP session, in particular the reception reports, and estimates the current quality of service for distribution monitoring, fault diagnosis, and long-term statistics.

- Non-RTP: Protocols and mechanisms that may be needed in

addition to RTP to provide a usable service. In particular, for multimedia conferences, a conference control application may distribute multicast addresses and keys for encryption, negotiate the encryption algorithm to be used, and define dynamic mappings between RTP payload type values and the payload formats they represent for formats that do not have a predefined payload type value.

Given the above functional definitions, we will now consider the RTP message header format that is shown in Figure 9-10. The first twelve octets are present in every RTP packet, while the list of CSRC identifiers is present only when inserted by a mixer.

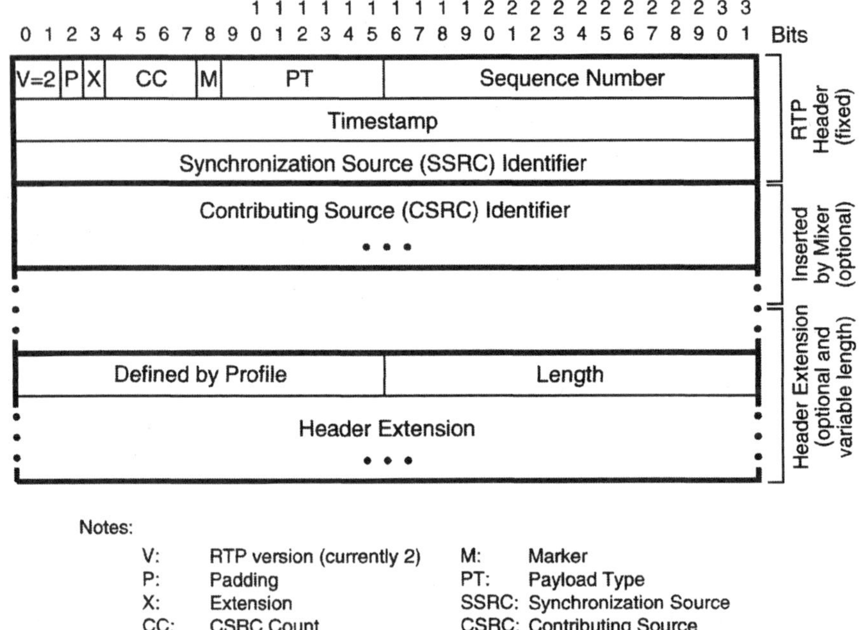

Figure 9-10. Real-Time Transport Protocol (RTP) Message Header.

The fields of the RTP header are:

- Version (V, 2 bits): Identifies the version of RTP, currently 2.
- Padding (P, 1 bit): If the padding bit is set, the packet contains one or more additional padding octets at the end that are not part of the payload. The last octet of the padding contains a count of how many padding octets should be ignored. Padding may be needed by some encryption algorithms with fixed block sizes or for carrying several RTP packets in a lower layer protocol data unit.
- Extension (X, 1 bit): If the extension bit is set, the fixed header is followed by exactly one header extension, with a format defined in RFC 3550, Section 5.3.1.
- CSRC count (CC, 4 bits): The CSRC count contains the number of CSRC identifiers that follow the fixed header.
- Marker (M, 1 bit): The interpretation of the marker is defined by a profile. It is intended to allow significant events such as frame boundaries to be marked in the packet stream.
- Payload type (PT, 7 bits): Identifies the format of the RTP payload and determines its interpretation by the application.
- Sequence number (16 bits): The sequence number increments by one for each RTP data packet sent, and may be used by the receiver to detect packet loss and to restore packet sequence. The initial value of the sequence number is random (unpredictable) to make known-plaintext attacks on encryption more difficult, even if the source itself does not encrypt (because the packets may flow through a translator that does).
- Timestamp (32 bits): The timestamp reflects the sampling instant of the first octet in the RTP data packet. The sampling instant must be derived from a clock that increments monotonically and linearly in time to allow synchronization and jitter calculations.
- SSRC (32 bits): The SSRC field identifies the synchronization source. This identifier is chosen randomly, with the intent that no two synchronization sources within the same RTP session will have the same SSRC identifier.
- CSRC list (0 to 15 items, 32 bits each): The CSRC list iden-

tifies the contributing sources for the payload contained in this packet. The number of identifiers is given by the CC field. If there are more than 15 contributing sources, only 15 may be identified. CSRC identifiers are inserted by mixers, using the SSRC identifiers of contributing sources. For example, for audio packets the SSRC identifiers of all sources that were mixed together to create a packet are listed, allowing correct talker indication at the receiver.

- RTP Header Extension (variable length): An optional extension mechanism is provided with RTP to allow individual implementations to experiment with new functions that require additional information in the RTP header.

Specific details regarding the use of these header fields are given in RFC 3550.

9.3.3 Real-Time Control Protocol

The RTP control protocol (RTCP) is also defined in RFC 3550 and is based on the periodic transmission of control packets to all participants in the session, using the same distribution mechanism as the data packets. The underlying protocol must provide multiplexing of the data and control packets, such as using separate port numbers with UDP. The following RTCP functions are identified in RFC 3550:

1. Providing feedback on the quality of the data distribution. This is an integral part of RTP's role as a transport protocol and is related to the flow and congestion control functions of other transport protocols. The feedback may be directly useful for control of adaptive encodings, but experiments with IP multicasting have shown that it is also critical to get feedback from the receivers to diagnose faults in the distribution. Sending reception feedback reports to all participants allows one who is observing problems to evaluate whether those problems are local or global. With a distribution mechanism like IP multicast, it is also possible for an entity such as a network service provider that is not otherwise involved in the session to receive the feedback information and act as a third-party monitor to diagnose network problems. This feedback function is performed by the RTCP sender and receiver reports, described in RFC 3550, Section 6.3.
2. Carrying a persistent transport-level identifier for an RTP source

called the canonical name or CNAME. Since the SSRC identifier may change if a conflict is discovered or a program is restarted, receivers require the CNAME to keep track of each participant. Receivers also require the CNAME to associate multiple data streams from a given participant in a set of related RTP sessions, for example to synchronize audio and video.

3. The first two functions require that all participants send RTCP packets; therefore, the rate must be controlled in order for RTP to scale up to a large number of participants. By having each participant send its control packets to all the others, each can independently observe the number of participants.
4. An optional function of conveying minimal session control information, for example participant identification to be displayed in the user interface. This is most likely to be useful in "loosely controlled" sessions where participants enter and leave without membership control or parameter negotiation. RTCP serves as a convenient channel to reach all the participants, but it is not necessarily expected to support all the control communication requirements of an application.

RFC 3550 defines five different RTCP packet formats:

- SR or Sender Report: for transmission and reception statistics from participants that are active senders
- RR or Receiver Report: for reception statistics from participants that are not active senders
- SDES or Source Description Items: includes CNAME
- BYE: indicates the end of participation
- APP: application-specific functions

Each RTCP packet begins with a fixed part similar to that of RTP data packets, followed by structured elements that may be of variable length according to the packet type but always end on a 32-bit boundary. The alignment requirement and a length field in the fixed part are included to make RTCP packets "stackable." Multiple RTCP packets may be linked together without any intervening separators to form a compound RTCP packet that is sent in a single packet of the lower layer protocol, for example UDP. There is no explicit count of individual RTCP packets in the compound packet since the lower layer protocols are expected to provide an overall length to determine the end of the compound packet.

The formats of these RTCP packets vary by function and are rather complex. Interested readers are referred to RFC 3550, section 6, for these details.

Details on the operation of RTP can be found in RFC 3550 [9-10], RFC 3551 [9-11], and Reference [9-12].

9.3.4 Resource Reservation Protocol

As more time-sensitive applications have been developed for the Internet, the need to define a Quality of Service (QoS), and the mechanisms to provide that QoS, have become requirements. The Resource Reservation Protocol (RSVP), defined in RFC 2205 [9-13] and updated in RFC 2750 [9-14], is designed to address those requirements. When a host has an application such as real-time video or multimedia, it may use RSVP to request an appropriate level of service from the network in support of that application. But for RSVP to be effective, every router in that path must support that protocol — something that some router vendors and ISPs are not yet equipped to do. In addition, RSVP is a control protocol, and therefore works in collaboration with — not instead of — traditional routing protocols. In other words, the routing protocol, such as RIP or OSPF, determines which datagrams are forwarded, while RSVP is concerned with the QoS of those datagrams that are forwarded.

RSVP requests that network resources be reserved to support data flowing on a simplex path, and that reservation is initiated and maintained by the receiver of the information. Using this model, RSVP can support both unicast and multicast applications. RSVP messages may be sent directly inside IP datagrams (using IP Protocol = 46) or encapsulated inside UDP datagrams, using Ports 1698 and 1699.

RSVP defines two basic message types: Reservation Request (or Resv) messages and Path messages (Figure 9-11). A receiver transmits Resv messages upstream toward the senders. These Resv messages create and maintain reservation state information in each node along the path or paths. A sender transmits Path messages downstream toward the receivers, following the paths prescribed by the routing protocols that follow the paths of the data. These Path messages store path state information in each node along the way.

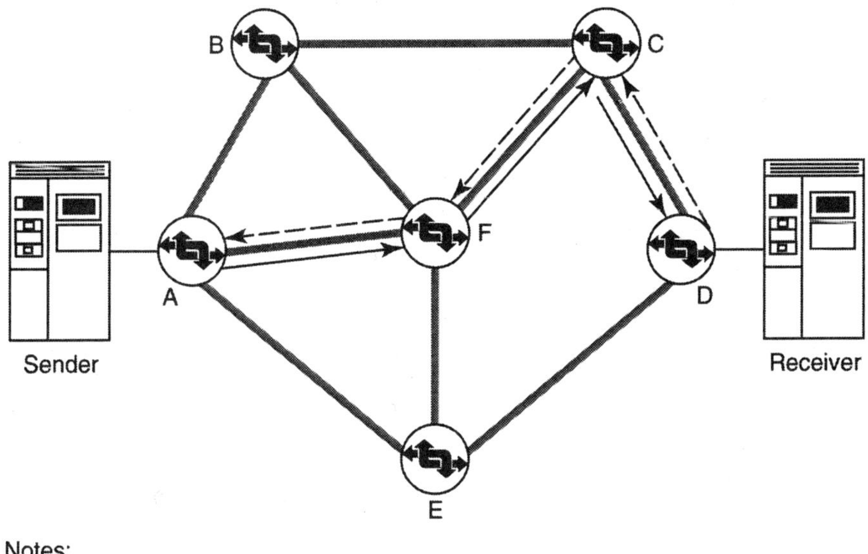

Figure 9-11. Resource Reservation Protocol (RSVP) Operation.

The RSVP message consists of three sections: a Common Header (8 octets), an Object Header (4 octets), and Object Contents (variable length), as shown in Figure 9-12. The Version field (4 bits) contains the protocol version (currently 1). The Flags field (4 bits) is reserved for future definition. The Message Type field (1 octet) defines one of seven currently defined RSVP messages:

Message Type	Message Name	Function
1	Path	Path message, from sender to receiver, along the same path used for the data packets.
2	Resv	Reservation message with reservation requests, carried hop-by-hop from receiver to senders.
3	PathErr	Path error message; reports errors

		in processing Path messages, and travels upstream toward senders.
4	ResvErr	Reservation error message; reports errors in processing Resv messages, and sent downstream toward receivers.
5	PathTear	Path teardown message, initiated by senders or by a timeout, and sent downstream to all receivers.
6	ResvTear	Reservation teardown message, initiated by receivers or by a timeout, and sent upstream to all matching senders.
7	ResvConf	Reservation confirmation message, acknowledges reservation requests.

The RSVP Checksum field (2 octets) provides error control. The Send_TTL field (1 octet) is the IP Time to Live field value with which the message was sent. The RSVP length field (2 octets) is the total length of the RSVP message, given in octets.

Every object consists of one or more 32-bit words with a 1-octet Object Header. This second header includes a Length field (2 octets), which is the total length of the object (with a minimum of 4 octets, and also a multiple of 4 octets); a Class Number (1 octet), which defines the object class; and a C-Type field (1 octet), which is an object type that is unique with the Class Number. The Object Contents complete the message.

Details on the operation of RSVP can be found in RFC 2205 [9-13], RFC 2750 [9-14], and Reference [9-15].

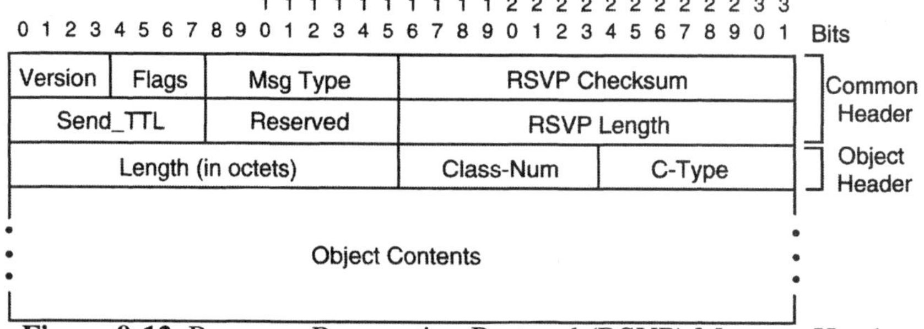

Figure 9-12. Resource Reservation Protocol (RSVP) Message Header.

9.3.5 Real-Time Streaming Protocol

The Real-Time Streaming Protocol, defined in RFC 2326 [9-16], is an Application Layer protocol that controls the delivery of data that has real-time properties, such as audio and video. RTSP does not typically deliver the continuous streams itself, although interleaving of the continuous media stream with the control stream (such as RTCP) is possible. In effect, RTSP acts as a "network remote control" for multimedia servers.

There is no notion of an RTSP connection; instead, a server maintains a session labeled by an identifier. An RTSP session is in no way tied to a transport-level connection such as a TCP connection. During an RTSP session, an RTSP client may open and close many reliable transport connections to the server to issue RTSP requests. Alternatively, it may use a connectionless transport protocol such as UDP.

The streams of information controlled by RTSP may use RTP; however, RTSP's operation is independent of the underlying transport mechanism. RTSP is intentionally similar in syntax and operation to the Hypertext Transfer Protocol (HTTP), version 1.1, defined in RFC 2616 [9-17], and updated by RFC 2817 [9-18], so that extension mechanisms to HTTP can in most cases also be added to RTSP. However, RTSP differs in a number of important aspects from HTTP, including its client/server operation, protocol identifier, and typical out-of-band data transport.

RFC 2326 defines the following operations that RTSP supports:

- Retrieval of media from media server: The client can request a presentation description via HTTP or some other method. If the presentation is being multicast, the presentation description contains the multicast addresses and ports to be used for the continuous media. If the presentation is to be sent only to the client via unicast, the client provides the destination for security reasons.
- Invitation of a media server to a conference: A media server can be "invited" to join an existing conference, either to play back media into the presentation or to record all or a subset of the media in a presentation. This mode is useful for distributed teaching applications. Several parties in the conference may take turns "pushing the remote control buttons."
- Addition of media to an existing presentation: Particularly useful for live presentations, the server can tell the client about additional media becoming available.

Details of RTSP operation can be found in RFC 2326.

9.3.6 Session Description and Session Announcement Protocols

The Session Description Protocol (SDP), defined in RFC 2327 [9-19], is used to describe a multimedia session for the purposes of session announcement, session invitation, and other forms of session-related initiation. SDP conveys information about media streams in multimedia sessions to allow the recipients of a session description to participate in that session. One of the applications for SDP is the session directory tool used on the MBONE, which advertises multimedia conferences on that network. Thus, SDP provides a means to communicate the existence of the session, and also conveys sufficient information to enable another station to join and participate in that session.

A conference session is announced by periodically multicasting an announcement packet to a well-known address and port using the Session Announcement Protocol, (SAP), defined in RFC 2974 [9-20], which is transmitted using UDP. The payload of the SAP packet may contain an SDP packet that describes the session of interest. That description includes:

- Session name and purpose.
- Time(s) the session is active.
- The media comprising the session.
- Information to receive those media, such as addresses, ports, formats, and so on.
- Information about the bandwidth to be used by the conference.
- Contact information for the person responsible for the session.

As noted in RFC 2327, the purpose of SDP is to define a format for describing session information, while allowing for different transports to convey that information. As such, SDP information may be transported using SAP (as described above, and in Figure 9-8), or by using the Session Initiation Protocol (SIP), the Real-time Streaming Protocol (RTSP), using electronic mail with Multimedia Internet Mail Extensions (MIME) format, or the Hypertext Transfer Protocol (HTTP).

9.3.7 Session Initiation Protocol

Thus far, we have considered Multicast IP, RTP, RTCP, RSVP, RTSP, SDP, and SAP as adjuncts to our familiar TCP/IP and/or UDP/IP protocol infrastructures. These protocols enhance the functions of IP to allow real-time information to be effectively communicated. (Recall that IP was developed in the 1970s to support data communication, which is non-real-time information, in contrast to the current requirements to support voice and video, which are real-time.) But before that information can be transmitted, a communication session between the two endpoints must be established. This is the purpose of the Session Initiation Protocol (SIP), defined in RFCs 3261 through 3265 ([9-21] through [9-25]). SIP is a control or signaling protocol used for creating, modifying, and terminating sessions between participants. These sessions may include multimedia conferences, telephone calls, distance learning, or other types of multimedia distribution. The parties to that session may or may not be human end users — robots, such as media storage devices, could also be participants in that session.

Thus, SIP is a part of the Internet multimedia protocol architecture, which includes the other protocols like RTP, SAP, and SDP that have already been discussed. Reviewing Figure 9-8, note that SIP can carry SDP data, and that the SIP information can use either TCP or UDP for transport. SIP's call control functionality is somewhat similar to control protocols defined by the ITU-T, including H.323 and others such as H.225.0 or H.245. Because of these functional parallels, we will postpone a more detailed discussion of SIP functions until our signaling discussion in Section 9.4. At that time we will consider call control in the larger context of VoIP architectures, and will look at gateways, gatekeepers, and the protocols such as the Media Gateway Control Protocol (MGCP) used for those communication functions.

9.3.8 Electronic Numbers

Thus far, we have considered a large number of protocols that are required for the proper operation of a voice, fax, or video over IP network. But one more challenge remains. Telephone, fax, or videoconferencing systems are identified by telephone numbers, such as +1.303.555.1212. In contrast, IP-connected devices are identified by an IP address, such as [10.222.18.66]. So what number do you enter to make a voice over IP call?

That challenge is addressed by a process known as ENUM, which is an abbreviation for electronic number. The ENUM work is based upon DNS and documented in RFC 2916 [9-26]. ENUM bridges the gap between telephone numbers, defined by ITU-T E.164, and IP addresses. These services support all forms of IP-based communication, including voice, fax, voicemail retrieval, messaging, and others. When an IP-based call is made from an end user, the connected system will query the ENUM system to determine if an IP address has been registered for the destination party. If such an IP address exists, the call is completed using IP on an end-to-end basis. If the IP address is not registered, the call is then redirected to the PSTN for completion. Thus, the telephone number supports both PSTN and IP-based network communication. The ENUM process utilizes the DNS for much of its communication activities, thus retaining compatibility with IP network systems in current use. Reference [9-27] details further resources for those readers wishing to dig deeper.

9.4 Signaling Processes and Protocols

Suppose you wish to have a voice call with a colleague over an IP-based infrastructure. On your end, the communications path includes a Voice over IP client application, a local network that supports IP, and a wide area network that supports IP, such as an ISDN or a T1 line. Likewise, your colleague requires a similar connection on their end. But there is one fundamental question that must be answered: how does the communication path get established? The answer to that question is found in a process known as signaling. Signaling is defined as the procedure or procedures undertaken to establish (or setup), manage (or supervise), and terminate (or disconnect) a communication session between two endpoints.

There are several assumptions built into these procedures. First, it must be assumed that the two endpoints have a need to communicate, and that need is being driven by some higher layer protocol or process (such as an application, like an email client that needs to communicate with the email server, or a human who needs to call home). Second, it must be assumed that these two endpoints have the ability to reach each other, share a common addressing scheme, and have some way of determining each other's address. For example, it would do me no good to try to contact you on my amateur radio station if you don't own a compatible transmitter and receiver, or if you are not presently on the air, or if I don't know your identifying address or call sign. Third, the signaling procedures may have to traverse several networks, possibly using different protocols, in order to reach the destination. For example, if the call is initiated on an IP-based network, but terminated on an analog telephone attached to the PSTN, both IP-based and PSTN-based signaling will be involved.

Thus, we can divide the signaling functions into two broad categories: signaling that occurs within the telephone network, and signaling that occurs within the packet network. Signaling within the PSTN is defined by a protocol known as Signaling System 7 (SS7), defined by the ITU-T for international calls [9-28], and by ANSI for calls within the United States [9-29]. PSTN signaling is divided into two elements. The first part occurs between the telephone set and the local central office switch and is known as subscriber loop signaling. Loop signaling conveys indications of on/off hook status, ringing signals, and so on to the subscriber's terminal. The second part is known as inter-exchange signaling, which conveys call information between central office switches. SS7 is a complex protocol beyond our discussion of enterprise-based VoIP networks.

Signaling within the packet network can be accomplished using some combination of four protocols: the H.323, developed by the ITU-T; the Session Initiation Protocol (SIP), developed by the IETF; the Media Gateway Control Protocol (MGCP), developed by the IETF; and the MEGACO/H.248 protocol, jointly developed by the IETF and the ITU-T. Depending on the network architecture and the components involved, one or more of these protocols may be deployed, although, as we will see, some provide very similar functions and would not necessarily be used simultaneously.

The next three sections will summarize the functions of these four signaling protocols, beginning with H.323. A few words of caution, however — all of these protocols are quite complex, and the documentation very extensive (for example, the baseline specifications for both H.323 and SIP are hundreds of pages in length). Therefore, what is presented in this section is intended as a summary of the key functions of these protocols, not a complete summary of all the possible operational conditions. Secondly, standards are always subject to change, so make sure that you have the most current document available for your research.

9.4.1 Comparing Protocol Flows for PSTN and VoIP Calls

Let's compare the processes used in both PSTN (circuit-switched) and VoIP (packet-switched) voice call scenarios. A circuit-switched telephone call through the PSTN involves the five components shown in Figure 9-13. These include: end-user equipment, such as telephones and fax machines; connections to the local exchange central office, typically provided on copper pairs; local switching offices; a signaling network; and a transport network. A typical telephone call through the PSTN would involve five major steps:

1. An analog telephone goes off-hook. This signal is recognized at the local central office, which returns dial tone and accepts the destination telephone number, typically transmitted using Dual Tone Multi-frequency (DTMF) tones.
2. The local central office generates a call setup packet that passes through the signaling network, identifying a path to the desired destination. Once that path has been identified, transport network resources are reserved in support of the connection.
3. The destination central office signals the destination telephone of an incoming call using ringing tone.
4. When the destination party answers, the signaling network starts billing, and the voice network connection becomes active.
5. The two parties carry on a conversation and, when completed, hang up their telephones. The on-hook signals are recognized at the local and destination central offices. Billing functions are then completed, and the reservations of the transport network facilities are released so that those resources can be devoted to another call.

Placing a call via a packet network is similar to the PSTN case. Figure 9-14 illustrates an example of a call using H.323 protocols, and the involvement of other processors such as gatekeepers and gateways (note that processes for other protocol architectures, such as SIP, MGCP, or Cisco SCCP would be different). Five major steps would also be involved in this scenario:

1. An analog telephone goes off hook and places a call to a remote telephone connected to an IP network. The signaling network within the PSTN receives the calling user's destination number, which is then passed to the VoIP gateway.
2. Call signaling information is passed from the gateway to the gatekeeper requesting admission to the network. This signaling information is sent using TCP/IP for greater reliability.
3. Since the destination telephone resides on another network, the respective gatekeepers communicate signaling information requesting call completion.
4. The destination client and remote gatekeeper exchange signaling information.
5. Once all call signaling is completed, the two end stations exchange media information (voice, video, etc. samples) using RTP/UDP/IP for greater efficiency. When the information transfer is completed, additional signaling messages are used to disconnect the call.

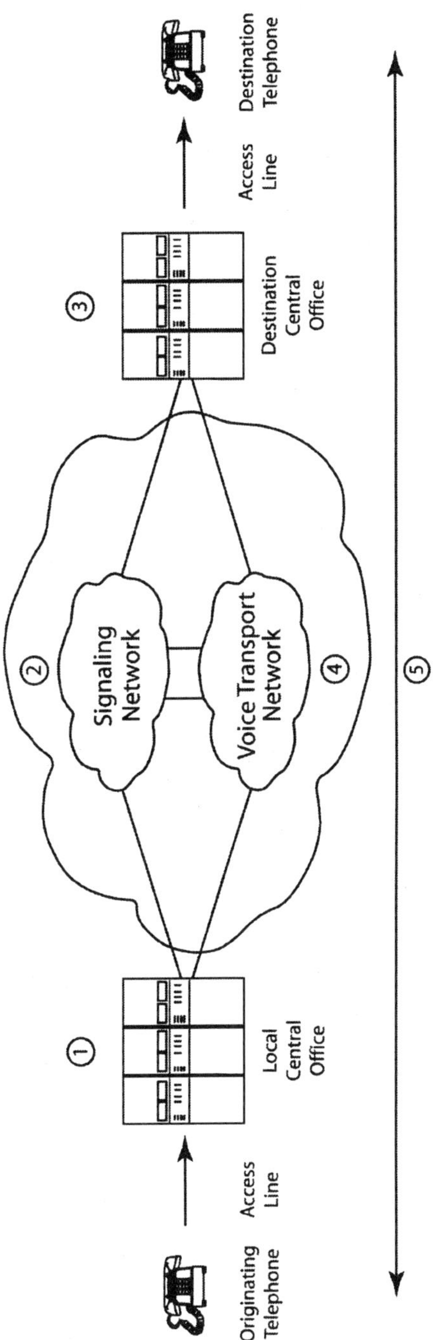

Figure 9-13. PSTN Call Processing and Protocol Flows.

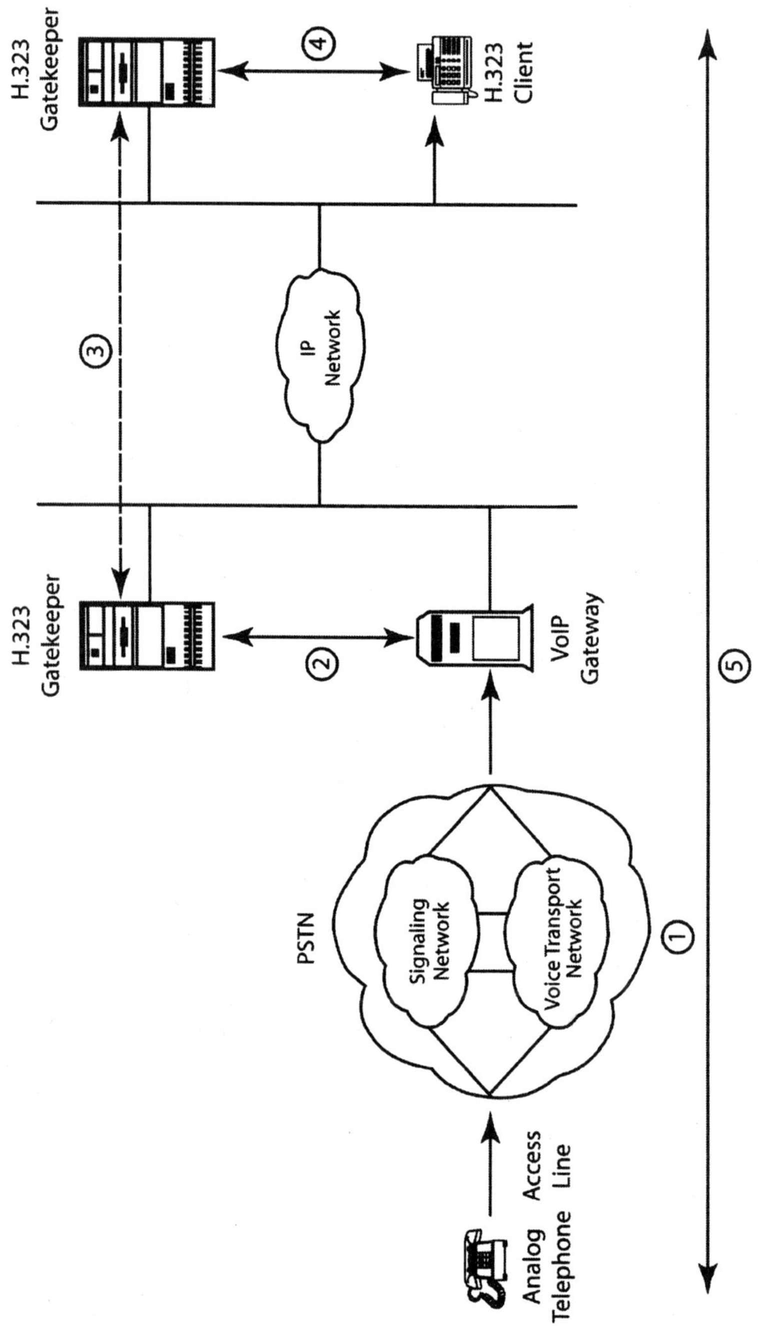

Figure 9-14. VoIP Call Processing and Protocol Flows.

Comparing the PSTN and VoIP call scenarios, note that two network operations are involved in each case: signaling, for call establishment, management, and termination, plus information transport from source to destination. We will look at the details of two of these signaling protocols, H.323 and SIP, in the next sections.

9.4.2 The H.323 Multimedia Standard

The first of the VoIP signaling standards that we will study, H.323, was developed by the ITU-T, and is currently in its fifth revision [9-30]. To better understand the scope of this standard, it helps to put it in context with other ITU-T Series H Recommendations that deal with audiovisual and multimedia systems. These other standards include:

- H.320: Narrowband visual telephone systems and terminal equipment (used with narrowband ISDN services).
- H.321: Adaptation of H.320 terminals to broadband ISDN (ATM) environments.
- H.322: Visual telephone systems and equipment for local area networks that provide a guaranteed quality of service.
- H.323: Packet-Based multimedia communications systems.
- H.324: Terminal for low bit rate multimedia communication (used for PTSN and wireless applications).

H.323 assumes that the transmission media is a LAN that does not provide guaranteed packet delivery. A typical Ethernet would be a good example — if two Ethernet workstations transmit at the same time, a collision occurs. Since the probability of such a collision is difficult to predict, defining a specific quality of service is also difficult. Other standards in this family address other network types, such as H.320 (ISDN), H.321 (ATM), and H.324 (low bit rate connections), or, in the case of H.324, networks that provide QoS guarantees. Thus, the H.323 standard is designed to work with the local and wide area network types that are most commonly found — those that do not provide guarantees on the QoS provided.

In addition to the network implementation standards noted above, there are other standards that fall within the umbrella of the H.323 recommendation. These include:

- H.225.0: Terminal to Gatekeeper signaling functions, as defined in Reference [9-31].
- H.245: Terminal control functions that are used to negotiate channel usage, capabilities, and other functions, as defined in Reference [9-32].
- Q.931: Call signaling functions to establish and terminate the call, as defined in Reference [9-33].
- T.120: Data conferencing, which might include shared whiteboarding and still image transfer applications, as defined in Reference [9-34].

9.4.2.1 H.323 Terms and Definitions

Before delving into the H.323 standard in greater detail, a few terms, as defined in that standard, should be mentioned:

- Call: A point-to-point multimedia communication between two H.323 endpoints, which begins with the call set-up procedure and ends with the call termination procedure.
- Endpoint: An H.323 terminal, Gateway (GW), Gatekeeper (GK), or Multipoint Control Unit (MCU). The endpoint can call and be called, and generates and/or terminates streams of information.
- Gatekeeper (GK): An entity on the LAN that provides address translation and controls access to the LAN for other devices, such as terminals, Gateways, and MCUs.
- Gateway (GW): An endpoint on the LAN that provides real-time, two-way communications between H.323 terminals on the LAN and other ITU terminals on a WAN, or to another H.323 Gateway.
- H.323 Entity: Any H.323 component, which includes terminals, Gateways, Gatekeepers, Multipoint Controllers (MCs), Multipoint Processors (MPs), and Multipoint Control Units (MCUs).
- Multipoint Control Unit (MCU): An endpoint on the LAN that provides the capability for three or more terminals and Gateways to participate in a multipoint conference. The MCU includes a mandatory MC and optional MPs.

- Multipoint Controller (MC): An entity on the LAN that provides for the control of three or more terminals participating in a multipoint conference.
- Multipoint Processor (MP): An H.323 entity on the LAN that provides for the centralized processing of audio, video, and/or data streams in a multipoint conference.
- Terminal (Tx): An endpoint on the LAN that provides for real-time, two-way communications with another H.323 terminal, Gateway, or Multipoint Control Unit.
- Zone: A collection of all Terminals, Gateways, and Multipoint Control Units managed by a single Gatekeeper.

9.4.2.2 H.323 Components and Information Streams

The various building blocks that comprise a typical H.323 network are illustrated in Figure 9-15. Note that the building blocks imply that each of these functions is in a distinct box. However, one physical box often contains more than one functional element. For example, both MCU and Gatekeeper functions could be located in the same physical device. A typical H.323 environment might include Gateways to other networks, such as the PSTN or ISDN.

Multimedia networks must support a number of different end-user applications; therefore, H.323 must support different streams of information. These are:

- Audio: Digitized and coded speech signals. The audio signal is accompanied by an audio control signal.
- Video: Digitized and coded motion video. The video signal is accompanied by a video control signal.
- Data: Still pictures, facsimile, computer files, and so on.
- Communications Control: Information that passes control data between like functional elements (such as terminal-to-terminal control) to exchange capabilities between these devices, to open and close logical channels, to control transmission modes, and to perform other functions.
- Call Control: Information that includes call establishment and call disconnect functions.

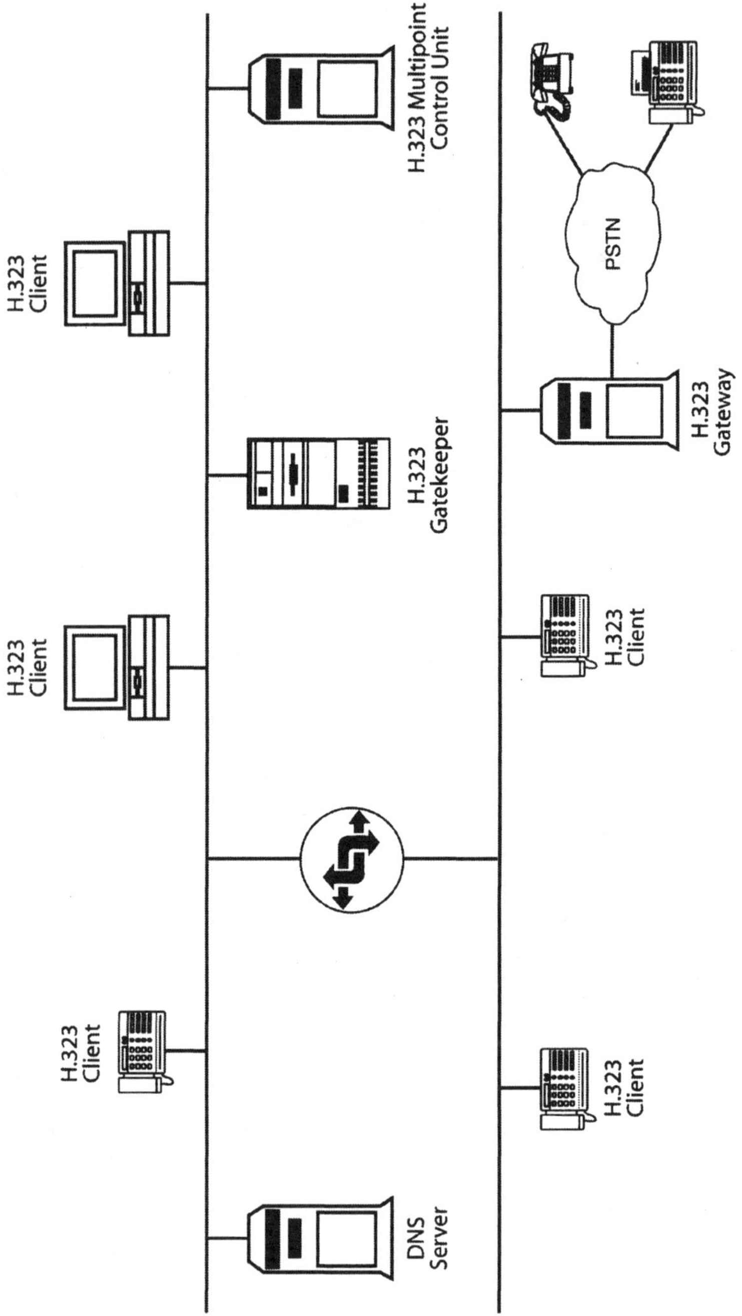

Figure 9-15. H.323 Components and Architecture.

9.4.2.3 H.323 Signaling Processes

Call signaling is defined in H.323 environments to establish a call, request changes in the bandwidth of that call, determine the status of endpoints associated with the call, and terminate or disconnect the call. The call signaling messages are specified in H.225.0, noted above. H.323 entities are identified using two levels of addressing structures, a Network address and a TSAP identifier. The Network address is a unique identifier for that entity on the network, and is specific to that network environment in which the entity resides. For the case of IP-based networks, the Network address would be the IP address assigned to the device. The TSAP identifier is used to multiplex several Transport Layer connections into an entity that has a single Network address. (The term TSAP comes from the Transport Service Access Point — an addressing scheme used at the OSI Transport Layer, and defined as part of the Open Systems Interconnection Reference Model.) For IP applications, the TSAP could be the UDP or TCP port number. H.225.0, Appendix IV, defines the transmission of H.225.0 messages over various transport protocol stacks, including UDP/IP and TCP/IP, and gives examples of addresses to be used. For example: [224.0.1.41] is the UDP Address, and 1718 is the UDP port, for multicast communication with gatekeepers.

Recall from the definitions above that the Gatekeeper is the device that controls access to the network. When endpoints are initialized on networks that contain a gatekeeper, they register their presence with that gatekeeper. The logical channel that is used to carry that type of communication between endpoints and gatekeepers is called the Registration, Admissions, and Status channel, or simply RAS for short. The RAS channel is known as an unreliable channel, meaning that, for IP-based networks, it would use IP/UDP (the more efficient connectionless transport) instead of the more rigorous TCP/IP transport (called a reliable channel). Note that the Transmission Control Protocol (TCP) and User Datagram Protocol (UDP) are used in support of these protocols, as shown in Figure 9-16.

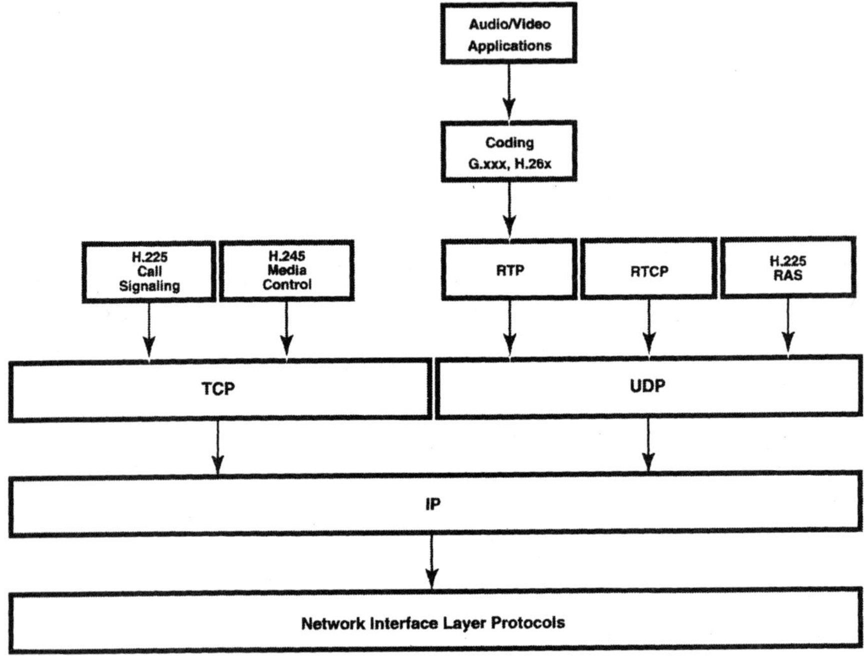

Figure 9-16. H.323 Related Protocols.

The H.323 signaling processes vary based on the active components on the network and are quite complex. In general, however, most signaling operations follow a sequence of four functions that are found with most voice and data communication processes. These functions utilize some of the other related protocols that were mentioned above:

1. Connection establishment using TCP and H.225.
2. Parameter exchange using TCP and H.245.
3. Information transfer using RTP and RTCP.
4. Connection termination using H.245, H.225, and TCP.

For further details, consult the H.323 standard [9-30].

9.4.3 Session Initiation Protocol

The second VoIP signaling protocol that we will study is the Session Initiation Protocol, or SIP, which has been developed by the IETF and documented in RFC 3261 [9-21]. SIP is an Application Layer control protocol used to create, modify, and terminate sessions between participants. These sessions could take on many forms, including multimedia conferences, Internet telephony, media distribution, and others. SIP has been developed using many of the constructs that are found in other IETF-developed protocols, including the Simple Mail Transfer Protocol (SMTP) for electronic messaging, and the Hypertext Transfer Protocol (HTTP) for web pages.

Much of this similarity will be found in the architecture of SIP, which is based on a straightforward client-server model, much like that found in other systems and protocols, like web browsers (or clients) and web servers. In addition, SIP is part of the IETF's multimedia architecture, which includes many of the protocols that were discussed previously, including RTP, RTSP, SAP, and SDP. In contrast to H.323 — which is, as we discovered in the previous section, really an umbrella for a number of standards (H.225.0, H.245, and so on) — SIP is based on IP and IP-derived technologies, such as IP addressing schemes, URLs, and so on. As a result, many feel that SIP is much simpler to implement, and therefore advantageous for certain applications. We will begin our discussion of SIP by considering some of the key terms and definitions.

9.4.3.1 SIP Terms and Definitions

RFC 3261 defines a number of terms related to the functions of SIP. Following are some of the most important:

- Call: a call consists of all participants in a conference invited by a common source. The SIP call is identified by a globally unique call ID.
- Conference: a multimedia session that is identified by a common session description. A conference may take on many different forms, including a multicast conference, a full-mesh conference, or a two-party telephone call or connection.
- Initiator, calling party, or caller: the party initiating a conference invitation, which may or may not be the same as the one creating the conference.
- Invitation: a request sent to a user (or service) requesting participation in a session. A successful SIP invitation consists of two transactions: an INVITE request followed by an ACK request.

- Invitee, invited user, called party, or callee: the person or service that the calling party is trying to invite to a conference.
- Location server: a service used by a SIP Redirect or Proxy server to obtain information regarding a callee's possible location(s). Location servers may be co-located with a SIP server.
- Proxy or proxy server: an intermediary program that acts as both a server and a client for the purpose of making requests on behalf of other clients.
- Redirect server: a server that accepts a SIP request, maps the address into zero or more new addresses, and returns these addresses to the client.
- Registrar: a server that accepts REGISTER requests.
- Server: an application program that accepts requests in order to service those requests, and sends back responses to those requests. Servers are either proxy, redirect, or user agent servers or registrars.
- Session: a multimedia session is a set of multimedia senders and receivers and the data streams flowing from senders to receivers.
- Transaction: a SIP transaction occurs between a client and a server, and comprises all messages from the first request sent from the client to the server, up to a final response sent from the server to the client.
- User agent client (UAC) or calling user agent: a client application that initiates the SIP request.
- User agent server (UAS) or called user agent: a server application that contacts the user when a SIP request is received, and that returns a response on behalf of the user.
- User agent (UA): an application that contains both a user agent client and a user agent server.

The next section will discuss how these various terms and definitions are realized within the SIP architecture.

9.4.3.2 SIP Components and Information Streams

There are two key components used in a SIP-based network, User Agents and Servers, which yield a straightforward, simplistic operation. The SIP User Agents contain a User Agent Client (UAC) and a User Agent Server (UAS), as shown in Figure 9-17. The UAC function initiates the SIP requests, while the UAS function contacts the user when a request is received and returns a response on behalf of that user. The response may be an acceptance, rejection, or redirection of the request.

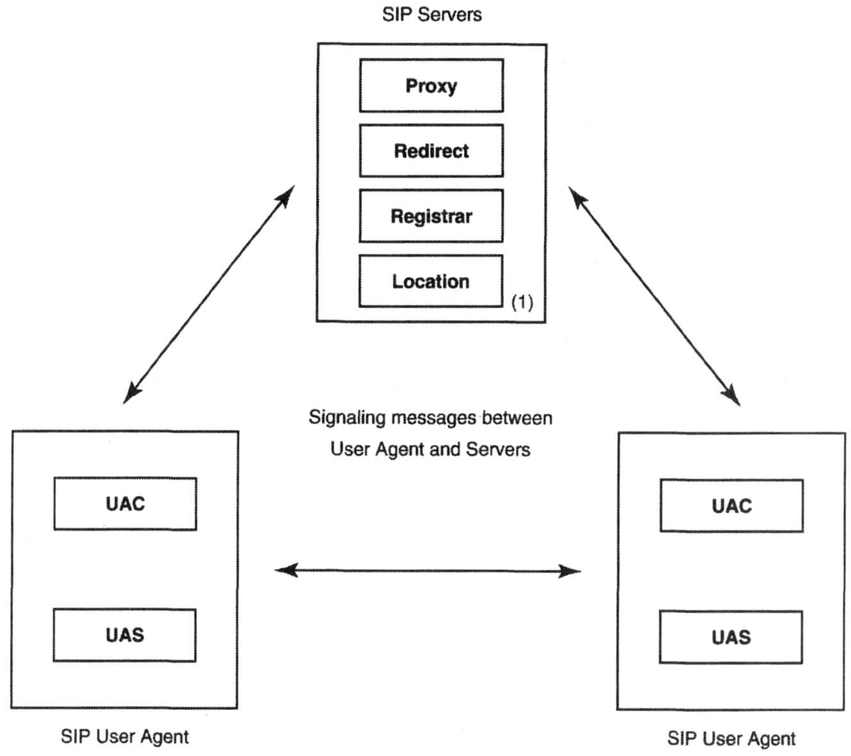

Figure 9-17. SIP Agent and Server Communication.

A SIP Server may also be deployed to perform specific functions within the network (illustrated at the top portion of Figure 9-17). A Proxy server would make requests on behalf of other clients, and possibly translate the message, as necessary, before forwarding. A Redirect Server accepts a SIP request, maps the address into another address, and then returns the new address to the Client. The Registrar Server accepts REGISTER requests, and may be co-located with the Proxy or Redirect Servers. Finally, the Location Server provides a service to the Proxy or Redirect Servers to obtain information regarding the callee's possible location. Location Servers may also be co-located with another SIP server.

SIP components are identified with a SIP Uniform Resource Locator, or URL, and take a form similar to that used with email systems, such as user@host. The user part could consist of a user name or telephone number, while the host part could consist of a domain name or a numeric network address. Thus, boomer@diginet.com, 3035551212@diginet.com, baron@10.3.4.91, and 3035551212@10.3.4.91 are all possible examples of SIP addresses.

A SIP message can either be a request from a client to a server, or a response from a server to a client. The request message defines the operation requested by the client, while the response message provides information from the server to the client indicating the status of that request.

There are six types of request messages, distinguished by what is called a method:

- INVITE: indicates that the user or service is being invited to participate in a session. The body of this message would include a description of the session to which the callee is being invited.
- ACK: confirms that the client has received a final response to an INVITE request, and is only used with INVITE requests.
- OPTIONS: is used to query a server about its capabilities.
- BYE: is sent by a User Agent Client to indicate to the server that it wishes to release the call.
- CANCEL: is used to cancel a pending request.
- REGISTER: is used by a client to register an address with a SIP server.

The response messages contain Status Codes and Reason Phrases that indicate the current condition of this request. The status code values are divided into six general categories:

- 1xx: Informational, the request has been received and processing is continuing.
- 2xx: Success, an ACK, to indicate that the action was successfully received, understood, and accepted.
- 3xx: Redirection, further action is required to process this request.
- 4xx: Client Error, the request contains bad syntax, and cannot be fulfilled at this server.
- 5xx: Server Error, the server failed (for internal reasons) to fulfill an apparently valid request.
- 6xx: Global Failure: the request cannot be fulfilled at any server.

Specific details on the SIP message formats, status codes, and other parameters are specified in RFC 3261.

SIP incorporates elements from several other IETF and ITU-T standards:

- IETF Session Description Protocol (SDP), which defines a standard method of describing the characteristics of a multimedia session.
- IETF Session Announcement Protocol (SAP), which periodically announces parameters of a conference session.
- IETF Real-Time Transport Protocol (RTP) and Real-Time Control Protocol (RTCP) provide information on transport and session management. RTP is the protocol within the H.323 suite that carries the digitally encoded voice or video packets between the end stations. RTCP manages the sessions by periodically transmitting packets containing feedback on the quality of the data distribution.
- ITU-T recommended encoding algorithms, such as G.723.1, G.728, and G.729 for audio encoding, or H.261 or H.263 for video encoding.

Note that the Transmission Control Protocol (TCP) and User Datagram Protocol (UDP) are used in support of these protocols, as shown in Figure 9-18.

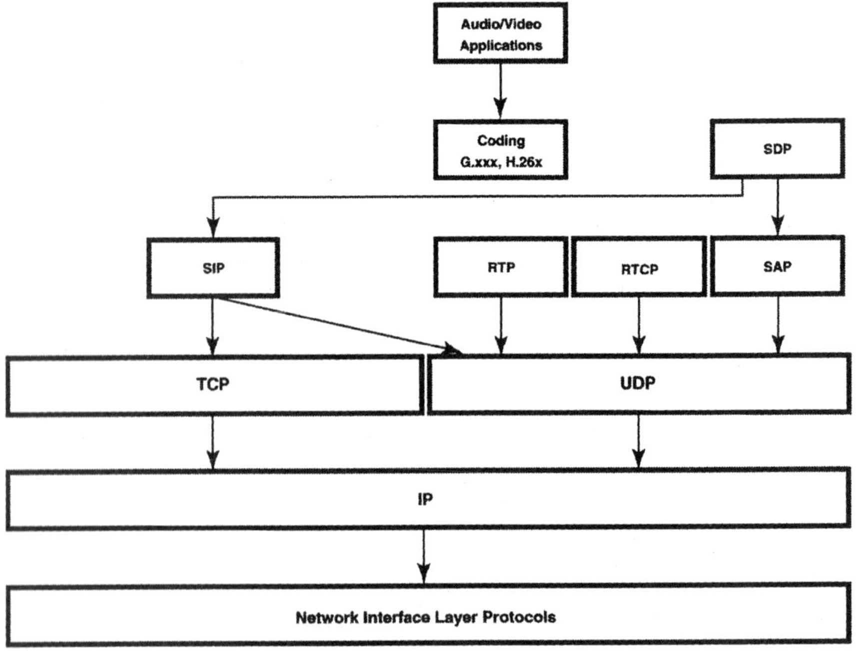

Figure 9-18. SIP-related Protocols.

9.4.3.3 SIP Signaling Processes

SIP is not as complex as H.323 (for example, SIP does not use Gatekeepers), and, as a result, the signaling processes are not as complex. A simple case of a SIP session between two user agents is illustrated in Figure 9-19. This process begins with the calling user agent sending an INVITE message to the called user agent. Included with that INVITE are parameters that indicate the calling party address, a description of the session, typically using the Session Description Protocol (SDP), and a call identifier. The called user agent would respond with an OK message (Response Code = 200), which would be followed by an ACK from the calling user agent as the final step in the call setup process. The two agents would then exchange the audio and/or video information, and, when finished, one of the agents would send the BYE message, indicating that it wishes to release the call. An OK message response would be returned, and then the call termination would be complete.

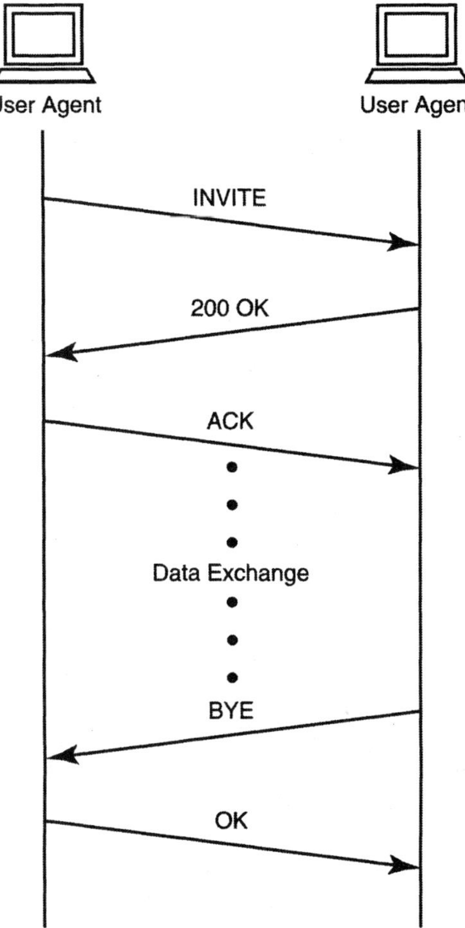

Figure 9-19. SIP Call Signaling.

RFC 3261 details a SIP connection when a Proxy server is required to complete the connection (Figure 9-20). The Proxy server accepts the INVITE request and queries the Location server for more complete addressing information. When this addressing information is returned, the Proxy server issues an INVITE to the address(es) returned from the Location server, which triggers the called user agent to indicate an incoming call (ringing). The called user agent accepts the call with a SIP Response Code of 200 (OK) message, which is passed via the Proxy server to the calling user agent. The calling user agent returns an ACK message, which, in turn, is passed via the Proxy server to the called user agent, thus completing the call setup.

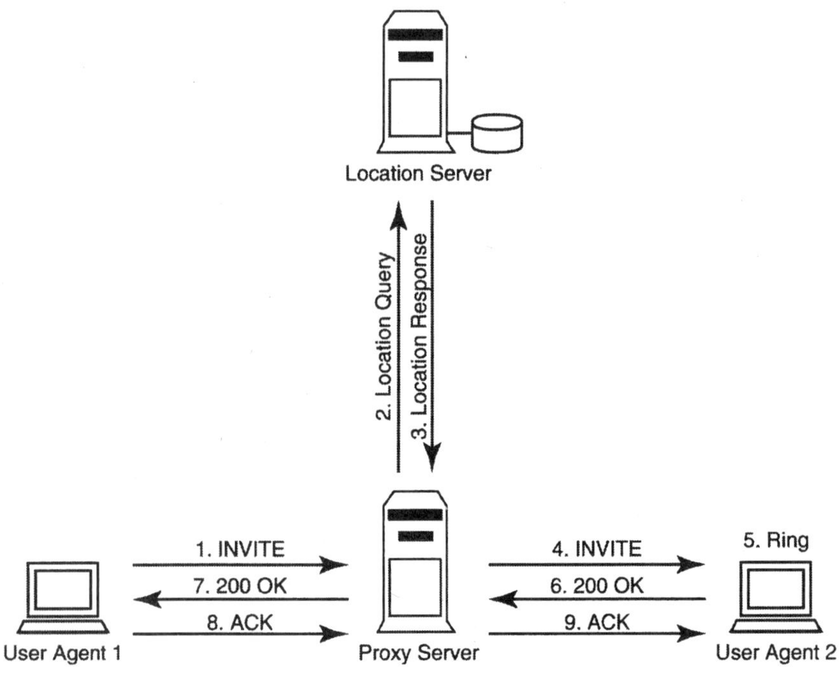

Figure 9-20. SIP Call Setup Using Proxy Server.

The Proxy server was able to identify the called user agent because it had registered its location with the Registrar server (Figure 9-21). Recall from the earlier definitions that the Registrar server accepts REGISTER requests, is typically co-located with a Proxy or Redirect server, and may also offer Location server functions.

RFC 3261 also details the procedures when the end user must be located and a Redirect server employed (Figure 9-22). In this case, the called user agent sends the INVITE message to the Redirect server, which consults with the Location server for complete address information, as before. However, in this case, the called user agent has moved, which is indicated with a SIP Response Code of 302 (Moved temporarily). This information is ACK'd to the Redirect server, and then a second INVITE message, this time to the appropriate address of the called user agent, is issued. The INVITE causes the called user agent to ring, and then a SIP Response Code of 200 (OK) message, which is then ACK'd by the calling user agent.

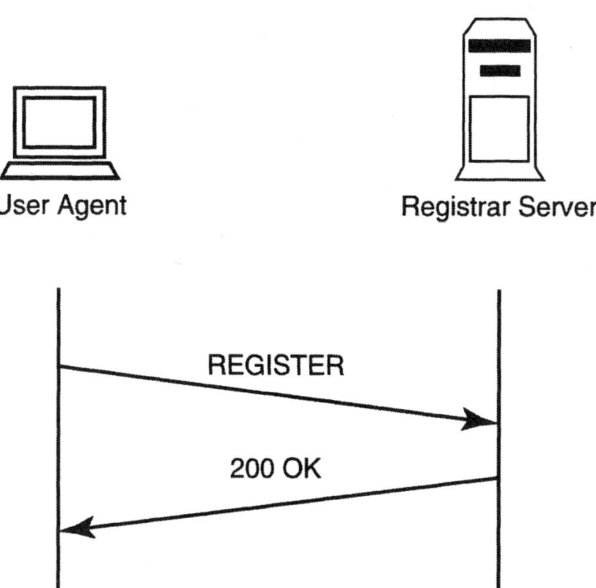

Figure 9-21. SIP Registration.

There are more scenarios that could be drawn, including support for mobile user agents. The interested reader is referred to RFC 3261 for details. Our next section will consider protocols that are used for signaling between gateways.

9.4.4 Gateway Control Protocols: MGCP and MEGACO/H.248

Thus far, we have considered signaling protocols, H.323 and SIP, that are primarily deployed to control end stations, such as a multimedia terminal or videoconferencing system. These protocols may be resident within the applications or operating systems of these end stations, enabling those end stations to initiate, control, and terminate a call. However, when dissimilar networks are involved in the communication path, such as a workstation attached to an IP network calling an analog telephone connected to the PSTN, gateways between those networks must get involved. A second category of signaling protocols is thus required between the telephony gateways to control their operation and to establish paths between those dissimilar networking environments.

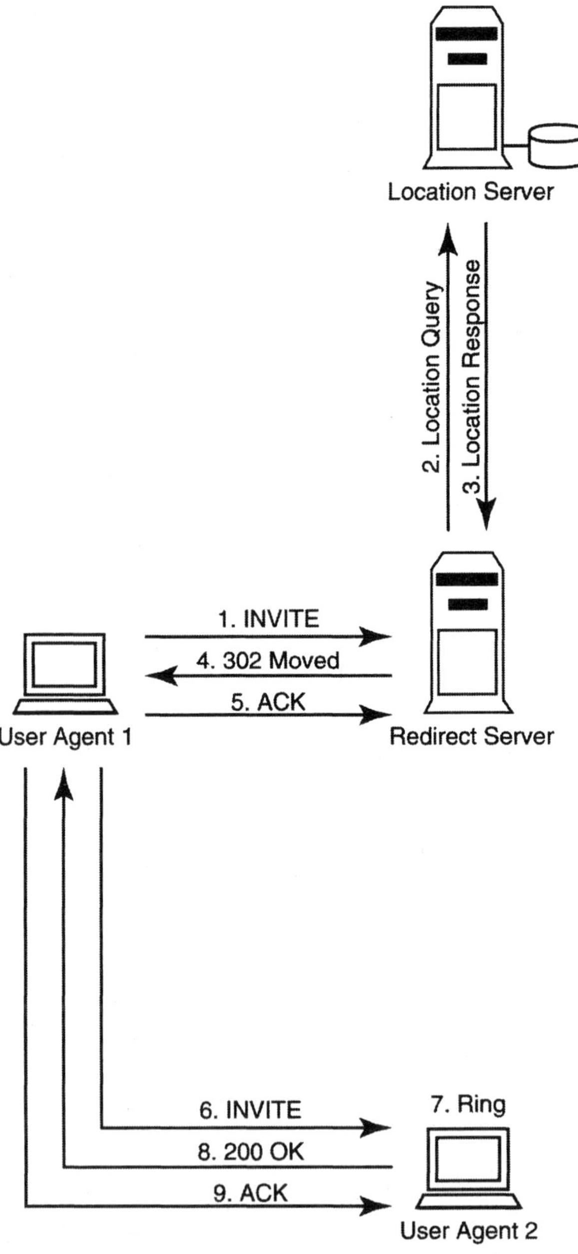

Figure 9-22. SIP Call Setup Using Redirect Server.

A model for communication between gateways has been developed by the European Telecommunications Standards Institute (ETSI) with a research initiative known as the TIPHON project, which stands for Telecommunications and Internet Protocol Harmonization Over Networks. The objective of the project is to support worldwide voice communication and to assure that users connected to IP networks will be able to communicate with those on switched circuit networks (SCNs), such as the PSTN or ISDN environments. ETSI is a European-based organization, but it cooperates with the ITU-T, the IETF, and also various trade organizations in an effort to produce standards that are applicable on a global basis. The TIPHON project is quite broad and covers a number of telecommunications activities, including network architectures, call control procedures and protocols, addressing, service interoperability, and quality of service.

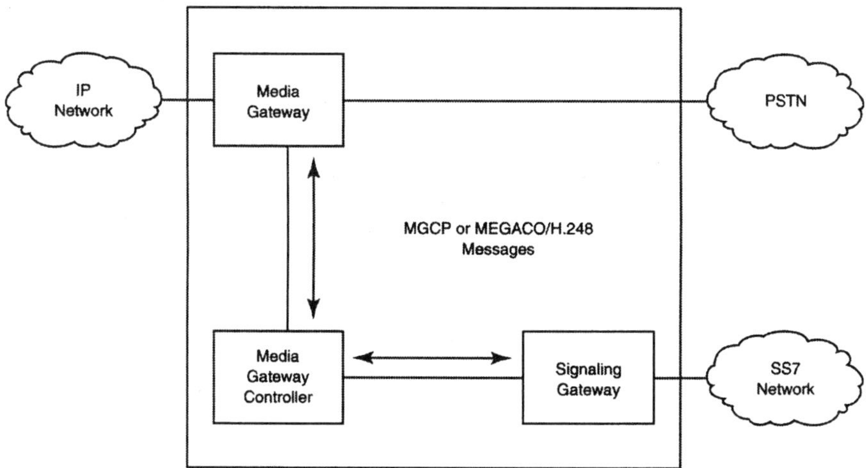

Figure 9-23. Decomposed Gateway Architecture.

The gateway model developed by ETSI divides the functions of a gateway into three key elements: a Signaling Gateway (SG), a Media Gateway (MG), and a Media Gateway Controller (MGC), as shown in Figure 9-23 and documented in Reference [9-35]. (As an aside, the ITU-T has developed a similar gateway model that is documented in H.323 section 6.3.1.) Note that, to the outside world, these three elements appear as a single gateway. When implemented, however, they may be distinct and may also be provided and/or administered by different organizations, such as carriers and enterprise network equipment vendors, thus realizing a distributed gateway architecture. The SG mediates the signaling functions between the IP network and the SCN. For example, it may provide correlation between the H.323 signaling on the packet network side and the SS7 signaling on the PSTN side. The MG mediates the media signals between the IP network and the SCN. For example, it may convert information transported on the IP network using RTP/UDP/IP packet formats to PCM encoded voice on the PSTN side. Finally, the MGC communicates with both the SG and the MG, providing the call processing functions required. The MGC uses either Media Gateway Control Protocol (MGCP) or MEGACO/H.248 protocols for these inter-gateway communication functions. MGCP was developed by the IETF and is documented in RFC 3435 [9-36]. MEGACO/H.248 was jointly developed by the IETF and the ITU-T, and is documented in RFC 3525 [9-37] and H.248.1 [9-38].

9.5 Quality of Service

ITU-T Recommendation E.800 [9-39] defines Quality of Service (QoS) as follows:

> *"The collective effect of service performance, which determines the degree of satisfaction of a user of the service."*

In other words, QoS will be significantly influenced by the end users and their perceptions of how well one network service performs in comparison with other network services. Almost everyone has experience with telephone calls over the PSTN and knows what level of quality to expect from that network. Therefore, one of the significant challenges that must be addressed is ensuring that the QoS the end user experiences with the converged network is on par with their previous experience with the PSTN.

Multimedia applications, such as voice and video, are connection-oriented. In contrast, IP-based internetworks are connectionless. Thus, one would expect some difficulties when transmitting voice or video signals over a network that was not originally designed for that application. As we discussed in Section 9.3.2, the use of the Real-Time Transport Protocol (RTP) and the User Datagram Protocol (UDP) provide some error and sequence control functions that IP lacks. However, other factors inherent in packet data transmission can affect the QoS that the end user perceives. And since the satisfaction of that community of end users may significantly impact the career of the network manager, a discussion of these factors is in order. But first we should discuss the measurement of QoS.

9.5.1 Measuring Quality of Service

For many years, the telephone industry has employed a very subjective rating system, known as the Mean Opinion Score, or MOS, to measure the quality of telephone connections. These measurement techniques are defined in ITU-T P.800 [9-40] and are based on the opinions of many testing volunteers who listen to a sample of voice traffic and rate the quality of that transmission. In doing so, they consider a number of factors that could degrade the quality of transmission, including loss, circuit noise, sidetone, talker echo, distortion, propagation time (or delay), and other transmission problems.

The most well known test, described in Annex A of P.800, is called the Conversation Opinion Test. The volunteer subjects are asked to provide their opinion of the connection they have just been using, based on a five-point scale:

Quality	Rating
Excellent	5
Good	4
Fair	3
Poor	2
Bad	1

Since the test subjects are human, some variation in the scores is expected. For that reason, a large number of people are used in the test, and their individual scores are averaged (hence the term "Mean" in Mean Opinion Score). A MOS of 4 is considered "toll quality" within the telephone industry.

Other tests are also defined in P.800. The listening test rates the ability of the tester to understand speech over the connection. The noise, fading, and disturbances test rates these problems from inaudible to intolerable. The comparison category rating compares a nonprocessed speech sample with a processed speech sample with ratings from "much better" to "much worse."

Readers needing further information can also consult two companion standards, P.861 [9-41] and P.862 [9-42], which define other quality measurements. The first method is called Perceptual Speech Quality Measurement (PSQM) and is defined in P.861. This measurement is used for assessing the quality of voice encoders on a scale of 0 (no degradation) to 6.5 (extensive distortion). The second method is called Perceptual Evaluation of Speech Quality (PESQ) and is defined in P.862. PESQ also addresses the effects of filters, variable delay, and coding distortions, and is applicable for both speech codec evaluation and end-to-end measurements, so is therefore preferred over PSQM for most applications. In summary, the P.800 MOS test is a subjective evaluation, while those tests defined in P.861 and P.862 are objective measurements implementable in hardware for greater accuracy.

9.5.2 Factors Affecting the Quality of Service

For most applications, voice and fax over IP signals will move through telephone company central offices, PBXs, gateways, and end stations before reaching their final destination. Along the way, multiple analog-to-digital conversions may be required. In an ideal case, where transmission bandwidth is either plentiful or has been reserved using some mechanism such as the Resource Reservation Protocol (RSVP), this packet-based (connectionless) transmission proceeds as if it were a connection-oriented network operation.

Unfortunately, not all packet transmission follows the ideal model. Some packets may be lost due to difficulties in the routing infrastructure, collisions on the local Ethernet, transmission impairments such as noisy lines, and so on. The result is a received message that is missing some of its components.

Delay, or latency, can also affect the quality of the transmission. The delay is the difference in time between when the signal is transmitted and when it is received. Delay is typically characterized by two components: a fixed delay and a variable delay. The fixed delays are found within the signal processing elements, such as the processing delays inherent in the voice codecs and also within the physical transmission systems, such as the copper pairs. The variable delays result from queuing times at packet processing points, such as switches and routers, as well as transmission variables, such as the specific route that a particular packet took and any difficulties, such as congestion, that might be experienced on that route. Some of these delays can be controlled by the network manager with thoughtful network engineering decisions. For example, voice gateways containing codecs with lower processing delays could be selected, or additional bandwidth could be provisioned on the wide area network to lower the likelihood of network congestion. There are delay elements that cannot be optimized, such as the physical propagation delay. Electromagnetic signals propagate at approximately 1 nanosecond per foot on copper transmission facilities, while optical fiber systems, including the repeaters, have approximately 10 milliseconds of delay per 1,000 miles. Both of these are principles of physics that must be dealt with.

The maximum recommended one-way delay for most user applications, as specified in ITU-T G.114, is 150 milliseconds, or a round-trip delay of 300 milliseconds [9-43]. However, since voice quality is a very subjective measurement, some end users find variations above and below these standard specifications to be acceptable; therefore, round-trip objectives between 200–400 milliseconds are often quoted. Other delays include switching, routing, and other packet processing delays (typically a few milliseconds); transmission delays (dependent on the speed of the transmission link, and typically less than 10 milliseconds); signal propagation delays (dependent on the physical length of transmission link); jitter buffer delays (typically a settable parameter between 20–40 milliseconds); and decoding delays (typically half of the encoding delay). These sources of typical delays are summarized in Figure 9-24.

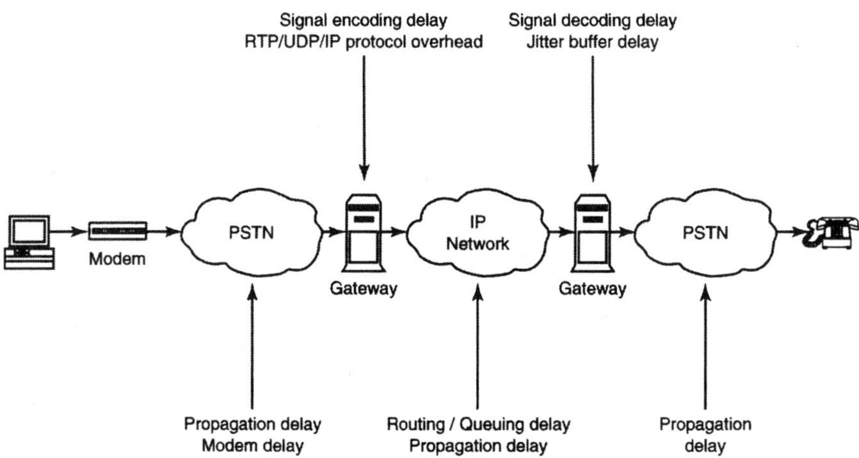

Figure 9-24. Sources of Delays with the VoIP Network.

Another possible factor is packet jitter, which is the variation in arrival rates between individual packets. With packet networks, it is possible that a sequence of packets that enter the network at a constant rate will reach the intended destination by a number of routes. Since each of these routes may have unique delay characteristics, it is possible for the arrival rates of the packets to vary. For non-real-time signals, such as an email transmission, this signal would not be an issue. However, for real-time signals such as voice, which are dependent upon a continuous flow of data, jitter affects the signal quality. To correct for jitter, the packets are buffered, or delayed, at the receiver, and then played out at a continuous rate. Unfortunately, this buffering adds to the overall latency of the packet transmission.

Echo can also cause signal degradation. Both the transmitter and the receiver have connections from the telephone sets to a transformer called a hybrid. The hybrid performs 2-wire to 4-wire conversions, such that the connection from the central office to your premises (2-wire) can be divided into both transmit and receive paths (total of 4 wires). Most transformers have some degree of inefficiency; in other words, they do not pass all of the energy that is applied to their primary winding onto the secondary winding(s). Some of that energy is reflected back to the source. That reflected energy is called an echo. An echo canceller circuit is required which subtracts (or cancels) out the energy that was reflected from the distant end of the connection prior to delivering that signal to the other client processes, such as the speech encoding, packet transmission, and other functions.

9.5.3 Implementing Quality of Service

Having discussed the various factors that affect QoS, such as latency within the network, packet delays and loss, and so on, the next question that must be addressed is the industry response — what steps are being taken to improve QoS within IP-based internetworks?

The traditional solution to improving service performance — throwing more bandwidth at the problem — can be expensive, especially on the WAN side of the equation. As a result, net managers are looking for ways to optimize the performance of their network infrastructures. A number of alternatives have been proposed, which are summarized in this section.

9.5.3.1 Integrated Services

Integrated Services (int-serv) is a model developed by the IETF and characterized by the reservation of network resources prior to the transmission of any data. The Resource Reservation Protocol (RSVP), which is defined in RFC 2205 and discussed in Section 9.3.4, is the signaling protocol that is used to reserve the bandwidth on that transmission path. An end station that supports RSVP may request a particular level of network performance from the network, and at each downstream node the protocol attempts to reserve network resources on behalf of that end station. RSVP is designed to operate in conjunction with current routing protocols, such as the Open Shortest Path First (OSPF) and Border Gateway Protocols (BGP), and relies on these other protocols to determine where reservation requests should be sent. The int-serv model is implemented by four components: the Signaling Protocol (RSVP), an Admission Control Routine (which determines if network resources are available), a Classifier (which puts packets in specific queues), and a Packet Scheduler (which schedules packets to meet QoS Requirements). Details regarding int-serv are found in RFC 1633 [9-44].

9.5.3.2 Differentiated Services

Differentiated Services (diff-serv) was also developed by the IETF and it distinguishes packets that require different network services into different classes. In effect, diff-serv provides a relative priority scheme, whereby the packets are handled differently based on the values of an 8-bit Differentiated Services (DS) field. That DS field replaces the IPv4 Type of Service (TOS) field or the IPv6 Traffic Class field (these fields are fortunately also 8 bits in length). Packets are classified at the network ingress node according to some Service Level Agreement (SLA) criteria established between the Internet Service Provider (ISP) and the customer. It is assumed that the ISP would offer different levels of service, and also different cost structures, for the various options defined by the DS field. Details regarding diff-serv are found in RFC 2474 [9-45].

9.5.3.3 Multiprotocol Label Switching

Multiprotocol Label Switching (MPLS) is a forwarding scheme that was based on tag switching mechanisms developed by router vendors over the past few years. With MPLS, a 32-bit header (sometimes called the "tag") is placed between the local network and IP headers of the packet when it enters the MPLS domain. This header determines how that packet will be handled by the intervening routers. Because these routers must be able to interpret and act upon this new label, the MPLS-capable routers are called Label Switching Routers, or LSRs. The packets are classified at the ingress LSR, and only that 32-bit header is examined to determine how the packet should be handled within the network. Since the LSR only examines the MPLS header and not the entire IP header, this QoS mechanism can operate independently of the Network Layer protocol (such as IP) that is currently in use. Hence the word "multiprotocol" in the term MPLS. Details regarding MPLS are found in RFC 2702 [9-46].

9.5.3.4 Queuing and Congestion Avoidance Mechanisms

Many router vendors have developed mechanisms to improve QoS for time-sensitive applications such as voice and video. Most of these techniques fall into two general categories: priority queuing and congestion avoidance mechanisms.

Queuing mechanisms create multiple signal queues within the router and allocate the available bandwidth to these queues according to some algorithm and/or network administrator-supplied rules. Cisco Systems' technique, called Weighted Fair Queuing (WFQ), classifies various incoming data flows according to source/destination address, protocol in use, or application/port, and then grants each one of these flows a specific amount of outgoing bandwidth based on some fairness criteria.

Congestion avoidance mechanisms augment the flow control properties that exist within protocols. With many of these protocols, such as TCP, data flows are retarded only after the network becomes congested, which causes packets to be dropped. The Random Early Detection (RED) and Weighted Random Early Detection (WRED) mechanisms attempt to anticipate when these congestion points will occur, and discard packets based on criteria such as buffer length or the Precedence bits within the IP header. Thus, RED or WRED minimizes the likelihood of large numbers of packets being sent to already congested queues, which would only result in further packet discard.

9.6 The Voice over IP Packet

As we have seen in this chapter, a number of protocols are involved in the transport of multimedia information through an IP network. In summary, the resulting packet format is shown in Figure 9-25, which would include the IP and UDP headers, the RTP header, and finally a digitized voice sample, which typically represents 20–24 milliseconds of information content.

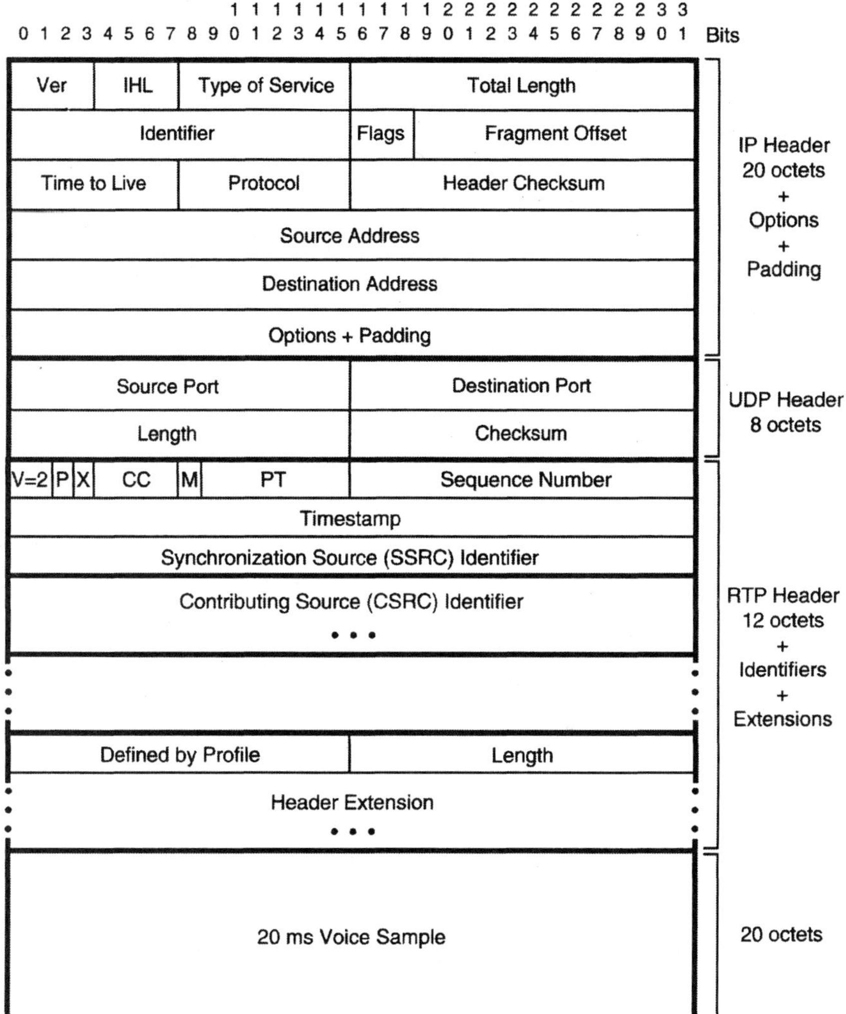
1 1 1 1 1 1 1 1 1 1 2 2 2 2 2 2 2 2 2 2 3 3
0 1 2 3 4 5 6 7 8 9 0 1 2 3 4 5 6 7 8 9 0 1 2 3 4 5 6 7 8 9 0 1 Bits
Ver
IHL
Type of Service
Total Length
Identifier
Flags
Fragment Offset
Time to Live
Protocol
Header Checksum
Source Address
Destination Address
Options + Padding
IP Header 20 octets + Options + Padding
Source Port
Destination Port
Length
Checksum
UDP Header 8 octets
V=2
P
X
CC
M
PT
Sequence Number
Timestamp
Synchronization Source (SSRC) Identifier
Contributing Source (CSRC) Identifier
Defined by Profile
Length
Header Extension
RTP Header 12 octets + Identifiers + Extensions
20 ms Voice Sample
20 octets

Figure 9-25. Voice over IP Packet Format.

9.7 Looking Ahead

In this chapter, we have considered the principles of network convergence, the differences between voice and data networking infrastructures, the projections for growth in these respective areas, and the various players in this marketplace. And, independent of these growth factors, today's networking environments are moving towards a convergence of the voice and data worlds — a trend that would be difficult to ignore. In the next chapter, we will examine case studies that illustrate the operation of both data and multimedia application protocols.

9.8 References

[9-1] Adapted from *Voice over IP Technologies: Building the Converged Network, 2nd Edition* by Mark A. Miller. Copyright © 2002 by Mark A. Miller. All rights reserved. Reproduced here by permission of Wiley Publishing, Inc.

[9-2] The Voice over IP Forum, part of the International Multimedia Teleconferencing Consortium, provides multimedia network information at http://www.imtc.org.

[9-3] Information regarding the European Telecommunications Standards Institute (ETSI) based in Sophia Antipolis, France, and the TIPHON project can be found at http://www.etsi.org.

[9-4] The International Telecommunications Union (ITU) is affiliated with the United Nations and based in Geneva, Switzerland. It can be reached at http://www.itu.int.

[9-5] Deering, S. "Host Extensions for IP Multicasting." RFC 1112, August 1989.

[9-6] Fenner, W. "Internet Group Management Protocol, version 2." RFC 2236, November 1997.

[9-7] A complete listing of IP multicast addresses can be found at: www.iana.org/assignments/multicast-addresses.

[9-8] Waitzman, D., et al. "Distance Vector Multicast Routing Protocol." RFC 1075, November 1988.

[9-9] Moy, J. "Multicast Extensions to OSPF." RFC 1584, March 1994.

[9-10] Schulzrinne, H., et al. "A Transport Protocol for Real-Time Applications." RFC 3550, July 2003.

[9-11] Schulzrinne, H. "RTP Profile for Audio and Video Conferences with Minimal Control." RFC 3551, July 2003.

[9-12] A good resource for RTP information is http://www.cs.columbia.edu/~hgs/rtp/.

[9-13] Braden, R., editor. "Resource Reservation Protocol (RSVP) Version 1 Functional Specification." RFC 2205, September 1997.

[9-14] Herzog, S. "RSVP Extensions for Policy Control." RFC 2750, January 2000.

[9-15] A good resource for RSVP information is the RSVP Project at http://www.isi.edu/div7/rsvp/rsvp-home.html.

[9-16] Schulzrinne, H., et al. "Real Time Streaming Protocol (RTSP)." RFC 2326, April 1998.

[9-17] Fielding, R., et al. "Hypertext Transfer Protocol — HTTP/1.1." RFC 2616, June 1999.

[9-18] Khare, R., and S. Lawrence. "Upgrading to Transport Layer Security (TLS) within HTTP/1.1." RFC 2817, May 2000.

[9-19] Handley, M., and V. Jacobson. "SDP: Session Description Protocol." RFC 2327, April 1998.

[9-20] Handley, M., et al. "Session Announcement Protocol." RFC 2974, October 2000.

[9-21] Rosenberg, J., et al. "SIP: Session Initiation Protocol." RFC 3261, June 2002.

phone-band (300-3,400 Hz) Speech Codecs. ITU-T Recommendation P.861, February 1998.

[9-42] International Telecommunications Union — Telecommunication Standardization Sector. *Perceptual Evaluation of Speech Quality (PESQ), an Objective Method for End-to-End Speech Quality Assessment of Narrow-Band Telephone Networks and Speech Codecs.* ITU-T Recommendation P.862, February 2001.

[9-43] International Telecommunications Union — Telecommunication Standardization Sector. *General Characteristics of International Telephone Connections and International Telephone Circuits — One-Way Transmission Time.* ITU-T Recommendation G.114, February 1996.

[9-44] Braden, R., et al. "Integrated Services in the Internet Architecture: An Overview." RFC 1633, June 1994.

[9-45] Nichols, K., et al. "Definition of the Differentiated Services Field (DS Field) in the IPv4 and IPv6 Headers." RFC 2474, December 1998.

[9-46] Awduche, D., et al. "Requirements for Traffic Engineering Over MPLS." RFC 2702, September 1999.

10

Case Studies in Application Support

To summarize our discussion of the applications that support end-user requirements, this chapter presents ten case studies, complete with network analyzer information captured from live networks, that illustrate how these application protocols operate (or in some cases, how they are *supposed* to operate!).

10.1 Using TFTP

In our first case study, a Sun workstation (shown as Sun IPX in Trace 10.1a) and a PC running TCP/IP workstation software attempt to use TFTP to read and write small source files.

In Frame 1, the Sun IPX initiates a connection to Station XT with a Read Request (RRQ) for file "TFTP_test" in netascii mode (Figure 10-1). Details of that frame (Trace 10.1b) indicate that the Sun IPX has selected Source Transaction ID 1167 (Source port = 1167) with Destination port = 69, which is defined as the TFTP port. The connection is accepted in Frame 2. Station XT selects TID 1183 (Source port = 1183) and returns the Sun IPX TID as the destination (Destination port = 1167). The TFTP header indicates a DATA packet, Block number = 1, with 128 octets of data. Frame 3 acknowledges receipt of the data and terminates that connection.

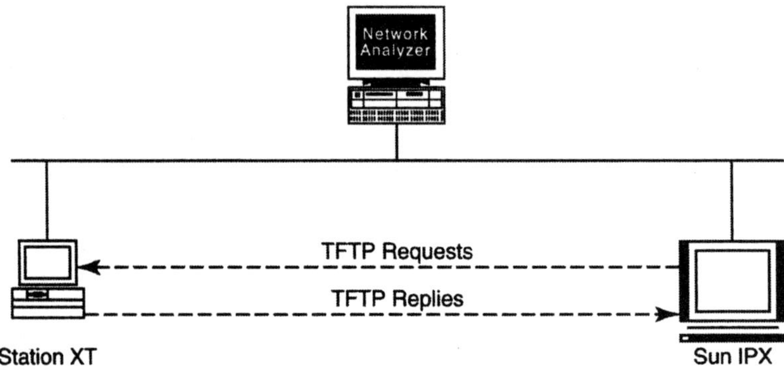

Figure 10-1. TFTP File Transfers.

Given that success, Sun IPX next attempts to write the same file using the Write Request (WRQ) from TID = 1167 in Frame 6. Station XT sends an ACK, selecting TID 1184 (Source port = 1184) and designating Block number = 0. The 128 octets of data are successfully written from Port 1167 (Sun IPX) to Port 1184 in Frame 8. An ACK from Station XT is sent in Frame 9, terminating the connection.

The next two operations are not so successful. In Frame 10, the Sun IPX sends an RRQ for file "\test" in netascii mode. Unfortunately, this file does not exist on Station XT. An ERROR packet, indicating an access violation, is returned in Frame 11. The Sun acknowledges its mistake and terminates the connection in Frame 12.

The Sun IPX makes one more attempt in Frame 16, this time a WRQ of the same file ("\test"), which still does not exist. The Sun IPX selects TID = 1167 again, while Station XT assigns TID = 1186 in Frame 17. Note that the ACK sent in response to the WRQ uses Block number = 0 (Frame 17). The Sun IPX then attempts to write the file. But it is empty (nonexistent) and no data is transferred. Station XT acknowledges the previous frame and terminates the connection.

TFTP thus performs as advertised — it transfers files with simplicity. It cannot overcome the faulty memory of the user, however, who can't seem to remember which files do and do not exist!

Trace 10.1a. TFTP Operation Summary

Sniffer Network Analyzer data 10-Mar at 13:08:44, TFTPTEST.TRC, Pg 1

SUMMARY	Delta T	Destination	Source	Summary
M 1		Station XT	Sun IPX	TFTP Read request File=TFTP_test
2	0.108	Sun IPX	Station XT	TFTP Data packet NS=1 (Last)
3	0.001	Station XT	Sun IPX	TFTP Ack NR=1
6	1.236	Station XT	Sun IPX	TFTP Write request File=TFTP_test
7	0.109	Sun IPX	Station XT	TFTP Ack NR=0
8	0.002	Station XT	Sun IPX	TFTP Data packet NS=1 (Last)
9	0.021	Sun IPX	Station XT	TFTP Ack NR=1
10	4.024	Station XT	Sun IPX	TFTP Read request File=\test
11	0.432	Sun IPX	Station XT	TFTP Error response (Access violation)
12	0.014	Station XT	Sun IPX	TFTP Ack NR=1
16	5.124	Station XT	Sun IPX	TFTP Write request File=\test
17	0.141	Sun IPX	Station XT	TFTP Ack NR=0
18	0.001	Station XT	Sun IPX	TFTP Data packet NS=1 (Last)
19	0.020	Sun IPX	Station XT	TFTP Ack NR=1

Trace 10.1b. TFTP Operation Details

Sniffer Network Analyzer data 10-Mar at 13:08:44, TFTPTEST.TRC, Pg 1

```
- - - - - - - - - - - - - - - - Frame 1 - - - - - - - - - - - - - - - - - -
UDP:        — UDP Header —
UDP:
UDP:        Source port = 1167 (TFTP)
UDP:        Destination port = 69
UDP:        Length = 29
UDP:        No checksum
UDP:
TFTP:       — Trivial file transfer —
TFTP:
TFTP:       Opcode = 1 (Read request)
TFTP:       File name = "TFTP_test"
TFTP:       Mode = "netascii"
TFTP:
TFTP:       [Normal end of "Trivial file transfer".]
TFTP:
```

```
- - - - - - - - - - - - - - - Frame 2 - - - - - - - - - - - - - - - - -
UDP:        — UDP Header —
UDP:
UDP:        Source port = 1183 (TFTP)
UDP:        Destination port = 1167
UDP:        Length = 140
UDP:        Checksum = F0DC (correct)
UDP:
TFTP:       — Trivial file transfer —
TFTP:
TFTP:       Opcode = 3 (Data packet)
TFTP:       Block number = 1
TFTP:       [128 bytes of data] (Last frame)
TFTP:
TFTP:       [Normal end of "Trivial file transfer".]
TFTP:

- - - - - - - - - - - - - - - Frame 3 - - - - - - - - - - - - - - - - -
UDP:        — UDP Header —
UDP:
UDP:        Source port = 1167 (TFTP)
UDP:        Destination port = 1183
UDP:        Length = 12
UDP:        No checksum
UDP:
TFTP:       — Trivial file transfer —
TFTP:
TFTP:       Opcode = 4 (Ack)
TFTP:       Block number = 1
TFTP:
TFTP:       [Normal end of "Trivial file transfer".]
TFTP:

- - - - - - - - - - - - - - - Frame 6 - - - - - - - - - - - - - - - - -
UDP:        — UDP Header —
UDP:
UDP:        Source port = 1167 (TFTP)
UDP:        Destination port = 69
UDP:        Length = 29
UDP:        No checksum
UDP:
TFTP:       — Trivial file transfer —
TFTP:
TFTP:       Opcode = 2 (Write request)
```

```
TFTP:       File name = "TFTP_test"
TFTP:       Mode = "netascii"
TFTP:
TFTP:       [Normal end of "Trivial file transfer".]
TFTP:

--------------- Frame 7 -----------------
UDP:        — UDP Header —
UDP:
UDP:        Source port = 1184 (TFTP)
UDP:        Destination port = 1167
UDP:        Length = 12
UDP:        Checksum = B7CB (correct)
UDP:
TFTP:       — Trivial file transfer —
TFTP:
TFTP:       Opcode = 4 (Ack)
TFTP:       Block number = 0
TFTP:
TFTP:       [Normal end of "Trivial file transfer".]
TFTP:

--------------- Frame 8 -----------------
UDP:        — UDP Header —
UDP:
UDP:        Source port = 1167 (TFTP)
UDP:        Destination port = 1184
UDP:        Length = 140
UDP:        No checksum
UDP:
TFTP:       — Trivial file transfer —
TFTP:
TFTP:       Opcode = 3 (Data packet)
TFTP:       Block number = 1
TFTP:       [128 bytes of data] (Last frame)
TFTP:
TFTP:       [Normal end of "Trivial file transfer".]
TFTP:

--------------- Frame 9 -----------------
UDP:        — UDP Header —
UDP:
UDP:        Source port = 1184 (TFTP)
UDP:        Destination port = 1167
```

```
UDP:        Length = 12
UDP:        Checksum = B7CA (correct)
UDP:
TFTP:       — Trivial file transfer —
TFTP:
TFTP:       Opcode = 4 (Ack)
TFTP:       Block number = 1
TFTP:
TFTP:       [Normal end of "Trivial file transfer".]
TFTP:

- - - - - - - - - - - - - - - Frame 10 - - - - - - - - - - - - - - - - -
UDP:        — UDP Header —
UDP:
UDP:        Source port = 1167 (TFTP)
UDP:        Destination port = 69
UDP:        Length = 25
UDP:        No checksum
UDP:
TFTP:       — Trivial file transfer —
TFTP:
TFTP:       Opcode = 1 (Read request)
TFTP:       File name = "\test"
TFTP:       Mode = "netascii"
TFTP:
TFTP:       [Normal end of "Trivial file transfer".]
TFTP:

- - - - - - - - - - - - - - - Frame 11 - - - - - - - - - - - - - - - - -
UDP:        — UDP Header —
UDP:
UDP:        Source port = 1185 (TFTP)
UDP:        Destination port = 1167
UDP:        Length = 41
UDP:        Checksum = 7497 (correct)
UDP:
TFTP:       — Trivial file transfer —
TFTP:
TFTP:       Opcode = 5 (Error response)
TFTP:       Error code = 2 (Access violation)
TFTP:       Error message = "unable to open file for read"
TFTP:
TFTP:       [Normal end of "Trivial file transfer".]
TFTP:
```

```
--------------- Frame 12 ----------------
UDP:          — UDP Header —
UDP:
UDP:          Source port = 1167 (TFTP)
UDP:          Destination port = 1185
UDP:          Length = 12
UDP:          No checksum
UDP:
TFTP:         — Trivial file transfer —
TFTP:
TFTP:         Opcode = 4 (Ack)
TFTP:         Block number = 1
TFTP:
TFTP:         [Normal end of "Trivial file transfer".]
TFTP:

--------------- Frame 16 ----------------
UDP:          — UDP Header —
UDP:
UDP:          Source port = 1167 (TFTP)
UDP:          Destination port = 69
UDP:          Length = 25
UDP:          No checksum
UDP:
TFTP:         — Trivial file transfer —
TFTP:
TFTP:         Opcode = 2 (Write request)
TFTP:         File name = "\test"
TFTP:         Mode = "netascii"
TFTP:
TFTP:         [Normal end of "Trivial file transfer".]
TFTP:

--------------- Frame 17 ----------------
UDP:          — UDP Header —
UDP:
UDP:          Source port = 1186 (TFTP)
UDP:          Destination port = 1167
UDP:          Length = 12
UDP:          Checksum = B7C9 (correct)
UDP:
TFTP:         — Trivial file transfer —
TFTP:
TFTP:         Opcode = 4 (Ack)
```

```
TFTP:          Block number = 0
TFTP:
TFTP:          [Normal end of "Trivial file transfer".]
TFTP:

--------------- Frame 18 -----------------
UDP:           — UDP Header —
UDP:
UDP:           Source port = 1167 (TFTP)
UDP:           Destination port = 1186
UDP:           Length = 12
UDP:           No checksum
UDP:
TFTP:          — Trivial file transfer —
TFTP:
TFTP:          Opcode = 3 (Data packet)
TFTP:          Block number = 1
TFTP:          [0 bytes of data] (Last frame)
TFTP:
TFTP:          [Normal end of "Trivial file transfer".]
TFTP:

--------------- Frame 19 -----------------
UDP:           — UDP Header —
UDP:
UDP:           Source port = 1186 (TFTP)
UDP:           Destination port = 1167
UDP:           Length = 12
UDP:           Checksum = B7C8 (correct)
UDP:
TFTP:          — Trivial file transfer —
TFTP:
TFTP:          Opcode = 4 (Ack)
TFTP:          Block number = 1
TFTP:
TFTP:          [Normal end of "Trivial file transfer".]
TFTP:
```

10.2 Collaborative Efforts of FTP, ARP, and TFTP

In the case study in Section 7.8, we examined the process by which a device can obtain its boot and configuration files (known as load and parameter images) from the internetwork. The following case study shows a similar process, using the ARP and TFTP protocols. ARP determines the device's (in this case, a bridge's) Internet (IP) address, and TFTP transfers the load and parameter files to the bridge. We'll also use FTP to transfer a new version of the bridge's load image to the TFTP server prior to the start of the ARP/TFTP sequence. Let's see what happens when these three processes must interact.

The topology of this internetwork consists of local and remote Ethernet segments connected via bridges (see Figure 10-2). The network manager, Ross, has received a new version of the bridge software from the manufacturer on a PC-formatted floppy. Ross uses a software utility to transfer the new bridge software from a PC (known as Snoopy) to the TFTP Server, which is a Sun SPARCstation (known as Maestro). Ross establishes a TCP control connection to Port number 21 on Maestro (Frames 2 through 4 of Trace 10.2a) and logs in (Frames 5 through 12). Next, Ross stores the new bridge boot image (filename rb1w1.sys) on Maestro in subdirectory/home/tftpboot (Frame 13). Maestro initiates an FTP data connection to Snoopy's Port number 20 in Frames 15 through 17, and the ASCII file transfer begins in Frame 19. Since the file is rather large, it does not complete until Frame 305, at which time both the FTP and TCP connections are terminated (Frames 306 through 314 of Trace 10.2b).

Next, Ross resets the local bridge (known as MX-3510) so that it can obtain and boot from the new load image. The bridges Ross uses are designed to get their boot and configuration files from a number of different servers that might exist on a network. In Frame 315, the bridge broadcasts for a proprietary boot server; in Frame 316, the bridge broadcasts for a BOOTP server; in Frame 317, the bridge broadcasts for a DEC file server using DEC's Maintenance Operations Protocol (MOP); finally, in Frame 318, the bridge broadcasts a RARP Request looking for its Internet address. This operation is successful and the bridge uses its IP address to broadcast a TFTP Read request (Frame 320). The SPARCstation (Maestro) responds, triggering the bridge to request a file transfer of the boot image using TFTP (Frame 327). Unfortunately, the file transfer is not successful (see Frame 1393). Details of the error response (Trace 10.2c) indicate a Transfer Size Error, with "22 bytes of additional data present." This error message prompts Ross to speculate that the boot image file should have been transferred in FTP Binary mode instead of ASCII mode.

Ross begins again, this time setting the FTP transfer mode to Binary when he loads the file from Snoopy to Maestro. A TCP connection between Snoopy and Maestro is established (Frames 1408 through 1410 of Trace 10.2d), and the FTP file transfer begins. But this time the File Type is set for Image (Type I in Frame 1417). The Binary (or Image) data connection is opened in Frame 1426 and completes in Frame 1692 (Trace 10.2e). The local bridge is again reset and its boot sequence started (Frames 1696 through 1707). The boot image (filename x004db3.img) is transferred from Maestro to the local bridge, using TFTP in Frames 1708 through 2774. A second file (x004db3.prm) is transferred in Frames 2787 through 2819. The bridge boot sequence is now complete. The lesson learned: check the data representation parameter, or file type, to avoid corrupting the transferred file.

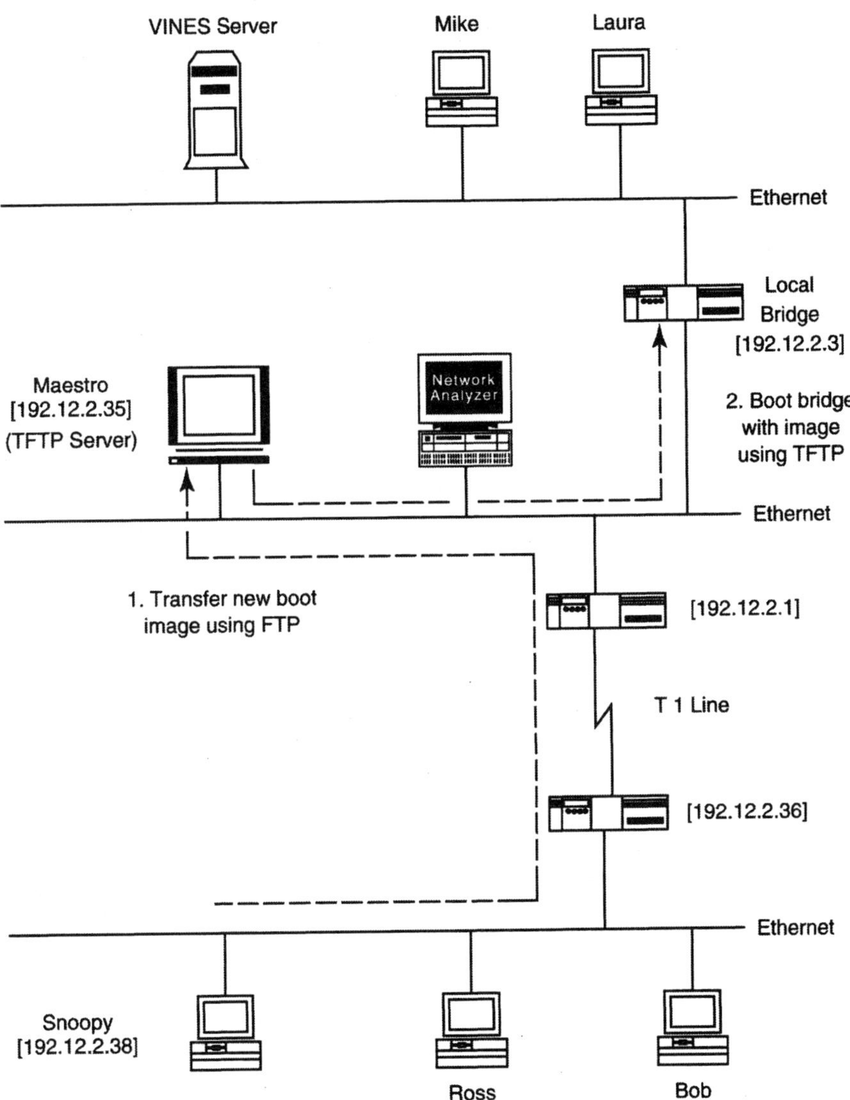

Figure 10-2. File Transfer of Bridge Boot Image.

Trace 10.2a. Boot Image Transfer Using FTP ASCII Connection
Sniffer Network Analyzer data 30-Aug at 14:07:42, BINARY.ENC, Pg 1

SUMMARY	Delta T	Destination	Source	Summary
M 1	AB0000020000	Local Bridge	MOP	RC System ID Receipt=0
2	1.2376	Maestro	Snoopy	TCP D=21 S=20294 SYN SEQ=19468734 LEN=0 WIN=2048
3	0.0009	Snoopy	Maestro	TCP D=20294 S=21 SYN ACK=19468735 SEQ=716544001 LEN=0 WIN=4096
4	0.0016	Maestro	Snoopy	TCP D=21 S=20294 ACK=716544002 WIN=2048
5	0.1115	Snoopy	Maestro	FTP R PORT=20294 220 maestro FTP server (SunOS 4.1) ready.<0D><0A>
6	0.0374	Maestro	Snoopy	FTP C PORT=20294 USER root<0D><0A>
7	0.0073	Snoopy	Maestro	FTP R PORT=20294 331 Password required 230 User root logged in.<0D><0A>
10	0.1621	Maestro	Snoopy	TCP D=21 S=20294 ACK=716544104 WIN=1946
11	21.4760	Maestro	Snoopy	FTP C PORT=20294 PORT 192,12,2, 38,79,71<0D><0A>
12	0.0127	Snoopy	Maestro	FTP R PORT=20294 200 PORT command successful <0D><0A>
13	0.0105	Maestro	Snoopy	FTP C PORT=20294 STOR /home/tftpboot /rb1w1.sys_v2.0 <0D><0A>
14	0.0169	Snoopy	Maestro	TCP D=20294 S=21 ACK=19468819 WIN=4096
15	0.0601	Snoopy	Maestro	TCP D=20295 S=20 SYN SEQ=719936001 LEN=0 WIN=24576

16	0.0020	Maestro	Snoopy	TCP D=20 S=20295 SYN ACK=719936002 SEQ=19468825 LEN=0 WIN=2048
17	0.0006	Snoopy	Maestro	TCP D=20295 S=20 ACK=19468826 WIN=24576
18	0.0018	Snoopy	Maestro	FTP R PORT=20294 150 ASCII data connection for /home/tftpboot/rb1w1...
19	0.1108	Maestro	Snoopy	TCP D=20 S=20295 ACK=719936002 SEQ=19468826
	0.0065	Maestro	Snoopy	TCP D=20 S=20295 ACK=719936002 SEQ=19470015 LEN=1460 WIN=2048
21	0.0058	Maestro	Snoopy	TCP D=20 S=20295 ACK=719936002 SEQ=19471475 LEN=1460 WIN=2048
22	0.0024	Snoopy	Maestro	TCP D=20295 S=20 ACK=19472935 WIN=24576

Trace 10.2b. ASCII File Transfer Completion and Unsuccessful Bridge Boot

Sniffer Network Analyzer data 30-Aug at 14:07:42, BINARY.ENC, Pg 1

SUMMARY	Delta T	Destination	Source	Summary
. . .				
301	0.0059	Maestro	Snoopy	TCP D=20 S=20295 ACK=719936002 SEQ=19739400 LEN=1123 WIN=2048
302	0.0053	Maestro	Snoopy	TCP D=20 S=20295 ACK=719936002 SEQ=19740523 LEN=1460 WIN=2048

303	0.0039	Maestro	Snoopy	TCP D=20 S=20295 FIN ACK=719936002 SEQ=19741983 LEN=627 WIN=2048
304	0.0007	Snoopy	Maestro	TCP D=20295 S=20 ACK=19742611 WIN=23949
305	0.0098	Snoopy	Maestro	FTP R PORT=20294 226 ASCII Transfer complete.<0D><0A>
306	0.0003	Snoopy	Maestro	TCP D=20295 S=20 FIN ACK=19742611 SEQ=719936002 LEN=0 WIN=24576
307	0.0020	Maestro	Snoopy	TCP D=21 S=20294 ACK=716544246 WIN=1804
308	0.0015	Maestro	Snoopy	TCP D=20 S=20295 ACK=719936003 WIN=2048
309	45.9027	Maestro	Snoopy	FTP C PORT=20294 QUIT<0D><0A>
310	0.0091	Snoopy	Maestro	FTP R PORT=20294 221 Goodbye. <0D><0A>
311	0.0060	Maestro	Snoopy	TCP D=21 S=20294 FIN ACK=716544260 SEQ=19468825 LEN=0 WIN=1790
312	0.0006	Snoopy	Maestro	TCP D=20294 S=21 ACK=19468826 WIN=4096
313	0.0009	Snoopy	Maestro	TCP D=20294 S=21 FIN ACK=19468826 SEQ=716544260 LEN=0 WIN=4096
314	0.0020	Maestro	Snoopy	TCP D=21 S=20294 ACK=716544261 WIN=1790
315	9.3205	09008780FFFF	Local Bridge	Ethertype=0889 (Unknown)
316	4.0209	Broadcast	Local Bridge	BOOTP Request

317	4.0194	AB0000010000	Local Bridge	MOP DL Request Program Device Type=UNA Program Type=System
318	8.0203	Broadcast	Local Bridge	ARP C HA=080087004DB3 PRO=IP
319	3.0660	Local Bridge	Maestro	ARP R PA=[192.12.2.3] HA=080087004DB3 PRO=IP
320	0.9553	Broadcast	Local Bridge	TFTP Read request File=x004db3.img
321	0.0168	Local Bridge	Maestro	UDP D=2001 S=1416 LEN=524
322	0.0037	Maestro	Local Bridge	UDP D=1416 S=2001 LEN=70
323	3.9820	Broadcast	Local Bridge	ARP C PA=[192.12.2.35] PRO=IP
324	0.0004	Local Bridge	Maestro	ARP R PA=[192.12.2.35] HA=080020103553 PRO=IP
325	7.9964	Maestro	Local Bridge	ARP C PA=[192.12.2.35] PRO=IP
326	0.0004	Local Bridge	Maestro	ARP R PA=[192.12.2.35] HA=080020103553 PRO=IP
327	0.0010	Maestro	Local Bridge	TFTP Read request File=x004db3.img
328	0.0167	Local Bridge	Maestro	TFTP Data packet NS=1
329	0.1344	Maestro	Local Bridge	TFTP Ack NR=1
330	0.0015	Local Bridge	Maestro	TFTP Data packet NS=2
331	0.0021	Maestro	Local Bridge	TFTP Ack NR=2

.
.
.

1390	0.0013	Local Bridge	Maestro	TFTP Data packet NS=532
1391	0.0022	Maestro	Local Bridge	TFTP Ack NR=532
1392	0.0012	Local Bridge	Maestro	TFTP Data packet NS=533
1393	0.0012	Maestro	Local Bridge	TFTP Error response (Not defined)
1394	0.0018	Maestro	Local Bridge	TFTP Write request File=loaderr.dmp
1395	3.0193	Local Bridge	Maestro	TFTP Error response (File not found)
1396	0.0047	Broadcast	Local Bridge	TFTP Read request File=type57.img
1397	3.0045	Local Bridge	Maestro	UDP D=2002 S=1421 LEN=27
1398	0.9889	09008780FFFF	Local Bridge	Ethertype=0889 (Unknown)

Trace 10.2c. TFTP Error Response Details

Sniffer Network Analyzer data 30-Aug at 14:07:42, BINARY.ENC, Pg 1

```
- - - - - - - - - - - - - - - Frame 1393 - - - - - - - - - - - - - - -
DLC:        — DLC Header —
DLC:
DLC:        Frame 1393 arrived at 14:09:49.7524; frame size is 88 (0058 hex) bytes.
DLC:        Destination = Station Sun   103553, Maestro
DLC:        Source      = Station Xyplex004DB3, Local Bridge
DLC:        Ethertype   = 0800 (IP)
DLC:
IP:         — IP Header —
IP:
IP:         Version = 4, header length = 20 bytes
IP:         Type of service = 00
IP:                 000. .... = routine
IP:                 ...0 .... = normal delay
IP:                 .... 0... = normal throughput
IP:                 .... .0.. = normal reliability
IP:         Total length = 74 bytes
IP:         Identification = 539
IP:         Flags = 0X
IP:         .0.. .... = may fragment
IP:         ..0. .... = last fragment
IP:         Fragment offset = 0 bytes
```

```
IP:        Time to live = 100 seconds/hops
IP:        Protocol = 17 (UDP)
IP:        Header checksum = D049 (correct)
IP:        Source address = [192.12.2.3]
IP:        Destination address = [192.12.2.35]
IP:         No options
IP:
UDP:       — UDP Header —
UDP:
UDP:       Source port = 2003 (TFTP)
UDP:       Destination port = 1417
UDP:       Length = 54
UDP:       No checksum
UDP:
TFTP:      — Trivial file transfer —
TFTP:
TFTP:      Opcode = 5 (Error response)
TFTP:      Error code = 0 (Not defined)
TFTP:      Error message = "Transfer size error"
TFTP:
TFTP:      *** 22 byte(s) of additional data present ***
TFTP:
TFTP:      [Abnormal end of "Trivial file transfer".]
TFTP:
```

Trace 10.2d. Boot Image Transfer Using FTP Binary Connection
Sniffer Network Analyzer data 30-Aug at 14:07:42, BINARY.ENC, Pg 1

SUMMARY	Delta T	Destination	Source	Summary
.				
.				
.				
.				
1399	4.0200	Broadcast	Local Bridge	BOOTP Request
1400	4.0185	AB0000010000	Local Bridge	MOP DL Request Program Device Type=UNA Program Type=System
1401	8.0202	Broadcast	Local Bridge	ARP C HA=080087004DB3 PRO=IP
1402	3.0631	Local Bridge	Maestro	ARP R PA=[192.12.2.3] HA=080087004DB3 PRO=IP

1403	0.9579	Broadcast	Local Bridge	TFTP Read request File=x004db3.img
1404	0.0180	Local Bridge	Maestro	UDP D=2001 S=1423 LEN=524
1405	0.0037	Maestro	Local Bridge	UDP D=1423 S=2001 LEN=70
1406	3.9809	Broadcast	Local Bridge	ARP C PA=[192.12.2.35] PRO=IP
1407	0.0004	Local Bridge	Maestro	ARP R PA=[192.12.2.35] HA=080020103553 PRO=IP
1408	25.2853	Maestro	Snoopy	TCP D=21 S=20296 SYN SEQ=19468830 LEN=0 WIN=2048
1409	0.0010	Snoopy	Maestro	TCP D=20296 S=21 SYN ACK=19468831 SEQ=739648001 LEN=0 WIN=4096
1410	0.0016	Maestro	Snoopy	TCP D=21 S=20296 ACK=739648002 WIN=2048
1411	0.1149	Snoopy	Maestro	FTP R PORT=20296220 maestro FTP server (SunOS 4.1) ready.<0D><0A>
1412	0.0377	Maestro	Snoopy	FTP C PORT=20296 USER root<0D><0A>
1413	0.0075	Snoopy	Maestro	FTP R PORT=20296331 Password required for root.<0D><0A>
1414	0.0249	Maestro	Snoopy	FTP C PORT=20296PASS bugoff<0D><0A>
1415	0.0480	Snoopy	Maestro	FTP R PORT=20296 230 User root logged in.<0D><0A>

1416	0.1375	Maestro	Snoopy	TCP D=21 S=20296 ACK=739648104 WIN=1946
1417	6.2568	Maestro	Snoopy	FTP C PORT=20296TYPE I<0D><0A>
1418	0.0015	Snoopy	Maestro	FTP R PORT=20296 200 Type set to I.<0D><0A>
1419	0.1605	Maestro	Snoopy	TCP D=21 S=20296 ACK=739648124 WIN=1926
1420	23.3762	Maestro	Snoopy	FTP C PORT=20296 PORT 192,12,2,38, 79,73<0D><0A>
1421	0.0126	Snoopy	Maestro	FTP R PORT=20296 200 PORT command successful <0D><0A>
1422	0.0118	Maestro	Snoopy	FTP C PORT=2029 STOR /home/tftpboot /rb1w1.sys_v2.0 <0D><0A>
1423	0.0919	Snoopy	Maestro	TCP D=20297 S=20 SYN SEQ=744128001 LEN=0 WIN=24576
1424	0.0015	Maestro	Snoopy	TCP D=20 S=20297 SYN ACK=744128002 SEQ=19468841 LEN=0 WIN=2048
1425	0.0006	Snoopy	Maestro	TCP D=20297 S=20 ACK=19468842 WIN=24576
1426	0.0018	Snoopy	Maestro	FTP R PORT=20296150 Binary data connection for /home/tftpboot/ rb1w...

1427	0.1065	Maestro	Snoopy	TCP D=20 S=2029 ACK=744128002 SEQ=19468842 LEN=1460 WIN=2048
1428	0.0031	Maestro	Snoopy	TCP D=20 S=20297 ACK=744128002 SEQ=19470302 LEN=1460 WIN=2048
1429	0.0013	Snoopy	Maestro	TCP D=20297 S=20 ACK=19471762 WIN=24576

.
.

Trace 10.2e. Binary File Transfer Completion and Successful Bridge Boot
Sniffer Network Analyzer data 30-Aug at 14:07:42, BINARY.ENC, Pg 1

SUMMARY	Delta T	Destination	Source	Summary
.				
.				
.				
1687	0.0087	Maestro	Snoopy	TCP D=20 S=20297 ACK=744128002 SEQ=19738450 LEN=1460 WIN=2048
1688	0.0031	Maestro	Snoopy	TCP D=20 S=20297 ACK=744128002 SEQ=19739910 LEN=1460 WIN=2048
1689	0.0002	Snoopy	Maestro	TCP D=20297 S=20 ACK=19739910 WIN=24576
1690	0.0026	Maestro	Snoopy	TCP D=20 S=20297 FIN ACK=744128002 SEQ=19741370 LEN=240 WIN=2048
1691	0.0005	Snoopy	Maestro	TCP D=20297 S=20 ACK=19741611 WIN=22876

1692	0.0092	Snoopy	Maestro	FTP R PORT=20296 226 Binary Transfer complete.<0D><0A>
1693	0.0003	Snoopy	Maestro	TCP D=20297 S=20 FIN ACK=19741611 SEQ=744128002 LEN=0 WIN=24576
1694	0.0020	Maestro	Snoopy	TCP D=21 S=20296 ACK=739648268 WIN=1782
1695	0.0008	Maestro	Snoopy	TCP D=20 S=20297 ACK=744128003 WIN=2048
1696	31.6613	09008780FFFF	Local Bridge	Ethertype=0889 (Unknown)
1697	4.0209	Broadcast	Local Bridge	BOOTP Request
1698	4.0194	AB0000010000	Local Bridge	MOP DL Request Program Device Type=UNA Program Type=System
1699	8.0204	Broadcast	Local Bridge	ARP C HA=080087004DB3 PRO=IP
1700	3.0628	Local Bridge	Maestro	ARP R PA=[192.12.2.3] HA=080087004DB3 PRO=IP
1701	0.9584	Broadcast	Local Bridge	TFTP Read request File=x004db3.img
1702	0.0168	Local Bridge	Maestro	UDP D=2001 S=1426 LEN=524
1703	0.0037	Maestro	Local Bridge	UDP D=1426 S=2001 LEN=70
1704	3.9820	Broadcast	Local Bridge	ARP C PA=[192.12.2.35] PRO=IP
1705	0.0004	Local Bridge	Maestro	ARP R PA=[192.12.2.35] HA=080020103553 PRO=IP

1706	7.9964	Maestro	Local Bridge	ARP C PA=[192.12.2.35] PRO=IP
1707	0.0004	Local Bridge	Maestro	ARP R PA=[192.12.2.35] HA=080020103553 PRO=IP
1708	0.0010	Maestro	Local Bridge	TFTP Read request File=x004db3.img
1709	0.0168	Local Bridge	Maestro	TFTP Data packet NS=1
1710	0.1344	Maestro	Local Bridge	TFTP Ack NR=1
1711	0.0015	Local Bridge	Maestro	TFTP Data packet NS=2
1712	0.0021	Maestro	Local Bridge	TFTP Ack NR=2
.				
.				
.				
2771	0.0012	Local Bridge	Maestro	TFTP Data packet NS=532
2772	0.0022	Maestro	Local Bridge	TFTP Ack NR=532
2773	0.0011	Local Bridge	Maestro	TFTP Data packet NS=533 (Last)
2774	0.0019	Maestro	Local Bridge	TFTP Ack NR=533
2775	11.3936	09008780FFFF	Local Bridge	Ethertype=0889 (Unknown)
2776	4.0129	Broadcast	Local Bridge	BOOTP Request
2777	4.0195	AB0000010000	Local Bridge	MOP DL Request Program Device Type=UNA Program Type=System
2778	8.0205	Broadcast	Local Bridge	ARP C HA=080087004DB3 PRO=IP
2779	3.0610	Local Bridge	Maestro	ARP R PA=[192.12.2.3] HA=080087004DB3 PRO=IP
2780	0.9599	Broadcast	Local Bridge	TFTP Read request File=x004db3.prm
2781	0.0159	Local Bridge	Maestro	UDP D=2001 S=1440 LEN=524

2782	0.0016	Maestro	Local Bridge	UDP D=1440 S=2001 LEN=70
2783	3.9846	Broadcast	Local BridgeARP	C PA=[192.12.2.35] PRO=IP
2784	0.0004	Local Bridge	Maestro	ARP R PA=[192.12.2.35] HA=080020103553 PRO=IP
2785	7.9971	Maestro	Local BridgeA	RP C PA=[192.12.2.35] PRO=IP
2786	0.0004	Local Bridge	Maestro	ARP R PA=[192.12.2.35] HA=080020103553 PRO=IP
2787	0.0010	Maestro	Local Bridge	TFTP Read request File=x004db3.prm
2788	0.0163	Local Bridge	Maestro	TFTP Data packet NS=1
2789	0.0147	Maestro	Local Bridge	TFTP Ack NR=1
2790	0.0015	Local Bridge	Maestro	TFTP Data packet NS=2
2791	0.0021	Maestro	Local Bridge	TFTP Ack NR=2
.				
.				
.				
2814	0.0012	Local Bridge	Maestro	TFTP Data packet NS=14
2815	0.0021	Maestro	Local Bridge	TFTP Ack NR=14
2816	0.0013	Local Bridge	Maestro	TFTP Data packet NS=15
2817	0.0022	Maestro	Local Bridge	TFTP Ack NR=15
2818	0.0012	Local Bridge	Maestro	TFTP Data packet NS=16 (Last)
2819	0.0021	Maestro	Local Bridge	TFTP Ack NR=16

10.3 TCP/IP Incompatible Terminal Type

The ARPA architecture groups all the upper layer functions into the Process/Application Layer, which includes the OSI Session, Presentation, and Application Layer services. The middle layer (Presentation) is responsible for any code conversion required to assure that the sending station and receiving station are speaking the same "language." For example, if my machine spoke

ASCII (the American Standard Code for Information Interchange), and your machine spoke EBCDIC (the Extended Binary Coded Decimal Interchange Code), some process would be required to perform a code conversion, or the two of us could never communicate. An excellent example of the requirement for code conversion is a PC that is emulating an IBM host terminal, such as a 3278 display station. The PC speaks ASCII, the host requires EBCDIC, and therefore a conversion is required. In most cases, the PC does the conversion so as to not burden the host with this seemingly trivial issue.

In this case study, a PC on an Ethernet needs to establish communication with a remote IBM host (see Figure 10-3). The PC establishes a connection, but then the connection abruptly fails. A network analyzer is placed on the Ethernet segment to determine the source of the failure. The initial analysis (see Trace 10.3a) indicates normal operation. The Emulator issues a query to the Domain Name Server to find the host's internet address (Frame 1), and the server responds with the address in Frame 2. Since the host is located on another network (network 129.6.x.y), a connection via the router is required. The TCP three-way handshake starts in Frame 3 and completes in Frames 4 and 6. (Note that any subsequent information to or from the remote host will go via the router. If our analyzer was placed on the same segment as the host, we could observe the host response directly.)

Trace 10.3a. TCP/IP Terminal Emulation Incompatibility Summary
Sniffer Network Analyzer data 21-Jun at 09:06:50 file TERMTYPE.ENC Pg 1

SUMMARY	Delta T	Destination	Source	Summary
M 1		Name Server	Emulator	DNS C ID=17 OP=QUERY
2	0.0028	Emulator	Name Server	DNS R ID=17 STAT=OK
3	0.1023	Router	Emulator	TCP D=23 S=3158 SYN SEQ=4660 LEN=O WIN=1024
4	4.0014	Router	Emulator	TCP D=23 S=3158 SYN SEQ=4660 LEN=0 WIN=1024
5	0.1316	Router	Emulator	ARP R PA=[132.163.200.15] HA=02608C1AB9BE PRO=IP

6	3.0033	Emulator	Router	TCP D=3158 S=23 SYN ACK=4661 SEQ=1051902676 LEN=0 WIN=8192
7	0.0033	Router	Emulator	TCP D=23 S=3158 ACK=1051902677 WIN=1024
8	0.2502	Emulator	Router	Telnet R PORT=3158 IAC Do Terminal type
9	0.1262	Router	Emulator	TCP D=23 S=3158 ACK=1051902680 WIN=1021
10	2.7604	Router	Emulator	Telnet C PORT=3158 IAC Will Terminal type
11	0.1311	Emulator	Router	TCP D=315 S=23 ACK=4664 WIN=8189
12	0.0122	Emulator	Router	Telnet R PORT=3158 IAC SB ...
13	0.0048	Router	Emulator	Telnet C PORT =3158 IAC SB ...
14	0.1418	Emulator	Router	TCP D=3158 S=23 ACK=4675 WIN=8178
15	0.0142	Emulator	Router	Telnet R PORT=3158 IAC SB ...
16	0.0049	Router	Emulator	Telnet C PORT =3158 IAC SB ...
17	0.1404	Emulator	Router	TCP D=3158 S=23 ACK=4686 WIN=8167
18	0.0142	Emulator	Router	TCP D=3158 S=23 FIN ACK=4686 SEQ=1051902692 LEN=0 WIN=8167
19	0.0029	Router	Emulator	TCP D=23 S=3158 ACK=1051902693 WIN=1009

20	0.0021	Router	Emulator	TCP D=23 S=3158 FIN ACK=1051902693 SEQ=4686 LEN=0 WIN=1009
21	0.1492	Emulator	Router	TCP D=31585=23 ACK=4687 WIN=8166

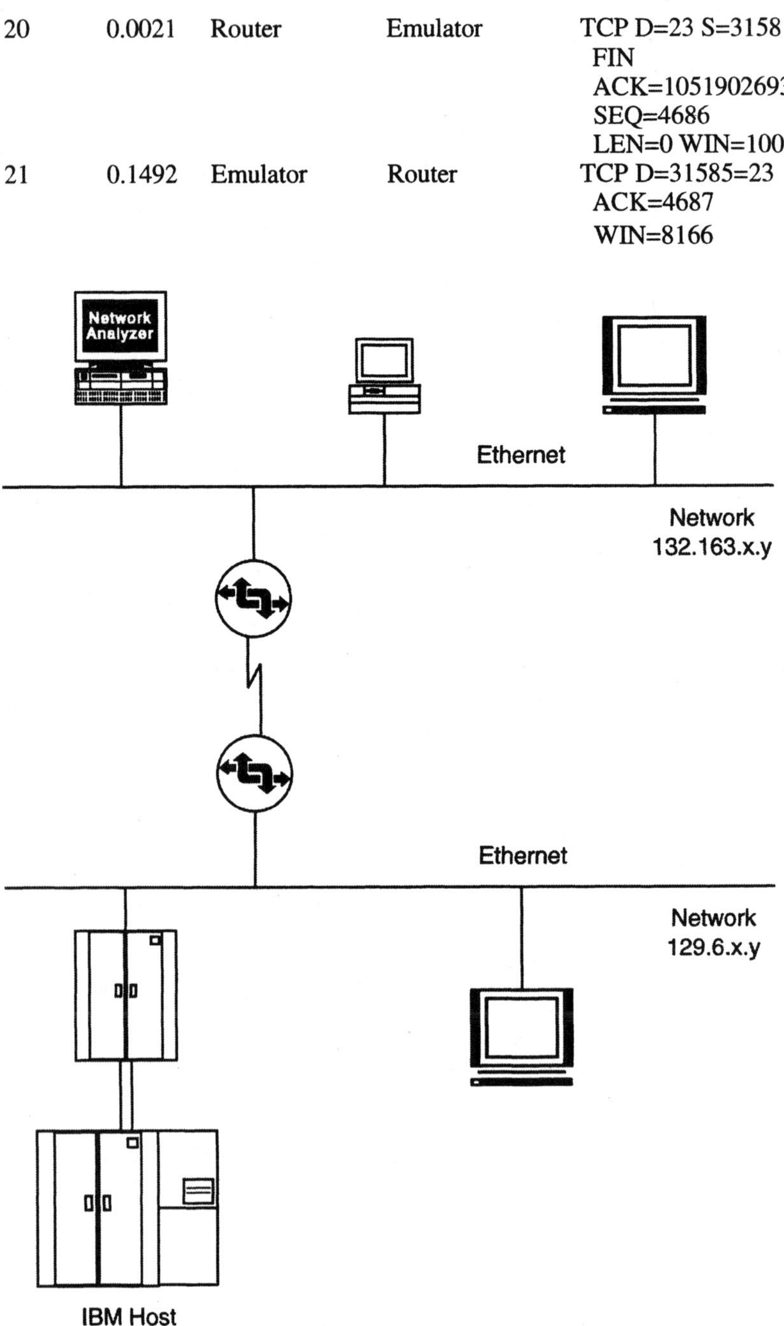

Figure 10-3. TCP/IP Incompatible Terminal Type.

Next, the emulator initiates the TELNET session with the remote host. All appears to be going well when the remote host terminates the connection in Frame 18. Frames 19 and 20 complete the TCP disconnect. The question remains: why did the session abruptly terminate before data could be transferred?

The answer is found in the details of the TELNET parameters that were passed from the emulator to the host (Trace 10.3b). Note that the source and destination stations are on different networks (see the IP address fields) and that the TELNET part of the host is being used (see the TCP header of Frame 12). The answer to the interoperability problem is found in the details of Frame 13, within the ASCII decode of that data. The emulator has selected a Terminal Type = DUMB, indicating that it will emulate a dumb terminal. This selection might work well for ASCII-based hosts, such as a DEC or Hewlett–Packard minicomputer, but it is unacceptable to the IBM host. The remote host wants to speak to a terminal of its own type (e.g. 3278 or 3279 display station) and rebels when asked to speak to a dumb terminal. The host's dissatisfaction is therefore demonstrated by the TCP disconnect in Frames 18–20. In Frame 12, the host (via the router) negotiates a parameter option with the terminal. The designation "IAC SB..." means "interpret as command, start of option subnegotiation." In other words, the host has a particular parameter option that must be negotiated. The emulator's response is given in Frame 13.

Trace 10.3b. TCP/IP Terminal Emulation Incompatibility Details
Sniffer Network Analyzer data 21-Jun-90 at 09:06:50 file TERMTYPE.ENC Pg 1

```
------- ------------- Frame 12 -------------------
IP:        — IP Header —
IP:
IP:        Version = 4, header length = 20 bytes
IP:        Type of service = 00
IP:              000. .... = routine
IP:              ...0 .... = normal delay
IP:              .... 0... = normal throughput
IP:              .... .0.. = normal reliability
IP:              .... ..0. = ECT bit - transport protocol will ignore the CE bit
IP:              .... ...0 = CE bit - no congestion
IP:        Total length    = 46 bytes
IP:        Identification  = 47266
IP:        Flags           = 0X
IP:              .0.. .... = may fragment
```

```
Q2931:                424 bps
Q2931:        Best effort indicator =  190
Q2931:
Q2931:        Info element id      = 5E (Broadband bearer capability)
Q2931:        Coding Standard/Action = 80
Q2931:                1... .... =        ext
Q2931:                .00. .... =         code stand (ITU-T standardized)
Q2931:                ...0 0000 =          IE field (not significant)
Q2931:        Length of info element = 3 byte(s)
Q2931:        Bearer class         = BCOB-X
Q2931:        Traffic type         = No indication
Q2931:        Timing requirements   = No indication
Q2931:        Susceptibility to clipping = Not susceptible to clipping
Q2931:        User plane conn config    = Point-to-point
Q2931:
Q2931:        Info element id      = 70 (Called party number)
Q2931:        Coding Standard/Action = 80
Q2931:                1... .... =        ext
Q2931:                .00. .... =         code stand (ITU-T standardized)
Q2931:                ...0 0000 =          IE field (not significant)
Q2931:        Length of info element = 21 byte(s)
Q2931:                1... .... =        ext
Q2931:                .000 .... =         type of num (Unknown)
Q2931:                .... 0010 =          addressing/num plan id (ATM Endsystem Address)
Q2931:        Authority and format id = DCC ATM Format
Q2931:        Data country code  = 0
Q2931:        HO_DSP             = 00000000000000000000
Q2931:        End system id      = 0000A145705A
Q2931:        selector           = 0
Q2931:
Q2931:        Info element id      = 6C (Calling party number)
Q2931:        Coding Standard/Action = 80
Q2931:                1... .... =        ext
Q2931:                .00. .... =         code stand (ITU-T standardized)
Q2931:                ...0 0000 =          IE field (not significant)
Q2931:        Length of info element = 22 byte(s)
Q2931:                0... .... =        ext
Q2931:                .000 .... =         type of num (Unknown)
Q2931:                .... 0010 =          addressing/num plan id (ATM Endsystem Address)
Q2931:                1... .... =        ext
Q2931:                .00. .... =         Presentation indicator (Presentation allowed)
Q2931:                .... ..00 =         Screening indicator (User-provided, not screened)
Q2931:        Authority and format id = DCC ATM Format
```

```
Q2931:      Data country code  = 0
Q2931:      HO_DSP             = 00000000000000000000
Q2931:      End system id      = 0000A14570FA
Q2931:      selector           = 0
Q2931:
Q2931:      Info element id        = 5C (Quality of service)
Q2931:      Coding Standard/Action = E0
Q2931:                1... .... =        ext
Q2931:                .11. .... =        code stand (Standard defined for network)
Q2931:                ...0 0000 =        IE field (not significant)
Q2931:      Length of info element = 2 byte(s)
Q2931:      QoS class
Q2931:      forward   =  QoS class 0 - Unspecified QoS class
Q2931:      backward  =  QoS class 0 - Unspecified QoS class
Q2931:
Q2931       :Info element id       = 58 (ATM adaptation layer parameters)
Q2931:      Coding Standard/Action = 80
Q2931:                1... .... =        ext
Q2931:                .00. .... =        code stand (ITU-T standardized)
Q2931:                ...0 0000 =        IE field (not significant)
Q2931:      Length of info element = 11 byte(s)
Q2931:      AAL type               = AAL type 5
Q2931:      Forward max CPCS-SDU
Q2931:                id   = 140
Q2931:                size  = 4478
Q2931:      Backward max CPCS-SDU
Q2931:                id   = 129
Q2931:                size  = 4478
Q2931:      Mode
Q2931:                id  = 131
Q2931:                mode  = Message mode
Q2931:      UNI 3.0 signaling
Q2931:      SSCS type
Q2931:        id  = 132
Q2931:        type = Null
Q2931:
Q2931:      Info element id        = 5F (Broadband low layer information)
Q2931:      Coding Standard/Action = 80
Q2931:                1... .... =        ext
Q2931:                .00. .... =        code stand (ITU-T standardized)
Q2931:                ...0 0000 =        IE field (not significant)
Q2931:      Length of info element = 1 byte(s)
Q2931:      Layer 2 protocol
Q2931:                1... .... = ext
```

```
Q2931:              .10. .... = layer 2 id
Q2931:              ...0 1100 = User info layer 2 protocol (LAN logical link
                    control (ISO 8802/2))
Q2931:

- - - - - - - - - - - - - - - Frame 2 - - - - - - - - - - - - - - - - -
ATM:          — ATM Cell Header —
ATM:
ATM:          Frame 2 arrived at  14:48:45.38319; frame size is 24 (0018 hex) bytes.
ATM:          Link = DCE
ATM:          Virtual path id = 0
ATM:          Virtual channel id = 5
ATM:
SSCOP:        — SSCOP trailer —
SSCOP:
SSCOP:        Sequenced Data PDU
SSCOP:        SD send seq num N(S) = 94
SSCOP:
Q2931:-       UNI 3.x Signaling -
Q2931:
Q2931:        Protocol discriminator   = 09
Q2931:        Length of call reference = 3 bytes
Q2931:        Call reference value     = 8003D8
Q2931:        Message type             = 02 (Call proceeding)
Q2931:        Message type Flag/Action = 80
Q2931:              ...0 .... = flag
Q2931:              .... ..00 = action (Clear call)
Q2931:        Message Length           = 9
Q2931:
Q2931:        Info element id          = 5A (Connection identifier)
Q2931:        Coding Standard/Action = 80
Q2931:              1... .... =     ext
Q2931:              .00. .... =     code stand (ITU-T standardized)
Q2931:              ...0 0000 =     IE field (not significant)
Q2931:        Length of info element = 5 byte(s)
Q2931:              1... .... = ext
Q2931:              .00. .... = spare
Q2931:              ...0 1... = VP assoc signaling
Q2931:              .... .000 = Preferred/exclusive
Q2931:        VPCI = 0
Q2931:        VCI  = 34
Q2931
```

```
- - - - - - - - - - - - - - - Frame 3 - - - - - - - - - - - - - - - - -
ATM:        — ATM Cell Header —
ATM:
ATM:        Frame 3 arrived at 14:48:45.44969; frame size is 24 (0018 hex) bytes.
ATM:        Link = DCE
ATM:        Virtual path id = 0
ATM:        Virtual channel id = 5
ATM:
SSCOP:      — SSCOP trailer —
SSCOP:
SSCOP:      Sequenced Data PDU
SSCOP:      SD send seq num N(S) = 95
SSCOP:
Q2931:      — UNI 3.x Signaling —
Q2931:
Q2931:      Protocol discriminator   = 09
Q2931:      Length of call reference = 3 bytes
Q2931:      Call reference value     = 8003D8
Q2931:      Message type             = 07 (Connect)
Q2931:      Message type Flag/Action = 80
Q2931:                ...0 .... = flag
Q2931:                .... ..00 = action (Clear call)
Q2931:      Message Length           = 9
Q2931:
Q2931       :Info element id         = 5A (Connection identifier)
Q2931:      Coding Standard/Action = 80
Q2931:                1... .... =        ext
Q2931:                .00. .... =        code stand (ITU-T standardized)
Q2931:                ...0 0000 =        IE field (not significant)
Q2931:      Length of info element = 5 byte(s)
Q2931:      1... .... = ext
Q2931:      .00. .... = spare
Q2931:      ...0 1... = VP assoc signaling
Q2931:      .... .000 = Preferred/exclusive
Q2931:      VPCI = 0
Q2931:      VCI  = 34
Q2931

- - - - - - - - - - - - - - - Frame 4 - - - - - - - - - - - - - - - - -
ATM:        — ATM Cell Header —
ATM:
ATM:        Frame 4 arrived at 14:48:45.45191; frame size is 48 (0030 hex) bytes.
ATM:        Link = DTE
ATM:        Virtual path id = 0
```

```
ATM:        Virtual channel id = 34
ATM:
LLC:        — LLC Header —
LLC:
LLC:        DSAP Address = AA, DSAP IG Bit = 00 (Individual Address)
LLC:        SSAP Address = AA, SSAP CR Bit = 00 (Command)
LLC:        Unnumbered frame: UI
LLC:
SNAP:       — SNAP Header —
SNAP:
SNAP:       Type = 0806 (ARP)
SNAP:
ARP:        — ARP/RARP frame —
ARP:
ARP:        Hardware type = 19 (ATM)
ARP:        Protocol type = 0800 (IP)
ARP:        Source ATM num
ARP:                type = ATM Forum NSAPA format
ARP:                length = 20
ARP:        Target ATM subaddr
ARP:                type = ATM Forum NSAPA format
ARP:                length = 0
ARP:        Opcode 8 (InARP_REQUEST)
ARP:        Length of source prot addr    = 4
ARP:        Target ATM num
ARP:                type = ATM Forum NSAPA format
ARP:                length = 0
ARP:        Target ATM num
ARP:                type = ATM Forum NSAPA format
ARP:                length = 0
ARP:        Length of target prot addr    = 4
ARP:        source ATM num = 390000000000000000000000000000A14570FA00
ARP:        source prot addr  = [XXX.YYY.112.250]
ARP:        target prot addr  = [0.0.0.0]
ARP:

- - - - - - - - - - - - - - - Frame 5 - - - - - - - - - - - - - - - - -
ATM:        — ATM Cell Header —
ATM:
ATM:        Frame 5 arrived at  14:48:45.45272; frame size is 16 (0010 hex) bytes.
ATM:        Link = DTE
ATM:        Virtual path id = 0
ATM:        Virtual channel id = 5
ATM:
```

```
SSCOP:      — SSCOP trailer —
SSCOP:
SSCOP:      Sequenced Data PDU
SSCOP:      SD send seq num N(S) = 95
SSCOP:
Q2931:      — UNI 3.x Signaling —
Q2931:
Q2931:      Protocol discriminator   = 09
Q2931:      Length of call reference = 3 bytes
Q2931:      Call reference value     = 0003D8
Q2931:      Message type             = 0F (Connect acknowledge)
Q2931:      Message type Flag/Action = 80
Q2931:                ...0 .... = flag
Q2931:                .... ..00 = action (Clear call)
Q2931:      Message Length           = 0

- - - - - - - - - - - - - - - Frame 6 - - - - - - - - - - - - - - - - -
ATM:        — ATM Cell Header —
ATM:
ATM:        Frame 6 arrived at  14:48:45.45296; frame size is 68 (0044 hex) bytes.
ATM:        Link = DCE
ATM:        Virtual path id = 0
ATM:        Virtual channel id = 34
ATM:
LLC:        — LLC Header —
LLC:
LLC:        DSAP Address = AA, DSAP IG Bit = 00 (Individual Address)
LLC:        SSAP Address = AA, SSAP CR Bit = 00 (Command)
LLC:        Unnumbered frame: UI
LLC:
SNAP:       — SNAP Header —
SNAP:
SNAP:       Type = 0806 (ARP)
SNAP:
ARP:        — ARP/RARP frame —
ARP:
ARP:        Hardware type = 19 (ATM)
ARP:        Protocol type = 0800 (IP)
ARP:        Source ATM num
ARP:                  type = ATM Forum NSAPA format
ARP:                  length = 20
ARP:        Target ATM subaddr
ARP:                  type = ATM Forum NSAPA format
```

```
ARP:                    length = 0
ARP:            Opcode 9 (InARP_REPLY)
ARP:            Length of source prot addr    = 4
ARP:            Target ATM num
ARP:                    type = ATM Forum NSAPA format
ARP:                    length = 20
ARP:            Target ATM num
ARP:                    type = ATM Forum NSAPA format
ARP:                    length = 0
ARP:            Length of target prot addr    = 4
ARP:            source ATM num = 390000000000000000000000000A145705A00
ARP:            source prot addr = [XXX.YYY.112.90]
ARP:            target ATM num  = 390000000000000000000000000A14570FA00
ARP:            target prot addr = [XXX.YYY.112.250]
ARP:
```

Now that the ATM connection is established between the router and the ATM switch, the TCP connection between the TELNET client (the PC) and the TELNET server (the Sun) may proceed. This communication is illustrated in Frames 7–9 (Trace 10.5c). In Frame 7, the client initiates the three-way handshake with SYN, ISN = 0. The Sun responds in Frame 8 with SYN, ISN = 2339991040, ACK = 1. The client then confirms the connection in Frame 9 with ACK, SEQ = 1, ACK = 2339991041. The workstation and server are now logically configured for a connection to Port 23 (TELNET). As a final note, look at the ATM Cell Header and observe that the virtual circuit that was established between the ATM switch and the router is now in use (VPI = 0, VCI = 34).

Trace 10.5c. TELNET over ATM TCP Three-Way Handshake
Sniffer Network Analyzer data from 23-Jan at 14:46:34, JIM02.ATC, Pg 1

```
- - - - - - - - - - - - - - - Frame 7 - - - - - - - - - - - - - - - - -
ATM:            — ATM Cell Header —
ATM:
ATM:            Frame 7 arrived at  14:48:45.92383; frame size is 52 (0034 hex) bytes.
ATM:            Link = DTE
ATM:            Virtual path id = 0
ATM:            Virtual channel id = 34
ATM:
LLC:            — LLC Header —
LLC:
LLC:            DSAP Address = AA, DSAP IG Bit = 00 (Individual Address)
```

```
LLC:       SSAP Address = AA, SSAP CR Bit = 00 (Command)
LLC:       Unnumbered frame: UI
LLC:
SNAP:      — SNAP Header —
SNAP:
SNAP:      Type = 0800 (IP)
SNAP:
IP:        — IP Header —
IP:
IP:        Version = 4, header length = 20 bytes
IP:        Type of service = 00
IP:                 000. .... = routine
IP:                 ...0 .... = normal delay
IP:                 .... 0... = normal throughput
IP:                 .... .0.. = normal reliability
IP:        Total length    = 44 bytes
IP:        Identification  = 60067
IP:        Flags           = 0X
IP:                 .0.. .... = may fragment
IP:                 ..0. .... = last fragment
IP:        Fragment offset = 0 bytes
IP:        Time to live    = 31 seconds/hops
IP:        Protocol        = 6 (TCP)
IP:        Header checksum = 8CE4 (correct)
IP:        Source address      = [XXX.YYY.113.95]
IP:        Destination address = [XXX.YYY.112.90]
IP:        No options
IP:
TCP:       — TCP header —
TCP:
TCP:       Source port          = 1405
TCP:       Destination port     = 23 (Telnet)
TCP:       Initial sequence number = 0
TCP:       Data offset          = 24 bytes
TCP:       Flags                = 02
TCP:                ..0. .... = (No urgent pointer)
TCP:                ...0 .... = (No acknowledgment)
TCP:                .... 0... = (No push)
TCP:                .... .0.. = (No reset)
TCP:                .... ..1. = SYN
TCP:                .... ...0 = (No FIN)
TCP:       Window               = 512
TCP:       Checksum             = 6C4E (correct)
TCP:
```

```
TCP:        Options follow
TCP:        Maximum segment size   = 1460
TCP:

--------------- Frame 8 ----------------
ATM:        — ATM Cell Header —
ATM:
ATM:        Frame 8 arrived at 14:48:45.92584; frame size is 52 (0034 hex) bytes.
ATM:        Link = DCE
ATM:        Virtual path id = 0
ATM:        Virtual channel id = 34
ATM:
LLC:        — LLC Header —
LLC:
LLC:        DSAP Address = AA, DSAP IG Bit = 00 (Individual Address)
LLC:        SSAP Address = AA, SSAP CR Bit = 00 (Command)
LLC:        Unnumbered frame: UI
LLC:
SNAP:       — SNAP Header —
SNAP:
SNAP:       Type = 0800 (IP)
SNAP:
IP:         — IP Header —
IP:
IP:         Version = 4, header length = 20 bytes
IP:         Type of service = 00
IP:                 000. .... = routine
IP:                 ...0 .... = normal delay
IP:                 .... 0... = normal throughput
IP:                 .... .0.. = normal reliability
IP:         Total length    = 44 bytes
IP:         Identification  = 22919
IP:         Flags           = 4X
IP:                 .1.. .... = don't fragment
IP:                 ..0. .... = last fragment
IP:         Fragment offset = 0 bytes
IP:         Time to live    = 255 seconds/hops
IP:         Protocol        = 6 (TCP)
IP:         Header checksum = FDFF (correct)
IP:         Source address      = [XXX.YYY.112.90]
IP:         Destination address = [XXX.YYY.113.95]
IP          No options
IP:
```

```
TCP      — TCP header —
TCP:
TCP:     Source port            = 23 (Telnet)
TCP:     Destination port       = 1405
TCP:     Initial sequence number = 2339991040
TCP:     Acknowledgment number  = 1
TCP:     Data offset            = 24 bytes
TCP:     Flags                  = 12
TCP:               ..0. .... = (No urgent pointer)
TCP:               ...1 .... = Acknowledgment
TCP:               .... 0... = (No push)
TCP:               .... .0.. = (No reset)
TCP:               .... ..1. = SYN
TCP:               .... ...0 = (No FIN)
TCP:     Window                 = 64240
TCP:     Checksum               = 79D2 (correct)
TCP:
TCP:     Options follow
TCP:     Maximum segment size   = 1460
TCP:

- - - - - - - - - - - - - - - Frame 9 - - - - - - - - - - - - - - - - -
ATM:     — ATM Cell Header —
ATM:
ATM:     Frame 9 arrived at 14:48:45.92857; frame size is 54 (0036 hex) bytes.
ATM:     Link = DTE
ATM:     Virtual path id = 0
ATM:     Virtual channel id = 34
ATM:
LLC:     — LLC Header —
LLC:
LLC:     DSAP Address = AA, DSAP IG Bit = 00 (Individual Address)
LLC:     SSAP Address = AA, SSAP CR Bit = 00 (Command)
LLC:     Unnumbered frame: UI
LLC:
SNAP:    — SNAP Header —
SNAP:
SNAP:    Type = 0800 (IP)
SNAP:
IP:      — IP Header —
```

```
IP:
IP:        Version = 4, header length = 20 bytes
IP:        Type of service = 00
IP:              000. .... = routine
IP:              ...0 .... = normal delay
IP:              .... 0... = normal throughput
IP:              .... .0.. = normal reliability
IP:        Total length    = 40 bytes
IP:        Identification  = 60068
IP:        Flags           = 0X
IP:              .0.. .... = may fragment
IP:              ..0. .... = last fragment
IP:        Fragment offset = 0 bytes
IP:        Time to live    = 31 seconds/hops
IP:        Protocol        = 6 (TCP)
IP:        Header checksum = 8CE7 (correct)
IP:        Source address      = [XXX.YYY.113.95]
IP:        Destination address = [XXX.YYY.112.90]
IP:        No options
IP:
TCP:       — TCP header —
TCP:
TCP:       Source port            = 1405
TCP:       Destination port       = 23 (Telnet)
TCP:       Sequence number        = 1
TCP:       Acknowledgment number  = 2339991041
TCP:       Data offset            = 20 bytes
TCP:       Flags                  = 18
TCP:             ..0. .... = (No urgent pointer)
TCP:             ...1 .... = Acknowledgment
TCP:             .... 1... = Push
TCP:             .... .0.. = (No reset)
TCP:             .... ..0. = (No SYN)
TCP:             .... ...0 = (No FIN)
TCP:       Window                 = 512
TCP:       Checksum               = 8A78 (correct)
TCP:       No TCP options
TCP:
```

The TELNET session then continues until the end users have completed their business. At that time, the TELNET session is terminated, the TCP connection is also terminated, and the ATM connection is then ready to be taken down. The ATM disconnect sequence is shown in Frames 132 and 135 (Trace 10.5d). Note that in Frame 132, the call reference (0003D8) matches the call reference of the initial connection in Frame 1, and that the Q.2931 message type is a RELEASE. A single IE specifies the cause of the release (normal). Frame 135 confirms the release with a RELEASE COMPLETE message. As with the previous frame, this message is sent over the signaling channel (VPI = 0 and VCI = 5).

In summary, TCP/IP communication over a connection-oriented network such as ATM requires steps for call setup and disconnect that a connectionless network such as an Ethernet does not require.

Trace 10.5d. TELNET over ATM Connection Release

Sniffer Network Analyzer data from 23-Jan at 14:46:34, file JIM02.ATC, Pg 1

```
--------------- Frame 132 -----------------
ATM:        — ATM Cell Header —
ATM:
ATM:        Frame 132 arrived at  14:49:11.93324; frame size is 20 (0014 hex)
bytes.
ATM:        Link = DTE
ATM:        Virtual path id = 0
ATM:        Virtual channel id = 5
ATM:
SSCOP:      — SSCOP trailer —
SSCOP:
SSCOP:      Sequenced Data PDU
SSCOP:      SD send seq num N(S) = 98
SSCOP:
Q2931:      — UNI 3.x Signaling —
Q2931:
Q2931:      Protocol discriminator   = 09
Q2931:      Length of call reference = 3 bytes
Q2931:      Call reference value     = 0003D8
Q2931:      Message type             = 4D (Release)
Q2931:      Message type Flag/Action = 80
Q2931:                  ...0 .... = flag
Q2931:                  .... ..00 = action (Clear call)
Q2931:      Message Length           = 6
Q2931:
Q2931:      Info element id       = 08 (Cause)
Q2931:      Coding Standard/Action = 80
Q2931:      1... .... =        ext
Q2931:      .00. .... =        code stand(ITU-T standardized)
Q2931:      ...0 0000 =        IE field(not significant)
Q2931:      Length of info element = 2 byte(s)
Q2931:      Octet 5
Q2931:      1... .... = ext
```

```
Q2931:     .000 .... = spare
Q2931:     .... 0000 = location(User)
Q2931:     cause value = normal, unspecified
Q2931

- - - - - - - - - - - - - - - Frame 135 - - - - - - - - - - - - - - - - -
ATM:       — ATM Cell Header —
ATM:
ATM:       Frame 135 arrived at 14:49:11.94152; frame size is 20 (0014 hex) bytes.
ATM:       Link = DCE
ATM:       Virtual path id = 0
ATM:       Virtual channel id = 5
ATM:
SSCOP:     — SSCOP trailer —
SSCOP:
SSCOP:     Sequenced Data PDU
SSCOP:     SD send seq num N(S) = 98
SSCOP:
Q2931:     — UNI 3.x Signaling —
Q2931:
Q2931:     Protocol discriminator  = 09
Q2931:     Length of call reference = 3 bytes
Q2931:     Call reference value    = 8003D8
Q2931:     Message type            = 5A (Release complete)
Q2931:     Message type Flag/Action = 80
Q2931:              ...0 .... = flag
Q2931:              .... ..00 = action (Clear call)
Q2931:     Message Length          = 6
Q2931:
Q2931:     Info element id      = 08 (Cause)
Q2931:     Coding Standard/Action = 80
Q2931:     1... .... =      ext
Q2931:     .00. .... =      code stand(ITU-T standardized)
Q2931:     ...0 0000 =      IE field(not significant)
Q2931:     Length of info element = 2 byte(s)
Q2931:     Octet 5
Q2931:     1... .... = ext
Q2931:     .000 .... = spare
Q2931:     .... 0000 = location(User)
Q2931:     cause value = normal, unspecified
Q2931:
```

10.6 SMTP Interoperability Problems

One of the premises of any mail system — be it the postal service, voice messaging, or an electronic text system — is that it must be a duplex, not a simplex, operation. Duplex means that if I send you a message, you should be able to reply to me. Let's see what happens when a system failure prevents this two-way communication.

The network in this case study is a single token ring to which a number of dissimilar workstations are attached (see Figure 10-6). The underlying operating systems are all variants of UNIX running on IBM RS/6000 and Sun SPARCstation II workstations. All of these workstations are implementing an SMTP package and theoretically should be interoperable. Unfortunately, this theory proves incorrect. When the RS/6000 sends a message to the Sun SPARCstation II (SS2), the message is delivered properly. But when the SS2 replies to the message, the delivery fails. The process for each workstation is as follows:

1. The user invokes the workstation's native mail program (Sendmail) to send a message.
2. The Sendmail program invokes SMTP for delivery via the network.
3. The recipient workstation obtains the SMTP message from the network and sends it to its native mail program (also Sendmail in this case).
4. The Sendmail program deposits the message in the recipient user's mailbox.

The process thus involves two significant operations. One — Steps 1 and 4 — is visible to the workstation but invisible to the network, while the other — Steps 2 and 3 — is visible to the network but invisible to the workstation.

In Trace 10.6, we see the portion that is visible to the network and thus available for capture by the network analyzer (note that this trace was filtered to remove non-SMTP frames, so not all of the frames are present in the Trace). The first scenario involves the RS/6000 sending a message to the SS2. The following steps are involved:

1. Establishing the SMTP connection:
   HELO <SP> <source mailbox> <CRLF>
   shown in Frames 535–537

2. Identifying the message originator:
   MAIL <SP> FROM: <source mailbox> <CRLF>
   shown in Frames 538–539
3. Identifying the message recipient:
   RCPT <SP> TO: <receiving mailbox> <CRLF>
   shown in Frames 540–541
4. Transferring the text of the message:
   DATA <CRLF>
   shown in Frames 542–548
5. Closing the SMTP connection:
   QUIT <CRLF>
   shown in Frames 549–550

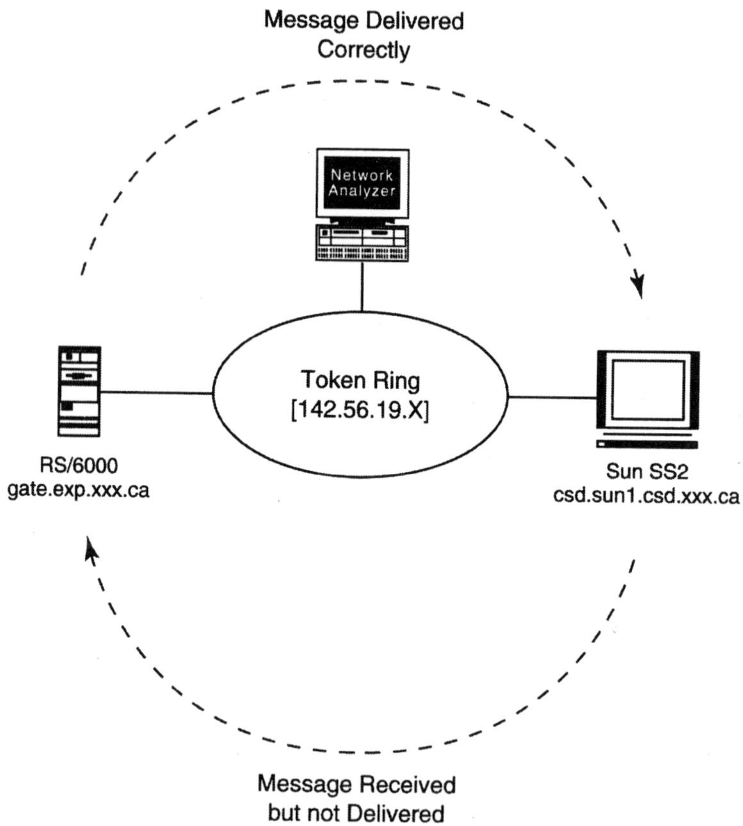

Figure 10-6. Mail Delivery Incompatibilities.

To verify that the message was delivered, the recipient host (SS2) transmits a "delivering mail..." message in Frame 550. The host process (invisible to the network) then takes over, delivering the mail to the recipient's mailbox. The message appears at the SS2, and all is well.

To verify the connectivity, the SS2 then attempts to reply to the message from the RS/6000. Unfortunately, this operation is unsuccessful and the failure is not readily apparent from the network's point of view. Reviewing Trace 10.6, we see a similar mail scenario:

1. Establishing connection: Frames 580–582
2. Identifying originator: Frames 583–594
3. Identifying recipient: Frames 595–597
4. Transferring text: Frames 598–603
5. Closing connection: Frames 604–605

This message was never delivered from the SS2 to the RS/6000, although that isn't readily apparent from the trace file. The clue that a problem existed was found by comparing the acknowledgment messages from the recipients. Recall that the message from the RS/6000 to the SS2 was received and delivered to the end user. The SS2 returned the following message in Frame 548:

```
250 Mail accepted <0D><0A>
```

The message from the SS2 to the RS/6000 was received but not delivered to the end user. The RS/6000 returned the following message in Frame 603:

```
250 Ok <0D><0A>
```

The differences in the two host acknowledgments prompted the network manager, Chris, to check the error log on the RS/6000 host. This is what he found:

```
  - Transcript of session follows -
>>> HELO gate.exp.xxx.ca
<<< 553 Local configuration error, hostname not rec-
ognized as local
554 <cdutchyn@gate.exp.xxx.ca.>... Service unavail-
able: Bad file number
```

```
  - Unsent message follows -
Received: from csd_sun1.csd.xxx.ca by gate.exp.xxx.ca
(AIX 3.1/UCB
5.61/4.03)
      id AA28142; 6 Apr 02 11:30:44 -0600
Received: by csd_sun1.csd.xxx.ca.csd.xxx.ca
      id AA00372; 6 Apr 02 11:37:42 MDT
Date: 16 Apr 02 11:37:42 MDT
From: cdutchyn@csd.xxx.ca (Christopher J. Dutchyn)
Message-Id:
<9204161737.AA00372@csd_sun1.csd.xxx.ca.csd.xxx.ca>
To: cdutchyn@gate.exp.xxx.ca.
Subject: This is a test.
This is a test response.  If it were a real response,
it would contain more information about what you
should know.
Chris
```

The message indicates that the SS2 placed an extra period (.) at the end of the intended recipient's address. Therefore, the recipient was noted as <cdutchyn@gate.exp.xxx.ca.> instead of <cdutchyn@gate.exp.xxx.ca>. When the address was delivered to the RS/6000 (gate.exp.xxx.ca), the RS/6000 was unable to recognize its own address because of the additional period. As a result, the connection failed at the HELO message and the following message was returned:

```
553 Local Configuration error,
     hostname not recognized as local
```

Since this error message occurred in the SMTP process within the RS/6000 host and was never transmitted via the network to the SS2, the message transfer appeared fine from a protocol point of view. Only when the error log of the RS/6000 host was examined did the actual source of the problem (the extraneous period after "ca") surface. Chris then studied the configuration of the SS2 host in greater detail, discovered a configuration error in the name server, and corrected the problem.

Trace 10.6. SMTP Mail Service Summary
Sniffer Network Analyzer data 16-Apr at 10:35:48, file SMTP.TRC, Pg 1

SUMMARY	Delta T	Destination	Source	Summary
535		RS/6000	SUN SS2	SMTP R PORT=2047220 csd_sun1.csd.xxx.ca. csd.xxx.ca
536	0.004	SUN SS2	RS/6000	SMTP C PORT=2047 HELO gate.exp.xxx.ca <0D><0A>
537	0.007	RS/6000	SUN SS2	SMTP R PORT=2047 250 csd_sun1.csd.xxx.ca. csd.xxx.ca Hello gate.exp....
538	0.003	SUN SS2	RS/6000	SMTP C PORT=2047 MAIL From:<cdutchyn @gate.exp.xxx.ca> <0D><0A>
539	0.143	RS/6000	SUN SS2	SMTP R PORT=2047 250 postmaster Sender ok<0D><0A>
540	0.012	SUN SS2	RS/6000	SMTP C PORT=2047 RCPT To:<cdutchyn @csd_sun1.csd.xxx.ca> <0D><0A>
541	0.111	RS/6000	SUN SS2	SMTP R PORT=204 250 <cdutchyn @csd_sun1.csd.xxx.ca> ... Recipient ok...
542	0.004	SUN SS2	RS/6000	SMTP C PORT=2047 DATA<0D><0A>
543	0.055	RS/6000	SUN SS2	SMTP R PORT=2047 354 Enter mail, end with "." on a line by itself<0D>...
545	0.026	SUN SS2	RS/6000	SMTP C PORT=2047 Received: by gate.exp.xxx.ca

547	0.200	SUN SS2	RS/6000	SMTP C PORT=2047 <0D><0A>
548	0.008	RS/6000	SUN SS2	SMTP R PORT=2047 250 Mail accepted <0D><0A>
549	0.004	SUN SS2	RS/6000	SMTP C PORT=2047 QUIT<0D><0A>
550	0.003	RS/6000	SUN SS2	SMTP R PORT=2047 221 csd_sun1.csd.xxx.ca. csd.xxx.ca delivering mail...
580	63.636	SUN SS2	RS/6000	SMTP R PORT=1041 220 gate.exp.xxx.ca
581	0.006	RS/6000	SUN SS2	SMTP C PORT=1041 HELO csd_sun1.csd.xxx.ca. csd.xxx.ca <0D><0A>
582	0.007	SUN SS2	RS/6000	SMTP R PORT=1041 250 gate.exp.xxx.ca Hello csd_sun1.csd.xxx.ca. csd....
583	0.004	RS/6000	SUN SS2	SMTP C PORT=1041 MAIL From:<cdutchyn @csd.xxx.ca> <0D><0A>
594	1.229	SUN SS2	RS/6000	SMTP R PORT=1041 250 <cdutchyn @csd.xxx.ca>... Sender ok<0D><0A>
595	0.006	RS/6000	SUN SS2	SMTP C PORT=1041 RCPT To:<cdutchyn @gate.exp.xxx.ca.> <0D><0A>
597	1.029	SUN SS2	RS/6000	SMTP R PORT=1041 250 <cdutchyn @gate.exp.xxx.ca.> Recipient ok<0D>...
598	0.004	RS/6000	SUN SS2	SMTP C PORT=1041 DATA<0D><0A>

599	0.018	SUN SS2	RS/6000	SMTP R PORT=1041 354 Enter mail, end with "." on a line by itself<0D>...
600	0.016	RS/6000	SUN SS2	SMTP C PORT=104 Received: by csd_sun1.csd.xxx.ca. csd.xxx.ca.
602	0.035	RS/6000	SUN SS2	SMTP C PORT=1041 <0D><0A>
603	0.117	SUN SS2	RS/6000	SMTP R PORT=1041 250 Ok<0D><0A>
604	0.006	RS/6000	SUN SS2	SMTP C PORT=1041 QUIT<0D><0A>
605	0.022	SUN SS2	RS/6000	SMTP R PORT=1041 221 gate.exp.xxx.ca closing connection <0D><0A>

10.7 NetBIOS and TCP Interactions

The following example of NetBIOS service illustrates how all of the ARPA layers must cooperate in order to ensure proper protocol operation. The objective in this case study is for one workstation (shown as Art in Trace 10.7a) to use the resources on another workstation, known as Robert (Figure 10-7). The trace begins with Art broadcasting a NetBIOS Name Service (DNS) Query looking for Robert. Details of the query in Frame 1 of Trace 10.7b show a Source port of 275 assigned to NetBIOS service within the UDP header. The query consists of a single question providing the name, type, and class of the object. The response returned in Frame 2 provides necessary details about Robert. These details include the uniqueness of the name and that Robert is a B-type (broadcast) node with an IP address of [XXX.YYY.200.85]. Now that Art knows more about Robert, he can use Robert's resources.

Frames 3–5 show a TCP connection initiated by Art and accepted by Robert. The NetBIOS Session Service begins in Frame 6 and is shown in detail in Trace 10.7c. Art begins with a NetBIOS Session Request message (Type = 81) having a length of 74 octets. The data consists of the called and calling names, ROBERT and ARTL, respectively. Robert then responds with a Positive Response (or Session Confirm) message, Type = 82. The logical session is now established.

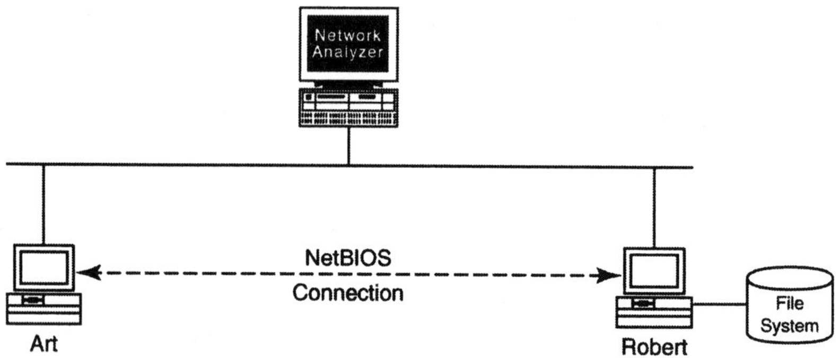

Figure 10-7. Workstation Communication Using NetBIOS.

Art must now negotiate the language he will use to communicate with Robert. To do so, he uses the Server Message Block (SMB) protocol. SMB was developed by Microsoft Corp. to enable a DOS-based client workstation to communicate with its server. SMB information is carried as data inside a NetBIOS message. Speaking in OSI terms, NetBIOS is providing the Session Layer (i.e., logical) connection while SMB information defines the format of the data to be transmitted and thus provides Presentation Layer functions. We see the protocol being negotiated in Frames 9 and 11.

The connection that Art is really after — access to Robert's disk drive — is shown in Frames 13 and 15 (see Trace 10.7d). This process is completed in Frame 19. Art may then search the specified directory on Robert's disk (Frame 22), finding one file in Frame 24. A subsequent search (Frame 26) returns ten additional filenames (Frame 28). No more files are noted in Frame 33, prompting Art to his next task: finding the disk attributes (Frames 35 and 38). Finished with his work, Art issues an SMB Disconnection message (Frame 40), which is acknowledged in Frame 42. The TCP connection is then torn down, using the FIN messages from both Art and Robert in Frames 44 and 45, respectively.

Notice the pattern of protocol operation: TCP establishes the end-to-end connection before NetBIOS and SMB undertake the logical session. When there is no longer a need for communication, the higher layer protocol (SMB) terminates the logical connection and TCP then terminates the end-to-end connection.

Trace 10.7a. NetBIOS over UDP and TCP (Summary)
Sniffer Network Analyzer data 6-Nov at 11:04:32, TCPNETB.ENC, Pg 1

SUMMARY	Delta T	Destination	Source	Summary
M 1		Broadcast	Art	DNS C ID=41128OP= QUERY NAME=ROBERT
2	0.0018	Art	Robert	DNS R ID=41128 STAT=OK
3	0.0243	Robert	Art	TCP D=139 S=257 SYN SEQ=172249 LEN=0 WIN=1024
4	0.0028	Art	Robert	TCP D=257 S=139 SYN ACK=172250 SEQ=257461 LEN=0 WIN=2152
5	0.0160	Robert	Art	TCP D=139 S=257 ACK=257462 WIN=1024
6	0.0046	Robert	Art	NETB D=ROBERT S=ARTL<00> Session request
7	0.0041	Art	Robert	NETB Session confirm
8	0.0204	Robert	Art	TCP D=139 S=257 ACK=257466 WIN=1020
9	0.0106	Robert	Art	SMB C PC NET WORK PROGRAM 1.0
10	0.0043	Art	Robert	TCP D=257 S=139 ACK=172391 WIN=2011
11	0.0036	Art	Robert	SMB R Negotiated Protocol 0
12	0.0191	Robert	Art	TCP D=139 S=257 ACK=257507 WIN=979
13	0.0096	Robert	Art	SMB C Connect A:\\ROBERT\DISK

14	0.0048	Art	Robert	TCP D=257 S=139 ACK=172451 WIN=1951
15	0.0077	Art	Robert	SMB R T=8F15 Connected
16	0.0138	Robert	Art	TCP D=139 S=257 ACK=257550 WIN=936
17	0.0347	Robert	Art	SMB C End of Process
18	0.0057	Art	Robert	TCP D=257 S=139 ACK=172490 WIN=1912
19	0.0057	Art	Robert	SMB R OK
20	0.0163	Robert	Art	TCP D=139 S=257 ACK=257589 WIN=897
21	0.1437	Art	Robert	TCP D=257 S=139 ACK=172490 WIN=2152
22	19.4487	Robert	Art	SMB C Search \????????.???
23	0.0054	Art	Robert	TCP D=257 S=139 ACK=172551 WIN=2091
24	0.0157	Art	Robert	SMB R 1 entry found (done)
25	0.0157	Robert	Art	TCP D=139 S=257 ACK=257676 WIN=810
26	0.0676	Robert	Art	SMB C Search \????????.???
27	0.0054	Art	Robert	TCP D=257 S=139 ACK=172612 WIN=2030
28	0.0276	Art	Robert	SMB R 10 entries found (done)
29	0.0188	Robert	Art	TCP D=139 S=257 ACK=258150 WIN=1024
30	0.1490	Art	Robert	TCP D=257 S=139 ACK=172612 WIN=2152

31	0.0852	Robert	Art	SMB C Continue Search
32	0.0052	Art	Robert	TCP D=257 S=139 ACK=172681 WIN=2083
33	0.0098	Art	Robert	SMB R No more files
34	0.0135	Robert	Art	TCP D=139 S=257 ACK=258194 WIN=980
35	0.0127	Robert	Art	SMB C Get Disk Attributes
36	0.0051	Art	Robert	TCP D=257 S=139 ACK=172720 WIN=2044
37	0.2589	Art	Robert	TCP D=257 S=139 ACK=172720 WIN=2152
38	0.9522	Art	Robert	SMB R Got Disk Attributes
39	0.0144	Robert	Art	TCP D=139 S=257 ACK=258243 WIN=931
40	4.5501	Robert	Art	SMB C T=8F15 Disconnect
41	0.0059	Art	Robert	TCP D=257 S=139 ACK=172759 WIN=2113
42	0.0057	Art	Robert	SMB R OK
43	0.0173	Robert	Art	TCP D=139 S=257 ACK=258282 WIN=892
44	0.0146	Robert	Art	TCP D=139 S=257 FIN ACK=258282 SEQ=172759 LEN=0 WIN=892
45	0.0029	Art	Robert	TCP D=257 S=139 FIN ACK=172760 SEQ=258282 LEN=0 WIN=2113
46	0.0116	Robert	Art	TCP D=139 S=257 ACK=258283 WIN=892

47	2.5477	Robert	Art	TCP D=139 S=257 ACK=258283 WIN=892
48	0.0018	Art	Robert	TCP D=257 S=139 RST WIN=892

Trace 10.7b. NetBIOS Name Service Query/Response
Sniffer Network Analyzer data 6-Nov at 11:04:32, TCPNETB.ENC, Pg 1

```
DLC:     — DLC Header —
DLC:
DLC:     Frame 1 arrived at 11:04:51.3527; frame size is 92 (005C hex) bytes.
DLC:     Destination = BROADCAST FFFFFFFFFFFF, Broadcast
DLC:     Source      = Station Bridge011084, Art
DLC:     Ethertype  = 0800 (IP)
DLC:
IP:      — IP Header —
IP:
IP:      Version = 4, header length = 20 bytes
IP:      Type of service = 00
IP:                000. .... = routine
IP:                ...0 .... = normal delay
IP:                .... 0... = normal throughput
IP:                .... .0.. = normal reliability
IP:      Total length = 78 bytes
IP:      Identification = 19
IP:      Flags = 0X
IP:      .0.. .... = may fragment
IP:      ..0. .... = last fragment
IP:      Fragment offset = 0 bytes
IP:      Time to live = 30 seconds/hops
IP:      Protocol = 17 (UDP)
IP:      Header checksum = 1420 (correct)
IP:      Source address = [XXX.YYY.200.99]
IP:      Destination address = [255.255.255.255]
IP:      No options
IP:
UDP:     — UDP Header —
UDP:
UDP:     Source port = 275 (NetBIOS)
UDP:     Destination port = 137
UDP:     Length = 58
UDP:     Checksum = 8153 (correct)
```

```
UDP:
DNS:        — Internet Domain Name Service header —
DNS:
DNS:        ID = 41128
DNS:        Flags = 01
DNS:        0... .... = Command
DNS:        .000 0... = Query
DNS:        .... ..0. = Not truncated
DNS:        .... ...1 = Recursion desired
DNS:        Flags = 1X
DNS:        ...1 .... = Broadcast packet
DNS:        Question count = 1, Answer count = 0
DNS:        Authority count = 0, Additional record count = 0
DNS:
DNS:        Question section:
DNS:                Name = ROBERT
DNS:                Type = NetBIOS name service (NetBIOS name,32)
DNS:                Class = Internet (IN,1)
DNS:
DNS:        [Normal end of "Internet Domain Name Service header".]
DNS:

- - - - - - - - - - - - - - - Frame 2 - - - - - - - - - - - - - - - -
DLC:        — DLC Header —
DLC:
DLC:        Frame 2 arrived at  11:04:51.3545; frame size is 104 (0068 hex) bytes.
DLC:        Destination = Station Bridge011084, Art
DLC:        Source      = Station U-B   F5D800, Robert
DLC:        Ethertype  = 0800 (IP)
DLC:
IP:         — IP Header —
IP:
IP:         Version = 4, header length = 20 bytes
IP:         Type of service = 00
IP:                 000. .... = routine
IP:                 ...0 .... = normal delay
IP:                 .... 0... = normal throughput
IP:                 .... .0.. = normal reliability
IP:         Total length = 90 bytes
IP:         Identification = 18756
IP:         Flags = 0X
IP:         .0.. .... = may fragment
IP:         ..0. .... = last fragment
IP:         Fragment offset = 0 bytes
```

```
IP:     Time to live = 60 seconds/hops
IP:     Protocol = 17 (UDP)
IP:     Header checksum = 2483 (correct)
IP:     Source address = [XXX.YYY.200.85]
IP:     Destination address = [XXX.YYY.200.99]
IP:     No options
IP:
UDP:    — UDP Header —
UDP:
UDP:    Source port = 137 (NetBIOS)
UDP:    Destination port = 275
UDP:    Length = 70
UDP:    Checksum = EC85 (correct)
UDP:
DNS:    — Internet Domain Name Service header —
DNS:
DNS:    ID = 41128
DNS:    Flags = 85
DNS:    1... .... = Response
DNS:    .... .1.. = Authoritative answer
DNS:    .000 0... = Query
DNS:    .... ..0. = Not truncated
DNS:    Flags = 0X
DNS:    ...0 .... = Unicast packet
DNS:    0... .... = Recursion not available
DNS:    Response code = OK (0)
DNS:    Question count = 0, Answer count = 1
DNS:    Authority count = 0, Additional record count = 0
DNS:    Answer section:
DNS:            Name = ROBERT
DNS:            Type = NetBIOS name service (NetBIOS name,32)
DNS:            Class = Internet (IN,1)
DNS:            Time-to-live = 0 (seconds)
DNS:    Node flags = 00
DNS:    0... .... = Unique NetBIOS name
DNS:    .00. .... = B-type node
DNS:    Node address = [XXX.YYY.200.85]
DNS:
DNS:    [Normal end of "Internet Domain Name Service header".]
DNS:
```

Trace 10.7c. NetBIOS Session Request/Confirm

Sniffer Network Analyzer data 6-Nov at 11:04:32, TCPNETB.ENC, Pg 1

```
- - - - - - - - - - - - - - - Frame 6 - - - - - - - - - - - - - - - - -
DLC:     — DLC Header —
DLC:
DLC:     Frame 6 arrived at 11:04:51.4023; frame size is 132 (0084 hex) bytes.
DLC:     Destination = Station U-B   F5D800, Robert
DLC:     Source      = Station Bridge011084, Art
DLC:     Ethertype  = 0800 (IP)
DLC:
IP:      — IP Header —
IP:
IP:      Version = 4, header length = 20 bytes
IP:      Type of service = 00
IP:              000. .... = routine
IP:              ...0 .... = normal delay
IP:              .... 0... = normal throughput
IP:              .... .0.. = normal reliability
IP:      Total length = 118 bytes
IP:      Identification = 281
IP:      Flags = 0X
IP:      .0.. .... = may fragment
IP:      ..0. .... = last fragment
IP:      Fragment offset = 0 bytes
IP:      Time to live = 30 seconds/hops
IP:      Protocol = 6 (TCP)
IP:      Header checksum = 8A9D (correct)
IP:      Source address = [XXX.YYY.200.99]
IP:      Destination address = [XXX.YYY.200.85]
IP:      No options
IP:
TCP:     — TCP header —
TCP:
TCP:     Source port = 257
TCP:     Destination port = 139
TCP:     Sequence number = 172250
TCP:     Acknowledgment number = 257462
TCP:     Data offset = 20 bytes
TCP:     Flags = 10
TCP:     ..0. .... = (No urgent pointer)
TCP:     ...1 .... = Acknowledgment
TCP:     .... 0... = (No push)
TCP:     .... .0.. = (No reset)
```

```
TCP:        .... ..0. = (No SYN)
TCP:        .... ...0 = (No FIN)
TCP:        Window = 1024
TCP:        Checksum = EFCD (correct)
TCP:        No TCP options
TCP:        [78 byte(s) of data]
TCP:
NETB:       — NetBIOS Session protocol —
NETB:
NETB:       Type = 81 (Session request)
NETB:       Flags = 00
NETB:       Total session packet length = 74
NETB:       Called NetBIOS name = ROBERT
NETB:       Calling NetBIOS name = ARTL<00>
NETB:

--------------- Frame 7 -----------------
DLC:        — DLC Header —
DLC:
DLC:        Frame 7 arrived at 11:04:51.4065; frame size is 60 (003C hex) bytes.
DLC:        Destination = Station Bridge011084, Art
DLC:        Source      = Station U-B   F5D800, Robert
DLC:        Ethertype  = 0800 (IP)
DLC:
IP:         — IP Header —
IP:
IP:         Version = 4, header length = 20 bytes
IP:         Type of service = 00
IP:                   000. .... = routine
IP:                   ...0 .... = normal delay
IP:                   .... 0... = normal throughput
IP:                   .... .0.. = normal reliability
IP:         Total length = 44 bytes
IP:         Identification = 2
IP:         Flags = 0X
IP:         .0.. .... = may fragment
IP:         ..0. .... = last fragment
IP:         Fragment offset = 0 bytes
IP:         Time to live = 60 seconds/hops
IP:         Protocol = 6 (TCP)
IP:         Header checksum = 6DFE (correct)
IP:         Source address = [XXX.YYY.200.85]
IP:         Destination address = [XXX.YYY.200.99]
IP:         No options
```

```
IP:
TCP:        — TCP header —
TCP:
TCP:        Source port = 139
TCP:        Destination port = 257
TCP:        Sequence number = 257462
TCP:        Acknowledgment number = 172328
TCP:        Data offset = 20 bytes
TCP:        Flags = 18
TCP:        ..0. .... = (No urgent pointer)
TCP:        ...1 .... = Acknowledgment
TCP:        .... 1... = Push
TCP:        .... .0.. = (No reset)
TCP:        .... ..0. = (No SYN)
TCP:        .... ...0 = (No FIN)
TCP:        Window = 2074
TCP:        Checksum = 8471 (correct)
TCP:        No TCP options
TCP:        [4 byte(s) of data]
TCP:
NETB:       — NetBIOS Session protocol —
NETB:
NETB:       Type = 82 (Positive response)
NETB:       Flags = 00
NETB:       Total session packet length = 0
NETB:
```

Trace 10.7d. NetBIOS with Server Message Block Information
Sniffer Network Analyzer data 6-Nov at 11:04:32, TCPNETB.ENC, Pg 1

```
- - - - - - - - - - - - - - - Frame 13 - - - - - - - - - - - - - - - - -
DLC:        — DLC Header —
DLC:
DLC:        Frame 13 arrived at 11:04:51.4744; frame size is 114 (0072 hex) bytes.
DLC:        Destination = Station U-B   F5D800, Robert
DLC:        Source      = Station Bridge011084, Art
DLC:        Ethertype  = 0800 (IP)
DLC:
IP:         — IP Header —
IP:
IP:         Version = 4, header length = 20 bytes
IP:         Type of service = 00
IP:                   000. .... = routine
IP:                   ...0 .... = normal delay
```

```
IP:                  .... 0... = normal throughput
IP:                  .... .0.. = normal reliability
IP:        Total length = 100 bytes
IP:        Identification = 285
IP:        Flags = 0X
IP:        .0.. .... = may fragment
IP:        ..0. .... = last fragment
IP:        Fragment offset = 0 bytes
IP:        Time to live = 30 seconds/hops
IP:        Protocol = 6 (TCP)
IP:        Header checksum = 8AAB (correct)
IP:        Source address = [XXX.YYY.200.99]
IP:        Destination address = [XXX.YYY.200.85]
IP:        No options
IP:
TCP:       — TCP header —
TCP:
TCP:       Source port = 257
TCP:       Destination port = 139
TCP:       Sequence number = 172391
TCP:       Acknowledgment number = 257507
TCP:       Data offset = 20 bytes
TCP:       Flags = 10
TCP:       ..0. .... = (No urgent pointer)
TCP:       ...1 .... = Acknowledgment
TCP:       .... 0... = (No push)
TCP:       .... .0.. = (No reset)
TCP:       .... ..0. = (No SYN)
TCP:       .... ...0 = (No FIN)
TCP:       Window = 979
TCP:       Checksum = D716 (correct)
TCP:       No TCP options
TCP:       [60 byte(s) of data]
TCP:
NETB:      — NetBIOS Session protocol —
NETB:
NETB:      Type = 00 (Session data)
NETB:      Flags = 00
NETB:      Total session packet length = 56
NETB:
SMB:       — SMB Tree Connect Command —
SMB:
SMB:       Function = 70 (Tree Connect)
SMB:       Tree id     (TID) = 0000
```

```
SMB:        Process id   (PID) = 0000
SMB:        File pathname = "\\ROBERT\DISK"
SMB:        Password = ""
SMB:        Device name = "A:"
SMB:

- - - - - - - - - - - - - - - Frame 15 - - - - - - - - - - - - - - - - -
DLC:        — DLC Header —
DLC:
DLC:        Frame 15 arrived at  11:04:51.4870; frame size is 97 (0061 hex) bytes.
DLC:        Destination = Station Bridge011084, Art
DLC:        Source      = Station U-B   F5D800, Robert
DLC:        Ethertype  = 0800 (IP)
DLC:
IP:         — IP Header —
IP:
IP:         Version = 4, header length = 20 bytes
IP:         Type of service = 00
IP:                  000. .... = routine
IP:                  ...0 .... = normal delay
IP:                  .... 0... = normal throughput
IP:                  .... .0.. = normal reliability
IP:         Total length = 83 bytes
IP:         Identification = 4
IP:         Flags = 0X
IP:         .0.. .... = may fragment
IP:         ..0. .... = last fragment
IP:         Fragment offset = 0 bytes
IP:         Time to live = 60 seconds/hops
IP:         Protocol = 6 (TCP)
IP:         Header checksum = 6DD5 (correct)
IP:         Source address = [XXX.YYY.200.85]
IP:         Destination address = [XXX.YYY.200.99]
IP:         No options
IP:
TCP:        — TCP header —
TCP:
TCP:        Source port = 139
TCP:        Destination port = 257
TCP:        Sequence number = 257507
TCP:        Acknowledgment number = 172451
TCP:        Data offset = 20 bytesTCP:Flags = 10
TCP:        ..0. .... = (No urgent pointer)
TCP:        ...1 .... = Acknowledgment
```

```
TCP:        .... 0... = (No push)TCP:..... .0.. = (No reset)
TCP:        .... ..0. = (No SYN)TCP:..... ...0 = (No FIN)
TCP:        Window = 1951
TCP:        Checksum = 97A2 (correct)
TCP:        No TCP options
TCP:        [43 byte(s) of data]
TCP:
NETB:       — NetBIOS Session protocol —
NETB:
NETB:       Type = 00 (Session data)
NETB:       Flags = 00
NETB:       Total session packet length = 39
NETB:
SMB:        — SMB Tree Connect Response —
SMB:
SMB:        Function = 70 (Tree Connect)
SMB:        Tree id     (TID) = 0000
SMB:        Process id  (PID) = 0000
SMB:        Return code = 0,0 (OK)
SMB:        Maximum transmit size = 8240
SMB:        TID = 8F15
SMB:
```

10.8 Web Page Access Using HTTP

In this case study we will examine how HTTP is used to access a web page. For this example, we will access our company web server, *www.diginet.com*, from a remote workstation via the Internet (Figure 10-8). We will first access the home page, then access a listing of the networking tutorials, and, finally, access information about the *Troubleshooting TCP/IP* tutorial.

A summary of the Remote Client's activities is shown in Trace 10.8a. There are several significant events that occur in this trace. First, the Remote Client establishes TCP connections with the www.diginet.com web server. Note that the Client opens multiple logical connections, using different Source Port numbers, over the same physical link. All of the Destination Port numbers are Port = 80, the HTTP port. These connections are:

Frames	Port Number
9–11	34438
17–19	34439
21–23	34440

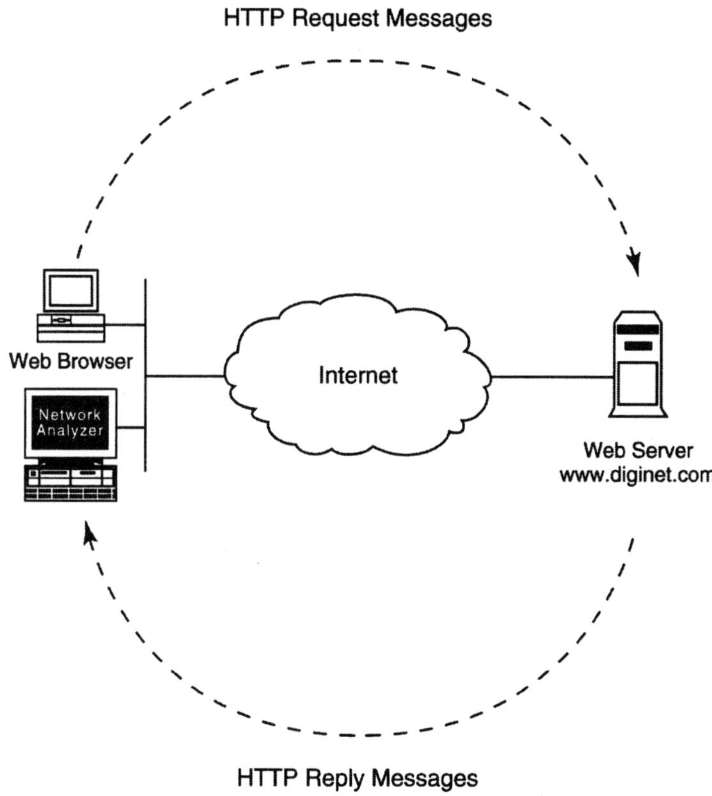

Figure 10-8. Web Page Access Using HTTP.

(Note that this trace has been filtered to show only the most relevant information, and that, even with filtering, some additional communication, which does not pertain to this case, is included.)

Also throughout this communication session, the Client is sending Request messages to the www.diginet.com web server and receiving Response messages in return. These messages are:

Frames	Port Number	File Requested	Information Requested
12–13	34438	*www.diginet.com*	Home Page
95–96	34440	*tutorial.html*	Tutorial page
121–122	34438	*troubletcp.html*	*Troubleshooting TCP/IP* tutorial outline

The responses from the web server include HTML and graphics files (such as Skidsblue.gif in Frame 24, Homepage1.gif in Frame 25, tutorial.html in Frame 95, and image54.gif in Frame 98) that eventually appear on the Client's workstation. After the requested information has been obtained, the user closes the HTTP application, which in turn initiates a closing of the TCP logical connections, signaled by the TCP FIN command:

Frame	Port Number
238	34440
240	34439
263	34438

Trace 10.8a. DigiNet Corporation Web Site Access Summary
Sniffer Network Analyzer data from 30-Apr at 09:58:22, file DNC.ENC, Pg 1

SUMMARY	Delta T	Destination	Source	Summary
9	0.0088	www.diginet.com	Remote Client	TCP D=80 S=34438 SYN SEQ=3089805603 LEN=0 WIN=8760
10	0.0083	Remote Client	www.diginet.com	TCP D=34438 S=80 SYN ACK=3089805604 SEQ=1037380740 LEN=0 WIN=61320
11	0.0004	www.diginet.com	Remote Client	TCP D=80 S=34438 ACK=1037380741 WIN=8760
12	0.0018	www.diginet.com	Remote Client	HTTP C Port=34438 GET / HTTP/1.0
13	0.0311	Remote Client	www.diginet.com	HTTP R Port=34438 OK
14	0.0020	www.diginet.com	Remote Client	TCP D=80 S=34438 ACK=1037381875 WIN=8760
15	0.0156	Remote Client	[XXX.YYY.73.73]	TCP D=34437 S=80 ACK=3074884629 WIN=8760
16	0.2158	[XXX.YYY.73.73]	Remote Client	TCP D=80 S=34435 FIN ACK=587129866 SEQ=3074501490 LEN=0 WIN=8760

17	0.0062	www.diginet.com	Remote Client	TCP D=80 S=34439 SYN SEQ=3089845435 LEN=0 WIN=8760
18	0.0034	Remote Client	www.diginet.com	TCP D=34439 S=80 SYN ACK=3089845436 SEQ=1037450141 LEN=0 WIN=61320
19	0.0004	www.diginet.com	Remote Client	TCP D=80 S=34439 ACK=1037450142 WIN=8760
20	0.0028	[XXX.YYY.73.73]	Remote Client	TCP D=80 S=34434 FIN ACK=587055032 SEQ=3074419387 LEN=0 WIN=8760
21	0.0014	www.diginet.com	Remote Client	TCP D=80 S=34440 SYN SEQ=3089875978 LEN=0 WIN=8760
22	0.0049	Remote Client	www.diginet.com	TCP D=34440 S=80 SYN ACK=3089875979 SEQ=1037577189 LEN=0 WIN=61320
23	0.0005	www.diginet.com	Remote Client	TCP D=80 S=34440 ACK=1037577190 WIN=8760
24	0.0397	www.diginet.com	Remote Client	HTTP C Port=34439 GET /Skidsblue.gif HTTP/1.0
25	0.0067	www.diginet.com	Remote Client	HTTP C Port=34440 GET /Homepage1.gif HTTP/1.0
26	0.0115	Remote Client	[XXX.YYY.73.73]	TCP D=34435 S=80 ACK=3074501491 WIN=8760

27	0.0067	Remote Client	www.diginet.com	HTTP R Port=34439 OK
28	0.0012	Remote Client	www.diginet.com	HTTP R Port=34440 OK
32	0.0091	Remote Client	www.diginet.com	HTTP R Port=34440 Graphics Data
33	0.0072	Remote Client	www.diginet.com	HTTP R Port=34439 Graphics Data
34	0.0096	Remote Client	www.diginet.com	HTTP R Port=34439 Graphics Data
35	0.0112	www.diginet.com	Remote Client	TCP D=80 S=34439 ACK=1037453062 WIN=8760
36	0.0099	www.diginet.com	Remote Client	TCP D=80 S=34440 ACK=1037578650 WIN=8760
37	0.0001	www.diginet.com	Remote Client	TCP D=80 S=34440 ACK=1037580110 WIN=8760
38	0.0071	Remote Client	www.diginet.com	HTTP R Port=34439 Graphics Data
39	0.0123	Remote Client	www.diginet.com	HTTP R Port=34440 Graphics Data
.				
.				
.				
91	0.0083	Remote Client	www.diginet.com	HTTP R Port=34440 Graphics Data
92	0.0046	Remote Client	www.diginet.com	HTTP R Port=34440 Graphics Data
93	0.0172	www.diginet.com	Remote Client	TCP D=80 S=34440 ACK=1037628290 WIN=8760
94	0.0714	www.diginet.com	Remote Client	TCP D=80 S=34440 ACK=1037630466 WIN=8760
95	2.7206	www.diginet.com	Remote Client	HTTP C Port=34440 GET /tutorial.html HTTP/1.0
96	0.0208	Remote Client	www.diginet.com	HTTP R Port=34440 OK
97	0.0485	www.diginet.com	Remote Client	TCP D=80 S=34440 ACK=1037631432 WIN=8760

98	0.1167	www.diginet.com	Remote Client	HTTP C Port=34439 GET /image54.gif HTTP/1.0
99	0.0267	Remote Client	www.diginet.com	HTTP R Port=34439 OK
100	0.0097	Remote Client	www.diginet.com	HTTP R Port=34439 Graphics Data
101	0.0067	Remote Client	www.diginet.com	HTTP R Port=34439 Graphics Data
102	0.0097	Remote Client	www.diginet.com	HTTP R Port=34439 Graphics Data
103	0.0050	www.diginet.com	Remote Client	TCP D=80 S=34439 ACK=1037458151 WIN=8760
104	0.0046	Remote Client	www.diginet.com	HTTP R Port=34439 Graphics Data
105	0.0088	Remote Client	www.diginet.com	HTTP R Port=34439 Graphics Data
106	0.0097	Remote Client	www.diginet.com	HTTP R Port=34439 Graphics Data
107	0.0245	www.diginet.com	Remote Client	TCP D=80 S=34439 ACK=1037462247 WIN=8760
108	0.0001	www.diginet.com	Remote Client	TCP D=80 S=34439 ACK=1037466623 WIN=8760
109	0.0208	Remote Client	www.diginet.com	HTTP R Port=34439 Graphics Data
114	0.0085	Remote Client	www.diginet.com	HTTP R Port=34439Graphics Data
115	0.0014	www.diginet.com	Remote Client	TCP D=80 S=34439 ACK=1037469543 WIN=8760
116	0.0001	www.diginet.com	Remote Client	TCP D=80 S=34439 ACK=1037472463 WIN=8760
117	0.0001	www.diginet.com	Remote Client	TCP D=80 S=34439 ACK=1037475383 WIN=8760
118	0.0231	Remote Client	www.diginet.com	HTTP R Port=34439 Graphics Data
119	0.0066	Remote Client	www.diginet.com	HTTP R Port=34439 Graphics Data

120	0.0900	www.diginet.com	Remote Client	TCP D=80 S=34439 ACK=1037477863 WIN=6280
121	2.0799	www.diginet.com	Remote Client	HTTP C Port=34438 GET /troubletcp.html HTTP/1.0
122	0.0357	Remote Client	www.diginet.com	HTTP R Port=34438 OK
123	0.0445	www.diginet.com	Remote Client	TCP D=80 S=34438 ACK=1037382456 WIN=8760
124	0.1261	www.diginet.com	Remote Client	HTTP C Port=34440 GET /image9.gif HTTP/1.0
125	0.0405	Remote Client	www.diginet.com	HTTP R Port=34440 OK
126	0.0101	Remote Client	www.diginet.com	HTTP R Port=34440 Graphics Data
127	0.0069	Remote Client	www.diginet.com	HTTP R Port=34440 Graphics Data
128	0.0012	www.diginet.com	Remote Client	TCP D=80 S=34440 ACK=1037632892 WIN=8760
129	0.0102	Remote Client	www.diginet.com	HTTP R Port=34440 Graphics Data
130	0.0082	Remote Client	www.diginet.com	HTTP R Port=34440 Graphics Data
131	0.0095	Remote Client	www.diginet.com	HTTP R Port=34440 Graphics Data
.				
.				
.				
232	7.1343	Remote Client	www.diginet.com	TCP D=34438 S=80 FIN ACK=3089806269 SEQ=1037407993 LEN=0 WIN=61320
233	0.0004	www.diginet.com	Remote Client	TCP D=80 S=34438 ACK=1037407994 WIN=8760

234	3.0438	Remote Client	[XXX.YYY.1.4]	DNS C ID=65073 OP=QUERY NAME= 117.37.172.140 .in-addr.arpa
235	0.0022	[XXX.YYY.1.4]	Remote Client	DNS R ID=65073 STAT=OK NAME= 117.37.172.140 in-addr.arpa
236	0.9173	Remote Client	[XXX.YYY.30.2]	NTP/SNTP Version 3
237	0.0012	[XXX.YYY.30.2]	Remote Client	NTP/SNTP Version 3
238	1.8867	Remote Client	www.diginet.com	TCP D=34440 S=80 FIN ACK=3089876905 SEQ=1037659772 LEN=0 WIN=61320
239	0.0004	www.diginet.com	Remote Client	TCP D=80 S=34440 ACK=1037659773 WIN=8760
240	0.1867	Remote Client	www.diginet.com	TCP D=34439 S=80 FIN ACK=3089846382 SEQ=1037500743 LEN=0 WIN=61320
241	0.0001	www.diginet.com	Remote Client	TCP D=80 S=34439 ACK=1037500744 WIN=8760
242	0.2287	Remote Client	[XXX.YYY.35.1..	NTP/SNTP Version 3
243	0.0011	[XXX.YYY.35.1..	Remote Client	NTP/SNTP Version 3
244	0.9355	[XXX.YYY.105...	Remote Client	TCP D=1090 S=80 FIN ACK=13563200 SEQ=3092005668 LEN=0 WIN=8760
245	0.0925	Remote Client	[XXX.YYY.105...	TCP D=80 S=1090 ACK=3092005669 WIN=7878
246	3.5564	[XXX.YYY.0.1]	Remote Client	IGMP Version 1, Un known

```
247   0.3732  [XXX.YYY.73.73]  Remote Client    TCP D=80 S=34436
                                                 FIN
                                                 ACK=587147383
                                                 SEQ=3074601632
                                                 LEN=0
                                                 WIN=8760
248   0.0644  Remote Client    [XXX.YYY.73.73]  TCP D=34436 S=80
                                                 ACK=3074601633
                                                 WIN=8760
249   0.0326  [XXX.YYY.130.17] Remote Client    TCP D=80 S=34441
                                                 SYN
                                                 SEQ=3094428755
                                                 LEN=0
                                                 WIN=8760
250   0.0688  Remote Client    [XXX.YYY.130.17] TCP D=34441 S=80
                                                 SYN
                                                 ACK=3094428756
                                                 SEQ=27046812
                                                 LEN=0
                                                 WIN=8760
```

Looking at the details of the home page request (Trace 10.8b), note that the HTTP Request message specifies the type of request (GET), protocol version (HTTP/1.0), host name (www.diginet.com), and the acceptable types of the response (image/gif, etc.). Note that Line 6 in Frame 12 is the blank line (CRLF), and that there is no message body in the Response.

In the Reply message given in Frame 13, note the status line, which contains the protocol (HTTP/1.1) and numeric status code (200 OK), plus various headers that indicate information about the server, file, date, and so on. Line 11 in Frame 13 is the blank line (CRLF); this is followed by the message body, which contains the HTML file describing the DigiNet Corporation home page, other hyperlinks, and so on.

Trace 10.8b. HTTP DigiNet Corporation Home Page Details
Sniffer Network Analyzer data from 30-Apr at 09:58:22, file DNC.ENC, Pg 1

```
- - - - - - - - - - - - - - - - Frame 12 - - - - - - - - - - - - - - - - -
HTTP:     — Hypertext Transfer Protocol —
HTTP:
HTTP:     Line  1:  GET / HTTP/1.0
HTTP:     Line  2:  Connection: Keep-Alive
HTTP:     Line  3:  User-Agent: Mozilla/3.01 (X11; I; SunOS 5.6 sun4m)
```

```
HTTP:      Line 4: Host: www.diginet.com
HTTP:      Line 5: Accept: image/gif, image/x-xbitmap, image/jpeg, image/pjpeg,
HTTP:     */*
HTTP:      Line 6:
HTTP:

- - - - - - - - - - - - - - - - Frame 13 - - - - - - - - - - - - - - - - -
HTTP:      — Hypertext Transfer Protocol —
HTTP:
HTTP:      Line 1: HTTP/1.1 200 OK
HTTP:      Line 2: Date: 30 Apr 17:02:15 GMT
HTTP:      Line 3: Server: Apache/1.2.4
HTTP:      Line 4: Last-Modified: 03 Dec 22:35:23 GMT
HTTP:      Line 5: ETag: "901390-358-3485deab"
HTTP:      Line 6: Content-Length: 856
HTTP:      Line 7: Accept-Ranges: bytes
HTTP:      Line 8: Keep-Alive: timeout=15, max=100
HTTP:      Line 9: Connection: Keep-Alive
HTTP:      Line 10: Content-Type: text/html
HTTP:      Line 11:
HTTP:      Line 12: <HTML>
HTTP:      Line 13: <HEAD>
HTTP:      Line 14: <META NAME="GENERATOR" CONTENT="Adobe PageMill 2.0 Mac">
HTTP:      Line 15:   <TITLE>DigiNet Home Page</TITLE>
HTTP:      Line 16: </HEAD>
HTTP:      Line 17: <BODY BGCOLOR="#f8faff" LINK="#f720ff" ALINK="#ff0af1" BACKG
HTTP:         ROUND="Skidsblue.gif">
HTTP:      Line 18:
HTTP:      Line 19: <P ALIGN=CENTER><MAP NAME="Homepage1">
HTTP:      Line 20:  <AREA SHAPE="circle" COORDS="276,104,12" HREF="brutus.html">
HTTP:      Line 21:  <AREA SHAPE="circle" COORDS="320,126,13" HREF="boomer.html">
HTTP:      Line 22:  <AREA SHAPE="circle" COORDS="371,136,13" HREF="publications.html">
HTTP:      Line 23:  <AREA SHAPE="circle" COORDS="454,115,12" HREF="tutorial.html">
HTTP:      Line 24:  <AREA SHAPE="circle" COORDS="427,70,13" HREF="principal.html">
HTTP:      Line 25:  <AREA SHAPE="circle" COORDS="353,44,12" HREF="previousas
HTTP:         signments.html">
HTTP:      Line 26:  <AREA SHAPE="circle" COORDS="284,44,12" HREF="capabilities.html">
HTTP:      Line 27: </MAP><IMG SRC="Homepage1.gif" WIDTH="493" HEIGHT="360" ALIG
HTTP:         N="BOTTOM"
HTTP:      Line 28: NATURALSIZEFLAG="3" USEMAP="#Homepage1" ISMAP>
HTTP:      Line 29: </BODY>
HTTP:      Line 30: </HTML>
HTTP:
```

After perusing the DigiNet Corporation home page, the user clicks on a hyperlink that points to the various tutorials. This issues an HTTP GET message, sent in Frame 95, and the resulting data transfer that begins in Frame 96 (Trace 10.8c). In Frame 95 you see the URI (tutorial.html), the host (www.diginet.com), and other information required to accurately communicate the request to the web server. The Response begins in Frame 96 with the HTML coding of the tutorials page.

Trace 10.8c. HTTP DigiNet Corporation Tutorial Page Details
Sniffer Network Analyzer data from 30-Apr at 09:58:22, file DNC.ENC, Pg 1

```
- - - - - - - - - - - - - - - - Frame 95 - - - - - - - - - - - - - - - - -
HTTP:     — Hypertext Transfer Protocol —
HTTP:
HTTP:     Line 1: GET /tutorial.html HTTP/1.0
HTTP:     Line 2: Referer: http://www.diginet.com/
HTTP:     Line 3: Connection: Keep-Alive
HTTP:     Line 4: User-Agent: Mozilla/3.01 (X11; I; SunOS 5.6 sun4m)
HTTP:     Line 5: Host: www.diginet.com
HTTP:     Line 6: Accept: image/gif, image/x-xbitmap, image/jpeg, image/pjpeg, */*
HTTP:     Line 7:
HTTP:

- - - - - - - - - - - - - - - - Frame 96 - - - - - - - - - - - - - - - - -
HTTP:     — Hypertext Transfer Protocol —
HTTP:
HTTP:     Line 1: HTTP/1.1 200 OK
HTTP:     Line 2: Date: 30 Apr 17:02:19 GMT
HTTP:     Line 3: Server: Apache/1.2.4
HTTP:     Line 4: Last-Modified: 02 Dec 20:39:30 GMT
HTTP:     Line 5: ETag: "80f432-2b1-34847202"
HTTP:     Line 6: Content-Length: 689
HTTP:     Line 7: Accept-Ranges: bytes
HTTP:     Line 8: Keep-Alive: timeout=15, max=99
HTTP:     Line 9: Connection: Keep-Alive
HTTP:     Line 10: Content-Type: text/html
HTTP:     Line 11:
HTTP:     Line 12: <HTML>
HTTP:     Line 13: <HEAD>
HTTP:     Line 14: <META NAME="GENERATOR" CONTENT="Adobe PageMill 2.0 Mac">
HTTP:     Line 15:   <TITLE>Tutorials</TITLE>
HTTP:     Line 16: </HEAD>
HTTP:     Line 17: <BODY BGCOLOR="#fffdfd" LINK="#ff17f7" BACKGROUND="Skidsblue.gif">
HTTP:     Line 18:
HTTP:     Line 19: <P ALIGN=CENTER><MAP NAME="image54">
```

```
HTTP:     Line 20: <AREASHAPE=”circle” COORDS=”373,135,14" HREF=”managinter.html”>
HTTP:     Line 21: <AREASHAPE=”circle” COORDS=”456,114,14" HREF=”implemipv6.html”>
HTTP:     Line 22: <AREA SHAPE=”circle” COORDS=”429,70,14" HREF=”troubletcp.html”>
HTTP:     Line 23: <AREA SHAPE=”circle” COORDS=”355,44,14" HREF=”analizbroad.html”>
HTTP:     Line 24: <AREA SHAPE=”circle” COORDS=”285,44,15" HREF=”intnetdesign.html”>
HTTP:     Line 25: </MAP><IMG SRC=”image54.gif” WIDTH=”530" HEIGHT=”339" ALIGN=
HTTP:     “BOTTOM” NATURALSIZEFLAG=
HTTP:     Line 26: “3” USEMAP=”#image54" ISMAP>
HTTP:     Line 27: </BODY>
HTTP:     Line 28: </HTML>
HTTP:
```

The user next clicks on the hyperlink that points to the Troubleshooting TCP/IP tutorial. This action generates the HTTP Request message shown in Frame 121, and the subsequent Response message and the file transfer that begins in Frame 122 (Trace 10.8d). These messages are similar to the two that were described above, except that the Port number used for this transmission (34438) is different from the Port number of the previous transmission (34440). Recall from our discussion above that the Client's web browser application opened multiple logical connections when it first accessed the web server, and that we are now seeing the use of those various connections.

Trace 10.8d. HTTP DigiNet Corporation Troubleshooting TCP/IP Tutorial Page Details
Sniffer Network Analyzer data from 30-Apr at 09:58:22, file DNC.ENC, Pg 1

```
---------------- Frame 121 -----------------
HTTP:     — Hypertext Transfer Protocol —
HTTP:
HTTP:     Line 1: GET /troubletcp.html HTTP/1.0
HTTP:     Line 2: Referer: http://www.diginet.com/tutorial.html
HTTP:     Line 3: Connection: Keep-Alive
HTTP:     Line 4: User-Agent: Mozilla/3.01 (X11; I; SunOS 5.6 sun4m)
HTTP:     Line 5: Host: www.diginet.com
HTTP:     Line 6: Accept: image/gif, image/x-xbitmap, image/jpeg, image/pjpeg,*/*
HTTP:     Line 7:
HTTP:

---------------- Frame 122 -----------------
HTTP:     — Hypertext Transfer Protocol —
HTTP:
HTTP:     Line 1: HTTP/1.1 200 OK
HTTP:     Line 2: Date: 30 Apr 17:02:22 GMT
HTTP:     Line 3: Server: Apache/1.2.4
```

```
HTTP:    Line 4: Last-Modified: 02 Dec 20:38:38 GMT
HTTP:    Line 5: ETag: "80f430-130-348471ce"
HTTP:    Line 6: Content-Length: 304
HTTP:    Line 7: Accept-Ranges: bytes
HTTP:    Line 8: Keep-Alive: timeout=15, max=99
HTTP:    Line 9: Connection: Keep-Alive
HTTP:    Line 10: Content-Type: text/html
HTTP:    Line 11:
HTTP:    Line 12: <HTML>
HTTP:    Line 13: <HEAD>
HTTP:    Line 14: <META NAME="GENERATOR" CONTENT="Adobe PageMill 2.0 Mac">
HTTP:    Line 15:   <TITLE>Troubleshooting TCP/IP</TITLE>
HTTP:    Line 16: </HEAD>
HTTP:    Line 17: <BODY BGCOLOR="#fdfffc" LINK="#ff22f2" BACKGROUND="Skidsblue.gif>
HTTP:    Line 18:
HTTP:    Line 19: <P ALIGN=CENTER><IMG SRC="image9.gif" WIDTH="576" HEIGHT="67
HTTP:    0" ALIGN="BOTTOM"
HTTP:    Line 20: NATURALSIZEFLAG="3">
HTTP:    Line 21: </BODY>
HTTP:    Line 22: </HTML>
HTTP:
```

10.9 Measuring WAN Response Times

Wide Area Network circuits contain propagation delays that can affect the performance of the protocols operating over those circuits. This case study will compare the response time of a LAN connection via fast Ethernet with that of a WAN connection via ATM (see Figure 10-9) to better understand the impacts of the WAN propagation delays.

To compare the LAN and WAN response times, a file transfer is initiated over both connections and the various results are compared. In the case of the LAN transfer, the connection is first initialized, and then the file transfer is initiated in Frame 91 of Trace 10.9a. Note that Frame 91 shows an FTP "Opening BINARY mode data connection" command to open a file called 100megb.zip. The timestamp (Absolute Time column) at the start of this file operation is 09:29:15. Since the file is quite large, the file transfer takes some time. It finally completes in Frame 7693, with the message "Transfer complete - file 100megb.zip sent successfully." The timestamp at the end of the file operation is 09:29:16, or approximately 1 second after it started. Note from Frame 91 that the file size was 6,811,648 bytes (or 54,493,184 bits), which yields a data throughput rate for this transaction of 54.4 Mbps.

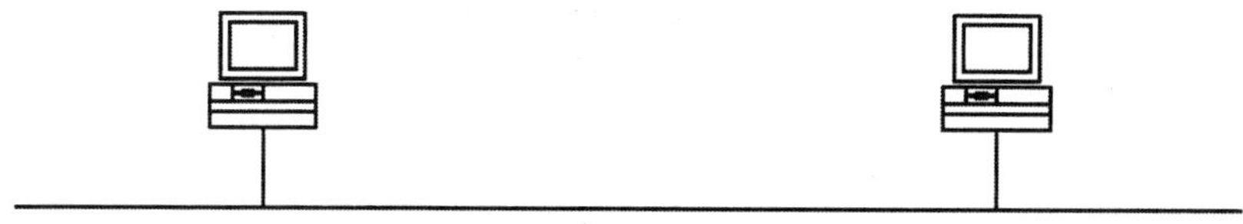

Local Connection via Fast Ethernet

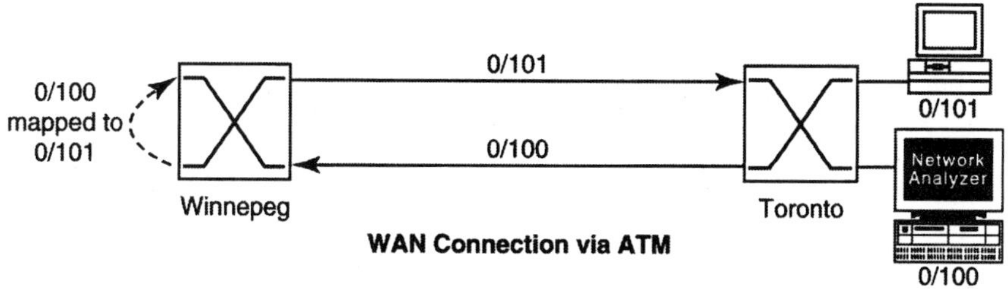

Figure 10-9. Local and WAN Connections.

Trace 10.9a. File Transfer using Local Network Connection
Sniffer Network Analyzer data from 01-Nov at 21:27:00, file Snif5_Local_Connect.cap, Pg 1

Frame	Destination	Source	Abs. Time	Summary
56	FTP Client	FTP Server	09:28:54	TCP: D=21 S=1032 SYN SEQ=84214 LEN=0 WIN=34064
57	FTP Server	FTP Client	09:28:54	TCP: D=1032 S=21 SYN ACK=84215 SEQ=1716569711 LEN=0 WIN=34064
58	FTP Client	FTP Server	09:28:54	TCP: D=21 S=1032 ACK=1716569712 WIN=34064
59	Broadcast	FTP Client	09:28:54	ARP: C PA=[172.16.0.253] PRO=IP
60	FTP Server	FTP Client	09:28:55	FTP: R PORT=1032 220 Serv-U FTP-Server v2.2 for WinSock ready...
61	FTP Client	FTP Server	09:28:55	TCP: D=21 S=1032 ACK=1716569762 WIN=34014

62	Broadcast	FTP Client	09:28:55	WINS: C ID=32859 OP=REGISTER NAME=TOR-ZSHIVJI<00>
63	Broadcast	FTP Client	09:28:55	ARP: C PA=[172.16.0.253] PRO=IP
64	Broadcast	FTP Client	09:28:57	ARP: C PA=[172.16.0.253] PRO=IP
65	FTP Client	FTP Server	09:28:58	FTP: C PORT=1032 USER admin
66	FTP Server	FTP Client	09:28:58	FTP: R PORT=1032 331 User name OK, send password
67	FTP Client	FTP Server	09:28:58	TCP: D=21 S=1032 ACK=1716569795 WIN=33981
68	Broadcast	FTP Client	09:28:59	ARP: C PA=[172.16.0.253] PRO=IP
69	Broadcast	FTP Client	09:29:00	WINS: C ID=32861 OP=REGISTER NAME=C1<00>
70	Broadcast	FTP Client	09:29:01	WINS: C ID=32861 OP=REGISTER NAME=C1<00>
71	FTP Client	FTP Server	09:29:02	FTP: C PORT=1032 PASS mainstreet
72	FTP Server	FTP Client	09:29:02	FTP: R PORT=1032 230 User ADMIN logged in
73	FTP Client	FTP Server	09:29:02	TCP: D=21 S=1032 ACK=1716569821 WIN=33955
74	Broadcast	FTP Client	09:29:02	WINS: C ID=32861 OP=REGISTER NAME=C1<00>
75	Broadcast	FTP Client	09:29:03	WINS: C ID=32861 OP=REGISTER NAME=C1<00>
76	FTP Client	FTP Server	09:29:03	FTP: C PORT=1032 TYPE I

77	FTP Server	FTP Client	09:29:03	FTP: R PORT=1032 200 TYPE set to IMAGE (binary)
78	Broadcast	FTP Client 0	9:29:03	ARP: C PA=[172.16.0.253] PRO=IP
79	FTP Client	FTP Server	09:29:03	TCP: D=21 S=1032 ACK=1716569853 WIN=33923
80	Broadcast	FTP Client	09:29:11	ARP: C PA=[172.16.0.253] PRO=IP
81	Broadcast	FTP Client	09:29:12	ARP: C PA=[172.16.0.253] PRO=IP
82	Broadcast	FTP Client	09:29:13	ARP: C PA=[172.16.0.253] PRO=IP
83	FTP Server	FTP Client	09:29:14	WINS: C ID=32863 OP=QUERY NAME= *<000000000000 0000000000000000<00>
84	FTP Client	FTP Server	09:29:14	WINS: R ID=32863 OP=QUERY STAT=OK
85	FTP Client	FTP Server	09:29:15	FTP: C PORT=1032 PORT 172,16,0,152,4,9
86	FTP Server	FTP Client	09:29:15	FTP: R PORT=1032 200 PORT Command OK
87	FTP Client	FTP Server	09:29:15	FTP: C PORT=1032 RETR 100megb.zip
88	FTP Server	FTP Client	09:29:15	TCP: D=1033 S=20 SYN SEQ=1721690893 LEN=0 WIN=34064
89	FTP Client	FTP Server	09:29:15	TCP: D=20 S=1033 SYN ACK=1721690894 SEQ=84219 LEN=0 WIN=34064
90	FTP Server	FTP Client	09:29:15	TCP: D=1033 S=20 ACK=84220 WIN=34064

91	FTP Server	FTP Client	09:29:15	FTP: R PORT=1032 150 Opening BINARY mode data connection for 100megb.zip (6811648 bytes).
92	FTP Server	FTP Client	09:29:15	FTP: R PORT=1033 Text Data
93	FTP Server	FTP Client	09:29:15	FTP: R PORT=1033 Text Data
94	FTP Client	FTP Server	09:29:15	TCP: D=20 S=1033 ACK=1721693814 WIN=34064
95	FTP Server	FTP Client	09:29:15	FTP: R PORT=1033 Text Data
96	FTP Client	FTP Server	09:29:15	TCP: D=20 S=1033 ACK=1721695274 WIN=34064
97	FTP Server	FTP Client	09:29:15	FTP: R PORT=1033 Text Data
98	FTP Server	FTP Client	09:29:15	FTP: R PORT=1033 Text Data
99	FTP Client	FTP Server	09:29:15	TCP: D=20 S=1033 ACK=1721698194 WIN=34064
100	FTP Server	FTP Client	09:29:15	FTP: R PORT=1033 Text Data
101	FTP Client	FTP Server	09:29:15	TCP: D=20 S=1033 ACK=1721699086 WIN=33172
102	FTP Server	FTP Client	09:29:15	FTP: R PORT=1033 Text Data
103	FTP Server	FTP Client	09:29:15	FTP: R PORT=1033 Text Data
104	FTP Client	FTP Server	09:29:15	TCP: D=20 S=1033 ACK=1721702006 WIN=30252
105	FTP Server	FTP Client	09:29:15	FTP: R PORT=1033 Text Data
106	FTP Server	FTP Client	09:29:15	FTP: R PORT=1033 Text Data
107	FTP Client	FTP Server	09:29:15	TCP: D=20 S=1033 ACK=1721704642 WIN=27616

108	FTP Server	FTP Client	09:29:15	FTP: R PORT=1033 Text Data
109	FTP Server	FTP Client	09:29:15	FTP: R PORT=1033 Text Data
110	FTP Client	FTP Server	09:29:15	TCP: D=20 S=1033 ACK=1721707278 WIN=24980
111	FTP Client	FTP Server	09:29:15	TCP: D=20 S=1033 ACK=1721707278 WIN=29540
112	FTP Client	FTP Server	09:29:15	TCP: D=20 S=1033 ACK=1721707278 WIN=34064
113	FTP Server	FTP Client	09:29:15	FTP: R PORT=1033 Text Data
114	FTP Client	FTP Server	09:29:15	TCP: D=20 S=1033 ACK=1721708738 WIN=34064
115	FTP Server	FTP Client	09:29:15	FTP: R PORT=1033 Text Data
116	FTP Server	FTP Client	09:29:15	FTP: R PORT=1033 Text Data
117	FTP Client	FTP Server	09:29:15	TCP: D=20 S=1033 ACK=1721711374 WIN=34064
118	FTP Server	FTP Client	09:29:15	FTP: R PORT=1033 Text Data
119	FTP Server	FTP Client	09:29:15	FTP: R PORT=1033
120	FTP Client	FTP Server	09:29:15	TCP: D=20 S=1033 ACK=1721714294 WIN=34064
121	FTP Server	FTP Client	09:29:15	FTP: R PORT=1033 Text Data
122	FTP Client	FTP Server	09:29:15	TCP: D=20 S=1033 ACK=1721715470 WIN=32888
123	FTP Server	FTP Client	09:29:15	FTP: R PORT=1033 Text Data
124	FTP Server	FTP Client	09:29:15	FTP: R PORT=1033 Text Data
125	FTP Client	FTP Server	09:29:15	TCP: D=20 S=1033 ACK=1721718390 WIN=34064

126	FTP Server	FTP Client	09:29:15	FTP: R PORT=1033 Text Data
127	FTP Client	FTP Server	09:29:15	TCP: D=20 S=1033 ACK=1721719566 WIN=32888
128	FTP Server	FTP Client	09:29:15	FTP: R PORT=1033 Text Data
129	FTP Server	FTP Client	09:29:15	FTP: R PORT=1033 Text Data
130	FTP Client	FTP Server	09:29:15	TCP: D=20 S=1033 ACK=1721722486 WIN=34064
131	FTP Server	FTP Client	09:29:15	FTP: R PORT=1033 Text Data
132	FTP Client	FTP Server	09:29:15	TCP: D=20 S=1033 ACK=1721723662 WIN=32888
133	FTP Server	FTP Client	09:29:15	FTP: R PORT=1033 Text Data
134	FTP Server	FTP Client	09:29:15	FTP: R PORT=1033 Text Data
135	FTP Client	FTP Server	09:29:15	TCP: D=20 S=1033 ACK=1721726582 WIN=34064
136	FTP Server	FTP Client	09:29:15	FTP: R PORT=1033 Text Data
.				
.				
.				
7654	FTP Client	FTP Server	09:29:16	TCP: D=20 S=1033 ACK=1728471234 WIN=34064
7655	FTP Server	FTP Client	09:29:16	FTP: R PORT=1033 Text Data
7656	FTP Server	FTP Client	09:29:16	FTP: R PORT=1033 Text Data
7657	FTP Client	FTP Server	09:29:16	TCP: D=20 S=1033 ACK=1728473870 WIN=34064
7658	FTP Server	FTP Client	09:29:16	FTP: R PORT=1033 Text Data
7659	FTP Server	FTP Client	09:29:16	FTP: R PORT=1033 Text Data

7660	FTP Client	FTP Server	09:29:16	TCP: D=20 S=1033 ACK=1728476790 WIN=34064
7661	FTP Server	FTP Client	09:29:16	FTP: R PORT=1033 Text Data
7662	FTP Server	FTP Client	09:29:16	FTP: R PORT=1033 Text Data
7663	FTP Client	FTP Server	09:29:16	TCP: D=20 S=1033 ACK=1728479426 WIN=34064
7664	FTP Server	FTP Client	09:29:16	FTP: R PORT=1033 Text Data
7665	FTP Server	FTP Client	09:29:16	FTP: R PORT=1033 Text Data
7666	FTP Client	FTP Server	09:29:16	TCP: D=20 S=1033 ACK=1728482062 WIN=34064
7667	FTP Server	FTP Client	09:29:16	FTP: R PORT=1033 Text Data
7668	FTP Server	FTP Client	09:29:16	FTP: R PORT=1033 Text Data
7669	FTP Client	FTP Server	09:29:16	TCP: D=20 S=1033 ACK=1728484982 WIN=34064
7670	FTP Server	FTP Client	09:29:16	FTP: R PORT=1033 Text Data
7671	FTP Server	FTP Client	09:29:16	FTP: R PORT=1033 Text Data
7672	FTP Client	FTP Server	09:29:16	TCP: D=20 S=1033 ACK=1728487618 WIN=34064
7673	FTP Server	FTP Client	09:29:16	FTP: R PORT=1033 Text Data
7674	FTP Server	FTP Client	09:29:16	FTP: R PORT=1033 Text Data
7675	FTP Client	FTP Server	09:29:16	TCP: D=20 S=1033 ACK=1728490254 WIN=34064
7676	FTP Server	FTP Client	09:29:16	FTP: R PORT=1033 Text Data
7677	FTP Server	FTP Client	09:29:16	FTP: R PORT=1033 Text Data
7678	FTP Client	FTP Server	09:29:16	TCP: D=20 S=1033 ACK=1728493174 WIN=34064

7679	FTP Server	FTP Client	09:29:16	FTP: R PORT=1033 Text Data
7680	FTP Server	FTP Client	09:29:16	FTP: R PORT=1033 Text Data
7681	FTP Client	FTP Server	09:29:16	TCP: D=20 S=1033 ACK=1728495810 WIN=34064
7682	FTP Server	FTP Client	09:29:16	FTP: R PORT=1033 Text Data
7683	FTP Server	FTP Client	09:29:16	FTP: R PORT=1033 Text Data
7684	FTP Client	FTP Server	09:29:16	TCP: D=20 S=1033 ACK=1728498446 WIN=34064
7685	FTP Server	FTP Client	09:29:16	FTP: R PORT=1033 Text Data
7686	FTP Server	FTP Client	09:29:16	TCP: R PORT=1033 Text Data
7687	FTP Client	FTP Server	09:29:16	TCP: D=20 S=1033 ACK=1728501366 WIN=34064
7688	FTP Server	FTP Client	09:29:16	FTP: R PORT=1033 Text Data
7689	FTP Server	FTP Client	09:29:16	TCP: D=1033 S=20 FIN ACK=84220 SEQ=1728502542 LEN=0 WIN=34064
7690	FTP Client	FTP Server	09:29:16	TCP: D=20 S=1033 ACK=1728502543 WIN=32888
7691	FTP Client	FTP Server	09:29:16	TCP: D=20 S=1033 FIN ACK=1728502543 SEQ=84220 LEN=0 WIN=32888
7692	FTP Server	FTP Client	09:29:16	TCP: D=1033 S=20 ACK=84221 WIN=34064
7693	FTP Server	FTP Client	09:29:16	FTP: R PORT=1032 226 Transfer complete - file 100megb.zip sent successfully
7694	FTP Client	FTP Server	09:29:16	TCP: D=21 S=1032 ACK=1716570008 WIN=33768

To compare the local network and wide area network response, the same experiment is repeated, but this time via a WAN connection from Toronto to Winnipeg and back, a roundtrip distance of 1,868 air miles (note that the circuit distance may not be the same as the air miles, but this measurement will give us an approximate number for discussion purposes). For the WAN connection, the file transfer is initialized in Frame 140 of Trace 10.9b and successfully concluded in Frame 7652. The timestamp at the beginning of the transfer is 08:55:12, and the timestamp at the end of the transfer is 08:56:01, making for an approximate time of 49 seconds. Since the file size has not changed (6,811,648 bytes or 54,493,184 bits), the throughput for the WAN connection is approximately 1.1 Mbps. By comparing the LAN and WAN file transfer times, we can conclude that the WAN circuit introduced approximately 48 seconds of delay into the overall file transfer operation.

Trace 10.9b. File Transfer using Wide Area Network Connection
Sniffer Network Analyzer data from 01-Nov at 20:53:00, file Snif3_64kWin_Wan.cap, Pg 1

Frame	Destination	Source	Abs. Time	Summary
71	FTP Server	FTP Client	08:54:50	TCP:D=21 S=1087 SYN SEQ=959503913 LEN=0 WIN=64240
72	FTP Client	FTP Server	08:54:50	TCP:D=1087 S=21 SYN ACK=959503914 SEQ=84497 LEN=0 WIN=65535
73	FTP Server	FTP Client	08:54:50	TCP:D=21 S=1087 ACK=84498 WIN=64240
74	FTP Client	FTP Server	08:54:50	FTP: R PORT=1087 220 Serv-U FTP-Server v2.2 for WinSock ready...
75	FTP Client	FTP Server	08:54:50	WINS: C ID=32886 OP=QUERY NAME=*<000000 0000000000000 000000000<00>

76	FTP Server	FTP Client	08:54:50	WINS: R ID=32886 OP=QUERY STAT=OK
77	FTP Server	FTP Client	08:54:50	FTP: C PORT=1087 USER admin
78	FTP Client	FTP Server	08:54:50	FTP: R PORT=1087 331 User name OK, send password
79	FTP Server	FTP Client	08:54:50	FTP: C PORT=1087 PASS mainstreet
80	FTP Client	FTP Server	08:54:50	FTP: R PORT=1087 230 User ADMIN logged in
81	0180C2000000	00027D363E86	08:54:50	BPDU: S: Pri=8000 Port=8012 Root: Pri=8000 Addr=00027D363E8C Cost=14
82	FTP Server	FTP Client	08:54:50	FTP: C PORT=1087 PWD
83	FTP Client	FTP Server	08:54:50	FTP: R PORT=1087 257 "/c:/ftp" is working directory
84	FTP Server	FTP Client	08:54:50	FTP: C PORT=1087 PASV
85	FTP Client	FTP Server	08:54:50	FTP: R PORT=1087 227 Entering Passive Mode (172,16,0,152,4,12)
86	FTP Server	FTP Client	08:54:50	TCP:D=1036 S=1088 SYN SEQ=959668720 LEN=0 WIN=64240

87	FTP Client	FTP Server	08:54:50	TCP:D=1088 S=1036 SYN ACK=959668721 SEQ=84504 LEN=0 WIN=65535
88	FTP Server	FTP Client	08:54:50	TCP:D=1036 S=1088 ACK=84505 WIN=64240
89	FTP Server	FTP Client	08:54:50	FTP: C PORT=1087 LIST
90	FTP Client	FTP Server	08:54:50	TCP:D=1088 S=1036 ACK=959668721 SEQ=84505 LEN=9 WIN=65535
91	FTP Client	FTP Server	08:54:50	FTP: R PORT=1087 150 Data connection open for PASV transfer
92	FTP Client	FTP Server	08:54:50	TCP:D=1088 S=1036 FIN ACK=959668721 SEQ=84514 LEN=0 WIN=65535
93	FTP Server	FTP Client	08:54:50	TCP:D=1036 S=1088 ACK=84515 WIN=64231
94	FTP Server	FTP Client	08:54:50	TCP:D=1036 S=1088 FIN ACK=84515 SEQ=959668721 LEN=0 WIN=64231
95	FTP Client	FTP Server	08:54:50	TCP:D=1088 S=1036 ACK=959668722 WIN=65535
96	FTP Server	FTP Client	08:54:50	TCP:D=21 S=1087 ACK=84734 WIN=64004
97	FTP Client	FTP Server	08:54:50	FTP: R PORT=1087 226 Transfer complete

98	FTP Server	FTP Client	08:54:51	TCP:D=21 S=1087 ACK=84757 WIN=63981
99	0180C2000000	00027D363E86	08:54:52	BPDU: S: Pri=8000 Port=8012 Root:Pri=8000 Addr=00027D363E8C Cost=14
100	0180C2000000	00027D363E86	08:54:54	BPDU: S: Pri=8000 Port=8012 Root:Pri=8000 Addr=00027D363E8C Cost=14
101	0180C2000000	00027D363E86	08:54:56	BPDU: S: Pri=8000 Port=8012 Root:Pri=8000 Addr=00027D363E8C Cost=14
102	FTP Server	FTP Client	08:54:56	FTP: C PORT=1087 CWD /e:/ftp
103	FTP Client	FTP Server	08:54:56	FTP: R PORT=1087 250 Directory changed to /e:/ftp
104	FTP Server	FTP Client	08:54:56	FTP: C PORT=1087 PWD
105	FTP Client	FTP Server	08:54:56	FTP: R PORT=1087 257 "/e:/ftp" is working directory
106	FTP Server	FTP Client	08:54:56	FTP: C PORT=1087 PASV
107	FTP Client	FTP Server	08:54:56	FTP: R PORT=1087 227 Entering Passive Mode (172,16,0,152,4,13)

108	FTP Server	FTP Client	08:54:56	TCP:D=1037 S=1089 SYN SEQ=961318082 LEN=0 WIN=64240
109	FTP Client	FTP Server	08:54:56	TCP:D=1089 S=1037 SYN ACK=961318083 SEQ=84513 LEN=0 WIN=65535
110	FTP Server	FTP Client	08:54:56	TCP:D=1037 S=1089 ACK=84514 WIN=64240
111	FTP Server	FTP Client	08:54:56	FTP: C PORT=1087 LIST
112	FTP Client	FTP Server	08:54:56	TCP:D=1089 S=1037 ACK=961318083 SEQ=84514 LEN=1411 WIN=65535
113	FTP Client	FTP Server	08:54:56	FTP: R PORT=1087 150 Data connection open for PASV transfer
114	FTP Client	FTP Server	08:54:56	TCP:D=1089 S=1037 FIN ACK=961318083 SEQ=85925 LEN=0 WIN=65535
115	FTP Server	FTP Client	08:54:57	TCP:D=1037 S=1089 ACK=85926 WIN=62829
116	FTP Server	FTP Client	08:54:57	TCP:D=1037 S=1089 FIN ACK=85926 SEQ=961318083 LEN=0 WIN=62829
117	FTP Client	FTP Server	08:54:57	TCP:D=1089 S=1037 ACK=961318084 WIN=65535

118	FTP Server	FTP Client	08:54:57	TCP:D=21 S=1087 ACK=84918 WIN=63820
119	FTP Client	FTP Server	08:54:57	FTP: R PORT=1087 226 Transfer complete
120	01000CCCCCCC	00027D363E86	08:54:57	CDP: Type=0x0001 (Device ID); Type=0x0002 (Address); Type=0x0003 (Port ID); Type=0x0004 (Capabilities); Type=0x0005 (Version); Type=0x0006 (Platform); Type=0x0008 (Unknown); Type=0x0009 (Unknown)
121	FTP Server	FTP Client	08:54:57	TCP:D=21 S=1087 ACK=84941 WIN=63797
122	0180C2000000	00027D363E86	08:54:58	BPDU: S: Pri=8000 Port=8012 Root:Pri=8000 Addr=00027D363E8C Cost=14
123	0180C2000000	00027D363E86	08:55:00	BPDU: S: Pri=8000 Port=8012 Root:Pri=8000 Addr=00027D363E8C Cost=14
124	0180C2000000	00027D363E86	08:55:02	BPDU: S: Pri=8000 Port=8012 Root:Pri=8000 Addr=00027D363E8C Cost=14

125	0180C2000000	00027D363E86	08:55:04	BPDU: S: Pri=8000 Port=8012 Root:Pri=8000 Addr=00027D363E8C Cost=14
126	0180C2000000	00027D363E86	08:55:06	BPDU: S: Pri=8000 Port=8012 Root:Pri=8000 Addr=00027D363E8C Cost=14
127	0180C2000000	00027D363E86	08:55:08	BPDU: S: Pri=8000 Port=8012 Root:Pri=8000 Addr=00027D363E8C Cost=14
128	0180C2000000	00027D363E86	08:55:10	BPDU: S: Pri=8000 Port=8012 Root:Pri=8000 Addr=00027D363E8C Cost=14
129	0180C2000000	00027D363E86	08:55:12	BPDU: S: Pri=8000 Port=8012 Root:Pri=8000 Addr=00027D363E8C Cost=14
130	FTP Server	FTP Client	08:55:12	FTP: C PORT=1087 TYPE I
131	FTP Client	FTP Server	08:55:12	FTP: R PORT=1087 200 TYPE set to IMAGE (binary)
132	FTP Server	FTP Client	08:55:12	FTP: C PORT=1087 PASV

133	FTP Client	FTP Server	08:55:12	FTP: R PORT=1087 227 Entering Passive Mode (172,16,0,152,4,14)
134	FTP Server	FTP Client	08:55:12	TCP:D=1038 S=1090 SYN SEQ=965280926 LEN=0 WIN=64240
135	FTP Client	FTP Server	08:55:12	TCP:D=1090 S=1038 SYN ACK=965280927 SEQ=84515 LEN=0 WIN=65535
136	FTP Server	FTP Client	08:55:12	TCP:D=1038 S=1090 ACK=84516 WIN=64240
137	FTP Server	FTP Client	08:55:12	FTP: C PORT=1087 RETR 100megb.zip
138	FTP Client	FTP Server	08:55:12	TCP:D=1090 S=1038 ACK=965280927 SEQ=84516 LEN=1460 WIN=65535
139	FTP Client	FTP Server	08:55:12	TCP:D=1090 S=1038 ACK=965280927 SEQ=85976 LEN=1460 WIN=65535
140	FTP Client	FTP Server	08:55:12	FTP: R PORT=1087 150 Opening BINARY mode data connection for 100megb.zip (6811648 bytes).
141	FTP Server	FTP Client	08:55:12	TCP:D=1038 S=1090 ACK=87436 WIN=64240
142	FTP Client	FTP Server	08:55:12	TCP:D=1090 S=1038 ACK=965280927 SEQ=87436 LEN=1460 WIN=65535

143	FTP Client	FTP Server	08:55:12	TCP:D=1090 S=1038 ACK=965280927 SEQ=88896 LEN=1460 WIN=65535
144	FTP Client	FTP Server	08:55:12	TCP:D=1090 S=1038 ACK=965280927 SEQ=90356 LEN=1460 WIN=65535
145	FTP Server	FTP Client	08:55:12	TCP:D=1038 S=1090 ACK=88896 WIN=64240
146	FTP Client	FTP Server	08:55:12	TCP:D=1090 S=1038 ACK=965280927 SEQ=91816 LEN=892 WIN=65535
147	FTP Client	FTP Server	08:55:12	TCP:D=1090 S=1038 ACK=965280927 SEQ=92708 LEN=1460 WIN=65535
148	FTP Server	FTP Client	08:55:12	TCP:D=1038 S=1090 ACK=91816 WIN=64240
.				
.				
.				
7617	FTP Client	FTP Server	08:56:01	TCP:D=1090 S=1038 ACK=965280927 SEQ=6866316 LEN=1176 WIN=65535
7618	FTP Server	FTP Client	08:56:01	TCP:D=1038 S=1090 ACK=6862220 WIN=64240
7619	FTP Server	FTP Client	08:56:01	TCP:D=1038 S=1090 ACK=6864856 WIN=64240
7620	FTP Client	FTP Server	08:56:01	TCP:D=1090 S=1038 ACK=965280927 SEQ=6867492 LEN=1460 WIN=65535

7621	FTP Client	FTP Server	08:56:01	TCPD=1090 S=1038 ACK=965280927 SEQ=6868952 LEN=1460 WIN=65535
7622	FTP Client	FTP Server	08:56:01	TCP:D=1090 S=1038 ACK=965280927 SEQ=6870412 LEN=1176 WIN=65535
7623	FTP Server	FTP Client	08:56:01	TCP:D=1038 S=1090 ACK=6867492 WIN=64240
7624	FTP Client	FTP Server	08:56:01	TCP:D=1090 S=1038 ACK=965280927 SEQ=6871588 LEN=1460 WIN=65535
7625	FTP Client	FTP Server	08:56:01	TCP:D=1090 S=1038 ACK=965280927 SEQ=6873048 LEN=1460 WIN=65535
7626	FTP Client	FTP Server	08:56:01	TCP:D=1090S=1038 ACK=965280927 SEQ=6874508 LEN=1176 WIN=65535
7627	FTP Server	FTP Client	08:56:01	TCP:D=1038 S=1090 ACK=6870412 WIN=64240
7628	FTP Server	FTP Client	08:56:01	TCP:D=1038 S=1090 ACK=6873048 WIN=64240
7629	FTP Client	FTP Server	08:56:01	TCP:D=1090 S=1038 ACK=965280927 SEQ=6875684 LEN=1460 WIN=65535
7630	FTP Client	FTP Server	08:56:01	TCP:D=1090 S=1038 ACK=965280927 SEQ=6877144 LEN=1460 WIN=65535

7631	FTP Client	FTP Server	08:56:01	TCP:D=1090 S=1038 ACK=965280927 SEQ=6878604 LEN=1176 WIN=65535
7632	FTP Server	FTP Client	08:56:01	TCP:D=1038 S=1090 ACK=6875684 WIN=64240
7633	FTP Client	FTP Server	08:56:01	TCP:D=1090 S=1038 ACK=965280927 SEQ=6879780 LEN=1460 WIN=65535
7634	FTP Client	FTP Server	08:56:01	TCP:D=1090 S=1038 ACK=965280927 SEQ=6881240 LEN=1460 WIN=65535
7635	FTP Client	FTP Server	08:56:01	TCP:D=1090 S=1038 ACK=965280927 SEQ=6882700 LEN=1176 WIN=65535
7636	FTP Server	FTP Client	08:56:01	TCP:D=1038 S=1090 ACK=6878604 WIN=64240
7637	FTP Server	FTP Client	08:56:01	TCP:D=1038 S=1090 ACK=6881240 WIN=64240
7638	FTP Client	FTP Server	08:56:01	TCP:D=1090 S=1038 ACK=965280927 SEQ=6883876 LEN=1460 WIN=65535
7639	FTP Client	FTP Server	08:56:01	TCP:D=1090 S=1038 ACK=965280927 SEQ=6885336 LEN=1460 WIN=65535
7640	FTP Client	FTP Server	08:56:01	TCP:D=1090 S=1038 ACK=965280927 SEQ=6886796 LEN=1176 WIN=65535

7641	FTP Server	FTP Client	08:56:01	TCP:D=1038 S=1090 ACK=6883876 WIN=64240
7642	FTP Client	FTP Server	08:56:01	TCP:D=1090 S=1038 ACK=965280927 SEQ=6887972 LEN=1460 WIN=65535
7643	FTP Client	FTP Server	08:56:01	TCP:D=1090 S=1038 ACK=965280927 SEQ=6889432 LEN=1460 WIN=65535
7644	FTP Client	FTP Server	08:56:01	TCP:D=1090 S=1038 ACK=965280927 SEQ=6890892 LEN=1176 WIN=65535
7645	FTP Server	FTP Client	08:56:01	TCP:D=1038 S=1090 ACK=6886796 WIN=64240
7646	FTP Server	FTP Client	08:56:01	TCP:D=1038 S=1090 ACK=6889432 WIN=64240
7647	FTP Client	FTP Server	08:56:01	TCP:D=1090 S=1038 ACK=965280927 SEQ=6892068 LEN=1460 WIN=65535
7648	FTP Client	FTP Server	08:56:01	TCP:D=1090 S=1038 ACK=965280927 SEQ=6893528 LEN=1460 WIN=65535
7649	FTP Client	FTP Server	08:56:01	TCP:D=1090 S=1038 ACK=965280927 SEQ=6894988 LEN=1176 WIN=65535
7650	FTP Client	FTP Server	08:56:01	TCP:D=1090 S=1038 FIN ACK=965280927 SEQ=6896164 LEN=0 WIN=65535

7651	FTP Server	FTP Client	08:56:01	TCP:D=1038 S=1090 ACK=6892068 WIN=64240
7652	FTP Client	FTP Server	08:56:01	FTP: R PORT=1087 226 Transfer completc - file 100megb.zip sent successfully
7653	FTP Server	FTP Client	08:56:01	TCP:D=1038 S=1090 ACK=6894988 WIN=64240
7654	FTP Server	FTP Client	08:56:01	TCP:D=1038 S=1090 ACK=6896165 WIN=63064
7655	FTP Server	FTP Client	08:56:02	TCP:D=21 S=1087 ACK=85154 WIN=63584
7656	FTP Server	FTP Client	08:56:02	TCP:D=1038 S=1090 FIN ACK=6896165 SEQ=965280927 LEN=0 WIN=63064
7657	FTP Client	FTP Server	08:56:02	TCP:D=1090 S=1038 ACK=965280928 WIN=65535

10.10 Analyzing SIP Phone Connections

The Session Initiation Protocol, or SIP, provides the signaling functions necessary to establish, manage, and terminate a connection between multimedia endpoints. As an example of SIP protocol analysis, consider the network shown in Figure 10-10a, in which an analog telephone connected to a Cisco Systems, Inc., SIP gateway initiates a call to a Cisco SIP telephone connected to the same network.

Note that a proxy server exists on this network and will be involved in the call establishment and termination procedures. The SIP control messages will flow between the SIP devices and the proxy server. The actual voice information will be transported using the Real Time Protocol, or RTP, which will flow between the SIP gateway and the SIP telephone (Figure 10-10b).

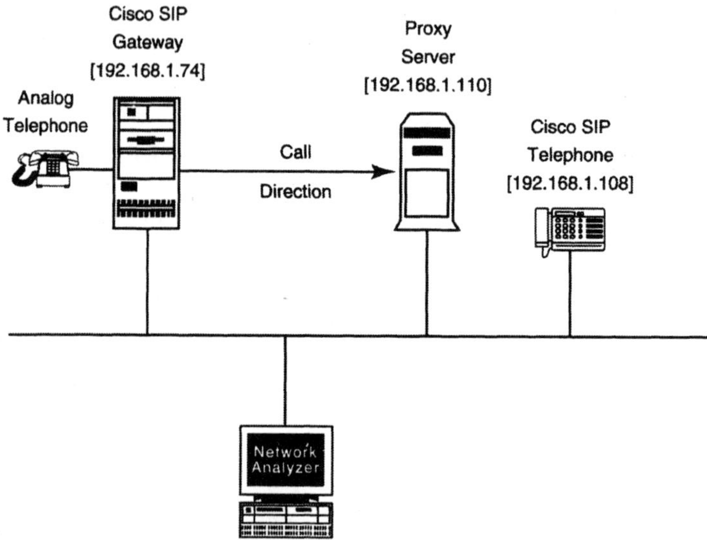

Figure 10-10a. SIP Phone-to-Phone Connection.

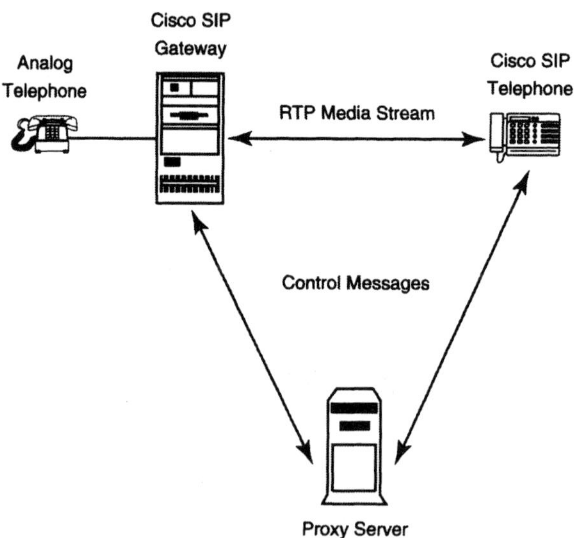

Figure 10-10b. SIP Control Message Flow.

The SIP control messages can be broken down into three general categories. These functions are detailed in Figure 10-10c and include the direction of the information flow:

1. Connection establishment using SIP (Frames 1–6, 129–131)
2. Information transfer using RTP (Frames 131–225)
3. Connection termination using SIP (Frames 226–229)

For brevity, some of the RTP voice traffic (Frames 131–225) has been filtered out and is not shown in either Figure 10-10c or the accompanying Trace 10.10a.

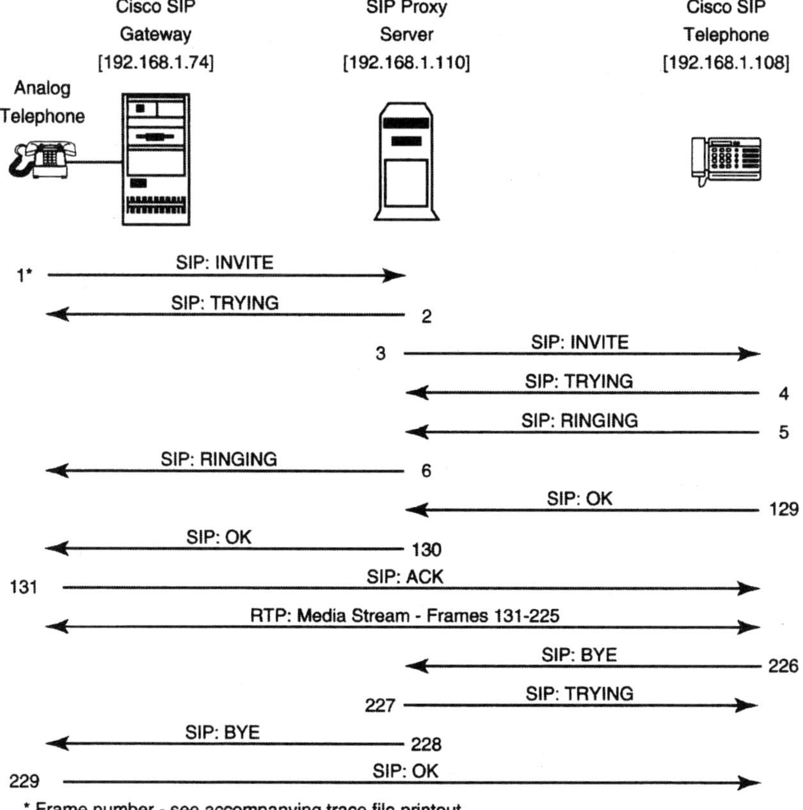

Figure 10-10c. SIP Control Message Details.

The details of these three functions are shown in Trace 10.10a, with traffic not relevant to our discussion filtered out (and therefore yielding some gaps in frame numbers).

Note that the inclusion of the proxy server in the network makes for a two-step communication process. For example, the SIP INVITE message is generated in Frame 1 and then passed to the desired destination in Frame 3. Similarly, the SIP RINGING response is sent to the proxy server in Frame 5 and then passed to the gateway in Frame 6. After the connection establishment is confirmed in Frame 130, then direct communication between the two endpoints begins.

Trace 10.10a. SIP Phone-to-Phone Connection Summary
Sniffer Network Analyzer data from 29-Apr at 15:01:10, file SIP2.cap, Pg 1

Frame	Source	Destination	Summary
1	Cisco SIP Gateway	SIP Proxy Server	SIP: CINVITE sip:108@192.168.1.110; user=phone;phone context=national SIP/2.0
2	SIP Proxy Server	Cisco SIP Gateway	SIP: R SIP/2.0 Response Status Code=100 (Informational) Response-Phrase=Trying
3	SIP Proxy Server	Cisco SIP Phone 108	SIP: C INVITE sip:108@192.168.1.108 SIP/2.0
4	Cisco SIP Phone 108	SIP Proxy Server	SIP: R SIP/2.0 Response Status Code=100 (Informational) Response-Phrase=Trying
5	Cisco SIP Phone 108	SIP Proxy Server	SIP: R SIP/2.0 Response Status Code=180 (Informational) Response-Phrase=Ringing
6	SIP Proxy Server	Cisco SIP Gateway	SIP: R SIP/2.0 Response Status Code=180 (Informational) Response-Phrase=Ringing

.
.
.

129	Cisco SIP Phone 108	SIP Proxy Server	SIP: R SIP/2.0 Response Status Code=200 (Success) Response-Phrase=OK
130	SIP Proxy Server	Cisco SIP Gateway	SIP: R SIP/2.0 Response Status Code=200 (Success) Response-Phrase=OK
131	Cisco SIP Gateway	Cisco SIP Phone 108	SIP: C ACK sip:108 @192.168.1.108:5060 SIP/2.0
132	Cisco SIP Gateway	Cisco SIP Phone 108	RTP: Payload=PCMU audio SEQ=699 SSRC=552665418
133	Cisco SIP Gateway	Cisco SIP Phone 108	RTP: Payload=PCMU audio SEQ=700 SSRC=552665418
134	Cisco SIP Phone 108	Cisco SIP Gateway	RTP: Payload=PCMU audio SEQ=4067 SSRC=490861163
135	Cisco SIP Gateway	Cisco SIP Phone 108	RTP: Payload=PCMU audio SEQ=701 SSRC=552665418
136	Cisco SIP Phone 108	Cisco SIP Gateway	RTP: Payload=PCMU audio SEQ=4068 SSRC=490861163
137	Cisco SIP Gateway	Cisco SIP Phone 108	RTP: Payload=PCMU audio SEQ=702 SSRC=552665418
138	Cisco SIP Phone 108	Cisco SIP Gateway	RTP: Payload=PCMU audio SEQ=4069 SSRC=490861163
139	Cisco SIP Gateway	Cisco SIP Phone 108	RTP: Payload=PCMU audio SEQ=703 SSRC=552665418

140	Cisco SIP Phone 108	Cisco SIP Gateway	RTP: Payload=PCMU audio SEQ=4070 SSRC=490861163
.			
.			
.			
220	Cisco SIP Phone 108	Cisco SIP Gateway	RTP: Payload=PCMU audio SEQ=4109 SSRC=490861163
221	Cisco SIP Gateway	Cisco SIP Phone 108	RTP: Payload=PCMU audio SEQ=743 SSRC=552665418
222	Cisco SIP Phone 108	Cisco SIP Gateway	RTP: Payload=PCMU audio SEQ=4110 SSRC=490861163
223	Cisco SIP Gateway	Cisco SIP Phone 108	RTP: Payload=PCMU audio SEQ=744 SSRC=552665418
224	Cisco SIP Phone 108	Cisco SIP Gateway	RTP: Payload=PCMU audio SEQ=4111 SSRC=490861163
225	Cisco SIP Gateway	Cisco SIP Phone 108	RTP: Payload=PCMU audio SEQ=745 SSRC=552665418
226	Cisco SIP Phone 108	SIP Proxy Server	SIP: C BYE sip:14083465118 @192.168.1.74:5060; user=phone SIP/2.0
227	SIP Proxy Server	Cisco SIP Phone 108	SIP: R SIP/2.0 Response Status Code=100 (Informational) Response-Phrase=Trying
228	SIP Proxy Server	Cisco SIP Gateway	SIP: C BYE sip:14083465118 @192.168.1.74; user=phone SIP/2.0

```
229    Cisco SIP Gateway    Cisco SIP Phone 108  SIP: R SIP/2.0 Response
                                                   Status Code=200
                                                   (Success)
                                                   Response-Phrase=OK
```

The details of the SIP call establishment frames are shown in Trace 10.10b. Frame 1 is the SIP INVITE message, which identifies the Cisco SIP telephone (extension 108) that is being called. Also included in the INVITE message is the Session Description Protocol (SDP) information, which provides details regarding the session owner/creator (Cisco Systems SIP GW), session name (SIP Call), connection information (Internet, IPv4 192.168.1.74), start and stop times (0 0, representing a permanent, unbounded call), media type (audio), and Audio/Video Profile (AVP 0, representing Pulse Code Modulation mu-Law (G.711) audio information). The INVITE message that is passed to the destination from the gateway includes a Via parameter, identifying that gateway as an intermediate message point.

Other SIP messages involved in the connection establishment sequence include the TRYING (Frames 2 and 4), RINGING (Frames 5 and 6), and successful operation acknowledgments (Frames 129, 130, and 131) that function similarly to processes on the more familiar analog telephone network.

Trace 10.10b. SIP Call Establishment Message Details

Sniffer Network Analyzer data from 29-Apr at 15:01:10, file SIP2.cap, Pg 1

```
- - - - - - - - - - - - - - - - - - - - Frame 1 - - - - - - - - - - - - - - - - - - - -
SIP: — Session Initiation Protocol —
      SIP:
      SIP: INVITE sip =108@192.168.1.110; user=phone;phone-context=national SIP/2.0
      SIP: Via = SIP/2.0/UDP  192.168.1.74:58440
      SIP: From = "14083465118" <sip:14083465118@192.168.1.74>
      SIP: To = <sip:108@192.168.1.110;user=phone;phone-context=national>
      SIP: Date = Mon, 03 Mar 2003 22:15:45 UTC
      SIP: Call-ID = 4E43092B-AD0ECA15-0-F13A2A8@192.168.1.74
      SIP: Cisco-Guid = 1313016107-2903427603-0-252945060
      SIP: User-Agent = Cisco VoIP Gateway/ IOS 12.x/ SIP enabled
      SIP: CSeq = 101 INVITE
      SIP: Max-Forwards = 6
      SIP: Timestamp = 731196945
      SIP: Contact  = <sip:14083465118@192.168.1.74:5060;user=phone>
      SIP: Expires = 180
      SIP: Content-Type = application/sdp
      SIP: Content-Length = 134
      SIP:
```

```
SDP: — SDP Header —
    SDP:
    SDP: v(sdp protocol version) =0
    SDP: o(owner/creator and session identifier) =CiscoSystemsSIP-GW-UserAgent 7802 3827 IN IP4 192.168.1.74
    SDP: s(session name) =SIP Call
    SDP: c(connection information) =IN IP4 192.168.1.74
    SDP: t(<start time> <stop time>) =0 0
    SDP: m(media name and transport address) =audio 20080 RTP/AVP 0
    SDP:

- - - - - - - - - - - - - - - - - - - - Frame 2 - - - - - - - - - - - - - - - - - - - -
SIP: — Session Initiation Protocol —
    SIP:
    SIP: R SIP/2.0 Response Status Code=100(Informational) Response-Phrase=Trying
    SIP: Via = SIP/2.0/UDP  192.168.1.74:58440
    SIP: From = "14083465118" <sip:14083465118@192.168.1.74>
    SIP: To = <sip:108@192.168.1.110;user=phone;phone-context=national>
    SIP: Call-ID = 4E43092B-AD0ECA15-0-F13A2A8@192.168.1.74
    SIP: CSeq = 101 INVITE
    SIP: Content-Length = 0
    SIP: Server = IndigoSipServer/4.0
    SIP:

- - - - - - - - - - - - - - - - - - - - Frame 3 - - - - - - - - - - - - - - - - - - - -
SIP: — Session Initiation Protocol —
    SIP:
    SIP: INVITE sip =108@192.168.1.108 SIP/2.0
    SIP: Via  = SIP/2.0/UDP
            192.168.1.110;branch=c45115398e3b9314d91685a56ec22d
            ce531ecb88f58c3ad59dbb84c3c31e9
    SIP: Via = SIP/2.0/UDP  192.168.1.74:58440
    SIP: From = "14083465118" <sip:14083465118@192.168.1.74>
    SIP: To = <sip:108@192.168.1.110;user=phone;phone-context=national>
    SIP: Date = Mon, 03 Mar 2003 22:15:45 UTC
    SIP: Call-ID = 4E43092B-AD0ECA15-0-F13A2A8@192.168.1.74
    SIP: User-Agent = Cisco VoIP Gateway/ IOS 12.x/ SIP enabled
    SIP: CSeq = 101 INVITE
    SIP: Max-Forwards = 5
    SIP: Timestamp = 731196945
    SIP: Contact = <sip:14083465118@192.168.1.74:5060;user=phone>
    SIP: Expires  = 180
    SIP: Content-Type = application/sdp
    SIP: Content-Length  = 134
```

```
      SIP:Record-Route
               <sip:14083465118@192.168.1.74;user=phone;maddr=192.168.1.110;
               ForkingID=c61532549d38a692ffcae34698f21c8>
      SIP:
SDP: — SDP Header —
      SDP:
      SDP: v(sdp protocol version) =0
      SDP: o(owner/creator and session identifier) =CiscoSystemsSIP-GW-UserAgent
               7802 3827 IN IP4 192.168.1.74
      SDP: s(session name) =SIP Call
      SDP: c(connection information)  =IN IP4 192.168.1.74
      SDP: t(<start time> <stop time>) =0 0
      SDP: m(media name and transport address) =audio 20080 RTP/AVP 0
      SDP:

- - - - - - - - - - - - - - - - - - - - Frame 4 - - - - - - - - - - - - - - - - - - - -
SIP: — Session Initiation Protocol —
      SIP:
      SIP: R SIP/2.0 Response Status Code=100(Informational) Response-
               Phrase=Trying
      SIP: Via  = SIP/2.0/UDP
               192.168.1.110;branch=c45115398e3b9314d91685a56ec22dc-
               e531ecb88f58c3ad59dbb84c3c31e9,
      SIP/2.0/UDP  192.168.1.74:58440
      SIP: From = "14083465118" <sip:14083465118@192.168.1.74>
      SIP: To = <sip:108@192.168.1.110;user=phone;phone-
               context=national>;tag=f26b030083a1d410-0
      SIP: Call-ID = 4E43092B-AD0ECA15-0-F13A2A8@192.168.1.74
      SIP: Server = Cisco IP Phone/ Rev. 1/ SIP enabled
      SIP: CSeq = 101 INVITE
      SIP: Content-Length = 0
      SIP:

- - - - - - - - - - - - - - - - - - - - Frame 5 - - - - - - - - - - - - - - - - - - - -
SIP: — Session Initiation Protocol —
      SIP:
      SIP: R SIP/2.0 Response Status Code=180(Informational) Response-
               Phrase=Ringing
      SIP: Via = SIP/2.0/UDP
               192.168.1.110;branch=c45115398e3b9314d91685a56ec22dc-
               e531ecb88f58c3ad59dbb84c3c31e9,
      SIP/2.0/UDP  192.168.1.74:58440
      SIP: From = "14083465118" <sip:14083465118@192.168.1.74>
```

```
SIP: To  = <sip:108@192.168.1.110;user=phone;phone-
        context=national>;tag=f26b030083a1d410-0
SIP: Call-ID = 4E43092B-AD0ECA15-0-F13A2A8@192.168.1.74
SIP: Server = Cisco IP Phone/ Rev. 1/ SIP enabled
SIP: CSeq = 101 INVITE
SIP: Content-Length    = 0
SIP:

- - - - - - - - - - - - - - - - - - - - Frame 6 - - - - - - - - - - - - - - - - - - - -
SIP: — Session Initiation Protocol —
    SIP:
    SIP: R SIP/2.0 Response Status Code=180(Informational) Response-Phrase=Ringing
    SIP: Via = SIP/2.0/UDP  192.168.1.74:58440
    SIP: From = "14083465118" <sip:14083465118@192.168.1.74>
    SIP: To = <sip:108@192.168.1.110;user=phone;phone-
            context=national>;tag=f26b030083a1d410-0
    SIP: Call-ID = 4E43092B-AD0ECA15-0-F13A2A8@192.168.1.74
    SIP: Server = Cisco IP Phone/ Rev. 1/ SIP enabled
    SIP: CSeq = 101 INVITE
    SIP: Content-Length    = 0
    SIP:

- - - - - - - - - - - - - - - - - - - - Frame 129 - - - - - - - - - - - - - - - - - - - -
SIP: — Session Initiation Protocol —
    SIP:
    SIP: R SIP/2.0 Response Status Code=200(Success) Response-Phrase=OK
    SIP: Via = SIP/2.0/UDP
            192.168.1.110;branch=c45115398e3b9314d91685a56ec22dc-
            e531ecb88f58c3ad59dbb84c3c31e9,
    SIP/2.0/UDP  192.168.1.74:58440
    SIP: From = "14083465118" <sip:14083465118@192.168.1.74>
    SIP: To = <sip:108@192.168.1.110;user=phone;phone-
            context=national>;tag=f26b030083a1d410-0
    SIP: Call-ID = 4E43092B-AD0ECA15-0-F13A2A8@192.168.1.74
    SIP: Server = Cisco IP Phone/ Rev. 1/ SIP enabled
    SIP: Contact = sip:108@192.168.1.108:5060
    SIP: Record-Route =
            <sip:14083465118@192.168.1.74;user=phone;maddr=192.168.1.110;
            ForkingID=c61532549d38a692ffcae34698f21c8>
    SIP: CSeq = 101 INVITE
    SIP: Content-Type = application/sdp
    SIP: Content-Length = 220
    SIP:
```

```
SDP: — SDP Header —
    SDP:
    SDP: v(sdp protocol version) =0
    SDP: o(owner/creator and session identifier) =CiscoSystemsSIP-IPPhone-
            UserAgent 4439 20135 IN IP4 192.168.1.108
    SDP: s(session name) =SIP Call
    SDP: c(connection information)  =IN IP4 192.168.1.108
    SDP: t(<start time> <stop time>) =0 0
    SDP: m(media name and transport address) =audio 18692 RTP/AVP 0 101
    SDP: a(media attributes) =rtpmap:0 pcmu/8000
    SDP: a(media attributes) =rtpmap:101 telephone-event/8000
    SDP: a(media attributes) =fmtp:101 0-11
    SDP:

- - - - - - - - - - - - - - - - - - - - Frame 130 - - - - - - - - - - - - - - - - - - - -
SIP: — Session Initiation Protocol —
    SIP:
    SIP: R SIP/2.0 Response Status Code=200(Success) Response-Phrase=OK
    SIP: Via = SIP/2.0/UDP  192.168.1.74:58440
    SIP: From = "14083465118" <sip:14083465118@192.168.1.74>
    SIP: To = <sip:108@192.168.1.110;user=phone;phone-
            context=national>;tag=f26b030083a1d410-0
    SIP: Call-ID = 4E43092B-AD0ECA15-0-F13A2A8@192.168.1.74
    SIP: Server = Cisco IP Phone/ Rev. 1/ SIP enabled
    SIP: Contact = sip:108@192.168.1.108:5060
    SIP: Record-Route = <sip:108@192.168.1.108;maddr=192.168.1.110;
            ForkingID=c61532549d38a692ffcae34698f21c8>
    SIP: CSeq = 101 INVITE
    SIP: Content-Type = application/sdp
    SIP: Content-Length = 220
    SIP:
SDP: — SDP Header —
    SDP:
    SDP: v(sdp protocol version) =0
    SDP: o(owner/creator and session identifier) =CiscoSystemsSIP-IPPhone-
            UserAgent 4439 20135 IN IP4 192.168.1.108
    SDP: s(session name) =SIP Call
    SDP: c(connection information) =IN IP4 192.168.1.108
    SDP: t(<start time> <stop time>) =0 0
    SDP: m(media name and transport address) =audio 18692 RTP/AVP 0 101
    SDP: a(media attributes) =rtpmap:0 pcmu/8000
    SDP: a(media attributes) =rtpmap:101 telephone-event/8000
    SDP: a(media attributes) =fmtp:101 0-11
    SDP:
```

```
- - - - - - - - - - - - - - - - - - - - - Frame 131 - - - - - - - - - - - - - - - - - - - -
SIP: — Session Initiation Protocol —
      SIP:
      SIP: ACK sip =108@192.168.1.108:5060 SIP/2.0
      SIP: Via = SIP/2.0/UDP  192.168.1.74:58440
      SIP: From = "14083465118" <sip:14083465118@192.168.1.74>
      SIP: To = <sip:108@192.168.1.110; user=phone; phone-context=national>;
             tag=f26b030083a1d410-0
      SIP: Date = Mon, 03 Mar 2003 22:15:45 UTC
      SIP: Call-ID = 4E43092B-AD0ECA15-0-F13A2A8@192.168.1.74
      SIP: Route = <sip:108@192.168.1.108:5060>
      SIP: Max-Forwards = 6
      SIP: Content-Type = application/sdp
      SIP: Content-Length = 134
      SIP: CSeq = 101 ACK
      SIP:
SDP: — SDP Header —
      SDP:
      SDP: v(sdp protocol version) =0
      SDP: o(owner/creator and session identifier) =CiscoSystemsSIP-GW-UserAgent
             7802 3827 IN IP4 192.168.1.74
      SDP: s(session name) =SIP Call
      SDP: c(connection information) =IN IP4 192.168.1.74
      SDP: t(<start time> <stop time>) =0 0
      SDP: m(media name and transport address) =audio 20080 RTP/AVP 0
      SDP:
```

Once the connection has been established, audio packets can flow between the two endpoints. The audio information is encapsulated using an Internet Protocol (IP) header, a User Datagram Protocol (UDP) header, and finally a Real-time Transport Protocol (RTP) header, as illustrated in Figure 10-10d.

From Trace 10.10c, note that, within the IP header, the source and destination IP addresses indicate the end-to-end nature of this communication, without the ongoing involvement of the proxy server (review Figure 10-10c). The RTP header indicates a number of parameters, including the RTP version (2, from RFC 1889), a timestamp for this sample (668592, indicating 83,574.000 milliseconds), and the Payload Type (0, indicating PCM mu-Law encoded audio). Figure 10-10d illustrates each of these fields, which can be compared with the analyzer details shown in Trace 10.10c.

Trace 10.10c. RTP Packet with G.711 Audio Payload Details
Sniffer Network Analyzer data from 29-Apr at 15:01:10, file SIP2.cap, Pg 1

```
- - - - - - - - - - - - - - - - - - - - Frame 138 - - - - - - - - - - - - - - - - - - - -
DLC: — DLC Header —
     DLC:
     DLC: Frame 138 arrived at 14:02:02.4409; frame size is 214 (00D6 hex) bytes.
     DLC: Destination = Station Cisco C5D202
     DLC: Source = Station Cisco F2411D
     DLC: Ethertype = 0800 (IP)
     DLC:
IP: — IP Header —
     IP:
     IP: Version = 4, header length = 20 bytes
     IP: Type of service = A0
     IP:      101. ....  = CRITIC/ECP
     IP:      ...0 .... = normal delay
     IP:      .... 0... = normal throughput
     IP:      .... .0.. = normal reliability
     IP:      .... ..0. = ECT bit - transport protocol will ignore the CE bit
     IP:      .... ...0 = CE bit - no congestion
     IP: Total length    = 200 bytes
     IP: Identification  = 14029
     IP: Flags           = 0X
     IP:      .0.. .... = may fragment
     IP:      ..0. .... = last fragment
     IP: Fragment offset = 0 bytes
     IP: Time to live    = 64 seconds/hops
     IP: Protocol        = 17 (UDP)
     IP: Header checksum = BEB1 (correct)
     IP: Source address      = [192.168.1.108]
     IP: Destination address = [192.168.1.74]
     IP: No options
     IP:
UDP: — UDP Header —
     UDP:
     UDP: Source port      = 18692
     UDP: Destination port = 20080
     UDP: Length           = 180
     UDP: No checksum
     UDP: [172 byte(s) of data]
     UDP:
```

```
RTP: — Real-Time Transport Protocol —
    RTP:
    RTP: Ver, Pad, Ext, CC:      = 80
    RTP:              10.. .... = Version = 2  (RFC 1889)
    RTP:              ..0. .... = Padding = 0  (Zero bytes of Padding at the End)
    RTP:              ...0 .... = Header Extension Bit = 0, (No Header Extension after Fixed Header)
    RTP:              .... 0000 = Contributor Count = 0
    RTP: Marker, Payload Type: = 00
    RTP:           0... .... = Marker 0
    RTP:           .000 0000 = Payload Type 0 (PCMU audio)
    RTP: Sequence Number      = 4069
    RTP: Time Stamp           = 668592 (83574.000 ms)
    RTP: SSRC                 = 490861163
    RTP:
    RTP: Payload Type = 0 (PCMU audio)
    RTP:
```

When the two parties have completed their conversation, the call is terminated, beginning with the BYE message sent from the Cisco Telephone in Frame 226 (review Figure 10-10c).

Note in Trace 10.10d that the same Call ID that was used in the INVITE message is referenced in the BYE message (4E43092B-AD0ECA15-0-F13A2A8@192.168.1.74). Two other confirmation messages, TRYING (Frame 227) and BYE (Frame 228), are passed between the proxy server and the end stations, followed by an end-to-end OK as the final confirmation in Frame 229. The previously mentioned Call ID is contained in these messages as well.

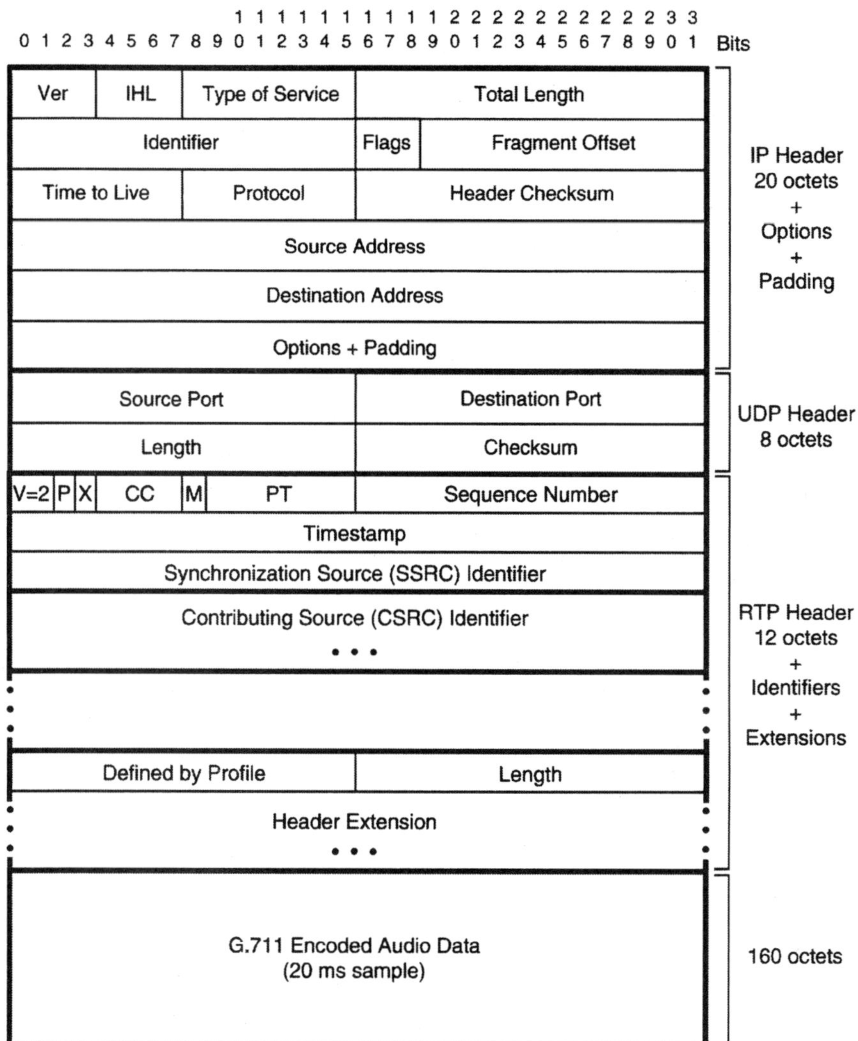

Figure 10-10d. Voice over IP Packet Format with G.711 Encoded Audio Data.

Trace 10.10d. SIP Call Disconnect Message Details

Sniffer Network Analyzer data from 29-Apr at 15:01:10, file SIP2.cap, Pg 1

```
- - - - - - - - - - - - - - - - - - - - Frame 226 - - - - - - - - - - - - - - - - - - - -
SIP: — Session Initiation Protocol —
      SIP:
      SIP: BYE sip =14083465118@192.168.1.74:5060;user=phone SIP/2.0
      SIP: Via = SIP/2.0/UDP 192.168.1.108:5060
      SIP: From = <sip:108@192.168.1.110;user=phone;phone-
                context=national>;tag=f26b030083a1d410-0
      SIP: To = "14083465118" <sip:14083465118@192.168.1.74>
      SIP: Call-ID = 4E43092B-AD0ECA15-0-F13A2A8@192.168.1.74
      SIP: User-Agent = Cisco IP Phone/ Rev. 1/ SIP enabled
      SIP: CSeq = 101 BYE
      SIP: Route = <sip:14083465118@192.168.1.74:5060;user=phone>
      SIP: Content-Length  = 0
      SIP:

- - - - - - - - - - - - - - - - - - - - Frame 227 - - - - - - - - - - - - - - - - - - - -
SIP: — Session Initiation Protocol —
      SIP:
      SIP: R SIP/2.0 Response Status Code=100(Informational) Response-Phrase=Trying
      SIP: Via = SIP/2.0/UDP 192.168.1.108:5060
      SIP: From = <sip>
      SIP:

- - - - - - - - - - - - - - - - - - - - Frame 228 - - - - - - - - - - - - - - - - - - - -
SIP: — Session Initiation Protocol —
      SIP:
      SIP: BYE sip =14083465118@192.168.1.74;user=phone SIP/2.0
      SIP: Via = SIP/2.0/UDP
                192.168.1.110;branch=f059ddc9dc161e2c7cd53bb1fc4d45a-
                a5ca34313ad49789bf3de2774b2f7e7
      SIP: Via = SIP/2.0/UDP 192.168.1.108:5060
      SIP: From = <sip:108@192.168.1.110;user=phone;phone-
                context=national>;tag=f26b030083a1d410-0
      SIP: To = "14083465118" <sip:14083465118@192.168.1.74>
      SIP: Call-ID = 4E43092B-AD0ECA15-0-F13A2A8@192.168.1.74
      SIP: User-Agent = Cisco IP Phone/ Rev. 1/ SIP enabled
      SIP: CSeq = 101 BYE
      SIP: Content-Length = 0
      SIP:
```

```
-------------------- Frame 229 --------------------
SIP: — Session Initiation Protocol —
    SIP:
    SIP: R SIP/2.0 Response Status Code=200(Success) Response-Phrase=OK
    SIP: Via = SIP/2.0/UDP
            192.168.1.110;branch=f059ddc9dc161e2c7cd53bb1fc4d45a-
            a5ca34313ad49789bf3de2774b2f7e7,
    SIP/2.0/UDP 192.168.1.108:5060
    SIP: From = <sip:108@192.168.1.110;user=phone;phone-
            context=national>;tag=f26b030083a1d410-0
    SIP: To = "14083465118" <sip:14083465118@192.168.1.74>
    SIP: Date = Wed, 03 Mar 1993 22:15:56 UTC
    SIP: Call-ID = 4E43092B-AD0ECA15-0-F13A2A8@192.168.1.74
    SIP: Server = Cisco VoIP Gateway/ IOS 12.x/ SIP enabled
    SIP: Content-Length = 0
    SIP: CSeq  = 101 BYE
    SIP:
```

10.11 Looking Ahead

This chapter concludes our study of the ARPA protocols that are used for data and multimedia transport. We will consider network management issues in Chapters 11, 12, and 13, and discuss how the Simple Network Management Protocol (SNMP) can effectively manage TCP/IP-based internetworks.

11

Network Management Architectures

As we have discovered in the first ten chapters of this text, computer and communications networks that are based on the TCP/IP suite of protocols, or in some other way tied to Internet technologies, can be quite complex. Quite often, a number of LAN and WAN connections, different hardware elements, protocols, and end-user applications are involved in the end-to-end communication system. For these individual elements to work together effectively requires an overall management process to oversee that entire operation. We refer to that management process as a *network management system*, or NMS, which is the subject of this and the next two chapters of this text. This chapter will consider the architecture of an NMS; Chapter 12 will look at the logical components, such as management databases and management protocols that are the building blocks for the NMS; and Chapter 13 will discuss case studies that illustrate the capabilities and benefits of the NMS.

11.1 The Challenge of Network Management

To briefly review Chapter 2, the 1970s was the decade of the centralized network. In a decade dominated by mainframe processing, data communication allowed terminals to talk to the mainframe, and low speed, asynchronous transmission (typically 300–9,600 bps) was the norm. Mainframe providers such as IBM and communication circuit providers such as AT&T or the local telephone company managed the network for those systems, leaving little responsibility for the on-site network manager. The 1980s saw significant changes in data communications, including microprocessors offering significant price and performance advantages over mainframes, the advent of the LAN, and the emergence of T-carrier circuits for WAN communication. In the 1990s, the proliferation of LANs gave rise to distributed processing and moved applications off the mainframe and onto the desktop, plus the Internet became a viable business communication medium.

As data communication shifted to distributed networks, network management became distributed as well (see Figure 11-1). Thus, the centralized management console used with mainframe environments gave way to a network management station for distributed LANs. Further shifts are resulting from the use of World Wide Web–based technologies, which use widely available web browsers to access network management information. As the information transport requirements increase in both complexity and speed, network management capabilities must mature as well.

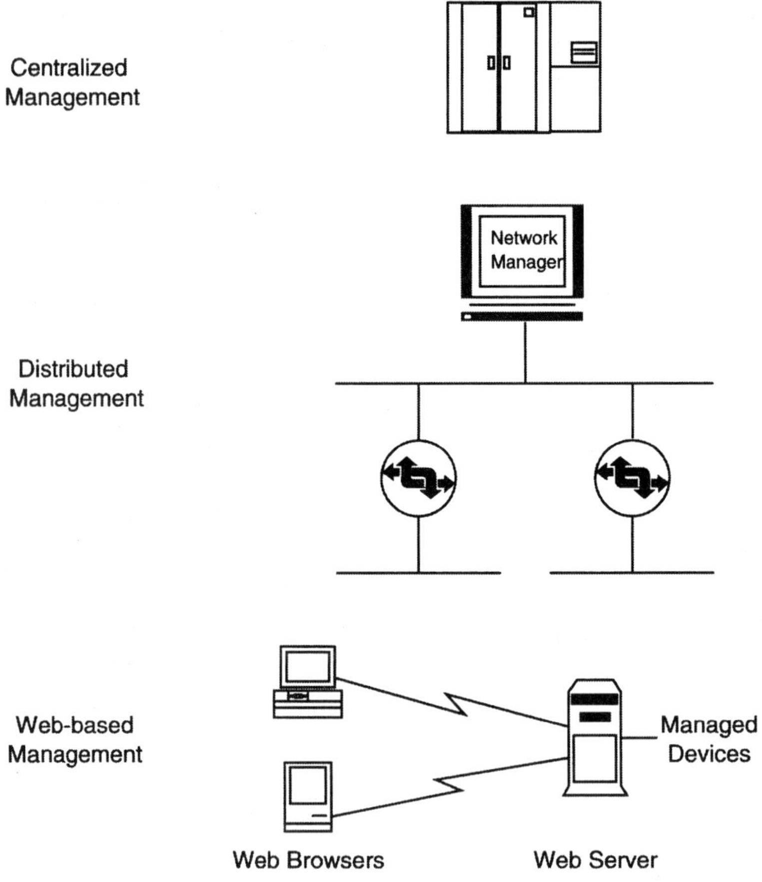

Figure 11-1. Evolution in Distributed Systems.

However, sometimes people forget that network management has two parts: the network and the management. To manage a network properly, all of the people involved must agree on the meaning of network management and on its objectives.

Network management can mean different things to the different individuals in an organization, such as the chief financial officer (CFO), the chief information officer (CIO), and the end users. The CFO tends to view the network (and its manager) as a line item on the expense budget. CFOs consider computing and data communications as a way to manage orders, inventory, accounting information, and so on. As long as overall corporate revenues hit their target, these budget items are likely to remain intact. Therefore, the CFO would define network management as the financial management of the corporate communications network.

The CIO must look at network management from the theoretical perspective of the CFO and the corporate budget as well as from the practical perspective of the end users. The goal is to keep the corporate network running 99.999 percent of the time (the often-quoted "5 nines"), and to schedule periods of downtime on weekends and holidays when few are around to notice. The CIO would, therefore, define network management as the ability to balance increasing end-user requirements with decreasing resources — that is, the ability to provide more service with less money.

End users spend their days in the network trenches, designing airplanes, writing dissertations, and attending less than inspiring meetings. Their jobs depend on the network remaining operational. Thus, end users would define network management as something that keeps the data communication infrastructure, which they depend upon, working at all times. A network failure could threaten their livelihood.

From the standpoint of the financial health of the corporation, its customers, and its employees, an all-encompassing rationale of network management would be something like this: The communications network is the vital link between customers and products. Our objective is to keep that link operating at all times, because when it fails our financial health suffers.

11.2 The System Being Managed

Now, let's shift to a systems-engineering perspective on network management. Most networks contain three key elements: centralized processing systems, distributed processing systems, and a communication infrastructure connecting the two (Figure 11-2). For example, the centralized applications could be an inventory control system or the corporate financial database. The distributed applications could be client-server or web-based applications running on LANs. In the middle is the glue that connects the different types of systems — the wide area transport. This transport may consist of public networks provided by Inter-Exchange Carriers (IXCs), Local Exchange Carriers (LECs), cellular telephone carriers, Internet Service Providers (ISPs), and Internet Telephony Service Providers (ITSPs). Private networks may include point-to-point lines, backbone connections, virtual private network (VPN) technologies, and so on.

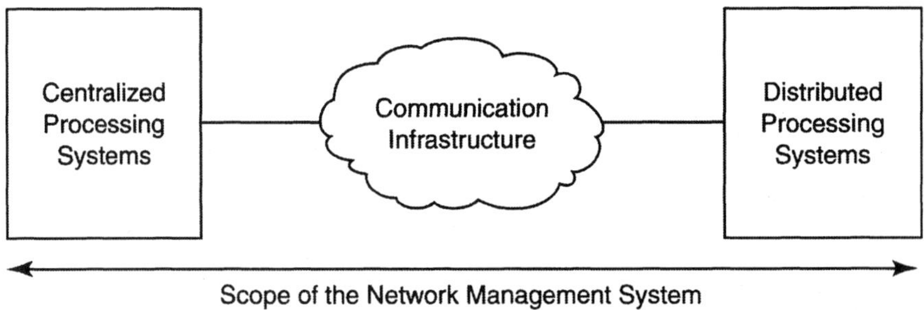

Figure 11-2. The Scope of Network Management Systems.

Managing the communications infrastructure can be especially challenging because of its cost structure. Most LANs have a relatively low cost (or a correspondingly efficient price/performance) per user, since a significant portion of the hardware is either integrated into the workstation (such as a built-in network interface card on a laptop computer) or is already provided within the building (such as the twisted pair cabling). However, the WAN is a completely different story, as the cost is typically based on transmission rate and distance traversed, also referred to as a "pay-as-you-go" pricing model. So when you try to connect two LANs across the wide area, you have a tradeoff to consider: how much bandwidth between the two endpoints are you willing to pay for, since the more you consume, the more you pay? If your economic analysis is like most, you will end up with a WAN bandwidth that is less than your LAN capacity, which is often referred to as the "WAN bottleneck" (Figure 11-3). Fortunately, this disparity in bandwidth capacities between LANs and WAN may not cause serious technical difficulties such as response time delays, because traffic patterns may dictate that a significant amount of the LAN traffic stays on that LAN (and does not traverse the WAN). Nevertheless, the potential for bottlenecks during periods of high network traffic does exist.

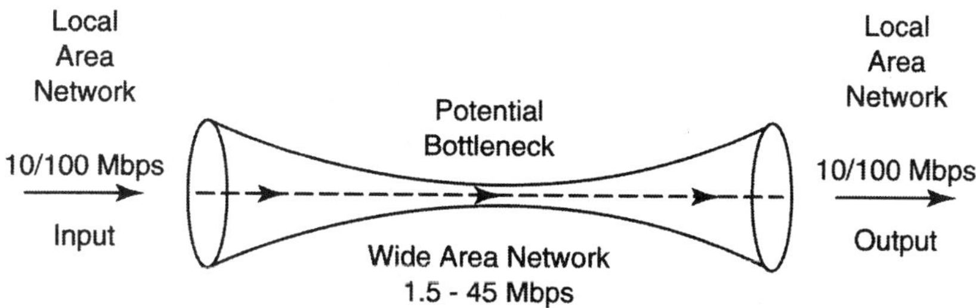

Figure 11-3. The WAN Bottleneck Challenge.

So when all the LAN and WAN elements plus their various cost factors are considered, the remaining key question to be considered is: "what part of this network are you responsible for?" In decades past, the network management model aligned budgets and personnel with the three elements: one management team for the mainframe, another for the local networks, and a third for the wide area networks and communication infrastructure. The present economic environment demands greater efficiencies, meaning that most organizations have one network management team that is responsible for all three of these elements, whether that team is experientially and technically equipped for that integrated task or not. Thus, the scope of your network management responsibilities is very likely to include all three subsystems: centralized and distributed processing, plus all communication infrastructures in between. And to meet this challenge, we must deploy a network management architecture that is equally adept at addressing each of these three elements.

11.3 The OSI Network Management Model

The ISO/OSI model has been a benchmark for computer networking since it was first published in 1978 (review Section 2.1.2 and Figures 2-5 and 2-6). An adjunct document, known as ISO 7498-4 [11-1] (and also ITU-T Recommendation X.700) breaks down the network management process into five smaller areas. These are called the OSI specific management functional areas (SMFAs): fault management, accounting management, configuration management, performance management, and security management (Figure 11-4). Let's look at these five areas from the perspective of the standard, and then from the perspective of practical applications.

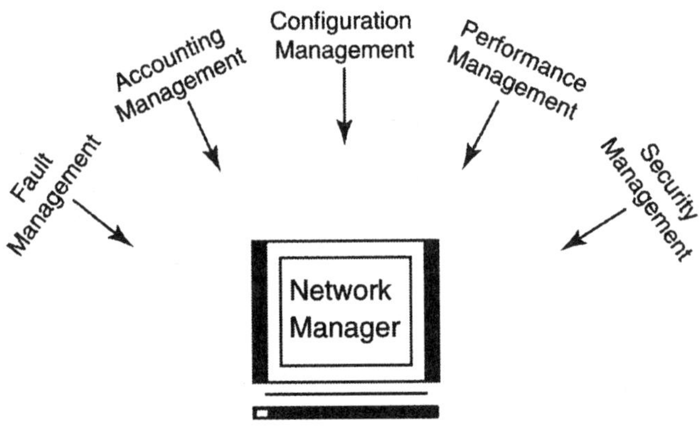

Figure 11-4. OSI Network Management Functional Areas.

11.3.1 Fault Management

The standard says that fault management "encompasses fault detection, isolation, and the correction of the abnormal operation of the OSI environment." The standard goes on to consider error logs, fault identification, and diagnostic testing.

In plain English, fault management means that you need to first identify, then repair, network faults. There are two ways to manage faults: reactively or proactively. A reactive manager waits for a problem and then troubleshoots it. A proactive manager examines the manager and agents to determine whether they are exceeding critical operational thresholds, such as network utilization. If excesses occur, the proactive administrator determines their source and reduces them accordingly.

11.3.2 Accounting Management

The standard says that accounting management "enables charges to be established for the use of resources in the OSI environment, and for cost to be identified for the use of those resources." Other considerations include informing the users of the costs and resources consumed, setting accounting limits, and incorporating tariff information into the overall accounting process.

In the real world, accounting means dealing with real people using real network resources with real operating expenses. Examples of these costs include disk usage and data archiving, telecommunication expenses for access to remote data, and charges for sending electronic mail messages. You can also use accounting management to determine whether network resource utilization is increasing because of growth, which might indicate the need for additions or rearrangements in the near future.

11.3.3 Configuration Management

The standard says that configuration management "identifies, exercises control over, collects data from, and provides data to open systems for the purpose of preparing for, initializing, starting, providing for the continuous operation of, and terminating interconnection services." These services might include the collection of information regarding the system, alerts regarding system changes, and changes to the configuration of the system.

In the real world, the acronym MAC, which stands for moves, adds, and changes, typifies the majority of configuration management activities. Networks are dynamic systems, and network administrators need to move personnel and rearrange their processing needs. This aspect of network management may be as simple as rearranging modular connectors at a wiring hub, or as complex as installing a LAN and its associated servers, communication circuits, and so on, at a remote location. Therefore, a significant aspect of the network management function involves keeping track of all these changes by using some type of database.

11.3.4 Performance Management

In the standard, performance management "enables the behavior of resources in the OSI environment and the effectiveness of communication activities to be evaluated." These functions include gathering statistical and historical information and evaluating the system's performance under a variety of real and hypothetical conditions.

Practically, performance management ensures that the administrator satisfies the end users' needs at all times. To do this, the administrator must select hardware and software systems according to the needs of the internetwork, then exercise these systems to their maximum potential. In some cases, administrative policies are developed which delineate these network performance requirements. Hence the term *policy management* is often used to further describe these performance management functions. In addition, performance and fault management are closely related, since you need to eliminate, or at least minimize, faults to obtain optimum performance. Many tools are available to measure performance. These include protocol analyzers, network monitoring software, and various utilities that come with the console programs of network operating systems.

11.3.5 Security Management

Academically, "the purpose of security management is to support the application of security policies by means of functions which include the creation, deletion, and control of security services and mechanisms; the distribution of security-relevant information; and the reporting of security related events."

In other words, security protects the network. It defends against viruses, ensures that remote and local users are authenticated, and installs encryption systems on any communication circuits that connect to a remote site.

Thus, a comprehensive network management system must address a multitude of elements — all the way from the physical configuration to network security — and do so in a cost-effective, efficient manner. Let's next see how such a system can be designed and deployed.

11.4 Elements of a Network Management Architecture

A network management system must have the flexibility to cover all three elements of the communications architecture (centralized and distributed processing, plus the interconnecting communications infrastructure), while addressing the five SMFAs described above. Most network management solutions accomplish this using a paradigm called the manager/agent model, which consists of a manager, a managed system, a database of management information, and the network protocol (see Figure 11-5).

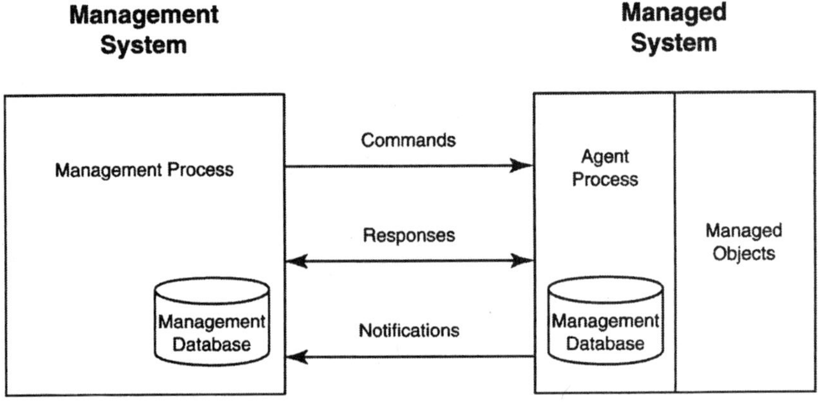

Figure 11-5. Network Manager/Agent Relationships.

The manager provides the interface between the human network manager and the devices being managed. It also provides the network management process. The management process performs tasks such as measuring traffic on a remote LAN segment or recording the transmission speed and physical address of a router's LAN interface. The manager also includes some type of output, usually graphical, to display management data, historical statistics, and so on. A common example of a graphical display is a map of the internetwork topology showing the locations of the LAN segments; selecting a particular segment might display its current operational status.

As Figure 11-5 shows, the managed system consists of the agent process and the managed objects. The agent process performs network management operations such as setting configuration parameters and current operational statistics for a router on a given segment. The managed objects could include workstations, servers, wiring hubs, communication circuits, and so on, or a subsystem of one of those devices, such as the processor type, amount of available memory on the workstation or server, or applications deployed. Associated with the managed objects are attributes, which may be statically defined (such as the speed of the interface), dynamic (such as entries in a routing table), or require ongoing measurement (such as the number of packets transmitted without errors in a given time period). A database of network management information, called the management information base (MIB), is associated with both the manager and the managed system. Just as a numerical database has a structure for storing and retrieving data, a MIB has a defined organization.

The network management protocol provides a way for the manager, the managed objects, and their agents to communicate. To structure the communication process, the protocol defines specific messages, referred to as commands, responses, and notifications. The manager uses these messages to request specific management information, and the agent uses them to respond. The building blocks of the messages are called protocol data units (PDUs). For example, a manager sends a *GetRequest PDU* to retrieve information, and the agent responds with a *GetResponse PDU*.

How does the manager/agent model relate to the network you need to manage? As you can see in Figure 11-6, a console, which typically has abundant processing power and memory, performs the network manager functions. The devices on the internetwork, such as routers and host computers, contain network management agents. A MIB is associated with both the manager and the various agents, but the router's view of the MIB and the host's view of the MIB are unlikely to be the same. First, these devices usually come from different manufacturers who have implemented network management functions in different, but complementary, ways. Second, routers and hosts perform different internetworking functions and may not need to store the same information. For example, the host may not require routing tables, and thus won't need to store routing table–related parameters such as the next hop to a particular destination in its MIB. Conversely, a router's MIB wouldn't contain a statistic such as CPU utilization that may be significant to a host.

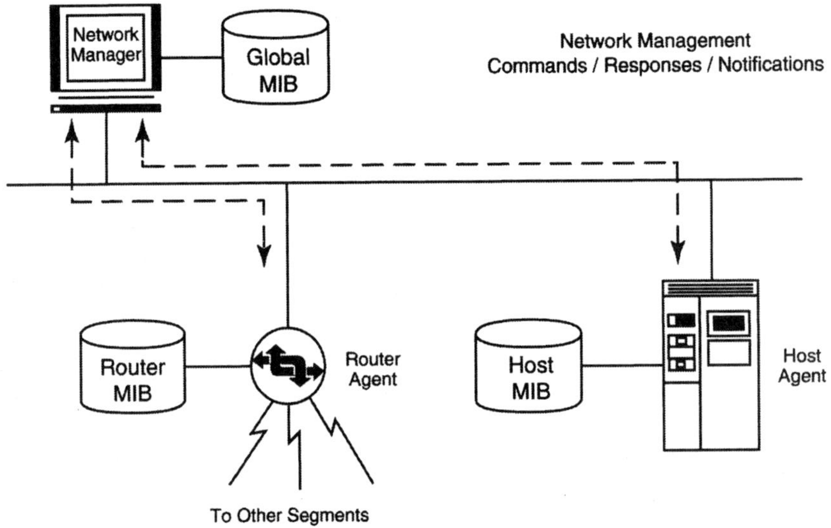

Figure 11-6. Network Manager/Agent Realization.

A protocol such as the Simple Network Management Protocol (SNMP) allows the manager and the agents to communicate. This protocol provides the structure for commands from the manager, notifies the manager of significant events from the agent, and responds to either the manager or the agent. SNMP is only part of what is known as the Internet Network Management Framework. Let's examine that framework before delving into the intricacies of SNMP.

11.5 The Internet Network Management Framework

Recall from our previous discussions that the Internet was derived from U.S. Government research dating back to 1969. The objective was to develop communication technologies and protocols so that government organizations, defense contractors, and academic researchers using dissimilar computer systems could collaborate on projects. The result of the government's research project was the Advanced Research Projects Agency network (ARPANET), which used packet switching technology to connect dissimilar systems. ARPANET came online in 1969 with nodes in four locations in the United States. From that humble beginning, the Internet now connects millions of host computers worldwide.

By the late 1980s, the Internet Architecture Board (IAB) realized that it needed a method to manage the growing Internet and the other attached networks. The board considered several proposals and decided to take a two-step approach to Internet management. Enhancements to the existing Simple Gateway Monitoring Protocol (SGMP), which the regional networks that made up the Internet were using, would provide a short-term solution. The long-term solution would be based on an ISO protocol architecture called the Common Management Information Service/Common Management Information Protocol (CMIS/CMIP) architecture. This proposed longer-term solution used TCP/IP instead of ISO protocols for the communications infrastructure, and was therefore called CMOT (CMIP over TCP/IP). RFC 1052 [11-2] summarizes these earlier network management directives.

The CMOT blend of ISO and Internet protocols made a lot of sense at the time, as the world was eagerly awaiting the advent of the computer architectures that were fully compliant with the OSI model and protocols (the CMIP part), while acknowledging that the vast majority of infrastructures were TCP/IP based (the TCP/IP part). Architecturally, CMOT fits the manager/agent paradigm. However, unlike SNMP, which provides connectionless service using UDP/IP, CMOT uses an association-oriented communication mechanism and the TCP/IP protocol to ensure reliable transport of data. To guarantee reliable transport, CMOT systems establish Application layer connections prior to transmitting management information. CMOT's Application layer services are built on three OSI services: the Common Management Information Service Element (CMISE), the Remote Operation Service Element (ROSE), and the Association Control Service Element (ACSE). A Lightweight Presentation Protocol (LPP) provides Presentation layer services. The long-term CMOT solution, however, never received the widespread acceptance of SNMP. In addition, dwindling support for ISO protocols in general has diminished interest in CMIP and CMOT, although these protocols are still mentioned in conjunction with some vendors' network management architectures. CMOT is documented in RFC 1189 [11-3].

The current version of the Internet-standard Network Management Framework (version 3, documented in RFC 3410 [11-4]) provides a bridge between the earlier work done in the late 1980s and the network management solutions that are required to support more current technologies. In making that bridge, the framework also documents the three versions of the SNMP: version 1, released in 1990; version 2, released in 1996; and version 3, released in 1999. We will study the specifics of these protocol versions in Chapter 12.

An enterprise deploying a management framework would contain four different components: managed nodes, called *agents*, which provide remote access to management information; a management process, called the *manager*, which runs management applications; a management protocol (such as SNMP), used to convey information between the entities; and the management information itself. The network management framework provides the architecture necessary for managing these four components. That framework consists of: a data definition language, known as the Structure of Management Information, or SMI; definitions of management information, which is called the Management Information Base, or MIB; a protocol definition (SNMP); plus security and administration.

As the three versions of the Internet-standard Network Management Framework have evolved, the SMI, MIB, protocol, and security/administration functions have progressed as well. As an introduction, consider a model of the SNMPv1 architecture illustrated in Figure 11-7.

SNMP is based on the manager/agent model. It is referred to as "simple" because the agent requires minimal software — most of the processing power and data storage resides on the management system, while a complementary subset of those functions resides on the managed system. This was done intentionally to encourage vendors to include agent capabilities in the hardware or software products, with a minimal cost and processing overhead. To achieve its goal of being simple, SNMP includes a limited set of management commands and responses (also shown in Figure 11-7). The management system issues Get, GetNext, and Set messages to retrieve single or multiple object variables or to establish the value of a single variable. The managed system sends a Response message to complete the Get, GetNext, or Set. The managed system sends an event notification, called a Trap, to the management system to identify the occurrence of conditions such as a threshold that exceeds a predetermined value.

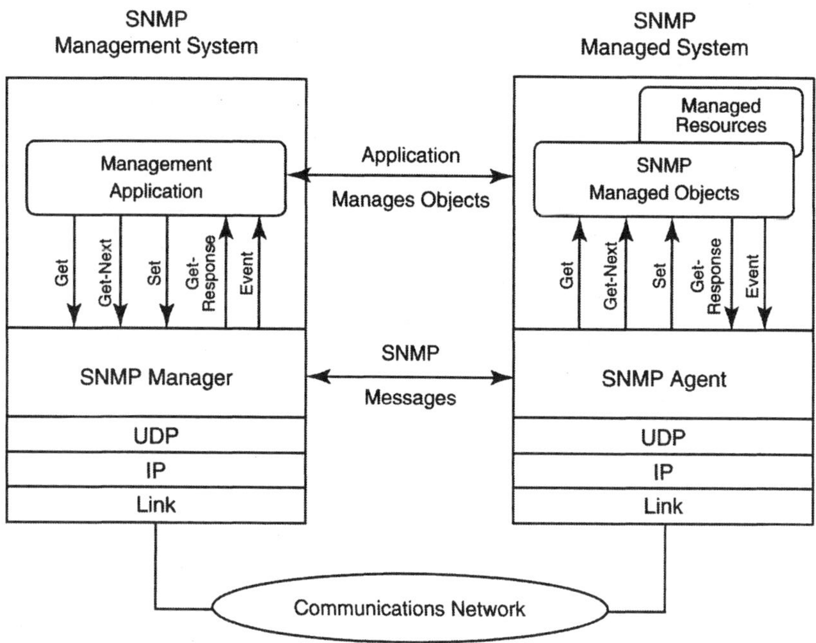

Figure 11-7. SNMP Architecture (© 1990, IEEE).

SNMP assumes that the communication path is a connectionless communication subnetwork. In other words, no prearranged communication path is established prior to the transmission of data. As a result, SNMP makes no guarantees about the reliable delivery of the data; however, in practice most messages get through, and those that don't can be retransmitted. From Figure 11-7, note that the transport protocols that this architecture implements are the User Datagram Protocol (UDP) and the Internet Protocol (IP). An SNMP-based management system also requires Data Link layer protocols, such as an Ethernet LAN or frame relay WAN, to implement the communication channel from the management system (or console) to the managed system (or agent).

SNMP's simplicity and connectionless communication also produce a degree of robustness. Neither the manager nor the agent relies on the other for its operation. Thus, a manager may continue to function even if a remote agent fails. When the agent resumes functioning, it can send a trap to the manager, notifying it of its change in operational status. SNMPv1 is defined in RFC 1157 [11-5].

11.6 Web-based Network Management

Network management platforms have evolved in the last few years, as client systems have become more distributed and complex. Most of these improvements have focused on the network management system and its interaction with agents and applications, rather than on the user interface to the network management information itself. As World Wide Web–based applications have proliferated through networking technologies, network management systems have adopted this same technology so that network management information stored on a web server can be accessed and disseminated to distributed users on a worldwide basis and in a platform-independent fashion. Web-based network management can take on one of several forms (see Figure 11-8).

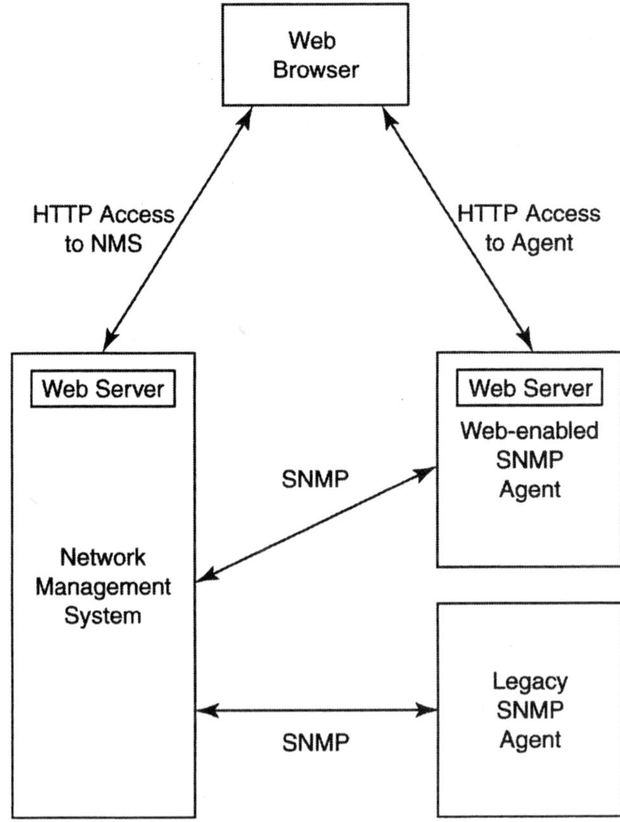

Figure 11-8. Web-based Management Architecture.

First, web-enabled agents can be managed through a browser using the HyperText Transfer Protocol (HTTP) for communication with that agent. Second, web-enabled managers, which may include a web server front end to an existing platform or a stand-alone manager running on a web server, enable access to their management data from any compatible web browser.

One standardization effort that has been undertaken to address web-based network management systems is the Web-based Enterprise Management (WBEM), which is sponsored by the Distributed Management Task Force (DMTF) [11-6]. The DMTF is an industry consortium of computer and networking manufacturers that includes Cisco Systems, Dell Computer, Hewlett-Packard, IBM, Intel, Microsoft, Novell, Oracle, and Sun Microsystems among its membership. The underlying objective of the DMTF is to extend network management capabilities down to the hardware and software components that make up a desktop or networking system. To meet this objective, the organization has developed a number of standard formats and protocols: the Alert Standard Format (ASF), Common Information Model (CIM), Desktop Management Interface (DMI), Directory Enabled Networking (DEN), System Management BIOS (SMBIOS), and the Web-Based Enterprise Management (WBEM).

The ASF provides remote control and alerting interfaces for computers that lack operating system–based systems management functions. The CIM is a model that describes networking environments, including the managed entities, their states, configuration, relationships, and so on. The DEN extends the CIM and describes how to use a CIM and a directory to locate management information and to access management data. The DMI established a standard framework for managing networked desktop systems and servers, including the ability to send management information across a network. The SMBIOS specification describes a standard format for motherboard and system vendors to present management information about their products. Finally, the WBEM is an initiative that couples the CIM with Internet-standard protocols, such as the Extensible Markup Language (XML) and the Hypertext Markup Language (HTML). The WEBM architecture includes a CIM server, which acts as an information broker between the providers of the instrumentation data and the management application. This CIM server defines the schemas that are used to represent the real-world objects that are being managed, using an object-oriented paradigm. It describes a common method of structuring management data across different management systems, and facilitates the integration of management information from different sources. Thus, while SNMP-based network management systems address managed resources on a component basis, the DMTF model also considers the *relationships* between the various resources.

11.7 Looking Ahead

In this chapter, we have considered the challenges that come from managing a distributed, heterogeneous network, and the solution that is provided by the Internet Network Management Framework. In the next chapter, we will focus on that framework in greater detail, and discuss the three versions of SNMP that have been developed to address that challenge.

11.8 References

[11-1] International Organization for Standardization. *Information Processing Systems: Open Systems Interconnection, Basic Reference Model Part 4: Management Framework.* ISO 7498-4, 1989.

[11-2] Cerf, V. "IAB Recommendations for the Development of Internet Network Management Standards." RFC 1052, April 1988.

[11-3] Warrier, U. S., et al. "Common Management Information Services and Protocols for the Internet (CMOT and CMIP)." RFC 1189, October 1990.

[11-4] Case, J., et al. "Introduction and Applicability Statements for Internet-Standard Management Framework." RFC 3410, December 2002.

[11-5] Case, J. D., M. Fedor, M. L. Schoffstall, and C. Davin. "Simple Network Management Protocol (SNMP)." RFC 1157, May 1990.

[11-6] Information on the Web-based Enterprise Management (WBEM) Initiative is available from the Distributed Management Task Force, Inc., at http://www.dmtf.org.

12

Network Management System Components

This chapter explores the details of SNMP-based network management and the Internet Network Management Framework. It discusses the structure of management information (SMI), which defines the rules for identifying managed objects; the management information bases (MIBs), which describe the managed objects themselves; plus SNMP versions 1, 2, and 3, which define the mechanism by which managers and agents communicate. For clarity, we will use the term SNMP to refer to SNMP Version 1, the term SNMPv2 to refer to SNMP Version 2, and SNMPv3 to refer to SNMP Version 3.

Three versions of this work have been produced, all of which build upon, revise, and/or enhance the efforts of the previous development. In other words, SNMPv3 is based upon SNMPv2, which in turn is based upon SNMPv1. However, if we just jumped immediately into a study of SNMPv3, we would miss some of the development progressions from one version to the next, and would find ourselves in an area of technology that becomes quite complex very quickly. For this reason, we are going to take our investigation one step at a time beginning with SNMPv1, describing the changes between versions as we proceed. Therefore, Sections 12.1 through 12.4 will consider the SMI, MIBs, and protocol structures associated with SNMPv1; Section 12.5 will consider SNMPv2; and Section 12.6 will study SNMPv3. Since SNMPv1 provides the foundation, we will spend the majority of our detailed discussion with that version. SNMPv1 is also the most prevalent implementation. As we will explore, the complexities associated with SNMPv2 did not receive widespread acceptance, and current deployments are just now catching up with SNMPv3, which leaves SNMPv1 as the network management protocol of choice (or default, depending upon your perspective) within many enterprises.

12.1 The SMI Defined for SNMPv1

In the manager/agent paradigm for network management, managed network objects must be physically and logically accessible. Physical accessibility means that some entity must physically check the address, count the packets, or otherwise quantify the network management information. Logical accessibility means that management information must be stored somewhere and, therefore, that the information must be retrievable and modifiable. (SNMP actually performs the retrieval and modification.) The structure of management information (SMI) (RFC 1155 [12-1]) organizes, names, and describes information so that logical access can occur.

The SMI states that each managed object must have a name, a syntax, and an encoding. The name, an object identifier (OID), uniquely identifies the object. The syntax defines the data type, such as an integer or a string of octets. The encoding describes how the information associated with the managed objects is serialized for transmission between machines. This section discusses how the SMI applies to SNMP. It begins by looking at the syntax (the Abstract Syntax Notation One, ASN.1), the encoding (the Basic Encoding Rules, or BER), and finally the names (the object identifiers, or OIDs).

12.1.1 Presenting Management Information

In terms of the ISO/OSI model, the ASN.1 syntax is a Presentation-layer (layer 6) function. Recall that the Presentation layer defines the format of the data stored within a host computer system.

In order for managers and agents to exchange data, both must understand it, regardless of the way either machine represents data internally. For this to occur, two items must be standardized: the abstract syntax and the transfer syntax. The abstract syntax defines specifications for data notation. The transfer syntax defines (transmittable) encodings for the elements in the abstract syntax.

The Internet SMI specifies that ASN.1 defines the abstract syntax for messages; that is, ASN.1 defines the basic language elements and provides rules for combining elements into messages. The Basic Encoding Rules (BER) provide the transfer syntax. The BER are associated with the abstract syntax and provide bit-level communication between machines. Thus the SMI and SNMP use the ASN.1 formalizations (ISO 8824-1 [12-2]) and BER (ISO 8825-1 [12-3]) to define various aspects of the Internet Network Management Framework. The specifics of ASN.1 are discussed in the next section.

12.1.2 ASN.1 Elements

Network administrators often criticize ASN.1 for its complexity. Some of their criticisms are fair: it is quite difficult to interpret the standard. However, ASN.1 has a straightforward objective. It is designed to define structured information (messages) in a machine-independent (or host-independent) fashion. To do this, ASN.1 defines basic data types, such as integers and strings, and new data types that are based on combinations of the basic ones. The BER then define the way the data is serialized for transmission.

ASN.1 defines data as a pattern of bits in computer memory, just as any high-level computer programming language defines data that the language manipulates as variables. The BER define a standard way to convert ASN.1 definitions into bit patterns for transmission, and then they actually transfer the data between computers. The BER are necessary because the ASN.1 description is "human-readable" and must be translated differently for each type of computer. The BER representation, however, is always the same for any ASN.1 description, regardless of the computers that send or receive that information. This ensures communication between machines, regardless of their internal architecture.

The objective here is to describe ASN.1 to the level of detail necessary to apply it to network management and SNMP. (SNMP uses a subset of ASN.1 for the sake of simplicity.) For additional information, refer to Douglas Steedman's text, *Abstract Syntax Notation One (ASN.1), the Tutorial and Reference* [12-4].

ASN.1 uses some unique terms to define its procedures, including type definitions, value assignments, macro definitions and evocations, and module definitions. You need to understand these terms before the discussion can proceed. Moreover, ASN.1 specifies some words as keywords, or reserved character sequences. Keywords, such as INTEGER, OBJECT, and NULL, have special meanings and appear in uppercase letters.

12.1.2.1 Types and Values

A *type* is a class of data. It defines the data structure that the machine needs in order to understand and process information. The SMI defines three types: Primitive, Constructor, and Defined. ASN.1 defines several Primitive types (also known as Simple types), including INTEGER, OCTET STRING, OBJECT IDENTIFIER, and NULL. By convention, types begin with an uppercase letter. (ASN.1 also defines the four types listed here as reserved character sequences and therefore represents them entirely in uppercase.) Constructor types (also known as Aggregate types) generate lists and tables. Defined types are alternate names for either simple or complex ASN.1 types and are usually more descriptive. Examples of SNMP-defined types include IpAddress, which represents a 32-bit Internet address, and TimeTicks, which is a time-stamp.

The *value* quantifies the type. In other words, once you know the type, such as INTEGER or OCTET STRING, the value provides a specific instance for that type. For example, a value could be an entry in a routing table. By convention, values begin with lowercase letters.

Some applications allow only a subset of the possible type values. A subtype specification indicates such a constraint. The subtype specification appears after the type and shows the permissible value or values, called the subtype values, in parentheses. For example, if an application uses an INTEGER type and the permissible values must fit within an 8-bit field, the possible range of values must be between 0 and 255. You would express this as:

```
INTEGER (0..255)
```

The two periods (..) are the range separator and indicate the validity of any integer value between 0 and 255.

12.1.2.2 Macros

Annex J of ISO 8824-1 defines a macro notation that allows you to extend the ASN.1 language. By convention, a macro reference (or macro name) appears entirely in uppercase letters.

For example, MIB definitions make extensive use of the ASN.1 macro, OBJECT-TYPE (originally defined in RFC 1155 and further described in RFC 1212 [12-5]). The first object in MIB-II is a system description (*sysDescr*). RFC 1213 uses the OBJECT-TYPE macro to define *sysDescr*, as follows:

```
SysDescr        OBJECT-TYPE
    SYNTAX      DisplayString (SIZE (0..255))
    ACCESS      read-only
    STATUS      mandatory
    DESCRIPTION

                "A textual description of the en-
                tity. This value should include the
                full name and version identifica-
                tion of the system's hardware type,
                software operating-system, and net
                working software. This must contain
                only printable ASCII characters."
    ::= { system 1 }
```

Thus, one concise package defines the object *sysDescr*. Section 12.1.3.1 explores the details of the OBJECT-TYPE macro (SYNTAX, ACCESS, and so on). Note that the range notation (0..255) specifies the permissible size of the DisplayString type, which in this case is between 0 and 255.

12.1.2.3 Modules

ASN.1 also collects descriptions into convenient groups called modules. For example, the remote monitoring (RMON) MIB is a discrete unit that is also part of MIB-II.

The module starts with a module name, such as RMON-MIB. Module names must begin with an uppercase letter. The BEGIN and END statements enclose the body of the module. The body may contain IMPORTS, which are the names of types, values, and macros, and the modules in which they are declared. In the following example, the first line after IMPORTS specifies that the Counter type, which will be used in this MIB module, is from another MIB module, RFC1155-SMI.

Following is the header section from RFC 1757 (an earlier version of the RMON MIB, which is now updated in RFC 2819), which represents a MIB module. Comment lines within ASN.1 syntax begin with a double hyphen (--):

```
RMON-MIB DEFINITIONS ::= BEGIN
      IMPORTS
            Counter               FROM RFC1155-SMI
            DisplayString         FROM RFC1158-MIB
            mib-2                 FROM RFC1213-MIB
            OBJECT-TYPE           FROM RFC-1212
            TRAP-TYPE             FROM RFC-1215;
-- Remote Network Monitoring MIB
   rmon      OBJECT IDENTIFIER ::= { mib-2 16 }

     - textual conventions
            .
            .
            .
END
```

In the preceding example, you can see the OBJECT IDENTIFIER value notation for RMON. Section 12.1.5 discusses this notation in detail, but for now simply note that the value of RMON is the sixteenth defined object under the mib-2 object tree. The curly brackets ({ }) indicate the beginning and end of a list — in this case a list of the OBJECT IDENTIFIER values defining RMON.

12.1.2.4 Summary of ASN.1 Conventions

In summary, ASN.1 makes distinctions between uppercase and lowercase letters, as follows:

Item	Convention
Types	Initial uppercase letter
Values	Initial lowercase letter
Macros	All uppercase letters
Modules	Initial uppercase letter
ASN.1 keywords	All uppercase letters

The ASN.1 keywords that are frequently used within SNMP are BEGIN, CHOICE, DEFINED, DEFINITIONS, END, EXPORTS, IDENTIFIER, IMPORTS, INTEGER, NULL, OBJECT, OCTET, OF, SEQUENCE, and STRING.

ASN.1 also gives special meanings to certain characters:

Item	Name
-	Signed number
—	Comment
::=	Assignment (defined as)
|	Alternation (options of a list)
{ }	Starts and ends a list
[ ]	Starts and ends a tag
( )	Starts and ends a subtype expression
..	Indicates a range

The sections that follow emphasize some of these special characters.

12.1.3 Details of ASN.1— Objects and Types

The previous discussion provided an overview of ASN.1. This section focuses on the ASN.1 objects and data types used within the Internet Network Management Framework. For ease of reference, examples will be derived from the SMI (RFC 1155, Reference [12-1]), the Concise MIB Definitions (RFC 1212, Reference [12-5]), and MIB-II (RFC 1213, Reference [12-6]) documents, which were all part of the SNMPv1 development effort.

12.1.3.1 Defining Objects in the MIBs

A MIB contains the objects to be managed. The OBJECT-TYPE macro defines these objects in a standard format that is consistent across various public and private MIBs. (Section 12.2 and 12.3 will discuss MIBs in greater detail.) The MIB-II ASN.1 definitions (RFC 1213, page 48) appear as follows:

```
tcpInSegs               OBJECT-TYPE
    SYNTAX              Counter
    ACCESS              read-only
    STATUS              mandatory
     ::= {tcp 10}
```

In English, this ASN.1 definition means: This defines an object named *tcpInSegs* that contains Counter information. The Counter type is a nonnegative number that increases monotonically. (Section 12.1.3.4 discusses counters, which is a defined type.) This object is read-only and is mandatory for all managed devices that support its parent, mib-2.tcp. When a management protocol accesses this object, it uses the name {tcp 10}, which identifies the tenth defined object within the tcp group. (Section 12.1.5 provides more detail on how the SMI names manage objects.)

12.1.3.2 Primitive (Simple) Types

To maintain SNMP's simplicity, the Internet SMI uses a subset of the ASN.1 data types. These are divided into two categories, the Primitive types and the Constructor types (see Section 12.1.3.3). Primitive data types (also called Simple types) include INTEGER, OCTET STRING, OBJECT IDENTIFIER, and NULL. The following examples come from MIB-II (RFC 1213).

INTEGER is a Primitive type with distinguished (or unique) values that are positive and negative whole numbers, including zero. The INTEGER type has two special cases. The first is the enumerated integer type, in which the objects have a specific, nonzero number such as 1, 2, or 3. The second, the integer-bitstring type, is used for short bit strings such as (0..127) and displays the value in hexadecimal. An example of INTEGER would be:

```
IpDefaultTTL            OBJECT-TYPE
    SYNTAX              INTEGER
    ACCESS              read-write
    STATUS              mandatory
    DESCRIPTION
```

```
                    "The default value inserted into the
                    Time-To-Live field of the IP header
                    of datagrams originating at this
                    entity, whenever a TTL value is not
                    supplied by the transport layer
                    protocol."
       ::= { ip 2 }
```

The OCTET STRING is a Primitive type whose distinguished values are an ordered sequence of zero, one, or more octets. SNMP uses three special cases of the OCTET STRING type: the DisplayString, the octetBitstring, and the PhysAddress. In the DisplayString, all of the octets are printable ASCII characters. The octetBitstring is used for bit strings that exceed 32 bits in length. (As we studied in the earlier chapters, TCP/IP frequently includes 32-bit fields. This quantity is a typical value for the internal word width of various processors — hosts and routers — within the Internet.) MIB-II defines the PhysAddress and uses it to represent media (or Physical layer) addresses.

An example of the use of a DisplayString would be:

```
sysContact        OBJECT-TYPE
    SYNTAX        DisplayString (SIZE (0..255))
    ACCESS        read-write
    STATUS        mandatory
    DESCRIPTION
                  "The textual identification of the
                  contact person for this managed node,
                  and information on how to contact
                  this person."
    ::= { system 4 }
```

Note that the subtype indicates that the permissible size of the DisplayString is between 0 and 255 octets. The OBJECT IDENTIFIER is a type whose distinguishing values are the set of all object identifiers allocated according to the rules of ISO 8824-1. The ObjectName type, a special case that SNMP uses, is restricted to the object identifiers of the objects and subtrees within the MIB, for example:

```
ipRouteInfo       OBJECT-TYPE
    SYNTAX        OBJECT IDENTIFIER
    ACCESS        read-only
    STATUS        mandatory
    DESCRIPTION
```

```
                    "A reference to MIB definitions spe
                    cific to the particular routing pro-
                    tocol responsible for this route,
                    as determined by the value speci-
                    fied in the route's ipRoutePro to
                    value. If this information is not
                    present, its value should be set to
                    the OBJECT IDENTIFIER { 0 0 }, which
                    is a syntactically valid object iden-
                    tifier, and any conforming imple-
                    mentation of ASN.1 and BER must be
                    able to generate and recognize this
                    value."
    ::= { ipRouteEntry 13 }
```

NULL is a type with a single value, also called null. The null serves as a placeholder, but is not currently used for SNMP objects. (You can see NULL used as a placeholder in the variable bindings field of the SNMP GetRequest PDU. The NULL is assigned to be the value of the unknown variable, that is, the value the GetRequest PDU seeks. We will see this in the case studies in Chapter 13.)

12.1.3.3 Constructor (Structured) Types

The Constructor types, SEQUENCE and SEQUENCE OF, define tables and rows (entries) within those tables. By convention, names for table objects end with the suffix Table, and names for rows end with the suffix Entry. The following discussion defines the Constructor types. The example comes from MIB-II.

SEQUENCE is a Constructor type defined by referencing a fixed, ordered list of types. Some of the types may be optional, and all may be different ASN.1 types. Each value of the new type consists of an ordered list of values, one from each component type. The SEQUENCE as a whole defines a row within a table. Each entry in the SEQUENCE specifies a column within the row.

SEQUENCE OF is a Constructor type that is defined by referencing a single existing type; each value in the new type is an ordered list of zero, one, or more values of that existing type. Like SEQUENCE, SEQUENCE OF defines the rows in a table; unlike SEQUENCE, SEQUENCE OF only uses elements of the same ASN.1 type.

The TCP connection table that follows illustrates both SEQUENCE and SEQUENCE OF:

```
tcpConnTable     OBJECT-TYPE
     SYNTAX      SEQUENCE OF TcpConnEntry
     ACCESS      not-accessible
     STATUS      mandatory
     DESCRIPTION
                 "A table containing TCP connection-
                 specific information."
     ::= { tcp 13 }
 tcpConnEntry    OBJECT-TYPE
     SYNTAX      TcpConnEntry
     ACCESS      not-accessible
     STATUS      mandatory
     DESCRIPTION
                 "Information about a particular cur-
                 rent TCP connection. An object of
                 this type is transient; it ceases
                 to exist when (or soon after) the
                 connection makes the transition to
                 the CLOSED state."
     INDEX { tcpConnLocalAddress,
           tcpConnLocalPort,
           tcpConnRemAddress,
           tcpConnRemPort }
     ::= { tcpConnTable 1 }
     TcpConnEntry ::=
     SEQUENCE {
     tcpConnState
           INTEGER,
     tcpConnLocalAddress
           IpAddress,
     tcpConnLocalPort
           INTEGER (0..65535),
     tcpConnRemAddress
           IpAddress,
     tcpConnRemPort
           INTEGER (0..65535)
     }
```

This example expands your ASN.1 grammar. The table name, *tcpConnTable*, ends with the suffix Table. The row name, *tcpConnEntry*, ends with the suffix Entry. The sequence name, *TcpConnEntry*, is the same as the row name, except that it begins with an uppercase letter. The INDEX clause defines the construction and order of the columns that make up the rows, and also defines how you address a row in the table.

12.1.3.4 Defined Types

The Internet Network Management Framework uses the Defined (or application-wide) types (described in RFC 1155). The Defined types include NetworkAddress, IpAddress, Counter, Gauge, TimeTicks, and Opaque. The examples that follow come from RFC 1213. The NetworkAddress type was designed to represent an address from one of several protocol families. A CHOICE is a Primitive type that provides alternatives between other types; it is found in several sections of the SMI definition given in RFC 1155. Currently, however, only one protocol family, the Internet family (called internet IpAddress in the SMI definition), has been defined for this CHOICE. Here is an example:

```
atNetAddress        OBJECT-TYPE
       SYNTAX       NetworkAddress
       ACCESS       read-write
       STATUS       deprecated
       DESCRIPTION
                    "The NetworkAddress (e.g., the
                    IP address) corresponding to
                    the media-dependent 'physical'
                    address."
       ::= { atEntry 3 }
```

Because it supports only IP addresses (hence the default "choice"), use of this type is discouraged. IpAddress is an application-wide type that represents a 32-bit Internet address. It is represented as an OCTET STRING of length 4 (octets) in network byte-order (the order bytes are transmitted over the network):

```
tcpConnRemAddress   OBJECT-TYPE
       SYNTAX       IpAddress
       ACCESS       read-only
       STATUS       mandatory
       DESCRIPTION
                    "The remote IP address for this
                    TCP connection."
       ::= {tcpConnEntry 4 }
```

The Counter is an application-wide type that represents a nonnegative integer that increases monotonically until it reaches a maximum value, then wraps around and increases again from zero. The maximum counter value is 2^{32}-1, or 4,294,967,295 decimal. In other words, the Counter is an unsigned 32-bit number. An INTEGER is a signed 32-bit value. By convention, you write the name of a Counter object as a plural; it ends in a lowercase *s*. Here is an example:

```
icmpInDestUnreachs OBJECT-TYPE
        SYNTAX     Counter
        ACCESS     read-only
        STATUS     mandatory
        DESCRIPTION
                   "The number of ICMP Destina-
                   tion Unreachable messages re-
                   ceived."
        ::= { icmp 3 }
```

A Gauge is an application-wide type that represents a nonnegative integer. It may increase or decrease, but it latches at a maximum value. The maximum counter value is 2^{32}-1 (4,294,967,295 decimal). Here is an example:

```
ifSpeed                OBJECT-TYPE
        SYNTAX         Gauge
        ACCESS         read-only
        STATUS         mandatory
        DESCRIPTION
                       "An estimate of the interface's
                       current bandwidth in bits per
                       second. For interfaces that do
                       not vary in bandwidth or for
                       those where no accurate esti-
                       mation can be made, this ob-
                       ject should contain the nomi-
                       nal bandwidth."
        ::= { ifEntry 5 }
```

TimeTicks is an application-wide type that represents a nonnegative integer that counts the time in hundredths of a second since some epoch or point in time. When the MIB defines object types that use this ASN.1 type, the description of the object type identifies the reference epoch. Here is an example:

```
sysUpTime              OBJECT-TYPE
        SYNTAX         TimeTicks
        ACCESS         read-only
        STATUS         mandatory
        DESCRIPTION
```

```
                     "The time (in hundredths of a
                     second) since the network man-
                     agement portion of the system
                     was last re-initialized."
        ::= { system 3 }
```

Opaque is an application-wide type that permits the passing of arbitrary ASN.1 syntax. The ASN.1 basic rules encode a value into a string of octets. This string, in turn, is encoded as an OCTET STRING, in effect "double-wrapping" the original ASN.1 value. SNMP does not currently use Opaque, although it may be found in some private MIBs.

12.1.3.5 Tagged Types

Tags distinguish between defined objects unequivocally. While a human reader might be able to distinguish defined objects through their names in ASN.1 notation, a machine can't do this without additional information. Therefore, the tagged types use a previously defined type as a base, and then add unique information. ASN.1 defines four classes of tags: universal, application, context specific, and private. ASN.1 (ISO 8824-1) defines universal tags. Other standards, such as the Internet standards, assign application class tags. The SNMP definition (RFC 1157) interprets context-specific class tags according to their context. Enterprise-specific applications use private class tags.

A number within square brackets ([ ]) identifies tagged types. For example, the concise SMI definition in RFC 1155 shows that:

```
TimeTicks ::=
   [APPLICATION 3]
      IMPLICIT INTEGER (0..4294967295)
```

Therefore, the TimeTicks type is a tagged type, designated APPLICATION 3. It is of the application class, and the tag number is 3. It may take on the range of values between 0 and 4294967295. The IMPLICIT keyword indicates that the tag associated with the INTEGER type is not transmitted, but the tag associated with TimeTicks is. This reduces the amount of data that must be encoded and transmitted.

12.1.4 Encoding Rules

The previous section considered the abstract syntax that represents management information. This section discusses the encoding rules that allow that information to be transmitted on a network. The Basic Encoding Rules (BER) define this transfer syntax, as specified in ISO 8825-1.

12.1.4.1 Encoding Management Information

Recall that each machine in the management system can have its own internal representation of the management information. The ASN.1 syntax describes that information in a standard form. The transfer syntax performs the bit-level communication (the external representation) between machines. For example, assume that the host needs management information from another device. The management application would generate an SNMP request, which the BER would encode and transmit on the network media. The destination machine would receive the information from the network, decode it using the BER rules, and interpret it as an SNMP command. The SNMP response would return in a similar, but reverse, manner. The encoding structure used for the external representation is called Type-Length-Value encoding (see Figure 12-1).

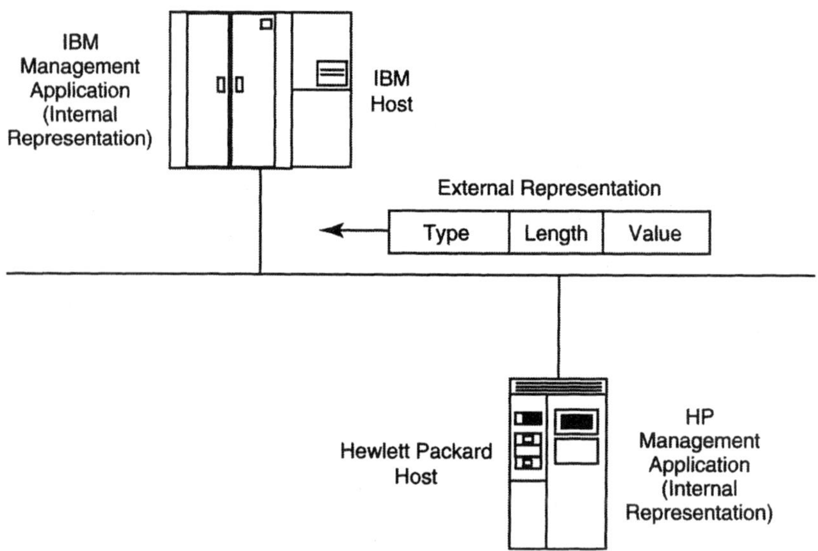

Figure 12-1. Internal and External Data Representations.

12.1.4.2 Type-Length-Value Encoding

To define the external data representation, the BER first specify the position of each bit within the octets being transmitted. Each octet transmits the most significant bit (MSB) first and defines it as bit 8 on the left-hand side of the octet (Figure 12-2). The octet defines the least significant bit (LSB) as bit 1 on the right-hand side. The data encoding structure itself has three components: Type, Length, and Value (TLV). Note that in the literature you will run across other names for Type-Length-Value, including Tag-Length-Value and Identifier-Length-Contents (from ISO 8825-1). The structure of a TLV encoding used with SNMP is shown in Figure 12-3.

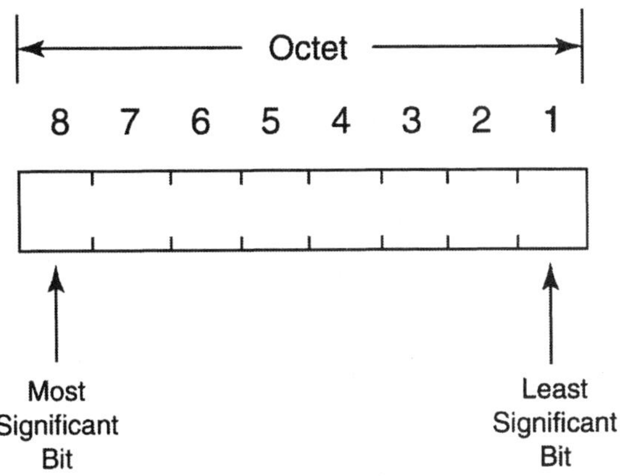

Figure 12-2. BER Bit Ordering, as Defined in ISO 8825-1.

By defining the order and structure of the bits, the BER guarantee that both ends of the communication channel interpret the bit stream consistently. The following sections examine the structure of each TLV field individually.

Type Field	Length Field	Value Field

Figure 12-3. Type-Length-Value (TLV) Encoding.

12.1.4.2.1 Type Field

The Type field comes first and alerts the destination to the structure that follows. Thus, the Type field contains an identification for the encoding structure; it encodes the ASN.1 tag (both the class and number) for the type of data contained in the Value field. A subfield within the Type field contains a bit designated as P/C that indicates whether the coding is Primitive (P/C = 0) or Constructed (P/C = 1), as shown in Figure 12-4.

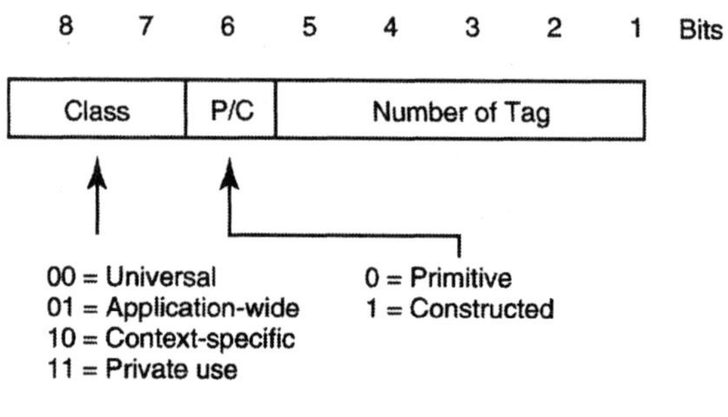

Figure 12-4. Type Field Encoding, as Defined in ISO 8825-1.

There are two types of Type fields; their use depends on the magnitude of the tag number. When the tag number is between 0 and 30, the Tag field contains a single octet, which is found with SNMP constructs, and is shown in Figure 12-4. When the tag number is 31 or greater, the Type field contains multiple octets. Since this format is not used with SNMP, it is not shown in Figure 12-4. ISO 8825-1 provides the details for those readers that may be interested.

The first octet of the Type field contains three subfields: the Class, P/C bit, and tag number. The Class subfield encodes the class of tag in use:

Class	Bit 8	Bit 7
Universal	0	0
Application	0	1
Context-specific	1	0
Private	1	1

SNMP applications use the first three classes: universal, application, and context-specific. The universal class encodes the INTEGER type, OCTET STRING type, and so on. The application class encodes the defined types (IpAddress, Counter, and so on). The context-specific class encodes the five SNMP protocol data units (PDUs), such as GetRequest, GetResponse, and so on.

The P/C subfield (bit 6) indicates the form of the data element. Primitive encoding (P/C = 0) means that the contents octets represent the value directly. A Constructor encoding (P/C = 1) means that the contents octets encode one or more additional data values, such as a SEQUENCE.

SNMP uses tag numbers between 0 and 30. The tag number appears in the third subfield and is represented in binary. Bit 5 is the tag's MSB; bit 1 is its LSB.

ISO 8824-1 contains tag numbers for the universal class (for example, UNIVERSAL 2 represents the INTEGER type). The SMI specification, RFC 1155, contains tag numbers for the application class (for example, IpAddress is a Primitive type with tag [0]). The SNMP specification, RFC 1157, contains tag numbers for the context-specific class (for example, GetRequest PDU is a Constructed type with tag [0]).

The following list summarizes the three classes of Type fields used with SNMP and the encodings for those fields: class, P/C, and tag number. These encodings appear in both binary and hexadecimal notation, where the H represents hexadecimal notation:

Universal Class	**Type Field Value**
INTEGER	00000010 = 02H
OCTET STRING	00000100 = 04H
NULL	00000101 = 05H
OBJECT IDENTIFIER	00000110 = 06H
SEQUENCE	00110000 = 30H
SEQUENCE-OF	00110000 = 30H

Application Class	**Type Field Value**
IpAddress	01000000 = 40H
Counter	01000001 = 41H
Gauge	01000010 = 42H
TimeTicks	01000011 = 43H
Opaque	01000100 = 44H

Context-Specific Class	**Type Field Value**
GetRequest	10100000 = A0H
GetNextRequest	10100001 = A1H
GetResponse	10100010 = A2H
SetRequest	10100011 = A3H
Trap	10100100 = A4H

12.1.4.2.2 Length Field

The Length field follows the Type field and determines the number of octets the Value field will contain. The Length field may take either the short definite or the long definite form, as shown in Figure 12-5. (Another form, called "indefinite," is not used with SNMP.) The definite indicates that the length of the encoding is known prior to transmission; the indefinite indicates otherwise.

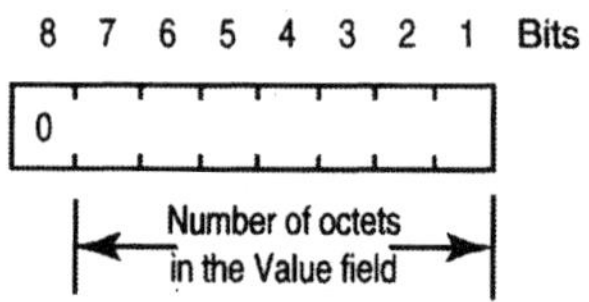

Short Definite Form: Length = 0-127 octets

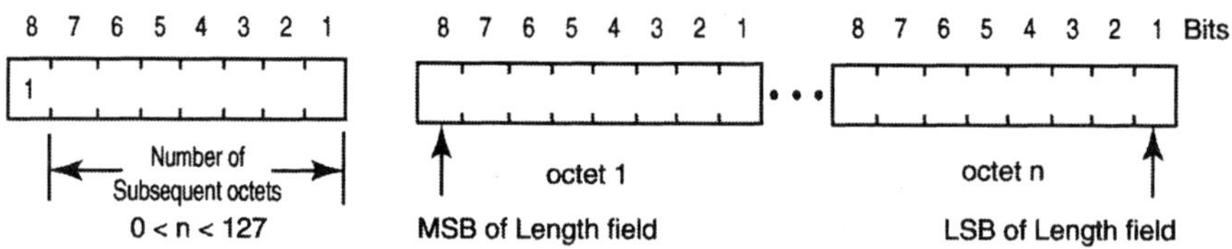

Long Definite Form: Length = 0 - (2^{1008} - 1) octets

Figure 12-5. Length Field Encoding.

The short definite form indicates a length of between 0 and 127 octets in the Contents field; the long definite form indicates 128 or more octets in the Contents field, although it can indicate shorter lengths.

The long form uses multiple octets to represent the total length. In the long form, the first octet of the Length field has Bit 8 = 1, followed by a binary number indicating the number of octets to follow. This number must be between 1 and 126; 127 is reserved for future extensions. Bit 8 of the second octet is considered the MSB of the Length field, and the following octets make up the rest of the length. Thus, the long definite form may represent a length up to 2^{1008}-1 octets. (The 1008 is derived from the product of 126 and 8: 126 subsequent octets times 8 bits per octet.)

12.1.4.2.3 Value Field

The Value field contains zero or more Contents octets, which convey the data values. Examples include an integer, ASCII character, or OBJECT IDENTIFIER, such as {1.3.6.1.2.}.

12.1.4.3 Encoding Examples

Section 12.1.3.2 mentioned that the Internet SMI defines a subset of the ASN.1 types. This subset includes the following universal Primitive types: INTEGER, OCTET STRING, OBJECT IDENTIFIER, and NULL. The universal Constructor types are SEQUENCE and SEQUENCE OF. The application Primitive types are IpAddress, Counter, Gauge, and TimeTicks. SNMPv1-related applications use only these ten types. We will use this information in the case studies presented in Chapter 13. For illustrations of the other types, consult ISO 8825-1.

12.1.4.3.1 INTEGER Type Encoding

The INTEGER type is a Simple type that has values of zero, positive, or negative whole numbers. It is a Primitive type encoded with a Value field containing one or more Contents octets. The Contents octets are a "two's-complement" binary number equal to the integer value, and they can use as many octets as necessary. For example, Boomer, our Labrador, weighs 75 pounds. The value of his weight would be encoded as: Type field = 02H, Length field = 01H, and Value field = 4BH (see Figure 12-6). Note that the value appears in quotes (Value = "75") to indicate that it represents a quantity, which can be numerical, ASCII characters, an IP address, and so on.

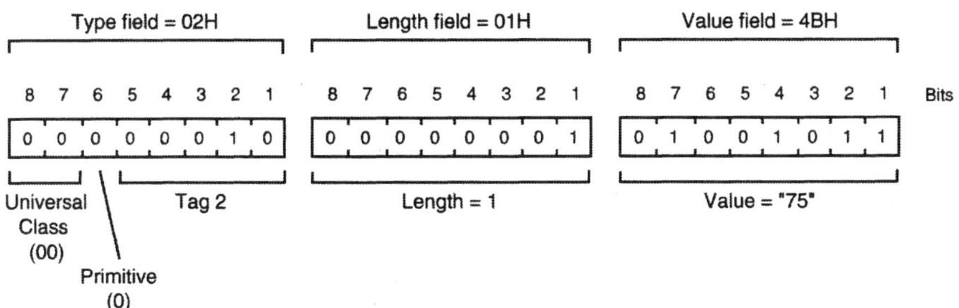

Figure 12-6. Encoding for the INTEGER Type, Value = "75."

12.1.4.3.2 OCTET STRING Type Encoding

The OCTET STRING is a Simple type whose distinguished values are an ordered sequence of zero, one, or more octets, each of which must be a multiple of 8 bits. Encoding for OCTET STRING values is Primitive, with the Type field = 04H. The Length field and Value field depend on the encoded information.

Let's again use Boomer as an example to show the OCTET STRING type encoding. Figure 12-7 shows how we have encoded the value for Boomer's initials (BBM, for Boomerang Buddy Miller). The Type field contains 04H, indicating a Primitive type, OCTET STRING (tag number 4). The Length field indicates 3 octets in the Value field. The Value field encodings come from the ASCII chart.

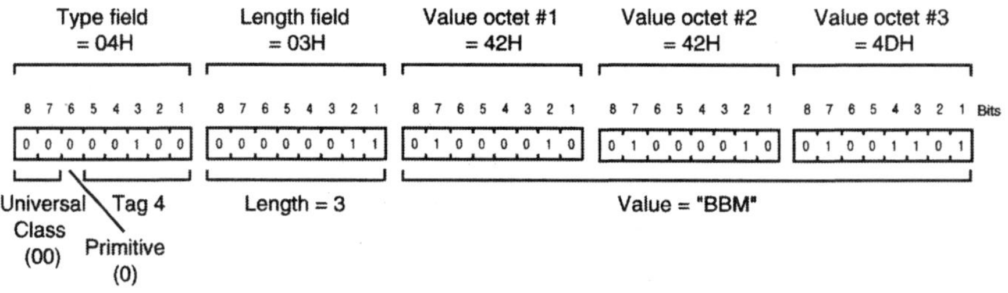

Figure 12-7. Encoding for the OCTET STRING Type, Value = "BBM."

12.1.4.3.3 OBJECT IDENTIFIER Type Encoding

The OBJECT IDENTIFIER names (or identifies) items. (In SNMP, these identify managed objects.) Its Value field contains an ordered list of subidentifiers. To save encoding and transmission effort, you can take advantage of the fact that the first subidentifier is a small number, such as 0, 1, or 2, and combine it mathematically with the second subidentifier, which may be larger. The total number of subidentifiers is, therefore, less than the number of object identifier components in the OID value being encoded. This reduced number (one less) results from a mathematical expression that uses the first two OID components to produce another expression:

```
Given X is the value of the first OID, and Y is the
second:
First subidentifier = (X * 40) + Y
```

The values for these subidentifiers are encoded and placed within the Value field. Bit 8 of each octet indicates whether or not that octet is the last in the series of octets required to fully describe the value. If Bit 8 = 1, at least one octet follows; Bit 8 = 0 indicates the last (or only) octet. Bits 7 through 1 of each octet encode subidentifiers. Using an example from the MIB-II object tree, the System group, assume that an OBJECT IDENTIFIER has a value of:

```
{iso org(3) dod(6) internet(1) mgmt(2) mib-2 (1) 1}
```

From the object tree (also discussed below), this is represented by:

```
{1.3.6.1.2.1.1}
```

Note that, by convention, the subidentifiers are separated by periods for clarification.

Using the values of X = 1 and Y = 3, and the expression above for the first subidentifier value,

```
(1 * 40) + 3 = 43
```

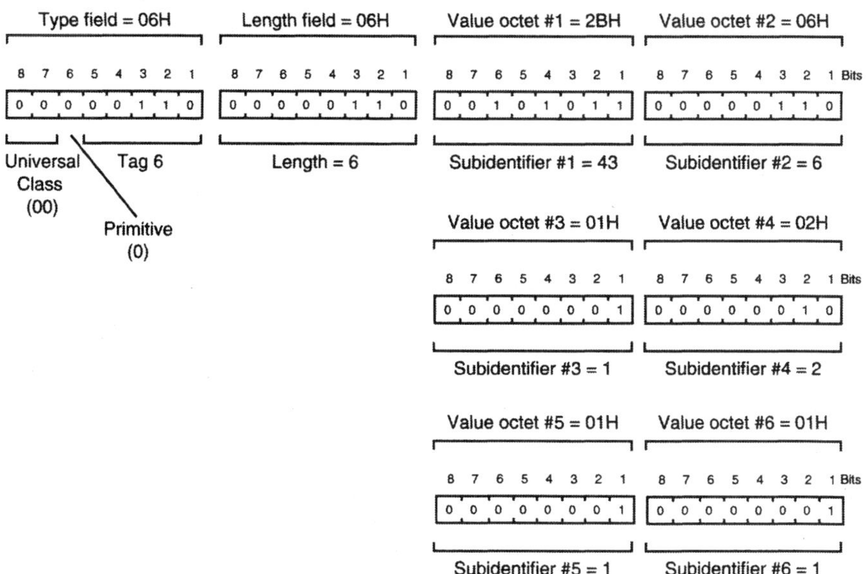

Figure 12-8. Encoding for OBJECT IDENTIFIER Type, Value = {1.3.6.1.2.1.1}.

This results in the first subidentifier value of 43, the second subidentifier value of 6, the third subidentifier value of 1, and so on. The first value (43) needs 6 bits, or one octet, for encoding (00101011). The second value (6) needs 3 bits for encoding (110), and requires only one octet. Subsequent values also require one octet. As you can see in Figure 12-8, the encoding becomes: Type field = 06H (OBJECT IDENTIFIER, tag = 6); Length field = 06H; and Value field = 2B 06 01 02 01 01 H.

12.1.4.3.4 NULL Type Encoding

The NULL type is a placeholder that communicates the absence of information. For example, when a manager requests the value of a variable, it uses the NULL type as a placeholder in the position where the agent will fill in the response.

Encoding for the NULL type is a Primitive. The Type field = 05H, and the Length field = 00H. The Value field is empty (no value octets), as shown in Figure 12-9.

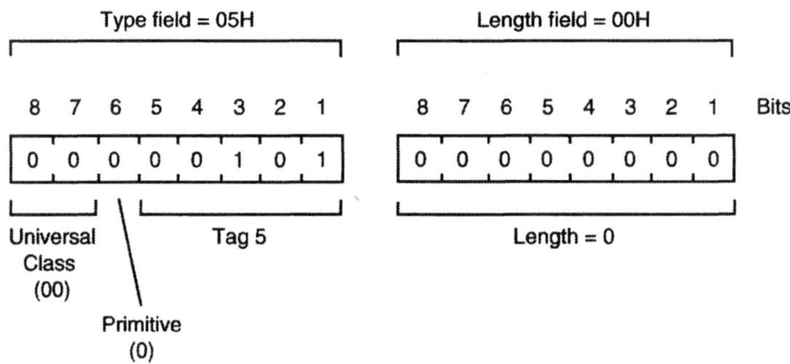

Figure 12-9. Encoding for the NULL Type, Value = NULL.

12.1.4.3.5 SEQUENCE Type Encoding

Recall from the discussion in Section 12.1.3.3 that the SEQUENCE type is a list of ASN.1 types. A SEQUENCE value is always encoded in Constructed form. The variable bindings used within the SNMP messages provide a good example of SEQUENCE. The variable bindings (or VarBind) pair an object name with its value, which is transmitted inside the Value field, as shown in Figure 12-3. SNMP (RFC 1157, page 32) defines the VarBind:

```
VarBind ::=
   SEQUENCE {
      name
         ObjectName,

      value
         ObjectSyntax,
}

VarBindList ::=
   SEQUENCE OF
      VarBind
END
```

As this syntax shows, the VarBind is a SEQUENCE (pairing) of a name and a value, and the VarBindList is a list of names and values.

Although this is getting ahead of the sequence of our SMI, MIB, and SNMP story, I'll provide an example. Suppose you need the system description for a particular object whose name is sysDescr. To obtain the system description, the manager transmits an SNMP GetRequest to the agent asking for the value of object sysDescr. The agent responds with an SNMP GetResponse message containing the value, such as "Retix Local Ethernet Bridge Model 2265M." The VarBind associates the object (sysDescr) and its value ("Retix ..."), as shown in Figure 12-10.

The first Type field (30H) indicates a Constructed type, with Tag = 16 (SEQUENCE). The first Length field contains 33H, indicating that 51 Value octets follow. The BER are then applied for every type in the SEQUENCE. The first sequence identifies a Primitive type with Tag = 6 (OBJECT IDENTIFIER) and Length = 08H. The Value field contains the numeric representation of the *sysDescr* object {1.3.6.1.2.1.1.1.0}. The second sequence identifies a Primitive type with Tag = 4 (OCTET STRING), and Length = 27H (39 decimal). The second Value field represents the value of the object *sysDescr* ("Retix Local Bridge ..."). If you've got a calculator, look at the total length of the encoding. Sequence #1 contains 10 octets (1 from the Type field + 1 from the Length + 8 from the Value). Sequence #2 contains 41 octets (1 + 1 + 39). Sequence #1 plus Sequence #2 (10 + 41) equals the value of the first Length field (51 octets).

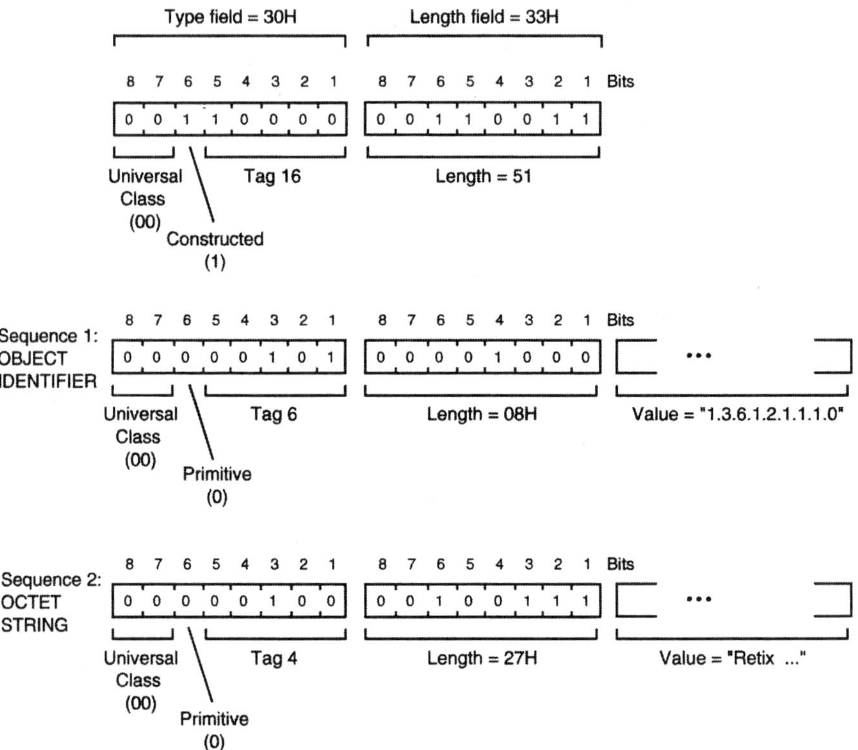

Figure 12-10. Encoding for the SEQUENCE Type, a Variable Binding (VarBind).

12.1.4.3.6 SEQUENCE-OF Type Encoding

The SEQUENCE-OF type value is encoded in Constructed form in the same way as the SEQUENCE type.

12.1.4.3.7 IpAddress Type Encoding

The discussion now moves to the Application class of encodings, which are tagged types. The Application class encodings begin with a binary 01, followed by a Primitive (P/C = 0) designation, and ending with a tag number between 0 and 4. Therefore, the Type fields will range from 40 to 44H, as shown in Figures 12-11 through 12-14.

The IpAddress type carries a 32-bit IP address, which is represented in four octets. Jumping ahead to the discussion of MIBs in Section 12.2, the IP group contains objects that relate to the IP process on a router or host. An object called *IpAdEntAddr* identifies the IP address that subsequent information is related to. To encode the *IpAdEntAddr* (see Figure 12-11), the Type field is set to 40H (application class, Primitive, Tag = 0). The Length field = 4, representing the four octets in the IP address. The Value field contains four Contents octets, which convey the IP address in dotted decimal notation. For the address shown in the example [128.150.161.8], the first octet in the Value field contains the binary equivalent of 128 (10000000), the second the binary equivalent of 150, and so on.

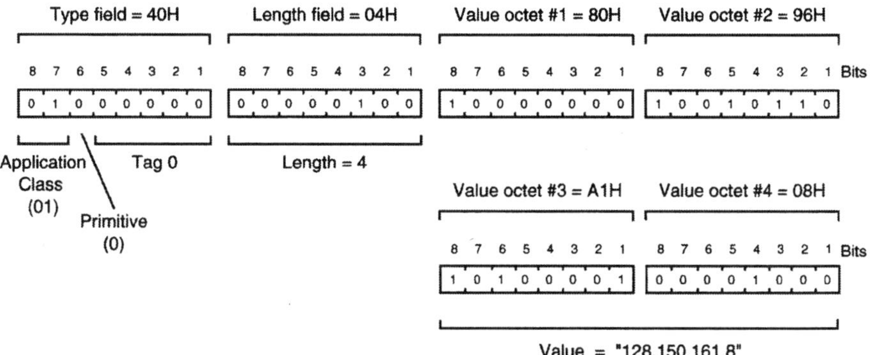

Figure 12-11. Encoding for the IpAddress Type, Value = "128.150.161.8."

12.1.4.3.8 Counter Type Encoding

A Counter type represents a nonnegative integer that increases monotonically to a maximum of 4,294,967,295, and then wraps around to zero. The ICMP Group uses many counters to record message statistics. One object, *icmpInMsgs*, records the number of messages that the ICMP process on a router or host has received. A sample encoding (see Figure 12-12) would have a Type field = 41H, representing application class, Primitive encoding, and Tag = 1. The Value (190,105) requires three octets. The Length field is, therefore, 03H, and the Value field contains 02 E6 99H, representing the 190,105 messages.

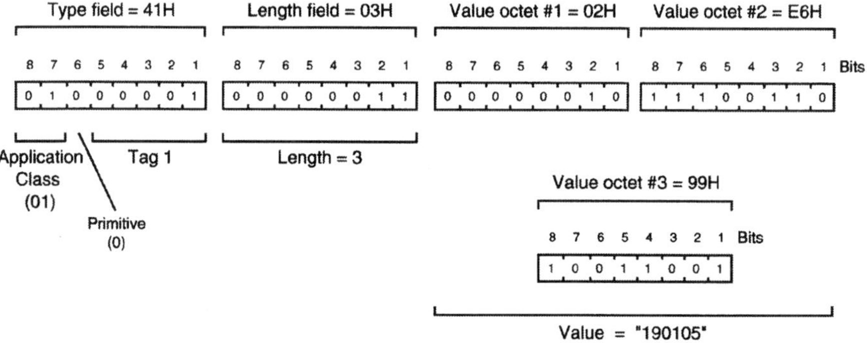

Figure 12-12. Encoding for the Counter Type, Value = "190105."

12.1.4.3.9 Gauge Type Encoding

A Gauge type is a nonnegative integer that may increase or decrease, but latches at a maximum value of 4,294,967,295. The Gauge type is not used frequently. MIB-II defines it for the *ifSpeed, ifOutQLen,* and *tcpCurrEstab* objects only. For example, Figure 12-13 assumes that the maximum output queue length of a particular interface is 32 packets. To encode this Gauge value, the Type field is set to 42H (application class, Primitive, Tag = 2). One octet encodes decimal 32; therefore, the Length field = 01H and the Value field contains 20H, the desired value of 32 decimal.

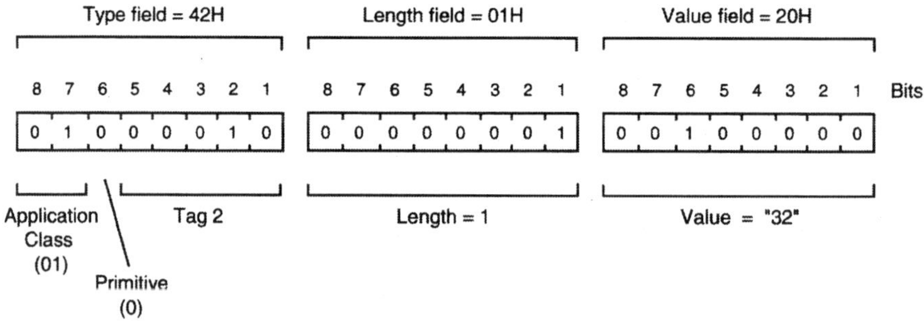

Figure 12-13. Encoding for the Gauge Type, Value = "32."

12.1.4.3.10 TimeTicks Type Encoding

The TimeTicks type contains a time-stamp that measures the elapsed time (in hundredths of a second) since some event. The *sysUpTime* object measures the time since the network management entity on a device was reinitialized. If the *sysUpTime* value for a particular device was 263,691,156 hundredths of a second (about 30 days), its value would be encoded as shown in Figure 12-14. The Type field would be set to 43H (application class, Primitive, Tag = 3). Four octets represent a Value equal to 263691156. Therefore, the Length field contains 04H. The four octets in the Value field contain the binary representation of the TimeTicks value.

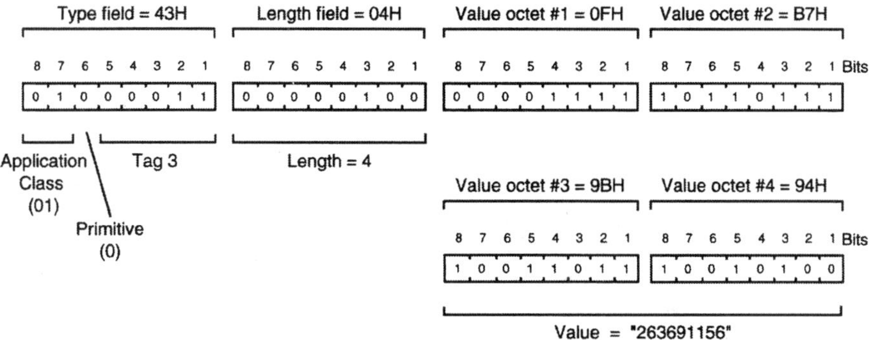

Figure 12-14. Encoding for the TimeTicks Type, Value = “263691156.”

12.1.4.3.11 Context-Specific Encodings for SNMP

The final class of encodings discussed in this section are the context-specific encodings, which are used within the context of SNMP. Five protocol data units (PDUs) convey SNMPv1 information. The PDUs are GetRequest, GetNextRequest, GetResponse, SetRequest, and Trap. These PDUs have tag numbers of 0 to 4, respectively. These encodings are all context-specific class (10) and Constructed (P/C = 1). The Type fields thus have values ranging from A0 to A4H (see Figure 12-15). The Length and Value fields depend on the information conveyed.

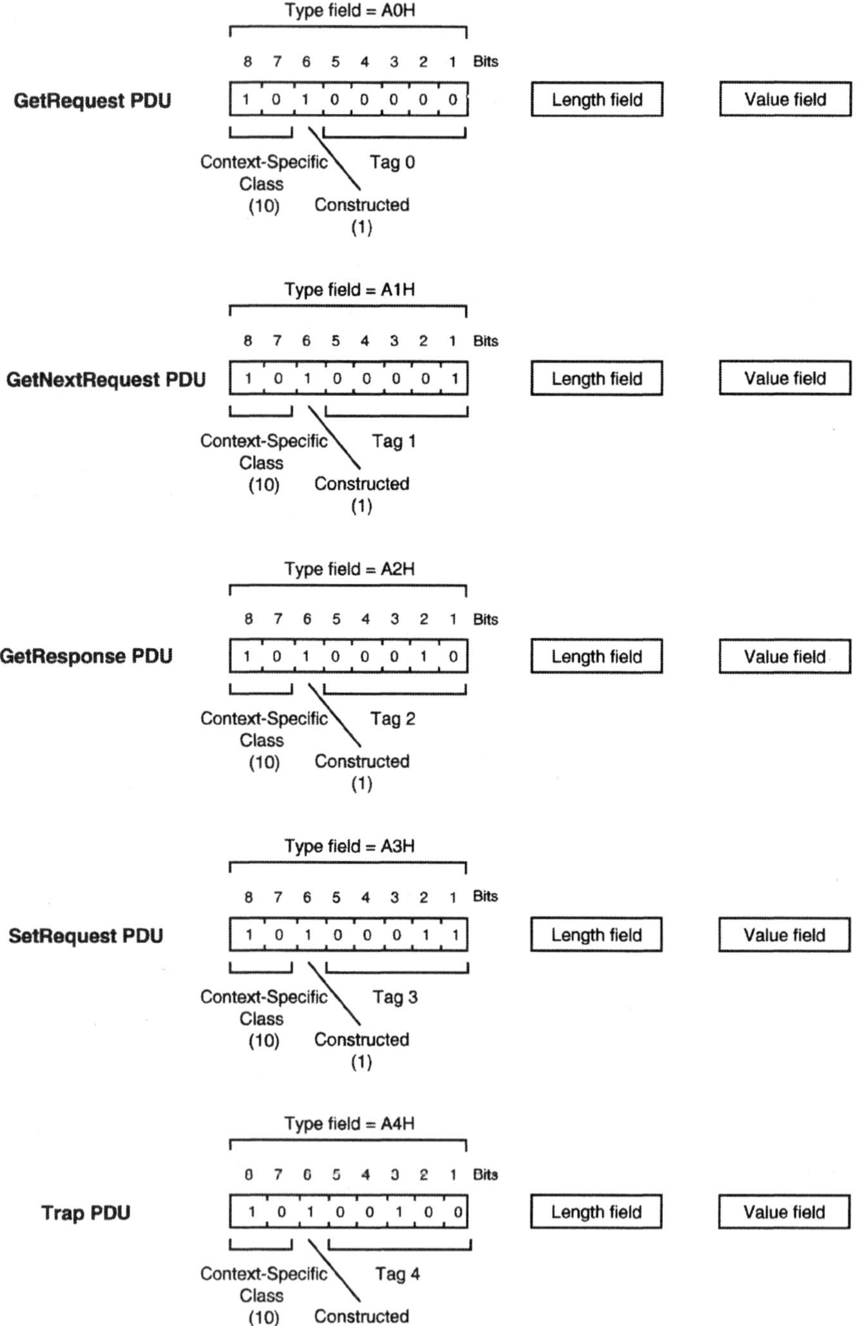

Figure 12-15. Encoding for the Context-specific Types Used with SNMP.

12.1.5 Object Names

Each object, whether it's a device or a characteristic of a device, must have a name by which it can be uniquely identified. That name is the object identifier. It is written as a sequence of integers separated by periods. For example, the sequence {1.3.6.1.2.1.1.1.0} specifies the system description, within the system group, of the mgmt subtree.

Annexes B, C, D, and E of ISO 8824-1 define the numerical sequences; they resemble a tree with a root and several directly attached branches, referred to as children (see Figure 12-16). These branches connect to other branches. You can use the structure of root, branches, subbranches, and leaves to diagram all of the objects within a particular MIB and their relationships.

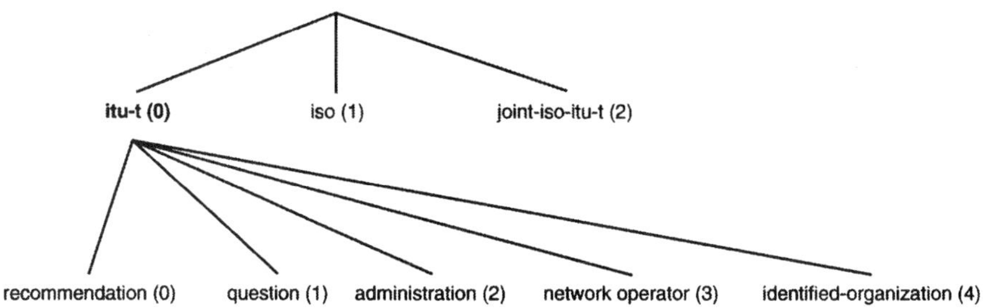

Figure 12-16. The Root Node and ITU-T-assigned OBJECT IDENTIFIER Component Values.

The root does not need a designation, but a specific numeric value designates the three connected arcs, or branches. The International Telecommunications Union — Telecommunications Standards Sector (ITU-T) administers the branch labeled 0, the International Organization for Standardization (ISO) administers the branch labeled 1, and ISO and ITU-T jointly administer the third branch, labeled 2.

The ITU-T branch has five children: recommendation (0) identifies ITU-T recommendations; question (1) is used for ITU-T study groups; administration (2) identifies the values of the X.121 DCCs (Data Country Codes); network-operator (3) identifies the values of the X.121 Data Network Identification codes (DNICs); and identified-organization (4) identifies values assigned by the ITU Telecommunication Standardization Bureau (TSB).

The ISO branch (Figure 12-17) has three children: standard (0) designates international standards; member-body (2) is a three-digit numeric country code that ISO 3166 assigns to each member of ISO/IEC; and identified-organizations (3) have values of an international code designator (ICD), defined in ISO 6523. (Branch (1) was previously assigned to registration-authority (1), but it is no longer in use, per ISO 8834-1.)

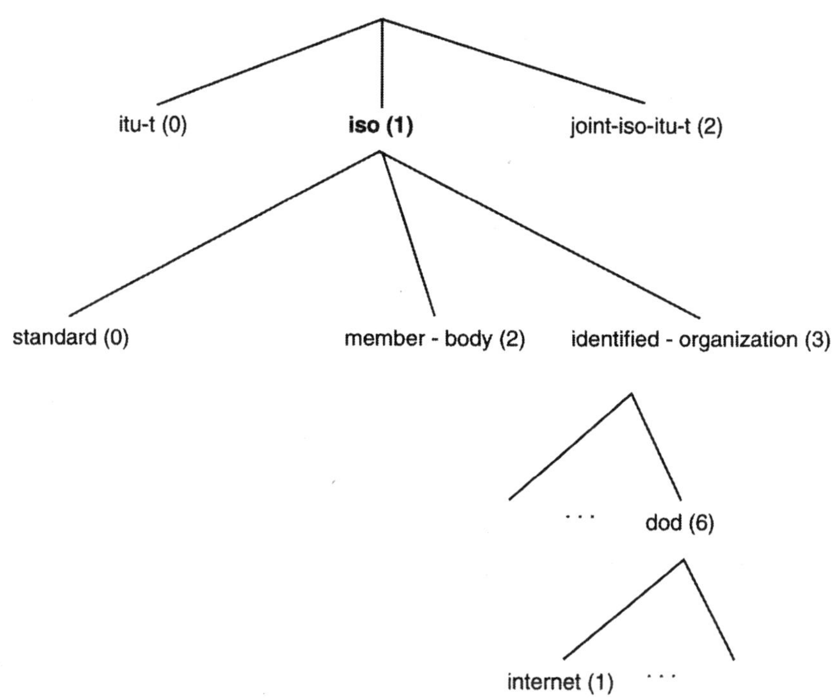

Figure 12-17. The Root Node and ISO-assigned OBJECT IDENTIFIER Component Values.

The U.S. Department of Defense is assigned to one of the children under 1.3, and is designated as 6. On this tree, the Internet community has designation 1.

To identify a particular position on the tree, you list the numeric values in a string, separated by periods. For example, to identify the position of the Internet subtree, you start at the root and move down until you reach position {1.3.6.1}.

At the Internet level (see Figure 12-18), you begin to see details germane to network management and SNMP. The Internet subtree has eight branches:

- The directory (1) subtree, {internet 1} or {1.3.6.1.1}, is reserved for future use by the OSI directory within the Internet.
- The mgmt (2) subtree, {internet 2} or {1.3.6.1.2}, is managed by the Internet Assigned Numbers Authority, and includes the standard MIBs.
- The experimental (3) subtree, {internet 3} or {1.3.6.1.3}, is used for Internet experiments.
- The private (4) subtree, {internet 4} or {1.3.6.1.4}, allows vendors to register objects.
- The security (5) subtree, {internet 5} or {1.3.6.1.5}, is used for security-related objects.
- The snmpV2 (6) subtree, {internet 6} or {1.3.6.1.6}, is used for SNMP Version 2 objects.
- The mail (7) subtree, {internet 7} or {1.3.6.1.7}, is used for mail objects.
- The features (8) subtree, {internet 8} or {1.3.6.1.8}, is used to describe features of devices.

These subtrees are documented along with other SMI-related network management parameters on the IANA website [12-7].

Let's now look at the structure of trees applicable to version 1 of the Internet Standard Network Management Framework. The Internet Standard MIB is defined by {mgmt 1} or {1.3.6.1.2.1}. Under this tree are objects defined by MIB-II (RFC 1213), such as the remote network monitoring (RMON) MIB, and many others.

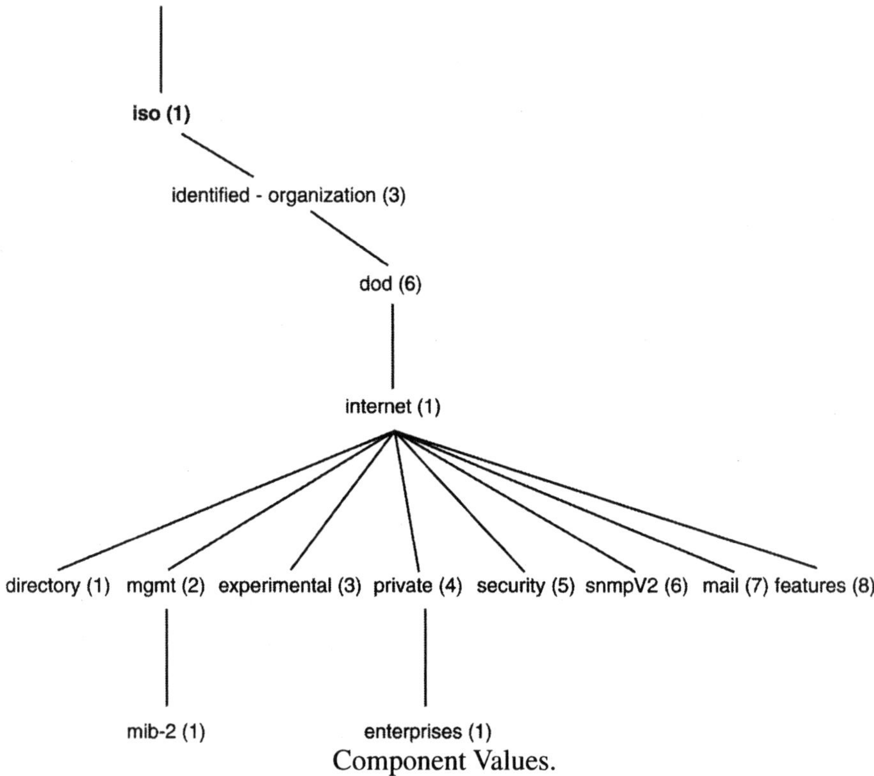

Component Values.

Reviewing the examples given at the beginning of this section, now that you know the identities of the individual tree structures, the following sequences can be constructed:

```
  internet OBJECT IDENTIFIER::={iso org(3) dod(6) 1}
mgmt OBJECT IDENTIFIER ::= {internet 2}
mib OBJECT IDENTIFIER ::= {mgmt 1}
system OBJECT IDENTIFIER ::= {mib-2 1}
sysDescr OBJECT IDENTIFIER ::= {system 1}
```

When these tree structures are combined, the result becomes:

```
sysDescr OBJECT IDENTIFIER ::= {1.3.6.1.2.1.1.1}
```

To the OID we need to add one last element — a suffix that identifies whether a particular variable occurs just once (a scalar) or whether the variable occurs multiple times (as in columnar entries).

Since *sysDescr* is a scalar, not columnar, object, there is only one instance of it. (In other words, you can have only one description of the system being managed.) Therefore, a .0 is added to the end of the OID:

```
{1.3.6.1.2.1.1.1.0}
```

If the object were a columnar entry, which could have multiple instances, an index plus a nonzero suffix (.1, .2, an IP address, and so on) would identify the object within the table. Section 12.4.2.2 discusses tables in greater detail.

Experimental codes, with prefix {1.3.6.1.3}, were assigned for many LAN and WAN objects and MIBs, such as the Synchronous Optical Network (SONET) objects, Asynchronous Transfer Mode (ATM) objects, or Fibre Channel objects, while the technologies and their MIBs were in the testing phase of development. Thus, as the Internet Engineering Task Force (IETF) working groups develop MIBs, they define them under a branch in the experimental tree. Once these MIBs are published and put on the standards track, they move to a branch on the Internet subtree.

Vendor-specific MIBs use private enterprise codes with prefix {1.3.6.1.4.1}. This is an area of rapid growth, because numerous vendors are developing structures to support their Internetworking devices, servers, and so on. Examples include 3Com Corporation (Santa Clara, CA) with code {1.3.6.1.4.1.43}; AT&T, with code {1.3.6.1.4.1.74}; Marconi Communications, with code {1.3.6.1.4.1.1012}; and OPNET Technologies, with code {1.3.6.1.4.1.4835}. Reference [12-8] is an online listing of current enterprise numbers.

12.2 The MIB Defined for SNMPv1

This is the second of four sections to discuss the Internet network management standards that are used with SNMPv1. Section 12.1 discussed the SMI, which defines the syntax that retrieves and communicates information; the ways information is placed into logical groups; and the naming mechanisms, known as the object identifiers, that identify each managed object.

This section extends the discussion of naming mechanisms to include management information bases (MIBs), which store management information. Section 12.3 will consider the Remote Monitoring (RMON) MIB, which facilitates the management of remote network segments. Section 12.4 will consider the SNMPv1 and its various message structures, which communicate management information. MIB details that are specific to SNMPv2 and SNMPv3 will be presented in Sections 12.5 and 12.6, respectively.

You can think of a MIB as a virtual information warehouse. Like a physical warehouse with specific floors, aisles, and bins, the MIB must implement an inventory control scheme. SMI defines the scheme for the MIBs. Just as a large company can have several warehouses, there are several different types of MIBs. Some, such as Internet standards, are for public use; specific organizations have developed others for private use by their products.

12.2.1 MIBs within the Internet Object Identifier Subtree

The previous section discussed the Internet network management naming structure. A tree represents the management structure, with branches and leaves representing the managed objects (see Figure 12-19). This discussion focuses on the Internet subtree, designated {1.3.6.1}. In the figure, you can see eight subtrees under Internet: directory (1), mgmt (2), experimental (3), private (4), security (5), snmpV2 (6), mail (7), and features (8).

The directory (1) subtree is reserved for future use of the OSI directory within the Internet. The mgmt (2) subtree handles Internet-approved documents, such as the Internet standard MIBs, which are MIB-I (see RFC 1156) and MIB-II (see RFC 1213). An object identifier (OID) with a prefix of {1.3.6.1.2.1} denotes managed objects within MIB-I and MIB-II.

Internet experiments use the experimental subtree (3). The Internet Assigned Numbers Authority (IANA) at the USC-Information Sciences Institute administers this subtree.

The private subtree (4) allows vendors to register a MIB for their equipment. The enterprise subtree, whose branches are private organizations, falls under the private subtree. The IANA assigns "enterprise codes" to branches representing private organizations and publishes them in an online document [12-8]. Enterprise OIDs begin with the prefix {1.3.6.1.4.1}.

This section focuses primarily on the mgmt MIBs. MIB-I was the first version of the mgmt MIBs and was defined in RFC 1156. MIB-II (RFC 1213) [12-6] replaced the earlier version. Section 12.3 discusses the remote network monitoring (RMON) MIBs for Ethernet and token ring networks, which are also under the mgmt subtree. Other subtrees contain objects for security, SNMPv2 and SNMPv3 (which will be discussed in subsequent sections), mail, and features.

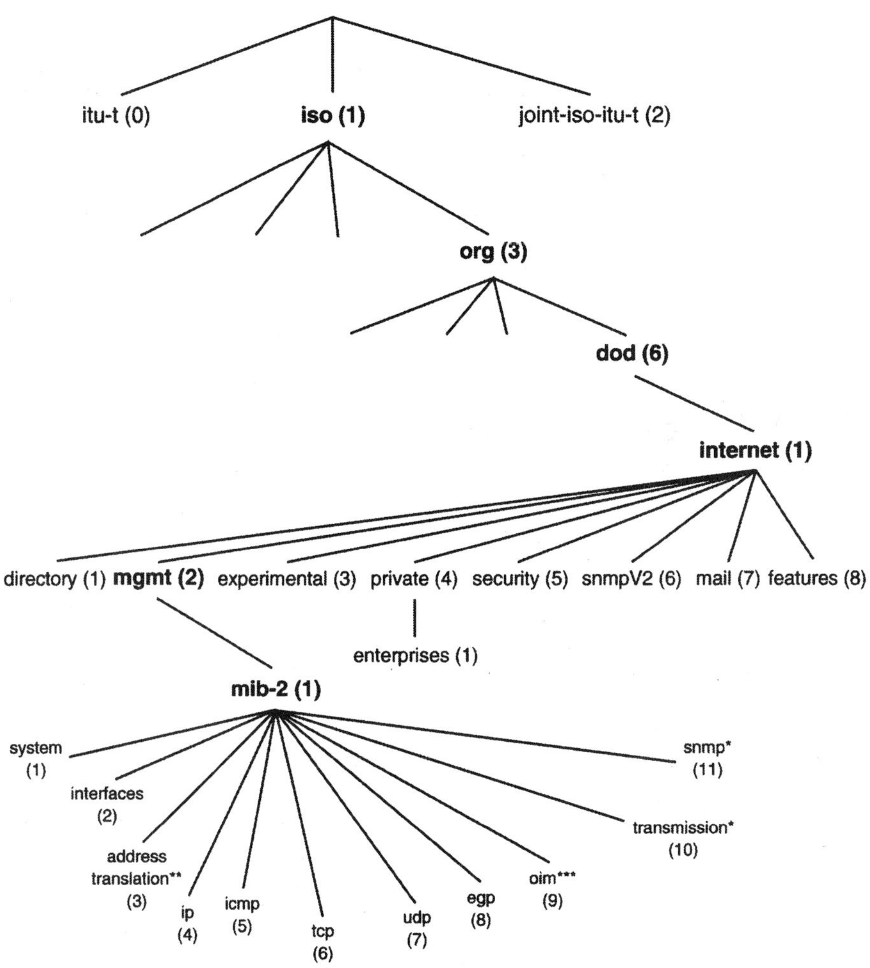

Figure 12-19. The Internet OID Tree.

12.2.2 MIB Development

As noted previously, MIBs address the need for a standard network management platform by the Internet as a whole and by private enterprises. These MIBs require a consistent objective and format to realize this objective. The first MIB, MIB-I (RFC 1156), was published in May 1990. MIB-I divided managed objects into eight groups in order to simplify OID assignment and implementation (that is, the SMI "structure"). Those groups were System, Interfaces, Address Translation, IP, ICMP, TCP, UDP, and EGP. With different private enterprises developing MIBs, it's necessary to develop a consistent format for MIB modules. RFC 1212, entitled "Concise MIB Definitions," addresses this issue [12-5]. Prior to the publication of RFC 1212, there were two ways to define objects: a textual definition and the ASN.1 OBJECT-TYPE macro. RFC 1212 embedded the textual definition within the OBJECT-TYPE macro, reducing the amount of documentation. Further work in this area produced RFC 1213 [12-6], known as "MIB-II," which was published in March 1991. MIB-II added the Transmission and SNMP groups.

12.2.3 The System Group

The System group provides a textural description of the entity in printable ASCII characters. This text includes a system description, OID, the length of time since the re-initialization of its network management entity, and other administrative details. Implementation of the System group is mandatory. The OID tree for the System group is designated {1.3.6.1.2.1.1}, as shown in Figure 12-20.

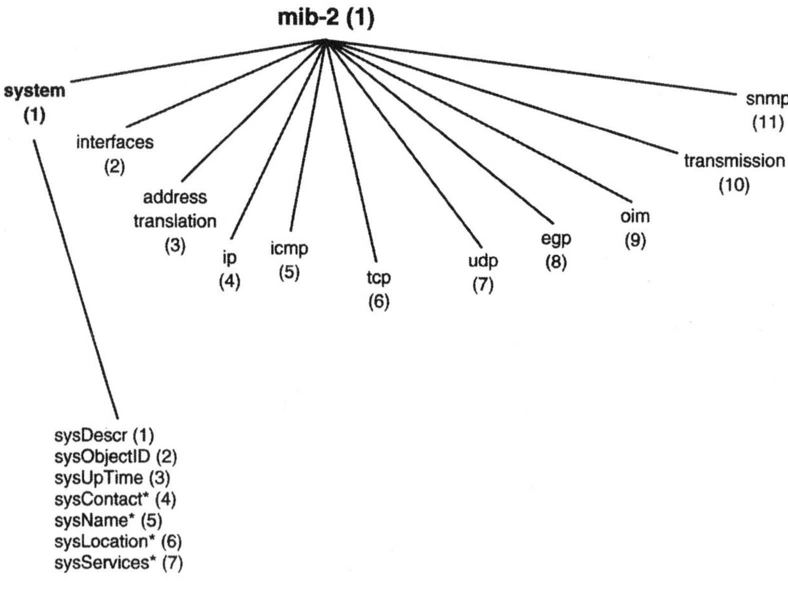

Figure 12-20. The System Group.

12.2.4 The Interfaces Group

The Interfaces group {1.3.6.1.2.1.2} provides information about the hardware interfaces on a managed device, as shown in Figure 12-21. This information is presented in a table. The first object (ifNumber) indicates the number of interfaces on the device. For each interface, a row entry is made into the table, with 22 column entries per row. The column entries provide information about the interfaces, such as the interface speed, physical (hardware) address, current operational state, and packet statistics.

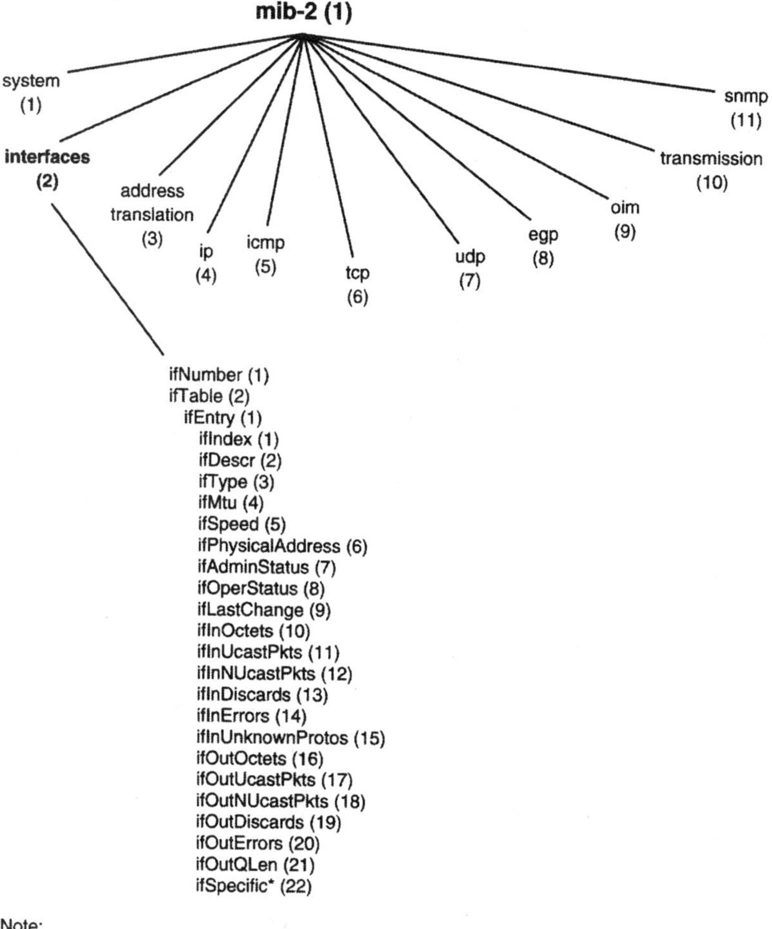

Figure 12-21. The Interfaces Group.

12.2.5 The Address Translation Group

MIB-I included the Address Translation group, shown in Figure 12-22, but it was deprecated in MIB-II. The "deprecated" status means that MIB-II includes the Address Translation group for compatibility with MIB-I, but will probably exclude it from future MIB releases. The Address Translation group provided a table that translated between IP addresses and physical (hardware) addresses. In MIB-II and future releases, each protocol group will contain its own translation tables. The Address Translation group is designated {1.3.6.1.2.1.3}. It contains one table with three columns per row.

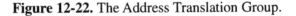

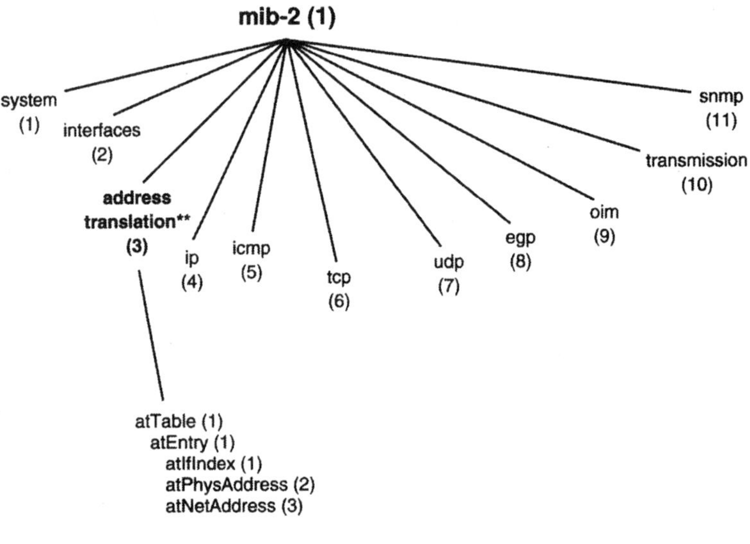

Figure 12-22. The Address Translation Group.

12.2.6 The IP Group

The Internet Protocol (IP) group, shown in Figure 12-23, is mandatory for all managed nodes and provides information on host and router use of the IP. This group includes a number of scalar objects that provide IP-related datagram statistics and the following three tables: an address table (*ipAddrTable*); an IP to physical address translation table (*ipNetToMediaTable*); and an IP forwarding table (*ipForwardTable*). Note that RFC 2096 [12-9] defined the *ipForwardTable*, which replaces and obsoletes the *ipRoutingTable* in MIB-II. The IP subtree is designated {1.3.6.1.2.1.4}.

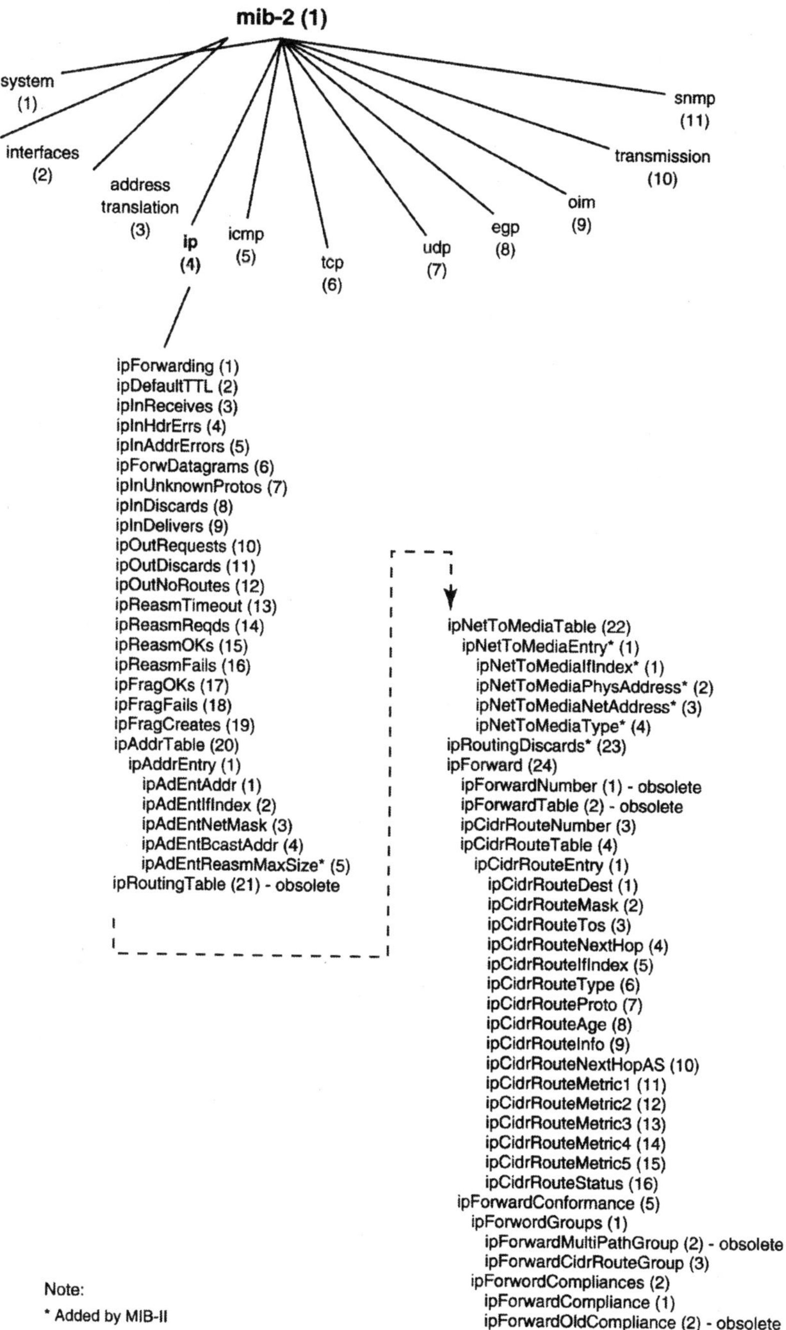

Figure 12-23. The IP Group.

12.2.7 The ICMP Group

The Internet Control Message Protocol (ICMP) group, shown in Figure 12-24, is a mandatory component of IP and is defined in RFC 792. The ICMP group provides intranetwork control messages and represents various ICMP operations within the managed entity. The ICMP group contains 26 scalar objects that maintain statistics for various ICMP messages, such as the number of ICMP Echo Request messages received or ICMP Redirect messages sent. This group is designated {1.3.6.1.2.1.5} on the OID tree.

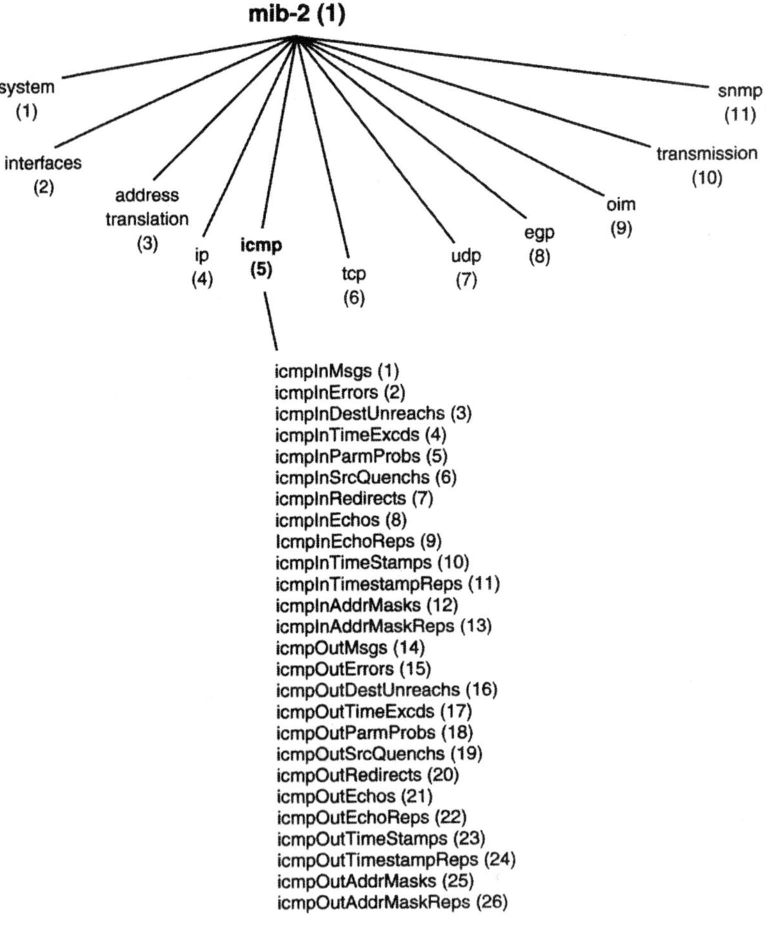

Figure 12-24. The ICMP Group.

12.2.8 The TCP Group

The Transmission Control Protocol (TCP) group, shown in Figure 12-25, is mandatory and provides information regarding TCP operation and connections. This group contains 14 scalar objects and one table. The scalar objects record various TCP parameters and statistics, such as the number of TCP connections that the device supports, or the total number of TCP segments transmitted. The table, *tcpConnTable*, contains information concerning a particular TCP connection. The OID for this group is {1.3.6.1.2.1.6}.

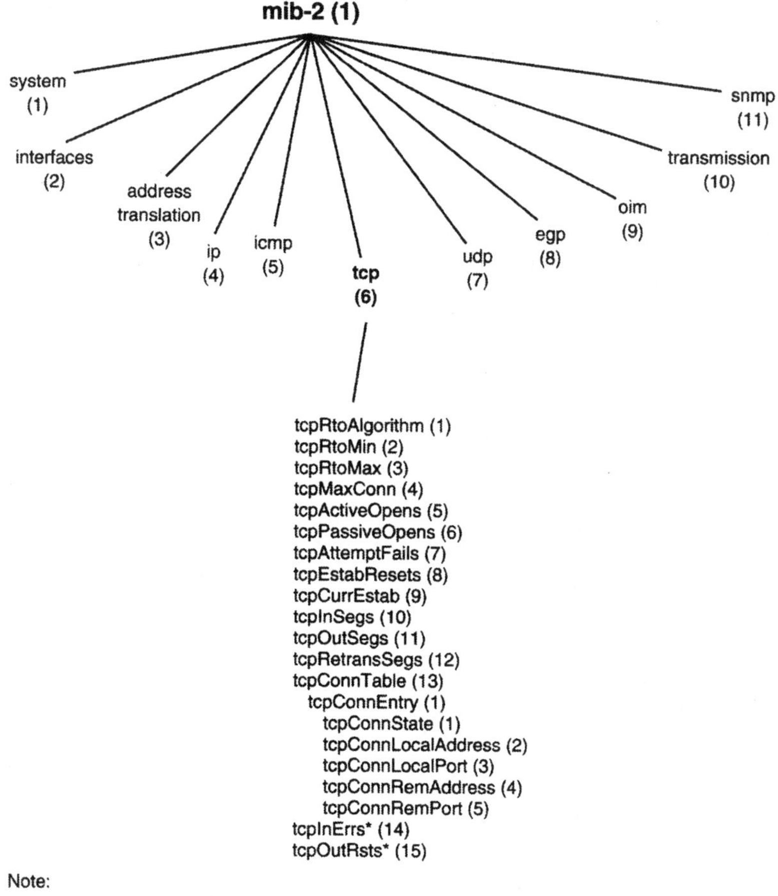

Figure 12-25. The TCP Group.

12.2.9 The UDP Group

The User Datagram Protocol (UDP) group, shown in Figure 12-26, is mandatory and provides information regarding UDP operation. Because UDP is connectionless, this group is much smaller than the connection-oriented TCP group. It does not have to compile information on connection attempts, establishment, reset, and so on. The UDP group contains four scalars and one table. The scalar objects maintain UDP-related datagram statistics, such as the number of datagrams sent from this entity. The table, *udpTable*, contains address and port information. The OID for this group is {1.3.6.1.2.1.7}.

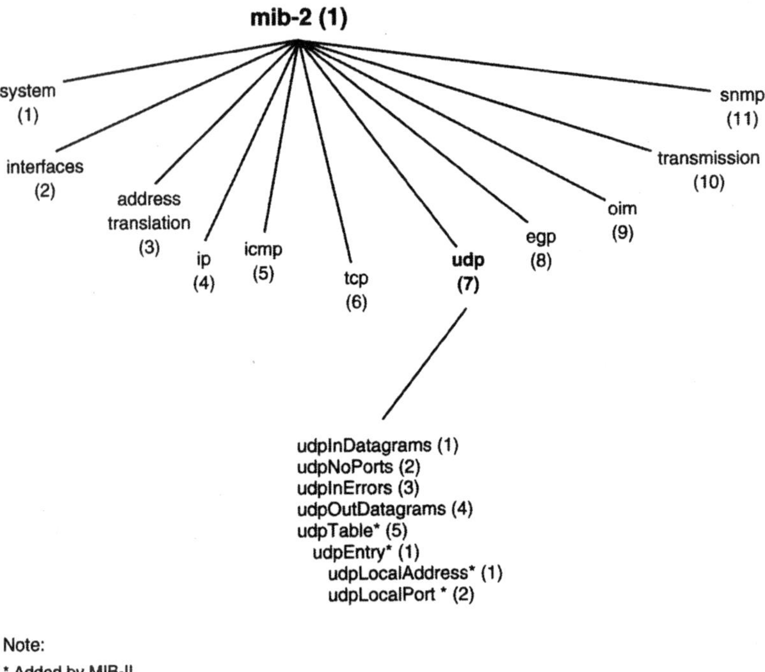

Figure 12-26. The UDP Group.

12.2.10 The EGP Group

The Exterior Gateway Protocol (EGP) group, shown in Figure 12-27, is mandatory for all systems that implement the EGP. The EGP communicates between autonomous (self-contained) systems; RFC 904 describes it in detail. The EGP group includes 5 scalar objects and one table. The scalars maintain EGP-related message statistics. The table, *egpNeighTable*, contains EGP neighbor information. The OID for this group is {1.3.6.1.2.1.8}.

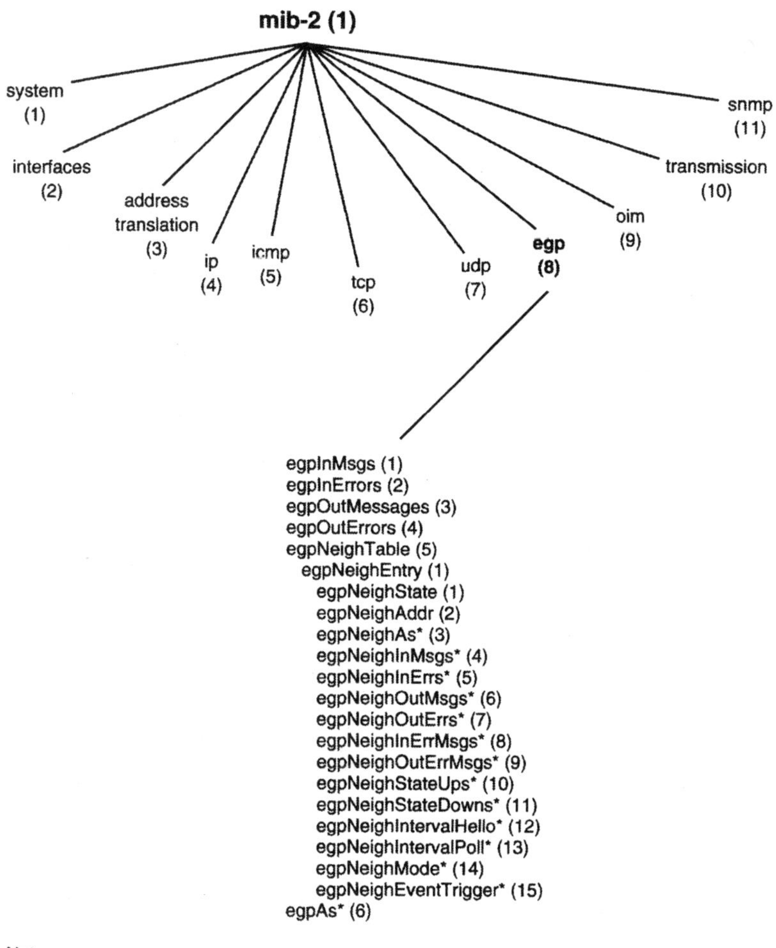

Figure 12-27. The EGP Group.

12.2.11 The CMOT (OIM) Group

At one time, during the development of the Internet Network Management Framework, there was an effort to use SNMP as an interim step in the push for a network management standard and to make the Common Management Information Protocol (CMIP) over TCP/IP (CMOT) the long-term, OSI-compliant solution. As a result, the CMOT group was placed within MIB-II. Experience has shown, however, that SNMP is not an interim solution, and that the OSI-related network management protocol requires unique MIBs. Therefore, it's unlikely that you'll encounter the OIM group within any commercially available SNMP managers or agents.

Nonetheless, the CMOT group was given a placeholder of {1.3.6.1.2.1.9} in MIB-II. RFC 1214 [12-10] details that subtree, which specifies the OSI Internet Management (OIM) MIB. At this time, RFC 1214 is classified as a "historic" protocol.

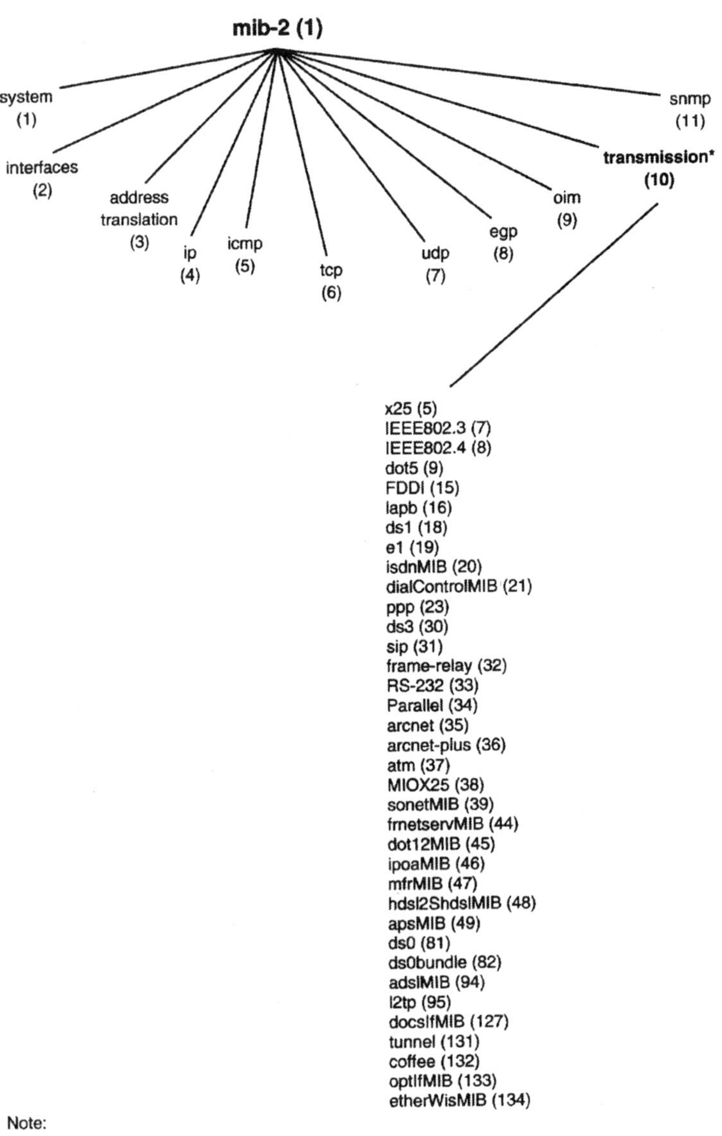

Figure 12-28. The Transmission Group.

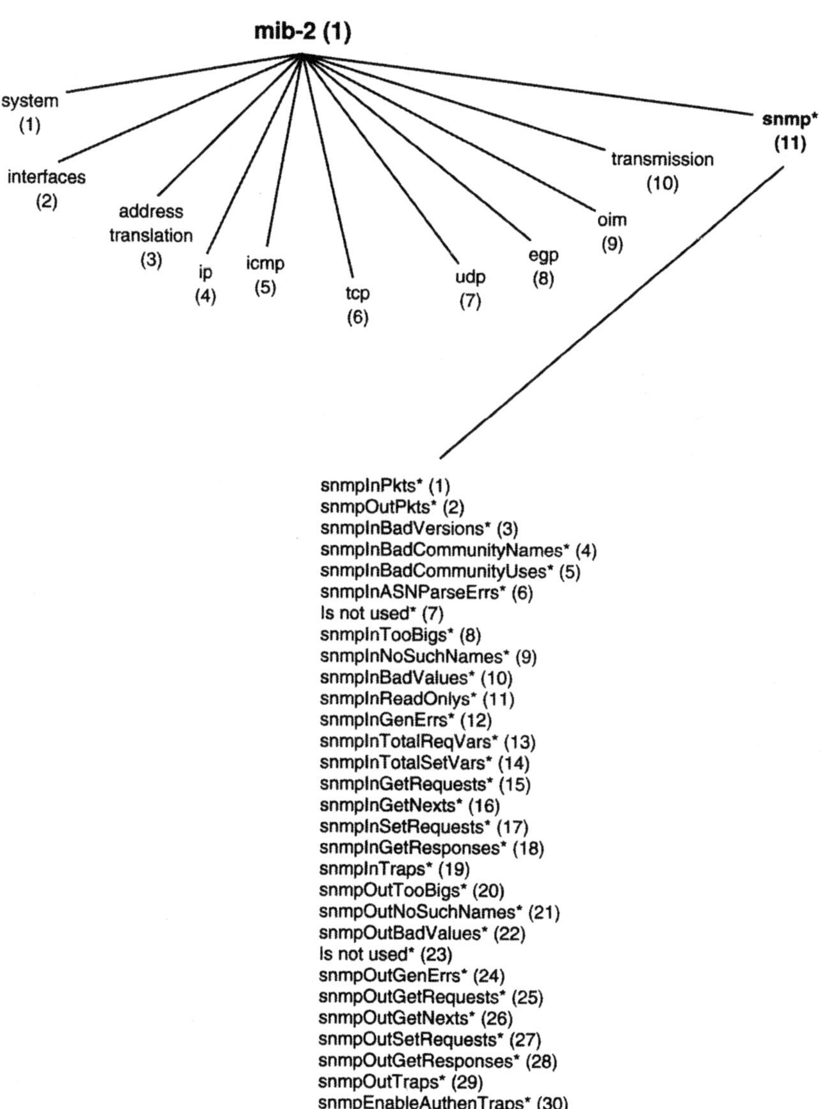
mib-2 (1)
system (1)
interfaces (2)
address translation (3)
ip (4)
icmp (5)
tcp (6)
udp (7)
egp (8)
oim (9)
transmission (10)
snmp* (11)
snmpInPkts* (1)
snmpOutPkts* (2)
snmpInBadVersions* (3)
snmpInBadCommunityNames* (4)
snmpInBadCommunityUses* (5)
snmpInASNParseErrs* (6)
Is not used* (7)
snmpInTooBigs* (8)
snmpInNoSuchNames* (9)
snmpInBadValues* (10)
snmpInReadOnlys* (11)
snmpInGenErrs* (12)
snmpInTotalReqVars* (13)
snmpInTotalSetVars* (14)
snmpInGetRequests* (15)
snmpInGetNexts* (16)
snmpInSetRequests* (17)
snmpInGetResponses* (18)
snmpInTraps* (19)
snmpOutTooBigs* (20)
snmpOutNoSuchNames* (21)
snmpOutBadValues* (22)
Is not used* (23)
snmpOutGenErrs* (24)
snmpOutGetRequests* (25)
snmpOutGetNexts* (26)
snmpOutSetRequests* (27)
snmpOutGetResponses* (28)
snmpOutTraps* (29)
snmpEnableAuthenTraps* (30)
Note:
* Added by MIB-II

Figure 12-29. The SNMP Group.

12.2.12 The Transmission Group

The Transmission group, shown in Figure 12-28 and designated {1.3.6.1.2.1.10}, contains objects that relate to the transmission of the data. RFC 1213 defines none of these objects explicitly. However, the document does say that these transmission objects will reside in the experimental subtree {1.3.6.1.3} until they are "proven."

The SMI Numbers document [12-7] lists the currently defined objects for the Transmission group.

12.2.13 The SNMP Group

The SNMP group provides information about SNMP objects (see Figure 12-29). There are a total of 30 scalar objects in this group, including SNMP message statistics, the number of MIB objects retrieved, and the number of SNMP traps sent. This group is designated {1.3.6.1.2.1.11}.

12.2.14 Other MIBs

The growing need for network management in general, along with SNMP's popularity, has led to the definition of a number of MIBs that support specific network architectures or platforms. Included are MIBs to support a specific architecture, protocol, or interface, and those that support specific transmission media types.

The following table lists the currently approved Internet-standard MIBs included under the MIB-II subtree that support a specific architecture, protocol, or interface:

MIB	OID	RFC
Reserved	1.3.6.1.2.1.0	IANA
System	1.3.6.1.2.1.1	1213
Interfaces	1.3.6.1.2.1.2	1213
Address Translation	1.3.6.1.2.1.3	1213
Internet Protocol	1.3.6.1.2.1.4	1213
Internet Control Message	1.3.6.1.2.1.5	1213
Transmission Control Protocol	1.3.6.1.2.1.6	1213
User Datagram Protocol	1.3.6.1.2.1.7	1213
Exterior Gateway Protocol	1.3.6.1.2.1.8	1213
CMIP over TCP	1.3.6.1.2.1.9	1213
Transmission	1.3.6.1.2.1.10	1213
Simple Network Management	1.3.6.1.2.1.11	1213
Generic Interface Extensions	1.3.6.1.2.1.12	1229, 1239
AppleTalk Protocols	1.3.6.1.2.1.13	1742

Open Shortest Path First	1.3.6.1.2.1.14	1253
Border Gateway Protocol	1.3.6.1.2.1.15	1657
Remote Monitoring	1.3.6.1.2.1.16	1271
Bridges	1.3.6.1.2.1.17	1286
DECnet Phase 4	1.3.6.1.2.1.18	1559
Character Streams	1.3.6.1.2.1.19	1658
SNMP Parties	1.3.6.1.2.1.20	1353
SNMP Secrets	1.3.6.1.2.1.21	1353
IEEE 802.3 Repeaters	1.3.6.1.2.1.22	2108
Routing Information Protocol	1.3.6.1.2.1.23	1389
Identification Protocol	1.3.6.1.2.1.24	1414
Host Resources	1.3.6.1.2.1.25	1514
IEEE 802.3 Medium Attachment Units	1.3.6.1.2.1.26	2668
Network Services Monitoring	1.3.6.1.2.1.27	2248
Mail Monitoring	1.3.6.1.2.1.28	2249
X.500 Directory Monitoring	1.3.6.1.2.1.29	1567
Interface Types	1.3.6.1.2.1.30	1573
Interface Types	1.3.6.1.2.1.31	1573
Domain Name System	1.3.6.1.2.1.32	1611
Uninterruptible Power Supplies	1.3.6.1.2.1.33	1628
SNA NAU	1.3.6.1.2.1.34	1666
Ethernet-like generic objects	1.3.6.1.2.1.35	2665
SMDS interface objects	1.3.6.1.2.1.36	1694
ATM objects	1.3.6.1.2.1.37	2515
Dial-up modem objects	1.3.6.1.2.1.38	1696
Relational database objects	1.3.6.1.2.1.39	1697
Traffic flow objects	1.3.6.1.2.1.40	2064
SNA SDLC	1.3.6.1.2.1.41	1747
Token Ring Station Source Route	1.3.6.1.2.1.42	1748
Printer	1.3.6.1.2.1.43	1759
Mobile IP	1.3.6.1.2.1.44	2006
IEEE 802.12	1.3.6.1.2.1.45	2020
Data Link Switch	1.3.6.1.2.1.46	2024
Entity	1.3.6.1.2.1.47	2037
Internet Protocol	1.3.6.1.2.1.48	2011
Transmission Control Protocol	1.3.6.1.2.1.49	2012
User Datagram Protocol	1.3.6.1.2.1.50	2013
Resource Reservation Protocol	1.3.6.1.2.1.51	2206
Integrated Services	1.3.6.1.2.1.52	2213
IEEE 802.12 Repeater	1.3.6.1.2.1.53	2266
System Application	1.3.6.1.2.1.54	2287
Internet Protocol version 6	1.3.6.1.2.1.55	2465
Internet Control Message Protocol version 6	1.3.6.1.2.1.56	2466

Multicast Address Resolution	1.3.6.1.2.1.57	2417
Performance History Textual Conventions	1.3.6.1.2.1.58	2493
ATM Accounting Information	1.3.6.1.2.1.59	2512
Accounting Control	1.3.6.1.2.1.60	2513
TN3270E Servers	1.3.6.1.2.1.61	2561
Application Management	1.3.6.1.2.1.62	2564
Distributed Management Schedule	1.3.6.1.2.1.63	2591
Distributed Management Script	1.3.6.1.2.1.64	3165
WWW Service	1.3.6.1.2.1.65	2594
Directory Server Monitoring	1.3.6.1.2.1.66	2605
Remote Authentication Dial-In User Service	1.3.6.1.2.1.67	2618
Virtual Router Redundancy Protocol	1.3.6.1.2.1.68	Internet Draft
Data Over Cable Service Interface Specifications Device	1.3.6.1.2.1.69	2669
Ethernet-Chipset	1.3.6.1.2.1.70	2666
NBMA Next Hop Resolution Protocol	1.3.6.1.2.1.71	2677
IANA Address Family Numbers	1.3.6.1.2.1.72	2677
IANA Language	1.3.6.1.2.1.73	3165
Agent Extensibility	1.3.6.1.2.1.74	2742
Fibre Channel Fabric Element	1.3.6.1.2.1.75	2837
Internet Network Address	1.3.6.1.2.1.76	2851
Interface Inverted Stack	1.3.6.1.2.1.77	2864
High Capacity Number Data Type Textual Conventions	1.3.6.1.2.1.78	2856
Physical Topology	1.3.6.1.2.1.79	2922
Distributed Management PING	1.3.6.1.2.1.80	2925
Distributed Management TRACEROUTE	1.3.6.1.2.1.81	2925
Distributed Management LOOKUP	1.3.6.1.2.1.82	2925
IPv4 Multicast Routing	1.3.6.1.2.1.83	2932
IANA-Route Protocol	1.3.6.1.2.1.84	2932
Internet Group Management Protocol	1.3.6.1.2.1.85	2933
Frame Relay/ATM PVC Service Interworking Function	1.3.6.1.2.1.86	2955
Real-Time Transport Protocol	1.3.6.1.2.1.87	2959
Distributed Management EVENT	1.3.6.1.2.1.88	2981
Common Open Policy Service CLIENT	1.3.6.1.2.1.89	2940
Distributed Management EXPRESSION	1.3.6.1.2.1.90	2982
IPv6 Multicast Listener Discovery	1.3.6.1.2.1.91	3019
Notification Log	1.3.6.1.2.1.92	3014
PSTN Services over the Internet	1.3.6.1.2.1.93	3055
Circuit-to-Interface Translation	1.3.6.1.2.1.94	3201
Frame Relay Service Level Definitions	1.3.6.1.2.1.95	3202
Differentiated Services Code Point	1.3.6.1.2.1.96	3289
Differentiated Services	1.3.6.1.2.1.97	3289
General Switch Management Protocol	1.3.6.1.2.1.98	3295
Entity Sensor	1.3.6.1.2.1.99	3433

Transport Address	1.3.6.1.2.1.100	3419
Multicast Address Allocation	1.3.6.1.2.1.101	3559
IANA Multicast Address Allocation	1.3.6.1.2.1.102	3559
IPv6 Flow Label	1.3.6.1.2.1.103	Internet Draft
Stream Control Transmission Protocol	1.3.6.1.2.1.104	Internet Draft
Power Ethernet	1.3.6.1.2.1.105	Internet Draft

12.2.15 Transmission Media MIBs

Other Internet-standard MIBs have been approved for specific transmission media, and are listed under the transmission subtree, {1.3.6.1.2.1.10}. These are defined in Reference [12-7], and include:

MIB	OID	RFC
X.25 Packet Layer objects	1.3.6.1.2.1.10.5	1382
CSMA/CD-like objects	1.3.6.1.2.1.10.7	2665
Token Bus-like objects	1.3.6.1.2.1.10.8	1230, 1239
Token Ring-like objects	1.3.6.1.2.1.10.9	1743
FDDI objects	1.3.6.1.2.1.10.15	12850
X.25 LAPB objects	1.3.6.1.2.1.10.16	1381
T1 Carrier objects	1.3.6.1.2.1.10.18	2495
E1 Carrier objects	1.3.6.1.2.1.10.19	2495
ISDN objects	1.3.6.1.2.1.10.20	2127
Dial Control objects	1.3.6.1.2.1.10.21	2128
PPP objects	1.3.6.1.2.1.10.23	1471
DS3/E3 Interface objects	1.3.6.1.2.1.10.30	2496
SMDS Interface objects	1.3.6.1.2.1.10.31	1694
Frame Relay objects	1.3.6.1.2.1.10.32	1315
RS-232 objects	1.3.6.1.2.1.10.33	1659
Parallel Printer objects	1.3.6.1.2.1.10.34	1660
ARCNET objects	1.3.6.1.2.1.10.35	N/A
ARCNETPLUS objects	1.3.6.1.2.1.10.36	N/A
ATM objects	1.3.6.1.2.1.10.37	N/A
Multiprotocol Interconnect over X.25 objects	1.3.6.1.2.1.10.38	1461
SONET objects	1.3.6.1.2.1.10.39	2558
Frame Relay Service objects	1.3.6.1.2.1.10.44	1604
IEEE 802.12 objects	1.3.6.1.2.1.10.45	2020
IP and ARP over ATM objects	1.3.6.1.2.1.10.46	2320
Multilink Frame Relay objects	1.3.6.1.2.1.10.47	3020
Digital Subscriber Line objects	1.3.6.1.2.1.10.48	3276
SONET Automatic Protection Switching objects	1.3.6.1.2.1.10.49	3498
Unassigned	1.3.6.1.2.1.10.50–80	N/A

DS0 objects	1.3.6.1.2.1.10.81	2494
DS0 Bundle objects	1.3.6.1.2.1.10.82	2494
Unassigned	Unassigned	N/A
ADSL objects	1.3.6.1.2.1.10.94	2662
Layer 2 Tunneling Protocol objects	1.3.6.1.2.1.10.95	3371
Unassigned	1.3.6.1.2.1.10.96–126	N/A
Data over Cable Service Interface objects	1.3.6.1.2.1.10.127	2670
Unassigned	1.3.6.1.2.1.10.128–130	N/A
IP Tunnel MIB	1.3.6.1.2.1.10.131	2667
Coffee pot	1.3.6.1.2.1.10.132	2325
Optical Interface objects	1.3.6.1.2.1.10.133	3591
Ethernet WAN Interface objects	1.3.6.1.2.1.10.134	Internet Draft

12.2.16 Private MIBs

Many vendors have developed private MIBs that support hubs, terminal servers, routers, and many other networking systems. You can find these MIBs under the enterprises subtree, {1.3.6.1.4.1.A}. The A indicates a private enterprise code, defined in the "enterprise-numbers" section of the IANA website [12-8]. In this file you can identify the vendor organization, plus a contact person at the organization. Another very helpful site is called the MIB Depot [12-11], which notes information and availability of several thousand vendor MIBs.

12.3 Remote Monitoring (RMON) MIB

This is the second section to examine Management Information Bases (MIBs) in detail. While Section 12.2 dealt with the concept of MIBs, the Internet Standard MIB defined in RFC 1213, and other MIBs, this chapter will extend those concepts and discuss a MIB module that has a very specific function. The Remote Monitoring (RMON) MIB is used to monitor and manage remote segments of distributed internetworks. We will discuss the general concepts of remote monitoring and the architecture of the RMON MIB modules.

In the last decade, LANs and internetworks have become more complex, and, with that, the methods and procedures for managing networks have matured as well. For example, a network consisting of a single segment, such as the topology shown in Figure 12-30, does not need much rigor in its network management strategy. First, there are relatively few devices to manage, and, more importantly, these devices are geographically located in a central location. So it should be a straightforward process to diagnose and repair a problem with this network.

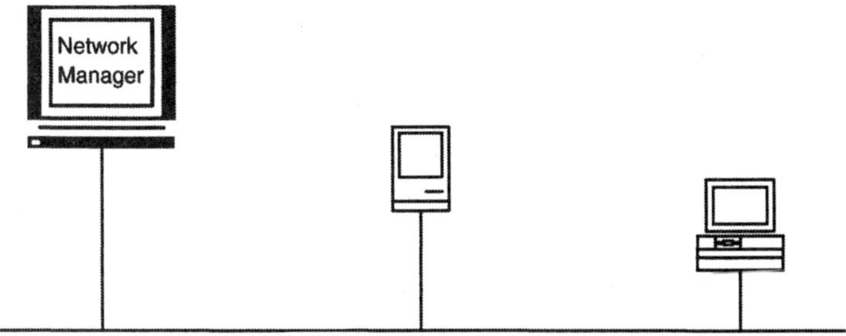

Figure 12-30. Single Segment Network Monitoring.

Distributed internetworks are a different story, however, for several reasons. First, there are more devices. Second, since there are multiple segments, the devices on those segments are more likely to have been acquired over time, to have come from different manufacturers, to have different software revision dates, and so on. (Not too many internetworks are built all at once. It is much more likely that they have evolved over time, with the typical challenges that network evolution brings.) Most importantly, when network difficulties occur, geography does not help the network manager. You can't be in multiple locations at the same time, and, if Murphy's Law holds true, the segment you are currently testing is not the segment you should be testing. And unless you have an unlimited staff and resources, it is not likely that you can afford to dispatch colleagues to every segment to assist with the problem. Thus, managing geographically dispersed internetwork segments can be a challenge.

The concepts of remote monitoring, or RMON, are designed to address those challenges. RMON places agents, called RMON probes, at strategic locations in the internetwork (see Figure 12-31). These probes consist of a network interface (either Ethernet or token ring), a processor, and memory, and are attached to the network much like any other device. The network management console determines the information that the probes are reporting on, such as the statistics to be monitored, time periods over which to collect historical information, and so on. In effect, the probes act as the eyes and ears of the network management system, providing managers with details regarding the operation of the distributed segments that would otherwise not be accessible without some type of direct human interaction.

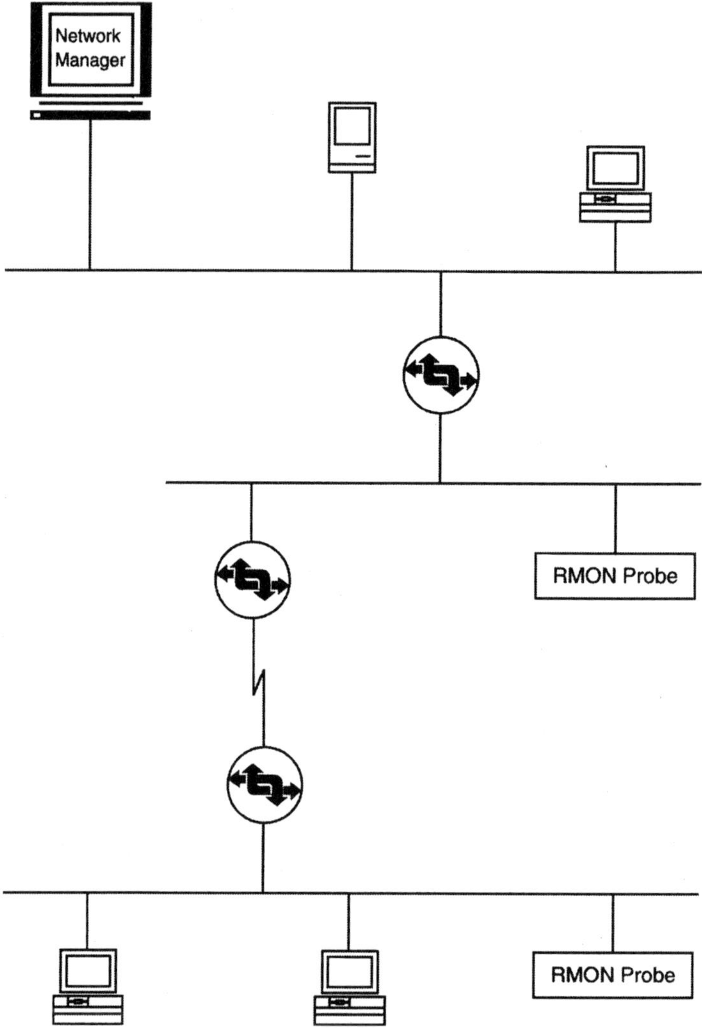

Figure 12-31. Distributed Segment Network Monitoring Using RMON.

The RMON MIB is assigned arc 16 under MIB-II, with a complete OID of {1.3.6.1.2.1.16}, as shown in Figure 12-32. RMON was developed in three stages. The first was the Ethernet RMON MIB shown in Figure 12-33. The second stage added token ring capabilities and was called the token ring RMON module, shown in Figure 12-34. The token ring module is not a separate MIB structure, but consists of additions to the original Ethernet module achieved by adding new objects in support of token ring networks. The third stage, known as RMON2, added ten more branches to the first ten, extending the capabilities of RMON to higher layer protocol monitoring. All three of these modules are complementary and frequently coexist in product implementations. These three elements will be discussed separately in the following sections.

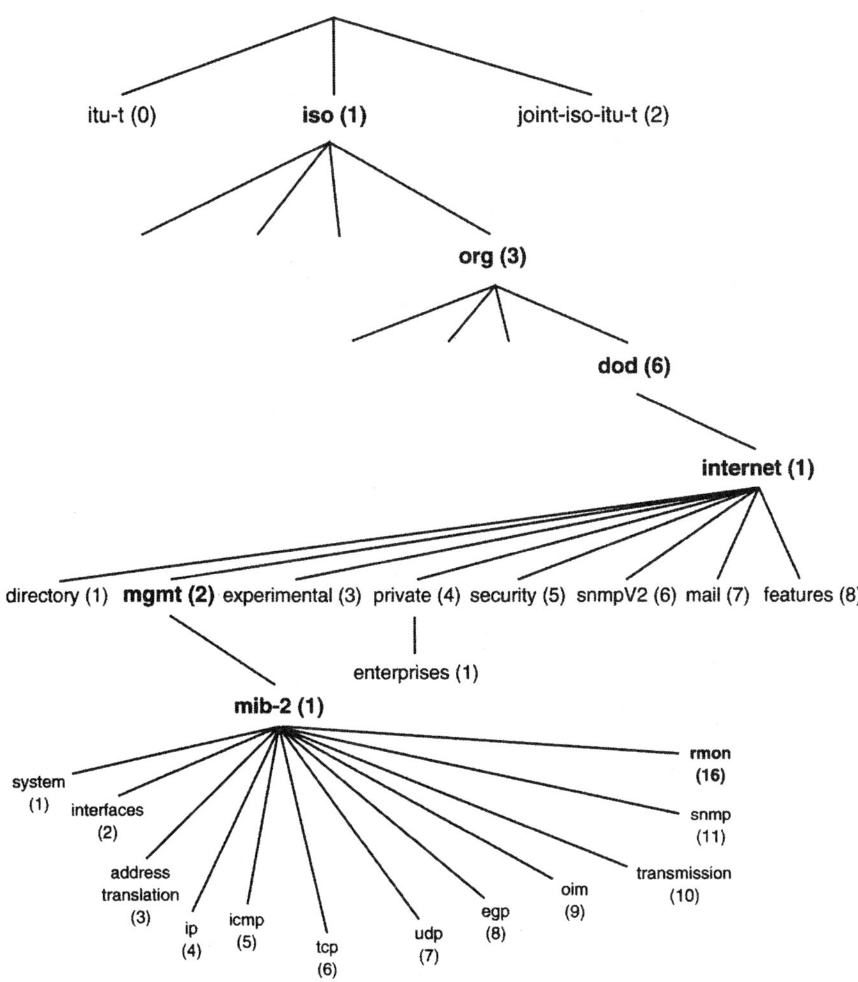

Figure 12-32. RMON MIB Position in the Internet OID Tree.

12.3.1 The Ethernet RMON MIB

The Ethernet RMON MIB module was the first to be developed. This module standardizes the management information sent to and from the Ethernet RMON probes and is presented in RFC 2819 [12-12].

The original RMON MIB is often referred to as the Ethernet RMON MIB. It is assigned OID {1.3.6.1.2.1.16} and contains nine groups. All of these groups are optional (not mandatory), but the implementation of some groups requires other groups. For example, the Filter group requires the Packet Capture group. The following table is a summary of the nine Ethernet groups shown in Figure 12-33:

Group	Description
statistics (1)	Provides probe-measured statistics, such as the number and sizes of packets, broadcasts, collisions, and so on.
history (2)	Records periodic statistical samples over time that you can use to analyze trends.
alarm (3)	Compares statistical samples with preset thresholds, generating alarms when a particular threshold is crossed.
host (4)	Maintains statistics of the hosts on the network, including the MAC addresses of the active hosts.
hostTopN (5)	Provides reports sorted by host table statistics, indicating which hosts are at the top of the list in a particular category.
matrix (6)	Stores statistics in a traffic matrix that tracks conversations between pairs of hosts.
filter (7)	Allows packets to be matched according to a filter equation.
capture (8)	Allows packets to be captured after they pass through a logical channel.
event (9)	Controls the generation and notification of events, which may include SNMP trap messages.

The current standard, RFC 2819, addresses only Ethernet network monitoring. The next section will discuss the token ring extensions.

12.3.2 The Token Ring RMON MIB

The token ring RMON MIB was created by adding tables to the Ethernet RMON MIB, and by adding a tenth group, all in support of token ring networks. Recall that the Ethernet RMON MIB defines nine groups, Statistics through Event. The token ring RMON MIB extends two of these groups, Statistics and History, and adds one unique group. This new group is called tokenRing, with object identifier {rmon 10} [12-13].

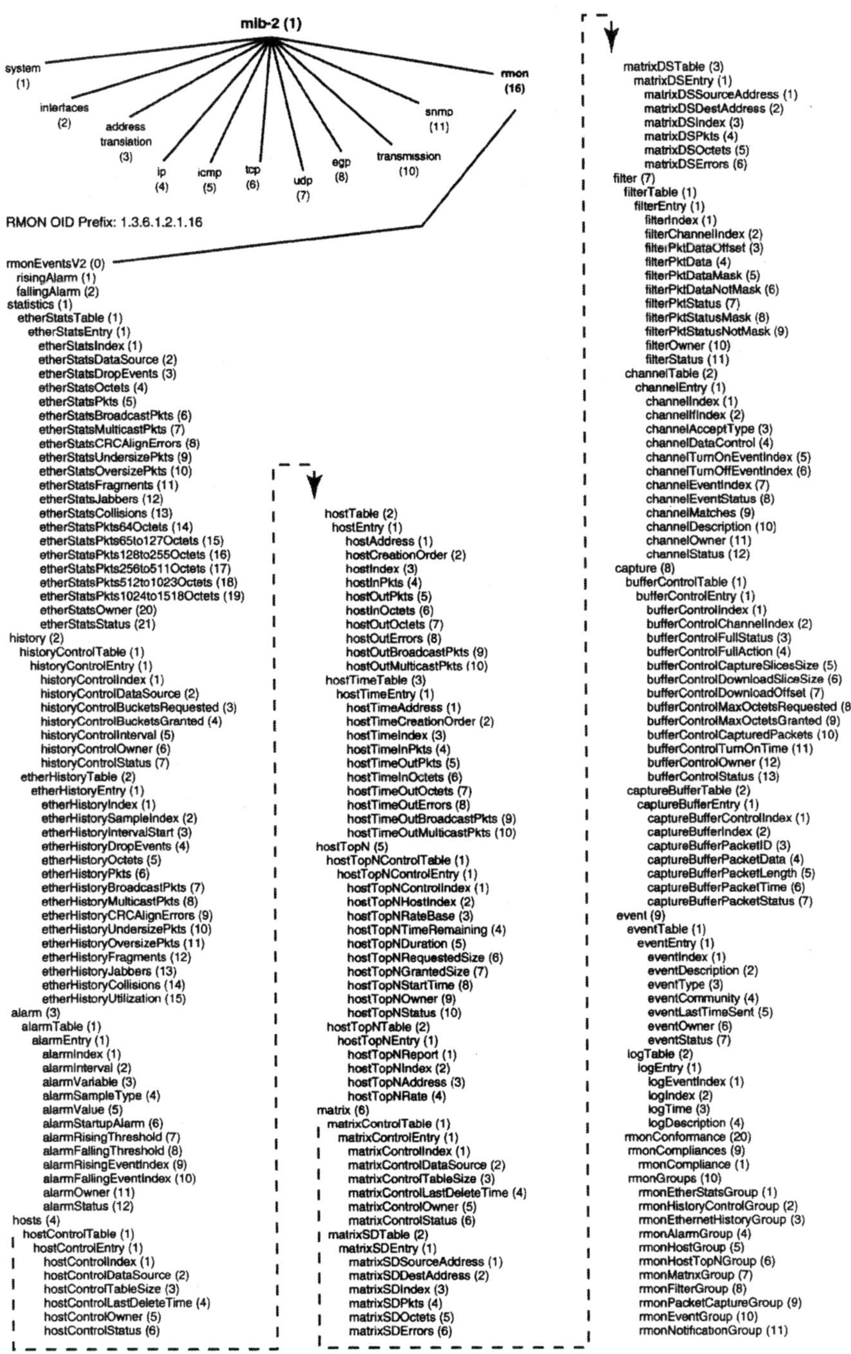

Figure 12-33. The Ethernet RMON MIB.

The Statistics extensions allow an RMON-compatible device to collect token ring MAC-Layer errors and promiscuous errors. The MAC-Layer errors, such as token errors and frame-copied errors, are specific to the token ring protocol; the promiscuous errors, such as counting the number of broadcast packages or data packets between 512 and 1023 octets in length, are more general. Similarly, the History information is divided into MAC-Layer and promiscuous details. The tokenRing group contains four subgroups: ring station, which monitors station- and ring-specific events; ring station order, which tracks the network topology; ring station configuration, which controls the removal and/or configuration of stations on the ring; and source routing, which details source routing bridging information. Figure 12-34 illustrates these token ring additions; it also shows the major Ethernet groups for clarification.

12.3.3 RMON2

The original RMON MIBs for Ethernet and token ring networks are primarily concerned with the operation and management of the Physical and Data Link Layers of a remote network. As such, they can compile statistics and historical information regarding Ethernet collisions, token ring frame copied errors, and so on, but they cannot look into the operation of the OSI Network through Application Layers of that remote network.

RMON2, defined in RFC 2021 [12-14], extends the RMON capabilities to those higher layers by adding ten new groups, designated {rmon 11} through {rmon 20}. Figure 12-35 illustrates the OID branches for both RMON and RMON2. Thus, RMON2 allows the higher layer protocols, such as TCP/IP or Novell's SPX/IPX, to be monitored for greater management visibility within the internetwork.

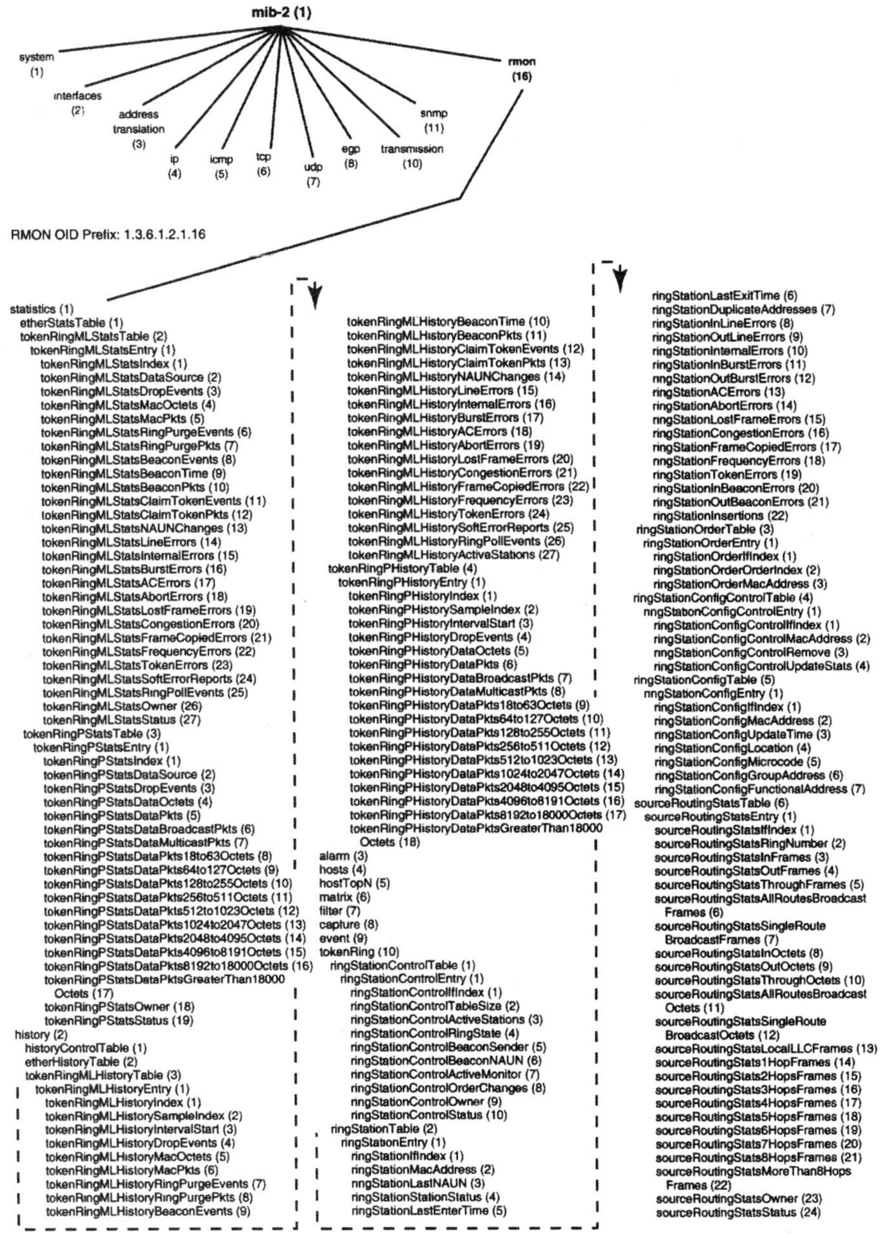

Figure 12-34. The Token Ring RMON MIB.

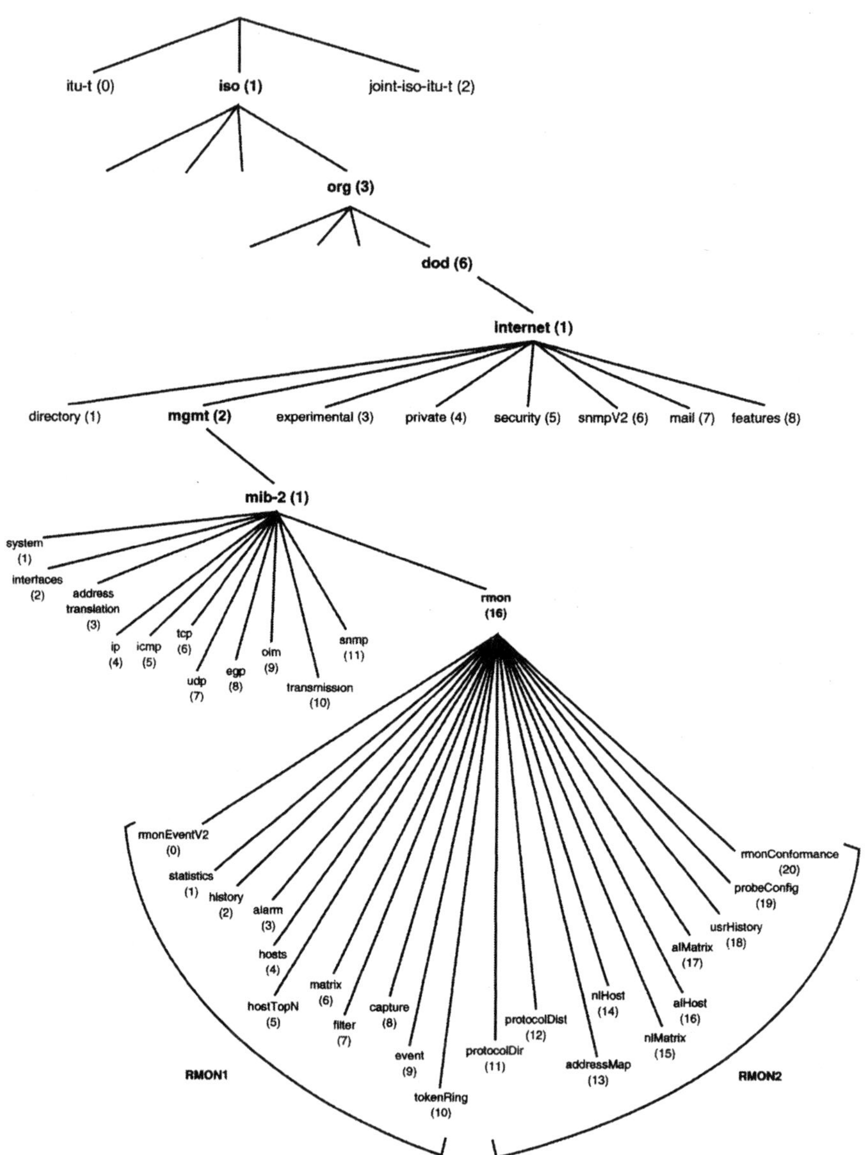

Figure 12-35. RMON1 and RMON2 Object Trees.

The ten groups within RMON2 are:

Group	Description
protocolDir (11)	Protocol Directory: lists, in a table, the inventory of protocols that the probe has the capability of monitoring. Each protocol is described by an entry in the table.
ProtocolDist (12)	Protocol Distribution: collects the relative amounts of octets and packets for the different protocols that are detected on a network segment. Each protocol is described by an entry in a table, and the network management station can easily determine the bandwidth consumed per protocol by accessing the information in that table.
addressMap (13)	Address Map: correlates Network Layer addresses and MAC Layer addresses, and stores the information in tables.
nlHost (14)	Network Layer Host: counts the amount of traffic sent from and to each Network Layer address discovered by the probe, and stores the information in tables.
nlMatrix (15)	Network Layer Matrix: counts the amount of traffic sent between each pair of network addresses discovered by the probe, and stores the information in tables from both source to destination and destination to source.
alHost (16)	Application Layer Host: counts the amount of traffic, by protocol and by host, that is sent from and to each network address discovered by the probe.
alMatrix (17)	Application Layer Matrix: counts the amount of traffic, by protocol, sent between each pair of network addresses discovered by the probe, and stores this information in tables. This group is similar to the nlMatrix group, but the focus is on the protocol in operation.
usrHistory (18)	User History: combines mechanisms seen in the alarm (3) and history (2) groups to provide user-specified history collection, and stores that information in tables.
probeConfig (19)	Probe Configuration: controls the configuration of various operational parameters by the probe, such as the Ethernet and token ring RMON groups that

	are supported by the probe, software and hardware revision numbers of the probe, a trap destination table, and so on.
rmonConformance (20)	RMON Conformance: describes the requirements for conformance to the RMON2 MIB.

Figures 12-36a, 12-36b, and 12-36c show the various RMON2 groups and objects.

12.4 SNMP version 1

So far, we have discussed the structure of management information (SMI) and management information bases (MIBs) defined for SNMPv1. This section completes the discussion of version 1 of the Internet Network Management Framework by looking at SNMP version 1, the protocol that communicates management information. In the next two sections we will consider SNMP versions 2 and 3, respectively. Note that we will adopt the convention of referring to SNMP version 1 as SNMP, SNMP version 2 as SNMPv2, and SNMP version 3 as SNMPv3.

12.4.1 SNMP Objectives and Architecture

RFC 1157 states that "SNMP explicitly minimizes the number and complexity of management functions realized by the management agent itself" [12-15]. In other words, SNMP is designed to be simple. SNMP does this in three ways. By reducing the development cost of the agent software, SNMP has decreased the burden on vendors who wish to support the protocol, thereby increasing the protocol's acceptance. Second, SNMP is extensible, allowing vendors to add network management functions. Third, it separates the management architecture from the architecture of hardware devices, such as hosts and routers, widening the base of multivendor support.

SNMP has a very straightforward architecture. Figure 12-37a compares the SNMP architecture to the ISO/OSI model and the Advanced Research Projects Agency (ARPA) model, around which the Internet protocols and TCP/IP were developed. Note that the four layers of the ARPA model do not map evenly to the seven layers of the OSI model.

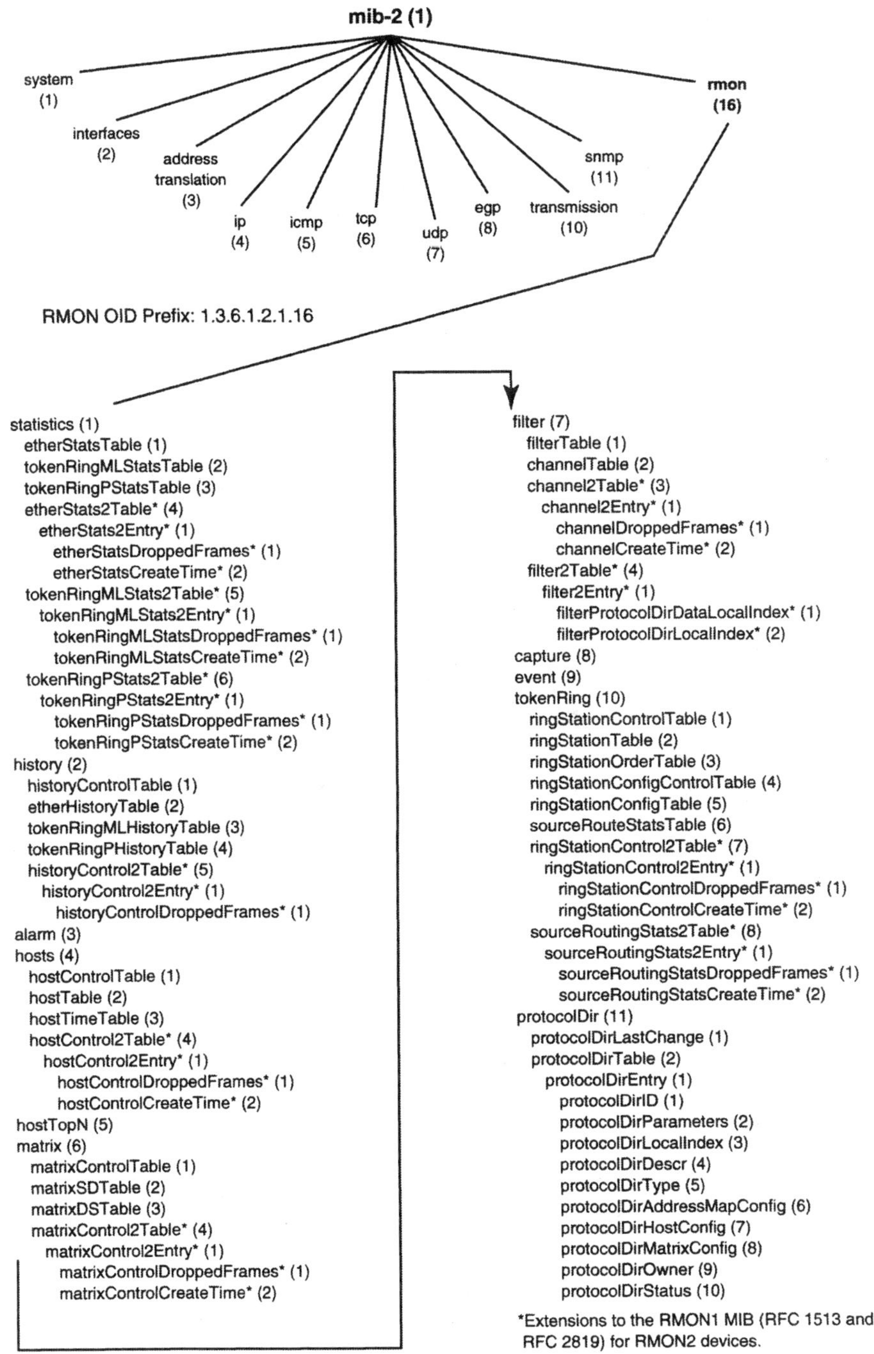

Figure 12-36a. RMON2 OID Tree.

```
protocolDist (12)
  protocolDistControlTable (1)
    protocolDistControlEntry (1)
      protocolDistControlIndex (1)
      protocolDistControlDataSource (2)
      protocolDistControlDroppedFrames (3)
      protocolDistControlCreateTime (4)
      protocolDistControlOwner (5)
      protocolDistControlStatus (6)
  protocolDistStatsTable (2)
    protocolDistStatsEntry (1)
      protocolDistStatsPkts (1)
      protocolDistStatsOctets (2)
addressMap (13)
  addressMapInserts (1)
  addressMapDeletes (2)
  addressMapMaxDesiredEntries (3)
  addressMapControlTable (4)
    addressMapControlEntry (1)
      addressMapControlIndex (1)
      addressMapControlDataSource (2)
      addressMapControlDroppedFrames (3)
      addressMapControlOwner (4)
      addressMapControlStatus (5)
  addressMapTable (5)
    addressMapEntry (1)
      addressMapTimeMark (1)
      addressMapNetworkAddress (2)
      addressMapSource (3)
      addressMapPhysicalAddress (4)
      addressMapLastChange (5)
nlhost (14)
  hlHostControlTable (1)
    hlHostControlEntry (1)
      hlHostControlIndex (1)
      hlHostControlDataSource (2)
      hlHostControlNlDroppedFrames (3)
      hlHostControlNlInserts (4)
      hlHostControlNlDeletes (5)
      hlHostControlNlMaxDesiredEntries (6)
      hlHostControlAlDroppedFrames (7)
      hlHostControlAlInserts (8)
      hlHostControlAlDeletes (9)
      hlHostControlAlMaxDesiredEntries (10)
      hlHostControlOwner (11)
      hlHostControlStatus (12)
  nlHostTable (2)
    nlHostEntry (1)
      nlHostTimeMark (1)
      nlHostAddress (2)
      nlHostInPkts (3)
      nlHostOutPkts (4)
      nlHostInOctets (5)
      nlHostOutOctets (6)
      nlHostOutMacNonUnicastPkts (7)
      nlHostCreateTime (8)
nlMatrix (15)
  hlMatrixControlTable (1)
    hlMatrixControlEntry (1)
      hlMatrixControlIndex (1)
      hlMatrixControlDataSource (2)
      hlMatrixControlNlDroppedFrames (3)
      hlMatrixControlNlInserts (4)
      hlMatrixControlNlDeletes (5)
      hlMatrixControlNlMaxDesiredEntries (6)
      hlMatrixControlAlDroppedFrames (7)
      hlMatrixControlAlInserts (8)
      hlMatrixControlAlDeletes (9)
      hlMatrixControlAlMaxDesiredEntries (10)
      hlMatrixControlOwner (11)
      hlMatrixControlStatus (12)
  nlMatrixSDTable (2)
    nlMatrixSDEntry (1)
      nlMatrixSDTimeMark (1)
      nlMatrixSDSourceAddress (2)
      nlMatrixSDDestAddress (3)
      nlMatrixSDPkts (4)
      nlMatrixSDOctets (5)
      nlMatrixSDCreateTime (6)
  nlMatrixDSTabl (3)
    nlMatrixDSEntry (1)
      nlMatrixDSTimeMark (1)
      nlMatrixDSSourceAddress (2)
      nlMatrixDSDestAddress (3)
      nlMatrixDSPkts (4)
      nlMatrixDSOctets (5)
      nlMatrixDSCreateTime (6)
  nlMatrixTopNControlTable (4)
    nlMatrixTopNControlEntry (1)
      nlMatrixTopNControlIndex (1)
      nlMatrixTopNControlMatrixIndex (2)
      nlMatrixTopNControlRateBase (3)
      nlMatrixTopNControlTimeRemaining (4)
      nlMatrixTopNControlGeneratedReports (5)
      nlMatrixTopNControlDuration (6)
      nlMatrixTopNControlRequestedSize (7)
      nlMatrixTopNControlGrantedSize (8)
      nlMatrixTopNControlStartTime (9)
      nlMatrixTopNControlOwner (10)
      nlMatrixTopNControlStatus (11)
  nlMatrixTopNTable (5)
    nlMatrixTopNEntry (1)
      nlMatrixTopNIndex (1)
      nlMatrixTopNProtocolDirLocalIndex (2)
      nlMatrixTopNSourceAddress (3)
      nlMatrixTopNDestAddress (4)
      nlMatrixTopNPktRate (5)
      nlMatrixTopNReversePktRate (6)
      nlMatrixTopNOctetRate (7)
      nlMatrixTopNReverseOctetRate (8)
```

alHost (16)
  alHostTable (1)
    alHostEntry (1)
      alHostTimeMark (1)
      alHostInPkts (2)
      alHostOutPkts (3)
      alHostInOctets (4)
      alHostOutOctets (5)
      alHostCreateTime (6)
alMatrix (17)
  alMatrixSDTable (1)
    alMatrixSDEntry (1)
      alMatrixSDTimeMark (1)
      alMatrixSDPkts (2)
      aMatrixSDOctets (3)
      alMatrixSDCreateTime (4)
  alMatrixDSTable (2)
    alMatrixDSEntry (1)
      alMatrixDSTimeMark (1)
      alMatrixDSPkts (2)
      alMatrixDSOctets (3)
      alMatrixDSCreateTime (4)
  alMatrixTopNControlTable (3)
    alMatrixTopNControlEntry (1)
      alMatrixTopNControlIndex (1)
      alMatrixTopNControlMatrixIndex (2)
      alMatrixTopNControlRateBase (3)
      alMatrixTopNControlTimeRemaining (4)
      alMatrixTopNControlGeneratedReports (5)
      alMatrixTopNControlDuration (6)
      alMatrixTopNControlRequestedSize (7)
      alMatrixTopNControlGrantedSize (8)
      alMatrixTopNControlStartTime (9)
      alMatrixTopNControlOwner (10)
      alMatrixTopNControlStatus (11)
  alMatrixTopNTable (5)
    alMatrixTopNEntry (1)
      alMatrixTopNIndex (1)
      alMatrixTopNProtocolDirLocalIndex (2)
      alMatrixTopNSourceAddress (3)
      alMatrixTopNDestAddress (4)
      alMatrixTopNAppProtocolDirLocalIndex (5)
      alMatrixTopNPktRate (6)
      alMatrixTopNReversePktRate (7)
      alMatrixTopNOctetRate (8)
      alMatrixTopNReverseOctetRate (9)
usrHistory (18)
  usrHistoryControlTable (1)
    usrHistoryControlEntry (1)
      usrHistoryControlIndex (1)
      usrHistoryControlObjects (2)
      usrHistoryControlBucketsRequested (3)
      usrHistoryControlBucketsGranted (4)
      usrHistoryControlInterval (5)
      usrHistoryControlOwner (6)
      usrHistoryControlStatus (7)
  usrHistoryObjectTable (2)
    usrHistoryObjectEntry (1)
      usrHistoryObjectIndex (1)
      usrHistoryObjectVariable (2)
      usrHistoryObjectSampleType (3)
  usrHistoryTable (3)
    usrHistoryEntry (1)
      usrHistorySampleIndex (1)
      usrHistoryIntervalStart (2)
      usrHistoryIntervalEnd (3)
      usrHistoryAbsValue (4)
      usrHistoryValStatus (5)
probeConfig (19)
  probeCapabilities (1)
  probeSoftwareRev (2)
  probeHardwareRev (3)
  probeDateTime (4)
  probeResetControl (5)
  probeDownloadFile (6)
  probeDownloadTFTPServer (7)
  probeDownloadAction (8)
  probeDownloadStatus (9)
  serialConfigTable (10)
    serialConfigEntry (1)
      serialMode (1)
      serialProtocol (2)
      serialTimeout (3)
      serialModeminitString (4)
      serialModemHangUpString (5)
      serialModemConnectResp (6)
      serialModemNoConnectResp (7)
      serialDialoutTimeout (8)
      serialStatus (9)
  netConfigTable (11)
    netConfigEntry (1)
      netConfigIPAddress (1)
      netConfigSubnetMask (2)
      netConfigStatus (3)
  netDefaultGateway (12)
  trapDestTable (13)
    trapDestEntry (1)
      trapDestIndex (1)
      trapDestCommunity (2)
      trapDestProtocol (3)
      trapDestAddress (4)
      trapDestOwner (5)
      trapDestStatus (6)
  serialConnectionTable (14)
    serialConnectionEntry (1)
      serialConnectIndex (1)
      serialConnectDestIpAddress (2)
      serialConnectType (3)
      serialConnectDialString (4)
      serialConnectSwitchConnectSeq (5)
      serialConnectSwitchDisconnectSeq (6)
      serialConnectSwitchResetSeq (7)
      serialConnectOwner (8)
      serialConnectStatus (9)
rmonConformance (20)
  rmon2MIBCompliances (1)
    rmon2MIBCompliance (1)
    rmon2MIBApplicationLayerCompliance (2)
  rmon2MIBGroups (2)
    protocolDirectoryGroup (1)
    protocolDistributionGroup (2)
    addressMapGroup (3)
    nlHostGroup (4)
    nlMatrixGroup (5)
    alHostGroup (6)
    alMatrixGroup (7)
    usrHistoryGroup (8)
    probeInformationGroup (9)
    probeConfigurationGroup (10)
    rmon1EnhancementGroup (11)
    rmon1EthernetEnhancementGroup (12)
    rmon1TokenRingEnhancementGroup (13)

OSI Layer	SNMP - Related Function	ARPA Layer
Application	Management Application (SNMP PDU)	Process / Application
Presentation	Structure of Management Information (ASN.1 & BER Encoding)	
Session	Authentication (SNMP Header)	
Transport	User Datagram Protocol (UDP)	Host-to-Host
Network	Internet Protocol (IP)	Internet
Data Link	LAN or WAN Interface Protocol	Network Interface
Physical		

Figure 12-37a. Comparing the SNMP Architecture with the OSI and ARPA Models.

Let's use an example to see how the processes within the SNMP architecture interact. Suppose a management console requests information about one of the managed nodes. The SNMP processes in both the manager and the agent respond to the console. The ASN.1 encoding at the Application layer provides the proper syntax for the SNMP message. The remaining functions authenticate the data (attach the SNMP header) and communicate the information request.

Because most management information does not demand the reliable delivery that connection-oriented systems provide, the communication channel between the SNMP manager and the agent is connectionless. When you compare the SNMP model to the ISO/OSI model, SNMP's connectionless communication mechanism removes some of the need for a Session layer and reduces the responsibilities of the lower four layers. For most implementations, the User Datagram Protocol (UDP) performs the Transport layer functions, the Internet Protocol (IP) provides the Network layer functions, and

LANs such as Ethernet or token ring or WANs such as a leased line or a frame relay connection provide the Data Link and Physical layer functions. (There are some exceptions to this rule. RFCs 1418, 1419, and 1420 describe implementations that use other transport mechanisms, such as OSI Apple Computer's AppleTalk, or Novell Inc.'s IPX protocols. However, RFC 1270, called "SNMP Communication Services" [12-16], states that UDP/IP are the protocols of choice for most circumstances.)

If you compare SNMP to the Internet (or ARPA) architectural model (see Figure 12-37a), you'll notice that the ARPA model uses four layers to describe the entire communication function. In the ARPA model, SNMP would reside at the Process/Application layer. However, while the ARPA Host-to-Host layer provides end-to-end communication reliability, SNMP's use of UDP ensures only proper port addressing and a checksum; it does not provide octet-by-octet error control. IP provides the Internet layer functions, such as addressing and fragmentation, that are necessary to deliver an SNMP message from the source to the destination. Finally, the Network Interface layer deals with the LAN or WAN hardware, such as an interface to an FDDI or Frame Relay network connection. Notice that Figure 12-37b also shows the relative complexities of the host and router functions. Hosts implement all four layers of the ARPA model, whereas routers implement only the lower two.

Comparing the SNMP architecture to the ISO/OSI and ARPA architectural models provides a theoretical basis for this discussion. But from a practical perspective, the SNMP model works as shown in Figure 12-38. This model includes a management system that uses the SNMP manager, an SNMP agent, and managed resources, and the SNMP messages communicate management information via five SNMP protocol data units (PDUs). The management application issues the Get, GetNext, or Set PDUs. The managed system returns a GetResponse PDU. The agent may initiate a Trap (sometimes called an Event) PDU when predefined conditions are met.

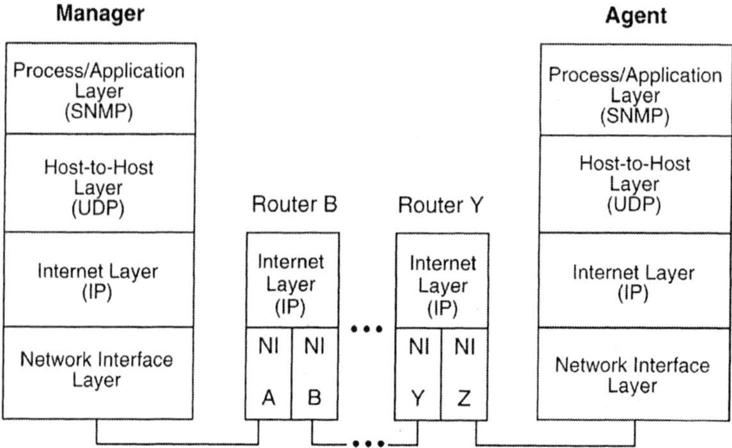

Figure 12-37b. Application-to-application Connection.

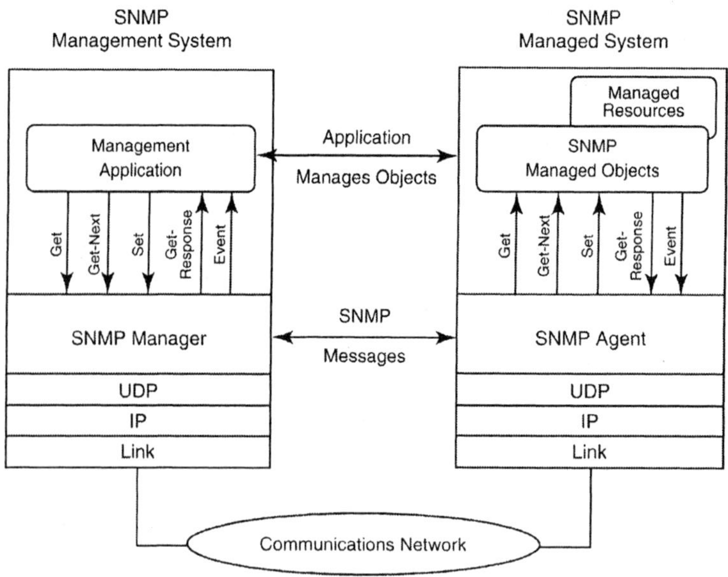

Figure 12-38. SNMP Architecture (© 1990, IEEE).

12.4.2 SNMP Operation

The SNMP processes described in the previous sections must occur in physical devices. For example, a router must have a physical processor that implements the software acting as an SNMP agent. Two sets of logical processes occur within those physical elements: the relationships that are specified between various network management entities, and the way network management information is communicated.

12.4.2.1 Network Management Relationships

The SNMP standard, RFC 1157, and the "SNMP Administrative Model," RFC 1351 [12-17], define a number of terms. Many of these definitions describe relationships between management entities:

- Network management stations are devices that execute the management applications that control and monitor the network elements.
- Network elements are devices such as hosts, switches, routers, and hubs that contain an agent and perform the network management functions that the network management stations request.
- The SNMP allows network management stations and the agents in the network elements to communicate.
- SNMP application entities reside at either a management station or a managed node, and use SNMP as a communication mechanism.
- Protocol entities are peer processes that implement SNMP, thus supporting the SNMP application entities.
- The SNMP community pairs an SNMP agent with an arbitrary set of SNMP application entities. The network administrator assigns the community a name (called the community name) that is essentially a password with associated rights and privileges. A management application with multiple community names may belong to multiple communities.
- Authentic SNMP messages are SNMP messages sent from an application entity to a specific SNMP community. The message contains the community name of interest.
- The authentication scheme is the method by which an SNMP message is identified as belonging to a specific SNMP community.
- The MIB View is the subset of MIB objects that may be contained within several subtrees that pertain to a network element.
- The SNMP access mode determines the level of access to

objects that a particular application entity is allowed. The choices are read-only and read-write.

- The community profile pairs the SNMP access mode with the SNMP MIB View. The community profile represents specific access privileges for the variables in a MIB View.
- The SNMP access policy pairs an SNMP community with an SNMP community profile. The access policy represents the specific community profile that an agent permits the other members of the community to have.
- The SNMP proxy agent provides management functions on behalf of network elements that would otherwise be inaccessible.

Figure 12-39 illustrates some of the definitions described above.

12.4.2.2 Identifying and Communicating Object Instances

SMI-managed object types have an object identifier (OID) that uniquely names them and locates their place on the object tree. An instance of an object type is an occurrence of that object type and has an assigned value. For example, the object *sysDescr* {1.3.6.1.2.1.1.1.0} might have a value of "Retix Remote Bridge Model 2265M."

Suppose a network management station wishes to retrieve an instance of a specific object. The management station must use SNMP to communicate its question to the agent.

Now, suppose multiple instances (or occurrences) of that object are possible. For example, say a router's routing table contains a number of entries. How would the network management station retrieve just the value of the third entry in the table?

RFC 1157, pages 12–15, specifies these tasks. For these SNMP operations, a variable name uniquely identifies each instance of an object type. This name consists of two parts of the form x.y. The x portion is the object type defined in the MIB, and the y portion is an OID fragment that identifies the desired instance. The following examples should clarify this.

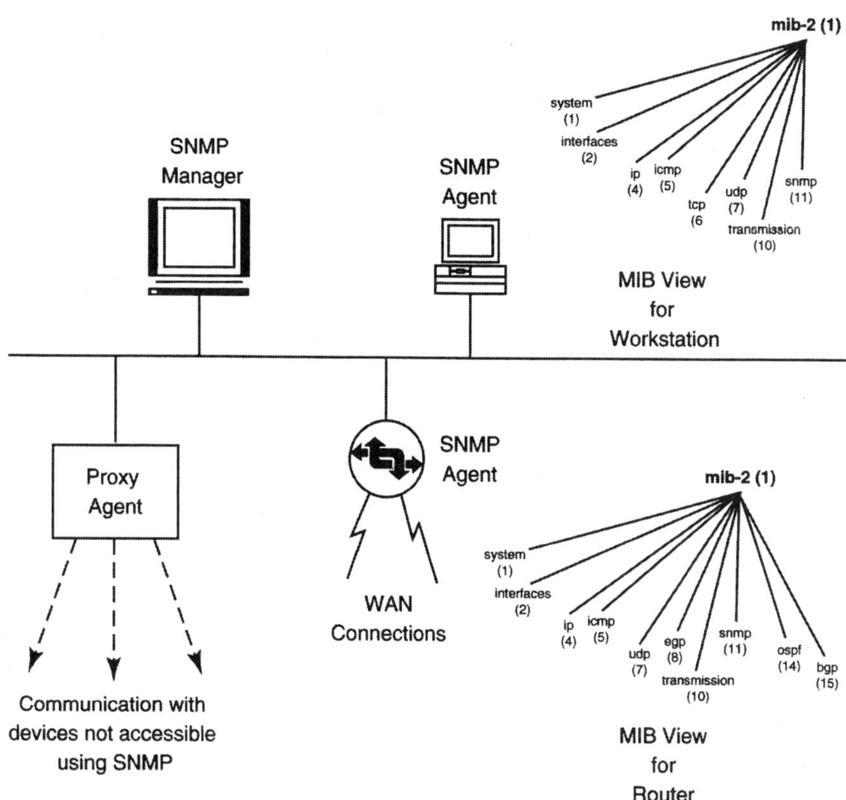

Consider a scalar object that has one instance. The objects contained in the System group are all scalar objects. For example, the *sysServices* object has an OID of {1.3.6.1.2.1.1.7} and occurs once. The x portion of the variable name is the OID, and the y portion has been assigned to 0. You can derive this by following the OID tree down to the object sysServices and adding the appropriate instance suffix (with the suffix, or y portion, shown in boldface type):

iso	org	dod	internet	mgmt	mib-2	system	sysServices	Instance
1	3	6	1	2	1	1	7	0

Thus, the variable name for *sysServices* is {1.3.6.1.2.1.1.7.0}.

The variable name for a columnar object is more complicated because it must identify the location of an object within a two-dimensional data structure, such as a table having both rows and columns. (Within the RMON MIB, three-dimensional data structures are added, making the identification even more complex.) Using the familiar spreadsheet as an example, the identification of a particular cell requires two coordinates, X and Y, which describe the horizontal and vertical positions, respectively. With columnar objects, an indexing scheme, specified in the INDEX clause in the ASN.1 definition for that object, provides a means for identifying the specific instance. The INDEX clause then further identifies the syntax to be used. And, as one might expect, some of the indexing schemes are more complicated than others. RFC 1212, pages 8–10, lists many of the INDEX clauses that are found within MIB-II. Let's look at some examples for further clarification.

Consider the IP Address Table object, *ipAdEntBcastAddr*, which specifies the value of the least significant bit (LSB) of the IP broadcast address (see Figures 12-40a and 12-40b). To begin, follow the OID tree down to *ipAdEntBcastAddr*:

iso	org	dod	internet	mgmt	mib-2	ip	ipAddrTable	ipAddrEntry	ipAdEntBcastAddr
1	3	6	1	2	1	4	20	1	4

The OID is {1.3.6.1.2.1.4.20.1.4}, consisting of the IP Group {1.3.6.1.2.1.4}, the IP Address Table (20), the *ipAddrEntry* (1), and the object *ipAdEntBcastAddr* (4), shown in Figure 12-40a.

Consulting MIB-II, RFC 1157, page 31, the ASN.1 definition for the object *ipAddrEntry* includes an INDEX clause that specifies the object *ipAdEntAddr*. Moving a few lines down in RFC 1157, we see that the object ipAdEntAddr has a SYNTAX of IpAddress. Thus, we would expect the index for this object to be defined by an IP address and shown in dotted decimal notation: a.b.c.d.

Returning to our example, to complete the variable name, the suffix (or y portion) is added, which consists of an IP address: a.b.c.d. The variable name for *ipAdEntBcastAddr* associated with IP Address a.b.c.d would therefore be {1.3.6.1.2.1.4.20.1.4.a.b.c.d}.

Figure 12-40b is a completed IP Address Table, built by retrieving all of the IP Address Table variables. The column headings show the five objects, *ipAdEntAddr* through *ipAdEntReasmMaxSize*. Each row contains the values of the five variables: *ipAdEntAddr* [XXX.YYY.150.2], *ipAdEntIfIndex* (1), and so on. A different index (2) identifies the second row, and it contains different values. Additional row entries are made, as necessary, until the table is completed.

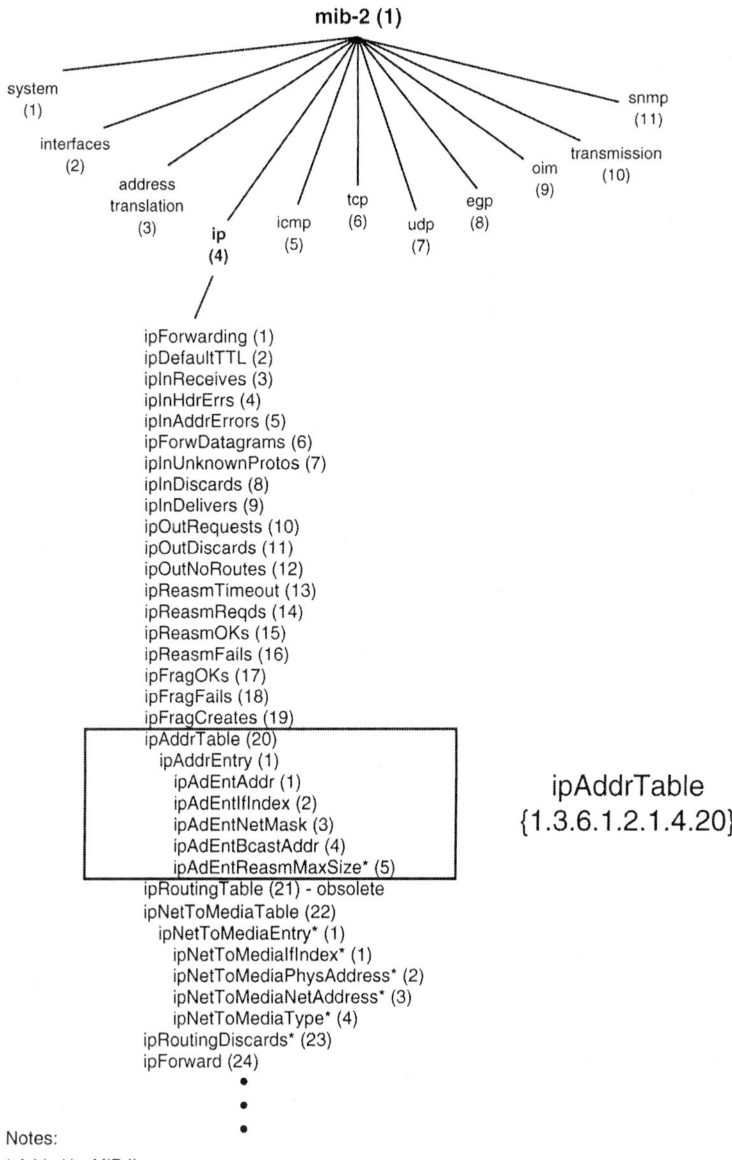

Figure 12-40a. The IP Address Table within the OID Tree.

ipAddrTable {1.3.6.1.2.1.4.20}

	ipAdEntAddr {1.3.6.1.2.1.4.20.1.1}	ipAdEntIfIndex {1.3.6.1.2.1.4.20.1.2}	ipAdEntNetMask {1.3.6.1.2.1.4.20.1.3}	ipAdEntBcastAddr {1.3.6.1.2.1.4.20.1.4}	ipAdEntReasmMaxSize {1.3.6.1.2.1.4.20.1.5}
Row 1	XXX.YYY.150.2	1	255.255.255.0	0	12000
Row 2	XXX.YYY.1.1	2	255.255.0.0	1	12000
• • •		• • •			
Row n					

Figure 12-40b. Object Instances in the IP Address Tables.

A final example (derived from RFC 1157, section 3.2.6.3) is from the TCP Connection Table, *tcpConnTable*. Suppose you wish to retrieve the state of the connection between port 575 on local address {a.b.c.d} and port 441 on remote address {w.x.y.z}. The OID for *tcpConnState* is {1.3.6.1.2.1.6.13.1.1}. The INDEX clause consists of four parts: *tcpConnLocalAddress*, *tcpConnLocalPort*, *tcpConnRemAddress*, and *tcpConnRemPort*. The y suffix would therefore be expressed as {a.b.c.d.575.w.x.y.z.441}. Therefore, the complete variable name would be:

```
{1.3.6.1.2.1.6.13.1.1.a.b.c.d.575.w.x.y.z.441}
```

The following examples show specific variable names for both scalar and columnar object types:

- The description of this system's services:

  ```
  sysServices ::=
  {1.3.6.1.2.1.1.7.0}
  ```

- The speed of interface 3:

  ```
  ifSpeed.3 ::=
  {1.3.6.1.2.1.2.2.1.5.3}
  ```

- The physical address associated with interface 2 and IP address {a.b.c.d} (Note that the first component is a .1, which indicates an IP address [see RFC 1157, page 13]):

```
atPhysAddress.2.1.a.b.c.d ::=
{1.3.6.1.2.1.3.1.1.2.2.1.a.b.c.d}
```

- The maximum IP datagram reassembly size associated with IP address {a.b.c.d}:

```
ipAdEntReasmMaxSize.a.b.c.d ::=
{1.3.6.1.2.1.4.20.1.5.a.b.c.d}
```

- The number of ICMP Echo (request) messages received at this device:

```
icmpInEchos ::=
{1.3.6.1.2.1.5.8.0}
```

- The state of a TCP connection between local port e, local address {a.b.c.d}, and remote port j, remote address {f.g.h.i}:

```
tcpConnState.a.b.c.d.e.f.g.h.i.j ::=
{1.3.6.1.2.1.6.13.1.1.a.b.c.d.e.f.g.h.i.j}
```

- Verification that a UDP listener is operational on port e of local IP address a.b.c.d:

```
udpLocalAddress.a.b.c.d.e ::=
{1.3.6.1.2.1.7.5.1.1.a.b.c.d.e}
```

- The neighbor state for the IP address a.b.c.d:

```
egpNeighState.a.b.c.d ::=
{1.3.6.1.2.1.8.5.1.1.a.b.c.d}
```

- The number of SNMP messages delivered to this device with unknown community names (a scalar):

```
snmpInBadCommNames ::=
{1.3.6.1.2.1.11.4.0}
```

RFC 1157, pages 12–15, and RFC 1212, pages 8–10, provide other examples that are worth further study.

With this background into the methods of identifying object instances, let's now discuss the SNMP protocol data units (PDUs) that carry the requests and responses for this information between manager and agent devices. The PDUs use the object instance examples shown here to identify the specific network management information that the manager is seeking.

12.4.3 SNMP Protocol Data Units (PDUs)

We will begin the discussion of PDUs by describing the position of the SNMP message within a transmitted frame. The frame is the unit of information transmitted between network nodes. For example, an IEEE 802.3 frame format defines the transmission between Ethernet/CSMA nodes, and an ANSI T1.617 format defines the transmission between Frame Relay nodes.

The local network header and trailers defined by the LAN or WAN protocol delimit the frame (see Figure 12-41). The transmitted data is called an Internet Protocol (IP) datagram. The IP datagram is a self-contained unit of information sent from the source host to its intended destination via the internetwork. Inside the datagram is a destination IP address that steers the datagram to the intended recipient. Next, the User Datagram Protocol (UDP) header identifies the higher layer protocol process (SNMP) that will process the datagram, and provides error control using a checksum. The SNMP message is the innermost part of the frame, carrying the actual data between the manager and the agent.

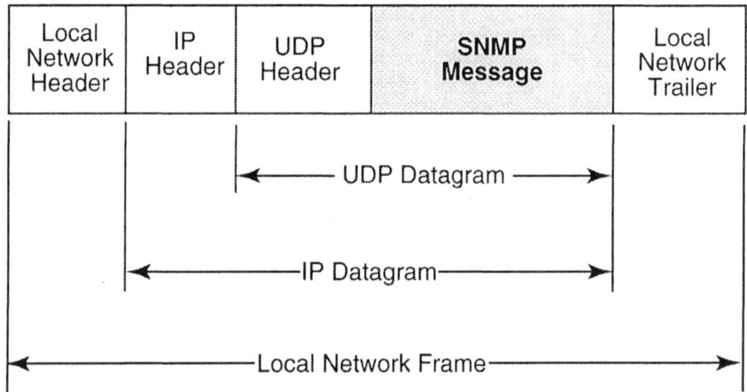

Figure 12-41. SNMP Message within a Transmission Frame.

When the IP is too long to fit inside one frame, it may be divided (or fragmented) into several frames for transmission on the LAN. For example, a datagram containing 2500 octets would require two Ethernet frames, each of which may contain a maximum of 1500 octets of higher layer data. The general structure of each frame, as shown in Figure 12-41, would remain the same.

The SNMP message itself is divided into two sections: a version identifier plus community name, and a PDU. The version identifier and community name are sometimes referred to as the SNMP authentication header. There are five different PDU types: GetRequest, GetNextRequest, GetResponse, SetRequest, and Trap. The Get, Set, and Response PDUs have a common format (see Figure 12-42), while the Trap PDU format is unique (Figure 12-46 later in this chapter illustrates the Trap PDU format).

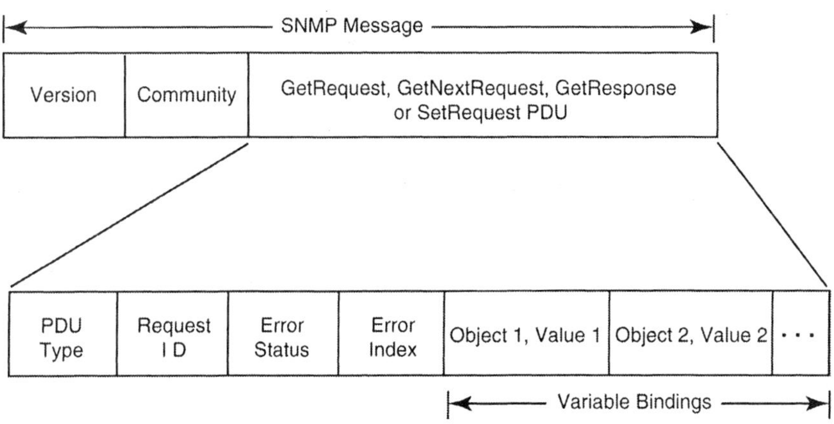

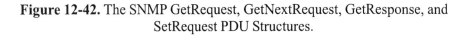

Figure 12-42. The SNMP GetRequest, GetNextRequest, GetResponse, and SetRequest PDU Structures.

The version number (an INTEGER type) ensures that both manager and agent are using the same version of the SNMP protocol. Messages between manager and agent containing different version numbers are discarded without further processing. The community name (an OCTET STRING type) authenticates the manager before allowing access to the agent. The community name, along with the manager's IP address, is stored in the agent's community profile. If there's a difference between the manager and agent values for the community name, the agent will send an authentication failure trap message to the manager. If both the version number and community name from the manager match the ones stored in the agent, the SNMP PDU begins processing.

In the following sections, we'll discuss the fields of the two PDU formats and the operation of the five PDUs.

12.4.3.1 Get, Set, and Response PDU Formats

The GetRequest, GetNextRequest, SetRequest, and GetResponse PDUs share a common format (see Figure 12-42). The first field, PDU Type, specifies the type of PDU the message contains:

PDU	PDU Type Field Value
GetRequest	0
GetNextRequest	1
GetResponse	2
SetRequest	3
Trap	4

The Request ID field is an INTEGER type that correlates the manager's request to the agent's response.

The Error Status field is an enumerated INTEGER type that indicates normal operation (noError) or one of five error conditions. The possible values are:

Error	Value	Meaning
noError	0	Proper manager/agent operation.
TooBig	1	The size of the required GetResponse PDU exceeds a local limitation.
noSuchName	2	The requested object name did not match the names available in the relevant MIB View.
badValue	3	A SetRequest contained an inconsistent type, length, and value for the variable.
readOnly	4	Not defined in RFC 1157. (Historical footnote: this error is listed, but the description of the SetRequest PDU processing does not describe how this error is generated. The standard interpretation is that this error should not be generated, although some vendors' agents nevertheless do.)
genErr	5	Other errors, not explicitly defined, have occurred.

When an error occurs, the Error Index field identifies the entry within the Variable Bindings list that caused the error. For example, if the fourth Variable Binding was misformatted, or otherwise not understandable by the receiver, an Error Index = 4 would be returned. To explain further, a Variable Binding (VarBind) pairs a variable name with its value. A VarBindList is a list of such pairings. Note that within the Variable Bindings fields of the SNMP PDUs (see Figures 12-42 through 12-47), the word Object identifies the variable name (OID encoding of object type plus the instance) for which a value is being communicated. Also note that GetRequest or GetNextRequest PDUs use a value of NULL, which is a special ASN.1 data type.

12.4.3.2 Using the GetRequest PDU

The manager uses the GetRequest PDU to retrieve the value of one or more object(s) from an agent. In most cases, these are scalar, not columnar, objects. To generate the GetRequest PDU, the manager assigns PDU Type = 0, specifies a locally defined Request ID, and sets both the Error Status and Error Index to 0. A VarBindList, containing the requested variables and corresponding NULL (placeholder) values, completes the PDU. Under error-free conditions, the agent generates a GetResponse PDU, which is assigned PDU Type = 2, has the same value of Request ID, Error Status = noError, and Error Index = 0. The Variable Bindings now contain the values associated with each of the variables noted in the GetRequest PDU (see Figure 12-43). Recall that the term variable refers to an instance of a managed object.

Four error conditions are possible:

- If a variable in the Variable Bindings field does not exactly match an available object, the agent returns a GetResponse PDU with Error Status = noSuchName, and with the Error Index indicating the index of the variable in question.
- If a variable is an aggregate type, such as a row object, the agent returns a GetResponse PDU with Error Status = noSuchName, and with the Error Index indicating the index of the variable in question.
- If the size of the appropriate GetResponse PDU would exceed a local limitation, then the agent returns a GetResponse PDU of identical form, with Error Status = tooBig, and Error Index = 0.
- If the value of a requested variable cannot be retrieved for any other reason, then the agent returns a GetResponse PDU with Error Status = genErr, and the Error Index indicating the index of the variable in question.

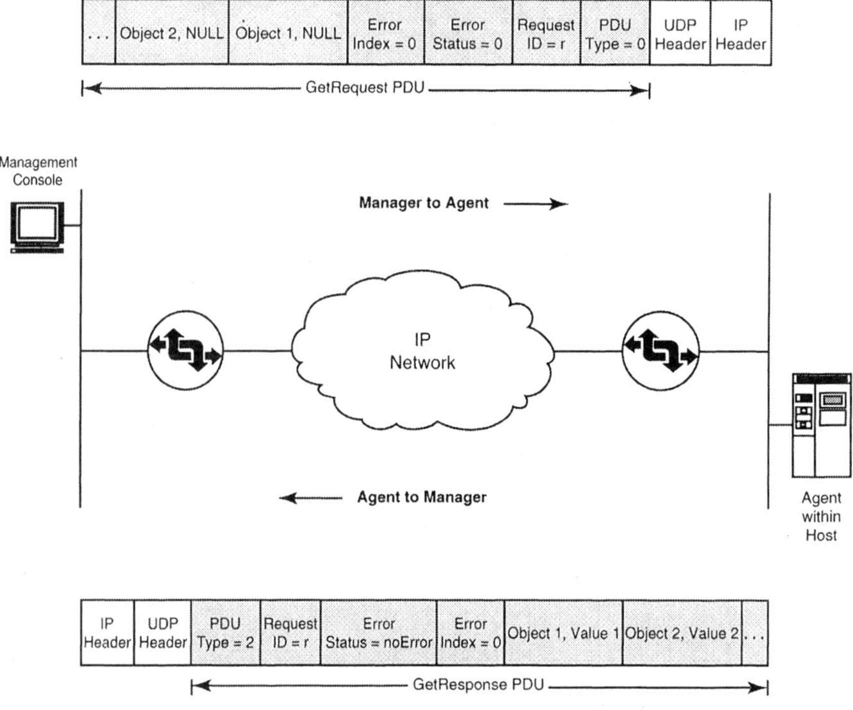

Figure 12-43. GetRequest/GetResponse PDU Transmission (with No Errors).

12.4.3.3 Using the GetNextRequest PDU

The manager uses the GetNextRequest PDU to retrieve one or more objects and their values from an agent. In most cases, these multiple objects will reside within a table. As you can see in Figure 12-44, to generate the GetNextRequest PDU the manager assigns PDU Type = 1, specifies a locally defined Request ID, and sets both the Error Status and the Error Index to 0. A VarBindList, containing the OIDs and corresponding NULL (placeholder) values, completes the PDU. These OIDs can be any OID (which may be a variable) that immediately precedes the variable and value returned. Under error-free conditions, the agent generates a GetResponse PDU, which is assigned PDU Type = 2, has the same value of Request ID, Error Status = noError, and Error Index = 0. The Variable Bindings contain the name and value associated with the lexicographical successor of each of the OIDs noted in the GetNextRequest PDU.

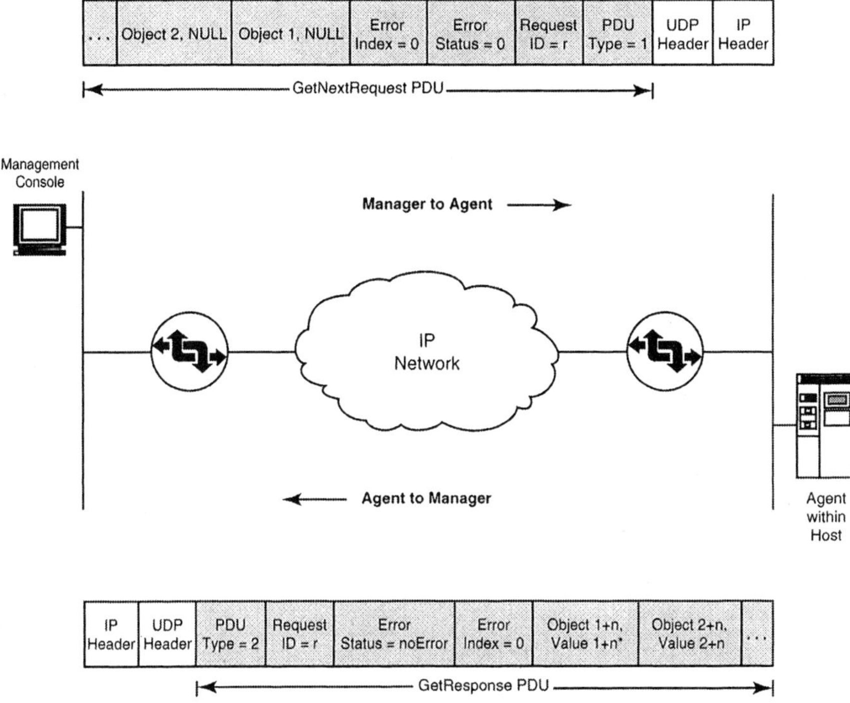

Figure 12-44. GetNextRequest/GetResponse PDU Transmission (with No Errors).

The key difference between the GetRequest and the GetNextRequest PDUs is the word lexicographical. This means that the GetNextRequest retrieves the value of the next object within the agent's MIB View.

Three error conditions are possible:

- If a variable in the Variable Bindings field does not lexicographically precede the name of an object that may be retrieved (that is, an object available for Get operations and within the relevant MIB View), the agent returns a GetResponse PDU with Error Status = noSuchName, and with the Error Index indicating the index of the variable in question. This condition is called "running off the end of the MIB View."

- If the size of the appropriate GetResponse PDU exceeds a local limitation, the agent returns a GetResponse PDU of identical form, with Error Status = tooBig and Error Index = 0.
- If the value of the lexicographical successor to a requested variable in the Variable Bindings field cannot be retrieved for any other reason, the agent returns a GetResponse PDU with Error Status = genErr, and the Error Index indicating the index of the variable in question.

12.4.3.4 Using the SetRequest PDU

The manager uses the SetRequest PDU to assign a value to an object residing in the agent. As you can see in Figure 12-45, to generate that PDU the manager assigns PDU Type = 3, specifies a locally defined Request ID, and sets both the Error Status and Error Index to 0. A VarBindList, containing the specified variables and their corresponding values, completes the PDU. When the agent receives the SetRequest PDU, it alters the values of the named objects to the values in the Variable Binding. Under error-free conditions, the agent generates a GetResponse PDU of identical form, except that the assigned PDU Type = 2, Error Status = noError, and Error Index = 0.

Four error conditions are possible:

- If a variable in the Variable Bindings field is not available for Set operations within the relevant MIB View, the agent returns a GetResponse PDU of identical form, with Error Status = noSuchName, and with the Error Index indicating the index of the object name in question. (Historical note: Some agent implementations return Error Status = readOnly if the object exists, but Access = read-only for that variable.)
- If the value of a variable named in the Variable Bindings field does not conform to the ASN.1 Type, Length, and Value required, the agent returns a GetResponse PDU of identical form, with Error Status = badValue and the Error Index indicating the index of the variable in question.
- If the size of the appropriate GetResponse PDU exceeds a local limitation, the agent returns a GetResponse PDU of identical form, with Error Status = tooBig and Error Index = 0.
- If the value of a variable cannot be altered for any other reason, the agent returns a GetResponse PDU of identical form, with Error Status = genErr and the Error Index indicating the index of the variable in question.

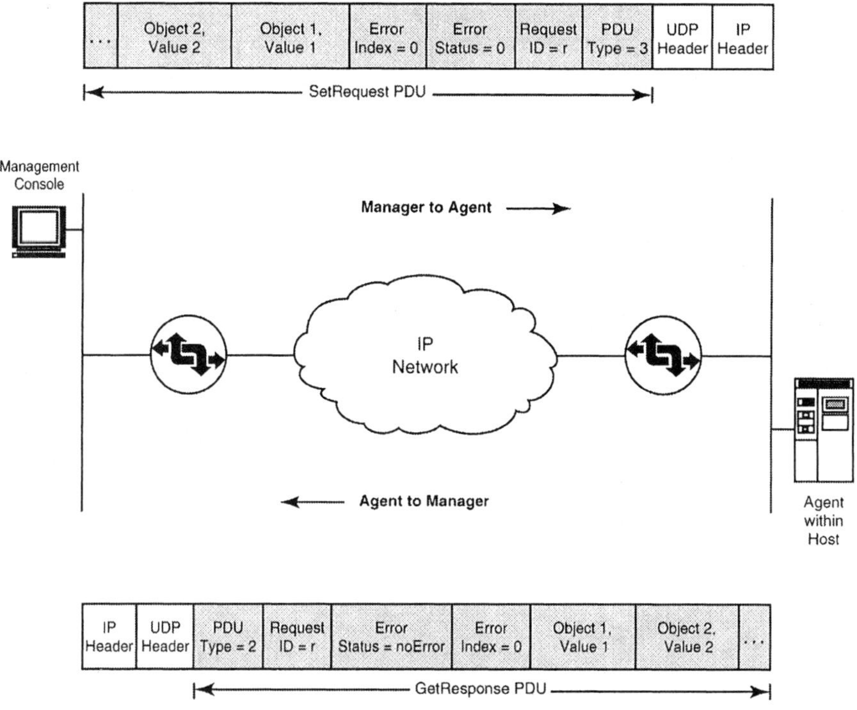

Figure 12-45. SetRequest/GetResponse PDU Transmission (with No Errors).

12.4.3.5 The Trap PDU Format

The Trap PDU has a format distinct from the other four SNMP PDUs, as you can see in Figure 12-46. The first field indicates the Trap PDU and contains PDU Type = 4. The Enterprise field identifies the management enterprise under whose registration authority the trap was defined. For example, the OID prefix {1.3.6.1.4.1.9} would identify Cisco Systems, Inc., as the Enterprise sending a trap. The Agent Address field, which contains the IP address of the agent, provides further identification. If a non-IP transport protocol is used, the value 0.0.0.0 is returned.

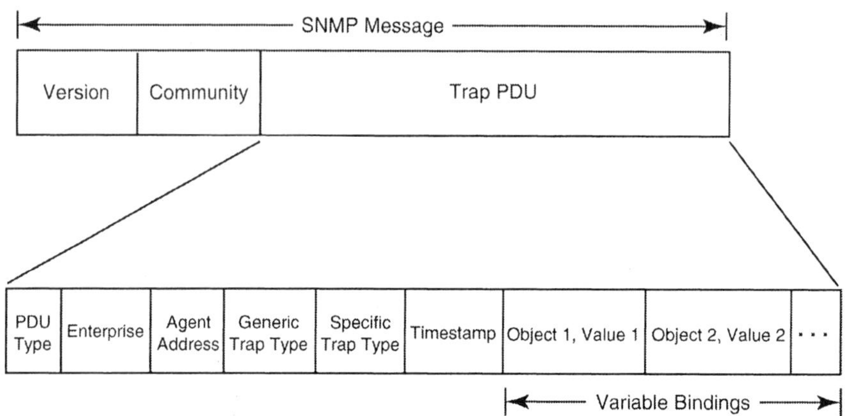

Figure 12-46. SNMP Trap PDU Structure.

The Generic Trap Type provides more specific information on the event being reported. There are seven defined values (enumerated INTEGER types) for this field:

Trap	Value	Meaning
coldStart	0	The sending protocol entity (higher layer network management) has reinitialized, indicating that the agent's configuration or entity implementation may be altered.
warmStart	1	The sending protocol has reinitialized, but neither the agent's configuration nor the protocol entity implementation has been altered.
linkDown	2	A communication link has failed. The affected interface is identified as the first element within the Variable Bindings field: name and value of the ifIndex instance.
linkUp	3	A communication link has come up. The affected interface is identified as the first element within the Variable Bindings field: name and value of the ifIndex instance.
authenticationFailure	4	The agent has received an improperly authenticated SNMP message from the manager; that is, the community name was incorrect.
egpNeighborLoss	5	An EGP peer neighbor is down.
enterpriseSpecific	6	A nongeneric trap has occurred, which is further identified by the Specific Trap Type field and the Enterprise field.

Two additional fields complete the Trap PDU. The Timestamp field contains the value of the sysUpTime object, representing the amount of time elapsed between the last (re-)initialization of the agent and the generation of that Trap. The last field contains the Variable Bindings.

The agent uses the Trap PDU to alert the manager that a predefined event has occurred. To generate the Trap PDU, the agent assigns PDU Type = 4 and fills in the Enterprise, Agent Address, Generic Trap Type, Specific Trap Type, and Timestamp fields, as well as the Variable Bindings list.

By definition (and convention), Traps are application specific. Therefore, it would be difficult to cover the range of uses for this PDU. RFC 1215, "A Convention for Defining Traps for Use with the SNMP" [12-18], offers some guidelines for their use. Figure 12-47 illustrates how an agent in a remote host could use a Trap to communicate a significant event to the manager.

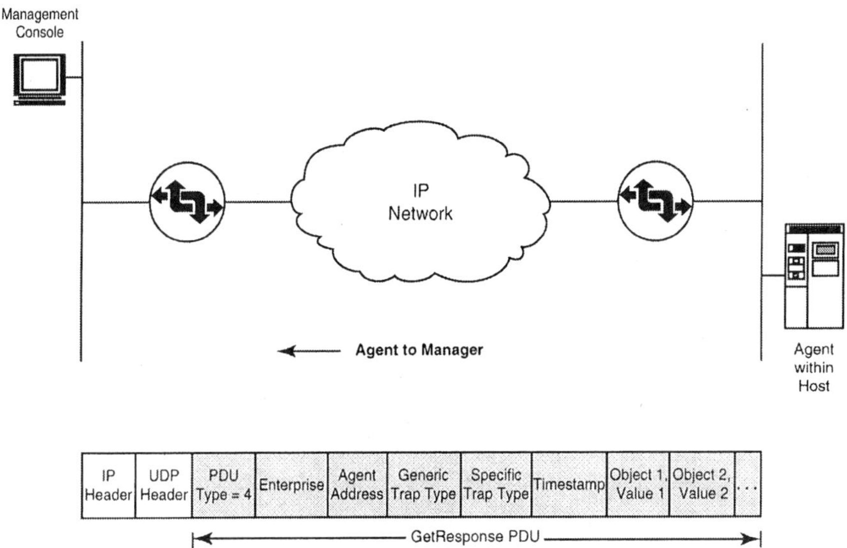

Figure 12-47. Trap PDU Operation.

12.4.3.6 SNMP PDU Encoding

Recall from our discussion in Section 12.1.4.3.11 that the SNMP PDUs are encoded using the context-specific class, with a tag that identifies the PDU (review Figure 12-15). The Length and Value fields are then constructed to convey a particular structure and quantity of information. Now that we have discussed the structure of the SNMP PDUs, we can revisit these encodings in more detail.

Figure 12-48 shows an example of a TLV encoding of a (non-Trap) SNMP PDU. Note that the entire encoding begins with a SEQUENCE OF type. The version is an INTEGER type, and the community name is an OCTET STRING type. A context-specific type then indicates the specific PDU and its length. Three INTEGER types provide the Request ID, Error Status, and Error Index. The VarBind list, consisting of multiple SEQUENCE OF encodings, completes the PDU.

12.5 SNMP version 2

The original version of SNMP (SNMPv1) was derived from the Simple Gateway Monitoring Protocol (SGMP) and published as RFC 1028 in 1988. At that time the industry agreed that SNMP would be an interim solution until OSI-based network management using CMIS/CMIP became more mature. After a few years of implementation experience, however, it was discovered that SNMP popularity and usage was dramatically increasing, while OSI-based solutions were experiencing the opposite effect. As a result, it became appropriate to revise and improve SNMPv1, which was undertaken in the early 1990s. This section will discuss the development of SNMPv2 and the resulting enhancements to SNMPv1.

One word of caution is in order, however. Very few implementations of SNMPv2 were ever realized in vendor products, and some of the RFCs that will be referenced in this section have been obsoleted by more current work and documentation on SNMPv3. Nevertheless, the research and development of SNMPv2 laid the groundwork for SNMPv3, and a thorough understanding of the current protocol (SNMPv3) would be difficult without a good grasp on its predecessor (SNMPv2). At the minimum, the reader is encouraged to briefly review this section to better understand this development progression, and therefore be better prepared for the SNMPv3 details presented in Section 12.6.

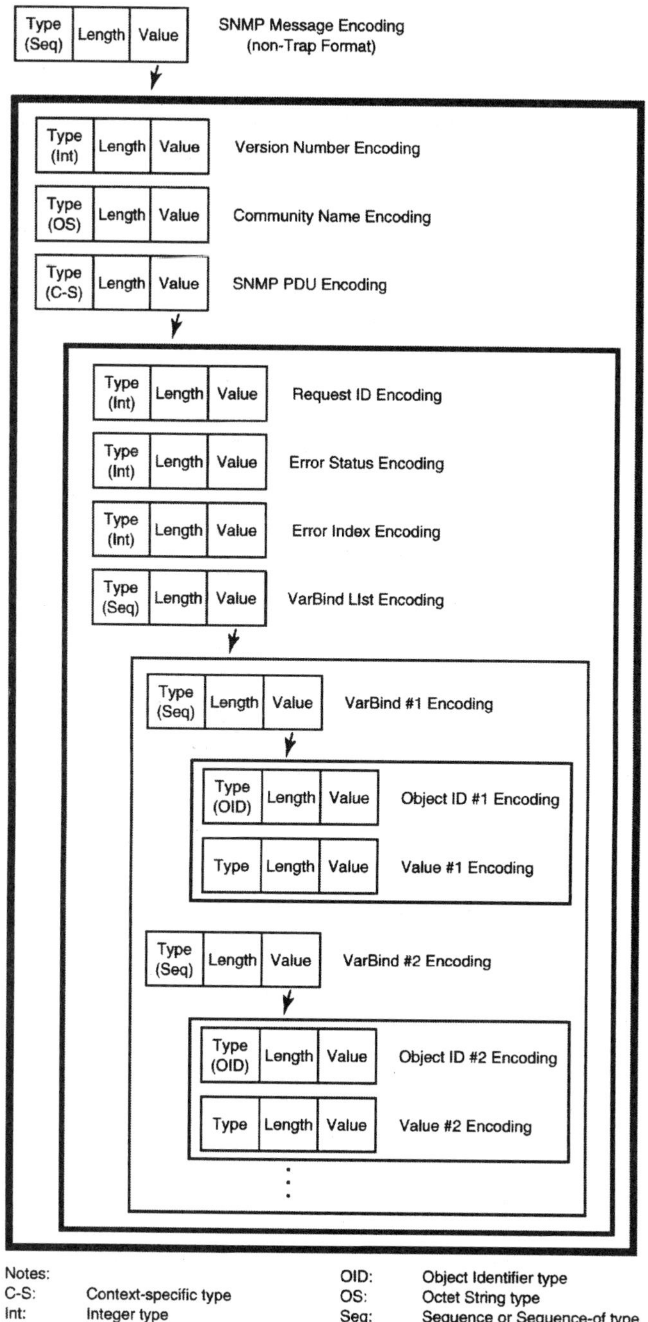

Figure 12-48. TLV Encoding of a (non-Trap) SNMP PDU.

12.5.1 SNMPv2 Development

In March of 1992, the IETF solicited proposals to enhance SNMPv1. A team consisting of Jeffrey Case, Keith McCloghrie, Marshall Rose, and Steven Waldbusser prepared a proposal called the Simple Management Protocol (SMP). At about the same time, the IETF initiated another effort aimed at enhancing SNMP security. These two research efforts merged and became known as version 2 of the Internet-standard Network Management Framework, or simply the SNMPv2 framework. The new documentation comprised more than 400 pages in twelve documents (RFCs 1441 to 1452). RFC 1441, "Introduction to version 2 of the Internet-standard Network Management Framework" [12-19], provided an overview of the remaining documents.

Unfortunately, the first SNMPv2 design (which is now referred to as Party-based SNMPv2) was not widely accepted in the marketplace. Frequently cited reasons were the complexities of the security enhancements and the administrative framework. The SNMPv2 Working Group reconvened in late 1994 to consider several simplified approaches for the administrative framework; however, no consensus was reached. As a result, three actions occurred:

1. Documents that had the consensus of the SNMPv2 Working Group were published in January 1996 as RFCs 1902–1908.
2. Minor modifications to the SNMPv2 Security and Administrative model, called Community-based SNMPv2 (or SNMPv2c), were published in January 1996 as RFC 1901 [12-20].
3. Work continued on the unfinished elements: security, administrative framework, a remote configuration MIB, and Manager-to-Manager communication. (Much of that work is now published as SNMPv3, which is the subject of Section 12.6.)

The major enhancements included in SNMPv2 were: new data types, new macros, textual conventions, protocol operations that facilitate bulk data transfers, richer error codes, and multiprotocol transport support. Section 12.5 will discuss these key enhancements in the SNMPv2c framework and provide a background for an understanding of the current protocol, SNMPv3.

12.5.2 SNMPv2 SMI

As sections 12.2 and 12.3 discussed, MIB modules provide a mechanism for grouping similar objects. The SMI for SNMPv2, RFC 1902 [12-21], defines the subset of the ASN.1 language that describes various MIB modules. SNMPv2 has two documents that support the SMI: the Conformance Statements and the Textual Conventions. The Textual Conventions, which are specified in RFC 1903 [12-22], define the data types used within these MIB modules and make it easier to read the modules. The Conformance Statements, which are specified in RFC 1904 [12-23], provide an implementation baseline and include, for example, a lower bound on what agents must support.

The SMI for SNMPv2 also defined two new branches of the Internet OID tree: security {1.3.6.1.5} and snmpV2 {1.3.6.1.6} (see Figure 12-49). Under snmpV2 are the Transport domains (snmpDomains), Transport proxies (snmpProxys), and Module identities (snmpModules). Defined under the snmpModules are the SNMPv2 MIB (snmpMIB, from RFC 1902) and a number of MIB modules that relate to work on SNMPv3 (we will describe SNMPv3-related enhancements to the SMI, and the SNMPv3 MIB modules, in Section 12.6).

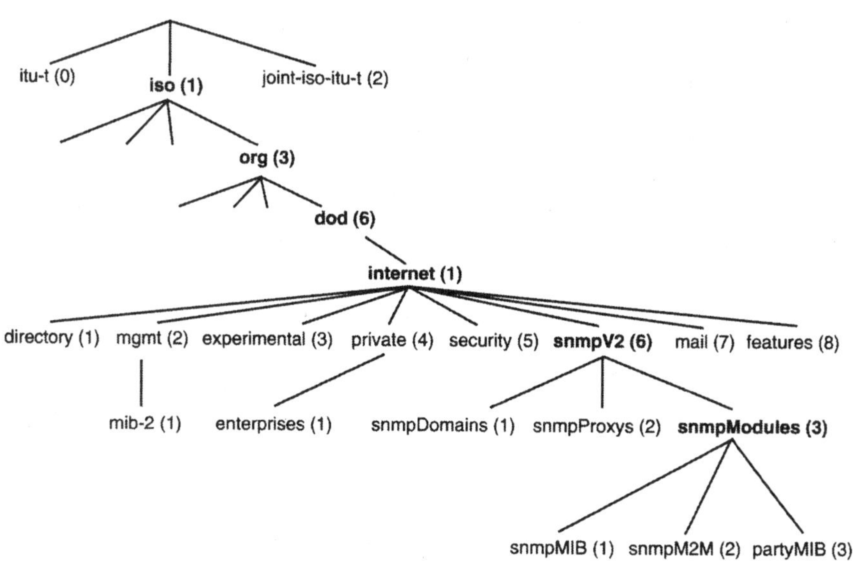

Figure 12-49. SNMPv2 Elements within the OID Tree.

According to RFC 1901, the SMI is divided into three parts:

- *Module definitions*, which are used to describe information modules, such as MIB modules, compliance statements for MIB modules, and capability statements for agent implementations.
- *Object definitions*, which are used to describe managed objects.
- *Notification definitions*, which are used to describe unsolicited transmissions of management information, such as traps.

The next three sections discuss these three elements of the SMI.

12.5.2.1 SNMPv2 SMI Module Definitions

As documented in RFC 1902 [12-21], module definitions are used when describing information modules. An ASN.1 macro, called MODULE-IDENTITY, is used to convey the semantics of an information module. More specifically, it conveys the contact and revision history for each information module, using the following clauses: LAST-UPDATED, ORGANIZATION, CONTACT-INFO, DESCRIPTION, and REVISION. The SMI definition in RFC 1902 explicitly defines the MODULE-IDENTITY macro. Other module definitions will be discussed in the following sections on textual conventions and conformance statements.

12.5.2.2 SNMPv2 Object Definitions

The object definitions in SNMPv2 are enhanced from SNMPv1 through the use of the new OBJECT-IDENTITY module, some new data types that were not used in SNMPv1, and a revised OBJECT-TYPE module. The OBJECT-IDENTITY module is used to define information about an OBJECT IDENTIFIER assignment. This module includes the following clauses: STATUS, DESCRIPTION, and REFERENCE.

The SNMPv2 SMI has added support for ASN.1 data types that SNMPv1 did not use, and it has also defined new data types. ASN.1 data types that remain unchanged from RFC 1155 include INTEGER, OCTET STRING, and OBJECT IDENTIFIER. Application-wide data types from RFC 1155 that have not changed are IpAddress, TimeTicks, and Opaque. New SNMPv2 data types include:

Data Type	Description
Integer32	A defined type that represents integer-valued information between -2^{31} and $2^{31}-1$ inclusive (-2147483648 and 2147483647 decimal). (Note: This type is indistinguishable from the INTEGER type, although the INTEGER type may have different numerical constraints.)
Counter32	A defined type that represents a nonnegative integer that monotonically increases until it reaches a maximum value of $2^{32}-1$ (4294967295 decimal), then wraps around and starts increasing again from zero.
Counter64	A defined type that represents a nonnegative integer that monotonically increases until it reaches a maximum value of $2^{64}-1$ (18446744073709551615 decimal), then wraps around and starts increasing again from zero. Counter64 is used for objects for which the 32-bit counter (Counter32) is too small or would wrap around too quickly. RFC 1902 states that the Counter64 type may be used only if the information being modeled would wrap in less than one hour using the Counter32 type.
Unsigned32	A defined type that represents integer-valued information between 0 and $2^{32}-1$ (4294967295 decimal), inclusive.
Gauge32	A defined type that represents a nonnegative integer, which may increase or decrease, but which never exceeds a maximum value ($2^{32}-1$, as above).
BITS	A construct that represents an enumeration of named bits.

It should be noted that three data types — BIT STRING, NsapAddress, and UInteger32 — were defined in the original version of the SNMPv2 SMI (RFC 1442), but were removed from the subsequent version of the SNMPv2 SMI (RFC 1902).

The SNMPv2 SMI also extends the OBJECT-TYPE macro, which conveys the syntax and semantics of a managed object. This macro has been enhanced from versions defined in RFC 1155 and RFC 1212. In the TYPE NOTATION, a new UNITS clause contains a textual definition of the units associated with that object. Examples include "packets," "messages," or "seconds." The MAX-ACCESS clause defines the maximum level of access for an object. In other words, this clause determines whether it makes sense within the proper operation of the protocol to read, write, and/or create an instance of an object. The values are ordered from least to greatest accessibility: not-accessible, accessible-for-notify, read-only, read-write, and read-create (where read-create is a superset of read-write). The revised STATUS clause of the OBJECT-TYPE macro eliminates the "optional" value in earlier versions of the macro. The DESCRIPTION clause is now mandatory, and the REFERENCE clause can provide a textual cross-reference to another module. The INDEX clause, which identifies instances of columnar objects within a table, can now be replaced with the AUGMENTS clause if the object corresponds to a conceptual row. The AUGMENTS clause thus augments (or extends) a conceptual row within a table. Finally, the DEFVAL clause defines a default value for the object.

12.5.2.3 SNMPv2 SMI Notification and Macro Definitions

The SNMPv2 SMI's new NOTIFICATION-TYPE macro defines the information contained within the unsolicited transmission of management information. This includes an SNMPv2-Trap-PDU or an Inform-Request-PDU. The SNMP Protocol Operations document, RFC 1905 [12-24], contains details on using the NOTIFICATION-TYPE macro. SMIv2 also references three other new ASN.1 macros: MODULE-COMPLIANCE, OBJECT-GROUP, and AGENT-CAPABILITIES, which are also described in detail in RFC 1904.

12.5.3 SNMPv2 Textual Conventions

The SMI defines a number of data types, such as INTEGER and OCTET STRING, which are used in defining managed objects. In some cases, it is useful to define new types, which are similar in syntax (form) but with a more precise semantics (meaning) than those types defined in the SMI. These newly defined types are called textual conventions, and they are specified in RFC 1903 [12-22]. The key advantage of these textual conventions is that they enable the human reader of a MIB module to more easily read and understand that module's intent.

The textual convention consists of a data type with a specific name and associated syntax and semantics. An ASN.1 macro, TEXTUAL-CONVENTION, also found in RFC 1903, is used to convey that syntax and semantics. For example, the textual convention MacAddress represents an IEEE 802 MAC address, which is an OCTET STRING of size 6. In other words, the MacAddress type is based on the existing OCTET STRING type, but with the restrictions that it is limited to six octets in length and is defined to represent an IEEE 802 MAC address. Within the TEXTUAL-CONVENTION macro, the DISPLAY-HINT clause is particularly useful because it describes how the value of the object will be displayed. It uses abbreviations such as x (hexadecimal), d (decimal), o (octal), b (binary), and a (ASCII) to more fully define the display format.

The following textual conventions have been defined in RFC 1903 for use with SNMPv2:

Convention	Description
DisplayString	Represents textual information taken from the NVT ASCII character set (see RFC 854).
PhysAddress	Represents Media- or Physical-level addresses. (Originally from RFC 1213.)
MacAddress	An 802 MAC address represented in the "canonical" order defined by IEEE 802.1a; that is, it is represented as if it were transmitted least significant bit first, even though 802.5 requires MAC addresses to be transmitted most significant bit first. (Originally from RFCs 1230 and 1231.)
TruthValue	Represents a boolean value, true or false. (Originally from RFC 1253.)
TestAndIncr	Represents integer-valued information for atomic operations. Atomic operations are self-contained but are performed in a specific order or sequence. The TestAndIncr ensures that these required sequences are maintained.
AutonomousType	Represents an independently extensible type identification value. It may, for example, indicate a particular subtree with further MIB definitions, or define a particular type of protocol or hardware. (Originally from RFC 1316.)
InstancePointer	A pointer to a row of a MIB table in the managed device. By convention, it is the name of the first columnar object in the conceptual row. Note that the term "conceptual row" defines all of the objects having the same instance value in a MIB table. The terms conceptual row and row are generally used interchangeably. (Originally from RFC 1316, but obsoleted in RFCs 1903 and 2578. The applications for InstancePointer are now replaced with the VariablePointer and RowPointer.)
VariablePointer	A pointer to a specific object instance (from the obsoleted InstancePointer).

RowPointer	A pointer to a conceptual row.
RowStatus	Creates and deletes conceptual rows, and is used as the value of the SYNTAX clause for the status column of a conceptual row. (See the SMI document for further details; originally from RFC 1271, the first version of the RMON MIB.)
TimeStamp	The value of the sysUpTime object at which a specific occurrence happened.
TimeInterval	A period of time, measured in hundredths of a second, between two events.
DateAndTime	A date-time specification, which can indicate the year, month, day, time, and so on.
StorageType	Describes the memory realization of a conceptual row, such as volatile, nonvolatile, read only, and so on. (Originally from RFC 1447.)
TDomain	Denotes a kind of transport service, such as those specified in the Transport Mappings document, RFC 1906 (obsoleted by RFC 3417).
TAddress	Denotes a transport service address, such as those specified in the Transport Mappings document, RFC 1906 (obsoleted by RFC 3417).

12.5.4 SNMPv2 Conformance Statements

The Conformance Statements are used to define acceptable lower bounds of implementation, along with the actual level of implementation for SNMPv2 that is achieved by the device. The Conformance Statements document, RFC 1904 [12-23], defines the notations, along with ASN.1 macros, that are used for these purposes. Two kinds of notations are used:

- Compliance statements: describe requirements for agents with respect to object definitions. The MODULE-COMPLIANCE macro is used to convey a minimum set of requirements with respect to implementation of one or more MIB modules. In other words, the MODULE-COMPLIANCE macro conveys a minimum conformance specification, including objects and groups required, which may come from different MIB modules.
- Capability statements: describe the capabilities of agents with respect to object definitions. The AGENT-

CAPABILITIES macro describes the capabilities of an SNMPv2 agent. It defines the MIB modules, objects, and values implemented within the agent. A description of the precise level of support that an agent claims is bound to the instance of the sysORID object. (See the SNMPv2 MIB, RFC 1907, for a complete definition of the sysORID object and other objects that convey object resource information.)

The Conformance Statements also define two other ASN.1 macros. The OBJECT-GROUP macro defines collections of related, managed objects. Similarly, collections of notifications may be grouped using the NOTIFICATION-GROUP macro.

12.5.5 SNMPv2 Protocol Operations and PDUs

Protocol operations and PDU formats for SNMPv2 are defined in RFC 1905 [12-24]. When it comes to processing protocol messages, an SNMPv2 entity may act as an agent, a manager, or both. The entity acts as an agent when it responds to protocol messages (other than the Inform notification, which is reserved for managers) or when it sends Trap notifications. The entity acts as a manager when it initiates protocol messages or responds to Trap or Inform notifications. The entity may also act as a proxy agent.

SNMPv2 provides three types of access to network management information: these types are determined by the network management entity's role and they relate to the Manager-to-Manager capabilities. The first type of interaction, called request-response, is where an SNMPv2 manager sends a request to an SNMPv2 agent, which responds. The second type of interaction is a request-response where both entities are SNMPv2 managers. The third type is an unconfirmed interaction, where an SNMPv2 agent sends an unsolicited message, or trap, to the manager and no response is returned.

SNMPv2 has significantly enhanced the PDUs that convey this management information (see Figure 12-50). SNMPv2 offers new PDUs and adds error codes and exception responses. The latter allows a management application to easily determine why a management operation failed.

RFC 1905 defines eight PDU types, of which three are new: the GetBulkRequest, the InformRequest, and the Report. In addition, the SNMPv2-Trap PDU format has been revised from the SNMPv1 Trap to conform to the format and structure of the other PDUs. (Recall that, in SNMPv1, the Trap PDU had a unique format. The fact that there are two PDU formats for SNMPv1, and one format for SNMPv2, renders SNMPv1 and SNMPv2 incompatible from a PDU perspective. Section 12.5.8 will discuss this issue further.)

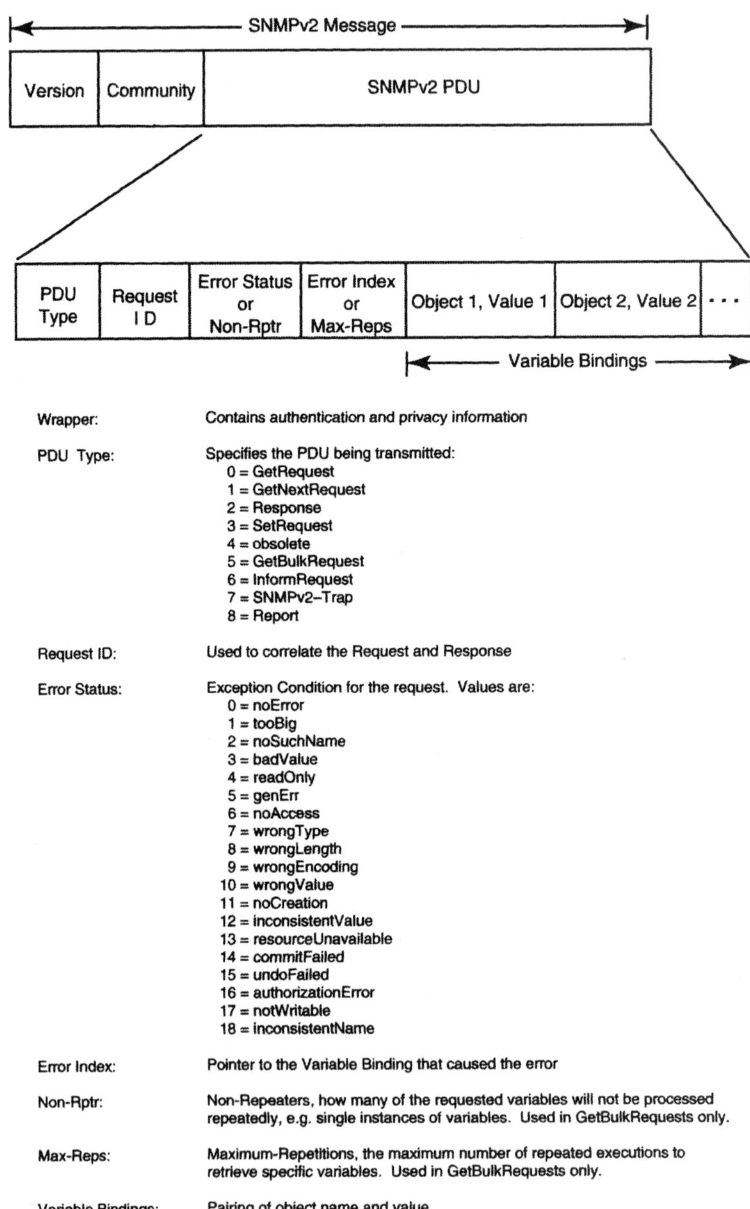

Figure 12-50. SNMPv2 PDU Structure.

The following is a list of all the SNMPv2 PDUs, along with their assigned tag numbers:

PDU/Tag Number	Description
GetRequest [0]	Retrieves values of objects listed within the Variable Bindings field.
GetNextRequest [1]	Retrieves values of objects that are the lexicographical successors of the variables, up to the end of the MIB View of this request.
Response [2]	Generated in response to a GetRequest, GetNextRequest, GetBulkRequest, SetRequest, or InformRequest PDU.
SetRequest [3]	Establishes the value of a variable.
GetBulkRequest [5]	Retrieves a large amount of data, such as the contents of a large table.
InformRequest [6]	Allows one manager to communicate information in its MIB View to another manager.
SNMPv2-Trap [7]	Used by an SNMPv2 agent to provide information regarding an exceptional condition. The Trap PDU, defined for SNMPv1 with tag [4], is now considered obsolete. The Coexistence document, RFC 1908, discusses conversion from the Trap PDU to the SNMPv2-Trap PDU.
Report [8]	Included in SNMPv2, but its usage is not defined in RFC 1905. It is expected that any Administrative Framework that makes use of this PDU would define its usage and semantics (see RFC 1905, page 6). The Report PDU is further defined with SNMPv3, and its usage is noted in RFC 2272, Section 7.1, paragraph (2).

The PDUs that the SNMPv2 entity generates or receives depend on the entity's role as an agent or manager:

SNMPv2 PDU	Agent Generate	Agent Receive	Manager Generate	Manager Receive
GetRequest		X	X	
GetNextRequest		X	X	
Response	X		X	X
SetRequest		X	X	
GetBulkRequest		X	X	
InformRequest			X	X
SNMPv2-Trap	X			X

12.5.6 SNMPv2 Transport Mappings

SNMP version 1 was originally defined for transmission over UDP and IP. Subsequent research explored the use of SNMP with other transport protocols, including OSI transport (RFC 1418), AppleTalk's Datagram Delivery Protocol (DDP) (RFC 1419), and Novell's Internetwork Packet Exchange (IPX) (RFC 1420). SNMPv2 formally defines implementations over these other transports in the Transport Mapping document, RFC 1906 [12-25]. The Transport Mapping document also includes instructions to provide proxy to SNMPv1 and for the use of the Basic Encoding Rules (BER).

12.5.7 SNMPv2 MIB

The April 1993 version of SNMPv2 (RFCs 1441–1452) provided three MIB documents. The first, RFC 1450, described a MIB module for SNMPv2 objects, which was identified by {snmpModules 1}. The second, RFC 1451, coordinated multiple management stations and was therefore called the Manager-to-Manager MIB, identified as {snmpModules 2}. The third module, RFC 1447, supported the SNMPv2 security protocols and was called the Party MIB {snmpModules 3}.

With the removal of the security-related aspects in the January 1996 version of SNMPv2 (RFCs 1901–1908), the MIB required revision as well. The fundamental structure is still the same, however (review Figure 12-49):

Branch	OID	RFC References
snmpV2	{1.3.6.1.6}	1442, 1902
snmpDomains	{1.3.6.1.6.1}	1442, 1902, 1906
snmpProxys	{1.3.6.1.6.2}	1442, 1902, 1906
snmpModules	{1.3.6.1.6.3}	1442, 1902
snmpMIB	{1.3.6.1.6.3.1}	1450, 1907
snmpM2M	{1.3.6.1.6.3.2}	1451
partyMIB	{1.3.6.1.6.3.3}	1447

Note that the last two branches (the Manager to Manager MIB and the Party MIB) are placeholders (with no currently defined objects). There are additional branches under the SNMPv2 branch that support the security and administrative functions of SNMPv3, and these additional branches will be explored in Section 12.6.

This section focuses on the structure of the snmpMIB branch, from RFC 1907 [12-26]. The changes include:

1. The inclusion of the *system* group from MIB-II; the addition of object resource information, which describes the SNMPv2 entity's support for various MIB modules, is also added to the *system* group (see Figure 12-51a).
2. The inclusion of the *snmp* group from MIB-II, making obsolete a number of objects and adding two new ones: *snmpSilentDrops* {snmp 31} and *snmpProxyDrops* {snmp 32} (see Figure 12-51a).
3. Changes to the *snmpMIB* group, making obsolete a number of objects and adding others (see Figure 12-51b). The objects within *snmpMIB* now fall into several categories: information for notifications and well-known traps; a set group, which allows managers to coordinate set operations; conformance information; and compliance statements.

Further details on these revisions are available in RFC 1907; use the other references in the table above as supplementary resources.

12.5.8 Coexistence of SNMPv1 and SNMPv2

The Coexistence document, RFC 1908 [12-27], presents a number of guidelines that outline the modifications necessary for successful coexistence of SNMPv1 and SNMPv2. Some of the issues noted in RFC 1908 deal with MIB structures — such as object definitions, trap definitions, compliance statements, and capabilities statements — that must be updated to conform to the specifications in SNMPv2.

From a practical point of view, two methods are defined to achieve coexistence: a proxy agent and a bilingual manager.

The proxy agent translates SNMPv1 messages to/from SNMPv2 (see Figure 12-52). When translating from SNMPv2 to SNMPv1, GetRequest, GetNextRequest, or SetRequest PDUs from the manager are passed directly to the SNMPv1 agent. GetBulkRequest PDUs are translated into GetNextPDUs. For translating from SNMPv1 to SNMPv2, the GetResponse PDU is passed unaltered to the manager. An SNMPv1 Trap PDU is mapped to an SNMPv2-Trap PDU, with the two new variable bindings, *sysUpTime.0* and *snmpTrapOID.0*, prepended to the Variable Bindings field.

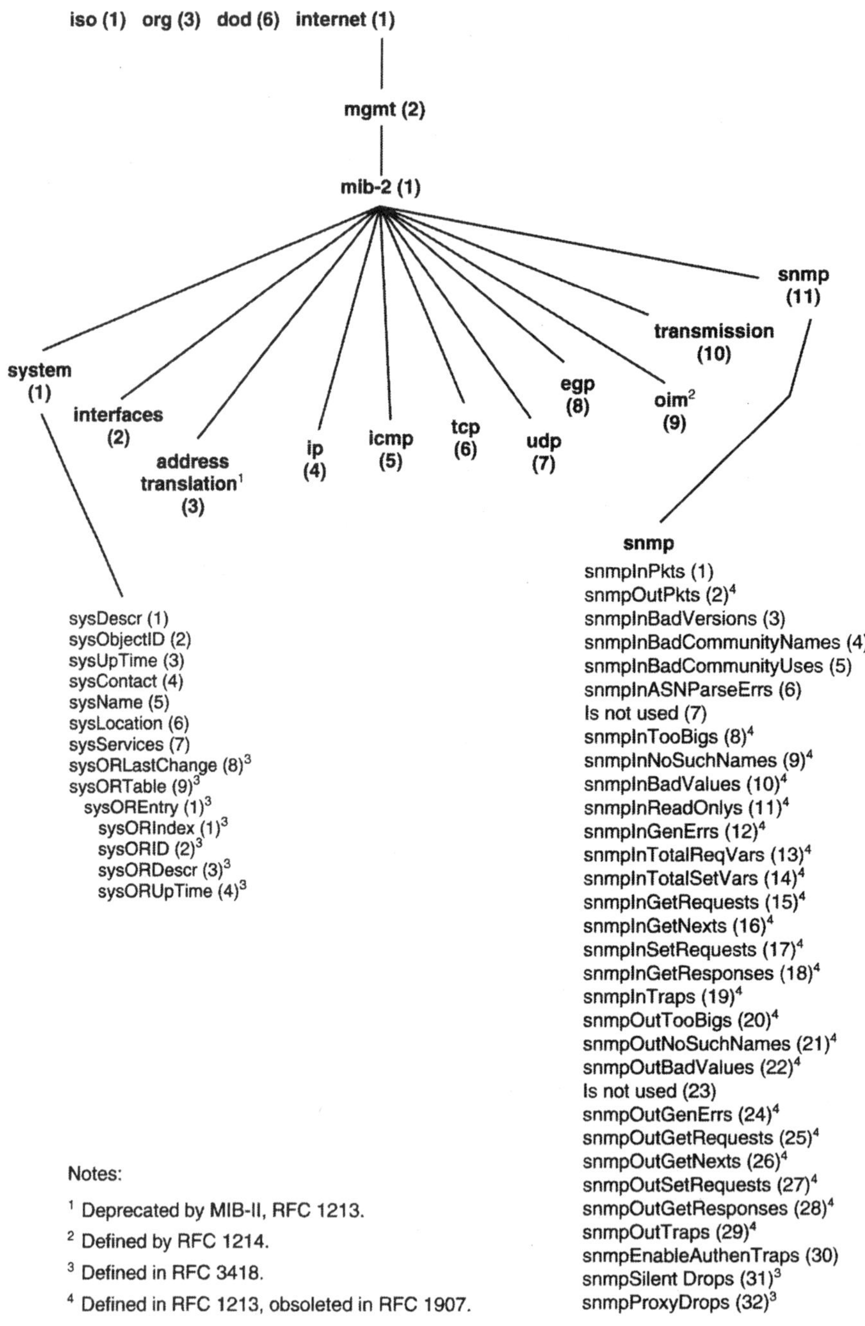

Figure 12-51a. The *system* and *snmp* Groups Implemented for SNMPv2.

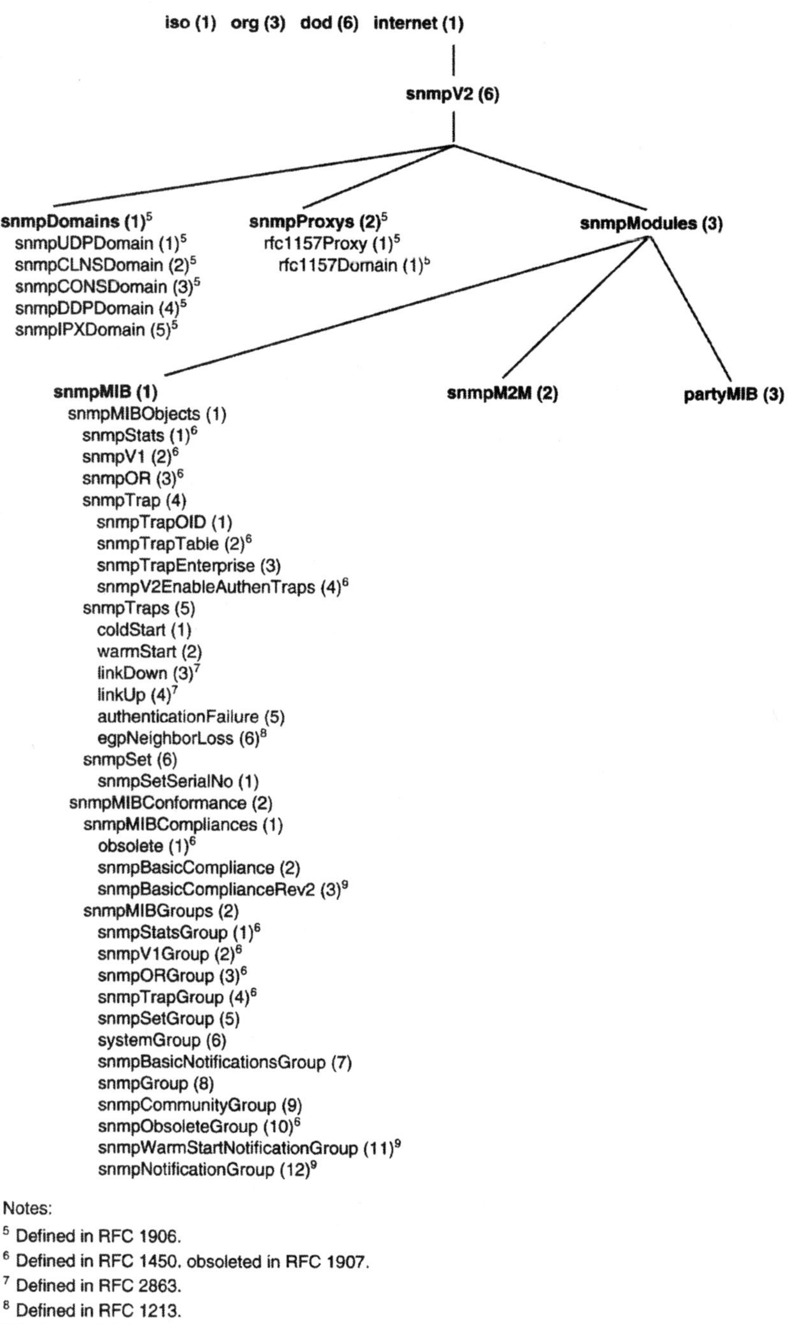
iso (1) org (3) dod (6) internet (1)
snmpV2 (6)
snmpDomains (1)[5]
snmpUDPDomain (1)[5]
snmpCLNSDomain (2)[5]
snmpCONSDomain (3)[5]
snmpDDPDomain (4)[5]
snmpIPXDomain (5)[5]
snmpProxys (2)[5]
rfc1157Proxy (1)[5]
rfc1157Domain (1)[5]
snmpModules (3)
snmpMIB (1)
snmpMIBObjects (1)
snmpStats (1)[6]
snmpV1 (2)[6]
snmpOR (3)[6]
snmpTrap (4)
snmpTrapOID (1)
snmpTrapTable (2)[6]
snmpTrapEnterprise (3)
snmpV2EnableAuthenTraps (4)[6]
snmpTraps (5)
coldStart (1)
warmStart (2)
linkDown (3)[7]
linkUp (4)[7]
authenticationFailure (5)
egpNeighborLoss (6)[8]
snmpSet (6)
snmpSetSerialNo (1)
snmpMIBConformance (2)
snmpMIBCompliances (1)
obsolete (1)[6]
snmpBasicCompliance (2)
snmpBasicComplianceRev2 (3)[9]
snmpMIBGroups (2)
snmpStatsGroup (1)[6]
snmpV1Group (2)[6]
snmpORGroup (3)[6]
snmpTrapGroup (4)[6]
snmpSetGroup (5)
systemGroup (6)
snmpBasicNotificationsGroup (7)
snmpGroup (8)
snmpCommunityGroup (9)
snmpObsoleteGroup (10)[6]
snmpWarmStartNotificationGroup (11)[9]
snmpNotificationGroup (12)[9]
snmpM2M (2)
partyMIB (3)
Notes:
[5] Defined in RFC 1906.
[6] Defined in RFC 1450. obsoleted in RFC 1907.
[7] Defined in RFC 2863.
[8] Defined in RFC 1213.
[9] Defined in RFC 3418.

Figure 12-51b. The *snmpMIB* Group for SNMPv2.

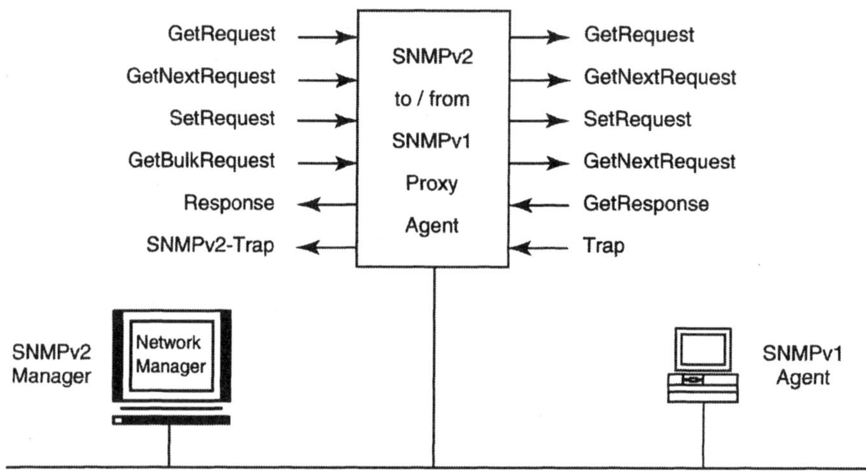

Figure 12-52. SNMPv1/SNMPv2 Proxy Agent Operation.

The second alternative is a bilingual manager, which incorporates both the SNMPv1 and SNMPv2 protocols. When the manager needs to communicate with an agent, it selects the protocol appropriate for the application.

12.5.9 SNMPv2 Security

When SNMPv1 was first published (circa 1988), the community name and the version number in the SNMP header provided the only message security capabilities. This provision, known as the trivial protocol, ensured that both agent and manager recognized the same community name before proceeding with network management operations.

Additional research into security issues yielded three documents on the subject, all released in July 1992:

RFC	Document Title
1351	SNMP Administrative Model
1352	SNMP Security Protocols
1353	Definitions for Managed Objects for Administration of SNMP Parties

Further refinements of the above yielded three additional documents, all released in April 1993:

RFC	Document Title
1445	Administrative Model for SNMPv2
1446	Security Protocols for SNMPv2
1447	Party MIB for SNMPv2

These RFCs were designed to address the authentication and privacy of network management communication. Authentication ensures the appropriate origin of the message, while privacy protects the messages from disclosure. Unfortunately, implementing these enhancements proved to be more complex than either vendors or network managers anticipated; consequently, few products containing these improvements were developed. The result, SNMPv2c, with Community-based security, has been the subject of this section.

Nevertheless, requirements for enhanced SNMP security exist. An administrative framework for SNMPv2, defined in the experimental RFC 1909 [12-28], described how network management can be effective in a variety of configurations and environments, using various techniques such as authorization, authentication, and privacy. In addition, two alternatives were proposed to address the security aspects in particular. The first was called SNMPv2U, which stands for a User-based security model; it is described in the experimental RFC 1910 [12-29]. The second was called SNMPv2* (pronounced SNMP vee-two-star), which was described in various trade publications. As might be expected, both proposals had very vocal proponents. An IETF working group, called the SNMPng (for SNMP next generation) working group, was chartered with proposing a resolution, which resulted in the protocol we now know as SNMPv3.

12.6 SNMP version 3

This section provides an introduction to the most up-to-date version of the protocol, SNMP version 3, which builds on the successes (and addresses the shortcomings) of the previous versions. At the time of this writing, however, this protocol is still in its early stages (even though it has been documented for several years), with many vendor implementations just getting underway. As a result, the objective of this section will be to point out the key architectural elements of the protocol rather than to go into fine details of the protocol enhancements. Readers who need deeper information will be directed to appropriate RFC documents.

We will begin by looking into the development process that led to SNMPv3.

12.6.1 SNMPv3 Development

Recall that SNMPv1 was designed with limited security (using only the plaintext or unencrypted Community string). This element of the SNMPv1 architecture was perceived as a serious shortcoming. Many managers of large enterprise internetworks demanded that any communication with key networking devices, such as backbone routers, be done in a secure fashion to prevent an imposter from misconfiguring the network under the guise of legitimate network management.

The first round of enhancements to SNMP, which became known as SNMPv2, included two distinct areas of improvement: protocol enhancements and security enhancements. These were published in April 1993 and documented in RFCs 1441–1452. The security portion of these enhancements was called "Party-based SNMPv2." The party was a logical relationship between an agent and a manager. The party defined a number of security parameters, which might include the security protocols to be used, encryption keys, and so on, that were required to ensure secure communication between the agent and the manager. Unfortunately, implementing Party-based SNMPv2 was a challenge, and very few vendors supported that version of the protocol.

The protocol enhancements that were part of SNMPv2, however, were considered quite sound. As a result, the SNMP Working Group kept those parts of the SNMPv2 effort, and went back to the drawing board for more work on the security portion. The dialect of SNMPv2 that included the v2 protocol enhancements, but only the limited security (the Community string from SNMPv1), became known as SNMPv2c (where the C stood for Community-based SNMP). This work was published as RFCs 1901–1908 in January 1996.

The work on the security area generated two major proposals. One, known as SNMPv2U (where the U stood for User-based security), was published as RFCs 1909 and 1910 in February 1996. The second proposal, known as SNMPv2* (pronounced "vee-two-star"), was documented in an Internet Draft. These two proposals, plus experience with Party-based SNMPv2, became the foundation for what we know today as SNMPv3.

Figure 12-53, from Dr. Jeff Case, one of the key developers of SNMPv1, SNMPv2, and SNMPv3, illustrates the relationships among the three different versions of the protocol. Note that there is overlap between SNMPv1 and SNMPv2, and between SNMPv2 and SNMPv3. Various dialects of SNMPv2 exist, such as Party-based SNMPv2, SNMPv2c, and so on; however, all four share some common ground (the protocol enhancements). SNMPv3 includes some new security functions, which will be explored in this chapter. Also note that the MIB definitions come from several sources, including the original RFC 1155 format, RFCs 1212 and 1215 (both from SNMPv1), plus RFCs 1442–1444 and RFCs 1902–1904 (from SNMPv2 and SNMPv2c, respectively).

From the perspective of this section, recall that the information on SNMPv1 (Sections 12.1 through 12.4) was enhanced with the information on SNMPv2 (Section 12.5). This section provides additional enhancements, and thus completes the process.

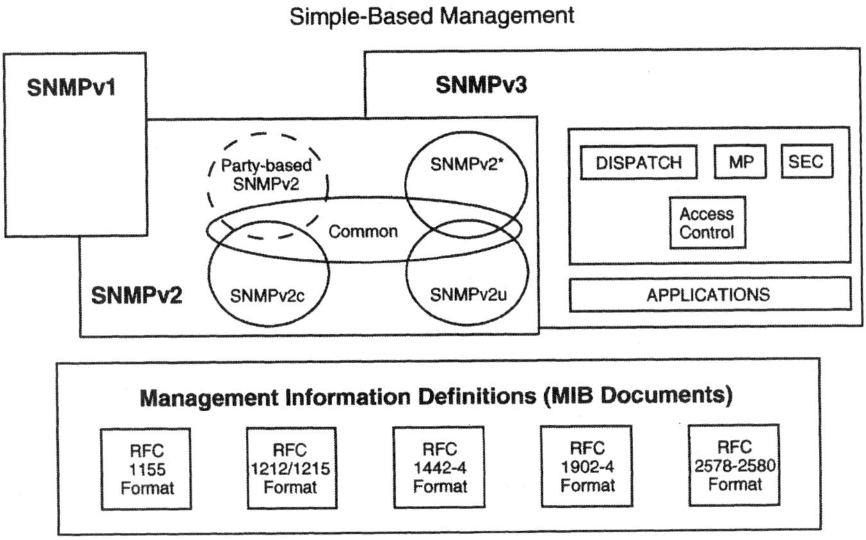

Figure 12-53. SNMPv1, SNMPv2, and SNMPv3 (Courtesy of SNMP Research International, Inc.).

12.6.2 SNMPv3 Objectives and Capabilities

In RFC 3410, the "Introduction and Applicability Statements for Internet-standard Network Management Framework" [12-30], the authors restate the four basic components of this framework that we have studied heretofore in this chapter:

- Several managed nodes, each with an SNMP entity (typically called the Agent) that provides remote access to that node's management information
- At least one SNMP entity (typically called the Manager) that provides the management applications

- A management protocol (SNMP) that conveys information among the SNMP entities
- Management information (defined by the SMI and the MIB modules)

As further stated in RFC 3410, this fundamental architecture has remained consistent as the generations of the Framework have evolved from SNMPv1 to SNMPv2 and now to SNMPv3. Thus, SNMPv3 builds on these four architectural components, but also defines new capabilities in the areas of security and network administration. Specifically, it addresses four key areas of network security that were found lacking in SNMPv2:

- Authentication: origin identification, message integrity, and some aspects of replay protection
- Privacy: confidentiality
- Authorization and access control
- Remote configuration and administration capabilities needed for the three capabilities above

Thus, the SNMPv3 development concentrated on the dual objectives of maintaining consistency with the Network Management Framework while enhancing that framework to provide a workable security platform that had been lacking in previous versions. Thus, SNMPv3 provides the additional security and administration capabilities that were not successfully included with SNMPv2.

12.6.3 SNMPv3 Documentation

The documentation set for SNMPv3 is described in RFC 3411 [12-31] and is derived from three sources: applicable and relevant information from SNMPv1 and SNMPv2, plus new information that documents the security aspects of the protocol (Figure 12-54). Since the SNMPv3 work is ongoing, some of these documents may be produced in the future.

At the top of Figure 12-54 are three documents that provide general direction for the SNMPv3 development: a Document Roadmap, which describes sets of documents, taken together, from specific implementation frameworks; an Applicability Statement, which describes the environments in which SNMP may be appropriately applied; and a Coexistence and Transition document, which describes evolutionary interactions among various versions of the SNMP.

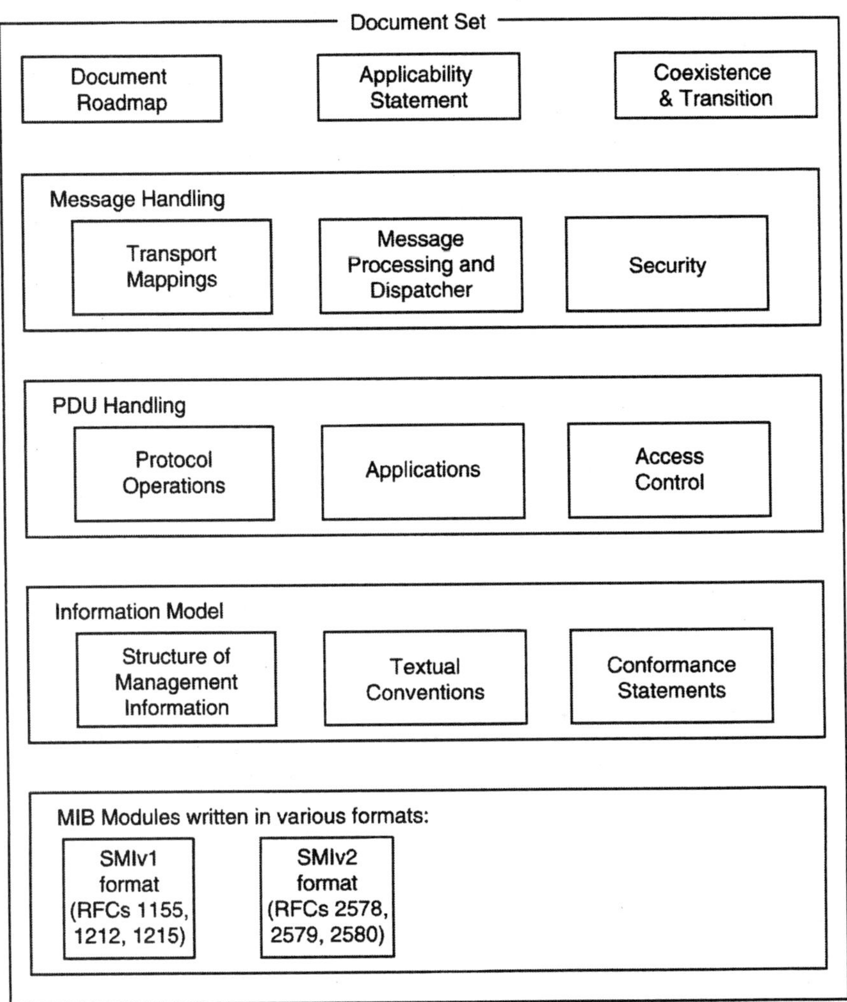

Figure 12-54. SNMPv3 Document Set.

The Message Handling documents define how the SNMP message is conveyed between the Manager and the Agent. The Transport Mappings document defines a number of lower layer options, such as the User Datagram Protocol (UDP) or Open Systems Interconnection (OSI) protocols that are used for data transport. The Message Processing and Dispatcher document defines the format of the SNMP message. The Security document defines support for protocol interactions that require security in the message transmission and reception, or in message processing.

The PDU Handling documents define how the Protocol Data Unit (PDU) is encapsulated inside the SNMP message. The Protocol Operations document defines the processing of the PDUs. The Applications document defines various management operations used to accomplish specific tasks, the various services required, and the protocol functions involved with those tasks. The Access Control document defines any constraints that may be required during processing to control access for operations to any of the managed objects.

The Information Model documents deal with the management information. The Structure of Management Information establishes the syntax for defining objects, modules, and other information elements. The Textual Conventions define new data types, similar to those in SMIv1 but with more precise, or special, semantics associated with them. The Conformance Statements document acceptable lower bounds for implementation, thus ensuring that all implementations of a particular function will meet some base level of compatibility.

The MIB Modules define the collection of managed objects. These MIBs may be based upon earlier work with SNMPv1 (typically called SMIv1, and specified in RFCs 1155, 1212, and 1215) or based upon more recent work from SNMPv2 and SNMPv3 (typically called SMIv2, and specified in RFCs 2578, 2579, and 2580).

The RFC documents that have been published thus far include:

- RFC 3410 [12-30]: "Introduction and Applicability Statements for Internet-Standard Network Management Framework," which outlines the development and migration from SNMPv1 to SNMPv2 and SNMPv3.
- RFC 3411 [12-31]: "An Architecture for Describing SNMP Management Frameworks," with emphasis on the architecture of security and administration.
- RFC 3412 [12-32]: "Message Processing and Dispatching for the SNMP," which describes the SNMP protocol engine and its functions.
- RFC 3413 [12-33]: "SNMP Applications," which describes the five types of applications that can be associated with an SNMP engine.
- RFC 3414 [12-34]: "User-based Security Model," which describes the threats, mechanisms, protocols, and supporting data used to provide SNMP message-level security.

- RFC 3415 [12-35]: "View-based Access Control Model," which describes how view-based access control can be applied to various applications.
- RFC 3416 [12-36]: "Protocol Operations for SNMP," which defines the syntax and elements of procedure for sending, receiving, and processing SNMP PDUs.
- RFC 3417 [12-37]: "Transport Mappings for SNMP," which defines the transport of SNMP messages over various protocols, including UDP, OSI, and IPX environments.
- RFC 3418 [12-38]: "Management Information Base for SNMP," which updates the MIB originally defined with SNMPv2.

In the following sections, we will examine the contents of these documents in greater detail. We will begin by looking at a revision to the Structure of Management Information specifications, known as SMIv2, which applies to SNMPv3 work in general, and is therefore referenced in many of the above-mentioned documents.

12.6.4 Structure of Management Information version 2 (SMIv2)

The work in conjunction with SNMPv2 that enhanced the original Structure of Management Information (RFCs 1155, 1212, and 1215) was published as RFC 1902. These efforts have been further refined with three documents that define what is now called SMIv2: RFC 2578 [12-39], "Structure of Management Information Version 2"; RFC 2579 [12-40], "Textual Conventions for SMIv2"; and RFC 2580 [12-41], "Conformance Statements for SMIv2."

12.6.4.1 SMIv2 Module Definitions

Much like the earlier work from RFC 1902, the SMIv2 is divided into three main parts: module definitions, object definitions, and notification definitions. The first part, the module definitions, is used to describe information modules, which could be of three kinds: MIB modules, compliance statements for MIB modules, or capability statements for agent implementations. All information modules begin with the MODULE-IDENTITY macro defined in RFC 2578:

```
MODULE-IDENTITY MACRO ::=
 BEGIN
      TYPE NOTATION ::=
      "LAST-UPDATED" value (Update ExtUTCTime)
      "ORGANIZATION" Text
      "CONTACT-INFO" Text
      "DESCRIPTION" Text
      RevisionPart

      VALUE NOTATION ::=
      Value (VALUE OBJECT IDENTIFIER)

      RevisionPart ::=
      Revisions
      | empty
      Revisions ::=
      Revision
      | Revisions Revision
      Revision ::=
      "REVISION" value (Update ExtUTCTime)
      "DESCRIPTION" Text
      Text ::= value(IA5String)
 END
```

The MODULE-IDENTITY macro provides contact information as well as a revision history, so that different versions of the same module can be uniquely identified. The macro includes several clauses, such as LAST UPDATED, which contains an editing date and time stamp; ORGANIZATION, which identifies the organization developing this module; CONTACT-INFO, for technical queries; and DESCRIPTION, which provides a textual description of this revision.

12.6.4.2 SMIv2 Object Definitions

RFC 2578 notes two general categories of base data types that are supported with SMIv1: the ASN.1 data types and the application-defined data types. The ASN.1 types include: INTEGER, OCTET STRING, and OBJECT IDENTIFIER. The application-defined types include: Integer32, IpAddress, Counter32, Gauge32, Unsigned32, TimeTicks, Opaque, and Counter64. In addition, there are three other data types noted: BITS, SEQUENCE, and SEQUENCE OF. These data types were previously defined for use with SNMPv2, as discussed in Section 12.5.2.2.

When objects are defined within a MIB, they are presented in a standard format and specified by the OBJECT-TYPE macro shown below:

```
OBJECT-TYPE MACRO ::=
BEGIN
TYPE NOTATION ::=
      "SYNTAX" Syntax
      UnitsPart
      "MAX-ACCESS" Access
      "STATUS" Status
      "DESCRIPTION" Text
      ReferPart
      IndexPart
      DefValPart

VALUE NOTATION ::=
      value(VALUE ObjectName)

Syntax ::= — Must be one of the following:
           — a base type (or its refinement),
           — a textual convention (or its re-
             finement), or
           — a BITS pseudo-type
      type
      | "BITS" "{" NamedBits "}"

NamedBits ::= NamedBit
      | NamedBits "," NamedBit

NamedBit ::=  identifier "(" number ")" —
number is nonnegative

UnitsPart ::=
      "UNITS" Text
      | empty

Access ::=
      "not-accessible"
      | "accessible-for-notify"
```

```
        | "read-only"
        | "read-write"
        | "read-create"

Status ::=
        "current"
        | "deprecated"
        | "obsolete"

ReferPart ::=
        "REFERENCE" Text
        | empty

IndexPart ::=
        "INDEX"    "{" IndexTypes "}"
        | "AUGMENTS" "{" Entry      "}"
        | empty
IndexTypes ::=
        IndexType
        | IndexTypes "," IndexType
IndexType ::=
        "IMPLIED" Index
        | Index
Index ::=
        - use the SYNTAX value of the
        - correspondent OBJECT-TYPE invocation
        value(ObjectName)
Entry ::=
        - use the INDEX value of the
        - correspondent OBJECT-TYPE invocation
        value(ObjectName)

DefValPart ::= "DEFVAL" "{" Defvalue "}"
        | empty

Defvalue ::=  - must be valid for the type
specified in
        - SYNTAX clause of same OBJECT-TYPE
        macro
```

```
        value(ObjectSyntax)
        | "{" BitsValue "}"

    BitsValue ::= BitNames
        | empty

    BitNames ::=  BitName
        | BitNames "," BitName

    BitName ::= identifier

    — a character string as defined in section
    3.1.1
    Text ::= value(IA5String)
  END
```

This macro also contains a number of clauses, with the options available for that clause separated by a vertical bar (|). For example, the Status clause has the options of "current," "deprecated," or "obsolete." The appropriate application of these clauses is detailed in RFC 2578.

12.6.4.3 SMIv2 Notation Definitions

The third part of SMIv2 is the notation definitions, which define the type of information that is contained within an unsolicited transmission of management information, such as an SNMPv2-Trap or InformRequest PDU. The NOTIFICATION-TYPE MACRO from RFC 2578 is shown below:

```
  NOTIFICATION-TYPE MACRO ::=
BEGIN
     TYPE NOTATION ::=
          ObjectsPart
          "STATUS" Status
          "DESCRIPTION" Text
          ReferPart

     VALUE NOTATION ::=
          value (VALUE NotificationName)

     ObjectsPart ::=
```

```
                "OBJECTS" "{" Objects "}"
                | empty
        Objects ::=
                Object
                | Objects "," Object
        Object ::=
                value (ObjectName)

        Status ::=
                "current"
                | "deprecated"
                | "obsolete"

        ReferPart ::=
                "REFERENCE" Text
                | empty

        Text ::= value (IA5String)
    END
```

Note that this macro specifies the formats for the data types and values that would be transmitted within a management notification message.

12.6.4.4 SMIv2 Textual Conventions

The SMIv2 Textual Conventions, documented in RFC 2579, are built upon earlier work from SNMPv2 that was published in RFC 1903. As defined in Section 12.5.3, the textual convention is a new form of data type that has similar syntax (form), but more precise semantics (meaning). For example, a MacAddress textual convention specifies an IEEE 802 address format used with LANs, and a specific bit ordering for the interpretation of that address. The textual conventions defined for SMIv2 include: DisplayString, PhysAddress, MacAddress, TruthValue, TestAndIncr, AutonomousType, InstancePointer, VariablePointer, RowPointer, RowStatus, Timestamp, TimeInterval, DateAndTime, StorageTime, TDomain, and TAddress. An ASN.1 macro, TEXTUAL-CONVENTION (which is somewhat similar to the OBJECT-TYPE macro discussed in Section 12.6.4.2), is used to concisely convey the syntax and semantics of a textual convention:

```
TEXTUAL-CONVENTION MACRO ::=

BEGIN
     TYPE NOTATION ::=
          DisplayPart
          "STATUS" Status
          "DESCRIPTION" Text
          ReferPart
          "SYNTAX" Syntax

     VALUE NOTATION ::=
          value (VALUE Syntax)
          — adapted ASN.1

     DisplayPart ::=
          "DISPLAY-HINT" Text
          | empty

     Status ::=
          "current"
          | "deprecated"
          | "obsolete"

     ReferPart ::=
          "REFERENCE" Text
          | empty

          — a character string
     Text ::= value (IA5String)

     Syntax ::=    — Must be one of the
     following:
          — a base type (or its refinement),
          or
          — a BITS pseudo-type
       type
       | "BITS" "{" NamedBits "}"
```

```
NamedBits ::= NamedBit
            | NamedBits "," NamedBit

     NamedBit ::=  identifier "(" number ")"
     — number is nonnegative

END
```

Further details on the usage of these textual conventions can be found in RFC 2579.

12.6.4.5 SMIv2 Conformance Statements

The SMIv2 Conformance Statements, documented in RFC 2580, are built upon earlier work from SNMPv2 that was published in RFC 1903 and discussed in Section 12.5.4. The Conformance Statements define thresholds to assure that an agent implements a minimum set of MIB modules (the compliance statements) or implements a minimum set of objects (the capability statements).

An example of a compliance statement, taken from RFC 2580, describes a hypothetical XYZv2-MIB:

```
xyzMIBCompliance MODULE-COMPLIANCE
     STATUS  current
     DESCRIPTION
         "The compliance statement for XYZv2
          entities which
          implement the XYZv2 MIB."
     MODULE  — compliance to the containing
     MIB module
          MANDATORY-GROUPS { xyzSystemGroup,
          xyzStatsGroup, xyzTrapGroup,
          xyzSetGroup,
          xyzBasicNotificationsGroup }

     GROUP   xyzV1Group
     DESCRIPTION
          "The xyzV1 group is mandatory only
          for those
          XYZv2 entities which also implement
          XYZv1."
::= { xyzMIBCompliances 1 }
```

Thus, to claim alignment with the compliance statement named {xyzMIBCompliances 1}, a system must implement the XYZv2-MIB's xyzSystemGroup, xyzStatsGroup, xyzTrapGroup, and xyzSetGroup object conformance groups, as well as the xyzBasicNotificationsGroup notifications group. Furthermore, if the XYZv2 entity also implements XYZv1, then it must also support the group XYZv1Group.

An example of a capability statement, also taken from RFC 2580, describes the capabilities of an agent:

```
exampleAgent AGENT-CAPABILITIES
     PRODUCT-RELEASE          "ACME Agent release 1.1
     for 4BSD."
     STATUS                   current
     DESCRIPTION              "ACME agent for 4BSD."

     SUPPORTS                 SNMPv2-MIB
          INCLUDES            { systemGroup, snmpGroup,
          snmpSetGroup, snmpBasicNotificationsGroup }

          VARIATION            coldStart
               DESCRIPTION  "A coldStart trap is
               generated on all reboots."

     SUPPORTS                 IF-MIB
               INCLUDES             { ifGeneralGroup,
               ifPacketGroup }

          VARIATION            ifAdminStatus
               SYNTAX      INTEGER { up(1), down(2)}
               DESCRIPTION  "Unable to set test
               mode on 4BSD."

          VARIATION            ifOperStatus
               SYNTAX      INTEGER { up(1), down(2)}
               DESCRIPTION  "Information limited
               on 4BSD."

     SUPPORTS                 IP-MIB
          INCLUDES             { ipGroup, icmpGroup }
```

```
        VARIATION           ipDefaultTTL
             SYNTAX          INTEGER (255..255)
             DESCRIPTION     "Hard-wired on 4BSD."

        VARIATION           ipInAddrErrors
             ACCESS          not-implemented
             DESCRIPTION     "Information not avail
             able on 4BSD."

        VARIATION           ipNetToMediaEntry
             CREATION-REQUIRES
             { ipNetToMediaPhysAddress }
             DESCRIPTION     "Address mappings on
             4BSD required both protocol and me
             dia addresses."

   SUPPORTS                TCP-MIB
        INCLUDES             { tcpGroup }
        VARIATION           tcpConnState
             ACCESS          read-only
             DESCRIPTION     "Unable to set this on
             4BSD."

   SUPPORTS                UDP-MIB
        INCLUDES             { udpGroup }

   SUPPORTS                EVAL-MIB
        INCLUDES             { functionsGroup,
        expressionsGroup }
        VARIATION           exprEntry
             CREATION-REQUIRES { evalString,
             evalStatus }
             DESCRIPTION     "Conceptual row cre
             ation is supported."

   ::= { acmeAgents 1 }
```

Thus, an agent with a sysORID value of {acmeAgents 1} supports objects defined in six MIB modules (SNMPv2 MIB, IF-MIB, IP-MIB, TCP-MIB, UDP-MIB, and EVAL-MIB). From SNMPv2-MIB, four conformance groups are supported. From IF-MIB, the *ifGeneralGroup* and *ifPacketGroup* groups are supported. However, the objects *ifAdminStatus* and *ifOperStatus* within IF-MIB have a restricted syntax.

From IP-MIB, all objects in the *ipGroup* and *icmpGroup* are supported except *ipInAddrErrors*, while *ipDefaultTTL* has a restricted range. In addition, when creating a new instance in the *ipNetToMediaTable*, the set-request must create an instance of *ipNetToMediaPhysAddress*.

From TCP-MIB, the *tcpGroup* is supported except that *tcpConnState* is available only for reading. From UDP-MIB, the udpGroup is fully supported.

From the EVAL-MIB, all the objects contained in the *functionsGroup* and *expressionsGroup* conformance groups are supported, without variation. In addition, creation of new instances in the *expr* table is supported and requires both of the objects, *evalString* and *evalStatus*, to be assigned a value.

12.6.5 SNMPv3 Architecture

In order to provide strong security, the architecture of SNMPv3 is more complex than that of previous versions. We will begin this section by considering the stated goals for the SNMPv3 architecture.

12.6.5.1 SNMPv3 Design Goals

The architecture of SNMPv3 is defined in RFC 3411 [12-31], which states the following goals:

- Use existing materials as much as possible. SNMPv3 is heavily based on previous work, informally known as SNMPv2U and SNMPv2*.
- Address the need for secure SET support, which is considered the most important deficiency in SNMPv1 and SNMPv2c.
- Make it possible to move portions of the architecture forward in the standards track, even if consensus has not been reached on all pieces.
- Define an architecture that allows for longevity of the SNMP Frameworks that have been and will be defined.
- Keep SNMP as simple as possible.
- Make it relatively inexpensive to deploy a minimal conforming implementation.
- Make it possible to upgrade portions of SNMP as new approaches become available, without disrupting an entire SNMP framework.

- Make it possible to support features required in large networks, but make the expense of supporting a feature directly related to the support of the feature.

12.6.5.2 The SNMPv3 Entity

The implementation of the SNMPv3 architecture is called an SNMP entity, as shown in Figure 12-55. Each entity is comprised of two major elements: an SNMP engine and one or more Applications.

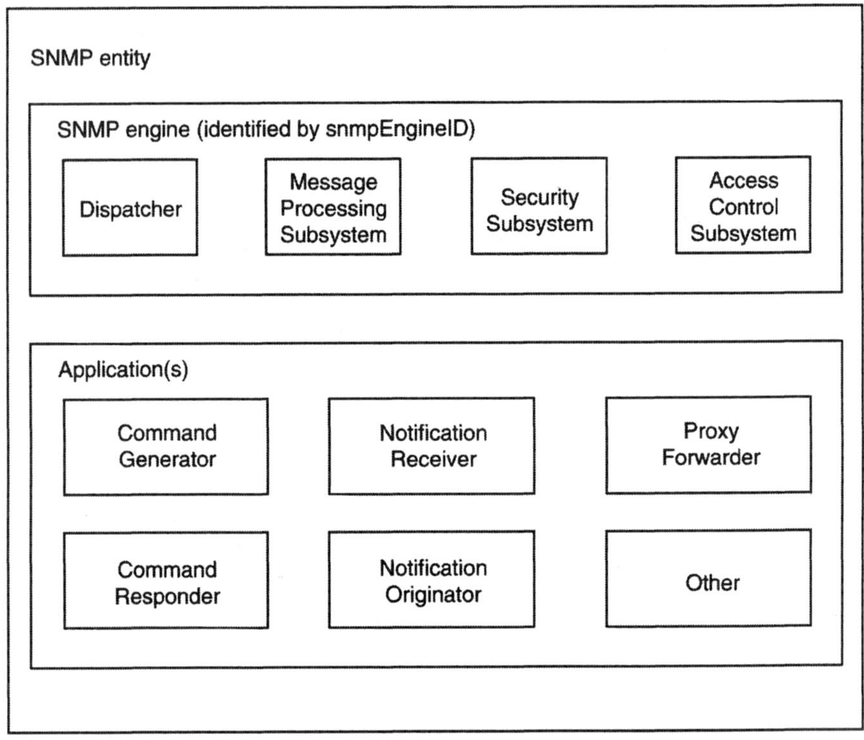

Figure 12-55. SNMPv3 Entity and its Components.

The SNMP engine sends and receives messages, authenticates and encrypts messages, and controls access to managed objects. There is a one-to-one association between an SNMP engine and the SNMP entity that contains that engine. The SNMP engine is identified by a unique and unambiguous identifier called the *snmpEngineID*. Since there is a one-to-one association between the SNMP entity and the SNMP engine, the *snmpEngineID* also identifies the SNMP entity.

The engine contains four elements: a Dispatcher, a Message Processing Subsystem, a Security Subsystem, and an Access Control Subsystem. Only one Dispatcher exists in an SNMP engine. The Dispatcher allows for concurrent support of multiple versions of SNMP messages (SNMPv1, SNMPv2c, etc.) in the SNMP engine. Its functions include sending and receiving SNMP messages to/from the network, determining the version of the SNMP message and interacting with the Message Processing Model that corresponds with that version, and interacting with SNMP applications for PDU delivery.

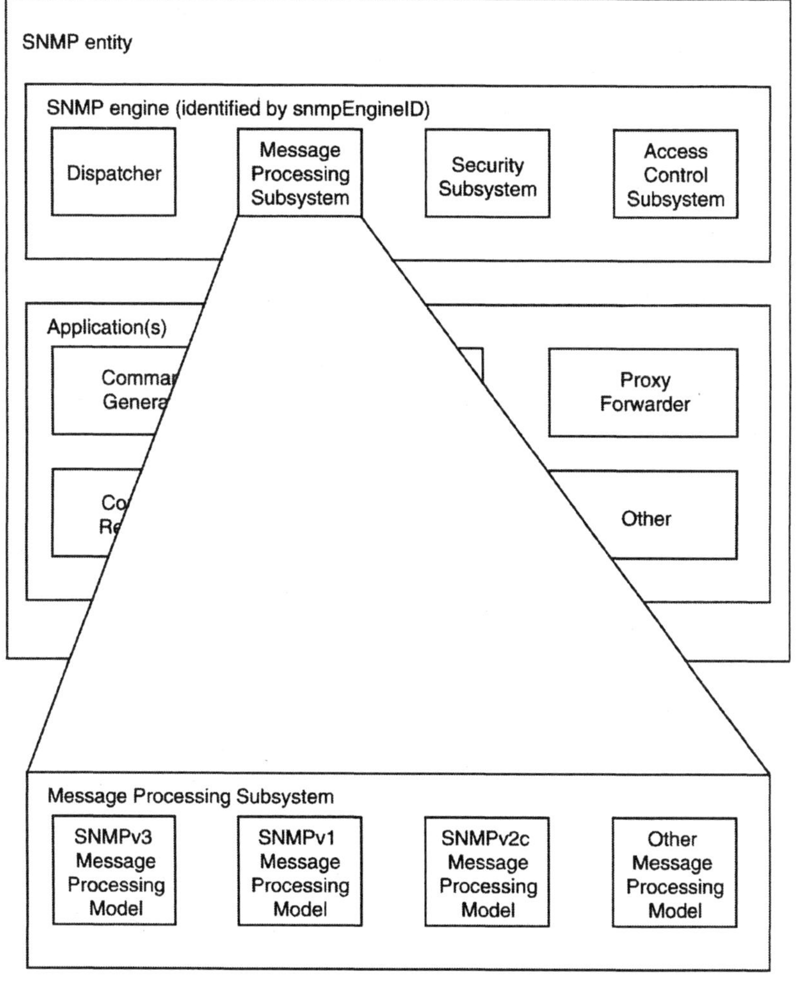

Figure 12-56. SNMPv3 Message Processing Subsystem.

The Message Processing Subsystem prepares messages for transmission and extracts data from received messages; it is further described in RFC 3412 [12-32]. The Message Processing Subsystem may contain one or more Message Processing Models, as shown in Figure 12-56. Each of the Message Processing Models defines the format of a particular SNMP message, and is responsible for formatting each of those messages as appropriate.

The Security Subsystem provides the security services, including authentication and privacy. Three levels of security are defined in RFC 3411:

- Without authentication and without privacy (noAuthNoPriv)
- With authentication but without privacy (authNoPriv)
- With authentication and with privacy (authPriv)

The Security Subsystem may contain multiple Security Models, as shown in Figure 12-57. Each of these Security Models may have different goals and associated services, and therefore protect against different security threats. Threats are categorized by their severity: principle, secondary, or those of lesser importance.

There are two principle threats:

- Modification of Information: an SNMP message is altered in transit by an unauthorized SNMP entity, thus causing unauthorized management operations or a false object value.
- Masquerade: an SNMP entity may attempt unauthorized management operations by appearing to have the identity of one that has the appropriate authorizations.

There are two secondary threats:

- Message Stream Modification: the message stream is maliciously reordered, delayed, or replayed in a manner that is not consistent with the normal operation of that network.
- Disclosure: eavesdropping on the exchanges between two SNMP engines.

There are also two threats that are of lesser importance, which need not be protected against:

- Denial of Service: where service to authorized users is denied.
- Traffic Analysis: where traffic patterns are examined in an attempt to derive sensitive information.

At the time of this writing, one security model, the User-based Security Model, has been documented in RFC 3414 [12-34]. The HMAC-MD5-96 and the HMAC-SHA-96 authentication protocols, plus the CBC-DES encryption protocol, are three security protocols defined for use within the various USM scenarios.

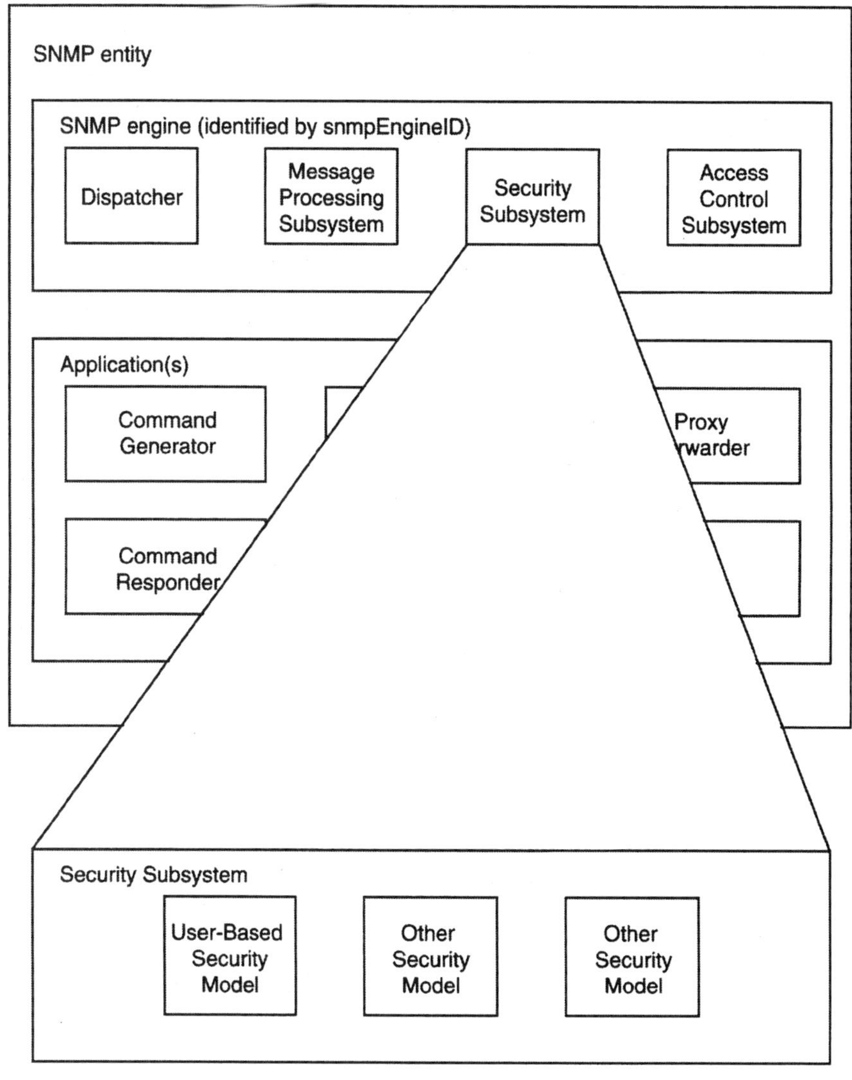

Figure 12-57. SNMPv3 Security Subsystem.

The Access Control Subsystem provides authorization services, using one or more Access Control Models (see Figure 12-58). The Access Control Model defines a specific access right decision function, which in turn supports the decisions regarding access rights. At the time of this writing, one Access Control Subsystem, the View-based Access Control Model (VACM), has been defined in RFC 3415 [12-35].

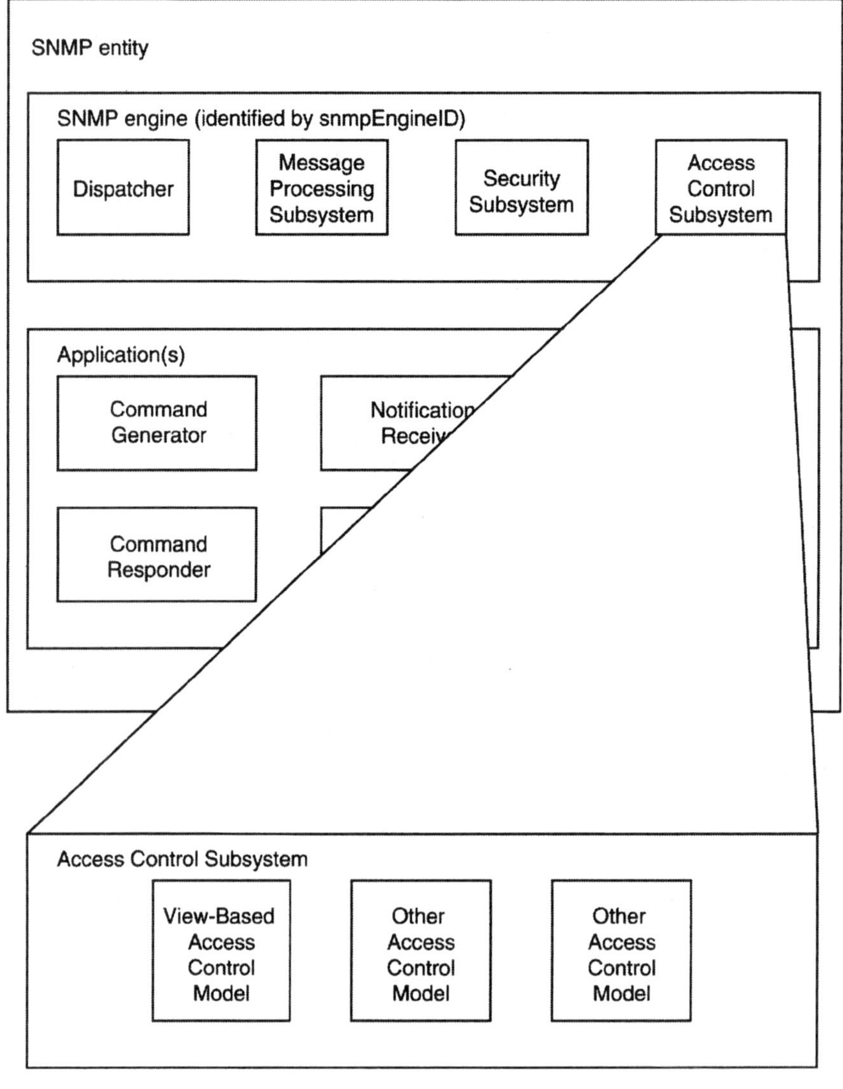

Figure 12-58. SNMPv3 Access Control Subsystem.

The final category of SNMP entity elements is the applications (review Figure 12-55). These include the following:

- Command Generators: monitor and manipulate management data
- Command Responders: provide access to management data
- Notification Originators: initiate asynchronous messages
- Notification Receivers: process asynchronous messages
- Proxy Forwarders: forward messages between entities

The applications make use of the services (Dispatcher, Message Processing Subsystem, Security Subsystem, and Access Control Subsystem) that are provided by the SNMP engine. The five applications listed above are further defined in RFC 3413 [12-33].

12.6.6 SNMPv3 Manager/Agent Communication

An SNMP entity that contains one or more Command Generator and/or Notification Receiver applications is called an SNMP Manager (Figure 12-59a). Likewise, an SNMP entity that contains one or more Command Responder and/or Notification Originator applications is called an SNMP Agent (Figure 12-59b).

A message originated from the SNMP Manager (Figure 12-59a) would use the various SNMP engine elements (Dispatcher, Message Processing Subsystem, etc.), access the appropriate transport protocol, such as UDP, and then be passed to the network (shown at the lower portion of Figure 12-59a) for transmission. That message would travel across the network and be received by the network interface of the SNMP Agent (shown at the upper portion of Figure 12-59b). After the transport protocol was decoded, the SNMP message would be passed to the Dispatcher, Message Processing Subsystem, and other elements. Note that, in this case, the applications also have access to the MIB instrumentation. This allows the Agent to respond to queries (such as a GetRequest from the Manager) with the appropriate MIB value, or to send a notification (such as a Trap) in response to exceptional conditions at the managed device.

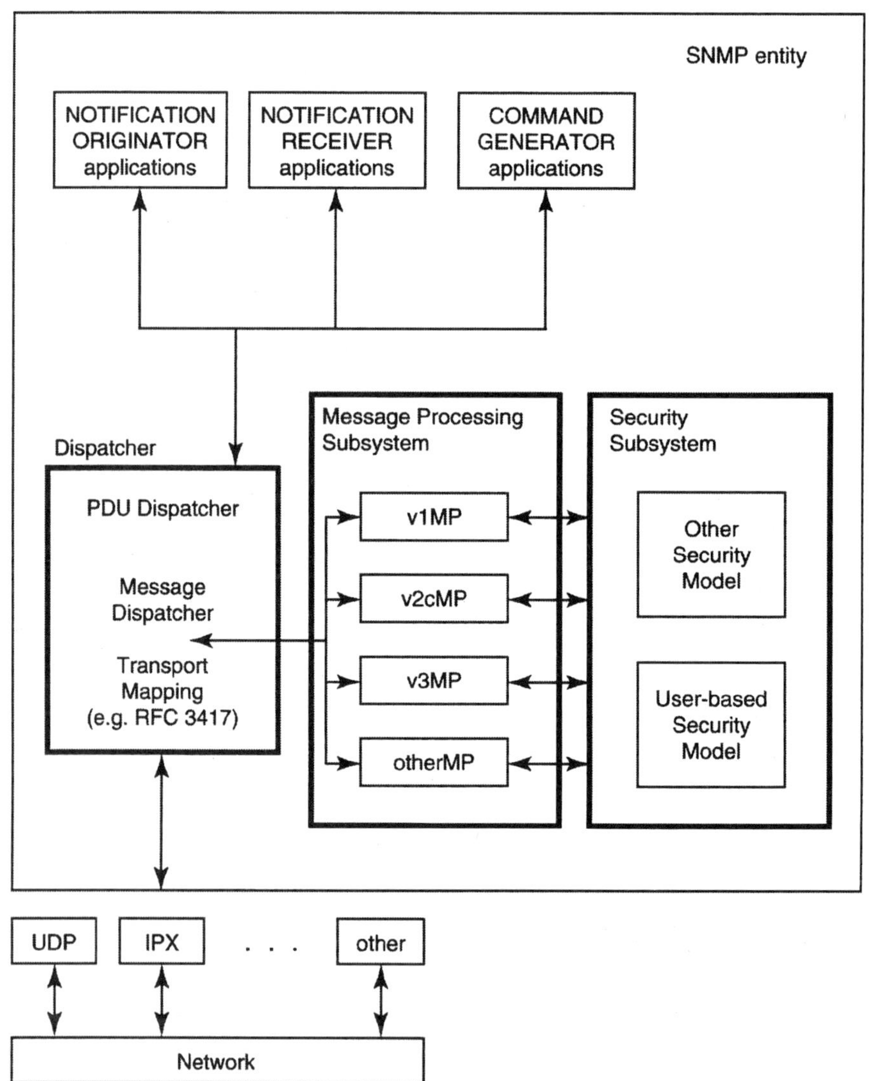

Figure 12-59a. SNMPv3 Manager.

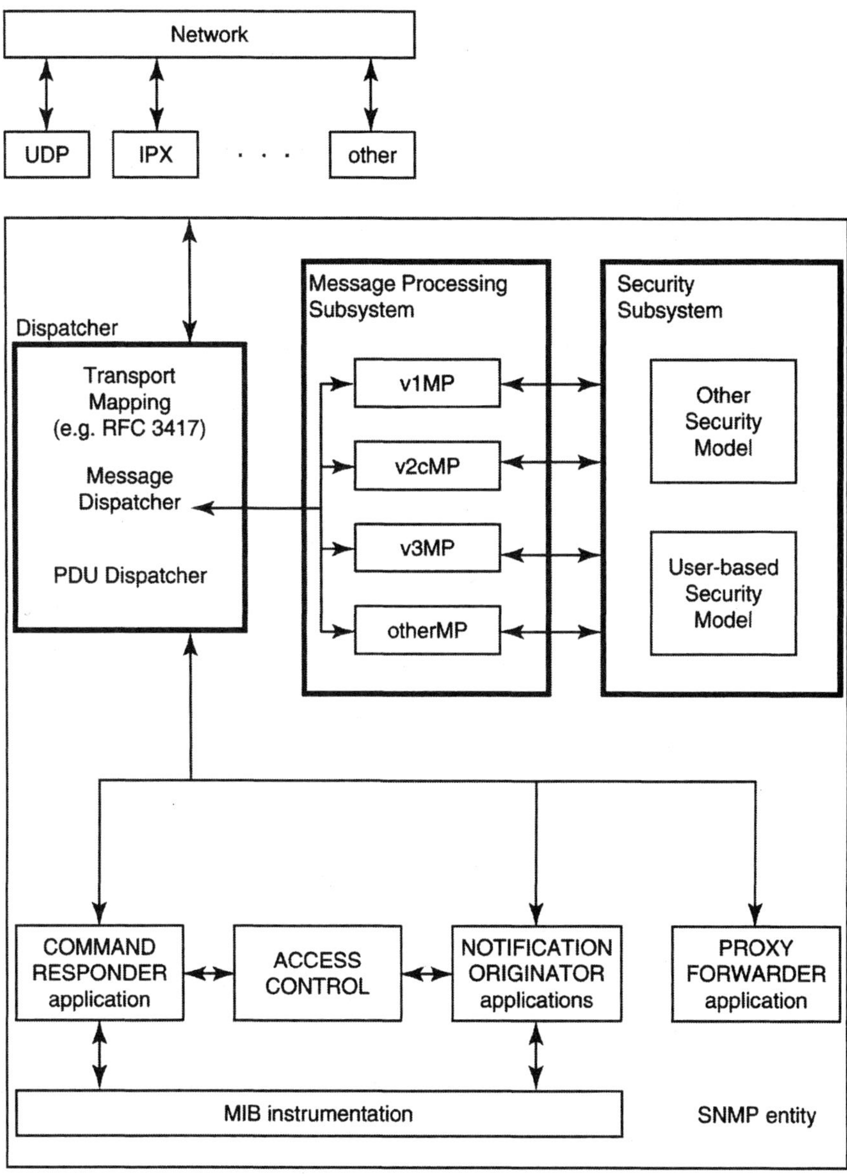

Figure 12-59b. SNMPv3 Agent.

12.6.7 SNMPv3 Message Processing

RFC 3411 also defines the internal communications activities among the SNMP engine subsystems: Dispatcher, Message Processing, Access Control, and Security. These internal communications elements are known as primitives; they include both functions to be performed and various parameters to be passed among these subsystems. RFC 3411 consolidates these interactions into two scenario diagrams, which are shown as Figure 12-60a (the Command Generator scenario) and Figure 12-60b (the Command Responder scenario).

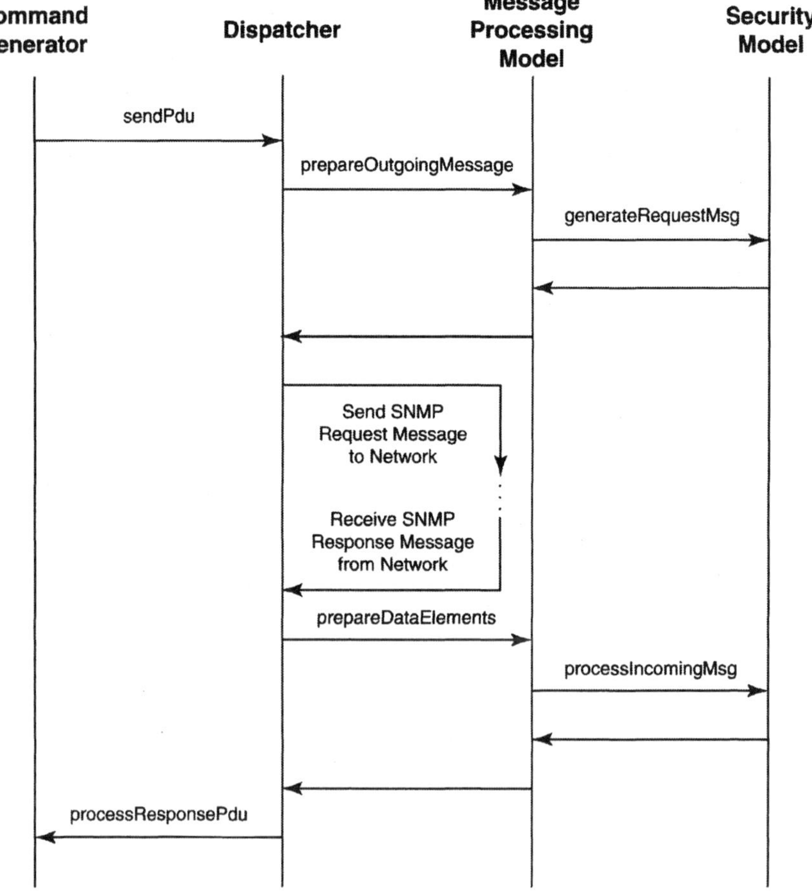

Figure 12-60a. Command Generator Scenario.

In the Command Generator (or Notification Originator) scenario, an application requests that a PDU be sent and passes this request to the Dispatcher. The Message Processing Model queries the Security Model and then sends the SNMP Request message to the network. When the Response message is received from the network, the incoming message is processed by the Dispatcher, with another query to the Security Model. The Response PDU is then passed to the originating application.

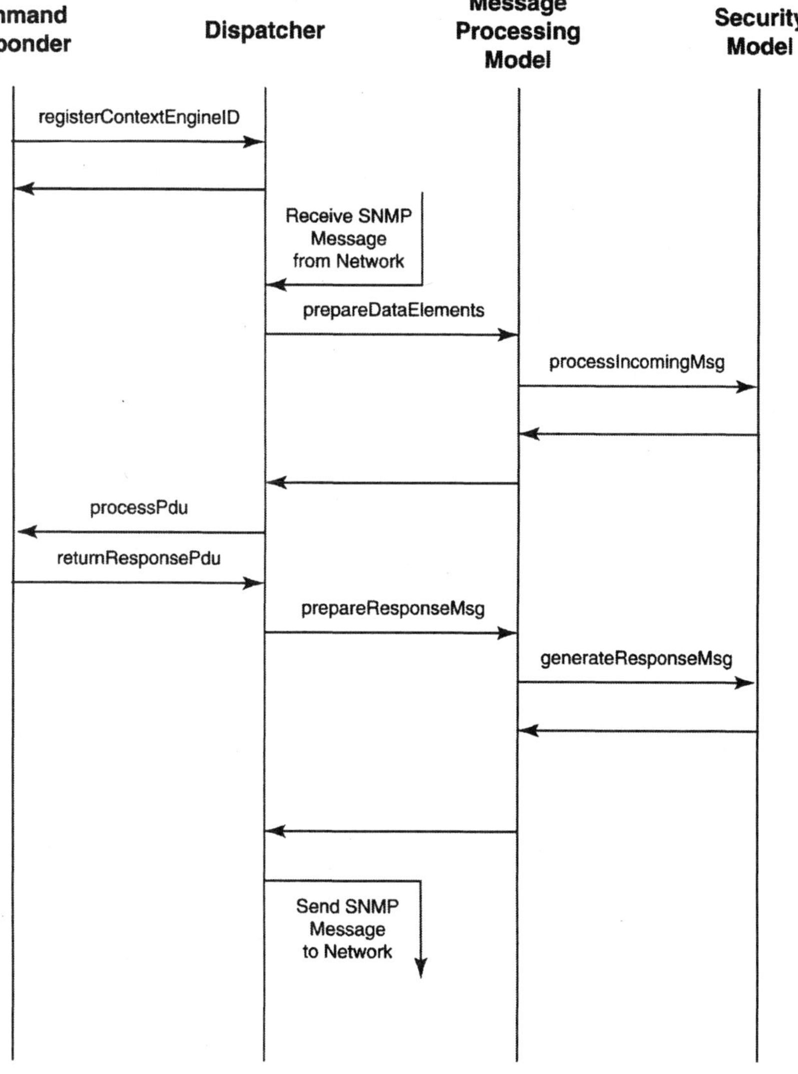

Figure 12-60b. Command Responder Scenario.

In a similar scenario, the Command Responder (or Notification Receiver) application also interacts with the various subsystems. The Command Responder first registers for handling a particular PDU type. Once a message is received from the network, it is processed by the Message Processing and Security Models, then the PDU is passed to the application (the Command Responder). The application returns a response PDU, then an outgoing response message is generated and is sent by the Dispatcher to the network.

12.6.8 SNMPv3 Message Formats

Recall from our earlier discussion that SNMPv3 consists of SNMPv2 plus additional security and administration capabilities. The SNMPv3 message illustrates this, as it consists of a message wrapper that encapsulates a Protocol Data Unit (PDU). The message wrapper provides the additional security and administration functions part, while the PDU represents the SNMPv2 part. The various PDUs defined for SNMPv2 in RFC 1905 are now used with SNMPv3 and defined in section 6 of the Message Processing and Dispatching document, RFC 3412 [12-32]. The various fields of the SNMPv3 message are illustrated in Figure 12-61.

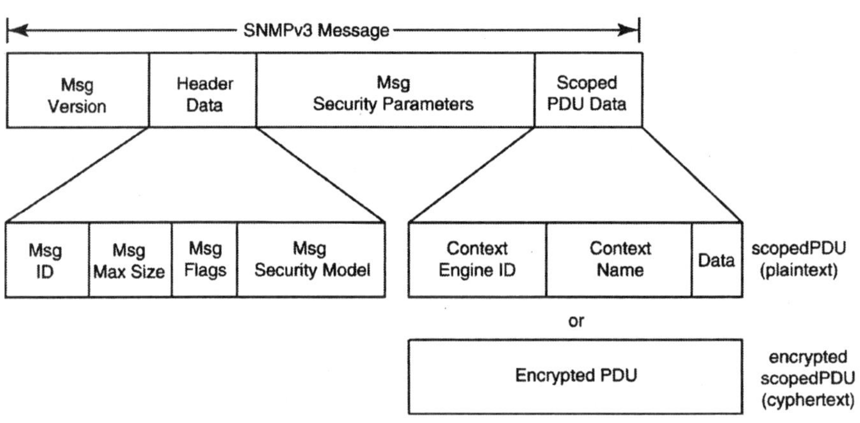

Figure 12-61. SNMPv3 Message Format.

The first field is the msgVersion field. A message is identified as an SNMPv3 message when msgVersion = 3.

Next, the HeaderData field is a sequence of four subfields: msgID, msgMaxSize, msgFlags, and msgSecurityModel, as described below:

- The msgID coordinates request messages and responses between two different SNMP entities. This field is also used by the version 3 Message Processor (or v3MP, as shown in Figures 12-59a and 12-59b) to coordinate message processing between different subsystem models within the architecture. (Note that while the PDU contains a Request ID field, which identifies the particular PDU in use, the msgID field is used by the engine to identify the message that carries a PDU.)
- The msgMaxSize field conveys the maximum message size that the sender (of this message) can accept. In other words, this field places an upper limit on the size of the message that this entity can receive, whether that message is a response or any other message.
- The msgFlags are single bit fields that control processing of the message:

Field	Meaning
....1	authFlag (when authFlag = 1, a process to authenticate the message is in use)
.... ..1.	privFlag (when privFlag = 1, a process to protect the message from disclosure is in use)
.... .1..	reportableFlag (when reportableFlag = 1, a Report PDU is returned to the sender under certain conditions)

Permissible values for the authFlag and privFlag are:

Field	Meaning
.... ..00	is OK, means noAuthNoPriv
.... ..01	is OK, means authNoPriv
.... ..10	reserved, must not be used
.... ..11	is OK, means authPriv

Note that various combinations of the last two bits indicate one of the three permissible security levels defined in Section 12.6.5.2, and that the combination of encryption without authentication is not allowed.

- Multiple Security Models may exist within the version 3 Message Processor. Therefore, the msgSecurityModel field identifies the Security Model used for message generation and reception. Similarly, the msgSecurityParameters field is used for communication between Security Model modules, and the format and contents of this field are defined by the Security Model in use. For example, the User-based Security Model specifies the msgSecurityParameters field in RFC 3414, section 2.4, as follows:

```
USMSecurityParametersSyntax DEFINITIONS IMPLICIT TAGS ::= BEGIN

 UsmSecurityParameters ::=
   SEQUENCE {
   —global User-based security parameters
      msgAuthoritativeEngineID          OCTET STRING,
      msgAuthoritativeEngineBoots       INTEGER
                                        (0..2147483647),
      msgAuthoritativeEngineTime        INTEGER
                                        (0..2147483647),
      msgUserName                       OCTET        STRING
                                        (SIZE(1..32)),
   —authentication protocol specific parameters
      msgAuthenticationParameters       OCTET STRING,
   —privacy protocol specific parameters
      msgPrivacyParameters              OCTET STRING
   }
 END
```

The scopedPduData contains either a plaintext scopedPDU if the privFlag in the msgFlags field is zero, or an encrypted PDU. A scopedPDU contains information to identify an administratively unique context, plus a PDU. An SNMP context is defined in RFC 3411, section 3.3.1, to be a collection of management information that is accessible by an SNMP entity. An item of management information may exist in more than one context, and, similarly, an SNMP entity may potentially access multiple contexts. A context may be a physical device or a logical device, may encompass multiple devices, be a subset of a single device, or be a subset of multiple devices, but it is always defined as a subset of a single SNMP entity. Four elements are used to identify an individual item of management information: the object type, the object's instance, the contextEngineID, and the contextName:

- The contextEngineID field uniquely identifies, within an administrative domain, an SNMP entity. For incoming messages, the contextEngineID is used to determine to which application the scopedPDU will be sent for processing. For outgoing messages, the version 3 Message Processor sets the contextEngineID to the value provided by the application in the request for the message to be sent.
- The contextName field is an identifier for the context that is unique within an SNMP entity.
- The Data field of the scopedPDU contains the SNMPv3 PDU, which must be one of the PDUs specified in RFC 3416. The SNMP PDU defines the operations that are performed by the receiving SNMP engine. RFC 3411, Section 2.8, indicates that the PDUs belong to one of five classes:
    - Read Class: for protocol operations that retrieve management information. These include the GetRequest PDU, GetNextRequest PDU, and GetBulkRequest PDU.
    - Write Class: for protocol operations that attempt to modify management information. This includes the SetRequest PDU.
    - Response Class: for protocol operations that are sent in response to a previous request. This includes the Response PDU.
    - Notification Class: for protocol operations that send a notification to a Notification Receiver application. These include the Trapv2 PDU and the InformRequest PDU.
    - Internal Class: for protocol operations that are exchanged internally between SNMP engines. This includes the Report PDU.

To summarize our discussion, the SNMPv3 message definition from Section 6 of RFC 3412 is given below.

```
SNMPv3MessageSyntax DEFINITIONS IMPLICIT TAGS ::=
BEGIN
     SNMPv3Message ::= SEQUENCE {
          – identify the layout of the SNMPv3Message
          – this element is in same position as in
          SNMPv1
          – and SNMPv2c, allowing recognition
          msgVersion INTEGER { snmpv3 (3) },
          – administrative parameters
          msgGlobalData HeaderData,
          – security model-specific parameters
          – format defined by Security Model
          msgSecurityParameters OCTET STRING,
          msgData  ScopedPduData
     }
     HeaderData ::= SEQUENCE {
          msgID      INTEGER (0..2147483647),
          msgMaxSize INTEGER (484..2147483647),
          msgFlags   OCTET STRING (SIZE(1)),
               –  .... ...1   authFlag
               –  .... ..1.   privFlag
               –  .... .1..   reportableFlag
               –              Please observe:
               –  .... ..00   is OK, means
               noAuthNoPriv
               – .... ..01  is OK, means authNoPriv
               – .... ..10  reserved, must NOT be
               used.
               – .... ..11  is OK, means authPriv
          msgSecurityModel INTEGER (0..2147483647)
     }
ScopedPduData ::= CHOICE {
          plaintext    ScopedPDU,
          encryptedPDU OCTET STRING  – encrypted
          scopedPDU value
     }
```

```
    ScopedPDU ::= SEQUENCE {
        contextEngineID  OCTET STRING,
        contextName      OCTET STRING,
        data           ANY – e.g., PDUs as defined
        in RFC3416
    }
END
```

12.6.9 SNMPv3 MIB Modules

The RFC documents that define SNMPv3 contain a number of MIB modules to support various functions of the protocol. These include:

MIB Module	OID	RFC Reference
snmpFrameworkMIB	{1.3.6.1.6.3.10}	3411
snmpMPDMIB	{1.3.6.1.6.3.11}	3412
snmpTargetMIB	{1.3.6.1.6.3.12}	3413
snmpNotificationMIB	{1.3.6.1.6.3.13}	3413
snmpProxyMIB	{1.3.6.1.6.3.14}	3413
snmpUsmMIB	{1.3.6.1.6.3.15}	3414
snmpVacmMIB	{1.3.6.1.6.3.16}	3415
snmpCommunityMIB	{1.3.6.1.6.3.18}	3584
snmpv2TMMIB	{1.3.6.1.6.3.19}	3417

OID {1.3.6.1.6.3.17} was previously assigned for multicast address resolution, but it has been obsoleted and therefore does not appear in the above list.

The SNMP Framework MIB (snmpFrameworkMIB) module is defined in RFC 3411. This MIB module is identified as {snmpModules 10}, or {1.3.6.1.6.3.10}, and includes objects for identifying and determining the configuration of an SNMP engine (Figure 12-62). One example is the *snmpEngineID*, which defines an SNMP engine's administratively unique identifier.

The SNMP Message Processing and Dispatching MIB (snmpMPDMIB) module is defined in RFC 3412. This MIB module is identified as {snmpModules 11}, or {1.3.6.1.6.3.11}, and includes objects for monitoring the SNMP message processing and dispatching process (Figure 12-63). For example, objects are defined in this module to count the SNMP packets received but not processed as a result of various syntax or security errors.

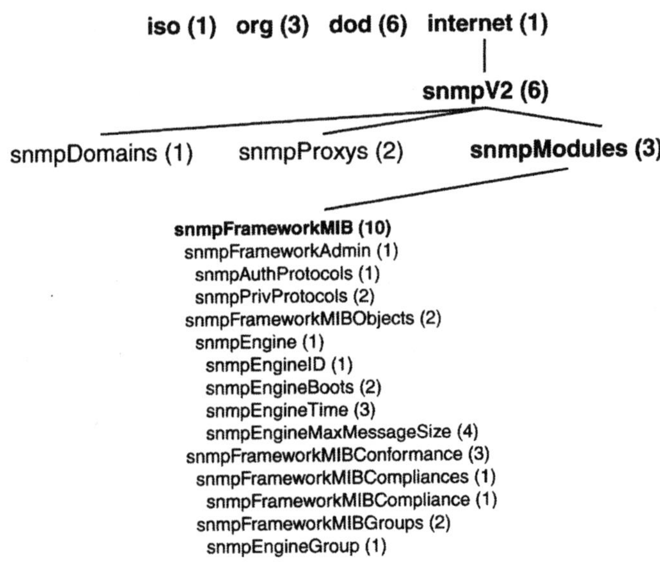

Figure 12-62. SNMP Framework MIB Module.

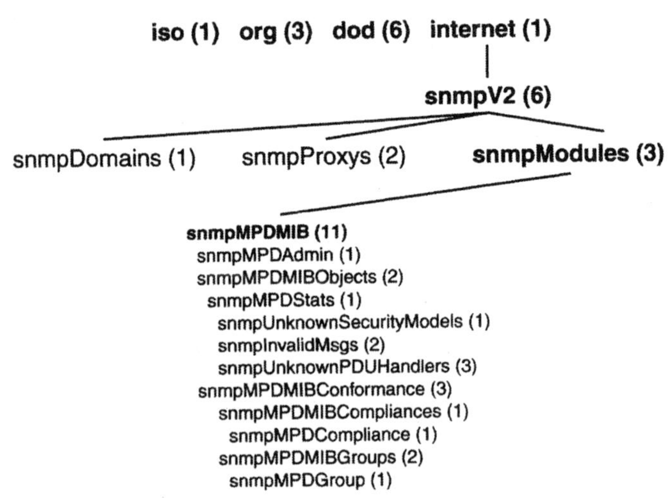

Figure 12-63. SNMP Message Processing and Dispatching MIB Module.

The SNMP Target MIB (snmpTargetMIB) module is defined in RFC 3413. This MIB module is identified as {snmpModules 12}, or {1.3.6.1.6.3.12}, and includes objects providing basic remote configuration of management targets (Figure 12-64). One example is the *snmpTargetAddrTable,* which defines a table of transport addresses to be used with message generation.

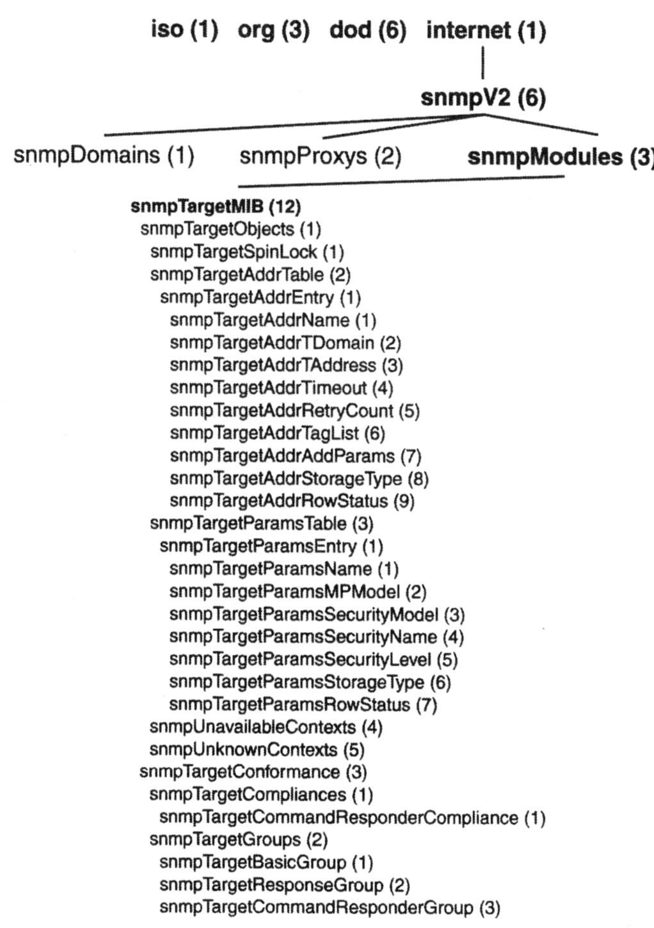

Figure 12-64. SNMP Target MIB Module.

The SNMP Notification MIB (snmpNotificationMIB) module is defined in RFC 3413. This MIB module is identified as {snmpModules 13}, or {1.3.6.1.6.3.13}, and includes objects for the remote configuration of parameters that are used in the process of generating notifications (Figure 12-65). One example is the *snmpNotifyTable*, which is used to select management targets that should receive notifications, as well as the type of notifications that these targets should receive.

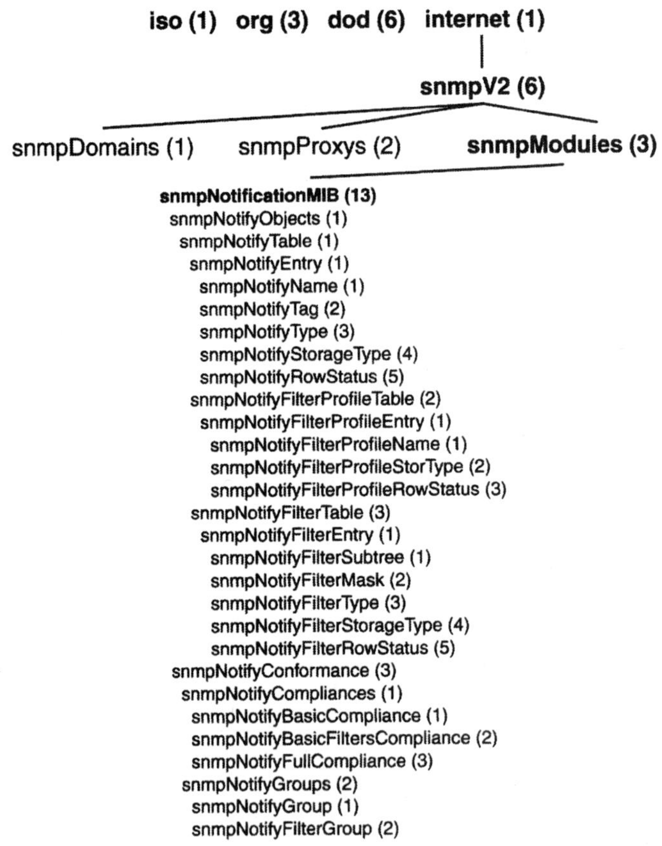

Figure 12-65. SNMP Notification MIB Module.

The SNMP Proxy MIB (snmpProxyMIB) module is defined in RFC 3413. This MIB module is identified as {snmpModules 14}, or {1.3.6.1.6.3.14}, and includes objects that provide mechanisms to remotely configure the parameters used in the process of proxy forwarding operations (Figure 12-66). One example is the *snmpProxyTable*, which contains translation parameters used by proxy forwarder applications for forwarding SNMP messages.

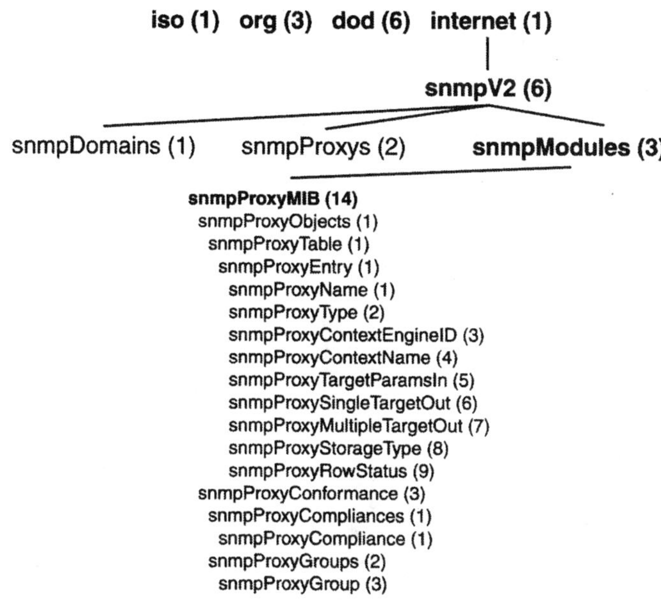

Figure 12-66. SNMP Proxy MIB Module.

The User-based Security Model MIB (snmpUsmMIB) module is defined in RFC 3414. This MIB module is identified as {snmpModules 15}, or {1.3.6.1.6.3.15}, and includes objects to configure an SNMP engine that implements the User-based Security Model (Figure 12-67). One example is usmUserAuthProtocol, which, in part, specifies the type of authentication protocol used.

The View-based Access Control Model MIB (snmpVacmMIB) module is defined in RFC 3415. This MIB module is identified as {snmpModules 16}, or {1.3.6.1.6.3.16}, and includes objects that provide for the remote configuration of an SNMP engine that implements the View-based Access Control Model (Figure 12-68). One example is *vacmAccessTable*, which is a table of access rights for groups.

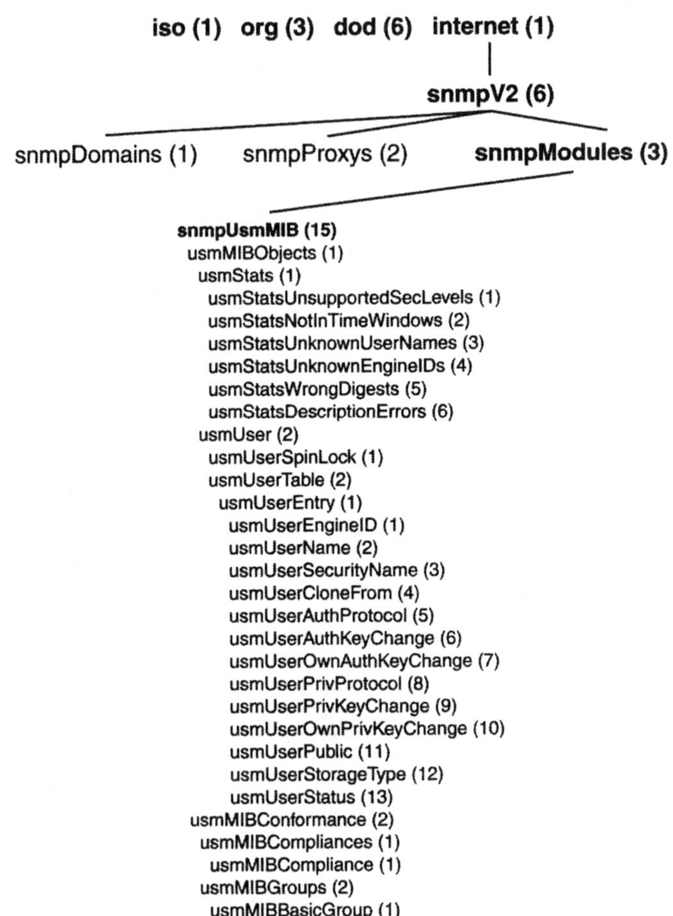

Figure 12-67. SNMP User-based Security Model MIB Module.

The SNMP Community MIB (snmpCommunityMIB) module is defined in RFC 3584 [12-42]. This MIB module is identified as {snmpModules 18}, or {1.3.6.1.6.3.18}, and includes objects for mapping between community strings and version-independent SNMP message parameters (Figure 12-69). In addition, the MIB has a mechanism to perform source address validation on incoming requests, and to select community strings based on target addresses for outgoing notifications. One example is *snmpCommunityTable*, which is a table of community strings that are configured in the SNMP engine's configuration parameters.

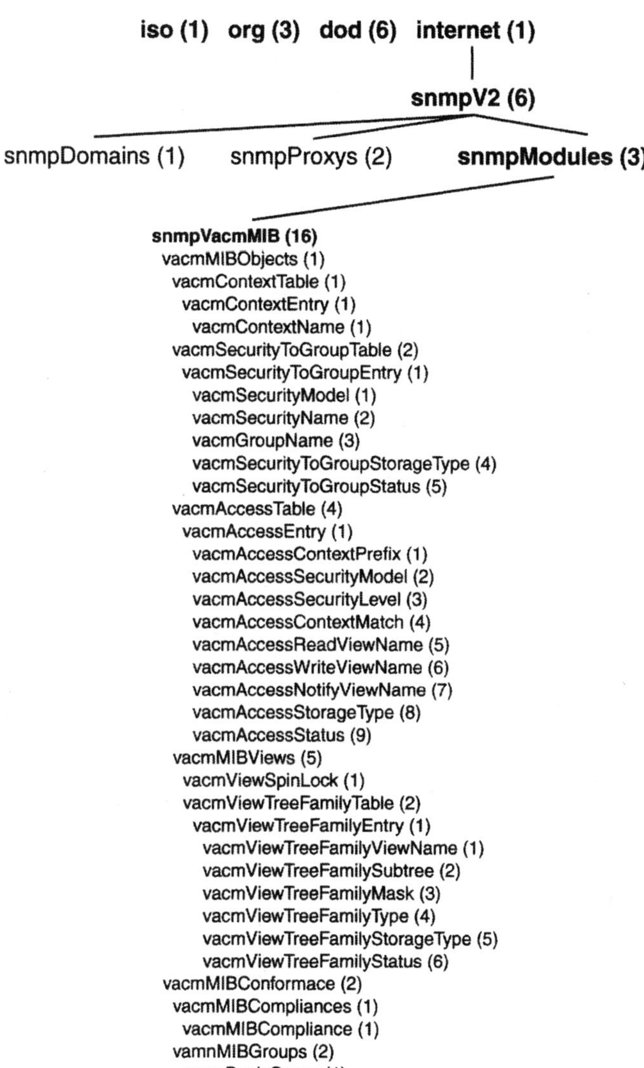
iso (1) org (3) dod (6) internet (1)
snmpV2 (6)
snmpDomains (1)
snmpProxys (2)
snmpModules (3)
snmpVacmMIB (16)
vacmMIBObjects (1)
vacmContextTable (1)
vacmContextEntry (1)
vacmContextName (1)
vacmSecurityToGroupTable (2)
vacmSecurityToGroupEntry (1)
vacmSecurityModel (1)
vacmSecurityName (2)
vacmGroupName (3)
vacmSecurityToGroupStorageType (4)
vacmSecurityToGroupStatus (5)
vacmAccessTable (4)
vacmAccessEntry (1)
vacmAccessContextPrefix (1)
vacmAccessSecurityModel (2)
vacmAccessSecurityLevel (3)
vacmAccessContextMatch (4)
vacmAccessReadViewName (5)
vacmAccessWriteViewName (6)
vacmAccessNotifyViewName (7)
vacmAccessStorageType (8)
vacmAccessStatus (9)
vacmMIBViews (5)
vacmViewSpinLock (1)
vacmViewTreeFamilyTable (2)
vacmViewTreeFamilyEntry (1)
vacmViewTreeFamilyViewName (1)
vacmViewTreeFamilySubtree (2)
vacmViewTreeFamilyMask (3)
vacmViewTreeFamilyType (4)
vacmViewTreeFamilyStorageType (5)
vacmViewTreeFamilyStatus (6)
vacmMIBConformace (2)
vacmMIBCompliances (1)
vacmMIBCompliance (1)
vamnMIBGroups (2)
vacmBasicGroup (1)

Figure 12-68. SNMP View-based Access Control Model MIB Module.

Figure 12-69. SNMP Community MIB Module.

12.6.10 Coexistence between SNMPv1, SNMPv2, and SNMPv3

With three versions of SNMP defined, developing ways for these different protocols to coexist is an essential element of the SNMP network management strategy. These issues are addressed in RFC 3584, titled "Coexistence between Version 1, Version 2 and Version 3 of the Internet-standard Network Management Framework" [12-42]. Four different aspects of this coexistence are described.

The conversion of MIB documents between SMIv1 (from SNMPv1) and SMIv2 (from SNMPv2) is required. SMIv2 is technically very close to a superset of SMIv1, with both versions based on ASN.1. Thus, SMIv1-based MIB modules may be used with protocol versions that support SMIv2 as long as certain syntactical changes are made. For example, an object defined with the Counter type from SMIv1 must be changed to use the Counter32 type, Gauge must be changed to Gauge32, and so on. Other syntactical changes may be required as well. In addition, changes in support of the compliance statements and capabilities statements may also be required.

Mapping of notification parameters from the SNMPv1 format to the SNMPv2 format is also required. For example, the first and second variable bindings in an SNMPv2-Trap or InformRequest PDU are the *sysUpTime* and *snmpTrapOID* parameters, respectively. Methods of translating these parameters, and others, must be defined.

Processing of protocol operations in a multilingual environment can take one of two forms: a multilingual implementation or a proxy implementation. For a multilingual implementation, the communicating elements must find an SNMP version that they both support. With the proxy implementation, a third-party network element provides a translation mechanism between the two (dissimilar) versions. For example, an SNMPv2 GetBulkRequest PDU must be translated into one (or more) GetNextRequest PDUs. A second example is the translation of error codes that may be required between the different versions.

Finally, the Message Processing Models for SNMPv1 and SNMPv2c vary, as do the Community-based Security Models for SNMPv1 and SNMPv2c. Differences in how these architectures handle incoming request processing, outgoing notification processing, and so on, must also be considered. All of these issues are considered in detail in RFC 3584.

A number of resources are available for those readers wanting to learn more about SNMPv3. A web page has been established as a focal point for SNMPv3 documentation and announcements regarding implementations [12-43], and an openly available newsletter, the *Simple Times*, chronicles SNMP development activities [12-44]. Finally, the website of one of the key software developers, SNMP Research International, Inc., provides information regarding interoperability testing and other current events [12-45].

12.7 Looking Ahead

This chapter has considered the technical intricacies of SNMPv1, SNMPv2, and SNMPv3. In the next chapter, we will consider case studies that illustrate the operation of the protocols.

12.8 References

[12-1] Rose, M. T., and K. McCloghrie. "Structure and Identification of Management Information for TCP/IP-based Internets." RFC 1155, May 1990.

[12-2] International Organization for Standardization. *Information Technology: Abstract Syntax Notation One (ASN.1): Specification of Basic Notation.* ISO/IEC 8824-1, 1998.

[12-3] International Organization for Standardization. *Information Technology: ASN.1 Encoding Rules: Specification of Basic Encoding Rules (BER), Canonical Encoding Rules (CER) and Distinguished Encoding Rules (DER).* ISO/IEC 8825-1, 1998.

[12-4] Steedman, Douglas. *Abstract Syntax Notation One (ASN.1), the Tutorial and Reference.* Isleworth, Middlesex, UK: Technology Appraisals, Ltd. ISBN 1-871802-06-7, 1990.

[12-5] Rose, M., and K. McCloghrie. "Concise MIB Definitions." RFC 1212, March 1991.

[12-6] McCloghrie, K., and M. T. Rose, Editors. "Management Information Base for Network Management of TCP/IP-based Internets: MIB-II." RFC 1213, March 1991.

[12-7] The Internet Assigned Numbers Authority (IANA) maintains an online database of SMI-related numbers at http://www.iana.org/assignments/smi-numbers.

[12-8] The Internet Assigned Numbers Authority (IANA) maintains an online database of private MIB assignment numbers at http://www.iana.org/assignments/enterprise-numbers.

[12-9] Baker, F. "IP Forwarding Table MIB." RFC 2096, January 1997.

[12-10] LaBarre, L. "OSI Internet Management: Management Information Base." RFC 1214, April 1991.

[12-11] Information regarding many vendor-developed private MIBs can be found at http://www.mibdepot.com.

[12-12] Waldbusser, S. "Remote Network Monitoring Management Information Base." RFC 2819, May 2000.

[12-13] Waldbusser, S. "Token Ring Extension to the Remote Network Monitoring MIB." RFC 1513, September 1993.

[12-14] Waldbusser, S. "Remote Network Monitoring Management Information Base version 2 using SMIv2." RFC 2021, January 1997.

[12-15] Case, J. D., M. Fedor, M. L. Schoffstall, and C. Davin. "Simple Network Management Protocol (SNMP)." RFC 1157, May 1990.

[12-16] Kastenholz, F., ed. "SNMP Communication Services." RFC 1270, October 1991.

[12-17] Davin, J., J. Galvin, and K. McCloghrie. "SNMP Administrative Model." RFC 1351, July 1992.

[12-18] Rose, M. T., Ed. "A Convention for Defining Traps for Use with the SNMP." RFC 1215, March 1991.

[12-19] Case, J. D., K. McCloghrie, M. T. Rose, and S. L. Waldbusser. "Introduction to version 2 of the Internet-standard Network Management Framework." RFC 1441, April 1993.

[12-20] Case, J., K. McCloghrie, M. Rose, and S. Waldbusser. "Introduction to Community-based SNMPv2." RFC 1901, January 1996.

[12-21] Case, J., K. McCloghrie, M. Rose, and S. Waldbusser. "Structure of Management Information for version 2 of the Simple Network Management Protocol (SNMPv2)." RFC 1902, January 1996.

[12-22] Case, J., K. McCloghrie, M. Rose, and S. Waldbusser. "Textual Conventions for version 2 of the Simple Network Management Protocol (SNMPv2)." RFC 1903, January 1996.

[12-23] Case, J., K. McCloghrie, M. Rose, and S. Waldbusser. "Conformance Statements for version 2 of the Simple Network Management Protocol (SNMPv2)." RFC 1904, January 1996.

[12-24] Case, J., K. McCloghrie, M. Rose, and S. Waldbusser. "Protocol Operations for Version 2 of the Simple Network Management Protocol (SNMPv2)." RFC 1905, January 1996.

[12-25] Case, J., K. McCloghrie, M. Rose, and S. Waldbusser. "Transport Mappings for version 2 of the Simple Network Management Protocol (SNMPv2)." RFC 1906, January 1996.

[12-26] Case, J., K. McCloghrie, M. Rose, and S. Waldbusser. "Management Information Base for version 2 of the Simple Network Management Protocol (SNMPv2)." RFC 1907, January 1996.

[12-27] Case, J. D., K. McCloghrie, M. T. Rose, and S. L. Waldbusser. "Coexistence between version 1 and version 2 of the Internet-standard Network Management Framework." RFC 1908, January 1996.

[12-28] McCloghrie, K., Editor. "An Administrated Infrastructure for SNMPv2." RFC 1909, February 1996.

[12-29] Waters, G., Editor. "User-based Security Model for SNMPv2." RFC 1910, February 1996.

[12-30] Case, J., et al. "Introduction and Applicability Statements for Internet-Standard Network Management Framework." RFC 3410, December 2002.

[12-31] Harrington, D., et al. "An Architecture for Describing Simple Network Management Protocol (SNMP) Management Frameworks." RFC 3411, December 2002.

[12-32] Case, J., et al. "Message Processing and Dispatching for the Simple Network Management Protocol (SNMP)." RFC 3412, December 2002.

[12-33] Levi, D., et al. "Simple Network Management Protocol (SNMP) Applications." RFC 3413, December 2002.

[12-34] Blumenthal, U., et al. "User-based Security Model (USM) for version 3 of the Simple Network Management Protocol (SNMPv3)." RFC 3414, December 2002.

[12-35] Wijnen, B., et al. "View-based Access Control Model (VACM) for the Simple Network Management Protocol (SNMP)." RFC 3415, December 2002.

[12-36] Presuhn, R. "Version 2 of the Protocol Operations for the Simple Network Management Protocol (SNMP)." RFC 3416, December 2002.

[12-37] Presuhn, R. "Transport Mappings for the Simple Network Management Protocol (SNMP)." RFC 3417, December 2002.

[12-38] Presuhn, R. "Management Information Base (MIB) for the Simple Network Management Protocol (SNMP)." RFC 3418, December 2002.

[12-39] McCloghrie, K., et al. "Structure of Management Information Version 2 (SMIv2)." RFC 2578, April 1999.

[12-40] McCloghrie, K., et al. "Textual Conventions for SMIv2." RFC 2579, April 1999.

[12-41] McCloghrie, K., et al. "Conformance Statements for SMIv2." RFC 2580, April 1999.

[12-42] Frye, R., et al. "Coexistence between Version 1, Version 2 and Version 3 of the Internet-standard Network Management Framework." RFC 3584, August 2003.

[12-43] The Computer Science Department of the Technical University of Braunschweig (Braunschweig, Germany) maintains an SNMPv3 home page that contains current information regarding SNMPv3 Working Group activities, documentation, vendor implementations, interoperability testing, and so on: http://www.ibr.cs.tu-bs.de/projects/snmpv3.

[12-44] The *Simple Times* is an openly available publication devoted to the promotion of the Simple Network Management Protocol (SNMP), which can be accessed at http://www.simple-times.org.

[12-45] SNMP Research International, Inc., provides information on SNMPv3 vendor implementations and interoperability testing at http://www.snmp.com.

13

Case Studies in Network and Performance Management

Chapter 11 considered some of the challenges of managing complex internetworks, and the advantages of a comprehensive network management architecture. Chapter 12 looked at the Internet Network Management Framework in general, plus the three versions of the Simple Network Management Protocol (SNMP) in particular. This chapter will illustrate topics from both of these areas by illustrating how the SNMP protocol addresses a number of network management challenges. We will concentrate our study on SNMPv1 and SNMPv3, as SNMPv2 was not widely implemented within vendors' products.

13.1 SNMPv1 PDUs

To illustrate the SNMP PDUs discussed in this chapter, this section presents four examples of SNMPv1 in use. The network analyzer captured each sample from an Ethernet backbone, which contained several other Ethernet segments connected by bridges and routers (see Figure 13-1). For these cases, the SNMP manager was a Sun workstation running SunNet Manager, and a Proteon (now OpenROUTE Networks) router contained the SNMP agent. In all of these examples, the traces are filtered to show only the SNMP protocol interaction.

13.1.1 SNMPv1 GetRequest Example

Recall from our earlier discussion that the GetRequest PDU retrieves one or more objects. Trace 13.1.1 illustrates how the UDP group (review Figure 12-26) does this.

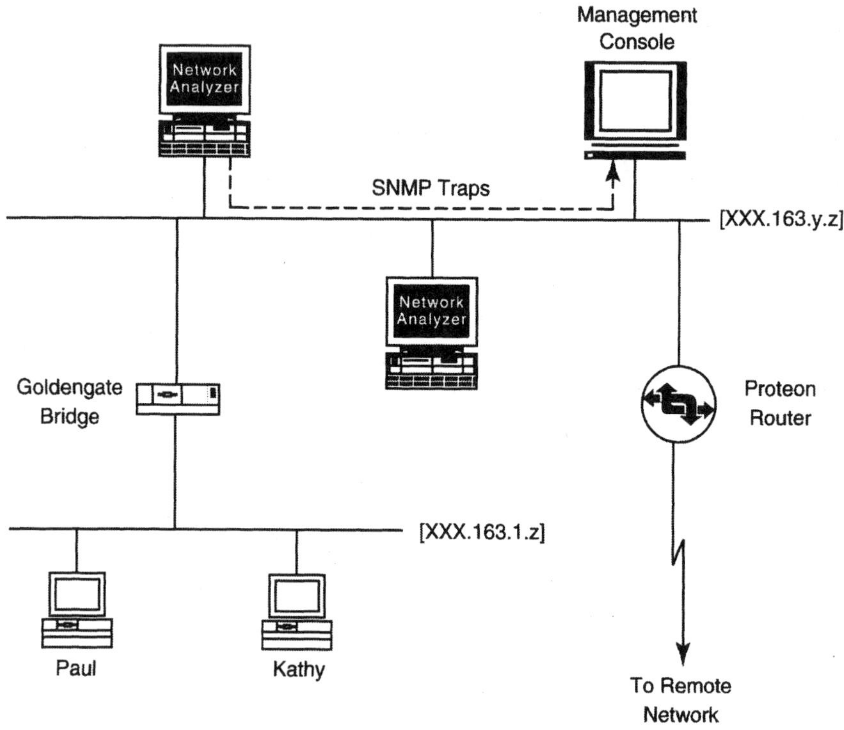

Figure 13-1. SNMPv1 Traps from a Network Analyzer.

Trace 13.1.1. Retrieving Scalar Data Using the GetRequest PDU: The UDP Group
Sniffer Network Analyzer data 10-Nov at 11:03:08, file UDP.ENC, Pg 1

```
- - - - - - - - - - - - - - - Frame 61 - - - - - - - - - - - - - - - -
SNMP:     — Simple Network Management Protocol —
SNMP:
SNMP:     Version = 0
SNMP:     Community = Brutus
SNMP:     Command = Get request
SNMP:     Request ID = 0
SNMP:     Error status = 0 (No error)
SNMP:     Error index = 0
SNMP:
SNMP:     Object = {1.3.6.1.2.1.1.3.0} (sysUpTime.0)
SNMP:     Value  = NULL
SNMP:
```

```
SNMP:        Object = {1.3.6.1.2.1.7.1.0} (udpInDatagrams.0)
SNMP:        Value  = NULL
SNMP:
SNMP:        Object = {1.3.6.1.2.1.7.2.0} (udpNoPorts.0)
SNMP:        Value  = NULL
SNMP:
SNMP:        Object = {1.3.6.1.2.1.7.3.0} (udpInErrors.0)
SNMP:        Value  = NULL
SNMP:
SNMP:        Object = {1.3.6.1.2.1.7.4.0} (udpOutDatagrams.0)
SNMP:        Value  = NULL
SNMP:

- - - - - - - - - - - - - - - Frame 62 - - - - - - - - - - - - - - - - -
SNMP:        — Simple Network Management Protocol —
SNMP:
SNMP:        Version = 0
SNMP:        Community = Brutus
SNMP:        Command = Get response
SNMP:        Request ID = 0
SNMP:        Error status = 0 (No error)
SNMP:        Error index = 0
SNMP:
SNMP:        Object = {1.3.6.1.2.1.1.3.0} (sysUpTime.0)
SNMP:        Value  = 263748621 hundredths of a second
SNMP:
SNMP:        Object = {1.3.6.1.2.1.7.1.0} (udpInDatagrams.0)
SNMP:        Value  = 573894 datagrams
SNMP:
SNMP:        Object = {1.3.6.1.2.1.7.2.0} (udpNoPorts.0)
SNMP:        Value  = 419103 datagrams
SNMP:
SNMP:        Object = {1.3.6.1.2.1.7.3.0} (udpInErrors.0)
SNMP:        Value  = 0 datagrams
SNMP:
SNMP:        Object = {1.3.6.1.2.1.7.4.0} (udpOutDatagrams.0)
SNMP:        Value  = 288892 datagrams
SNMP:
```

Trace 13.1.1 consists of two SNMP PDUs: the GetRequest (Frame 61) and the GetResponse (Frame 62). Both frames illustrate their respective PDU structures (as described in Figures 12-42 and 12-43): Version = 0, Community = Brutus, Command (PDU Type 0 or 2), Request ID = 0, Error Status = 0, and Error Index = 0.

Next, the VarBindList indicates the variables and associated values being requested or supplied. You can observe two things here. First, the SunNet Manager always asks for the value of the *sysUpTime* object before requesting other objects. (Other management consoles may construct the VarBindList in another fashion.) The *sysUpTime* provides a timestamp update for the Sun console, and is an input to the Sun graphical display of the network management statistics. Second, the values associated with the objects in the GetRequest have a Value = NULL. Recall that NULL is the ASN.1 type used as a placeholder in the data stream. When you look at the GetResponse in Frame 62, you'll see that each NULL value has been replaced with a measured value. For example, the number of UDP datagrams that have been delivered to UDP users, the object *udpInDatagrams* {1.3.6.1.2.1.7.1.0}, has a value of 573,894 datagrams. You can interpret the rest of the VarBindList in Frame 62 in a similar manner.

13.1.2 SNMPv1 GetNextRequest Example

The GetNextRequest PDU retrieves the lexicographical successor (or next) object, and is often used to retrieve objects from a table. Reviewing Figure 12-26, consider the UDP Table {1.3.6.1.2.7.5}. This table contains two columnar objects, *udpLocalAddress* and *udpLocalPort*. Together, these two entries associate a local IP address with a local port number, as follows:

udpTable

udpLocalAddress	udpLocalPort
0.0.0.0	1
a.b.c.d	2
.	.
.	.
.	.

Note that the address [0.0.0.0] indicates that a UDP listener is willing to accept datagrams for any interface on this node.

In Trace 13.1.2, the SunNet Manager wishes to retrieve all the values in the UDP Table. To do so, it issues the first GetNextRequest to specify OIDs that are lexicographically immediately after any *udpLocalAddress* objects and after *udpLocalPort* objects (see Frame 33). Frame 34 returns the contents of the first row, indicating a *udpLocalAddress* [0.0.0.0], *udpLocalPort* (69, the Trivial File Transfer Protocol (TFTP) port), and object {1.3.6.1.2.1.7.5.1.1.0.0.0.0.69}. [Reviewing Section 12.4.2.2, note that this table is indexed by the IP address (e.g. 0.0.0.0) and the local port (e.g. 69).]

The GetNextRequest in Frame 35 is identical to the one in Frame 33, except that it requests the object returned in Frame 34, that is, the next value after {1.3.6.1.2.1.7.5.1.1.0.0.0.0.69}. Frame 36 returns the response, which is the value of the second row: {1.3.6.1.2.1.7.5.1.1.0.0.0.0.161}, which identifies the SNMP port. Frame 37 continues the pattern, with the value of the third row identifying port 520 (a local routing port).

Note that Frame 38 reaches the end of the table, indicated by the GetResponse given in Frame 40, which moves to a different group within the OID tree. This final value is the lexicographical next item in the router's MIB, object *snmpInPkts*, the first object of the SNMP group. This means that the router did not have any item in the Transmission group {mib-2 10} in its MIB View. (These groups were in the router's MIB, however, so access to those groups using the same community name was not possible.)

The UDP Table constructed from the data in Trace 13.1.2 is as follows:

udpTable

udpLocalAddress	**udpLocalPort**
0.0.0.0	69 (TFTP)
0.0.0.0	161 (SNMP)
0.0.0.0	20 (Router)

Trace 13.1.2. Retrieving Tabular Data with the GetNextRequest PDU: The udpTable
Sniffer Network Analyzer data 10-Nov at 11:03:58, file UDT.ENC, Pg 1

```
- - - - - - - - - - - - - - - Frame 33 - - - - - - - - - - - - - - - -
SNMP:     — Simple Network Management Protocol —
SNMP:
SNMP:     Version = 0
SNMP:     Community = Brutus
SNMP:     Command = Get next request
SNMP:     Request ID = 0
SNMP:     Error status = 0 (No error)
SNMP:     Error index = 0
SNMP:
SNMP:     Object = {1.3.6.1.2.1.1.3} (sysUpTime)
SNMP:     Value  = NULL
SNMP:
SNMP:     Object = {1.3.6.1.2.1.7.5.1.1} (udpLocalAddress)
SNMP:     Value  = NULL
SNMP:
SNMP:     Object = {1.3.6.1.2.1.7.5.1.2} (udpLocalPort)
```

```
SNMP:        Value  = NULL
SNMP:

- - - - - - - - - - - - - - - Frame 34 - - - - - - - - - - - - - - - -
SNMP:        — Simple Network Management Protocol —
SNMP:
SNMP:        Version = 0
SNMP:        Community = Brutus
SNMP:        Command = Get response
SNMP:        Request ID = 0
SNMP:        Error status = 0 (No error)
SNMP:        Error index = 0
SNMP:
SNMP:        Object = {1.3.6.1.2.1.1.3.0} (sysUpTime.0)
SNMP:        Value  = 263753458 hundredths of a second
SNMP:
SNMP:        Object = {1.3.6.1.2.1.7.5.1.1.0.0.0.0.69} (udpLocalAddress.0.0.0.0.69)
SNMP:        Value  = [0.0.0.0]
SNMP:
SNMP:        Object = {1.3.6.1.2.1.7.5.1.2.0.0.0.0.69} (udpLocalPort.0.0.0.0.69)
SNMP:        Value  = 69
SNMP:

- - - - - - - - - - - - - - - Frame 35 - - - - - - - - - - - - - - - -
SNMP:        — Simple Network Management Protocol —
SNMP:
SNMP:        Version = 0
SNMP:        Community = Brutus
SNMP:        Command = Get next request
SNMP:        Request ID = 0
SNMP:        Error status = 0 (No error)
SNMP:        Error index = 0
SNMP:
SNMP:        Object = {1.3.6.1.2.1.1.3} (sysUpTime)
SNMP:        Value  = NULL
SNMP:
SNMP:        Object = {1.3.6.1.2.1.7.5.1.1.0.0.0.0.69} (udpLocalAddress.0.0.0.0.69)
SNMP:        Value  = NULL
SNMP:
SNMP:        Object = {1.3.6.1.2.1.7.5.1.2.0.0.0.0.69} (udpLocalPort.0.0.0.0.69)
SNMP:        Value  = NULL
SNMP:
```

```
- - - - - - - - - - - - - - - Frame 36 - - - - - - - - - - - - - - - - -
SNMP:     — Simple Network Management Protocol —
SNMP:
SNMP:     Version = 0
SNMP:     Community = Brutus
SNMP:     Command = Get response
SNMP:     Request ID = 0
SNMP:     Error status = 0 (No error)
SNMP:     Error index = 0
SNMP:
SNMP:     Object  = {1.3.6.1.2.1.1.3.0} (sysUpTime.0)
SNMP:     Value  = 263753461 hundredths of a second
SNMP:
SNMP:     Object = {1.3.6.1.2.1.7.5.1.1.0.0.0.0.161} (udpLocalAddress.0.0.0.0.161)
SNMP:     Value  = [0.0.0.0]
SNMP:
SNMP:     Object = {1.3.6.1.2.1.7.5.1.2.0.0.0.0.161}(udpLocalPort.0.0.0.0.161)
SNMP:     Value  = 161
SNMP:

- - - - - - - - - - - - - - - Frame 37 - - - - - - - - - - - - - - - - -
SNMP:     — Simple Network Management Protocol —
SNMP:
SNMP:     Version = 0
SNMP:     Community = Brutus
SNMP:     Command = Get next request
SNMP:     Request ID = 0
SNMP:     Error status = 0 (No error)
SNMP:     Error index = 0
SNMP:
SNMP:     Object = {1.3.6.1.2.1.1.3} (sysUpTime)
SNMP:     Value  = NULL
SNMP:
SNMP:     Object ={1.3.6.1.2.1.7.5.1.1.0.0.0.0.161} (udpLocalAddress.0.0.0.0.161)
SNMP:     Value  = NULL
SNMP:
```

```
SNMP:      Object = {1.3.6.1.2.1.7.5.1.2.0.0.0.0.161} (udpLocalPort.0.0.0.0.161)
SNMP:      Value  = NULL
SNMP:

--------------- Frame 38 ----------------
SNMP:      — Simple Network Management Protocol —
SNMP:
SNMP:      Version = 0
SNMP:      Community = Brutus
SNMP:      Command = Get response
SNMP:      Request ID = 0
SNMP:      Error status = 0 (No error)
SNMP:      Error index = 0
SNMP:
SNMP:      Object = {1.3.6.1.2.1.1.3.0} (sysUpTime.0)
SNMP:      Value  = 263753463 hundredths of a second
SNMP:
SNMP:      Object = {1.3.6.1.2.1.7.5.1.1.0.0.0.0.520}  (udpLocalAddress.0.0.0.0.520)
SNMP:      Value  = [0.0.0.0]
SNMP:
SNMP:      Object = {1.3.6.1.2.1.7.5.1.2.0.0.0.0.520}  (udpLocalPort.0.0.0.0.520)
SNMP:      Value  = 520
SNMP:

--------------- Frame 39 ----------------
SNMP:      — Simple Network Management Protocol —
SNMP:
SNMP:      Version = 0
SNMP:      Community = Brutus
SNMP:      Command = Get next request
SNMP:      Request ID = 0
SNMP:      Error status = 0 (No error)
SNMP:      Error index = 0
SNMP:
SNMP:      Object = {1.3.6.1.2.1.1.3} (sysUpTime)
SNMP:      Value  = NULL
SNMP:
SNMP:      Object = {1.3.6.1.2.1.7.5.1.1.0.0.0.0.520}  (udpLocalAddress.0.0.0.0.520)
SNMP:      Value  = NULL
SNMP:
SNMP:      Object = {1.3.6.1.2.1.7.5.1.2.0.0.0.0.520}  (udpLocalPort.0.0.0.0.520)
SNMP:      Value  = NULL
SNMP:
```

```
--------------- Frame 40 -----------------
SNMP:     — Simple Network Management Protocol —
SNMP:
SNMP:     Version = 0
SNMP:     Community = Brutus
SNMP:     Command = Get response
SNMP:     Request ID = 0
SNMP:     Error status = 0 (No error)
SNMP:     Error index = 0
SNMP:
SNMP:     Object = {1.3.6.1.2.1.1.3.0} (sysUpTime.0)
SNMP:     Value  = 263753466 hundredths of a second
SNMP:
SNMP:     Object = {1.3.6.1.2.1.7.5.1.2.0.0.0.0.69} (udpLocalPort.0.0.0.0.69)
SNMP:     Value  = 69
SNMP:
SNMP:     Object = {1.3.6.1.2.1.11.1.0} (snmpInPkts.0)
SNMP:     Value  = 116744 (counter)
SNMP:
```

13.1.3 SNMPv1 SetRequest Example

This example issues a SetRequest PDU for an object on the Proteon router, then issues a GetRequest for the same object (see Trace 13.1.3) to double check that the action was properly completed. Frames 1 and 2 retrieve the current value of the object *ipDefaultTTL*, Frames 3 and 4 set a new value for that object, Frames 5 and 6 verify the new value, Frames 7 and 8 set the value back to the original, and, finally, Frames 9 and 10 verify the previous operation.

Looking at the details, notice that in Frame 1 the Value = NULL, as in the previous example. The GetResponse PDU (Frame 2) contains the requested value (60) of *ipDefaultTTL* (the default value of the Time-to-Live field within the IP header). Frame 3 contains a SetRequest PDU, assigning Value = 64 to *ipDefaultTTL*. The router sends a confirming GetResponse PDU in Frame 4. Frame 5 issues a GetRequest PDU to verify that the SetRequest changed the value of *ipDefaultTTL* to 64 (Frame 6). Frame 7 issues a second SetRequest, this time with Value = 60, which is acknowledged in Frame 8. Frames 9 and 10 confirm that the operation was successful.

Trace 13.1.3. SNNP Set ipDefaultTTL Details

```
Sniffer Network Analyzer data 11-Dec at 15:16:52 file SETIPTTL.ENC Pg 1

- - - - - - - - - - - - - - - Frame 1 - - - - - - - - - - - - - - - - -
SNMP:     — Simple Network Management Protocol —
SNMP:
SNMP:     Version = 0
SNMP:     Community = Brutus
SNMP:     Command = Get request
SNMP:     Request ID = 0
SNMP:     Error status = 0 (No error)
SNMP:     Error index = 0
SNMP:
SNMP:     Object = {1.3.6.1.2.1.1.3.0} (sysUpTime.0)
SNMP:     Value  = NULL
SNMP:
SNMP:     Object = {1.3.6.1.2.1.4.1.0} (ipForwarding.0)
SNMP:     Value  = NULL
SNMP:
SNMP:     Object = {1.3.6.1.2.1.4.2.0} (ipDefaultTTL.0)
SNMP:     Value  = NULL
SNMP:

- - - - - - - - - - - - - - - Frame 2 - - - - - - - - - - - - - - - - -
SNMP:     — Simple Network Management Protocol —
SNMP:
SNMP:     Version = 0
SNMP:     Community = Brutus
SNMP:     Command = Get response
SNMP:     Request ID = 0
SNMP:     Error status = 0 (No error)
SNMP:     Error index = 0
SNMP:
SNMP:     Object = {1.3.6.1.2.1.1.3.0} (sysUpTime.0)
SNMP:     Value  = 16862273 hundredths of a second
SNMP:
SNMP:     Object = {1.3.6.1.2.1.4.1.0} (ipForwarding.0)
SNMP:     Value  = 1 (gateway)
SNMP:
SNMP:     Object = {1.3.6.1.2.1.4.2.0} (ipDefaultTTL.0)
SNMP:     Value  = 60
SNMP:
```

```
--------------- Frame 3 ----------------
SNMP:     — Simple Network Management Protocol —
SNMP:
SNMP:     Version = 0
SNMP:     Community = Brutus
SNMP:     Command = Set request
SNMP:     Request ID = 0
SNMP:     Error status = 0 (No error)
SNMP:     Error index = 0
SNMP:
SNMP:     Object = {1.3.6.1.2.1.4.2.0} (ipDefaultTTL.0)
SNMP:     Value  = 64
SNMP:

--------------- Frame 4 ----------------
SNMP:     — Simple Network Management Protocol —
SNMP:
SNMP:     Version = 0
SNMP:     Community = Brutus
SNMP:     Command = Get response
SNMP:     Request ID = 0
SNMP:     Error status = 0 (No error)
SNMP:     Error index = 0
SNMP:
SNMP:     Object = {1.3.6.1.2.1.4.2.0} (ipDefaultTTL.0)
SNMP:     Value  = 64
SNMP:

--------------- Frame 5 ----------------
SNMP:     — Simple Network Management Protocol —
SNMP:
SNMP:     Version = 0
SNMP:     Community = Brutus
SNMP:     Command = Get request
SNMP:     Request ID = 0
SNMP:     Error status = 0 (No error)
SNMP:     Error index = 0
SNMP:
SNMP:     Object = {1.3.6.1.2.1.1.3.0} (sysUpTime.0)
SNMP:     Value  = NULL
SNMP:
SNMP:     Object = {1.3.6.1.2.1.4.1.0} (ipForwarding.0)
SNMP:     Value  = NULL
SNMP:
```

```
SNMP:    Object = {1.3.6.1.2.1.4.2.0} (ipDefaultTTL.0)
SNMP:    Value  = NULL
SNMP:

--------------- Frame 6 -----------------
SNMP:    — Simple Network Management Protocol —
SNMP:
SNMP:    Version = 0
SNMP:    Community = Brutus
SNMP:    Command = Get response
SNMP:    Request ID = 0
SNMP:    Error status = 0 (No error)
SNMP:    Error index = 0
SNMP:
SNMP:    Object = {1.3.6.1.2.1.1.3.0} (sysUpTime.0)
SNMP:    Value  = 16863228 hundredths of a second
SNMP:
SNMP:    Object = {1.3.6.1.2.1.4.1.0} (ipForwarding.0)
SNMP:    Value  = 1 (gateway)
SNMP:
SNMP:    Object = {1.3.6.1.2.1.4.2.0} (ipDefaultTTL.0)
SNMP:    Value  = 64
SNMP:

--------------- Frame 7 -----------------
SNMP:    — Simple Network Management Protocol —
SNMP:
SNMP:    Version = 0
SNMP:    Community = Brutus
SNMP:    Command = Set request
SNMP:    Request ID = 0
SNMP:    Error status = 0 (No error)
SNMP:    Error index = 0
SNMP:
SNMP:    Object = {1.3.6.1.2.1.4.2.0} (ipDefaultTTL.0)
SNMP:    Value  = 60
SNMP:

--------------- Frame 8 -----------------
SNMP:    — Simple Network Management Protocol —
```

```
SNMP:
SNMP:        Version = 0
SNMP:        Community = Brutus
SNMP:        Command = Get response
SNMP:        Request ID = 0
SNMP:        Error status = 0 (No error)
SNMP:        Error index = 0
SNMP:
SNMP:        Object = {1.3.6.1.2.1.4.2.0} (ipDefaultTTL.0)
SNMP:        Value  = 60
SNMP:

--------------- Frame 9 ----------------
SNMP:        — Simple Network Management Protocol —
SNMP:
SNMP:        Version = 0
SNMP:        Community = Brutus
SNMP:        Command = Get request
SNMP:        Request ID = 0
SNMP:        Error status = 0 (No error)
SNMP:        Error index = 0
SNMP:
SNMP:        Object = {1.3.6.1.2.1.1.3.0} (sysUpTime.0)
SNMP:        Value  = NULL
SNMP:
SNMP:        Object = {1.3.6.1.2.1.4.1.0} (ipForwarding.0)
SNMP:        Value  = NULL
SNMP:
SNMP:        Object = {1.3.6.1.2.1.4.2.0} (ipDefaultTTL.0)
SNMP:        Value  = NULL
SNMP:

--------------- Frame 10 ----------------
SNMP:        — Simple Network Management Protocol —
SNMP:
SNMP:        Version = 0
SNMP:        Community = Brutus
SNMP:        Command = Get response
SNMP:        Request ID = 0
SNMP:        Error status = 0 (No error)
SNMP:        Error index = 0
SNMP:
SNMP:        Object = {1.3.6.1.2.1.1.3.0} (sysUpTime.0)
```

```
SNMP:       Value  = 16863846 hundredths of a second
SNMP:
SNMP:       Object = {1.3.6.1.2.1.4.1.0} (ipForwarding.0)
SNMP:       Value  = 1 (gateway)
SNMP:
SNMP:       Object = {1.3.6.1.2.1.4.2.0} (ipDefaultTTL.0)
SNMP:       Value  = 60
SNMP:
```

13.1.4 SNMPv1 Trap Example

The final example in this section shows how a Trap PDU indicates an alarm condition to the network manager. In this case, the agent generating the trap is a Network Associates, Inc. (formerly Network General Corp.) *Sniffer*® protocol analyzer (review Figure 13-1).

One set of network statistics is network utilization. Network utilization is a ratio between the total number of bits transmitted in a period of time (in this case five seconds) divided by the total number of bits that could theoretically be transmitted during the same period. A typical shared Ethernet segment in a network would have a network utilization in the 5 to 20 percent range. For this example, we set the threshold to the unrealistically low value of 1 percent over a five-second period. When the network reaches that threshold, the *Sniffer* generates a Trap PDU and sends it to the SunNet Manager. Another *Sniffer* analyzer captured the results.

This transmission follows the Trap PDU structure shown in Figure 12-46. The SNMP authentication header contains the version number and community string, and the PDU Type specifies a Trap (PDU Type = 4). The Enterprise field gives the OID for the authority that defined the trap. The prefix {1.3.6.1.4.1} identifies the Private Enterprises subtree, and the 110 identifies Network General Corp. (review Reference [12-8]). The Generic Trap field indicates an enterprise-specific trap (Trap = 6). This means that the value of the Enterprise field indicates the authority (Network General) that defined this trap.

The Specific Trap field has Type = 7, which Network General defined. The Variable Bindings also contain variables and values that Network General defined. The third object's numeric value was defined by Network General to mean "Abs usage exceeded 1 percent."

Trace 13.1.4. An Enterprise-specific Trap: Network Utilization Exceeded 1 Percent during a Five-second Period

Sniffer Network Analyzer data 11-Dec at 16:13:26 file SNIFTRAP.ENC Pg 1

```
- - - - - - - - - - - - - - - Frame 1 - - - - - - - - - - - - - - - - -
SNMP:   — Simple Network Management Protocol —
SNMP:
SNMP:   Version = 0
SNMP:   Community = public
SNMP:   Command = Trap
SNMP:   Enterprise = {1.3.6.1.4.1.110.1.1.1.0}
SNMP:   Network address = [132.163.128.102]
SNMP:   Generic trap = 6 (Enterprise specific)
SNMP:   Specific trap = 7
SNMP:   Time ticks = 244894900
SNMP:
SNMP:   Object = {1.3.6.1.4.1.110.1.1.1.1.1.1.1} (Network General Corp. 1.1.1.1.1.1.1)
SNMP:   Value  = 53 (counter)
SNMP:
SNMP:   Object = {1.3.6.1.4.1.110.1.1.1.1.1.2.1} (Network General Corp.1.1.1.1.1.2.1)
SNMP:   Value  = 1
SNMP:
SNMP:   Object = {1.3.6.1.4.1.110.1.1.1.1.1.3.1} (Network General Corp.1.1.1.1.1.3.1)
SNMP:   Value  = Abs usage exceeded 1%
SNMP:
SNMP:   Object = {1.3.6.1.4.1.110.1.1.1.1.1.4.1} (Network General Corp.1.1.1.1.1.4.1)
SNMP:   Value  = 5
SNMP:
SNMP:   Object = {1.3.6.1.4.1.110.1.1.1.1.1.5.1} (Network General Corp.1.1.1.1.1.5.1)
SNMP:   Value  = 0
SNMP:
SNMP:   Object = {1.3.6.1.4.1.110.1.1.1.1.1.6.1} (Network General Corp.1.1.1.1.1.6.1)
SNMP:   Value  = 7
SNMP:
SNMP:   Object = {1.3.6.1.4.1.110.1.1.1.1.1.7.1} (Network General Corp.1.1.1.1.1.7.1)
SNMP:   Value  = 724119640 (counter)
SNMP:
SNMP:   Object = {1.3.6.1.4.1.110.1.1.1.1.1.8.1}  (Network General Corp .1.1.1.1.1.8.1)
SNMP:   Value  = Global Network
SNMP:
```

13.2 Accessing a MIB

This section gives an example of an SNMP management console retrieving values for MIB objects from a remote SNMP agent. In this case, the manager is a Sun Microsystems' SunNet Manager, and the agent is located in a Proteon (now OpenROUTE Networks) router. Both devices connect to an Ethernet backbone.

This exchange between the manager and the agent (see Trace 13.2a) involves two frames of information. Frame 109 contains an SNMP GetRequest PDU (protocol data unit, the core of the SNMP message) and Frame 110 contains a GetResponse PDU.

The manager sends the GetRequest to the agent asking for the values of the objects within the system subtree, OID {1.3.6.1.2.1.1}. The PDU requests information about all seven of the objects: *sysDescr, sysObjectID, sysUpTime, sysContact, sysName, sysLocation,* and *sysServices*. On the trace, you can see two coding elements for each of these objects. First, the manager requests the sysUpTime object to determine whether the agent within the router has restarted (warm or cold boot). Second, the manager asks for the values of each individual object in order (review Figure 12-20). This trace also illustrates the use of the SEQUENCE type encoding of VarBinds discussed in Section 12.1.4.3.5. Each object is encoded with an OBJECT IDENTIFIER type, for example {1.3.6.1.2.1.1.2.0}. The Object Value field is encoded with a NULL type because the manager does not know this information.

Frame 110 gives the agent's GetResponse. The response returns each object and its associated value in the order that Frame 109 requested. The *sysDescr* provides a textual description of the device (Portable I80386 C Gateway ...). The *sysObjectID* has a value of {1.3.6.1.4.1.1.1.1.41}. From the prefix {1.3.6.1.4.1}, you know that this is a private enterprise subtree. The next digit (.1) is the enterprise code for Proteon, Inc. (see Reference [12-8]).

The *sysUpTime* object has a value of 263,621,778 hundredths of a second, which translates to roughly 30 days since the router's network management system was restarted. Two of the objects, *sysContact* {system 4} and *sysLocation* {system 6}, appear not to have a value. In reality, they have a value of a zero-length string, but the network manager entered no values for those objects in the router's configuration file. The *sysName* is the domain name of the node (boulder.org). Finally, the *sysServices* {system 7} is a calculated sum that indicates the services this node performs. In this case, the value is 72, indicating a host offering application services (see RFC 1213, page 15).

See Trace 13.2b for a quick review of the ASN.1-encoding discussed in Chapter 12. This data shows the details of Frame 110, but with the Sniffer's ASN.1 decoding option activated. You can trace each ASN.1 element, identifying the Type-Length-Value encodings as well as the hexadecimal display of those values. As a reference point, the first SNMP encoding (SEQUENCE [of], Length = 235) appears in bold type with the characters 30 81 EB (see the third line of the hexadecimal decode, that begins with Address = 0020). Reviewing Chapter 12, we know that the Type field = 30H (the SEQUENCE OF type, see Figure 12-10). The Length field is the Long Definite form (see Figure 12-5), with one subsequent octet having a value of 81 EBH. (Hexadecimal values of X are dummy characters to maintain the confidentiality of the trace.)

Trace 13.2a. Browsing the System Subtree (SNMP Protocol Decode)
Sniffer Network Analyzer data 10-Nov at 10:42:04 file ASAN_SYS.ENC Pg 1

```
- - - - - - - - - - - - - - - Frame 109 - - - - - - - - - - - - - - - - -
SNMP:    — Simple Network Management Protocol —
SNMP:
SNMP:    Version = 0
SNMP:    Community = boulder
SNMP:    Command = Get request
SNMP:    Request ID = 0
SNMP:    Error status = 0 (No error)
SNMP:    Error index = 0
SNMP:
SNMP:    Object = {1.3.6.1.2.1.1.3.0} (sysUpTime.0)
SNMP:    Value = NULL
SNMP:
SNMP:    Object = {1.3.6.1.2.1.1.1.0} (sysDescr.0)
SNMP:    Value = NULL
SNMP:
SNMP:    Object = {1.3.6.1.2.1.1.2.0} (sysObjectID.0)
SNMP:    Value = NULL
SNMP:
SNMP:    Object = {1.3.6.1.2.1.1.3.0} (sysUpTime.0)
SNMP:    Value = NULL
SNMP:
SNMP:    Object = {1.3.6.1.2.1.1.4.0} (system.4.0)
```

```
SNMP:        Value = NULL
SNMP:
SNMP:        Object = {1.3.6.1.2.1.1.5.0} (system.5.0)
SNMP:        Value = NULL
SNMP:
SNMP:        Object = {1.3.6.1.2.1.1.6.0} (system 6.0)
SNMP:        Value = NULL
SNMP:
SNMP:        Object = {1.3.6.1.2.1.1.7.0} (system 7.0)
SNMP:        Value = NULL
SNMP:

- - - - - - - - - - - - - - - Frame 110 - - - - - - - - - - - - - - - -
SNMP:        — Simple Network Management Protocol —
SNMP:
SNMP:        Version = 0
SNMP:        Community = boulder
SNMP:        Command = Get response
SNMP:        Request ID = 0
SNMP:        Error status = 0 (No error)
SNMP:        Error index = 0
SNMP:
SNMP:        Object = {1.3.6.1.2.1.1.3.0] (sysUpTime.0)
SNMP:        Value = 263621778 hundredths of a second
SNMP:
SNMP:        Object = {1.3.6.1.2.1.1.1.0} (sysDescr.0)
SNMP:        Value = Portable I80386 C Gateway BOULDER.ORG S/N XXX V12.0
SNMP:
SNMP:        Object = {1.3.6.1.2.1.1.2.0} (sysObjectID.0)
SNMP:        Value = {1.3.6.1.4.1.1.1.1.41}
SNMP:
SNMP:        Object = {1.3.6.1.2.1.1.3.0} (sysUpTime.0)
SNMP:        Value = 263621778 hundredths of a second
SNMP:
SNMP:        Object = {1.3.6.1.2.1.1.4.0} (system.4.0)
SNMP:        Value =
SNMP:
SNMP:        Object = {1.3.6.1.2.1.1.5.0} (system.5.0)
SNMP:        Value = BOULDER.ORG
SNMP:
```

```
SNMP:     Object = {1.3.6.1.2.1.1.6.0} (system.6.0)
SNMP:     Value =
SNMP:
SNMP:     Object = {1.3.6.1.2.1.1.7.0} (system.7.0)
SNMP:     Value = 72
SNMP:
```

Trace 13.2b. Browsing the System Subtree (ASN.1 and Hexadecimal Decode)
Sniffer Network Analyzer data 10-Nov at 10:42:04 file ASAN_SYS.ENC Pg 1

```
- - - - - - - - - - - - - - - Frame 110 - - - - - - - - - - - - - - - -
SNMP:     1.1 SEQUENCE [of], Length=235
SNMP:     2.1 INTEGER, Length=1, Value = "0"
SNMP:     2.2 OCTET STRING, Length=7, Value = "boulder"
SNMP:     2.3 Context-Specific Constructed [2], Length=220
SNMP:     3.1 INTEGER, Length=1, Value = "0"
SNMP:     3.2 INTEGER, Length=1, Value = "0"
SNMP:     3.3 INTEGER, Length=1, Value =  "0"
SNMP:     3.4 SEQUENCE [of], Length=208
SNMP:     4.1 SEQUENCE [of], Length=16
SNMP:     5.1 OBJECT IDENTIFIER, Length=8, Value = "{1.3.6.1.2.1.1.3.0}"
SNMP:     5.2 Application Primitive [3], Length=4, Data = "<0FB68C92>"
SNMP:     4.2 SEQUENCE [of], Length=74
SNMP:     5.1 OBJECT IDENTIFIER, Length=8, Value = "{1.3.6.1.2.1.1.1.0}"
SNMP:     5.2 OCTET STRING, Length=62, Value = "Portable I80386 C Gateway XXX. ..."
SNMP:     4.3 SEQUENCE [of], Length=21
SNMP:     5.1 OBJECT IDENTIFIER, Length=8,Value="{1.3.6.1.2.1.1.2.0}"
SNMP:     5.2 OBJECT IDENTIFIER, Length=9,Value="{1.3.6.1.4.1.1.1.1.41}"
SNMP:     4.4 SEQUENCE [of], Length=16
SNMP:     5.1 OBJECT IDENTIFIER, Length=8,Value="{1.3.6.1.2.1.1.3.0}"
SNMP:     5.2 Application Primitive [3], Length=4, Data = "<0FB68C92>"
SNMP:     4.5 SEQUENCE [of], Length=12
SNMP:     5.1 OBJECT IDENTIFIER, Length=8, Value="{1.3.6.1.2.1.1.4.0}"
SNMP:     5.2 OCTET STRING, Length=0, Value = ""
SNMP:     4.6 SEQUENCE [of], Length=28
SNMP:     5.1 OBJECT IDENTIFIER, Length=8, Value="{1.3.6.1.2.1.1.5.0}"
SNMP:     5.2 OCTET STRING, Length=16, Value="XXX.XXX.XXXX.XXX"
SNMP:     4.7 SEQUENCE [of], Length=12
```

```
SNMP:       5.1 OBJECT IDENTIFIER, Length=8, Value="{1.3.6.1.2.1.1.6.0}"
SNMP:       5.2 OCTET STRING, Length=0, Value = ""
SNMP:       4.8 SEQUENCE [of], Length=13
SNMP:       5.1 OBJECT IDENTIFIER, Length=8, Value = "{1.3.6.1.2.1.1.7.0}"
SNMP:       5.2 INTEGER, Length=1, Value = "72"
SNMP:
ADDR  HEX                                              ASCII
0000  08 00 20 09 00 C8 AA 00 04 00 44 86 08 00 45 00  .. .......D...E.
0010  01 0A 81 20 00 00 3B 11 73 77 84 A3 01 01 84 A3  .....;.sw......
0020  80 04 00 A1 0D 20 00 F6 C6 62 30 81 EB 02 01 00  ..... ...b0.....
0030  04 07 XX XX XX XX XX XX XX A2 81 DC 02 01 00 02  ..XXXXXXX........
0040  01 00 02 01 00 30 81 D0 30 10 06 08 2B 06 01     .....0..0...+...
0050  01 01 03 00 43 04 0F B6 8C 92 30 4A 06 08 2B 06  ...C.....0J..+.
0060  01 02 01 01 01 00 04 3E 50 6F 72 74 61 62 6C 65  .......>Portable
0070  20 49 38 30 33 38 36 20 43 20 47 61 74 65 77 61  I80386 C Gateway
0080  79 20 XX XX XX XX XX XX XX XX XX XX XX XX XX XX  XXX.XXX.XXXX.X
0090  XX XX 20 53 2F 4E 20 33 33 33 20 56 31 32 2E 30  XX S/N 333 V12.0
00A0  20 20 5B 20 20 5D 30 15 06 08 2B 06 01 02 01 01  [ ]0...+.....
00B0  02 00 06 09 2B 06 01 04 01 01 01 01 29 30 10 06  ....+.......)0..
00C0  08 2B 06 01 02 01 01 03 00 43 04 0F B6 8C 92 30  .+.......C.....0
00D0  0C 06 08 2B 06 01 02 01 01 04 00 04 00 30 1C 06  ...+.........0..
00E0  08 2B 06 01 02 01 01 05 00 04 10 XX XX XX XX XX  .+.........XXX.X
00F0  XX XX XX XX XX XX XX XX XX XX XX 30 0C 06 08 2B  XX.XXXX.XXX0...+
0100  06 01 02 01 01 06 00 04 00 30 0D 06 08 2B 06 01  .........0...+..
0110  02 01 01 07 00 02 01 48                          .......H
```

13.3 Using SNMPv1 with UDP and IP

This section shows how the SNMP GetRequest and GetResponse PDUs fit within the structure of an Ethernet frame (review Figure 12-41). In this section we will examine the Ethernet frame format, the IP header, the UDP header, the SNMP PDUs, and the ASN.1 encoding of the variable bindings.

Trace 13.3a shows four layers of protocol operating in two Frames, 7 and 8. You can easily identify the Data Link Control (DLC) layer as Ethernet because of the EtherType (or Type) field. The next field, the IP Header, is 20 octets long, has routine service, and is not fragmented. The Protocol field identifies the next higher layer (UDP), while the source and destination addresses identify the origin and destination of this datagram.

The UDP header gives the source and destination port numbers. Note that the SunMgr assigns port number 3234 for SNMP, while the Retix bridge (GoldGate) uses the standard port number of 161 for SNMP.

The SNMP authentication header containing the version number and community string precedes the SNMP PDU. We see the PDU identified (GetRequest) and the various error fields. Next come the variable bindings, which consist of an object name and its value. The GetRequest PDU uses NULL for all the Value fields (Frame 7), while the GetResponse (Frame 8) contains the actual values retrieved. To review, you could return to Figure 12-20 and trace the subtree for the System group, verifying the accuracy of the OID designation {1.3.6.1.2.1....}.

Trace 13.3a. Using SNMP with Ethernet, IP, and UDP

Sniffer Network Analyzer data 10-Nov at 10:29:36 file GOLD_SYS.ENC Pg 1

```
--------------- Frame 7 -----------------
DLC:    — DLC Header —
DLC:
DLC:    Frame 7 arrived at 10:29:37.30; frame size is 138 (008A hex) bytes
DLC:    Destination = Station Retix 034CF1, GoldGate
DLC:    Source      = Station Sun   0900C8, SunMgr
DLC:    Ethertype   = 0800 (IP)
DLC:
IP:     — IP Header —
IP:
IP:     Version = 4, header length = 20 bytes
IP:     Type of service = 00
IP:           000. .... = routine
IP:           ...0 .... = normal delay
IP:           .... 0... = normal throughput
IP:           .... .0.. = normal reliability
IP:     Total length = 124 bytes
IP:     Identification = 20055
IP:     Flags = 0X
IP:     .0.. .... = may fragment
IP:     ..0. .... = last fragment
IP:     Fragment offset = 0 bytes
IP:     Time to live = 60 seconds/hops
IP:     Protocol = 17 (UDP)
IP:     Header checksum = A5C5 (correct)
IP:     Source address = [XXX.YYY.128.4]
IP:     Destination address = [XXX.YYY.1.10]
IP:     No options
```

```
IP:
UDP:        — UDP Header —
UDP:
UDP:        Source port = 3234 (SNMP)
UDP:        Destination port = 161
UDP:        Length = 104
UDP:        No checksum
UDP:
SNMP:       — Simple Network Management Protocol (Version 1) —
SNMP:
SNMP:       Version = 0
SNMP:       Community = public
SNMP:       Command = Get request
SNMP:       Request ID = 0
SNMP:       Error status = 0 (No error)
SNMP:       Error index = 0
SNMP:
SNMP:       Object = {1.3.6.1.2.1.1.3.0} (sysUpTime.0)
SNMP:       Value  = NULL
SNMP:
SNMP:       Object = {1.3.6.1.2.1.1.1.0} (sysDescr.0)
SNMP:       Value  = NULL
SNMP:
SNMP:       Object = {1.3.6.1.2.1.1.2.0} (sysObjectID.0)
SNMP:       Value  = NULL
SNMP:
SNMP:       Object = {1.3.6.1.2.1.1.3.0} (sysUpTime.0)
SNMP:       Value  = NULL
SNMP:
SNMP:       Object = {1.3.6.1.2.1.1.6.0} (system.6.0)
SNMP:       Value  = NULL
SNMP:

- - - - - - - - - - - - - - - Frame 8 - - - - - - - - - - - - - - - - -
DLC:        — DLC Header —
DLC:
DLC:        Frame 8 arrived at 10:29:37.33; frame size is 195 (00C3 hex) bytes
DLC:        Destination = Station Sun   0900C8, SunMgr
DLC:        Source      = Station Retix 034CF1, GoldGate
DLC:        Ethertype   = 0800 (IP)
DLC:
IP:         — IP Header —
IP:
IP:         Version = 4, header length = 20 bytes
IP:         Type of service = 00
```

```
IP:                       000. .... = routine
IP:                       ...0 .... = normal delay
IP:                       .... 0... = normal throughput
IP:                       .... .0.. = normal reliability
IP:         Total length = 181 bytes
IP:         Identification = 0
IP:         Flags = 0X
IP:         .0.. .... = may fragment
IP:         ..0. .... = last fragment
IP:         Fragment offset = 0 bytes
IP:         Time to live = 16 seconds/hops
IP:         Protocol = 17 (UDP)
IP:         Header checksum = 1FE4 (correct)
IP:         Source address = [XXX.YYY.1.10]
IP:         Destination address = [XXX.YYY.128.4]
IP:         No options
IP:
UDP:        — UDP Header —
UDP:
UDP:        Source port = 161 (SNMP)
UDP:        Destination port = 3234
UDP:        Length = 161
UDP:        Checksum = 6417 (correct)
UDP:
SNMP:       — Simple Network Management Protocol (Version 1) —
SNMP:
SNMP:       Version = 0
SNMP:       Community = public
SNMP:       Command = Get response
SNMP:       Request ID = 0
SNMP:       Error status = 0 (No error)
SNMP:       Error index = 0
SNMP:
SNMP:       Object = {1.3.6.1.2.1.1.3.0} (sysUpTime.0)
SNMP:       Value  = 240267300 hundredths of a second
SNMP:
SNMP:       Object = {1.3.6.1.2.1.1.1.0} (sysDescr.0)
SNMP:       Value  = Retix Local Ethernet Bridge Model 2265M
SNMP:
SNMP:       Object = {1.3.6.1.2.1.1.2.0} (sysObjectID.0)
SNMP:       Value  = {1.3.6.1.4.1.72.8.3}
SNMP:
SNMP:       Object = {1.3.6.1.2.1.1.3.0} (sysUpTime.0)
SNMP:       Value  = 240267300 hundredths of a second
```

```
SNMP:
SNMP:        Object = {1.3.6.1.2.1.1.6.0} (system.6.0)
SNMP:        Value =
SNMP:
```

Trace 13.3b amplifies the first trace by looking at the actual ASN.1-encoded information included within the SNMP PDUs. Frame 7 begins with a SEQUENCE OF type, code 30H, followed by the length of the encoding (94 octets, or 5E). (These two octets are shown in bold within the third line of the hexadecimal display and are the 43rd and 44th octets transmitted.) The ASN.1 encoding continues within the SNMP PDU, using the TLV (Type-Length-Value) structure discussed in Chapter 12. You can identify the first OID Value requested (line 6.1) by looking for the preceding OID. That information is the 72nd octet transmitted (numbered starting from 0, as octet 71), with an OBJECT IDENTIFIER type (06H), a Length of 8 octets (08H), and a Value of 2B 06 01 02 01 01 03 00H. Recall that the 1.3 prefix is translated into a 43 decimal (or 2BH) through the expression we studied in Section 12.1.4.3.3. The rest of the PDU follows in a similar manner.

The final example (see Figure 13-2) shows how the SNMP GetRequest PDU is encapsulated within the Ethernet frame. Frame 7 illustrates this with the hexadecimal characters shown below their respective fields. The data capture begins with the destination address (08 00 90 03 4C F1), which identifies the Retix bridge. Likewise, the source address (08 00 20 09 00 C8) identifies the Sun Manager. The EtherType field (08 00) indicates that IP will be the next protocol in the Data field. You can decode the IP header, UDP header, SNMP authentication header, and SNMP GetRequest PDU (including the Variable Bindings) in a similar manner.

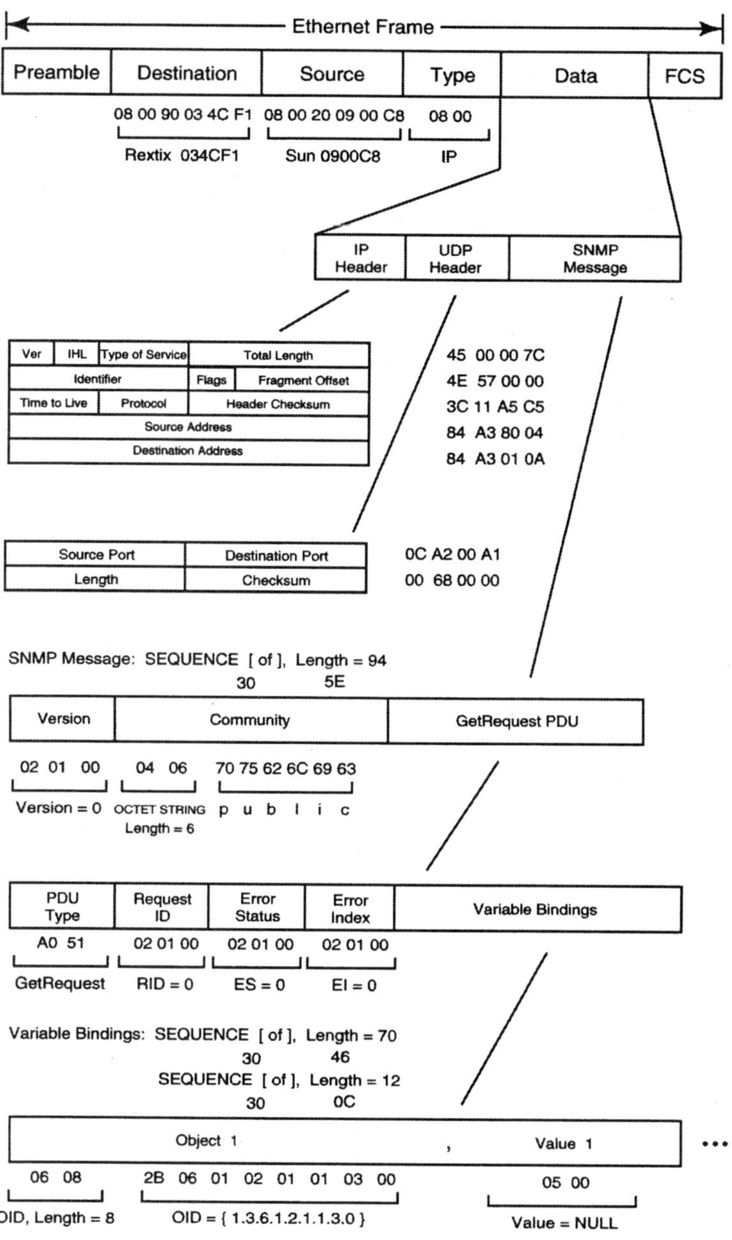

Figure 13-2. Expansion of the SNMP GetRequest PDU within an Ethernet Frame.

Trace 13.3b. ASN.1 Encoding of SNMP GetRequest and GetResponse PDUs
Sniffer Network Analyzer data 10-Nov at 10:29:36 file GOLD_SYS.ENC Pg 1

```
---------------- Frame 7 -----------------
SNMP:      1.1 SEQUENCE [of], Length=94
SNMP:      2.1 INTEGER, Length=1, Value = "0"
SNMP:      2.2 OCTET STRING, Length=6, Value = "public"
SNMP:      2.3 Context-Specific Constructed [0], Length=81
SNMP:      3.1 INTEGER, Length=1, Value = "0"
SNMP:      3.2 INTEGER, Length=1, Value = "0"
SNMP:      3.3 INTEGER, Length=1, Value = "0"
SNMP:      3.4 SEQUENCE [of], Length=70
SNMP:      4.1 SEQUENCE [of], Length=12
SNMP:      6.1 OBJECT IDENTIFIER, Length=8, Value = "{1.3.6.1.2.1.1.3.0}"
SNMP:      6.2 NULL, Length=0, Value = ""
SNMP:      4.2 SEQUENCE [of], Length=12
SNMP:      6.1 OBJECT IDENTIFIER, Length=8, Value = "{1.3.6.1.2.1.1.1.0}"
SNMP:      6.2 NULL, Length=0, Value = ""
SNMP:      4.3 SEQUENCE [of], Length=12
SNMP:      6.1 OBJECT IDENTIFIER, Length=8, Value = "{1.3.6.1.2.1.1.2.0}"
SNMP:      6.2 NULL, Length=0, Value = ""
SNMP:      4.4 SEQUENCE [of], Length=12
SNMP:      6.1 OBJECT IDENTIFIER, Length=8, Value = "{1.3.6.1.2.1.1.3.0}"
SNMP:      6.2 NULL, Length=0, Value = ""
SNMP:      4.5 SEQUENCE [of], Length=12
SNMP:      6.1 OBJECT IDENTIFIER, Length=8, Value = "{1.3.6.1.2.1.1.6.0}"
SNMP:      6.2 NULL, Length=0, Value = ""
SNMP:
ADDR  HEX                                              ASCII
0000  08 00 90 03 4C F1 08 00 20 09 00 C8 08 00 45 00  ....L... .....E.
0010  00 7C 4E 57 00 00 3C 11 A5 C5 84 A3 80 04 84 A3  .|NW..<.........
0020  01 0A 0C A2 00 A1 00 68 00 00 30 5E 02 01 00 04  .......h..0^....
0030  06 70 75 62 6C 69 63 A0 51 02 01 00 02 01 00 02  .public.Q.......
0040  01 00 30 46 30 0C 06 08 2B 06 01 02 01 01 03 00  ..0F0...+.......
0050  05 00 30 0C 06 08 2B 06 01 02 01 01 01 00 05 00  ..0...+.........
0060  30 0C 06 08 2B 06 01 02 01 01 02 00 05 00 30 0C  0...+.........0.
0070  06 08 2B 06 01 02 01 01 03 00 05 00 30 0C 06 08  ..+.........0...
0080  2B 06 01 02 01 01 06 00 05 00                    +.........
```

```
---------------- Frame 8 -----------------
SNMP:    1.1 SEQUENCE [of], Length=150
SNMP:    2.1 INTEGER, Length=1, Value = "0"
SNMP:    2.2 OCTET STRING, Length=6, Value = "public"
SNMP:    2.3 Context-Specific Constructed [2], Length=136
SNMP:    3.1 INTEGER, Length=1, Value = "0"
SNMP:    3.2 INTEGER, Length=1, Value = "0"
SNMP:    3.3 INTEGER, Length=1, Value = "0"
SNMP:    3.4 SEQUENCE [of], Length=125
SNMP:    4.1 SEQUENCE [of], Length=16
SNMP:    6.1 OBJECT IDENTIFIER, Length=8, Value = "{1.3.6.1.2.1.1.3.0}"
SNMP:    6.2 Application Primitive [3], Length=4, Data = "<0E>R0$"
SNMP:    4.2 SEQUENCE [of], Length=51
SNMP:    6.1 OBJECT IDENTIFIER, Length=8, Value = "{1.3.6.1.2.1.1.1.0}"
SNMP:    6.2 OCTET STRING, Length=39, Value = "Retix Local Ethernet Bridge Model "
SNMP:    4.3 SEQUENCE [of], Length=20
SNMP:    6.1 OBJECT IDENTIFIER, Length=8, Value = "{1.3.6.1.2.1.1.2.0}"
SNMP:    6.2 OBJECT IDENTIFIER, Length=8, Value = "{1.3.6.1.4.1.72.8.3}"
SNMP:    4.4 SEQUENCE [of], Length=16
SNMP:    6.1 OBJECT IDENTIFIER, Length=8, Value = "{1.3.6.1.2.1.1.3.0}"
SNMP:    6.2 Application Primitive [3], Length=4, Data = "<0E>R0$"
SNMP:    4.5 SEQUENCE [of], Length=12
SNMP:    6.1 OBJECT IDENTIFIER, Length=8, Value = "{1.3.6.1.2.1.1.6.0}"
SNMP:    6.2 OCTET STRING, Length=0, Value = ""
SNMP:
ADDR  HEX                                              ASCII
0000  08 00 20 09 00 C8 08 00 90 03 4C F1 08 00 45 00  .. .......L...E.
0010  00 B5 00 00 00 00 10 11 1F E4 84 A3 01 0A 84 A3  ................
0020  80 04 00 A1 0C A2 00 A1 64 17 30 81 96 02 01 00  ........d.0.....
0030  04 06 70 75 62 6C 69 63 A2 81 88 02 01 00 02 01  ..public........
0040  00 02 01 00 30 7D 30 10 06 08 2B 06 01 02 01 01  ....0}0...+.....
0050  03 00 43 04 0E 52 30 24 30 33 06 08 2B 06 01 02  ..C..R0$03..+...
0060  01 01 01 00 04 27 52 65 74 69 78 20 4C 6F 63 61  .....íRetix Loca
0070  6C 20 45 74 68 65 72 6E 65 74 20 42 72 69 64 67  l Ethernet Bridg
0080  65 20 4D 6F 64 65 6C 20 32 32 36 35 4D 30 14 06  eModel2265M0..
0090  08 2B 06 01 02 01 01 02 00 06 08 2B 06 01 04 01  .+.........+....
00A0  48 08 03 30 10 06 08 2B 06 01 02 01 01 03 00 43  H..0...+.......C
00B0  04 0E 52 30 24 30 0C 06 08 2B 06 01 02 01 01 06  ..R0$0...+......
00C0  00 04 00                                         ...
```

13.4 Communicating Device and Link Status with Traps

One of the most useful aspects of SNMP traps is their ability to communicate significant events to a remote network manager. This example illustrates how vendors embellish traps to provide additional information for their customers. The internetwork for this case study consists of more than 20,000 workstations, over 500 servers, and over 350 bridges and routers. Without SNMP, managing such extensive systems would be extremely difficult. In the example shown in Figure 13-3, a remote Router D and another serial link are having difficulties. This example shows how SNMP alerts the network manager to the problems.

Router D with IP address [XXX.YYY.250.1] has a power failure and then returns to normal operation. In Frame 1 of Trace 13.4a, it signals to the manager by sending a LinkUp trap. The SNMPv1 standard (RFC 1157) requires that the trap include the name and value of the *ifIndex* instance for the affected interface. The router's manufacturer, Cisco Systems, Inc., includes additional information to further identify the interface (see Trace 13.4b). For example, in Frame 1, the Enterprise = {1.3.6.1.4.1.9.1.1} identifies Cisco. The first three object values transmitted come from the *ifTable* under the Interfaces subtree {1.3.6.1.2.1.2}. These are the ifIndex (1 or 2), the *ifDescription* (Ethernet0 or Ethernet1), and the *ifType* (ethernet-csmacd). The last object value, taken from Cisco's private MIB, further identifies what happened (the link is now up).

In Frame 210, a second problem occurs on the serial link between Router B and Router C. This failure triggers the transmission of LinkDown traps from the router. As before, the Enterprise field identifies a Cisco device as the source of the traps and further identifies the failed router port by its IP address: [XXX.YYY.2.3]. The four object values transmitted to the manager communicate the link description (serial0), the type of link (proprietary point-to-point serial), and the reason for the trap (down).

Thus, if failures occur on other segments or communication links, which could even be across the country from each other, SNMP traps can alert the manager that a problem exists. Further troubleshooting using software utilities such as ICMP Echo (PING) messages, SNMP queries (such as the IP or ICMP groups), or test equipment (such as network analyzers), can then proceed.

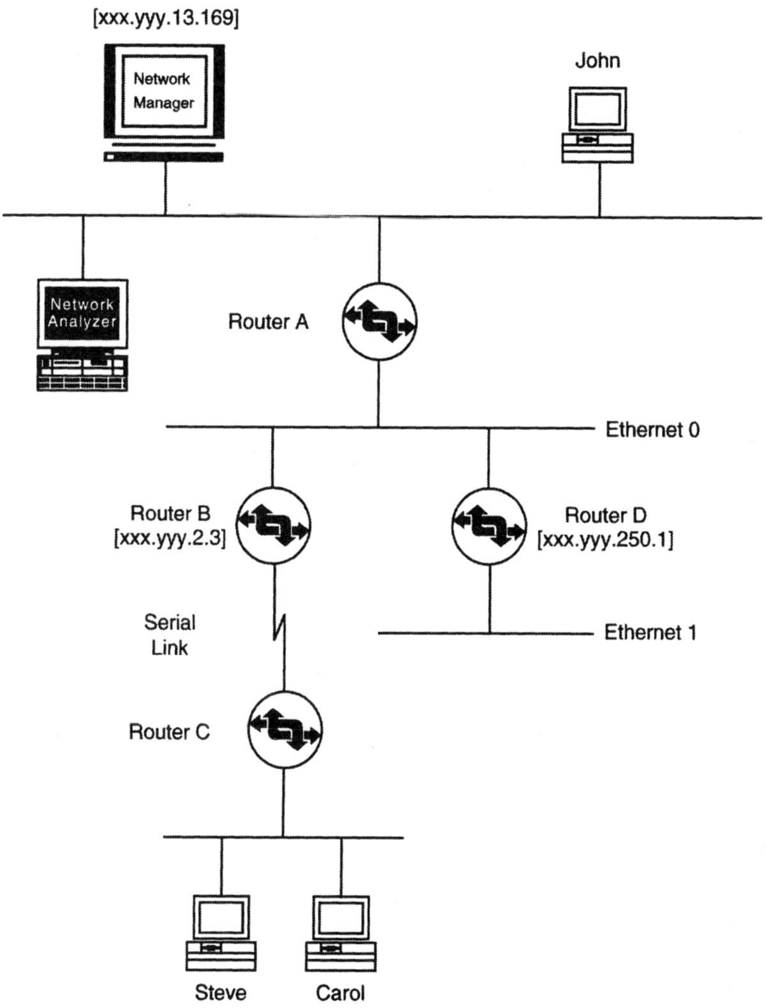

Figure 13-3. Communicating Device and Link Status Information.

Trace 13.4a. Link Up and Link Down Traps (Summary)
Sniffer Network Analyzer data 23-Mar at 13:08:58, file A:TRAP.ENC, Pg 1

SUMMARY	Delta T	Destination	Source	Summary
1		Manager	Router D	SNMP Trap -v1 Link up ifIndex .. cisco.2.2.1.1.20.1 (4 items)
2	0.4585	Manager	Router D	SNMP Trap -v1 Link up ifIndex .. cisco.2.2.1.1.20.2 (4 items)
.				
.				
.				
210	27.6608	Manager	Router B	SNMP Trap Link down ifIndex .. cisco.2.2.1.1.20.1 (4 items)

Trace 13.4b. Link Up and Link Down Traps (Details)
Sniffer Network Analyzer data 23-Mar at 13:08:58, file A:TRAP.ENC, Pg 1

```
- - - - - - - - - - - - - - - Frame 1 - - - - - - - - - - - - - - - - -
SNMP:     — Simple Network Management Protocol (Version 1) —
SNMP:
SNMP:     Version = 0
SNMP:     Community = public
SNMP:     Command = Trap
SNMP:     Enterprise = {1.3.6.1.4.1.9.1.1}
SNMP:     Network address = [XXX.YYY.12.250]
SNMP:     Generic trap = 3 (Link up)
SNMP:     Specific trap = 0
SNMP:     Time ticks = 797
SNMP:
SNMP:     Object = {1.3.6.1.2.1.2.2.1.1.1} (ifIndex.1)
SNMP:     Value = 1
SNMP:
SNMP:     Object = {1.3.6.1.2.1.2.2.1.2.1} (ifDescr.1)
SNMP:     Value = Ethernet0
```

```
SNMP:
SNMP:       Object = {1.3.6.1.2.1.2.2.1.3.1} (ifType.1)
SNMP:       Value = 6 (ethernet-csmacd)
SNMP:
SNMP:       Object = {1.3.6.1.4.1.9.2.2.1.1.20.1} (cisco.2.2.1.1.20.1)
SNMP:       Value = up
SNMP:

--------------- Frame 2 -----------------
SNMP:       — Simple Network Management Protocol (Version 1) —
SNMP:
SNMP:       Version = 0
SNMP:       Community = public
SNMP:       Command = Trap
SNMP:       Enterprise = {1.3.6.1.4.1.9.1.1}
SNMP:       Network address = [XXX.YYY.12.250]
SNMP:       Generic trap = 3 (Link up)
SNMP:       Specific trap = 0
SNMP:       Time ticks = 799
SNMP:
SNMP:       Object = {1.3.6.1.2.1.2.2.1.1.2} (ifIndex.2)
SNMP:       Value = 2
SNMP:
SNMP:       Object = {1.3.6.1.2.1.2.2.1.2.2} (ifDescr.2)
SNMP:       Value = Ethernet1
SNMP:
SNMP:       Object = {1.3.6.1.2.1.2.2.1.3.2} (ifType.2)
SNMP:       Value = 6 (ethernet-csmacd)
SNMP:
SNMP:       Object = {1.3.6.1.4.1.9.2.2.1.1.20.2} (cisco.2.2.1.1.20.2)
SNMP:       Value = up
SNMP:

--------------- Frame 210 -----------------
SNMP:       — Simple Network Management Protocol (Version 1) —
SNMP:
SNMP:       Version = 0
SNMP:       Community = public
SNMP:       Command = Trap
SNMP:       Enterprise = {1.3.6.1.4.1.9.1.1}
```

```
SNMP:    Network address = [XXX.YYY.2.3]
SNMP:    Generic trap = 2 (Link down)
SNMP:    Specific trap = 0
SNMP:    Time ticks = 45039280
SNMP:
SNMP:    Object = {1.3.6.1.2.1.2.2.1.1.1} (ifIndex.1)
SNMP:    Value = 1
SNMP:
SNMP:    Object = {1.3.6.1.2.1.2.2.1.2.1} (ifDescr.1)
SNMP:    Value = Serial0
SNMP:
SNMP:    Object = {1.3.6.1.2.1.2.2.1.3.1} (ifType.1)
SNMP:    Value = 22 (propPointToPointSerial)
SNMP:
SNMP:    Object = {1.3.6.1.4.1.9.2.2.1.1.20.1} (cisco.2.2.1.1.20.1)
SNMP:    Value = down
SNMP:
```

13.5 Incompatible Private Enterprise MIBs

SNMP's popularity has encouraged numerous vendors to incorporate the protocol and its functions into their products. Unfortunately, any time more than one vendor gets involved in a system, incompatibilities can arise, as illustrated in this example (see Figure 13-4).

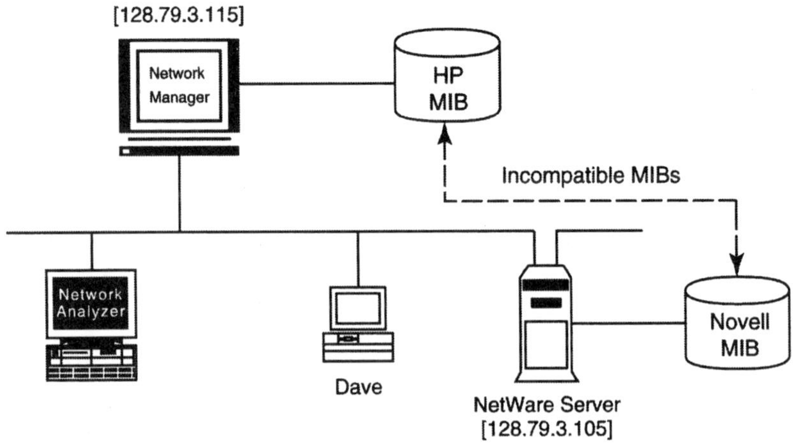

Figure 13-4. Incompatible Private Enterprise MIBs.

The manager in this case is a Hewlett-Packard OpenView console, and the agent is a Novell file server running NetWare and using TCP/IP and SNMP. The manager requests the value of a specific object from the agent in Frame 1; the response is "no such name." The details of the captured frames identify what happened (see Trace 13.5).

In the first IP datagram (Frame 1), the manager's SNMP process includes Identification = 29913 to identify the message. Note that the HP manager assigned Source port = 4837 on its host and used the Destination port = 161, the standard SNMP port, for the agent's host. Also note that the Community name = public. The SNMP GetRequest PDU has a Request ID = 1840, which is used to correlate this request with the agent's response. The object in question is from the private enterprise tree {1.3.6.1.4.1} and is further identified as belonging to the Hewlett-Packard enterprise {1.3.6.1.4.1.11}. The specific object is {1.3.6.1.4.1.11.2.3.1.1.3.0} or {HP 2.3.1.1.3.0}, which identifies the CPU utilization. (Note that the network analyzer is programmed to identify the HP subtree {1.3.6.1.4.1.11}, but not the exact object. We know that this object is CPU utilization from looking into the details of HP's private MIB.)

The agent responds in Frame 2. Note that the agent uses a different Identification (22662) for the IP datagram. This is not a problem, since IP processes of the manager and the agent are independent. The agent correctly designates the Destination port (4837) within its UDP header, which sends the SNMP reply to the manager's SNMP process. The agent's message contains the community name (public) and the Request ID (1840) that correlate with the manager's request. The last two fields provide a clue to the problem: the Error Status = 2 (no such name) and the Error Index = 1. These specify that the object name given in the GetRequest was unknown to the agent, and that the first object specified contained the error. When you consider that the HP manager was asking a Novell server for its CPU utilization, the confusion isn't surprising. Thus, while both manager and agent support MIB-II, their private enterprise MIBs are incompatible. This is another example of systems that are standards based, yet incompatible.

Trace 13.5. Inconsistent Private Enterprise MIBs

Sniffer Network Analyzer data 5-Oct at 09:42:54, file PRVMIB2.ENC Pg 1

```
- - - - - - - - - - - - - - - Frame 1 - - - - - - - - - - - - - - - - -
DLC:        — DLC Header —
DLC:
DLC:        Frame 1 arrived at 09:43:20.8073; Frame size is 87 (0057 hex) bytes.
DLC:        Destination = Station H-P 133ADE
```

```
DLC:        Source    = Station H-P  17B65F
DLC:        Ethertype = 0800 (IP)
DLC:
IP:         — IP Header —
IP:
IP:         Version = 4, header length = 20 bytes
IP:         Type of service = 00
IP:                   000. .... = routine
IP:                   ...0 .... = normal delay
IP:                   .... 0... = normal throughput
IP:                   .... .0.. = normal reliability
IP:         Total length = 73 bytes
IP:         Identification = 29913
IP:         Flags = 0X
IP:         .0.. .... = may fragment
IP:         ..0. .... = last fragment
IP:         Fragment offset = 0 bytes
IP:         Time to live = 30 seconds/hops
IP:         Protocol = 17 (UDP)
IP:         Header checksum = 2051 (correct)
IP:         Source address = [128.79.3.115], Manager
IP:         Destination address = [128.79.3.105], Agent
IP:         No options
IP:
UDP:        — UDP Header —
UDP:
UDP:        Source port = 4837 (SNMP)
UDP:        Destination port = 161
UDP:        Length = 53
UDP:        Checksum = CBCC (correct)
UDP:
SNMP:       — Simple Network Management Protocol (Version 1) —
SNMP:
SNMP:       Version = 0
SNMP:       Community = public
SNMP:       Command = Get request
SNMP:       Request ID = 1840
SNMP:       Error status = 0 (No error)
SNMP:       Error index = 0
SNMP:
SNMP:       Object = {1.3.6.1.4.1.11.2.3.1.1.3.0} (HP.2.3.1.1.3.0)
SNMP:       Value = NULL
SNMP:
```

```
--------------- Frame 2 -----------------
DLC:        — DLC Header —
DLC:
DLC:        Frame 2 arrived at 09:43:20.8093; Frame size is 88 (0058 hex) bytes.
DLC:        Destination = Station H-P  17B65F
DLC:        Source   = Station H-P  133ADE
DLC:        Ethertype = 0800 (IP)
DLC:
IP:         — IP Header —
IP:
IP:         Version = 4, header length = 20 bytes
IP:         Type of service = 00
IP:                 000. .... = routine
IP:                 ...0 .... = normal delay
IP:                 .... 0... = normal throughput
IP:                 .... .0.. = normal reliability
IP:         Total length = 73 bytes
IP:         Identification = 22662
IP:         Flags = 0X
IP:         .0.. .... = may fragment
IP:         ..0. .... = last fragment
IP:         Fragment offset = 0 bytes
IP:         Time to live = 128 seconds/hops
IP:         Protocol = 17 (UDP)
IP:         Header checksum = DAA3 (correct)
IP:         Source address = [128.79.3.105], Agent
IP:         Destination address = [128.79.3.115], Manager
IP:         No options
IP:
UDP:        — UDP Header —
UDP:
UDP:        Source port = 161 (SNMP)
UDP:        Destination port = 4837
UDP:        Length = 53
UDP:        No checksum
UDP:
SNMP:       — Simple Network Management Protocol (Version 1) —
SNMP:
SNMP:       Version = 0
SNMP:       Community = public
SNMP:       Command = Get response
SNMP:       Request ID = 1840
SNMP:       Error status = 2 (No such name)
```

```
SNMP:      Error index = 1
SNMP:
SNMP:      Object = {1.3.6.1.4.1.11.2.3.1.1.3.0} (HP.2.3.1.1.3.0)
SNMP:      Value = NULL
SNMP:
```

13.6 Proper Handling of an Invalid Object Identifier (OID)

Chapter 12 discussed the ASN.1 encodings for objects within the SNMP MIBs. These encodings are based on a tree structure, and specific object identifiers (OIDs) locate the positions of objects on the tree. Since these OIDs are sequences of numbers, a mistake of just one digit renders the sequence invalid. This example shows how an agent responds to a manager's mistake.

In this example, the manager wishes to obtain the value of the system description (see Figure 13-5 and Trace 13.6a). The first time the request is made (Frame 1), a correct response is returned (Frame 3). The second request (Frame 4) is unsuccessful (Frame 5). The details of the SNMP messages (Trace 13.6b) reveal that an invalid OID caused the problem.

In the first GetRequest (Frame 1 of Trace 13.6b), the OID given for sysDescr is {1.3.6.1.2.1.1.1.0}. This OID consists of the prefix {1.3.6.1.2.1.1.1} and an instance (.0). Recall that an instance of .0 indicates a scalar object, that is, one that occurs only once. (Columnar objects may have multiple instances, requiring a more complex suffix that incorporates the indexing scheme in order to identify the specific instance of interest.) In Frame 3, you can see that the GetResponse returns:

```
Value = /usr3/wf/wf.rel/v5.75/wf.pj/proto.ss/
ace_test.p./
```

So far, so good.

Now, as an experiment, the network administrator issues another GetRequest (Frame 4), which returns an error (Frame 5). The details show why this problem occurred. The GetRequest contains an invalid OID {1.3.6.1.3.1.1.1.1}, otherwise known as sysDescr.1. Since this is a scalar, not a tabular, object the ASN.1 syntax is invalid. The response returned in Frame 5 indicates this error: No such name. Thus, the agent provided a proper response for an OID that was not within its MIB.

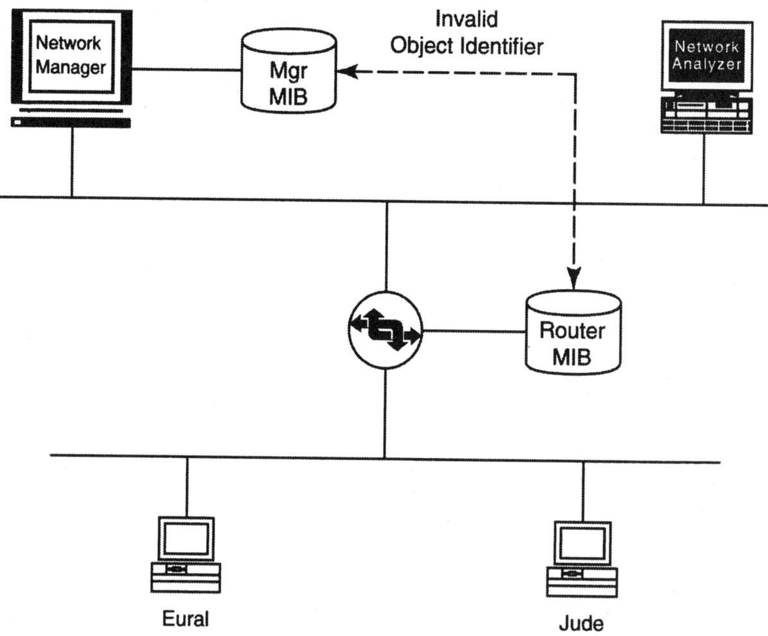

Figure 13-5. Invalid Object Identifier (OID).

Trace 13.6a. Handling an Invalid Object Identifier (Summary)
Sniffer Network Analyzer data 16-Nov at 17:43:32, file 7-15.ENC, Pg 1

SUMMARY	Delta T	Destination	Source	Summary
1		Router	Manager 146	SNMP Get sysDescr
2	0.0488	Router	SGI 020C5D	ARP R PA= [XXX.YYY.3.146] HA=080069020C5D PRO=IP
3	0.0031	Manager 146	Router	SNMP GetReply sysDescr = /usr3/wf/wf.rel/v5.75 /wf.pj/proto.ss /ace_test.p/
4	7.2153	Router	Manager 146	SNMP Get sysDescr
5	0.0421	Manager 146	Router	SNMP GetReply No such name sysDescr

Trace 13.6b. Handling an Invalid Object Identifier (Details)

```
--------------- Frame 1 ----------------
SNMP:    — Simple Network Management Protocol (Version 1) —
SNMP:
SNMP:    Version = 0
SNMP:    Community = xyzsnmp
SNMP:    Command = Get request
SNMP:    Request ID = 13227000
SNMP:    Error status = 0 (No error)
SNMP:    Error index = 0
SNMP:
SNMP:    Object = {1.3.6.1.2.1.1.1.0} (sysDescr.0)
SNMP:    Value = NULL
SNMP:

--------------- Frame 3 ----------------
SNMP:    — Simple Network Management Protocol (Version 1) —
SNMP:
SNMP:    Version = 0
SNMP:    Community = xyzsnmp
SNMP:    Command = Get response
SNMP:    Request ID = 13227000
SNMP:    Error status = 0 (No error)
SNMP:    Error index = 0
SNMP:
SNMP:    Object = {1.3.6.1.2.1.1.1.0} (sysDescr.0)
SNMP:    Value = /usr3/wf/wf.rel/v5.75/wf.pj/proto.ss/ace_test.p/
SNMP:

--------------- Frame 4 ----------------
SNMP:    — Simple Network Management Protocol (Version 1) —
SNMP:
SNMP:    Version = 0
SNMP:    Community = xyzsnmp
SNMP:    Command = Get request
SNMP:    Request ID = 1094416166
SNMP:    Error status = 0 (No error)
SNMP:    Error index = 0
SNMP:
SNMP:    Object = {1.3.6.1.2.1.1.1.1} (sysDescr.1)
SNMP:    Value = NULL
SNMP:
```

```
- - - - - - - - - - - - - - - Frame 5 - - - - - - - - - - - - - - - -
SNMP:      — Simple Network Management Protocol (Version 1) —
SNMP:
SNMP:      Version = 0
SNMP:      Community = xyzsnmp
SNMP:      Command = Get response
SNMP:      Request ID = 1094416166
SNMP:      Error status = 2 (No such name)
SNMP:      Error index = 1
SNMP:
SNMP:      Object = {1.3.6.1.2.1.1.1.1} (sysDescr.1)
SNMP:      Value = NULL
SNMP:
```

13.7 Comparing TELNET and SNMP for Network Management

This example looks at two alternatives for accessing the configuration parameters and operational statistics of a remote bridge. One alternative is to access the bridge with a workstation and then access the bridge's configuration menus using the Telecommunication Network Protocol (TELNET). TELNET allows a remote user to access a host or device as if it were a local terminal. The second alternative is to access the bridge with the management console using SNMP and retrieve the appropriate MIB information. Let's compare these two methods.

In the first method, a Sun workstation initiates a TELNET session with the 3Com bridge (see Figure 13-6). The network administrator uses commands defined by 3Com to retrieve the system parameters and statistics. Each of these commands is then sent from the workstation to the bridge in a TELNET message, and the bridge returns a corresponding response. For example, in Frame 1 (see Trace 13.7a), the system version is requested using the 3Com *show -sys ver* command. The bridge responds in Frame 2 (see Trace 13.7 b) with:

```
"SW/NBII-BR-5.0.1, booted on Mon Mar 22 12:05 from
local floppy"
```

Subsequent requests obtain the system contact (Frames 7 through 10), the system location (Frames 11 through 14), and the system name (Frames 15 through 18). The network administrator next accesses the IP Address Translation (ARP) table (Frames 19 through 53) and the IP table (Frames 54 through 65). The final operation retrieves the Path Statistics, such as the number of packets transmitted and received, collisions, and network utilization (Frames 66 through 99). The TELNET method requires the transmission of a total of 99 frames, including 8,599 octets of information.

The second method uses the network management console to access the bridge's MIB using SNMP (see Trace 13.7c). The administrator queries the System group (Frames 1 through 4), the Address Translation group (Frames 5 through 18), the IP group (Frames 19 through 22), and the Interfaces group (Frames 23 through 42). The details of these transactions reveal information almost identical to that discovered earlier (see Trace 13.7d). The SNMP method requires the transmission of a total of 42 frames and 8,910 octets of information.

The question, then, is which of these two methods is best? From a network traffic point of view, the results are almost identical: 8,599 octets are transmitted using TELNET, and 8,910 octets using SNMP.

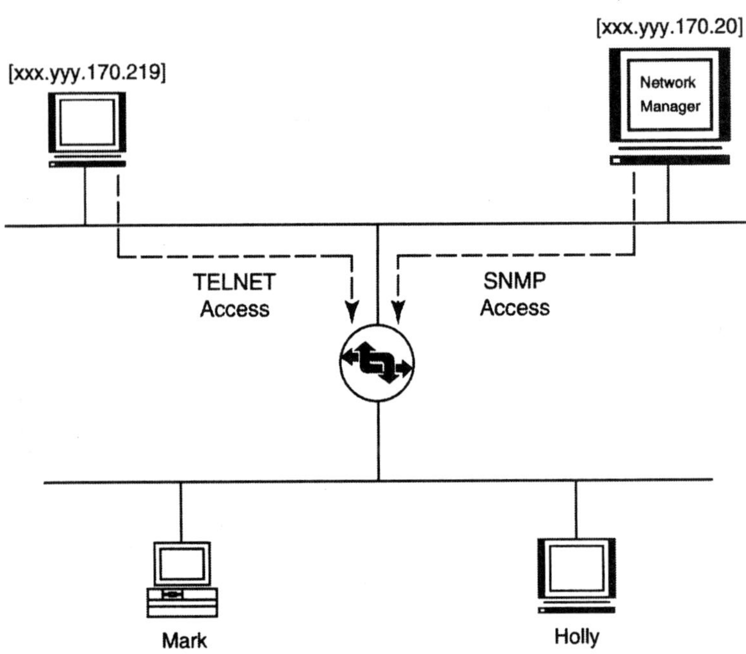

Figure 13-6. Remote Device Configuration Using TELNET and SNMP.

From a practical angle, however, SNMP has an advantage. TELNET requires the network manager to have a workstation available, to understand TELNET commands, and to understand the product-specific parameters, menus, configuration screens, and so on that are part of that managed system. SNMP does not require these details. The administrator simply goes to the management console, accesses the device in question (the 3Com bridge), and enters well-known SNMP commands. Few details of that bridge's internal configuration are necessary. What's more, SNMP may also use vendor-specific traps to alert the administrator to significant events.

In summary, great synergies can come from using an open network management platform instead of a multitude of vendor-specific solutions. Perhaps this is one reason that SNMP has achieved its great popularity!

Trace 13.7a. Accessing Bridge Parameters with a TELNET Session (Summary)
Sniffer Network Analyzer data 26-Mar at 13:52:54, file DIG1.ENC, Pg 1

SUMMARY	Delta T	Destination	Source	Summary
1		3Com Bridge	Sun Station	Telnet C PORT=2200 show -sys ver<0D0A>
2	0.0031	Sun Station	3Com Bridge	Telnet R PORT=2200 SW/NBII-BR-5.0.1, booted on Mon Mar 22
3	0.0005	Sun Station	3Com Bridge	Telnet R PORT=2200 nb2 REM: 1.0.2 <0D0A>
4	0.0007	Sun Station	3Com Bridge	Telnet R PORT=2200 Copyright 1985-1992 3Com Corporation <0D0A>
5	0.0015	Sun Station	3Com Bridge	Telnet R PORT=2200 [24]nbii650b#
6	0.1308	3Com Bridge	Sun Station	TCP D=23 S=2200 ACK=344321989 WIN=4096
7	9.9600	3Com Bridge	Sun Station	Telnet C PORT=2200 show -sys scon<0D0A>
8	0.0029	Sun Station	3Com Bridge	Telnet R PORT=2200 SysCONtact = "John Doe"<0D0A>

9	0.0014	Sun Station	3Com Bridge	Telnet R PORT=2200 [25]nbii650b#
10	0.0357	3Com Bridge	Sun Station	TCP D=23 S=2200 ACK=344322028 WIN=4096
11	4.8660	3Com Bridge	Sun Station	Telnet C PORT=2200 show -sys sloc<0D0A>
12	0.0030	Sun Station	3Com Bridge	Telnet R PORT=2200 SysLOCation = “650 Computer Room”<0D0A>
13	0.0015	Sun Station	3Com Bridge	Telnet R PORT=2200 [26]nbii650b#
14	0.1295	3Com Bridge	Sun Station	TCP D=23 S=2200 ACK=344322077 WIN=4096
15	4.8946	3Com Bridge	Sun Station	Telnet C PORT=2200 show -sys snam<0D0A>
16	0.0029	Sun Station	3Com Bridge	Telnet R PORT=2200 SysNAMe = “nbii650b”<0D0A>
17	0.0015	Sun Station	3Com Bridge	Telnet R PORT=2200 [27]nbii650b#
18	0.1011	3Com Bridge	Sun Station	TCP D=23 S=2200 ACK=344322113 WIN=4096
19	4.4364	3Com Bridge	Sun Station	Telnet C PORT=2200 show -ip addr<0D0A>
20	0.0029	Sun Station	3Com Bridge	Telnet R PORT=2200 —IP Address Trans... XXX.YYY.170.214
.				
.				
.				
54	5.2126	3Com Bridge	Sun Station	Telnet C PORT=2200 show -ip netaddr<0D0A>
55	0.0030	Sun Station	3Com Bridge	Telnet R PORT=2200 —IP Directly Conn..
56	0.0010	Sun Station	3Com Bridge	Telnet R PORT=2200 IP

				Address Port Subnet Mask Status...
.				
.				
.				
66	15.9375	3Com Bridge	Sun Station	Telnet C PORT=2200 show stat - path<0D0A>
67	0.0046	Sun Station	3Com Bridge	Telnet R PORT=2200 ACCUMULATED VALUES<0D0A>
68	0.0006	Sun Station	3Com Bridge	Telnet R PORT=2200 <0D0A>
69	0.0010	Sun Station	3Com Bridge	Telnet R PORT=2200 == PATH statistic...
70	0.0007	Sun Station	3Com Bridge	Telnet R PORT=2200
71	0.0010	Sun Station	3Com Bridge	Telnet R PORT=2200
.				
.				
.				
96	0.0010	Sun Station	3Com Bridge	Telnet R PORT=2200 Byte/Sec Rcv Good: Packets 29885492 2732272...
97	0.0009	Sun Station	3Com Bridge	Telnet R PORT=2200 Xmit Bad: Pkt/Sec
98	0.0014	Sun Station	3Com Bridge	Telnet R PORT=2200 [30]nbii650b#
99	0.1944	3Com Bridge	Sun Station	TCP D=23 S=2200 ACK=344324928 WIN=4096

Trace 13.7b. Accessing Bridge Parameters with a TELNET Session (Details)

Sniffer Network Analyzer data 26-Mar at 13:52:54, file DIG1.ENC, Pg 1

```
- - - - - - - - - - - - - - - Frame 1 - - - - - - - - - - - - - - - - -
Telnet:      — Telnet data —
Telnet:
Telnet:      show -sys ver<0D0A>
Telnet:
```

```
- - - - - - - - - - - - - - - Frame 2 - - - - - - - - - - - - - - - - -
Telnet:        — Telnet data —
Telnet:
Telnet:        SW/NBII-BR-5.0.1, booted on Mon Mar 22 12:05 from local
               floppy<0D0A>
Telnet:

- - - - - - - - - - - - - - - Frame 3 - - - - - - - - - - - - - - - - -
Telnet:        — Telnet data —
Telnet:
Telnet:        nb2 REM: 1.0.2 <0D0A>
Telnet:

- - - - - - - - - - - - - - - Frame 4 - - - - - - - - - - - - - - - - -
Telnet:        — Telnet data —
Telnet:
Telnet:        Copyright 1985-2002, 3Com Corporation<0D0A>
Telnet:

- - - - - - - - - - - - - - - Frame 5 - - - - - - - - - - - - - - - - -
Telnet:        — Telnet data —
Telnet:
Telnet:        [24]nbii650b#
Telnet:

- - - - - - - - - - - - - - - Frame 7 - - - - - - - - - - - - - - - - -
Telnet:        — Telnet data —
Telnet:
Telnet:        show -sys scon<0D0A>
Telnet:

- - - - - - - - - - - - - - - Frame 8 - - - - - - - - - - - - - - - - -
Telnet:        — Telnet data —
Telnet:
Telnet:        SysCONtact = “John Doe”<0D0A>
Telnet:

- - - - - - - - - - - - - - - Frame 9 - - - - - - - - - - - - - - - - -
Telnet:        — Telnet data —
Telnet:
Telnet:        [25]nbii650b#
Telnet:
```

Trace 13.7c. Accessing Remote Bridge Parameters with SNMP (Summary)
Sniffer Network Analyzer data 26-Mar at 13:55:46, file DIG2.ENC, Pg 1

SUMMARY	Delta T	Destination	Source	Summary
1		3Com Bridge	Manager	SNMP GetNext sysDescr .. sysServices (7 items)
2	0.0035	Manager	3Com Bridge	SNMP GetReply sysDescr .. sysServices (7 items)
.				
.				
.				
5	1.2927	3Com Bridge	Manager	SNMP GetNext atIfIndex .. atNetAddress (3 items)
6	0.0038	Manager	3Com Bridge	SNMP GetReply atIfIndex .. atNetAddress (3 items)
.				
.				
.				
19	2.6261	3Com Bridge	Manager	SNMP GetNext ipAdEntAddr .. ipAdEnt ReasmMaxSize (5 items)
20	0.0034	Manager	3Com Bridge	SNMP GetReply ipAdEntAddr .. ipAdEnt ReasmMaxSize (5 items)
.				
.				
.				
23	2.3142	3Com Bridge	Manager	SNMP GetNext ifIndex .. ifOutOctets (16 items)

24	0.0106	Manager	3Com Bridge	SNMP GetReply ifIndex .. ifOutOctets (16 items)
.				
.				
.				
41	0.0045	3Com Bridge	Manager	SNMP GetNext ifOutUcastPkts .. ifSpecific (6 items)
42	0.0058	Manager	3Com Bridge	SNMP GetReply ifOutNUcastPkts .. atIfIndex (6 items)

Trace 13.7d. Accessing Remote Bridge Parameters with SNMP (Details)
Sniffer Network Analyzer data 26-Mar at 13:55:46, file DIG2.ENC, Pg 1

```
---------------- Frame 1 -----------------
SNMP:     — Simple Network Management Protocol (Version 1) —
SNMP:
SNMP:     Version = 0
SNMP:     Community = public
SNMP:     Command = Get next request
SNMP:     Request ID = 733178616
SNMP:     Error status = 0 (No error)
SNMP:     Error index = 0
SNMP:
SNMP:     Object = {1.3.6.1.2.1.1.1} (sysDescr)
SNMP:     Value = NULL
SNMP:
SNMP:     Object = {1.3.6.1.2.1.1.2} (sysObjectID)
SNMP:     Value = NULL
SNMP:
SNMP:     Object = {1.3.6.1.2.1.1.3} (sysUpTime)
SNMP:     Value = NULL
SNMP:
SNMP:     Object = {1.3.6.1.2.1.1.4} (sysContact)
SNMP:     Value = NULL
SNMP:
SNMP:     Object = {1.3.6.1.2.1.1.5} (sysName)
SNMP:     Value = NULL
SNMP:
```

```
SNMP:     Object = {1.3.6.1.2.1.1.6} (sysLocation)
SNMP:     Value = NULL
SNMP:
SNMP:     Object = {1.3.6.1.2.1.1.7} (sysServices)
SNMP:     Value = NULL
SNMP:

--------------- Frame 2 -----------------
SNMP:      — Simple Network Management Protocol (Version 1) —
SNMP:
SNMP:     Version = 0
SNMP:     Community = public
SNMP:     Command = Get response
SNMP:     Request ID = 733178616
SNMP:     Error status = 0 (No error)
SNMP:     Error index = 0
SNMP:
SNMP:     Object = {1.3.6.1.2.1.1.1.0} (sysDescr.0)
SNMP:     Value = SW/NBII-BR-5.0.1
SNMP:
SNMP:     Object = {1.3.6.1.2.1.1.2.0} (sysObjectID.0)
SNMP:     Value = {1.3.6.1.4.1.43.1.4}
SNMP:
SNMP:     Object = {1.3.6.1.2.1.1.3.0} (sysUpTime.0)
SNMP:     Value = 35594870 hundredths of a second
SNMP:
SNMP:     Object = {1.3.6.1.2.1.1.4.0} (sysContact.0)
SNMP:     Value = John Doe
SNMP:
SNMP:     Object = {1.3.6.1.2.1.1.5.0} (sysName.0)
SNMP:     Value = nbii650b
SNMP:
SNMP:     Object = {1.3.6.1.2.1.1.6.0} (sysLocation.0)
SNMP:     Value = 650 Computer Room
SNMP:
SNMP:     Object = {1.3.6.1.2.1.1.7.0} (sysServices.0)
SNMP:     Value = 74
SNMP:
  .
  .
  .
```

```
---------------- Frame 5 ----------------
SNMP:     — Simple Network Management Protocol (Version 1) —
SNMP:
SNMP:     Version = 0
SNMP:     Community = public
SNMP:     Command = Get next request
SNMP:     Request ID = 733178619
SNMP:     Error status = 0 (No error)
SNMP:     Error index = 0
SNMP:
SNMP:     Object = {1.3.6.1.2.1.3.1.1.1} (atIfIndex)
SNMP:     Value = NULL
SNMP:
SNMP:     Object = {1.3.6.1.2.1.3.1.1.2} (atPhysAddress)
SNMP:     Value = NULL
SNMP:
SNMP:     Object = {1.3.6.1.2.1.3.1.1.3} (atNetAddress)
SNMP:     Value = NULL
SNMP:

---------------- Frame 6 ----------------
SNMP:     — Simple Network Management Protocol (Version 1) —
SNMP:
SNMP:     Version = 0
SNMP:     Community = public
SNMP:     Command = Get response
SNMP:     Request ID = 733178619
SNMP:     Error status = 0 (No error)
SNMP:     Error index = 0
SNMP:
SNMP:     Object = {1.3.6.1.2.1.3.1.1.1.0.1.XXX.YYY.170.0}
          (atIfIndex.0.1.XXX.YYY.170.0)
SNMP:     Value = 0
SNMP:
SNMP:     Object = {1.3.6.1.2.1.3.1.1.2.0.1.XXX.YYY.170.0}
          (atPhysAddress.0.1.XXX.YYY.170.0)
SNMP:     Value = FFFFFFFFFFFF, Broadcast
SNMP:
SNMP:     Object = {1.3.6.1.2.1.3.1.1.3.0.1.XXX.YYY.170.0}
          (atNetAddress.0.1.XXX.YYY.170.0)
SNMP:     Value = [XXX.YYY.170.0]
SNMP:
  .
  .
  .
```

```
--------------- Frame 19 ----------------
SNMP:    — Simple Network Management Protocol (Version 1) —
SNMP:
SNMP:    Version = 0
SNMP:    Community = public
SNMP:    Command = Get next request
SNMP:    Request ID = 733178620
SNMP:    Error status = 0 (No error)
SNMP:    Error index = 0
SNMP:
SNMP:    Object = {1.3.6.1.2.1.4.20.1.1} (ipAdEntAddr)
SNMP:    Value = NULL
SNMP:
SNMP:    Object = {1.3.6.1.2.1.4.20.1.2} (ipAdEntIfIndex)
SNMP:    Value = NULL
SNMP:
SNMP:    Object = {1.3.6.1.2.1.4.20.1.3} (ipAdEntNetMask)
SNMP:    Value = NULL
SNMP:
SNMP:    Object = {1.3.6.1.2.1.4.20.1.4} (ipAdEntBcastAddr)
SNMP:    Value = NULL
SNMP:
SNMP:    Object = {1.3.6.1.2.1.4.20.1.5} (ipAdEntReasmMaxSize)
SNMP:    Value = NULL
SNMP:

--------------- Frame 20 ----------------
SNMP:    — Simple Network Management Protocol (Version 1) —
SNMP:
SNMP:    Version = 0
SNMP:    Community = public
SNMP:    Command = Get response
SNMP:    Request ID = 733178620
SNMP:    Error status = 0 (No error)
SNMP:    Error index = 0
SNMP:
SNMP:    Object = {1.3.6.1.2.1.4.20.1.1.XXX.YYY.170.214}
         (ipAdEntAddr.XXX.YYY.170.214)
SNMP:    Value = [XXX.YYY.170.214], 3Com Bridge
SNMP:
```

```
SNMP:      Object = {1.3.6.1.2.1.4.20.1.2.XXX.YYY.170.214}
           (ipAdEntIfIndex.XXX.YYY.170.214)
SNMP:      Value = 0
SNMP:
SNMP:      Object = {1.3.6.1.2.1.4.20.1.3.XXX.YYY.170.214}
           (ipAdEntNetMask.XXX.YYY.170.214)
SNMP:      Value = [255.255.255.0]
SNMP:
SNMP:      Object = {1.3.6.1.2.1.4.20.1.4.XXX.YYY.170.214}
           (ipAdEntBcastAddr.XXX.YYY.170.214)
SNMP:      Value = 0
SNMP:
SNMP:      Object = {1.3.6.1.2.1.4.20.1.5.XXX.YYY.170.214}
           (ipAdEntReasmMaxSize.XXX.YYY.170.214)
SNMP:      Value = 65535
SNMP:
   .
   .
   .

---------------- Frame 23 -----------------
SNMP:      — Simple Network Management Protocol (Version 1) —
SNMP:
SNMP:      Version = 0
SNMP:      Community = public
SNMP:      Command = Get next request
SNMP:      Request ID = 733178622
SNMP:      Error status = 0 (No error)
SNMP:      Error index = 0
SNMP:
SNMP:      Object = {1.3.6.1.2.1.2.2.1.1} (ifIndex)
SNMP:      Value = NULL
SNMP:
SNMP:      Object = {1.3.6.1.2.1.2.2.1.2} (ifDescr)
SNMP:      Value = NULL
SNMP:
SNMP:      Object = {1.3.6.1.2.1.2.2.1.3} (ifType)
SNMP:      Value = NULL
SNMP:
SNMP:      Object = {1.3.6.1.2.1.2.2.1.4} (ifMtu)
SNMP:      Value = NULL
SNMP:
SNMP:      Object = {1.3.6.1.2.1.2.2.1.5} (ifSpeed)
```

```
SNMP:       Value = NULL
SNMP:
SNMP:       Object = {1.3.6.1.2.1.2.2.1.6} (ifPhysAddress)
SNMP:       Value = NULL
SNMP:
SNMP:       Object = {1.3.6.1.2.1.2.2.1.7} (ifAdminStatus)
SNMP:       Value = NULL
SNMP:
SNMP:       Object = {1.3.6.1.2.1.2.2.1.8} (ifOperStatus)
SNMP:       Value = NULL
SNMP:
SNMP:       Object = {1.3.6.1.2.1.2.2.1.9} (ifLastChange)
SNMP:       Value = NULL
SNMP:
SNMP:       Object = {1.3.6.1.2.1.2.2.1.10} (ifInOctets)
SNMP:       Value = NULL
SNMP:
SNMP:       Object = {1.3.6.1.2.1.2.2.1.11} (ifInUcastPkts)
SNMP:       Value = NULL
SNMP:
SNMP:       Object = {1.3.6.1.2.1.2.2.1.12} (ifInNUcastPkts)
SNMP:       Value = NULL
SNMP:
SNMP:       Object = {1.3.6.1.2.1.2.2.1.13} (ifInDiscards)
SNMP:       Value = NULL
SNMP:
SNMP:       Object = {1.3.6.1.2.1.2.2.1.14} (ifInErrors)
SNMP:       Value = NULL
SNMP:
SNMP:       Object = {1.3.6.1.2.1.2.2.1.15} (ifInUnknownProtos)
SNMP:       Value = NULL
SNMP:
SNMP:       Object = {1.3.6.1.2.1.2.2.1.16} (ifOutOctets)
SNMP:       Value = NULL
SNMP:

- - - - - - - - - - - - - - - Frame 24 - - - - - - - - - - - - - - - - -
SNMP:       — Simple Network Management Protocol (Version 1) —
SNMP:
SNMP:       Version = 0
SNMP:       Community = public
SNMP:       Command = Get response
SNMP:       Request ID = 733178622
SNMP:       Error status = 0 (No error)
```

```
SNMP:       Error index = 0
SNMP:
SNMP:       Object = {1.3.6.1.2.1.2.2.1.1.1} (ifIndex.1)
SNMP:       Value = 1
SNMP:
SNMP:       Object = {1.3.6.1.2.1.2.2.1.2.1} (ifDescr.1)
SNMP:       Value = 3Com NETBuilderETH/1-1
SNMP:
SNMP:       Object = {1.3.6.1.2.1.2.2.1.3.1} (ifType.1)
SNMP:       Value = 6 (ethernet-csmacd)
SNMP:
SNMP:       Object = {1.3.6.1.2.1.2.2.1.4.1} (ifMtu.1)
SNMP:       Value = 1500 octets
SNMP:
SNMP:       Object = {1.3.6.1.2.1.2.2.1.5.1} (ifSpeed.1)
SNMP:       Value = 10000000 bits per second
SNMP:
SNMP:       Object = {1.3.6.1.2.1.2.2.1.6.1} (ifPhysAddress.1)
SNMP:       Value = Bridge034ACD
SNMP:
SNMP:       Object = {1.3.6.1.2.1.2.2.1.7.1} (ifAdminStatus.1)
SNMP:       Value = 1 (up)
SNMP:
SNMP:       Object = {1.3.6.1.2.1.2.2.1.8.1} (ifOperStatus.1)
SNMP:       Value = 1 (up)
SNMP:
SNMP:       Object = {1.3.6.1.2.1.2.2.1.9.1} (ifLastChange.1)
SNMP:       Value = 1995 hundredths of a second
SNMP:
SNMP:       Object = {1.3.6.1.2.1.2.2.1.10.1} (ifInOctets.1)
SNMP:       Value = 1611173470 octets
SNMP:
SNMP:       Object = {1.3.6.1.2.1.2.2.1.11.1} (ifInUcastPkts.1)
SNMP:       Value = 13162801 packets
SNMP:
SNMP:       Object = {1.3.6.1.2.1.2.2.1.12.1} (ifInNUcastPkts.1)
SNMP:       Value = 8368939 packets
SNMP:
SNMP:       Object = {1.3.6.1.2.1.2.2.1.13.1} (ifInDiscards.1)
SNMP:       Value = 0 packets
SNMP:
SNMP:       Object = {1.3.6.1.2.1.2.2.1.14.1} (ifInErrors.1)
SNMP:       Value = 579 packets
SNMP:
```

```
SNMP:       Object = {1.3.6.1.2.1.2.2.1.15.1} (ifInUnknownProtos.1)
SNMP:       Value = 6620842 packets
SNMP:
SNMP:       Object = {1.3.6.1.2.1.2.2.1.16.1} (ifOutOctets.1)
SNMP:       Value = -1258921820 octets
SNMP:
  .
  .
  .

- - - - - - - - - - - - - - - Frame 41 - - - - - - - - - - - - - - - - -
SNMP:       — Simple Network Management Protocol (Version 1) —
SNMP:
SNMP:       Version = 0
SNMP:       Community = public
SNMP:       Command = Get next request
SNMP:       Request ID = 733178609
SNMP:       Error status = 0 (No error)
SNMP:       Error index = 0
SNMP:
SNMP:       Object = {1.3.6.1.2.1.2.2.1.17.4} (ifOutUcastPkts.4)
SNMP:       Value = NULL
SNMP:
SNMP:       Object = {1.3.6.1.2.1.2.2.1.18.4} (ifOutNUcastPkts.4)
SNMP:       Value = NULL
SNMP:
SNMP:       Object = {1.3.6.1.2.1.2.2.1.19.4} (ifOutDiscards.4)
SNMP:       Value = NULL
SNMP:
SNMP:       Object = {1.3.6.1.2.1.2.2.1.20.4} (ifOutErrors.4)
SNMP:       Value = NULL
SNMP:
SNMP:       Object = {1.3.6.1.2.1.2.2.1.21.4} (ifOutQLen.4)
SNMP:       Value = NULL
SNMP:
SNMP:       Object = {1.3.6.1.2.1.2.2.1.22.4} (ifSpecific.4)
SNMP:       Value = NULL
SNMP:
```

```
--------------- Frame 42 -----------------
SNMP:     — Simple Network Management Protocol (Version 1) —
SNMP:
SNMP:     Version = 0
SNMP:     Community = public
SNMP:     Command = Get response
SNMP:     Request ID = 733178609
SNMP:     Error status = 0 (No error)
SNMP:     Error index = 0
SNMP:
SNMP:     Object = {1.3.6.1.2.1.2.2.1.18.1} (ifOutNUcastPkts.1)
SNMP:     Value = 10078921 packets
SNMP:
SNMP:     Object = {1.3.6.1.2.1.2.2.1.19.1} (ifOutDiscards.1)
SNMP:     Value = 0 packets
SNMP:
SNMP:     Object = {1.3.6.1.2.1.2.2.1.20.1} (ifOutErrors.1)
SNMP:     Value = 409 packets
SNMP:
SNMP:     Object = {1.3.6.1.2.1.2.2.1.21.1} (ifOutQLen.1)
SNMP:     Value = 0 packets
SNMP:
SNMP:     Object = {1.3.6.1.2.1.2.2.1.22.1} (ifSpecific.1)
SNMP:     Value = {0}
SNMP:
SNMP:     Object = {1.3.6.1.2.1.3.1.1.1.0.1.XXX.YYY.170.0} (atIfIndex.0.1.XXX.YYY.170.0)
SNMP:     Value = 0
SNMP:
```

13.8 Supporting the RMON MIB with a Network Monitor

RMON extends the reach of the network manager to remote LAN segments located anywhere on that internetwork. The RMON agent can be a simple device that connects to a local LAN segment and gathers statistics on that segment's performance. RMON agents can also be built into internetworking devices such as bridges, routers, and intelligent hubs. The RMON manager can be a software application running on the network management console.

This example shows sample statistics that the Ethernet RMON agent can tabulate for the manager (see Figure 13-7).

The Ethernet RMON1 MIB contains nine groups (review Figure 12-33). The first group maintains a table of statistics, measured on the agent's segment. In Trace 13.8a, the manager wishes to retrieve the Statistics Table and check its values. Frames 86 and 88 transmit the manager's GetNextRequests, and Frames 87 and 89 contain the RMON agent's responses.

Recall that the GetNext Request is asking for the value of the object that is the lexicographical successor to the one in the variable binding. In Frame 86 of Trace 13.8b, the first object is *sysUpTime*, with an OID of {1.3.6.1.2.1.1.3}. Since *sysUpTime* is a scalar object, we would expect the response to be OID {1.3.6.1.2.1.1.3.0}, and this is confirmed in the details of Frame 87. (Recall that a .0 suffix indicates scalar objects.) The second OID in Frame 86 is {1.3.6.1.2.1.16.1.1.1.1}, which identifies the *etherStatsIndex* object of the *etherStatsEntry* in the *etherStatsTable*. Note that to actually reference the value of that index, we would have to know the value ahead of time, as required by the module definition for this table. But since we do not have prior knowledge of the value of this index, we use the GetNext PDU, which does not have this limitation. Thus, in Frame 87, we not only see the value we require for the first *etherStatsIndex*, Value = 1, but we are also given the full OID of that variable {1.3.6.1.2.1.16.1.1.1.1.1}. This OID is then used in the subsequent GetNextRequest in Frame 88. This same logic can be applied to the remaining variable bindings in Frames 86 and 87.

Frame 88 is an attempt to retrieve the next row of the Statistics Table. However, Frame 89 shows that this is not possible. Instead of returning the index of the second row, the GetResponse in Frame 89 returns the OID and value of the next object: *etherStatsDataSource*, with OID {1.3.6.1.2.1.16.1.1.1.2.1}. This means that there is no second row to the Statistics Table, and that the GetNext operation has moved on to the lexicographical successor, which is another object.

The details of the agent's response in Frame 87 show the statistics that are maintained (see Trace 13.8b). Of particular interest is the number of errored frames (fragments, jabbers, and collisions), which are counted by objects {1.3.6.1.2.1.16.1.1.1.11.1}, 12.1, and 13.1, respectively. Frame 87 reveals values of 21,389 fragments, 4 jabbers, and 163 collisions, respectively, for those objects. (Review the Statistics group of Figure 12-33, the RMON Ethernet MIB.)

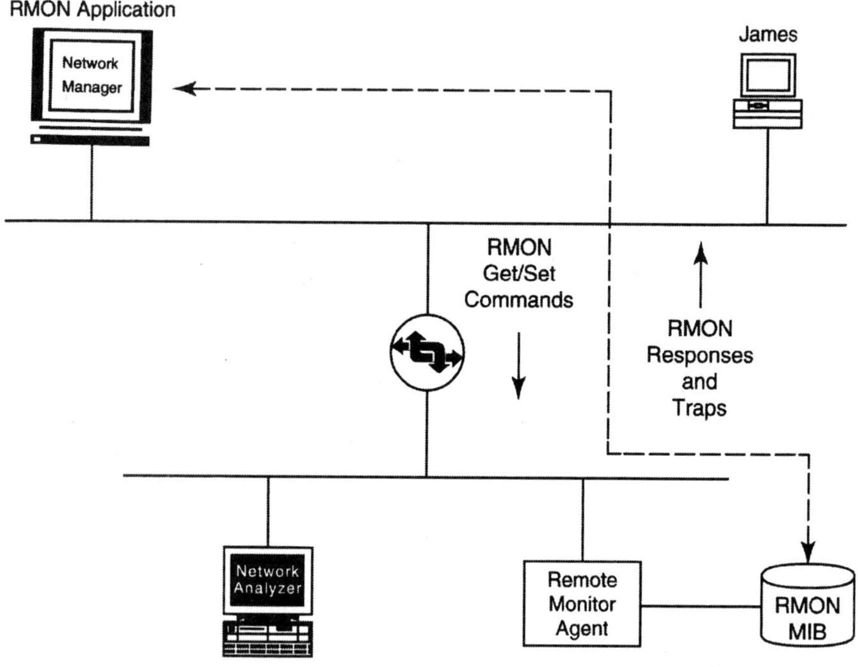

Figure 13-7. Retrieving Remote Information Using the RMON MIB.

Trace 13.8a. RMON MIB Objects (Summary)
Sniffer Network Analyzer data 30-Mar at 15:01:14, RMON3.ENC, Pg 1

SUMMARY	Delta T	Destination	Source	Summary
86	3.1904	RMON Agent	Manager	SNMP GetNext sysUpTime.. rmon.1.1.1.21 (22 items)
87	0.0079	Manager	RMON Agent	SNMP GetReply sysUpTime.. rmon.1.1.1.21.1 (22 items)
88	0.0813	RMON Agent	Manager	SNMP GetNext sysUpTime ..rmon.1.1.1.21.1 (22 items)
89	0.0301	Manager	RMON Agent	SNMP GetReply sysUpTime ..rmon.2.1.1.1.1 (22 items)

Trace 13.8b. RMON MIB Objects (Details)
Sniffer Network Analyzer data 30-Mar at 15:01:14, RMON3.ENC, Pg 1

```
- - - - - - - - - - - - - - - Frame 86 - - - - - - - - - - - - - - - - -
SNMP:     — Simple Network Management Protocol (Version 1) —
SNMP:
SNMP:     Version = 0
SNMP:     Community = armon
SNMP:     Command = Get next request
SNMP:     Request ID = 0
SNMP:     Error status = 0 (No error)
SNMP:     Error index = 0
SNMP:
SNMP:     Object = {1.3.6.1.2.1.1.3} (sysUpTime)
SNMP:     Value = NULL
SNMP:
SNMP:     Object = {1.3.6.1.2.1.16.1.1.1.1} (rmon.1.1.1.1)
SNMP:     Value = NULL
SNMP:
SNMP:     Object = {1.3.6.1.2.1.16.1.1.1.2} (rmon.1.1.1.2)
SNMP:     Value = NULL
SNMP:
SNMP:     Object = {1.3.6.1.2.1.16.1.1.1.3} (rmon.1.1.1.3)
SNMP:     Value = NULL
SNMP:
SNMP:     Object = {1.3.6.1.2.1.16.1.1.1.4} (rmon.1.1.1.4)
SNMP:     Value = NULL
SNMP:
SNMP:     Object = {1.3.6.1.2.1.16.1.1.1.5} (rmon.1.1.1.5)
SNMP:     Value = NULL
SNMP:
SNMP:     Object = {1.3.6.1.2.1.16.1.1.1.6} (rmon.1.1.1.6)
SNMP:     Value = NULL
SNMP:
SNMP:     Object = {1.3.6.1.2.1.16.1.1.1.7} (rmon.1.1.1.7)
SNMP:     Value = NULL
SNMP:
SNMP:     Object = {1.3.6.1.2.1.16.1.1.1.8} (rmon.1.1.1.8)
SNMP:     Value = NULL
SNMP:
SNMP:     Object = {1.3.6.1.2.1.16.1.1.1.9} (rmon.1.1.1.9)
SNMP:     Value = NULL
SNMP:
SNMP:     Object = {1.3.6.1.2.1.16.1.1.1.10} (rmon.1.1.1.10)
```

```
SNMP:      Value = NULL
SNMP:
SNMP:      Object = {1.3.6.1.2.1.16.1.1.1.11} (rmon.1.1.1.11)
SNMP:      Value = NULL
SNMP:
SNMP:      Object = {1.3.6.1.2.1.16.1.1.1.12} (rmon.1.1.1.12)
SNMP:      Value = NULL
SNMP:
SNMP:      Object = {1.3.6.1.2.1.16.1.1.1.13} (rmon.1.1.1.13)
SNMP:      Value = NULL
SNMP:
SNMP:      Object = {1.3.6.1.2.1.16.1.1.1.14} (rmon.1.1.1.14)
SNMP:      Value = NULL
SNMP:
SNMP:      Object = {1.3.6.1.2.1.16.1.1.1.15} (rmon.1.1.1.15)
SNMP:      Value = NULL
SNMP:
SNMP:      Object = {1.3.6.1.2.1.16.1.1.1.16} (rmon.1.1.1.16)
SNMP:      Value = NULL
SNMP:
SNMP:      Object = {1.3.6.1.2.1.16.1.1.1.17} (rmon.1.1.1.17)
SNMP:      Value = NULL
SNMP:
SNMP:      Object = {1.3.6.1.2.1.16.1.1.1.18} (rmon.1.1.1.18)
SNMP:      Value = NULL
SNMP:
SNMP:      Object = {1.3.6.1.2.1.16.1.1.1.19} (rmon.1.1.1.19)
SNMP:      Value = NULL
SNMP:
SNMP:      Object = {1.3.6.1.2.1.16.1.1.1.20} (rmon.1.1.1.20)
SNMP:      Value = NULL
SNMP:
SNMP:      Object = {1.3.6.1.2.1.16.1.1.1.21} (rmon.1.1.1.21)
SNMP:      Value = NULL
SNMP:

--------------- Frame 87 ----------------
SNMP:      — Simple Network Management Protocol (Version 1) —
SNMP:
SNMP:      Version = 0
SNMP:      Community = armon
SNMP:      Command = Get response
SNMP:      Request ID = 0
SNMP:      Error status = 0 (No error)
```

```
SNMP:        Error index = 0
SNMP:
SNMP:        Object = {1.3.6.1.2.1.1.3.0} (sysUpTime.0)
SNMP:        Value = 10132171 hundredths of a second
SNMP:
SNMP:        Object = {1.3.6.1.2.1.16.1.1.1.1.1} (rmon.1.1.1.1.1)
SNMP:        Value = 1
SNMP:
SNMP:        Object = {1.3.6.1.2.1.16.1.1.1.2.1} (rmon.1.1.1.2.1)
SNMP:        Value = {1.3.6.1.2.1.2.2.1.1.1}
SNMP:
SNMP:        Object = {1.3.6.1.2.1.16.1.1.1.3.1} (rmon.1.1.1.3.1)
SNMP:        Value = 0 (counter)
SNMP:
SNMP:        Object = {1.3.6.1.2.1.16.1.1.1.4.1} (rmon.1.1.1.4.1)
SNMP:        Value = 1557016926 (counter)
SNMP:
SNMP:        Object = {1.3.6.1.2.1.16.1.1.1.5.1} (rmon.1.1.1.5.1)
SNMP:        Value = 7705082 (counter)
SNMP:
SNMP:        Object = {1.3.6.1.2.1.16.1.1.1.6.1} (rmon.1.1.1.6.1)
SNMP:        Value = 272349 (counter)
SNMP:
SNMP:        Object = {1.3.6.1.2.1.16.1.1.1.7.1} (rmon.1.1.1.7.1)
SNMP:        Value = 20682 (counter)
SNMP:
SNMP:        Object = {1.3.6.1.2.1.16.1.1.1.8.1} (rmon.1.1.1.8.1)
SNMP:        Value = 2082 (counter)
SNMP:
SNMP:        Object = {1.3.6.1.2.1.16.1.1.1.9.1} (rmon.1.1.1.9.1)
SNMP:        Value = 0 (counter)
SNMP:
SNMP:        Object = {1.3.6.1.2.1.16.1.1.1.10.1} (rmon.1.1.1.10.1)
SNMP:        Value = 0 (counter)
SNMP:
SNMP:        Object = {1.3.6.1.2.1.16.1.1.1.11.1} (rmon.1.1.1.11.1)
SNMP:        Value = 21389 (counter)
SNMP:
SNMP:        Object = {1.3.6.1.2.1.16.1.1.1.12.1} (rmon.1.1.1.12.1)
SNMP:        Value = 4 (counter)
SNMP:
SNMP:        Object = {1.3.6.1.2.1.16.1.1.1.13.1} (rmon.1.1.1.13.1)
SNMP:        Value = 163 (counter)
SNMP:
```

```
SNMP:      Object = {1.3.6.1.2.1.16.1.1.1.14.1} (rmon.1.1.1.14.1)
SNMP:      Value = 2425300 (counter)
SNMP:
SNMP:      Object = {1.3.6.1.2.1.16.1.1.1.15.1} (rmon.1.1.1.15.1)
SNMP:      Value = 3309942 (counter)
SNMP:
SNMP:      Object = {1.3.6.1.2.1.16.1.1.1.16.1} (rmon.1.1.1.16.1)
SNMP:      Value = 448315 (counter)
SNMP:
SNMP:      Object = {1.3.6.1.2.1.16.1.1.1.17.1} (rmon.1.1.1.17.1)
SNMP:      Value = 94386 (counter)
SNMP:
SNMP:      Object = {1.3.6.1.2.1.16.1.1.1.18.1} (rmon.1.1.1.18.1)
SNMP:      Value = 1019509 (counter)
SNMP:
SNMP:      Object = {1.3.6.1.2.1.16.1.1.1.19.1} (rmon.1.1.1.19.1)
SNMP:      Value = 386237 (counter)
SNMP:
SNMP:      Object = {1.3.6.1.2.1.16.1.1.1.20.1} (rmon.1.1.1.20.1)
SNMP:      Value = monitor
SNMP:
SNMP:      Object = {1.3.6.1.2.1.16.1.1.1.21.1} (rmon.1.1.1.21.1)
SNMP:      Value = 1
SNMP:
--------------- Frame 88 -----------------
SNMP:      — Simple Network Management Protocol (Version 1) —
SNMP:
SNMP:      Version = 0
SNMP:      Community = armon
SNMP:      Command = Get next request
SNMP:      Request ID = 0
SNMP:      Error status = 0 (No error)
SNMP:      Error index = 0
SNMP:
SNMP:      Object = {1.3.6.1.2.1.1.3} (sysUpTime)
SNMP:      Value = NULL
SNMP:
SNMP:      Object = {1.3.6.1.2.1.16.1.1.1.1.1} (rmon.1.1.1.1.1)
SNMP:      Value = NULL
SNMP:
```

```
SNMP:     Object = {1.3.6.1.2.1.16.1.1.1.2.1} (rmon.1.1.1.2.1)
SNMP:     Value = NULL
SNMP:
SNMP:     Object = {1.3.6.1.2.1.16.1.1.1.3.1} (rmon.1.1.1.3.1)
SNMP:     Value = NULL
SNMP:
SNMP:     Object = {1.3.6.1.2.1.16.1.1.1.4.1} (rmon.1.1.1.4.1)
SNMP:     Value = NULL
SNMP:
SNMP:     Object = {1.3.6.1.2.1.16.1.1.1.5.1} (rmon.1.1.1.5.1)
SNMP:     Value = NULL
SNMP:
SNMP:     Object = {1.3.6.1.2.1.16.1.1.1.6.1} (rmon.1.1.1.6.1)
SNMP:     Value = NULL
SNMP:
SNMP:     Object = {1.3.6.1.2.1.16.1.1.1.7.1} (rmon.1.1.1.7.1)
SNMP:     Value = NULL
SNMP:
SNMP:     Object = {1.3.6.1.2.1.16.1.1.1.8.1} (rmon.1.1.1.8.1)
SNMP:     Value = NULL
SNMP:
SNMP:     Object = {1.3.6.1.2.1.16.1.1.1.9.1} (rmon.1.1.1.9.1)
SNMP:     Value = NULL
SNMP:
SNMP:     Object = {1.3.6.1.2.1.16.1.1.1.10.1} (rmon.1.1.1.10.1)
SNMP:     Value = NULL
SNMP:
SNMP:     Object = {1.3.6.1.2.1.16.1.1.1.11.1} (rmon.1.1.1.11.1)
SNMP:     Value = NULL
SNMP:
SNMP:     Object = {1.3.6.1.2.1.16.1.1.1.12.1} (rmon.1.1.1.12.1)
SNMP:     Value = NULL
SNMP:
SNMP:     Object = {1.3.6.1.2.1.16.1.1.1.13.1} (rmon.1.1.1.13.1)
SNMP:     Value = NULL
SNMP:
SNMP:     Object = {1.3.6.1.2.1.16.1.1.1.14.1} (rmon.1.1.1.14.1)
SNMP:     Value = NULL
SNMP:
```

```
SNMP:      Object = {1.3.6.1.2.1.16.1.1.1.15.1} (rmon.1.1.1.15.1)
SNMP:      Value = NULL
SNMP:
SNMP:      Object = {1.3.6.1.2.1.16.1.1.1.16.1} (rmon.1.1.1.16.1)
SNMP:      Value = NULL
SNMP:
SNMP:      Object = {1.3.6.1.2.1.16.1.1.1.17.1} (rmon.1.1.1.17.1)
SNMP:      Value = NULL
SNMP:
SNMP:      Object = {1.3.6.1.2.1.16.1.1.1.18.1} (rmon.1.1.1.18.1)
SNMP:      Value = NULL
SNMP:
SNMP:      Object = {1.3.6.1.2.1.16.1.1.1.19.1} (rmon.1.1.1.19.1)
SNMP:      Value = NULL
SNMP:
SNMP:      Object = {1.3.6.1.2.1.16.1.1.1.20.1} (rmon.1.1.1.20.1)
SNMP:      Value = NULL
SNMP:
SNMP:      Object = {1.3.6.1.2.1.16.1.1.1.21.1} (rmon.1.1.1.21.1)
SNMP:      Value = NULL
SNMP:

- - - - - - - - - - - - - - - Frame 89 - - - - - - - - - - - - - - - - -
SNMP:      — Simple Network Management Protocol (Version 1) —
SNMP:
SNMP:      Version = 0
SNMP:      Community = armon
SNMP:      Command = Get response
SNMP:      Request ID = 0
SNMP:      Error status = 0 (No error)
SNMP:      Error index = 0
SNMP:
SNMP:      Object = {1.3.6.1.2.1.1.3.0} (sysUpTime.0)
SNMP:      Value = 10132180 hundredths of a second
SNMP:
SNMP:      Object = {1.3.6.1.2.1.16.1.1.1.2.1} (rmon.1.1.1.2.1)
SNMP:      Value = {1.3.6.1.2.1.2.2.1.1.1}
SNMP:
SNMP:      Object = {1.3.6.1.2.1.16.1.1.1.3.1} (rmon.1.1.1.3.1)
SNMP:      Value = 0 (counter)
SNMP:
SNMP:      Object = {1.3.6.1.2.1.16.1.1.1.4.1} (rmon.1.1.1.4.1)
SNMP:      Value = 1557018586 (counter)
```

```
SNMP:
SNMP:        Object = {1.3.6.1.2.1.16.1.1.1.5.1} (rmon.1.1.1.5.1)
SNMP:        Value = 7705087 (counter)
SNMP:
SNMP:        Object = {1.3.6.1.2.1.16.1.1.1.6.1} (rmon.1.1.1.6.1)
SNMP:        Value = 272349 (counter)
SNMP:
SNMP:        Object = {1.3.6.1.2.1.16.1.1.1.7.1} (rmon.1.1.1.7.1)
SNMP:        Value = 20682 (counter)
SNMP:
SNMP:        Object = {1.3.6.1.2.1.16.1.1.1.8.1} (rmon.1.1.1.8.1)
SNMP:        Value = 2082 (counter)
SNMP:
SNMP:        Object = {1.3.6.1.2.1.16.1.1.1.9.1} (rmon.1.1.1.9.1)
SNMP:        Value = 0 (counter)
SNMP:
SNMP:        Object = {1.3.6.1.2.1.16.1.1.1.10.1} (rmon.1.1.1.10.1)
SNMP:        Value = 0 (counter)
SNMP:
SNMP:        Object = {1.3.6.1.2.1.16.1.1.1.11.1} (rmon.1.1.1.11.1)
SNMP:        Value = 21389 (counter)
SNMP:
SNMP:        Object = {1.3.6.1.2.1.16.1.1.1.12.1} (rmon.1.1.1.12.1)
SNMP:        Value = 4 (counter)
SNMP:
SNMP:        Object = {1.3.6.1.2.1.16.1.1.1.13.1} (rmon.1.1.1.13.1)
SNMP:        Value = 163 (counter)
SNMP:
SNMP:        Object = {1.3.6.1.2.1.16.1.1.1.14.1} (rmon.1.1.1.14.1)
SNMP:        Value = 2425301 (counter)
SNMP:
SNMP:        Object = {1.3.6.1.2.1.16.1.1.1.15.1} (rmon.1.1.1.15.1)
SNMP:        Value = 3309943 (counter)
SNMP:
SNMP:        Object = {1.3.6.1.2.1.16.1.1.1.16.1} (rmon.1.1.1.16.1)
SNMP:        Value = 448315 (counter)
SNMP:
SNMP:        Object = {1.3.6.1.2.1.16.1.1.1.17.1} (rmon.1.1.1.17.1)
SNMP:        Value = 94388 (counter)
SNMP:
SNMP:        Object = {1.3.6.1.2.1.16.1.1.1.18.1} (rmon.1.1.1.18.1)
SNMP:        Value = 1019510 (counter)
SNMP:
SNMP:        Object = {1.3.6.1.2.1.16.1.1.1.19.1} (rmon.1.1.1.19.1)
```

```
SNMP:       Value = 386237 (counter)
SNMP:
SNMP:       Object = {1.3.6.1.2.1.16.1.1.1.20.1} (rmon.1.1.1.20.1)
SNMP:       Value = monitor
SNMP:
SNMP:       Object = {1.3.6.1.2.1.16.1.1.1.21.1} (rmon.1.1.1.21.1)
SNMP:       Value = 1
SNMP:
SNMP:       Object = {1.3.6.1.2.1.16.2.1.1.1.1} (rmon.2.1.1.1.1)
SNMP:       Value = 1
SNMP:
```

In summary, RMON can provide a number of host-derived statistics and store these statistics in several different ways for the convenience of the end user.

13.9 Event Notification Using RMON

In Section 12.3, we studied the RMON MIB and presented an example of retrieving RMON statistics. In this case study, we will go beyond the retrieval of information and illustrate how RMON can send a notification if a particular network parameter has exceeded the range of acceptable operation. The network consists of a single Ethernet segment with an attached RMON network management console and remote monitoring agent, or Probe (Figure 13-8). In order for the Probe to monitor and respond to a particular network statistic, it must first be programmed by the Manager (see Trace 13.9a).

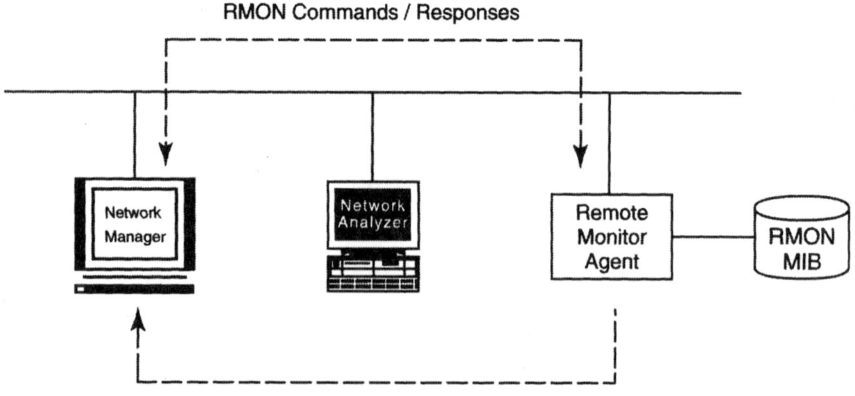

Figure 13-8. RMON Event Notification.

Trace 13.9a. RMON Threshold Analysis Summary

Sniffer Network Analyzer data from 21-Feb at 16:47:08, file TRAP.ENC, Page 1

SUMMARY	Delta T	Destination	Source	Summary
M 1	Probe	Manager	SNMP	Get sysUpTime .. ifSpeed (3 items)
2	0.0020	Manager	Probe	SNMP GetReply sysUpTime .. ifSpeed (3 items)
3	0.0396	Probe	Manager	SNMP Get sysUpTime
4	0.0007	Probe	Manager	SNMP Set Protools.2.1.2. XXX.YYY.135.195, Protools.2.1.3. XXX.YYY.135.195
5	0.0006	Manager	Probe	SNMP GetReply sysUpTime = 69284 ticks
6	0.0008	Manager	Probe	SNMP GetReply Protools.2.1.2. XXX.YYY.135.195, Protools.2.1.3. XXX.YYY.135.195
7	0.0041	Probe	Manager	SNMP GetNext rmon.4.1.1.5.0, rmon.4.1.1.2.0
8	0.0008	Probe	Manager	SNMP GetNext rmon.1.1.1.20.0, rmon.1.1.1.2.0
9	0.0008	Manager	Probe	SNMP GetReply rmon.4.1.1.5.1, rmon.4.1.1.2.1
10	0.0008	Manager	Probe	SNMP GetReply rmon.1.1.1.20.1, rmon.1.1.1.2.1
11	0.0155	Probe	Manager	SNMP Set rmon.9.1.1.7.31860 = 2
12	0.0006	Probe	Manager	SNMP Get rmon.1.1.1.5.1
13	0.0008	Manager	Probe	SNMP GetReply rmon.9.1.1.7.31860 = 2

14	0.0007	Manager	Probe	SNMP GetReply rmon.1.1.1.5.1 = 74126
15	0.0012	Probe	Manager	SNMP Set rmon.9.1.1.2.31860 .. rmon.9.1.1.6.31860 (4 items)
16	0.0014	Manager	Probe	SNMP GetReply rmon.9.1.1.2.31860 .. rmon.9.1.1.6.31860 (4 items)
17	0.0012	Probe	Manager	SNMP Set rmon.9.1.1.7.31860 = 1
18	0.0010	Manager	Probe	SNMP GetReply rmon.9.1.1.7.31860 = 1
19	0.0011	Probe	Manager	SNMP Set rmon.3.1.1.12.29148 = 2
20	0.0010	Manager	Probe	SNMP GetReply rmon.3.1.1.12.29148 = 2
21	0.0025	Probe	Manager	SNMP Set rmon.3.1.1.2.29148 .. rmon.3.1.1.11.29148 (9 items)
22	0.0019	Manager	Probe	SNMP GetReply rmon.3.1.1.2.29148 .. rmon.3.1.1.11.29148 (9 items)
23	0.0016	Probe	Manager	SNMP Set rmon.3.1.1.12.29148 = 1
24	0.0010	Manager	Probe	SNMP GetReply rmon.3.1.1.12.29148 = 1
25	0.0011	Probe	Manager	SNMP GetNext rmon.9.2.1.3.31860, rmon.9.2.1.4.31860
26	0.0014	Manager	Probe	SNMP GetReply rmon.11.1.0, rmon.11.1.0

27	0.0011	Probe	Manager	SNMP Get rmon.9.1.1.7.31860, rmon.3.1.1.12.29148
28	0.0010	Manager	Probe	SNMP GetReply rmon.9.1.1.7.31860, rmon.3.1.1.12.29148
29	4.9868	Probe	Manager	SNMP Get sysUpTime = 69284 ticks
30	0.0003	Probe	Manager	SNMP Get rmon.1.1.1.5.1 = 74126
31	0.0009	Manager	Probe	SNMP GetReply sysUpTime = 69787 ticks
32	0.0007	Manager	Probe	SNMP GetReply rmon.1.1.1.5.1 = 74582
33	0.0087	Manager	Probe	SNMP Trap-v1 Enterprise specific rmon.3.1.1.1.29148 .. rmon.3.1.1.7.29148 (5 items)
34	0.0003	Probe	Manager	ICMP Destination unreachable (Port unreachable)
35	0.0091	Probe	Manager	SNMP GetNext rmon.9.2.1.3.31860, rmon.9.2.1.4.31860
36	0.0012	Manager	Probe	SNMP GetReply rmon.9.2.1.3.31860.1, rmon.9.2.1.4.31860.1
37	0.1371	Probe	Manager	SNMP GetNext rmon.9.2.1.3.31860.1, rmon.9.2.1.4. 31860.1
38	0.0013	Manager	Probe	SNMP GetReply rmon.9.2.1.4.31860.1, rmon.11.1.0
39	0.0039	Probe	Manager	SNMP Get rmon.9.1.1.7.31860, rmon.3.1.1.12.29148
40	0.0010	Manager	Probe	SNMP GetReply rmon.9.1.1.7.31860, rmon.3.1.1.12.29148

41	1.8471	Probe	Manager	SNMP Set rmon.3.1.1.12.29148 = 4
42	0.0004	Probe	Manager	SNMP Set rmon.9.1.1.7.31860 = 4
43	0.0008	Manager	Probe	SNMP GetReply rmon.3.1.1.12.29148 = 4
44	0.0003	Probe	Manager	ICMP Destination unreachable (Port unreachable)
45	0.0011	Manager	Probe	SNMP GetReply rmon.9.1.1.7.31860 = 4

In Frame 4, the Manager sets the value of a private MIB object, called *etherUtilization*, to a threshold that is acceptable for the current network conditions. The Probe issues a confirmation in Frame 6. In Frames 11–18, the Manager initializes values in the *eventTable*, which creates a row entry that is indexed by a random number chosen by the Manager (31860). Note that in Frame 15 (Trace 13.9b), values of four of the objects from the *eventTable* {1.3.6.1.2.1.16.9.1} are specified:

Object	OID	Value
eventDescription	{rmon 9.1.1.2}	% Network usage
eventType	{rmon 9.1.1.3}	4
eventCommunity	{rmon 9.1.1.4}	public
eventOwner	{rmon 9.1.1.6}	FM/X (XXX.YYY.135.195): PID 3632 ID 0x008a0310

The *eventDescription* object provides a textual description that describes the event (% Network usage). The *eventType* object defines the type of notification that the probe will make as a result of this event. The RMON MIB (page 85 of RFC 2819) defines Type = 4 as a "log-and-trap" notification. The *eventCommunity* object defines the community where the SNMP trap will be sent. The *eventOwner* object defines the entity that configured this entry (the RMON console, with the IP address specified above). The PID value is a process identification assigned by the Manager.

Frame 16 is a GetResponse from the Probe to the Manager to confirm the parameters that were set. In Frame 17, the Manager issues another SetRequest, this time for the *eventStatus* object {1.3.6.1.2.1.16.9.1.1.7} with Value = 1, which validates this table entry. The Probe sends a confirming GetResponse in Frame 18.

Trace 13.9b. RMON EventTable Initialization

Sniffer Network Analyzer data from 21-Feb at 16:47:08, file TRAP.ENC, Page 1

```
--------------- Frame 15 -----------------
SNMP:      — Simple Network Management Protocol (Version 1) —
SNMP:
SNMP:      Version = 0
SNMP:      Community = public
SNMP:      Command = Set request
SNMP:      Request ID = 918004690
SNMP:      Error status = 0 (No error)
SNMP:      Error index = 0
SNMP:
SNMP:      Object = {1.3.6.1.2.1.16.9.1.1.2.31860} (rmon.9.1.1.2.31860)
SNMP:      Value  = % Network usage
SNMP:
SNMP:      Object = {1.3.6.1.2.1.16.9.1.1.3.31860} (rmon.9.1.1.3.31860)
SNMP:      Value  = 4
SNMP:
SNMP:      Object = {1.3.6.1.2.1.16.9.1.1.4.31860} (rmon.9.1.1.4.31860)
SNMP:      Value  = public
SNMP:
SNMP:      Object = {1.3.6.1.2.1.16.9.1.1.6.31860} (rmon.9.1.1.6.31860)
SNMP:      Value  = FM/X (XXX.YYY.135.195) : PID 3632 ID 0x008a0310
SNMP:

--------------- Frame 16 -----------------
SNMP:      — Simple Network Management Protocol (Version 1) —
SNMP:
SNMP:      Version = 0
SNMP:      Community = public
SNMP:      Command = Get response
SNMP:      Request ID = 918004690
SNMP:      Error status = 0 (No error)
```

```
SNMP:     Error index = 0
SNMP:
SNMP:     Object = {1.3.6.1.2.1.16.9.1.1.2.31860} (rmon.9.1.1.2.31860)
SNMP:     Value  = % Network usage
SNMP:
SNMP:     Object = {1.3.6.1.2.1.16.9.1.1.3.31860} (rmon.9.1.1.3.31860)
SNMP:     Value  = 4
SNMP:
SNMP:     Object = {1.3.6.1.2.1.16.9.1.1.4.31860} (rmon.9.1.1.4.31860)
SNMP:     Value  = public
SNMP:
SNMP:     Object = {1.3.6.1.2.1.16.9.1.1.6.31860} (rmon.9.1.1.6.31860)
SNMP:     Value  = FM/X (XXX.YYY.135.195) : PID 3632 ID 0x008a0310
SNMP:

- - - - - - - - - - - - - - - Frame 17 - - - - - - - - - - - - - - - -
SNMP:     — Simple Network Management Protocol (Version 1) —
SNMP:
SNMP:     Version = 0
SNMP:     Community = public
SNMP:     Command = Set request
SNMP:     Request ID = 918004691
SNMP:     Error status = 0 (No error)
SNMP:     Error index = 0
SNMP:
SNMP:     Object = {1.3.6.1.2.1.16.9.1.1.7.31860} (rmon.9.1.1.7.31860)
SNMP:     Value  = 1
SNMP:

- - - - - - - - - - - - - - - Frame 18 - - - - - - - - - - - - - - - -
SNMP:     — Simple Network Management Protocol (Version 1) —
SNMP:
SNMP:     Version = 0
SNMP:     Community = public
SNMP:     Command = Get response
SNMP:     Request ID = 918004691
SNMP:     Error status = 0 (No error)
SNMP:     Error index = 0
SNMP:
```

```
SNMP:      Object = {1.3.6.1.2.1.16.9.1.1.7.31860} (rmon.9.1.1.7.31860)
SNMP:      Value  = 1
SNMP:
```

Reviewing Frames 19–24 in Trace 13.9a, the Manager initializes the *alarmTable* {1.3.6.1.2.1.16.3.1}. Note that in Frame 19, the Manager sets the *alarmStatus* object {rmon.3.1.1.12} to a Value = 2 (defined as "createRequest" on page 15 of RFC 2819), and selects a random number for an index for this table entry (29148). Other values of the *alarmStatus* table are set in Frame 21 (Trace 13.9c), including:

Object	OID	Value
alarmInterval	{rmon 3.1.1.2}	5
alarmVariable	{rmon 3.1.1.3}	{1.3.6.1.4.1.209.2.2.1.2.1}
alarmSampleType	{rmon 3.1.1.4}	1
alarmRisingThreshold	{rmon 3.1.1.7}	1
alarmFallingThreshold	{rmon 3.1.1.8}	1
alarmFallingEventIndex	{rmon 3.1.1.10}	0
alarmRisingEventIndex	{rmon 3.1.1.9}	31860
alarmOwner	{rmon 3.1.1.11}	FM/X (XXX.YYY.135.195): PID 3632 ID 0x008a0310

Trace 13.9c. RMON alarmTable Initialization

Sniffer Network Analyzer data from 21-Feb at 16:47:08, file TRAP.ENC, Page 1

```
- - - - - - - - - - - - - - - - Frame 21 - - - - - - - - - - - - - - - - - -
SNMP:      — Simple Network Management Protocol (Version 1) —
SNMP:
SNMP:      Version = 0
SNMP:      Community = public
SNMP:      Command = Set request
SNMP:      Request ID = 918004693
SNMP:      Error status = 0 (No error)
SNMP:      Error index = 0
SNMP:
SNMP:      Object = {1.3.6.1.2.1.16.3.1.1.2.29148} (rmon.3.1.1.2.29148)
SNMP:      Value  = 5
SNMP:
SNMP:      Object = {1.3.6.1.2.1.16.3.1.1.3.29148} (rmon.3.1.1.3.29148)
SNMP:      Value  = {1.3.6.1.4.1.209.2.2.1.2.1}
SNMP:
SNMP:      Object = {1.3.6.1.2.1.16.3.1.1.4.29148} (rmon.3.1.1.4.29148)
SNMP:      Value  = 1
SNMP:
```

```
SNMP:        Object = {1.3.6.1.2.1.16.3.1.1.4.29148} (rmon.3.1.1.4.29148)
SNMP:        Value  = 1
SNMP:
SNMP:        Object = {1.3.6.1.2.1.16.3.1.1.7.29148} (rmon.3.1.1.7.29148)
SNMP:        Value  = 1
SNMP:
SNMP:        Object = {1.3.6.1.2.1.16.3.1.1.8.29148} (rmon.3.1.1.8.29148)
SNMP:        Value  = 1
SNMP:
SNMP:        Object = {1.3.6.1.2.1.16.3.1.1.10.29148} (rmon.3.1.1.10.29148)
SNMP:        Value  = 0
SNMP:
SNMP:        Object = {1.3.6.1.2.1.16.3.1.1.9.29148} (rmon.3.1.1.9.29148)
SNMP:        Value  = 31860
SNMP:
SNMP:        Object = {1.3.6.1.2.1.16.3.1.1.11.29148} (rmon.3.1.1.11.29148)
SNMP:        Value  = FM/X (XXX.YYY.135.195) : PID 3632 ID 0x008a0310
SNMP:
```

In Frame 23, the Manager sets the *alarmStatus* object to a value of 1 (defined as "valid" on page 15 of RFC 2819), which is confirmed in Frame 24.

The network does exceed the threshold of network utilization that was specified; as a result, the Probe issues a trap in Frame 33 (Trace 13.9d). Note that this is an enterprise-specific trap, and that the trap reports the values of some objects from the *alarmTable* {1.3.6.1.2.1.16.3.1}. These objects are:

Object	OID	Value
alarmIndex	{rmon.3.1.1.1}	29148
alarmVariable	{rmon.3.1.1.3}	{1.3.6.1.4.1.209.2.2.1.2.1}
alarmSampleType	{rmon.3.1.1.4}	1
alarmValue	{rmon.3.1.1.5}	3
alarmRisingThreshold	{rmon.3.1.1.7}	1

The *alarmIndex* object uniquely identifies an entry in the alarm table; its value is a random value that was originally determined by the Manager (review Frame 19). The *alarmVariable* object identifies the OID of the variable to be sampled. In this case, the variable is derived from a Private MIB (enterprise number 209, assigned to ProTools). The *alarmSampleType* object specifies the method of sampling the selected variable and calculating the value to be compared against the thresholds. The value of 1 specifies an absolute value, meaning that the value of the selected variable will be compared directly with the thresholds at the end of the sampling interval. The *alarmValue* object is the value of the statistic during the last sampling period. Finally, the *alarmRisingThreshold* object is a threshold for the sampled statistic. According to page 37 of RFC 2819, when the current sampled value is greater than or equal to this threshold, and the value at the last sampling interval was less than this threshold, a single event will be generated.

Trace 13.9d. Trap Sent in Response to an Excessive Network Condition
Sniffer Network Analyzer data from 21-Feb at 16:47:08, file TRAP.ENC, Page 1

```
- - - - - - - - - - - - - - Frame 33 - - - - - - - - - - - - - - - -
SNMP:         — Simple Network Management Protocol (Version 1) —
SNMP:
SNMP:         Version = 0
SNMP:         Community = public
SNMP:         Command = Trap
SNMP:         Enterprise = {1.3.6.1.2.1.16}
SNMP:         Network address = [XXX.YYY.135.11], Probe
SNMP:         Generic trap = 6 (Enterprise specific)
SNMP:         Specific trap = 1
SNMP:         Time ticks = 69788
SNMP:
SNMP:         Object = {1.3.6.1.2.1.16.3.1.1.1.29148} (rmon.3.1.1.1.29148)
SNMP:         Value  = 29148
SNMP:
SNMP:         Object = {1.3.6.1.2.1.16.3.1.1.3.29148} (rmon.3.1.1.3.29148)
SNMP:         Value  = {1.3.6.1.4.1.209.2.2.1.2.1}
SNMP:
SNMP:         Object = {1.3.6.1.2.1.16.3.1.1.4.29148} (rmon.3.1.1.4.29148)
SNMP:         Value  = 1
SNMP:
SNMP:         Object = {1.3.6.1.2.1.16.3.1.1.5.29148} (rmon.3.1.1.5.29148)
SNMP:         Value  = 3
SNMP:
SNMP:         Object = {1.3.6.1.2.1.16.3.1.1.7.29148} (rmon.3.1.1.7.29148)
SNMP:         Value  = 1
SNMP:
```

The next step is for the Manager to retrieve the values from the *logTable* (recall that the eventType object set in Frame 11 specified a "log-and-trap" event, event type = 4). Frames 35–36 read the values from this *logTable* (see Trace 13.9e), and Frames 37–38 (which are not shown in the trace) merely confirm that all entries in that table have been read. Frame 36 contains the entries of the *logTable*, indexed by the random number initially selected by the Manager (31860):

Object	OID	Value
logTime	{rmon.9.2.1.3}	69788
logDescription	{rmon.9.2.1.4}	% Network usage 3. Over threshold of (1)

Trace 13.9e. Reading the logTable
Sniffer Network Analyzer data from 21-Feb at 16:47:08, file: TRAP.ENC, Page 1

```
- - - - - - - - - - - - - - - Frame 35 - - - - - - - - - - - - - - - -
SNMP:      — Simple Network Management Protocol (Version 1) —
SNMP:
SNMP:      Version = 0
SNMP:      Community = public
SNMP:      Command = Get next request
SNMP:      Request ID = 918004699
SNMP:      Error status = 0 (No error)
SNMP:      Error index = 0
SNMP:
SNMP:      Object = {1.3.6.1.2.1.16.9.2.1.3.31860} (rmon.9.2.1.3.31860)
SNMP:      Value  = NULL
SNMP:
SNMP:      Object = {1.3.6.1.2.1.16.9.2.1.4.31860} (rmon.9.2.1.4.31860)
SNMP:      Value  = NULL
SNMP:

- - - - - - - - - - - - - - - Frame 36 - - - - - - - - - - - - - - - -
SNMP:      — Simple Network Management Protocol (Version 1) —
SNMP:
SNMP:      Version = 0
SNMP:      Community = public
SNMP:      Command = Get response
SNMP:      Request ID = 918004699
SNMP:      Error status = 0 (No error)
SNMP:      Error index = 0
SNMP:
SNMP:      Object = {1.3.6.1.2.1.16.9.2.1.3.31860.1} (rmon.9.2.1.3.31860.1)
SNMP:      Value  = 69788 (time ticks)
SNMP:
SNMP:      Object = {1.3.6.1.2.1.16.9.2.1.4.31860.1} (rmon.9.2.1.4.31860.1)
SNMP:      Value  = % Network usage   3. Over threshold of (1)
SNMP:
```

The balance of the events (review Trace 13.9a) reads and invalidates the *alarmTable* and *eventTable* by setting the values of the *alarmStatus* {rmon 3.1.1.12} and *eventStatus* {rmon 9.1.1.7} to a value of 4 (invalid).

In summary, this trace has illustrated how the RMON alarm and event groups interact to provide both monitoring and notification of significant network conditions.

13.10 Encrypting Management Information with SNMPv3

In our discussions regarding the various versions of SNMP, we discussed the matter of network security. When SNMP was becoming more widely deployed in the mid-1990s, many network managers recognized the need for stronger security in SNMPv1, which led to the development of SNMPv2. Unfortunately, the security solutions provided with SNMPv2 did not meet with rave reviews, which led to the development of SNMPv3. As this case study will demonstrate, it would appear that the SNMPv3 security enhancements are finally ready for "prime time."

In this example, an SNMPv3 management console wishes to query an SNMPv3 agent within an ATM switch, but without the danger of other systems or devices becoming privy to that communication (Figure 13-9). As we will discover, the SNMPv3 messages are transmitted in a secure manner to identify the source of the information (the authentication function) and to keep the message contents secret (the encryption function).

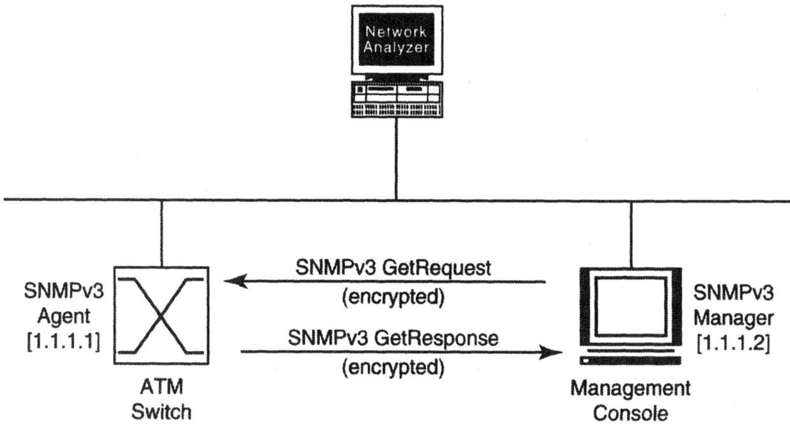

Figure 13-9. SNMPv3 Agent/Manager Interaction.

The devices in this network implement the User-Based Security Model (USM) defined in RFC 3414, and use a message format illustrated in Figure 13-10. (Note that this illustration is similar to Figure 12-61, but with the added parameters defined by the USM.) Of particular interest are the Message Security Parameters defined by the USM, which include the Authoritative Engine ID, Authoritative Engine Boots, Authoritative Engine Time, Message User Name, Authentication Parameters, and Privacy Parameters fields.

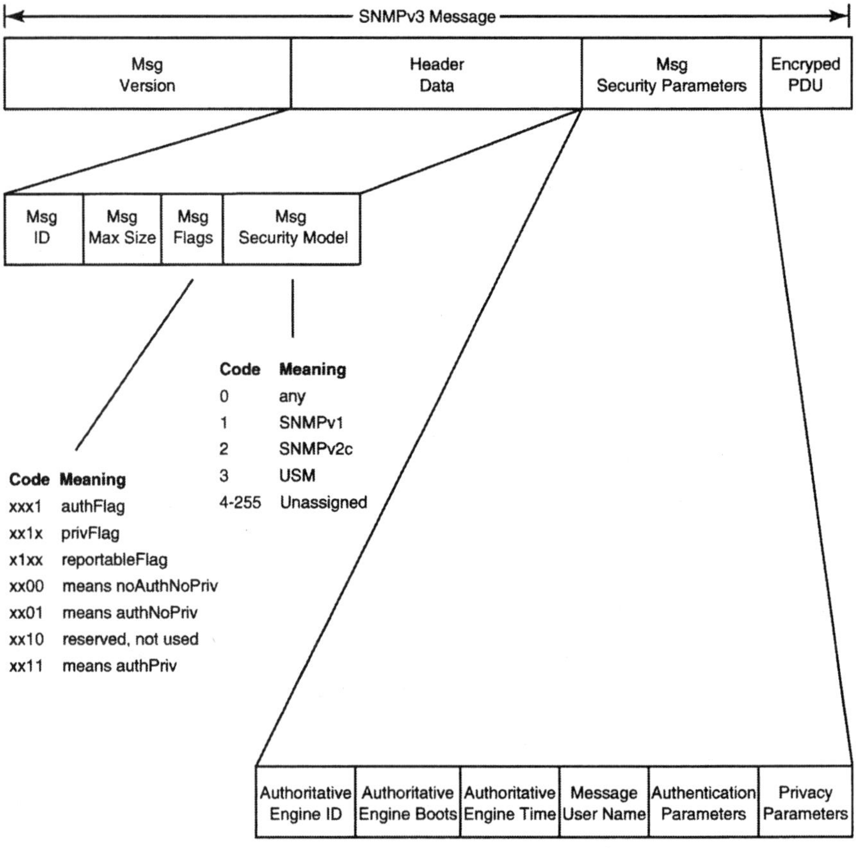

Figure 13-10. SNMPv3 Message Format for User-Based Security Model (USM) with Encryption.

From the details of Frame 5 shown in Trace 13.10a, note that the same UDP Port Number (161) that was used with SNMPv1 is also used to identify this message, and that the first field of the SNMP message specifies the version number of the protocol (SNMPv3). We also see a number of fields that are sent in clear (unencrypted) text: the Message ID (31081) that correlates sent and received messages; the Message Flags (0111), which indicate a reportable message using both authentication and encryption; the Security Model (3) identifying that the USM is being used for this communication; the Authoritative Engine ID (<800001>F<0300> H+<FDEF>), which specifies the snmpEngineID of the authoritative SNMP engine involved in this message exchange; the Authoritative Engine Boots (8), which specifies the snmpEngineBoots value at the authoritative SNMP engine involved in this message exchange; the Authoritative Engine Time (1286918), which specifies the snmpEngineTime value at the authoritative SNMP engine involved in this message exchange; the User Name (snmpv3), which specifies the user on whose behalf the message is being exchanged; the Authentication Parameters (<E00F>P<0FCF>i@o<8A>~\b), which are defined by the authentication protocol in use for the message and are found in the *usmUserAuthProtocol* column of the usmUserTable (review Figure 12-67); and finally the Privacy Parameters, which are defined by the privacy protocol in use for the message and are found in the *usmUserPrivProtocol* column in the *usmUserTable* (also shown in Figure 12-67).

In the hexidecimal display, note that the Privacy Parameters are shown in the line that begins with 0070 (shown in boldface type):

```
08 00 00 00 08 c0 77 7f 73
```

The encrypted PDU that follows these Privacy Parameters cannot be decoded by the ASCII display of the protocol analyzer — it just appears as a random sequence of hexadecimal characters. A similar condition occurs in Frame 6 (also shown in Trace 13.10a), in that the information following the Privacy Parameters is also not decodable by the network analyzer.

Trace 13.10a. Protocol Decode of Encrypted SNMPv3 Messages Between Manager and Agent

Sniffer Network Analyzer data from 15-May at 13:54:02, file SNMPv3_trace1.cap, Pg 1

```
- - - - - - - - - - - - - - - - Frame 5 - - - - - - - - - - - - - - - -
DLC:  —— DLC Header ——
    DLC:
    DLC:  Frame 5 arrived at 13:54:02; frame size is 243 bytes.
    DLC:  Destination = Station Marcni1A368C
    DLC:  Source      = Station MegHz 4E40B0
    DLC:  Ethertype   = 0800 (IP)
    DLC:
IP: —— IP Header ——
    IP:
    IP: Version = 4, header length = 20 bytes
    IP: Type of service = 00
    IP:       000. ....  = routine
    IP:       ...0 .... = normal delay
    IP:       .... 0... = normal throughput
    IP:       .... .0.. = normal reliability
    IP:       .... ..0. = ECT bit - trans proto will ignore CE bit
    IP:       .... ...0 = CE bit - no congestion
    IP: Total length    = 229 bytes
    IP: Identification  = 0
    IP: Flags           = 4X
    IP:       .1.. .... = don't fragment
    IP:       ..0. .... = last fragment
    IP: Fragment offset = 0 bytes
    IP: Time to live    = 64 seconds/hops
    IP: Protocol        = 17 (UDP)
    IP: Header checksum = 3604 (correct)
    IP: Source address      = [1.1.1.2]
    IP: Destination address = [1.1.1.1]
    IP: No options
    IP:
UDP: —— UDP Header ——
    UDP:
    UDP: Source port      = 32769 (Proteon Tknrng)
    UDP: Destination port =   161 (SNMP)
    UDP: Length           = 209
    UDP: Checksum         = 5F7C (correct)
    UDP: [201 byte(s) of data]
    UDP:
```

```
SNMP: —— Simple Network Management Protocol (Version 3) ——
    SNMP:
    SNMP: SNMP Version = 3
    SNMP: Id             = 31081
    SNMP: Maximum size   = 1472
    SNMP: Message flags  = 07
    SNMP:       .... ...1 = authFlag is on
    SNMP:       .... ..1. = privFlag is on
    SNMP:       .... .1.. = reportableFlag is on
    SNMP: Security model = 3
    SNMP: Authoritative engine id    = <800001>F<0300> H+<FDEF>
    SNMP: Authoritative engine boots = 8
    SNMP: Authoritative engine time  = 1286918
    SNMP: User name                  = snmpv3
    SNMP: Authentication parameters  = <E00F>P<0FCF>i@o<8A>~\b
    SNMP: Privacy parameters         = <00000008C0>w|s
```

```
ADDR  HEX                                              ASCII
0000: 00 20 48 1a 36 8c 00 00 86 4e 40 b0 08 00 45 00 | . H.6..N@..E.
0010: 00 e5 00 00 40 00 40 11 36 04 01 01 01 02 01 01 | ...@.@.6.......
0020: 01 01 80 01 00 a1 00 d1 5f 7c 30 81 c6 02 01 03 | ....?._|0?...
0030: 30 0e 02 02 79 69 02 02 05 c0 04 01 07 02 01 03 | 0...yi... .....
0040: 04 37 30 35 04 0b 80 00 01 46 03 00 20 48 2b fd | .705....F.. H+
0050: ef 02 01 08 02 03 13 a3 06 04 06 73 6e 6d 70 76 |  .........snmpv
0060: 33 04 0c e0 0f 50 0f cf 69 40 6f 8a 7e 5c 62 04 | 3...P. i@o~\b.
0070: 08 00 00 00 08 c0 77 7f 73 04 78 f9 2c db 2d 91 | .....w|s.x ,-?
0080: 6b 61 aa 8c 13 7d ab e8 a2 4b 6d 3f d2 06 f6 42 | ka.}Km? .B
0090: c0 8d f8 5f 69 31 ad e2 13 97 87 16 3c 0f 41 84 |  _i1 ..<.A
00a0: 87 a3 10 b3 41 6f b8 ed a0 fb c4 21 2b 3b b0 c8 | .Ao !+;??
00b0: 5a f1 58 d9 14 20 79 52 9f 74 fe b8 77 f8 ea 14 | Z X. yR?tw .
00c0: e9 5b aa 27 0d e1 e7 e3 44 86 74 68 f7 82 eb fe |  ['. Dth
00d0: 18 7a 92 19 32 e2 3d b8 d6 1b 05 97 e9 24 1a f1 | .z.2= ..$.
00e0: ef f0 c8 06 03 19 4a 87 c0 0c 82 5c cf 4b 0b f1 |  ...J?._\K.
00f0: 3d d1 56                                        | =V
```

```
- - - - - - - - - - - - - - - - Frame 6 - - - - - - - - - - - - - - - -
DLC: —— DLC Header ——
    DLC:
    DLC: Frame 6 arrived at  13:54:02; frame size is 329 bytes.
    DLC: Destination = Station MegHz 4E40B0
    DLC: Source      = Station Marcni1A368C
    DLC: Ethertype   = 0800 (IP)
    DLC:
IP: —— IP Header ——
    IP:
    IP: Version = 4, header length = 20 bytes
    IP: Type of service = 00
```

```
    IP:      000. ....  = routine
    IP:      ...0 .... = normal delay
    IP:      .... 0... = normal throughput
    IP:      .... .0.. = normal reliability
    IP:      .... ..0. = ECT bit - trans proto will ignore CE bit
    IP:      .... ...0 = CE bit - no congestion
    IP: Total length   = 315 bytes
    IP: Identification  = 30330
    IP: Flags          = 0X
    IP:      .0.. .... = may fragment
    IP:      ..0. .... = last fragment
    IP: Fragment offset = 0 bytes
    IP: Time to live   = 64 seconds/hops
    IP: Protocol       = 17 (UDP)
    IP: Header checksum = FF33 (correct)
    IP: Source address     = [1.1.1.1]
    IP: Destination address = [1.1.1.2]
    IP: No options
    IP:
UDP: —— UDP Header ——
    UDP:
    UDP: Source port      =  161 (SNMP)
    UDP: Destination port = 32769 (Proteon Tknrng)
    UDP: Length           = 295
    UDP: Checksum         = 933D (correct)
    UDP: [287 byte(s) of data]
    UDP:
SNMP: —— Simple Network Management Protocol (Version 3) ——
    SNMP:
    SNMP: SNMP Version = 3
    SNMP: Id           = 31081
    SNMP: Maximum size  = 2047
    SNMP: Message flags  = 03
    SNMP:      .... ...1 = authFlag is on
    SNMP:      .... ..1. = privFlag is on
    SNMP:      .... .0.. = reportableFlag is off
    SNMP: Security model = 3
    SNMP: Authoritative engine id    = <800001>F<0300> H+<FDEF>
    SNMP: Authoritative engine boots = 8
    SNMP: Authoritative engine time  = 1286918
    SNMP: User name                  = snmpv3
    SNMP: Authentication parameters  = <D7>$<80A7>n&<F6>4<CF>'<E7>2
    SNMP: Privacy parameters         = <0000008000000>&
```

```
ADDR  HEX                                                   ASCII
0000: 00 00 86 4e 40 b0 00 20 48 1a 36 8c 08 00 45 00  | ..N@. H.6..E.
0010: 01 3b 76 7a 00 00 40 11 ff 33 01 01 01 01 01 01  | .;vz..@.3......
0020: 01 02 00 a1 80 01 01 27 93 3d 30 82 01 1b 02 01  | .....'=0....
0030: 03 30 82 00 0e 02 02 79 69 02 02 07 ff 04 01 03  | .0....yi... ...
0040: 02 01 03 04 39 30 82 00 35 04 0b 80 00 01 46 03  | ....90.5....F.
0050: 00 20 48 2b fd ef 02 01 08 02 03 13 a3 06 04 06  | . H+ .........
0060: 73 6e 6d 70 76 33 04 0c d7 24 80 a7 6e 26 f6 34  | snmpv3..$?n&4
0070: cf 60 e7 32 04 08 00 00 00 08 00 00 00 26 04 81  |  `2.........&.
0080: c8 e9 ff 93 7b 9a e0 1f 2c 34 66 dc 1f ee 8f 09  |    {.,4f ..
0090: 51 9c fa d9 4f b5 e0 60 38 2e d2 18 a2 3a 61 a9  | Q?? O`8.  .:a
00a0: 21 fe 66 f8 44 14 76 3e f0 c3 01 8d 36 27 f9 37  | !f D.v> .6'7
00b0: 47 7c 57 a2 e5 29 d3 13 d3 5c 84 6a f6 ec 50 8d  | G|W ). \j P
00c0: 8c e0 4b fd 8f 8f dd 7a da a6 d2 33 9e 9d e3 ad  | K z 3??
00d0: a3 9b 4d 6e f4 fe f6 74 af db d8 43 38 b8 5d 44  | ?Mn t C8??]D
00e0: 30 8a e0 41 25 6a c7 ae d3 75 ee e1 52 ab 36 2e  | 0A%j u R6.
00f0: 28 9c 16 ac 47 94 82 7b 40 3c 8d 07 6d cb d1 90  | (?.G{@<.m
0100: a7 93 fe 1f 14 32 f7 fc 13 f8 2c 92 5f ee 26 74  | ..2 . ,_&t
0110: 5b 61 58 b2 53 d8 ab 12 f1 81 93 ad 4b ff ff d4  | [aXS .??K
0120: 7c cb ae a9 8f be 2c f5 89 6e b8 3a b6 73 f1 af  | | ,n:??s
0130: a1 f3 37 69 e5 b2 c7 d6 6e 9b 9a 2a 35 a8 98 5e  |  7i n??*5^
0140: 7f df 0b 17 9f 88 a5 72 82                       | |..?r
```

We can derive further information by looking into the software algorithms used at the management console and the agent at the ATM switch to send the SNMPv3 messages. (We are fortunate to have access to this information since this network was used for product development testing.) In Trace 13.10b, note that the USM parameters that were actually transmitted are shown in hex, and are displayed differently than those shown from the analyzer. For example, in Frame 5, the analyzer showed the following:

```
Authoritative engine id = <800001>F<0300> H+<FDEF>
```

The hexadecimal dump from the software algorithm is slightly different:

```
Authoritative Engine ID:0x80000146030020482bfdef
```

The differences can be easily explained. The "0x" is another way of indicating that hexadecimal notation follows, thus the first "0x" can be ignored. The next six characters (800001) are identical in both cases. The next sequence (an "F" in the analyzer display, and a "46" in the hexadecimal display) is actually the same — the 46 in hexadecimal is actually an uppercase F in the ASCII code, so the analyzer was just doing some decoding for us. The next "0300" is identical in both cases, and the "H+" from the analyzer translates to a "482B" in hexadecimal. The final four characters (FDEF) are identical. So, although the two formats look different, they actually represent the same sequence of information.

The balance of the message contains the encrypted packet, which is displayed in hexadecimal but protected from disclosure because of the encryption.

Trace 13.10b. Hexadecimal Dump of Encrypted SNMPv3 Messages Between Manager and Agent

```
=============== Packet 5 transmitted to [1.1.1.1] ===============
Message Length:                 198 bytes
Version:                        version-3(3)
Global Data Length:             14 bytes
Message ID:                     31081
Maximum Size:                   1472 octets
Message Flags:                  reportable,authPriv(7)
Security Model:                 USM(3)
USM Parameters Length:          53 bytes
Authoritative Engine ID:        0x800001460300204 82bfdef
Authoritative Engine Boots:     8
Authoritative Engine Time:      1286918
Message User Name:              snmpv3
Authentication Parameters:      0xe00f500fcf69406f8a7e5c62
Privacy Parameters:             0x00000008c0777f73
Encrypted PDU:                  120 bytes
   Hex Display [0081]:   f9 2c db 2d  91 6b 61 aa
   Hex Display [0089]:   8c 13 7d ab  e8 a2 4b 6d
   Hex Display [0097]:   3f d2 06 f6  42 c0 8d f8
   Hex Display [0105]:   5f 69 31 ad  e2 13 97 87
   Hex Display [0113]:   16 3c 0f 41  84 87 a3 10
   Hex Display [0121]:   b3 41 6f b8  ed a0 fb c4
   Hex Display [0129]:   21 2b 3b b0  c8 5a f1 58
   Hex Display [0137]:   d9 14 20 79  52 9f 74 fe
   Hex Display [0145]:   b8 77 f8 ea  14 e9 5b aa
   Hex Display [0153]:   27 0d e1 e7  e3 44 86 74
   Hex Display [0161]:   68 f7 82 eb  fe 18 7a 92
   Hex Display [0169]:   19 32 e2 3d  b8 d6 1b 05
   Hex Display [0177]:   97 e9 24 1a  f1 ef f0 c8
   Hex Display [0185]:   06 03 19 4a  87 c0 0c 82
   Hex Display [0193]:   5c cf 4b 0b  f1 3d d1 56
```

```
=============== Packet 6 received from [1.1.1.1] ===============
Message Length:                283 bytes
Version:                       version-3(3)
Global Data Length:            14 bytes
Message ID:                    31081
Maximum Size:                  2047 octets
Message Flags:                 authPriv(3)
Security Model:                USM(3)
USM Parameters Length:         53 bytes
Authoritative Engine ID:       0x800001460300204 82bfdef
Authoritative Engine Boots:    8
Authoritative Engine Time:     1286918
Message User Name:             snmpv3
Authentication Parameters:     0xd72480a76e26f634cf60e732
Privacy Parameters:            0x0000000800000026
Encrypted PDU:                 200 bytes
  Hex Display [0087]:    e9 ff 93 7b  9a e0 1f 2c
  Hex Display [0095]:    34 66 dc 1f  ee 8f 09 51
  Hex Display [0103]:    9c fa d9 4f  b5 e0 60 38
  Hex Display [0111]:    2e d2 18 a2  3a 61 a9 21
  Hex Display [0119]:    fe 66 f8 44  14 76 3e f0
  Hex Display [0127]:    c3 01 8d 36  27 f9 37 47
  Hex Display [0135]:    7c 57 a2 e5  29 d3 13 d3
  Hex Display [0143]:    5c 84 6a f6  ec 50 8d 8c
  Hex Display [0151]:    e0 4b fd 8f  8f dd 7a da
  Hex Display [0159]:    a6 d2 33 9e  9d e3 ad a3
  Hex Display [0167]:    9b 4d 6e f4  fe f6 74 af
  Hex Display [0175]:    db d8 43 38  b8 5d 44 30
  Hex Display [0183]:    8a e0 41 25  6a c7 ae d3
  Hex Display [0191]:    75 ee e1 52  ab 36 2e 28
  Hex Display [0199]:    9c 16 ac 47  94 82 7b 40
  Hex Display [0207]:    3c 8d 07 6d  cb d1 90 a7
  Hex Display [0215]:    93 fe 1f 14  32 f7 fc 13
  Hex Display [0223]:    f8 2c 92 5f  ee 26 74 5b
  Hex Display [0231]:    61 58 b2 53  d8 ab 12 f1
  Hex Display [0239]:    81 93 ad 4b  ff ff d4 7c
  Hex Display [0247]:    cb ae a9 8f  be 2c f5 89
  Hex Display [0255]:    6e b8 3a b6  73 f1 af a1
  Hex Display [0263]:    f3 37 69 e5  b2 c7 d6 6e
  Hex Display [0271]:    9b 9a 2a 35  a8 98 5e 7f
  Hex Display [0279]:    df 0b 17 9f  88 a5 72 82
```

We are also fortunate to have access to the decryption algorithm that can decode these two SNMPv3 messages (Trace 13.10c). Using this information, we find that Frame 5 was an SNMPv3 GetRequest message, and Frame 6 was the GetResponse. We can also determine that the GetRequest was walking through six of the objects of the system group from MIB-II: *sysDescr, sysObjectID, sysUpTime, sysContact, sysName,* and *sysLocation.* As would be expected, the object values are NULL in the GetRequest, and filled in with values (such as Marconi ASX-200BX) in the GetResponse. In addition, we can observe the presence of padding at the end of the GetResponse, which provides another form of security by disguising the actual length of the message.

Trace 13.10c. Hexadecimal Dump of Decrypted SNMPv3 Messages Between Manager and Agent

```
=============== Packet 5 transmitted to [1.1.1.1] ===============
Message Length:               198 bytes
Version:                      version-3(3)
Global Data Length:           14 bytes
Message ID:                   31081
Maximum Size:                 1472 octets
Message Flags:                reportable,authPriv(7)
Security Model:               USM(3)
USM Parameters Length:        53 bytes
Authoritative Engine ID:      0x80000146030020482bfdef
Authoritative Engine Boots:   8
Authoritative Engine Time:    1286918
Message User Name:            snmpv3
Authentication Parameters:    0xe00f500fcf69406f8a7e5c62
Privacy Parameters:           0x00000008c0777f73
Encrypted PDU:                120 bytes
Scoped PDU Length:            113 bytes
Context Engine ID:            0x80000146030020482bfdef
Context Name:                 <null>
PDU Type:                     get-request(0)
Request ID:                   31081
Error Status:                 noError(0)
Error Index:                  0
VarBindList Length:           84 bytes
VarBind Length:               12 bytes
Object Identifier:            1.3.6.1.2.1.1.1.0
  Name:                       sysDescr.0
Object Value:                 NULL
```

```
VarBind Length:              12 bytes
Object Identifier:           1.3.6.1.2.1.1.2.0
  Name:                      sysObjectID.0
Object Value:                NULL
VarBind Length:              12 bytes
Object Identifier:           1.3.6.1.2.1.1.3.0
  Name:                      sysUpTime.0
Object Value:                NULL
VarBind Length:              12 bytes
Object Identifier:           1.3.6.1.2.1.1.4.0
  Name:                      sysContact.0
Object Value:                NULL
VarBind Length:              12 bytes
Object Identifier:           1.3.6.1.2.1.1.5.0
  Name:                      sysName.0
Object Value:                NULL
VarBind Length:              12 bytes
Object Identifier:           1.3.6.1.2.1.1.6.0
  Name:                      sysLocation.0
Object Value:                NULL
Padding:                     05 05 05 05  05 .. .. ..

=============== Packet 6 received from [1.1.1.1] ===============
Message Length:              283 bytes
Version:                     version-3(3)
Global Data Length:          14 bytes
Message ID:                  31081
Maximum Size:                2047 octets
Message Flags:               authPriv(3)
Security Model:              USM(3)
USM Parameters Length:       53 bytes
Authoritative Engine ID:     0x80000146030020482bfdef
Authoritative Engine Boots:  8
Authoritative Engine Time:   1286918
Message User Name:           snmpv3
Authentication Parameters:   0xd72480a76e26f634cf60e732
Privacy Parameters:          0x0000000800000026
Encrypted PDU:               200 bytes
Scoped PDU Length:           189 bytes
Context Engine ID:           0x80000146030020482bfdef
Context Name:                <null>
PDU Type:                    get-response(2)
Request ID:                  31081
Error Status:                noError(0)
```

```
Error Index:               0
VarBindList Length:        156 bytes
VarBind Length:            29 bytes
Object Identifier:         1.3.6.1.2.1.1.1.0
 Name:                     sysDescr.0
Object Value:              Marconi ASX-200BX
VarBind Length:            22 bytes
Object Identifier:         1.3.6.1.2.1.1.2.0
 Name:                     sysObjectID.0
Object Value:              1.3.6.1.4.1.326.2.2.5
 Name:                     fore.2.2.5
VarBind Length:            16 bytes
Object Identifier:         1.3.6.1.2.1.1.3.0
 Name:                     sysUpTime.0
Object Value:              128691872
VarBind Length:            20 bytes
Object Identifier:         1.3.6.1.2.1.1.4.0
 Name:                     sysContact.0
Object Value:              sparsons
VarBind Length:            23 bytes
Object Identifier:         1.3.6.1.2.1.1.5.0
 Name:                     sysName.0
Object Value:              taclab-sw-9
VarBind Length:            22 bytes
Object Identifier:         1.3.6.1.2.1.1.6.0
 Name:                     sysLocation.0
Object Value:              Scott Quad
Padding:                   07 07 07 07  07 07 07 ..
```

13.11 Looking Back

This text has taken you on a journey through the technical details of the most exciting communications network known today — the Internet. We have considered the underlying communications infrastructure (LANs and WANs), the protocols that assure reliable packet delivery (such as TCP and IP), the applications that support the end users (such as FTP, TELNET, and HTTP), and, finally, the management systems that (should) keep it all running smoothly. I trust that you have found this information to be helpful. Thanks for coming along on the ride!

Appendix A

Acronyms and Abbreviations

A

A	Ampere
AAL	ATM Adaption Layer
AARP	AppleTalk Address Resolution Protocol
ABP	Alternate Bipolar
ACD	Automatic Call Distributor
ACELP	Algebraic-Code-Excited-Linear-Prediction
ACK	Acknowledgment
ACSE	Association Control Service Element
ACTLU	Activate Logical Unit
ACTPU	Activate Physical Unit
ADPCM	Adaptive Differential Pulse Code Modulation
AH	Authentication Header
AI	Artificial Intelligence
AIN	Advanced Intelligent Network
AIS	Alarm Indication Signal
AL	Alignment
AMI	Alternate Mark Inversion
ANSI	American National Standards Institute
API	Applications Program Interface
APPC	Advanced Program-to-Program Communication
ARP	Address Resolution Program
ARPA	Advanced Research Projects Agency

ARPANET	Advanced Research Projects Agency Network
ASCII	American Standard Code for Information Interchange
ASIC	Application-Specific Integrated Circuits
ASN.1	Abstract Syntax Notation One
ATIS	Alliance for Telecommunications Industry Solutions
ATM	Asynchronous Transfer Mode
ATM DXI	ATM Data Exchange Interface
ATM UNI	ATM User-Network Interface
AU	Access Unit
AUP	Acceptable Use Policy
AVT	Audio-Video Transport

B

B	B Channel
B	Broadband
B8ZS	Bipolar with 8 ZERO Substitution
BC	Bearer Capability
BC	Block Check
Bc	Committed Burst
BCD	Binary Coded Decimal
BCN	Backward Congestion Notification
Be	Excess Burst
BECN	Backward Explicit Congestion Notification
BER	Basic Encoding Rules
BER	Bit Error Ratio
B-ICI	Broadband Inter-Carrier Interface
BIOS	Basic Input/Output System
B-ISDN	Broadband Integrated Services Digital Network
B-ISSI	Broadband Inter-Switching System Interface
BITNET	Because It's Time NETwork
B-NT1	Network Termination 1 for B-ISDN

B-NT2	Network Termination 2 for B-ISDN
BOC	Bell Operating Company
BOF	Birds of a Feather
BOM	Beginning-of-Message
BOOTP	Bootstrap Protocol
BPDU	Bridge Protocol Data Unit
bps	Bits Per Second
BPV	Bipolar Violations
BRI	Basic Rate Interface
BSC	Binary Synchronous Communication
BSD	Berkeley Software Distribution
BSS	Broadband Switching System
B-TA	Terminal Adapter for B-ISDN
B-TE	B-ISDN Terminal Equipment
B-TE1	Terminal Equipment 1 for B-ISDN
B-TE2	Terminal Equipment 2 for B-ISDN
BW	Bandwidth

C

CAD/CAM	Computer-Aided Design and Manufacturing
CAS	Channel Associated Signaling
CATNIP	Common Architecture for Next Generation Internet Protocol
CBR	Constant Bit Rate
CCIS	Common Channel Interoffice Signaling
CCITT	International Telegraph and Telephone Consultative Committee
CCS	Common Channel Signaling
CD	Compact Disc
CDPD	Cellular Digital Packet Data
CDR	Call Detail Records
CDV	Cell Delay Variance
CDV	Constant Delay Value

CELP	Codebook Excited Predictive Linear Coding
CIDR	Classless Interdomain Routing
CIR	Committed Information Rate
CLNP	Connectionless Network Protocol
CLNS	Connectionless-Mode Network Services
CLS	Connectionless Service
CLTP	Connectionless Transport Protocol
CMIP	Common Management Information Protocol
CMIS	Common Management Information Service
CMISE	Common Management Information Service Element
CMOL	CMIP on IEEE 802.2 Logical Link Control
CMOT	Common Management Information Protocol over TCP/IP
CO	Central Office
Codec	Coder-Decoder
CONS	Connection-Mode Network Services
CORBA	Common Object Request Broker Architecture
CPCS	Common Part Convergence Sublayer
CPE	Convergence Protocol Entity
CPE	Customer Premises Equipment
CPI	Common Part Indicator
CPU	Central Processing Unit
C/R	Command/Response Bit
CRC	Cyclic Redundancy Check
CRS	Cell Relay Service
CS	Convergence Sublayer
CS ACELP	Conjugate-Structure Algebraic-Code-Excited-Linear-Prediction
CSI	Convergence Sublayer Indicator
CSMA/CD	Carrier Sense Multiple Access with Collision Detection
CSNET	Computer+Science Network
CSPDN	Circuit Switched Public Data Network
CSU	Channel Service Unit

CT	Computer Telephony
CTI	Computer Telephony Integration

D

D	D Channel
DA	Destination Address
DAP	Data Access Protocol
DARPA	Defense Advanced Research Projects Agency
DCC	Data Country Code
DCE	Data Circuit-Terminating Equipment
DCN	Data Communications Network
DCS-n	Digital Cross Connect System at Level n
DDCMP	Digital Data Communications Message Protocol
DDN	Defense Data Network
DDP	Datagram Delivery Protocol
DDS	Digital Data Service
DE	Discard Eligible
DES	Data Encryption Standard
DFI	DSP Format Identifier
DFT	Discrete Fourier Transform
DHCP	Dynamic Host Configuration Protocol
DIX	DEC, Intel, and Xerox
DL	Data Link
DLC	Data Link Control
DLCI	Data Link Connection Identifier
DMA	Direct Memory Access
DMI	Desktop Management Interface
DMTF	Desktop Management Task Force
DNIC	Data Network Identification Code
DNS	Domain Name System
DOD	Department of Defense

DPCM	Differential Pulse Code Modulation
DRP	DECnet Routing Protocol
DS0	Digital Signal Level 0 (64 Kbps)
DS1	Digital Signal Level 1 (1.544 Mbps)
DS3	Digital Signal Level 3 (44.736 Mbps)
DSAP	Destination Service Access Point
DSI	Digital Speech Interpolation
DSP	Digital Signal Processing
DSU	Data Service Unit
DSU/CSU	Data Service Unit/Channel Service Unit
DSX-n	Digital Signal Cross Connect Level n
DTE	Data Terminal Equipment
DTMF	Dual Tone Multifrequency
DTR	Data Terminal Ready
DXC	Digital Cross-Connect
DXI	Data Exchange Interface

E

EA	Extended Address
E and M	Ear and Mouth
EBCDIC	Extended Binary Coded Decimal Interchange Code
ECSA	Exchange Carriers Standards Association
EDI	Electronic Data Interchange
EFCN	Explicit Forward Congestion Notification
EGA	Enhanced Graphics Array
EGP	Exterior Gateway Protocol
EIA	Electronic Industries Association
EOM	End-of-Message
EOT	End of Transmission
ES	Errored Second
ESF	Extended Superframe Format

ES-IS	End System to Intermediate System Protocol
ET	Exchange Termination
ETSI	European Telecommunications Standards Institute
Ext	Extension

F

FAS	Frame Alignment Signal
FAT	File Access Table
FCC	Federal Communications Commission
FCN	Forward Congestion Notification
FCS	Frame Check Sequence
FDDI	Fiber Data Distributed Interface
FDM	Frequency Division Multiplexing
FEBE	Far End Block Error
FEC	Forward Error Correction
FECN	Forward Error Congestion Notification
FFS	For Further Study
FIFO	First-In First-Out
FIPS	Federal Information Processing Standard
FoIP	Fax over Internet Protocol
FOTS	Fiber Optic Transport System
FR	Frame Relay
FRAD	Frame Relay Assembler/Disassembler
FRND	Frame Relay Network Device
FRS	Frame Relay Service
FR-SSCS	Frame Relay Service Specific Convergence Sublayer
FR-UNI	Frame Relay User Network Interface
FT1	Fractional T1
FTAM	File Transfer Access and Management
FTP	File Transfer Protocol
FXO	Foreign Exchange Office

G

G	Giga
GB	Gigabyte
GCF	Gatekeeper Confirmation Message
GFC	Generic Flow Control
GHz	Gigahertz
GOSIP	Government OSI Profile
GRJ	Gatekeeper Reject Message
GRQ	Gatekeeper Request Message
GSTN	General Switched Telephone Network
GUI	Graphical User Interface

H

HA	Hardware Address
HCS	Header Check Sequence
HDLC	High Level Data Link Control
Hdr	Header
HEC	Header Error Control
HEL	Header Extension Length
HLLAPI	High Level Language API
HLPI	Higher Layer Protocol Identifier
HSSI	High Speed Serial Interface
HTML	Hypertext Markup Language
HTTP	Hypertext Transfer Protocol
Hz	Hertz

I

I	Information
IA5	International Alphabet No. 5
IAB	Internet Architecture Board
IANA	Internet Assigned Numbers Authority

ICD	International Code Designator
ICI	Interexchange Carrier Interface
ICMP	Internet Control Message Protocol
ICP	Internet Control Protocol
IDI	Initial Domain Identifier
IDP	Internetwork Datagram Protocol
IDRP	Inter Domain Routing Protocol
IE	Information Element
IEC	International Electrotechnical Commission
IEEE	Institute of Electrical and Electronics Engineers
IETF	Internet Engineering Task Force
I/G	Individual/Group
IGMP	Internet Group Management Protocol
IGP	Interior Gateway Protocol
IGRP	Internet Gateway Routing Protocol
ILMI	Interim Local Management Interface
IMTC	International Multimedia Teleconferencing Consortium
I/O	Input/Output
IOC	Inter-Office Channel
IP	Internet Protocol
IPC	Interprocess Communications Protocol
IPng	Internet Protocol, Next Generation
IPsec	Internet Protocol Security
IPv6	Internet Protocol, version 6
IPv6CP	Internet Protocol version 6 Control Protocol
IPX	Internetwork Packet Exchange
IR	Internet Router
IRTF	Internet Research Task Force
ISA	Industry Standard Architecture
ISAKMP	Internet Secure Association Key Management Protocol
ISDN	Integrated Services Digital Network

IS-IS	Intermediate System to Intermediate System Protocol
ISN	Initial Sequence Number
ISO	International Organization for Standardization
ISOC	Internet Society
ISP	Internet Service Provider
ISSI	Inter-Switching System Interface
ISSIP_CLS	Inter-Switching System Interface Protocol Connectionless Service
ISU	Isochronous Service User
IT	Information Type
ITSP	Internet Telephony Service Provider
ITU	International Telecommunication Union
ITU-D	International Telecommunication Union Development Sector
ITU-R	International Telecommunication Union Radiocommunications Sector
ITU-T	ITU Telecommunication Standardization Sector (formerly CCITT)
IVDLAN	Integrated Voice/Data Local Area Network
IVR	Interactive Voice Response
IWF	Interworking Function
IWU	Interworking Unit
IXC	Inter-Exchange Carrier

J

JPEG	Joint Photographic Experts Group

K

Kbps	Kilo Bits per Second
KHz	Kilohertz

L

LAA	Locally Administered Address

LAN	Local Area Network
LANE	LAN Emulation
LAP	Link Access Procedure
LAPB	Link Access Procedure Balanced
LAPD	Link Access Procedure D Channel
LAPF	Link Access Procedures to Frame Mode Bearer Services
LAT	Local Area Transport
LATA	Local Access Transport Area
LCP	Link Control Protocol
LDAP	Lightweight Directory Access Protocol
LD-CELP	Low Delay Code Excited Linear Prediction
LE	Local Exchange
LEC	Local Exchange Carrier
LEN	Length
LI	Length Indicator
LI	Link Identifier
LL	Logical Link
LLC	Logical Link Control
LLI	Logical Link Identifier
LME	Layer Management Entity
LMI	Layer Management Interface
LMI	Local Management Interface
LOC	Loss of Cell Delineation
LOF	Loss of Frame
LOH	Line Overhead
LOP	Loss of Pointer
LOS	Loss of Signal
LPAS	Linear Prediction Analysis-by-Synthesis
LPC	Linear Predictive Coding
LPP	Lightweight Presentation Protocol
LRQ	Location Request

LSB	Least Significant Bit
LSL	Link Support Layer
LSS	Link Status Signal
LT	Line Termination
LTE	Line Terminating Equipment

M

MAC	Medium Access Control
MAN	Metropolitan Area Network
MAP	Management Application Protocol
MBONE	Multicasting Backbone
Mbps	Mega Bits Per Second
MC	Multipoint Controller
MCS	Multipoint Communication Service
MCU	Multipoint Control Unit
MELP	Mixed-Excitation LPC
MG	Media Gateway
MGC	Media Gateway Controller
MGCP	Media Gateway Control Protocol
MHS	Message Handling Service
MHz	Megahertz
MIB	Management Information Base
MILNET	Military NETwork
MILSTD	Military Standard
MIOX	Multiprotocol Interconnect over X.25
MIPS	Millions Instructions Per Second
MIS	Management Information Systems
MMF	Multi Mode Fiber
M/O	Mandatory/Optional
MOP	Maintenance Operations Protocol
MOS	Mean Opinion Scores

MPEG	Motion Pictures Expert Group
MP-MLQ	Multipulse Maximum Likelihood Quantization
MPOA	Multiprotocol over ATM
ms	Milliseconds
MSAU	Multistation Access Unit
MSB	Most Significant Bit
MSN	Monitoring Cell Sequence Number
MSS	MAN Switching System
MTA	Message Transfer Agent
MTBF	Mean Time Between Failure
MTTR	Mean Time to Repair
MTU	Maximum Transmission Unit
MUX	Multiplex, Multiplexor

N

NACS	NetWare Asynchronous Communications Server
NAK	Negative Acknowledgment
NAPs	Network Access Points
NASI	NetWare Asynchronous Service Interface
NAU	Network Addressable Unit
NAUN	Nearest Active Upstream Neighbor
NBP	Name Binding Protocol
NCP	NetWare Core Protocol
NCP	Network Control Protocol
NCSI	Network Communications Services Interface
NDIS	Network Driver Interface Standard
NetBEUI	NetBIOS Extended User Interface
NetBIOS	Network Basic Input/Output System
NFS	Network File System
NIC	Network Information Center
NIC	Network Interface Card

N-ISDN	Narrowband-ISDN
NIST	National Institute of Standards and Technology
NLA	Next-Level Aggregation Identifier
NLM	NetWare Loadable Module
NMP	Network Management Process
NMS	Network Management Station
NNI	Network-Network Interface
NNI	Network-Node Interface
NOC	Network Operations Center
NRZ	Non-Return to Zero
NSAP	Network Service Access Point
NSF	National Science Foundation
NSP	Network Services Protocol
NT	Network Termination
NT1	Network Termination of Type 1
NT2	Network Termination of Type 2
NTSC	National Television Standards Committee

O

OAM	Operations and Maintenance
OBASC	Object-Based Analysis-Synthesis Coding
OBC	Object-Based Coding
OC1	Optical Carrier, Level 1 (51.84 Mbps)
OC3	Optical Carrier, Level 3 (155.52 Mbps)
OCn	Optical Carrier, Level n
ODI	Open Data Link Interface
OID	Object Identifier
OIM	OSI Internet Management
OS	Operating System
OSI	Open Systems Interconnection
OSI-RM	Open Systems Interconnection Reference Model

OSPF	Open Shortest Path First
OUI	Organizationally Unique Identifier

P

PA	Protocol Address
PABX	Private Automatic Branch Exchange
PAD	Packet Assembler and Disassembler
PAD	Padding
PAL	Phased Alternating Line
PAP	Printer Access Protocol
PBX	Private Branch Exchange
PC	Personal Computer
PCI	Protocol Control Information
PCM	Pulse Code Modulation
PDN	Public Data Network
PDU	Protocol Data Unit
PEL	Picture Element
PEP	Packet Exchange Protocol
PH	Packet Handler
PHY	Physical Layer Protocol
PI	Protocol Identification
PIC	Primary Interexchange Carrier
PID	Protocol Identifier
PL	PAD Length
PL	Physical Layer
PLCP	Physical Layer Convergence Protocol
PLEN	Protocol Length
PLP	Packet Layer Protocol
PM	Physical Medium
PMTU	Path Maximum Transmission Unit
POH	Path Overhead

POI	Path Overhead Identifier
POP	Point of Presence
POSIX	Portable Operating System Interface — UNIX
POTS	Plain Old Telephone Service
PPP	Point-to-Point Protocol
PRI	Primary Rate Interface
PSN	Packet Switch Node
PSPDN	Packet Switched Public Data Network
PSTN	Public Switched Telephone Network
PT	Payload Type
PTN	Personal Telecommunications Number
PTP	Point-to-Point
PUC	Public Utility Commission
PVC	Permanent Virtual Circuit
PVC	Permanent Virtual Connection

Q

QCIF	Quarter Common Intermediate Format
QOS	Quality of Service

R

RARP	Reverse Address Resolution Protocol
RAS	Registration Admissions Status
RBOC	Regional Bell Operating Company
RC	Routing Control
RD	Route Descriptor
RD	Routing Domain Identifier
RDT	Remote Digital Terminal
RDTD	Restricted Differential Time Delay
RED	Random Early Detection
REQ	Request

RFC	Request for Comments
RFH	Remote Frame Handler
RFS	Remote File System
RGB	Red-Green-Blue
RH	Request/Response Header
RI	Routing Information
RIP	Routing Information Protocol
RJE	Remote Job Entry
RMI	Remote Method Invocation
RMON	Remote Monitoring
ROSE	Remote Operations Service Element
RPC	Remote Procedure Call
RQ	Request Counter
RQM	Request Queue Machine
RRJ	Registration Reject Message
RSVP	Resource Reservation Protocol
RSX	Realtime Resource-Sharing Executive
RT	Routing Type
RTCP	Real Time Control Protocol
RTP	Real Time Protocol
RTSP	Real Time Stream Protocol
RTT	Round Trip Time
RU	Request/Response Unit
RX	Receive

S

SA	Security Association
SA	Source Address
SABME	Set Asynchronous Balanced Mode Extended
SAP	Service Access Point
SAP	Service Advertising Protocol

SAPI	Service Access Point Identifier
SAR	Segmentation and Reassembly
SAT	Subscriber Access Termination
SB-ADPCM	Subband Adaptive Differential Pulse Code Modulation
SCN	Switched Circuit Network
SCS	System Communication Services
SDH	Synchronous Digital Hierarchy
SDLC	Synchronous Data Link Control
SDN	Software Defined Network
SDU	Service Data Unit
SEL	NSAP Selector
SEQ	Sequence
SET	Switching System (SS) Exchange Termination
SG	Signaling Gateway
SG	Study Group
SGCP	Simple Gateway Control Protocol
SGMP	Simple Gateway Management Protocol
SIP	SMDS Interface Protocol
SIPP	Simple Internet Protocol Plus
SIR	Sustained Information Rate
SIU	Subscriber Interface Unit
SKIP	Simple Key Management for the Internet Protocol
S/L	Strict/Loose Bits
SLA	Site-Level Aggregation Identifier
SLIP	Serial Line IP
SMB	Server Message Block
SMDS	Switched Multimegabit Data Service
SMF	Single Mode Fiber
SMI	Structure of Management Information
SMT	Station Management
SMTP	Simple Mail Transfer Protocol

SN	Sequence Number
SNA	System Network Architecture
SNADS	System Network Architecture Distribution Services
SNAP	Sub Network Access Protocol
SNCF	Sub Network Configuration Field
SNI	Subscriber-Network Interface
SNMP	Simple Network Management Protocol
SOH	Section Overhead
SOH	Start of Header
SONET	Synchronous Optical Network
SPE	Synchronous Payload Envelope
SPI	Security Parameters Index
SPN	Subscriber Premises Network
SPP	Sequenced Packet Protocol
SPX	Sequenced Packet Exchange
SR	Source Routing
SRF	Specifically Routed Frame
SRI	Stanford Research Institute
SRT	Source Routing Transparent
SRTS	Synchronous Residual Time Stamp
SS	Switching System
SS7	Common Channel Signalling System No. 7
SSAP	Source Service Access Point
SSCF	Service Specific Coordination Function
SSCOP	Service Specific Connection Oriented Protocol
SSCS	Service Specific Convergence Sublayer
SSM	Single Segment Message
ST	Segment Type
STE	Section Terminating Equipment
STM	Synchronous Transfer Mode
STS-N	Synchronous Transport Signal Level-N

SUA	Stored Upstream Address
SVC	Switched Virtual Circuit
SVC	Switched Virtual Connection
SVGA	Super Visual Graphics Array

T

TA	Technical Advisory
TA	Terminal Adapter
TAG	Technology Advisory Group
TAPI	Telephony Application Programming Interface
TASI	Time-Assigned Speech Interpolation
TB	Terabyte
TCP	Transmission Control Protocol
TCP/IP	Transmission Control Protocol/Internet Protocol
TDM	Time Division Multiplexing
TE	Terminal Equipment
TE1	Terminal Equipment of Type 1
TE2	Terminal Equipment of Type 2
TEI	Terminal Endpoint Identifier
TELNET	Telecommunications Network
TFTP	Trivial File Transfer Protocol
TH	Transmission Header
TID	Terminal Identifier
TIPHON	Telecommunications and Internet Protocol Harmonization over Networks
TLA	Top-Level Aggregation Identifier
TLI	Transport Layer Interface
TLV	Type-Length-Value Encoding
TOS	Type of Service
TP	Transport Protocol
TR	Technical Reference

TS	Time Stamp
TSI	Time Slot Interchange
TSR	Terminate-and-Stay Resident
TTL	Time to Live
TTS	Text-to-Speech
TUBA	TCP/UDP with Bigger Addresses
TX	Transmit

U

UA	Unnumbered Acknowledgment
UA	User Agent
UAS	Unavailable Seconds
UDI	Unrestricted Digital Information
UDI-TA	Unrestricted Digital Information with Tones/Announcements
UDP	User Datagram Protocol
UI	Unnumbered Information Frame
U/L	Universal/Local
ULP	Upper Layer Protocols
UNI	User Network Interface
UPC	Usage Parameter Control
URJ	Unregister Reject Message
URQ	Unregister Request Message
UT	Universal Time
UTP	Unshielded Twisted Pair
UU	User-to-User
UUCP	UNIX to UNIX Copy Program

V

V	Volt
VAN	Value Added Network
VAP	Value Added Process

VBR	Variable Bit Rate
VC	Virtual Channel
VCC	Virtual Channel Connection
VCI	Virtual Channel Identifier
VF	Voice Frequency Services
VGA	Video Graphics Array
VLBV	Very Low Bit-Rate Video
VLC	Variable Length Code
VLSI	Very Large Scale Integration
VLSM	Variable Length Submask
VMS	Virtual Memory System
VoIP	Voice over IP
VP	Virtual Path
VPC	Virtual Path Connection
VPI	Virtual Path Identifier
VPI/VCI	Virtual Path Identifier/Virtual Channel Identifier
VPN	Virtual Private Network
VT	Virtual Terminal
VT	Virtual Tributary

W

WAN	Wide Area Network
WBEM	Web-Based Enterprise Management
WDM	Wavelength Division Multiplexing
WFQ	Weighted Fair Queuing
WIN	Window
WRED	Weighted RED

X

X	Unassigned Bit
XDR	External Data Representation

XID	Exchange Identification
XNS	Xerox Network System

Appendix B

Sources of Internet Information

Internet Organizations

A number of groups contribute to the management, operation, and proliferation of the Internet. These include (in alphabetical order):

CommerceNet
510 Logue Avenue
Mountain View, CA 94304
Tel: (650) 962-2600
(888) 255-1900
Fax: (650) 962-2601
Email: info@commerce.net
www.commerce.net

Internet Architecture Board (IAB)
Email: iab@isi.edu
www.iab.org

Internet Assigned Numbers Authority (IANA)
4676 Admiralty Way, Suite 330
Marina del Rey, CA 90292
Tel: (310) 823-9358
Fax: (310) 823-8649
Email: iana@iana.org
www.iana.org

Internet Corporation for Assigned Names and Numbers (ICANN)
4676 Admiralty Way, Suite 330
Marina del Rey, CA 90292
Tel: (310) 823-9358
Fax: (310) 823-8649
Email: icann@icann.org
www.icann.org

Internet Engineering Task Force (IETF)
Email: ietf-info@ietf.org
www.ietf.org

Internet Society (ISOC)
1775 Wiehle Avenue, Suite 102
Reston, VA 20190-5108
Tel: (703) 326-9880
Fax: (703) 326-9881
Email: isoc@isoc.org
www.isoc.org

InterNIC
www.internic.net

North American Network Operators Group (NANOG)
c/o Merit Network
4251 Plymouth Road, Suite 2000
Ann Arbor, MI 48105-2785
Tel: (734) 764-9430
Email: www@merit.edu
www.nanog.org

United States Internet Service Provider Association
1330 Connecticut Avenue, NW
Washington, DC 20036
Tel: (202) 862-3816
Fax: (202) 261-0604
Email: kdean@steptoe.com
www.cix.org

World Wide Web Consortium (W3C)
c/o MIT Laboratory for Computer Science
200 Technology Square, Bldg. NE43
Cambridge, MA 02139
Tel: (617) 253-5851
Fax: (617) 258-8682
Email: info@lcs.mit.edu
www.w3.org

Obtaining RFCs

The following is an excerpt from the file www.isi.edu/in-notes/rfc-retrieval.txt, which is available from many of the RFC repositories listed below. This information is subject to change. Obtain the current version of this file if problems occur. Also note that each RFC site may have instructions for file retrieval (such as a particular subdirectory) that are unique to that location.

RFCs may be obtained via EMAIL or FTP from many RFC Repositories. The Primary Repositories will have the RFC available when it is first announced, as will many Secondary Repositories. Some Secondary Repositories may take a few days to make available the most recent RFCs. Many of these repositories also now have World Wide Web servers. Try the following URL as a starting point:

http://www.rfc-editor.org/

Primary Repositories

RFCs can be obtained via FTP from NIS.NSF.NET, FTP.RFC-EDITOR.ORG, WUARCHIVE.WUSTL.EDU, SRC.DOC.IC.AC.UK, FTP.NCREN.NET, FTP.NIC.IT, FTP.IMAG.FR, FTP.IETF.RNP.BR, WWW.NORMOS.ORG, FTP.GIGABELL.NET, or OASISSTUDIOS.COM.

NIS.NSF.NET

To obtain RFCs from NIS.NSF.NET via FTP, login with username "anonymous" and password "name@host.domain"; then connect to the directory of RFCs with cd /internet/documents. The file name is of the form rfcnnnn.txt (where "nnnn" refers to the RFC number).

For sites without FTP capability, electronic mail query is available from NIS.NSF.NET. Address the request to NIS-INFO@NIS.NSF.NET and leave the subject field of the message blank. The first text line of the message must be "send rfcnnnn.txt" with nnnn the RFC number.

Contact: rfc-mgr@merit.edu

FTP.RFC-EDITOR.ORG

RFCs can be obtained via FTP from FTP.RFC-EDITOR.ORG, with the pathname in-notes/rfcnnnn.txt (where "nnnn" refers to the number of the RFC). Login with FTP username "anonymous" and password "name@host.domain".

RFCs can also be obtained via electronic mail from FTP.RFC-EDITOR.ORG by using the RFC-INFO service. Address the request to "rfc-info@rfc-editor" with a message body of:

Retrieve: RFC

Doc-ID: RFCnnnn

(Where "nnnn" refers to the number of the RFC (always use 4 digits — the DOC-ID of RFC 822 is "RFC0822").) The RFC-INFO@RFC-EDITOR.ORG server provides other ways of selecting RFCs based on key-words and such; for more information send a message to rfc-info@rfc-editor.org with the message body "help: help".

Contact: rfc-editor@rfc-editor.org

WUARCHIVE.WUSTL.EDU

RFCs can also be obtained via FTP from WUARCHIVE.WUSTL.EDU, with the pathname info/rfc/rfcnnnn.txt.Z (where "nnnn" refers to the number of the RFC and "Z" indicates that the document is in compressed form).

At WUARCHIVE.WUSTL.EDU the RFCs are in an "archive" file system and various archives can be mounted as part of an NFS file system. Please contact Chris Myers (chris@wugate.wustl.edu) if you want to mount this file system in your NFS.

WUArchive now keeps RFCs and STDs under ftp://wuarchive.wustl.edu./doc/ or http://wuarchive.wustl.edu./doc/.

Contact: chris@wugate.wustl.edu

SUNSITE.ORG.UK (also known as SRC.DOC.IC.AC.UK)

RFCs can be obtained via FTP from SUNSITE.ORG.UK with the pathname rfc/rfcnnnn.txt or rfc/rfcnnnn.ps (where "nnnn" refers to the number of the RFC). Login with FTP username "anonymous" and password "your-email-address". To obtain the RFC Index, use the pathname rfc/rfc-index.txt.

For users with good fast Internet connections, the whole archive is also available by NFS (readonly) and the RFC area can be mounted as, e.g.

```
mount -r sunsite.org.uk:/public/rfc /mnt
```

RFCs are also available via the web at http://sunsite.org.uk/rfc/

Contact: wizards@sunsite.org.uk

FTP.NCREN.NET

To obtain RFCs from FTP.NCREN.NET via FTP, login with username "anonymous" and your internet email address as password. The RFCs can be found in the directory /rfc, with file names of the form: rfcNNNN.txt or rfcNNNN.ps where NNNN refers to the RFC number.

This repository is also accessible via WAIS and the Internet Gopher.

Contact: rfc-mgr@ncren.net

FTP.NIC.IT

RFCs can be obtained from the ftp.nic.it FTP archive with the pathname rfc/rfcnnnn.txt (where "nnnn" refers to the number of the RFC). Login with FTP, username "anonymous" and password "name@host.domain".

The summary of ways to get RFCs from the Italian Network Information Center is the following:

Via ftp: ftp.nic.it directory rfc

Via email: send a message to listserv@nic.it whose body contains "get RFC/rfc<number>.[txt,ps]". For receiving a full list of the existing RFCs include in the body the command "index RFC/rfc".

Contact: D.Vannozzi@cnuce.cnr.it

FTP.IMAG.FR

RFCs can be obtained via FTP from ftp.imag.fr with the pathname /pub/archive/IETF/rfc/rfcnnnn.txt (where "nnnn" refers to the number of the RFC).

Login with FTP username "anonymous" and password "your-email-address". To obtain the RFC Index, use the pathname /pub/archive/IETF/rfc/rfc-index.txt.

Internet drafts and other IETF related documents are also mirrored in the /pub/archive/IETF directory.

Contact: rfc-adm@imag.fr

WWW.NORMOS.ORG

RFCs, STD, BCP, FYI, RTR, IEN, Internet-Drafts, RIPE, and other internet engineering documents can be found at http://www.normos.org and ftp://ftp.normos.org.

The RFCs are available as:
http://www.normos.org/ietf/rfc/rfcXXXX.txt
ftp://ftp.normos.org/ietf/rfc/rfcXXXX.txt

STD, BCP, FYI, RTR, IEN documents are available as:
http://www.normos.org/ietf/[std,bcp,fyi,rtr,ien][std,bcp,fyi,rtr,ien]XXXX.txt

ftp://ftp.normos.org/ietf/[std,bcp,fyi,rtr,ien]/[std,bcp,fyi,rtr,ien]XXXX.txt

Internet-drafts are available as:
http://www.normos.org/ietf/internet-drafts/draft-....txt
ftp://ftp.normos.org/ietf/internet-drafts/draft-....txt

Full-text search and database queries are available from the web interface.

Please send questions, comments, suggestions to info@normos.org.

FTP.IETF.RNP.BR

RFCs can be obtained via FTP from FTP.IETF.RNP.BR with the pathname rfc/rfcnnnn.txt (where "nnnn" refers to the number of the RFC). Login with FTP username "anonymous" and password "your-email-address". To obtain the RFC Index, use the pathname rfc/rfc-index.txt.

Internet-Drafts and other IETF related documents are also mirrored.
Contact: rfc-admin@ietf.rnp.br

FTP.GIGABELL.NET

To obtain RFCs from FTP.GIGABELL.NET via FTP, login with username "anonymous" and password "name@host.domain"; then connect to the directory of RFCs with cd /pub/rfc. The file name is of the form rfcnnnn.txt (where "nnnn" refers to the RFC number). An index can be obtained with the pathname pub/rfc/rfc-index.txt.

Contact: ftpadmin@gigabell.net

FTP.FCCN.PT

To obtain RFCs from Oasis Studios via FTP, login to FTP.FCCN.PT with username "anonymous" and password "name@host.domain"; then connect to the directory of RFCs with cd /pub/IETF/RFCs. The file name is of the form rfcnnnn.txt (where "nnnn" refers to the RFC number).

Contact: webmaster@fccn.pt

OASISSTUDIOS.COM

To obtain RFCs from Oasis Studios via FTP, login to FTP.OASISSTUDIOS.COM with username "anonymous" and password "name@host.domain"; then connect to the directory of RFCs with cd /pub/RFC. The file name is of the form rfcnnnn.txt (where "nnnn" refers to the RFC number).

For sites without FTP capability, electronic mail query is available from oasisstudios@OASISSTUDIOS.COM. Address the request to oasisstudios@OASISSTUDIOS.COM and leave the body of the message blank. The subject of the message must be "send rfcnnnn.txt" where nnnn is the RFC number.

Oasis Global Inc. also provides an HTTP interface to the RFC archive. To browse or search the archives via a web browser surf to: http://www.oasisstudios.com/RFC.

For more information send a message to rfc-info@OASISSTUDIOS.COM with the subject "help".

Contact: rfc-admin@oasisstudios.com

SUNSITE.DK

RFCs can be obtained via FTP from SUNSITE.DK with the pathname mirrors/rfc/rfcnnnn.txt or mirrors/rfc/rfcnnnn.ps (where "nnnn" refers to the number of the RFC). Login with FTP username "anonymous" and password "your-email-address". To obtain the RFC Index, use the pathname rfc/rfc-index.txt.

RFCs are also available via the web at: http://mirrors.sunsite.dk/rfc.

Contact: staff@sunsite.dk

Secondary Repositories:

Australia and Pacific Rim

Site: munnari

Contact: Robert Elz <kre@munnari.oz.au>

Host: munnari.oz.au

Directory: rfc (rfcs in compressed format rfcNNNN.Z postscript rfcs rfcNNNN.ps.Z)

Site: The Programmers' Society University of Technology, Sydney

Contact: ftp@progsoc.uts.edu.au

Host: ftp.progsoc.uts.edu.au

Directory: pub/internet

Both are stored uncompressed.

Chile

Site: OK Internet
Host: http://www.ok.cl/rfcs/
Directory: http://www.ok.cl/rfcs/

Denmark

Site: University of Copenhagen
Host: ftp.denet.dk
Directory: mirror/ftp.isi.edu/in-notes

Finland

Site: FUNET
Host: nic.funet.fi
Directory: index/RFC
Directory: /pub/netinfo/rfc
Notes: RFCs in compressed format. Also provides email access by sending mail to archive-server@nic.funet.fi.

France

Site: Centre d'Informatique Scientifique et Medicale (CISM)
Contact: ftpmaint@univ-lyon1.fr
Host: ftp.univ-lyon1.fr pub/mirrors/rfc Mirror of Internic
Notes: Files compressed with gzip. Online decompression done by the FTP server.

Indonesia

Site: UILA — University of Indonesia at Lenteng Agung
Contact: rfc-mirror at vlsm dot org
The RFC-Editor directory URL:
http://bebas.vLSM.org/v07/org/rfc-editor/rfc-ed-all/
ftp://bebas.vlsm.org/bebas/v07/org/rfc-editor/rfc-ed-all/
The Internet Drafts directory URL:
http://bebas.vLSM.org//v08/org/rfc-editor/internet-drafts/
ftp://bebas.vlsm.org/bebas/v08/org/rfc-editor/internet-drafts/
Notes: Not mirroring the ZIP files.

Romania

Site: SunSITE Romania at the Politehnica University of Bucharest
Contact: space@sunsite.pub.ro
Host: sunsite.pub.ro/pub/rfc
or via http:
sunsite.pub.ro/pub/mirrors/ds.internic.net

South Africa

Site: The Internet Solution
Contact: ftp-admin@is.co.za
Host: ftp.is.co.za
Directory: internet/in-notes/rfc

Sweden

Host: ftp.chalmers.se
Directory: rfc

United Kingdom

Site: rfc.net
Contact: Alaric Williams <webmaster@rfc.net>

United States

Site: uunet
Contact: James Revell <revell@uunet.uu.net>
Host: ftp.uu.net
Directory: inet/rfc

UUNET Archive

UUNET archive, which includes the RFCs, various IETF documents, and other information regarding the Internet, is available to the public via anonymous ftp (to ftp.uu.net/inet/rfc/) and anonymous uucp, and will be available via an anonymous kermit server soon. Get the file /archive/inet/ls-lR.Z for a listing of these documents.

Any site in the US running UUCP may call +1 900 GOT SRCS and use the login "uucp". There is no password. The phone company will bill you at $0.50 per minute for the call. The 900 number only works from within the US.

Requests for special distribution of RFCs should be addressed to either the author of the RFC in question or to RFC-EDITOR@RFC-EDITOR.ORG.

Submissions for Requests for Comments should be sent to RFC-EDITOR@RFC-EDITOR.ORG. Please consult "Instructions to RFC Authors," RFC 2223, for further information.

Requests to be added to or deleted from the RFC distribution list should be sent to RFC-REQUEST@ISI.EDU.

Users with .MIL addresses may send a request to MAJORDOMO@NIC.DDN.MIL with an empty Subject: line and a message: subscribe rfc [your email address].

Changes to this file "rfc-retrieval.txt" should be sent to RFC-EDITOR@RFC-EDITOR.ORG.

Internet Mailing Lists

A number of mailing lists are maintained on the Internet for the purposes of soliciting information and discussions on specific subjects. In addition, a number of the Internet Engineering Task Force (IETF) working groups maintain a list for the exchange of information that is specific to that group.

For example, the IETF maintains two lists: the IETF General Discussion list and the IETF Announcement list. To join the IETF Announcement list, send a request to:

ietf-announce-request@ietf.org, and enter the word subscribe in the subject line of the message and in the message body.

To join the IETF General Discussion, send a request to:

ietf-request@ietf.org, and enter the word subscribe in the subject line of the message and in the message body.

To unsubscribe from either list, send a request as described above, but containing unsubscribe in the subject line and message body.

A number of other mailing lists are available. To join a mailing list, send a message to the associated request list:

listname-request@listhost (for example, snmp-request@psi.com)

With the following as the message body:

subscribe listname (for example, subscribe snmp)

A complete listing of the current IETF working groups and their respective mailing lists is available at:

www.ietf.org/maillist.html

Active IETF Working Groups

Much of the Internet research is documented by the various IETF Working Groups. A listing of the active working groups, organized by subject matter, can be found at:

http://www.ietf.org/html.charters/wg-dir.html

Appendix C

Addresses of Standards Organizations

ANSI STANDARDS

American National Standards Institute
25 West 43rd Street, 4th Floor
New York, NY 10036
Telephone: (212) 642-4900
Customer Service: (212) 642-4980
Fax: 212-398-0023
www.ansi.org

ATIS PUBLICATIONS

Alliance for Telecommunications Industry Solutions
1200 G Street NW, Suite 500
Washington, DC 20005
Tel: (202) 628-6380
Fax: (202) 393-5453
www.atis.org

BELLCORE STANDARDS

Telcordia Technologies, Inc.
PYA3
8 Corporate Place
Piscataway, NJ 08854-4156
Tel: (732) 699-2000
Customer Service: (800) 521-2673
www.telcordia.com

CSA STANDARDS

Canadian Standards Association
5060 Spectrum Way
Mississauga, ONT L4W 5N6
Canada
Tel: (416) 747-4000
Customer Service: (800) 463-6727
Fax: (416) 747-2473
www.csa.ca

DISA STANDARDS

Defense Information Systems Agency
www.itsi.disa.mil

DOD STANDARDS

DoD Network Information Center
Boeing Corporation
7990 Boeing Court, M/S CV-50
Vienna, VA 22183-7000
Tel: (703) 821-6266
Help Desk: (800) 365-3642
Fax: (703) 821-6161
www.nic.mil

ECMA STANDARDS

European Computer Manufacturers Association
Rue de Rhône 114
CH-1204 Geneva
Switzerland
Tel: 41 22 849 6000
Fax: 41 22 849 6001
Email: helpdesk@ecma.ch
www.ecma-international.org

EIA STANDARDS

Electronic Industries Association
2500 Wilson Boulevard
Arlington, VA 22201
Tel: (703) 907-7500
www.eia.org

ETSI STANDARDS

European Telecommunications Standards Institute
ETSI Secretariat
650, route des Lucioles
06921 Sophia-Antipolis Cedex
France
Tel: 33 (0)4 92 94 42 00
Fax: 33 (0)4 93 65 47 16
Email: helpdesk@etsi.org
www.etsi.org

FEDERAL INFORMATION PROCESSING STANDARDS (FIPS)

U.S. Department of Commerce
National Technical Information Service (NTIS)
5285 Port Royal Road
Springfield, VA 22161
Tel: (703) 605-6585
www.ntis.gov

IEC STANDARDS

International Electrotechnical Commission
IEC Central Office
3, rue de Varembé
P.O. Box 131
CH-1211 Geneva 20
Switzerland
Phone: 41 22 919 02 11
Fax: 41 22 919 03 00
Email: info@iec.ch
www.iec.ch

IEEE STANDARDS

Institute of Electrical and Electronics Engineers
445 Hoes Lane
P.O. Box 1331
Piscataway, NJ 08855-1331
Tel: (732) 981-0060
Customer Service: (800) 678-4333
Fax: (732) 981-1721
www.ieee.org

INTERNET STANDARDS

Internet Society International
1775 Wiehle Ave., Suite 102
Reston, VA 20190-5108
Tel: (703) 326 9880
Fax: (703) 326 9881
Email: membership@isoc.org
www.isoc.org

ISO STANDARDS

International Organization for Standardization
1, rue de Varembé
Case postale 56
CH-1211 Geneva 20
Switzerland
Tel: 41 22 749 0111
Fax: 41 22 733 3430
Email: central@isocs.iso.ch
www.iso.ch

ITU STANDARDS

International Telecommunications Union
Information Services Department
Place des Nations
CH-1211 Geneva 20
Switzerland
Tel: 41 22 730 5115
Fax: 41 22 730 5137
Email: sgo@itu.ch
www.itu.ch

NATIONAL INSTITUTE OF STANDARDS AND TECHNOLOGY

NIST
100 Bureau Drive, Stop 3460
Gaithersburg, MD 20899-3460
Tel: (301) 975-6478
Email: inquiries@nist.gov
www.nist.gov

WWW STANDARDS

World Wide Web Consortium
Massachusetts Institute of Technology
Computer Science and Artificial Intelligence Laboratory (CSAIL)
200 Technology Square
Cambridge, MA 02139
Tel: (617) 253-2613
Fax: (617) 258-5999
Email: www-request@w3.org
www.w3.org

Many of the above standards may be purchased from:

Global Engineering Documents
15 Inverness Way East
Englewood, CO 80112
Tel: (303) 792-2181
Customer Service: (800) 624-3974
Fax: (303) 792-2192
Email: globalcustomerservice@ihs.com
www.global.ihs.com

Phillips Business Information, Inc.
7811 Montrose Road
Potomac, MD 20854
Tel: (301) 340-2100
Customer Service: (800) 777-5006
Email: clientservices.pbi@phillips.com
www.phillips.com

Appendix D

Trademarks

PostScript is a trademark of Adobe Systems.

Apple, the Apple logo, AppleTalk, EtherTalk, LocalTalk, Macintosh, and TokenTalk are registered trademarks of Apple Computer, Inc.

Cisco is a registered trademark of Cisco Systems, Inc.

DEC, DECmcc, DECnet, Digital, LAT, LAVC, Micro-VAX, MOP, POLYCENTER, ThinWire, Ultrix, VAX, and VAX Cluster are trademarks, and Ethernet is a registered trademark, of Compaq Computer Corporation.

DigiNet is a registered trademark of Digital Network Corporation.

Ethernet and Intel are registered trademarks of Intel Corporation.

IBM PC LAN, PC/AT, PC/XT, SNA, and NetBIOS are trademarks of International Business Machines Corporation, and AIX, AT, IBM, and NetView are registered trademarks of International Business Machines Corporation.

Microsoft, Microsoft NetMeeting, MS-DOS, Windows, and Windows NT are registered trademarks of Microsoft Corporation.

IPX, ManageWise, NetWare, NetWare 386, Novell, and SPX are trademarks, and Novell is a registered trademark, of Novell, Inc.

BSD is a trademark of the Regents of the University of California.

Java, Network File System, NFS, Sun, Sun Microsystems Inc., Sun Microsystems, SunNet, SunOS, and SunSoft are trademarks or registered trademarks of Sun Microsystems, Inc. SPARC is a registered trademark of SPARC International, Inc., licensed to Sun Microsystems, Inc.

Ethernet and Xerox are registered trademarks of Xerox Corporation.

UNIX is a registered trademark of The Open Group.

All other trademarks are the property of their respective owners.

Index